Spintronics Handbook: Spin Transport and Magnetism, Second Edition

Metallic Spintronics—Volume One

Spintronics Handbook: Spin Transport and Magnetism, Second Edition

Metallic Spintronics—Volume One

Edited by
Evgeny Y. Tsymbal and Igor Žutić

CRC Press
Taylor & Francis Group
Boca Raton London New York

CRC Press is an imprint of the
Taylor & Francis Group, an **informa** business

CRC Press
Taylor & Francis Group
6000 Broken Sound Parkway NW, Suite 300
Boca Raton, FL 33487-2742

First issued in paperback 2021

© 2019 by Taylor & Francis Group, LLC
CRC Press is an imprint of Taylor & Francis Group, an Informa business

No claim to original U.S. Government works

ISBN 13: 978-0-367-77977-1 (pbk)
ISBN 13: 978-1-4987-6952-5 (hbk)

Visit the Taylor & Francis Web site at
http://www.taylorandfrancis.com

and the CRC Press Web site at
http://www.crcpress.com

Cover illustration courtesy of Markus Lindemann, Nils C. Gerhardt, and Carsten Brenner.

The e-book of this title contains full colour figures and can be purchased here: http://www.crcpress.com/9780429423079. The figures can also be found under the 'Additional Resources' tab.

Contents

SECTION I—Introduction

SECTION II—Magnetic Metallic Multilayers

SECTION III—Magnetic Tunnel Junctions

Foreword

Spintronics is a field of research in which novel properties of materials, especially atomically engineered magnetic multilayers, are the result of the manipulation of currents of spin-polarized electrons. Spintronics, in its most recent incarnation, is a field of research that is almost 30 years old. To date, its most significant technological impact has been in the development of a new generation of ultra-sensitive magnetic recording read heads that have powered magnetic disk drives since late 1997. These magnetoresistive read head which use spin-valves, that are based on spin-dependent scattering at magnetic/non-magnetic interfaces, and since 2007, magnetic tunnel junctions (MTJs), that are based on spin-dependent tunneling across ultra-thin insulating layers, have a common thin film structure. These structures involve "spin engineering" to eliminate the influence of long-range magneto-dipole fields via the use of synthetic or artificial antiferromagnets, which are formed from thin magnetic layers coupled antiferromagnetically via the use of atomically thin layers of ruthenium. These structures involve the discoveries of spin-dependent tunneling in 1975, giant magnetoresistance at low temperatures in Fe/Cr in 1988, oscillatory interlayer coupling in 1989, the synthetic antiferromagnet in 1990, giant magnetoresistance at room temperature in Co/Cu and related multilayers in 1991, and the origin of giant magnetoresistance as being a result of predominant interface scattering in 1991–1993. Together these discoveries led to the spin-valve recording read head that was introduced by IBM in 1997 and led, within a few years, to a 1,000 fold increase in the storage capacity of magnetic disk drives. This rapid pace of improvement has stalled over the past years as the difficulty of stabilizing tiny magnetic bits against thermal fluctuations, whilst, at the same time, being able to generate large enough magnetic fields to write them, has proved intractable. The possibility of creating novel spintronic magnetic memory-storage devices to rival magnetic disk drives in capacity and to vastly exceed them in performance has emerged in the form of Racetrack Memory. This concept and the physics underlying it are discussed in this book together with a more conventional spintronic memory, magnetic random access memory (MRAM). MRAM is based on MTJ magnetic memory bits each one accessed in a two-dimensional cross point array via a transistor. The fundamental concept of MRAM was proposed in 1995 using local fields to write the MTJ elements. This basic concept was proven in 1999 with the subsequent demonstration of large scale fully integrated 64 Mbit memory chips in the following decade. Writing these same elements using spin angular momentum from sufficiently large spin-polarized currents passed through the tunnel junction elements emerged in the 1990s and is now key to the development of massive scale MRAM chips. The second edition of this book discusses these emerging spintronic technologies as well as other breakthroughs and key advances both fundamental and applied, in the field of spintronics.

Beyond MRAM and Racetrack Memory, this book elucidates other nascent opportunities in spintronics that do not rely directly on magneto-resistive effects, such as fault-tolerant quantum computing, non-Boolean spin-wave logic, and lasers that are enhanced by spin-polarized carriers. It is interesting that spintronics is a field of research that continues to surprise even though the fundamental property of spin was realized nearly a century ago, and the basic concept of spin-dependent scattering in magnetic materials was introduced by Neville Mott just shortly after the notion of "spin" was conceived.

Since the first edition of this book, spintronics has so much evolved that a new name of "spin-orbitronics" has been coined to describe these new discoveries and developments. In the first edition of this book spin-orbit coupling was regarded rather negatively as a property that leads to mixing between spin-channels and the loss of spin angular momentum from spin currents to the lattice, thereby limiting the persistence of these same spin currents, both temporally and spatially. In this edition several physical phenomena derived from spin-orbit coupling are shown to be key to the development of several new technologies, such as, in particular, the current induced motion of a series of magnetic domain walls that underlies Racetrack Memory. This relies especially on the generation of pure spin currents via the spin Hall effect (SHE). The magnitude of the SHE was thought for some time to be very small in conventional metals but over the past few years this has rather been shown to be incorrect. Significant and useful SHEs have been discovered in a number of heavy materials where spin-orbit coupling is large. These spin currents can be used to help move domain walls or to help switch the magnetization direction of nanoscale magnets. Whether they can be usefully used for MRAM, however, is still a matter of debate.

Another very interesting development since the first edition of this book is the explosive increase in our understanding and knowledge of topological insulators and their cousins including, most recently, Weyl semi-metals. The number of such materials has increased astronomically and, indeed, it is now understood that a significant fraction of all extant materials are "topological". What this means, in some cases, is that the spin of the carriers is locked to their momentum leading, for example, to the quantum spin Hall effect.

The very concept of these materials is derived from band inversion which, is often due to strong spin-orbit coupling. From a spintronics perspective the novel properties of these materials can lead to intrinsic spin currents and spin accumulations that are topologically "protected" to a greater or lesser degree. The concept of topological protection is itself evolving.

Distinct from electronic topological effects are topological spin textures such as skyrmions and anti-skyrmions. The latter were only experimentally found 2 years ago. These spin textures are nano-sized magnetic objects, related to magnetic bubbles that are also found in magnetic materials with perpendicular magnetic anisotropy, but which have boundaries or walls that are innately chiral. The chirality is determined by a vector magnetic exchange—a Dzyaloshinskii–Moriya interaction (DMI)—that is often derived from spin-orbit coupling. The DMI favors orthogonal alignment of neighboring magnetic moments in contrast to conventional ferromagnetic or antiferromagnetic exchange interactions that favors collinear magnetic arrangements. Skyrmion and anti-skyrmion spin textures have very interesting properties that could also be useful for Racetrack Memories. Typically, skyrmions and anti-skyrmions evolve from helical or conical spin textures. The magnetic phase of such systems can have complex dependences on temperature, magnetic field and strain. Some chiral antiferromagnetic spin textures have interesting properties such as an anomalous Hall effect (AHE), that is derived from their topological chiral spin texture in the absence of any net magnetization. In practice, however, a small unbalanced moment is needed to set the material in a magnetic state with domains of the same chirality in order to evidence the AHE. On the other hand, these same chiral textures can display an intrinsic spin Hall effect whose sign is independent of the chirality of the spin texture.

The DMI interaction can also result from interfaces particularly between heavy metals and magnetic layers. Such interfacial DMIs can give rise to chiral domain walls as well as magnetic bubbles with chiral domain walls – somewhat akin to skyrmions. The tunability of the interfacial DMI via materials engineering makes it of especial interest.

Thus, since the first edition of this book chiral spin phenomena, namely chiral spin textures and domain walls, and the spin Hall effect itself, which is innately chiral, have emerged as one of the most interesting developments in spintronics. The impact of these effects was largely unanticipated. It is not too strong to say that we are now in the age of "chiraltronics"!

Another topic that has considerably advanced since the first edition of this book is the field of what is often now termed spin caloritronics, namely the use of temperature gradients to create spin currents and the use of thermal excitations of magnetic systems, i.e. magnons, for magnonic devices. Indeed, magnons carry spin angular momentum and can propagate over long distances. Perhaps here it is worth mentioning the extraordinarily long propagation distances of spin currents via magnons in antiferromagnetic systems that has recently been realized.

Recently discovered atomically thin ferromagnets reveal how the presence of spin-orbit coupling overcomes the exclusion of two-dimensional ferromagnetism expected from the Mermin–Wagner theorem. These two-dimensional materials which are similar to graphene in that they can readily be exfoliated from bulk samples, provide a rich platform to study magnetic proximity effects and transform a rapidly growing class of van der Waals materials. Through studies of magnetic materials it is possible to reveal their peculiar quantum manifestations. Topological insulators can become magnetic by doping with 3d transition metals: the quantum anomalous Hall effect has been discovered in such materials and heterostructures that consist of magnetic and non-magnetic topological insulators have been used to demonstrate current induced control of magnetism.

Spintronics remains a vibrant research field that spans many disciplines ranging from materials science and chemistry to physics and engineering. Based on the rich developments and discoveries over the past thirty years one can anticipate a bountiful future.

Stuart Parkin
Director at the Max Planck Institute of Microstructure Physics
Halle (Saale), Germany
and
Alexander von Humboldt Professor, Martin-Luther-Universität Halle-Wittenberg, Germany
Max Planck Institute of Microstructure Physics
Halle (Saale), Germany

Preface

The second edition of this book continues the path from the foundations of spin transport and magnetism to their potential device applications, usually referred to as spintronics. Spintronics has already left its mark on several emerging technologies, e.g., in magnetic random access memories (MRAMs), where the fundamental properties of magnetic tunnel junctions are key for device performance. Further, many intricate fundamental phenomena featured in the first edition have since evolved from an academic curiosity into the potential basis for future spintronic devices. Often, as in the case of spin Hall effects, spin-orbit torques, and electrically controlled magnetism, the research has migrated from the initial low-temperature discovery in semiconductors to technologically more suitable room temperature manifestations in metallic systems. This path from exotic behavior to possible application continues to the present day and is reflected in the modified title of the book, which now explicitly highlights "spintronics," as its overarching scope. Exotic topics of today, for example, pertaining to topological properties, such as skyrmions, topological insulators, or even elusive Majorana fermions, may become suitable platforms for the spintronics of tomorrow. Impressive progress has been seen in the last decade in the field of spin caloritronics, which has evolved from a curious prediction 30 years ago to a vibrant field of research.

Since the first edition, there has been a significant evolution in material systems displaying spin-dependent phenomena, making it difficult to cover even the key developments in a single volume. The initially featured chapter on graphene spintronics is now complemented by a chapter on the spin-dependent properties of a broad range of two-dimensional materials that can form a myriad of heterostructures coupled by weak van der Waals forces and support superconductivity or ferromagnetism even in a single atomic layer. Exciting developments have also been seen in the field of complex oxide heterostuctures, where the non-trivial properties are driven by the interplay between the electronic, spin, and structural degrees of freedom. A particular example is the magnetism emerging in two-dimensional electron gases at oxide interfaces composed of otherwise non-magnetic constituents. The updated structure of a significantly expanded book reflects various materials developments and it is now thematically divided into three volumes, each based on broadly defined metallic and semiconductor systems or their nanoscale and applied aspects.

Spintronics becomes more and more attractive as a viable platform for propelling semiconducting technology beyond its current limits. Various schemes have been proposed to enhance the functionalities of the existing technologies based on the spin degree of freedom. Among them is the voltage control of magnetism, exploiting the non-volatile performance of ferromagnet-based devices in conjunction with their low-power operation. Another approach is utilizing spin currents carried by magnons to transport and process information. Magnon spintronics involves interesting fundamental physics and offers novel spin wave-based computing technologies and logic circuits. Optical control of magnetism is another approach, which has attracted a lot of attention due to the recent discovery of the all-optical switching of magnetization and its realization at the nanoscale. Chapters on these subjects are included in the new edition of the book.

Nearly nine decades after the discovery of superconducting proximity effects by Ragnar Holm and Walther Meissner, several new chapters now explore how a given material can be transformed through proximity effects whereby it acquires the properties of its neighbors, for example, becoming superconducting, magnetic, topologically non-trivial, or with an enhanced spin-orbit coupling. Such proximity effects not only complement the conventional methods of designing materials by doping or functionalization but can also overcome their various limitations and enable yet more unexplored spintronic applications.

We are grateful both to the authors who set aside their many priorities and contributed new chapters, which have significantly expanded the scope of this book, as well as to those who patiently provided valuable updates to their original chapters and kept this edition even more timely. The completion of the second edition was again greatly facilitated by Verona Skomski, who tirelessly collected authors' contributions and assisted their preparation for the submission to the publisher. We acknowledge the support of NSF-DMR, NSF-MRSEC, NSF-ECCS, SRC, DOE-BES, US ONR which, through the support of our research and involvement in spintronics, has also enabled our editorial work. We are thankful to our families for their support, patience, and understanding during extended periods of time when we remained focused on the completion of this edition.

Evgeny Y. Tsymbal
Department of Physics and Astronomy,
Nebraska Center for Materials and Nanoscience, University of Nebraska,
Lincoln, Nebraska 68588, USA

Igor Žutić
Department of Physics, University at Buffalo,
State University of New York, Buffalo, New York 14260, USA

About the Editors

Evgeny Y. Tsymbal is a George Holmes University Distinguished Professor at the Department of Physics and Astronomy of the University of Nebraska-Lincoln (UNL), and Director of the UNL's Materials Research Science and Engineering Center (MRSEC). He joined UNL in 2002 as an Associate Professor, was promoted to a Full Professor with Tenure in 2005 and named a Charles Bessey Professor of Physics in 2009 and George Holmes University Distinguished Professor in 2013. Prior to his appointment at UNL, he was a research scientist at University of Oxford, United Kingdom, a research fellow of the Alexander von Humboldt Foundation at the Research Center-Jülich, Germany, and a research scientist at the Russian Research Center "Kurchatov Institute," Moscow. Evgeny Tsymbal's research is focused on computational materials science aiming at the understanding of fundamental properties of advanced ferromagnetic and ferroelectric nanostructures and materials relevant to nanoelectronics and spintronics. He has published over 230 papers, review articles, and book chapters and presented over 180 invited presentations in the areas of spin transport, magnetoresistive phenomena, nanoscale magnetism, complex oxide heterostructures, interface magnetoelectric phenomena, and ferroelectric tunnel junctions. Evgeny Tsymbal is a fellow of the American Physical Society, a fellow of the Institute of Physics, UK, and a recipient of the UNL's College of Arts & Sciences Outstanding Research and Creativity Award (ORCA). His research has been supported by the National Science Foundation, Semiconductor Research Corporation, Office of Naval Research, Department of Energy, Seagate Technology, and the W. M. Keck Foundation.

Igor Žutić received his Ph.D. in theoretical physics at the University of Minnesota, after undergraduate studies at the University of Zagreb, Croatia. He was a postdoc at the University of Maryland and the Naval Research Lab. In 2005 he joined the State University of New York at Buffalo as an Assistant Professor of Physics and got promoted to an Associate Professor in 2009 and to a Full Professor in 2013. He proposed and chaired Spintronics 2001: International Conference on Novel Aspects of Spin-Polarized Transport and Spin Dynamics, at Washington DC. Work with his collaborators spans a range of topics from high-temperature superconductors, Majorana fermions, proximity effects, van der Waals materials, and unconventional magnetism, to the prediction and experimental realization of spin-based devices that are not limited to magnetoresistance. He has published over 100 refereed articles and given over 150 invited presentations on spin transport, magnetism, spintronics, and superconductivity. Igor Žutić is a recipient of the 2006 National Science Foundation CAREER Award, the 2019 State University of New York Chancellor's Award for Excellence in Scholarship and Creative Activities, the 2005 National Research Council/American Society for Engineering Education Postdoctoral Research Award, and the National Research Council Fellowship (2003–2005). His research is supported by the National Science Foundation, the Office of Naval Research, the Department of Energy, Office of Basic Energy Sciences, the Defense Advanced Research Project Agency, and the Airforce Office of Scientific Research. He is a fellow of the American Physical Society.

Contributors

Anthony S. Arrott
Department of Physics
Simon Fraser University
Burnaby, British Columbia, Canada

Agnès Barthélémy
Unité Mixte de Physique
Centre National de la Recherche
    Scientifique-Thales
Université Paris-Sud
Palaiseau, France

Jack Bass
Department of Physics and Astronomy
Michigan State University
East Lansing, Michigan

Kirill D. Belashchenko
Department of Physics and Astronomy
Nebraska Center for Materials and
    Nanoscience
University of Nebraska
Lincoln, Nebraska

Manuel Bibes
Unité Mixte de Physique
Centre National de la Recherche
    Scientifique-Thales
Université Paris-Sud
Palaiseau, France

Andrii V. Chumak
Fachbereich Physik
Technische Universität
    Kaiserslautern
Kaiserslautern, Germany

Matthias Eschrig
Department of Physics, Royal Holloway
University of London
Egham, United Kingdom

Albert Fert
Unité Mixte de Physique
Centre National de la Recherche
    Scientifique-Thales
Université Paris-Sud
Palaiseau, France

Erol Girt
Department of Physics
Simon Fraser University
Burnaby, British Columbia, Canada

Bretislav Heinrich
Department of Physics
Simon Fraser University
Burnaby, British Columbia, Canada

Mark Johnson
Materials Physics Division
Naval Research Laboratory
Washington, DC

Alexey V. Kimel
Radboud University
Institute for Molecules and
    Materials
Nijmegen, the Netherlands

Andrei Kirilyuk
Radboud University
Institute for Molecules and
    Materials
Nijmegen, the Netherlands

Hitoshi Kubota
National Institute of Advanced
    Industrial Science and
    Technology
Spintronics Research
Tsukuba, Japan

Patrick R. LeClair
Department of Physics and
    Astronomy and MINT
The University of Alabama
Tuscaloosa, Alabama

Aurélien Manchon
Physical Science and Engineering
    Division
King Abdullah University of Science
    and Technology (KAUST)
Thuwal, Saudi Arabia

Jagadeesh S. Moodera
Francis Bitter Magnet Laboratory
Massachusetts Institute of
    Technology
Cambridge, Massachusetts

Pavlo Omelchenko
Department of Physics
Simon Fraser University
Burnaby, British Columbia, Canada

Theo Rasing
Radboud University
Institute for Molecules and Materials
Nijmegen, the Netherlands

Tiffany S. Santos
Western Digital Corporation
San Jose, California

Yoshishige Suzuki
Department of Materials
Engineering Science
Osaka University
Osaka, Japan

Maxim Tsoi
Department of Physics
The University of Texas at Austin
Austin, Texas

Evgeny Y. Tsymbal
Department of Physics and
    Astronomy
Nebraska Center for Materials and
    Nanoscience
University of Nebraska
Lincoln, Nebraska

Hyunsoo Yang
Department of Electrical and
    Computer Engineering, and
    NUSNNI
National University of Singapore
Singapore

Shinji Yuasa
National Institute of Advanced
    Industrial Science and
    Technology (AIST)
Tsukuba, Japan

Shufeng Zhang
Department of Physics
University of Arizona
Tucson, Arizona

Section I
Introduction

1

Historical Overview
From Electron Transport in Magnetic Materials to Spintronics

Albert Fert

1.1 INTRODUCTION

Spintronics is now an important field of research with major applications in several technologies. Its development has been triggered by the discovery [1, 2] of giant magnetoresistance (GMR) in 1988. The basic concept of spintronics is the manipulation of spin-polarized currents, in contrast to mainstream electronics in which the spin of the electron is ignored. Adding the spin degree of freedom provides new effects, new capabilities, and new functionalities. Spin-polarized currents can be generated by exploiting the influence of the spin on the transport properties of the electrons in ferromagnetic conductors. This influence, first suggested by Mott [3], had been experimentally demonstrated and theoretically described in early works [4, 5] more than 10 years before the discovery of the GMR. The GMR was the first step on the road of the utilization of the spin degree of freedom in magnetic nanostructures. Its application to the read heads of hard disks greatly contributed to the fast rise in the density of stored information and led to the extension of hard disk technology to consumer electronics. Then, more development and intensive research revealed many other phenomena related to the control and manipulation of spin-polarized currents. Today, the field of spintronics is expanding considerably, with very promising new axes, such as the manipulation of magnetic moments and the generation of microwaves by spin transfer, spintronics with semiconductors, molecular spintronics, the spin Hall effect (SHE), the quantum spin Hall effect (QSHE), and single-electron spintronics for quantum computing. In this chapter, I will tell the story of this development from the early experiments on spin-dependent conduction in ferromagnets to the emerging directions of today.

1.2 SPIN-DEPENDENT CONDUCTION IN FERROMAGNETS AND EARLY EXAMPLES OF SPIN TRANSPORT EXPERIMENTS

GMR and spintronics take their roots from previous research on the influence of spin on electrical conduction in ferromagnetic metals [3–5]. The spin dependence of the conduction can be understood from the typical band structure of a ferromagnetic metal, which is shown in Figure 1.1a. Due to the splitting between the energies of the "majority spin" and "minority spin" directions (spin up and spin down in the usual notation), the electrons at the Fermi level, which carry the electrical current, are in different states and exhibit different conduction properties for opposite spin directions. This spin-dependent conduction was proposed by Mott [3] in 1936 to explain some features of the resistivity of ferromagnetic metals at the Curie temperature. However, in 1966, when I started my Ph.D. thesis, the subject

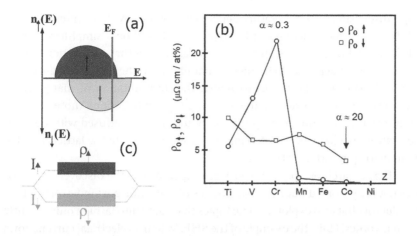

FIGURE 1.1 The basics of spintronics. (a) Schematic band structure of a ferromagnetic metal. (b) Schematic for spin-dependent conduction through independent spin-up and spin-down channels in the limit of negligible spin mixing ($\rho_{\uparrow\downarrow} = 0$ in the formalism of Ref. [4]). (c) Resistivities of the spin-up and spin-down conduction channels for nickel doped with 1% of several types of impurity (measurements at 4.2 K) [4]. The ratio α between the resistivities $\rho_{0\downarrow}$ and $\rho_{0\uparrow}$ can be as large as 20 (Co impurities) or smaller than 1 (Cr or V impurities). (After Fert, A. et al., *J. Phys. F Met. Phys.* 6, 849, 1976. With permission.)

was still almost completely unexplored. My supervisor, Ian Campbell, and I investigated the transport properties of Ni- and Fe-based alloys and, from the analysis of the temperature dependence of the resistivity and also from experiments on ternary alloys I describe in Section 1.3, we demonstrated the spin dependence of conduction in various metals and alloys. In particular, we showed that the resistivities of the two channels can be very different in metals doped with impurities presenting a strongly spin-dependent scattering cross-section [4]. In Figure 1.1c, I show the example of the spin-up and spin-down resistivities of nickel doped with 1% of different impurities. It can be seen that ratio α of the spin-down to spin-up resistivity can be as large as 20 for Co impurities or smaller than 1 for Cr or V impurities. This was consistent with the theoretical models developed at this time by Jacques Friedel for the electronic structures of impurities in metals. The two-current conduction was rapidly confirmed in other groups and, for example, extended to Co-based alloys by Loegel and Gautier [5].

The so-called two-current model [4] for conduction in ferromagnetic metals was worked out for the quantitative interpretation of the experiments described in the preceding paragraph. This model is based on a picture of spin-up and spin-down currents coupled with spin mixing, that is, by exchange of current between the spin-up and spin-down channels. Spin mixing comes from spin-flip scattering, mainly from electron-magnon scattering, which conserves the total spin of the electronic system but is a mechanism of current transfer between the two channels. It increases with temperature and tends to equalize partly the spin-up and spin-down currents at room temperature in most ferromagnetic metals [4]. The two-current model is the basis of spintronics today, but, except in very few publications

[6, 7] discussing the temperature dependence of the GMR, the interpretation of the spintronics phenomena is surprisingly based on a simplified version of the model neglecting spin mixing and assuming that conduction occurs through two independent channels in parallel, as in the sketch of Figure 1.1c. It should certainly be useful to revisit the interpretation of many recent experiments by considering the spin mixing contributions (note that the spin mixing mechanism by spin-flips should not be confused with the spin relaxation mechanism transferring spin accumulation to the lattice and due mainly to spin–orbit scattering).

The research on spin transport developed before the discovery of GMR has not only worked out the basic physics of spin transport in ferromagnetic conductors but also explored some topics that came into fashion only recently in spintronics. I take the example of the SHE. When an electrical current flows in a nonmagnetic conductor, the electrons of opposite spins are deflected in opposite transverse directions by spin–orbit interactions. With equal spin-up and spin-down currents, there is no charge accumulation and consequently no Hall voltage between edge contacts, but the deflections induce opposite spin accumulations on the edges of the conductor, as illustrated in Figure 1.2a. This is the SHE, already described in 1971 by D'yakonov and Perel' [8].

As spin accumulation by the SHE can be used to generate spin currents, the SHE is nowadays presented as a possible method for carrying out spintronics without magnetic materials, which explains the intense current research on this topic. In most recent experiments, the SHE is detected by breaking the symmetry between the opposite spin directions to obtain a transverse voltage, either by using a ferromagnetic metal for one of the Hall probes or by locally injecting a spin-polarized current (or even a pure spin current) from a ferromagnetic contact, the so-called inverse SHE detection. This can be done only in nanodevices fabricated by lithographic techniques [9–11]. Thirty years ago, it was not technically possible to fabricate such nanodevices, but, nevertheless, a precise determination of the SHE could be performed in other ways. In Figure 1.2b, we show the results published in 1981 [12] on the SHE induced by 5d impurities (Lu, Ta, Ir, Au) in Cu. The SHE is detected by adding 0.01% of Mn impurities and applying a magnetic field. It had previously been shown that applying a field with only Mn impurities in Cu does not give any significant contribution to the Hall effect but induces (by exchange scattering) a spin polarization of the current, which can be seen through the associated negative magnetoresistance (GMR-like effect). In brief, the spin polarization induced by exchange with dilute impurities of Mn replaces the spin injection of modern experiments. As the current polarization follows the polarization of the paramagnetic Mn impurities and is inversely proportional to the temperature (T), one obtains the variations of the Hall constant as $1/T$ seen in Figure 1.2b for different types of nonmagnetic impurities. The amplitude of the SHE can be characterized by the Hall angle, $\Phi_H = \rho_{xy}/\rho_{xx}$. From the results of Figure 1.2b, Fert et al. [12] derived Hall angles varying from -2.4×10^{-2} for CuLu to $+5.2 \times 10^{-2}$ for CuIr, with the typical change of sign between the beginning and the end of the 5d series predicted by a model of resonant scattering on impurity 5d states split by a spin–orbit interaction [12, 13]. Similar values of the SHE angle

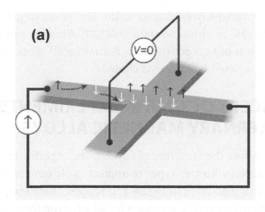

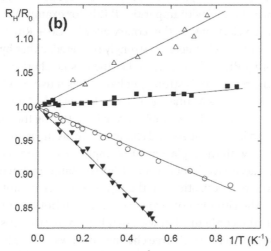

FIGURE 1.2 (a) Sketch illustrating the SHE in a nonmagnetic conductor. (b) Experimental results [12] on the SHE induced by resonant scattering on 5d levels (split by spin–orbit coupling) of nonmagnetic impurities in Cu (open triangles = Lu 0.013%, circles = Ta 0.023%, black triangle = Ir 0.19%). The Inverse SHE of the Cu alloys is revealed by adding ≈0.01% of Mn impurities and applying a field H to spin-polarize the current by exchange scattering on the spin-polarized Mn impurities. With a paramagnetic-like spin polarization proportional to H/T, the SHE is revealed through a contribution to the Hall constant proportional to T^{-1} with different slopes for different impurities (the squares represent measurement on Cu with only 0.015% of Mn and without 5d impurities, which shows the quasi-absence of $1/T$ contribution to the Hall effect from Mn alone). Note the characteristic change of sign of the SHE between the beginning (Lu) and the end (Ir) of the 5d series.

in Cu doped with Ir have again been found in "modern" SHE experiments with spin injection from lateral spin valves [13].

The SHE is not the only effect that had been already observed in the "prehistory" of spintronics before beginning the object of intense attention in the very recent years of the developments of spintronics. For example, the Dzyaloshinskii-Moriya Interactions (DMI) have been introduced for non-centrosymmetric magnetic compounds [14, 15], metallic spin glasses [16], and at the interface of magnetic films with heavy metals [17] well before being extensively exploited in the last years to generate magnetic skyrmions [18–19] or chiral domain wall [20]. Another example is the very early observation of current-induced domain wall motion by Berger [21], as it will be discussed later in this chapter.

1.3 CONCEPT OF GMR IN EXPERIMENTS ON TERNARY MAGNETIC ALLOYS

Twenty years before the discovery of GMR, some experiments with ferromagnetic metals doped with two types of impurities [4] were already anticipating the GMR. This is illustrated in Figure 1.3. Suppose, for example, that nickel is doped with impurities A (Co, for example), which strongly scatter the electrons of the spin-down channel with impurities B (Rh, for example), which strongly scatter the spin-up electrons. In the ternary alloy Ni(Co + Rh), that I call type 1, the electrons of both channels are strongly scattered, either by Co in one of the channels or by Rh in the other, so that there is no shorting by one of the channels and the resistivity is strongly enhanced, as illustrated in Figure 1.3a. In contrast, there is no such enhancement in alloys of type 2 doped with impurities A and B (Co and Au, for example) strongly scattering the electrons in the same channel and leaving the second channel open, as in Figure 1.3b.

GMR occurs with the replacement of impurities A and B in the ternary alloy with the successive layers A and B of the same magnetic metal in a multilayer. If the magnetizations of the layers A and B are antiparallel, this corresponds to the situation of strong scattering in both channels in alloys of type 1, while the configuration with parallel magnetizations corresponds to the situation with a relatively free channel in alloys of type 2. What is new with respect to the ternary alloys is the possibility of switching between high and low resistivity states applying a magnetic field and by simply changing the relative orientation of the magnetizations of layers A and B from antiparallel to parallel. However, the transport equations tell us that the relative orientation of layers A and B can be felt by the electrons only if their distance is smaller than the electron mean free path, that is, practically, if they are spaced by only a few nanometers. Unfortunately, in the 1970s, it was not technically possible to make multilayers with layers as thin as a few nanometers, and the discovery of the GMR had to wait until the development of sophisticated deposition techniques.

1.4 DISCOVERY OF THE GMR

In the mid-1980s, with the development of techniques such as molecular beam epitaxy (MBE), it became possible to fabricate multilayers composed of very thin individual layers and one could consider trying to extend the experiments on ternary alloys to multilayers. In addition, in 1986, I saw the beautiful Brillouin scattering experiments of Grünberg and coworkers [22] revealing the existence of antiferromagnetic interlayer exchange couplings in Fe/Cr multilayers. Fe/Cr appeared as a magnetic multilayered system in which it was possible to switch the relative orientation of the

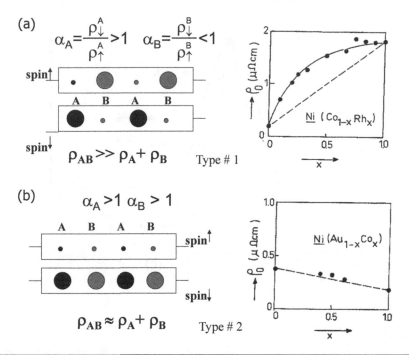

FIGURE 1.3 Experiments on ternary alloys based on the same concept as that of GMR [4]. The sketches illustrate the conduction by two channels in a ferromagnet doped with impurities A (black) and B (gray), the circles are at the scale of the scattering cross-sections of impurities A and B. (a) Schematic for the spin-dependent conduction in alloys with impurities of opposite scattering spin asymmetries ($\alpha_A = \rho_{A\downarrow}/\rho_{A\uparrow} > 1$, $\alpha_B = \rho_{B\downarrow}/\rho_{B\uparrow} < 1$, $\rho_{AB} \gg \rho_A + \rho_B$) and experimental results for Ni(Co$_{1-x}$Rh$_x$) alloys. (b) Same for alloys with impurities of similar scattering spin asymmetries ($\alpha_A = \rho_{A\downarrow}/\rho_{A\uparrow} > 1$, $\alpha_B = \rho_{B\downarrow}/\rho_{B\uparrow} > 1$, $\rho_{AB} \approx \rho_A + \rho_B$) and experimental results for Ni(Au$_{1-x}$Co$_x$) alloys. In GMR, the impurities A and B are replaced by multilayers, the situation of (a) corresponding to the antiparallel magnetic configurations of adjacent magnetic layers and (b) corresponding to parallel.

magnetization in adjacent magnetic layers from antiparallel to parallel by applying a magnetic field. In collaboration with the group headed by Alain Friederich at the Thomson-CSF company, I started the fabrication and investigation of Fe/Cr multilayers. In 1988, this led us to the discovery [1] of very large magnetoresistance effects that we called GMR (Figure 1.4a). Effects of the same type in Fe/Cr/Fe trilayers were obtained practically at the same time by Grünberg at Jülich [2] (Figure 1.4b). The interpretation of GMR is similar to that described above for the ternary alloys and is illustrated in Figure 1.4c. The first classical model of GMR was published in 1989 by Camley and Barnas [23], and I collaborated with Levy and Zhang on the first quantum model [24] in 1991.

I am often asked if I was expecting such large MR effects. My answer is yes and no: on one hand, a very large magnetoresistance could be expected from an extrapolation of the preceding results on ternary alloys, on the other hand, one could fear that the unavoidable structural defects of the multilayers, interface roughness, for example, might introduce spin-independent scatterings canceling the spin-dependent scattering inside the magnetic layers.

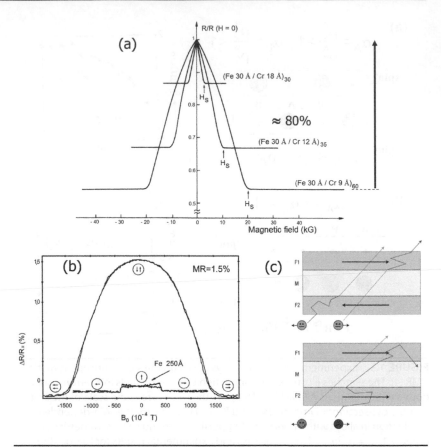

FIGURE 1.4   The first observations of giant magnetoresistance. (a) Fe/Cr(001) multilayers [1] (with the current definition of the magnetoresistance ratio, MR = $100[R_{AP} - R_P]/R_P$, MR = 85% for the Fe 3 nm/Cr 0.9 nm multilayer). (b) Fe/Cr/Fe trilayers [2]. (c) Schematic of the mechanism of GMR. In the parallel magnetic configuration (bottom), the electrons of one of the spin directions can easily go through all the magnetic layers, and the short circuit through this channel leads to a small resistance. In the antiparallel configuration (top), the electrons of each channel are slowed down on every second magnetic layer and the resistance is high. (After Chappert, C. et al., *Nat. Mater.* 6, 813, 2007. With permission.)

Fortunately, the scattering by the roughness of the interfaces was also spin dependent and added its contribution to the "bulk" (the "bulk" and interface contributions can be separately derived from GMR experiments with the current perpendicular to the layers).

1.5 GOLDEN AGE OF GMR

Rapidly, the papers reporting the discovery of GMR attracted attention for their fundamental interest, as well as for the many possibilities for application, and research on magnetic multilayers and GMR became a very hot topic. In my team, as well as in the small but rapidly increasing community working in the field, we had the exalting impression of exploring a wide virgin country, with so many amazing surprises in store. On the experimental

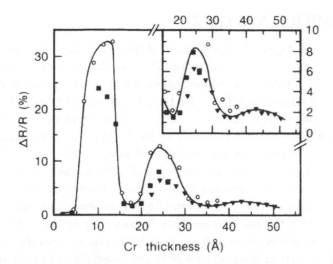

FIGURE 1.5 Oscillatory variation of the GMR ratio of Fe/Cr multilayers as a function of the thickness of the Cr layers. (After Parkin, S.S.P. et al., *Phys. Rev. Lett.* 64, 2304, 1990. With permission.)

side, two important results were published in 1990. Parkin et al. [25] demonstrated the existence of GMR in multilayers made by the simpler and faster technique of sputtering (Fe/Cr, Co/Ru, and Co/Cr) and found that the oscillatory behavior of the GMR was due to the oscillations of the interlayer exchange as a function of the thickness of the nonmagnetic layers (see Figure 1.5). Also in 1990, Shinjo and Yamamoto [26], as well as Dupas et al. [27], demonstrated that GMR effects can be found in multilayers without antiferromagnetic interlayer coupling but composed of magnetic layers of different coercivities. Another important result, in 1991, was the observation of large and oscillatory GMR effects in Co/Cu, which became an archetypical GMR system. The first observations were obtained at Orsay [28] with multilayers prepared by sputtering at Michigan State University and at about the same time at IBM [29].

Also in 1991, Dieny et al. [30] reported the first observation of GMR in spin valves, that is, trilayered structures in which the magnetization of one of the two magnetic layers is pinned by coupling with an antiferromagnetic layer, while the magnetization of the second one is free. The magnetization of the free layer can be reversed by very small magnetic fields, so the concept is now used in many devices. The various applications of GMR are described in other chapters of this book. Its application to the read heads of hard disks is certainly the most important [31, 32]. The GMR, by providing a sensitive and scalable read technique, has led to an increase of the areal recording density by more than two orders of magnitude (from ≈ 1 to ≈ 600 Gbit/in.^{2} in 2009). This increase opened the way both to unprecedented drive capacities (up to 1 terabyte) for video recording or backup and to smaller hard disk drive (HDD) sizes (down to 0.85 inch disk diameter) for mobile appliances like ultra-light laptops or portable multimedia players. GMR sensors are also used in many other types of application, mainly in the automotive industry and biomedical technology [33].

1.6 TMR RELAYS GMR

An important stage in the development of spintronics has been the research on the tunneling magnetoresistance (TMR) of the magnetic tunnel junctions (MTJs). The MTJs are tunnel junctions with ferromagnetic electrodes, and their resistance is different for the parallel and antiparallel magnetic configurations of their electrodes. Some early observations of small TMR effects had been already reported by Jullière [34] in 1975 and Maekawa and Gäfvert [35] in 1982, but they were found to be hardly reproducible and actually could not be really reproduced until 1995. It was at this time only that large (≈20%) and reproducible effects were obtained by Moodera's and Miyasaki's groups on MTJ with a tunnel barrier of amorphous alumina [36].

After 1995, the research on TMR became very active, and the most important step was the transition from MTJ with an amorphous tunnel barrier (alumina) to single-crystal MTJ and especially MTJ with an MgO barrier. In the first results with MgO, the TMR ratio was only slightly larger than that found with alumina barriers and similar electrodes [37]. The important breakthrough came in 2004 at Tsukuba [38] and IBM [39] where it was found that very large TMR ratios, up to 200% at room temperature, could be obtained with MgO MTJ of a very high structural quality, as illustrated in Figure 1.6. Since 2004, these results have been progressively improved [40], and TMR ratios up to 1000% have been now reached [41].

The large TMR of MTJ with single-crystal tunnel barriers, such as MgO(001), come from symmetry selection [42–45]. This is illustrated in Figure 1.6c where one sees the calculated density of states (DOS) of evanescent wave functions of different symmetries, Δ_1, Δ_5, etc., and the much slower decay of the symmetry Δ_1 in an MgO(001) barrier between Co electrodes [45]. The key point is that, at least for interfaces of high quality, an evanescent wave function of a given symmetry is connected to the Bloch functions of the same symmetry and same spin direction at the Fermi level of the electrodes. For Co electrodes, the Δ_1 symmetry is well represented at the Fermi level in the majority spin direction sub-band and not in the minority one. Consequently, a good connection between the majority spin direction sub-bands of the Co electrodes by the slowly decaying channel Δ_1 can be obtained only in their parallel magnetic configuration, which explains the very high TMR. Other types of barriers can select symmetries other than the symmetry Δ_1 selected by MgO(001). For example, a $SrTiO_3$ barrier predominantly selects evanescent wave functions of Δ_5 symmetry, which are well connected to minority spin states of cobalt [46]. This explains the negative effective spin polarization of cobalt observed in $SrTiO_3$-based MTJ [47].

The high spin polarization obtained by symmetry selection gives a very good illustration of what is under the word "spin polarization" in a spintronic experiment. There is no intrinsic spin polarization of a magnetic conductor. In an MTJ, the effective polarization is related to the symmetry selected by the barrier and, depending on the barrier, can be positive or negative, large or small. In the same way, as we have seen in Section 1.2, the spin polarization

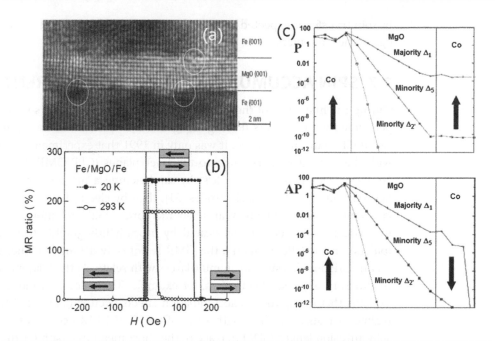

FIGURE 1.6   TMR of MTJ with MgO barrier. (a) Electron microscopy image of an Fe(001)/MgO(001)/Fe(001) MTJ [38]. (b) MR = $(R_{max} - R_{min})/R_{min}$ measured by Yuasa et al. [38] for a Fe(001)/MgO(001)/Fe(001) MTJ. (c) Physics of the TMR illustrated by the decay of evanescent electronic waves of different symmetries in an MgO(001) layer between cobalt electrodes calculated by Zhang and Butler [45]. The Δ_1 symmetry of the slow decay tunneling channel is well represented at the Fermi level of the spin conduction band of cobalt for the majority spin direction and not for the minority spin direction, so that a spin conserving connection through the channel Δ_1 is possible in the parallel configuration. This explains the very high TMR of this type of junction.

of metallic conduction depends strongly on the spin dependence of the scattering by impurities, as shown in Figure 1.1b.

The MTJ with MgO barriers are today the most efficient "spin polarizers" and are used in many experiments and devices. However, other directions are also explored to obtain even larger TMR. One of these directions is the research of half-metallic materials, in other words, materials of a metallic character for one of the spin directions and insulating for the other, which means a 100% polarization. In some oxides and Heusler alloys close to half-metallicity have been found [48, 49]. Spin filtering by a ferromagnetic barrier is another interesting concept [50].

From a technological point of view, the interest of the MTJ with respect to the metallic spin valves comes from the vertical direction of the current and from the resulting possibility of a reduction of the lateral size to a submicronic scale by lithographic techniques. The MTJs are at the basis of a new concept of magnetic memory called magnetic random access memory (MRAM), which is expected to combine the short access time of the semiconductor-based RAMs and the nonvolatile character of the magnetic memories [32]. In the first MRAMs, put onto the market in 2006, the memory cells are MTJs with an alumina barrier. The magnetic fields generated by "word" and "bit" lines are used to switch their magnetic configuration. The next generation of MRAM, based on MgO tunnel junctions and switching by spin transfer, is expected to have a much stronger impact on the technology of the computers.

1.7 SPIN ACCUMULATION AND SPIN CURRENTS

During the first years of the research on GMR, the experiments were only performed with currents flowing along the layer planes—in the geometry called CIP (current in plane). It was only in 1991 that experiments of GMR with the current perpendicular to the layer planes (CPP-GMR) begun to be performed. This was done first by sandwiching a magnetic multilayer between superconducting electrodes [51], then by electrodepositing multilayers into the pores of a polycarbonate membrane [52], and, more recently, in vertical nanostructures fabricated by e-beam lithographic techniques (pillars). In the CPP geometry, the GMR is not only definitely higher than that in CIP but also subsists in multilayers with relatively thick layers, up to the micron range, see Figure 1.7a, for example. The Valet–Fert model [53] explains that owing to spin accumulation effects occurring in the diffusive regime of transport, the length scale of the CPP-GMR becomes the long spin diffusion length (SDL) in place of the short mean free path for the CIP geometry.

Actually, the CPP-GMR clearly revealed the spin accumulation effects, which govern the propagation of a spin-polarized current through a succession of magnetic and nonmagnetic materials and play an important role in all the current developments of spintronics. The diffusion current induced by the accumulation of spins at the magnetic/nonmagnetic interface is the mechanism driving a spin-polarized current in a nonmagnetic conductor to a long distance from the interface, well beyond the ballistic range (i.e., well beyond the mean free path) up to the distance of the SDL. In carbon molecules, for example, the SDL largely exceeds the micron range and, as

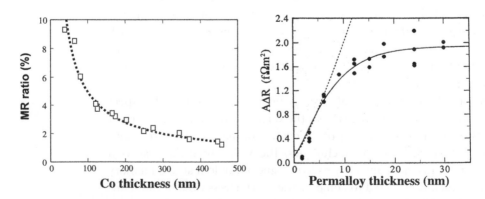

FIGURE 1.7   Experimental results illustrating the thickness dependence of the CPP-GMR and its relation to the SDL. (a) CPP-GMR of Co/Cu/Co multilayers electrodeposited into pores of membranes [52]. With the long SDL of Co (60 nm), the GMR can be seen up to Co thicknesses in the μm range. (b) CPP-GMR of Py/Cu/Py multilayers [51]. With a SDL of only 5 nm, the resistance change between P and AP configuration flattens out, for permalloy layers thicker than 10 nm and, as the resistance still increases, the GMR ratio decreases to zero.

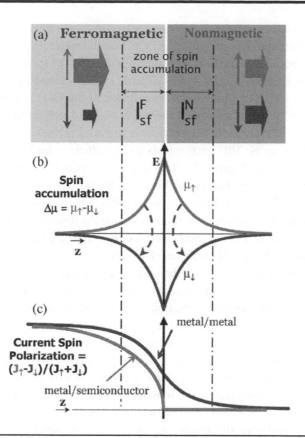

FIGURE 1.8 Schematic representation of the spin accumulation at an interface between a ferromagnetic metal and a nonmagnetic layer. (a) Incoming and outgoing spin-up and spin-down currents. (b) Splitting of the chemical potentials, $E_{F\uparrow}$ and $E_{F\downarrow}$, in the interface region with arrows symbolizing the spin-flips induced by the spin accumulation (out of equilibrium distribution), spin-flips controlling the progressive depolarization of the current. With an opposite direction of the current, the inversion of the spin accumulation and the opposite spin-flips progressively polarize the current. (c) Variation of the current spin polarization when there is an approximate balance between the spin-flips on both sides (metal/metal curve) and when the spin-flips on the right side are predominant (metal/semiconductor curve in the situation without spin-dependent interface resistance). (After Chappert, C. et al., *Nat. Mater.* 6, 813, 2007. With permission.)

we see in Section 1.11, strongly spin-polarized currents can be transported throughout long carbon nanotubes (CNTs) [54].

The physics of the spin accumulation occurring when an electron flux crosses an interface between a ferromagnetic and a nonmagnetic material is explained in Figure 1.8 for a simple case of single interface (no interface resistance, no band bending, and single polarity). To sum up, there is a broad zone of spin accumulation, which extends on both sides of an F/N interface and in which the spin polarization of the injected current subsists in N to a distance of the order of the SDL (or start at this distance for the opposite current direction). The spin polarization just at the interface depends on the proportion of the depolarizing (polarizing) spin-flips induced by the spin accumulation in F and N. If the DOS and the spin relaxation times are similar on both sides, there is a balanced proportion of spin-flips on the N and F sides, and the current is significantly spin-polarized in N (as for the metal/metal curve in Figure 1.8c). In contrast, if the DOS is much smaller or the spin relaxation time much shorter in N, the depolarization (or polarization for spin extraction) occurs mainly in F and the current is negligibly spin-polarized in N. This corresponds to the situation for the metal/semiconductor curve in Figure 1.8c. This difficulty, when F is a metal and N a semiconductor, was first described by Schmidt et al. [55] and called "conductivity mismatch." The problem can be solved by inserting a spin-dependent interface resistance (tunnel junction) to induce the discontinuity of the spin accumulation, which increases the Fermi energy spitting and the proportion of spin-flips on the side N of the interface [56, 57].

The physics of spin accumulation and spin-polarized currents, which plays an important role in most fields of spintronics today, can be described by a new type of transport equation [53, 57, 58], often called drift/diffusion equations, in which the electrical potential is replaced by a spin- and position-dependent electrochemical potential. These equations can also be extended to consider band bending and high current density effects [59]. They have been frequently applied to the general situation of multi-contact systems with the interplay of the spin accumulation effects at different contacts [57, 60, 61]. A standard structure is a nonmagnetic lateral channel, metal, semiconductor, or carbon-based conductor, between a spin-polarized source F1 and a spin-polarized drain F2. The output "spin signal" is the resistance or voltage difference between the different magnetic configurations of the contacts. Very generally, one distinguishes local geometries in which the output signal is measured between the source and the drain (Figure 1.9a), and nonlocal geometries in which the output signal is between two other contacts (Figure 1.9b). In the nonlocal geometry of Figure 1.9b, where the current flows to the left with a spin accumulation spreading to the right (Figure 1.9c), the output voltage, V, reflects the spin accumulation splitting at the magnetic contact, F2, and is inverted when one reverses the magnetization of F2. Note that, in the right part of the channel, the opposite gradients of spin-up and spin-down, chemical potentials generate opposite spin-up and spin-down currents, or, in other words, a pure spin current without charge current.

In Figure 1.9e, we show an example of the application of the drift-diffusion equations to the problem of spin transport in a nonmagnetic lateral channel between the spin-polarized source and drains connected to the channel by a spin-dependent interface resistance (MTJ, for example). The calculation is performed for the local and nonlocal detections of Figure 1.9a and b, and also for the local detection of Figure 1.9d in which the channel does not extend outside the length L between the contacts (confined spin relaxation). The output signal, $\Delta R = (V_{AP} - V_P)/I$, is plotted as a function of the ratio of the mean interface resistance R_T^* (defined in the caption and supposed to be the same for the contacts F1 and F2) to the spin resistance of the channel (R_N = product

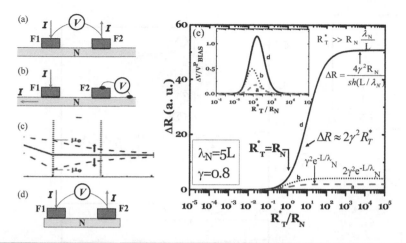

FIGURE 1.9 (a) and (b) Geometries for lateral spin transport between spin-polarized sources and drains with, respectively, local and nonlocal detection of the "spin signal." (c) Spin-up and spin-down electrochemical potential profiles for the nonlocal geometry of (b). The nonlocal signal, V, is related to the splitting (spin accumulation) under the nonlocal contact. (d) Geometry similar to (a) but with a lateral channel not extending outside the contacts. (e) Output spin signal, $\Delta R = [V_{AP} - V_P]/I$, for geometries of (a), (b), and (d) calculated from drift/diffusion equations as a function of R_T^*/R_N with the following definitions: L = length between F1 and F2, $R_{T\uparrow(\downarrow)} = 2[1 - (+)\gamma]$ R_T^* = tunnel resistances between the channel and F1 or F2, $R_N = \rho_N\lambda_N/S_N$ = channel spin resistance (ρ_N = resistivity, λ_N = SDL, S_N = section), $R_F = \rho_F^* \lambda_F/S_F \ll R_N$ ($\gamma = 0.8$, $\lambda_N = 5L$ in the calculation). ΔR flattens out at a value of the order of R_N for $R_T^* \gg R_N$ in the open structures of (a) and (b) (dotted and dashed lines). For the geometry of (d), ΔR does not flatten out for $R_T^* > R_N$ and goes on increasing proportionally to R_T^* before reaching saturation at the level $4\gamma^2 R_N/sh(L/\lambda_N)$, well above R_N, for $R_T^* \gg R_N\lambda_N / L$ (solid line).

of the resistivity by the SDL λ_N divided by the section; the spin relaxation per unit length of N is inversely proportional to R_N) for the three geometries of Figure 1.9a–d (other details on the parameters are in the caption of the figure). The calculation is performed in a situation of "conductivity mismatch," that is, $R_F \ll R_N$, so that, for the three curves, the spin signal turns out when R_T^* exceeds R_N. Then, the difference between the three curves illustrates an important rule: the scale of ΔR is the "resistance" controlling the spin relaxation. For a and b with $R_T^* \gg R_N$, the relaxation in N between $-$ and $+\lambda_N$ is larger ($\propto 1/R_N$) than the spin escape through R_T^* ($\propto 1R_T^*$) and ΔR flattens out at a level of the order of $R_N \ll R_T^*$ ($\approx 2\gamma^2 R_N \exp(-L/\lambda_N)$ for local detection and $\approx \gamma^2 R_N \exp(-L/\lambda_N)$ for nonlocal). In contrast for d with $R_N \ll R_T^* \ll R_N\lambda_N / L$, the relaxation is controlled by the spin escape through R_T^* (rather than by the relaxation inside L) and ΔR goes on increasing in the proportion of R_T^* well above the saturation value $\approx R_N$ of the open structures a and b. This regime, also described by a dwell time τ_n shorter than the spin lifetime τ_{sf}, is illustrated by the large MR observed in CNTs (60%–70%, ΔR in the MΩ range well above the spin resistance of the nanotube) [54]. Then, for $R_T^* \gg R_N\lambda_N/L$, the relaxation inside L becomes predominant and ΔR flattens out at the level $4\gamma^2 R_N/sh(\lambda_N/L) \approx 4\gamma^2 R_N\lambda_N/L$ (well above the saturation for curves a and b in

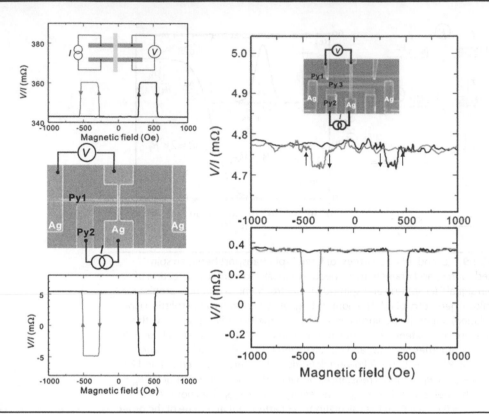

FIGURE 1.10   Experimental examples of lateral spin transport in permalloy/silver nanodevices [63]. (a) Local spin-valve measurements at 77 K. A sketch of the nanodevice is shown above the experimental curves. (b) Nonlocal spin-valve measurements at 77 K. An image of the nanodevice is shown on the top. (c) Nonlocal spin-valve measurements at 77 K (bottom) and room temperature (middle) on a nanodevice similar to that of (b) but with an additional permalloy contact (Py3 on the image) between the injection and detection contacts. The signal reduction with respect to devices without additional contact is due to the spin current escaping through this contact and relaxing in the additional channel. The reduction is inversely proportional to the spin resistance of the additional channel and, for example, becomes larger and larger as one goes from Cu to Au and Py. (After Kimura, T. et al., *Phys. Rev. B* 72, 014461, 2005. With permission.)

Figure 1.9e). For confined relaxation with nonlocal detection, $4\gamma^2 R_N/\text{sh}(L/\lambda_N)$ becomes $3\gamma^2 R_N/\text{sh}(L/\lambda_N)$, as shown by Jaffrès et al. [57]. The different rules described above can explain the main features of the results found in various structures with metals, semiconductor, or carbon-based materials.

In Figure 1.10, we show examples of lateral spin transport in the simple situation of a metallic channel and ohmic contacts with metallic electrodes [58, 60, 62, 63]. The output signal ΔR, at the typical scale of the channel spin resistance, is usually of the order of a few mΩ, which corresponds to local or nonlocal output voltages in the μV range for current densities in the 10^7 A/cm^2 range (the ratio to the bias voltage can reach 10%). With a metallic channel but with tunnel contacts, most experiments have been performed in the geometry of Figure 1.9a and b, so that the output signals are larger but do not significantly exceed the spin resistance of a metallic channel ($\Delta R \approx 10^{-3} - 10^{-4} R_{\text{bias}}$). For the experiments with a semiconducting channel, generally performed with tunnel or Schottky contacts to solve the "conductivity mismatch," the review article of Jonker and Flatté [64] concludes that the

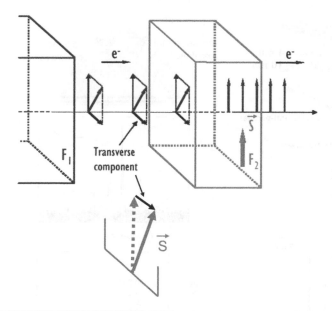

FIGURE 1.11   Illustration of the spin-transfer concept introduced by Slonczewski [67]. A spin-polarized current is generated by a first magnetic layer, F, with an obliquely oriented spin polarization with respect to the magnetization axis of a second layer F2. When it goes through F2, the exchange interaction aligns its spin polarization along the magnetization axis. The exchange interaction being spin conserving, the transverse spin polarization lost by the current is transferred to the total spin of F2, which can also be described by a spin-transfer torque acting on F2. (After Chappert, C. et al., *Nat. Mater.* 6, 813, 2007. With permission.)

contrast between the conductance of the parallel and antiparallel configuration never exceeds the range 0.1%–1%. With CNTs and tunnel contacts in the range of 100 MΩ with $La_{2/3}Sr_{1/3}MnO_3$ (LSMO) electrodes [54], some experiments in the geometry of Figure 1.9d have led to output signals $\Delta R \approx$ 90 MΩ (60% of the total resistance of the two contacts), which appears to correspond to the situation of the corresponding curve in Figure 1.9e with $\Delta R \approx R_T \gg R_{CNT}$. This shows the possibility of obtaining large spin signals at the level of tunnel resistances by preventing the spin relaxation in branches of small spin resistance. This also reflects the potential of carbon-based materials like CNT [54, 65] or graphene [66], which can combine a long spin lifetime, due to the small spin–orbit of carbon, with the advantage of their very high electron velocity for short dwell times.

1.8 SPIN TRANSFER

Spin transfer provides the means to manipulate the magnetic moment of a ferromagnetic body without applying any magnetic field but only through the transfer of the spin angular momentum from a spin-polarized current. The concept that was introduced by Slonczewski [67] and also appears in studies by Berger [68] is illustrated in Figure 1.11. As described in the caption of the figure, the transfer of the transverse component of a spin current to the "free" magnetic layer, F2, can be described by a torque acting on the

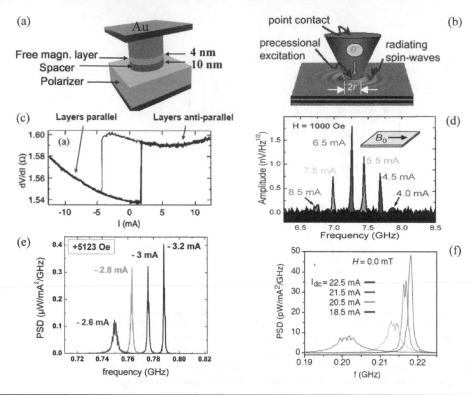

FIGURE 1.12   (a) and (b) Nanodevices for spin-transfer experiments ("pillar" and nanocontacts [72]). (c) Differential resistance vs. current for a Co/Cu/Co "pillar" with current-induced switching between parallel and antiparallel magnetic configurations [70]. (d) Microwave spectra obtained by current injection into a CoFe/Cu/NiFe trilayer through a nanocontact at different values of current [72]. (e) Microwave power spectra obtained at different values of current from vortex gyrations in the permalloy layer of a CoFeB/MgO/Permalloy MTJ [73]. (f) Progressive synchronization of the gyration of four vortices appearing at increasing current in the microwave spectrum of a device for current injection into a Co/Cu/permalloy trilayer through four nanocontacts. (After Ruotolo, A. et al., *Nat. Nanotechnol.* 4, 528, 2009. With permission.)

magnetic moment of F2. This torque can induce a switching or reorientation of this magnetic moment, which can be described as a transport of non-volatile magnetization by an electrical wire. In a second regime, generally in the presence of an applied field, the spin-transfer torque generates periodic motions of the magnetic moment and voltage oscillations in the microwave frequency range. Actually, the concept of spin transfer has opened a new direction in spintronics: GMR or TMR allowed us to electrically detect a magnetic configuration, while spin transfer now enables us to create a magnetic configuration with an electrical current.

The first evidence that spin transfer can work was indicated by the experiments on spin injection through point contacts by Tsoi et al. [69], but a clearer understanding came later from measurements [71, 72] performed on pillar-shaped metallic trilayers and tunnel junctions (Figure 1.12a) or by injecting a current into a magnetic trilayer through the type of nanocontact represented in Figure 1.12b [72]. In Figure 1.12c, we show an example of magnetic switching detected by GMR in a metallic pillar. Figure 1.12d represents the microwave power spectrum of the voltage oscillations generated by current-induced magnetic precessions in a structure of a nanocontact type. Actually, the spin-transfer torque can be used not only to induce switchings and precessions of magnetic moments but also more sophisticated magnetic excitations. Figure 1.12e shows the microwave power spectrum generated by the gyration of the core of a magnetic vortex in one of the electrodes of an MTJ [73]. Alternatively, the spin-transfer torque can also be used to move domain walls in nanowires. The many different spin-transfer experiments raise a great variety of interesting physical problems mixing spin transport, spin dynamics, and micromagnetism.

The spin-transfer phenomena will certainly have important applications. Switching by spin transfer will be used in the next generation of MRAMs (called STT-RAMs) for a more precise addressing and a smaller energy consumption. The generation of oscillations in the microwave frequency range will lead to the design of spin-transfer oscillators (STOs). One of the main interests of the STOs is their tunability, that is the possibility of rapidly changing their frequency by tuning a DC current. Their disadvantage is the very small microwave power of an individual STO. The microwave power obtained with MTJs is not too far from what is needed for applied devices, but the linewidth of their emission is still much too broad. Today it seems that the excitation of magnetic vortices in MTJs leads the race for application as they combine power and narrow linewidth, see Figure 1.12e [73]. One interesting challenge, to increase the emitted power and reduce the linewidth, is the synchronization of a set of STOs. Figure 1.12f shows an example of the synchronization of the gyration of a square of four vortices [74].

1.9 SPINTRONICS TODAY AND TOMORROW

Nowadays spintronics is expanding in so many directions that we will only list and summarize the main axes.

There is a lot of research [75] on the axis "spintronics with semiconductors" which aims at combining the potential of semiconductors (control of the current by the gate, coupling with optics, etc.) with the potential of the magnetic materials (control of the current by spin manipulation, non-volatility, etc.). It should be possible, for example, to gather storage, detection, logic, and communication capabilities on a single chip that could replace several components. One of the ways is with ferromagnetic semiconductors (FSs) based on the well-known semiconductors of today. Excellent results have been obtained with GaAs doped with Mn [76], but it has not been possible to increase its Curie temperature above 170 K and this rules out most possibilities of applied devices. Some announcements of FS at room temperature have been made, but the situation is still rather confused on this point today. The second way is the fabrication of hybrid structures associating ferromagnetic metals and conventional semiconductors. In a spin LED [77–80], for example, the injection of spin-polarized current into an LED leads to the emission of a circularly polarized light. In other experiments, it has now been possible to electrically detect the spin accumulation injected into a semiconductor

[80–83]. Intense research has been devoted to the implementation of the concept of a spin transistor proposed by Datta and Das [84], but only very modest advances [85] have been made on this topic up to now. Finally, a hot topic of today is the SHE and the QSHE [86, 87].

Spintronics with multiferroics is also a promising axis. Multiferroic materials combine ferromagnetic and ferroelectric properties, and the possible coupling between the ferromagnetic and ferroelectric orders could give the means to control a magnetic configuration with an electric field [88–90]. The recent results [90] with $BiFeO_3$ are promising, and the research on several other types of multiferroic oxides is now very active. Note that an alternative way for the electric field control of the magnetization direction turns out to be through direct effects on the band structure by spin–orbit interactions in FSs [91] or even in ultrathin layers of ferromagnetic metals [92].

Other interesting new properties of oxides have been revealed recently. For example, the existence of 2DEGs of high mobility at the interface between two insulating oxides [93–95] opens the way to spintronic devices exploiting these 2DEGs in "all oxide" nanostructures. Also, the extension of the technologies developed for MTJ to tunnel junctions with ferroelectric barriers has led to the observation of huge electroresistance effects [96] and to interesting prospects for the design of ferroelectric memories.

Another important field of research today is spintronics with carbon-based materials. Their general advantage comes from the long spin lifetime related to the small spin–orbit coupling of carbon. For metallic CNT and graphene, another advantage comes from their linear dispersion curves and very large velocities even for small carrier concentrations. In lateral spin-valve structures of CNT between LSMO electrodes (Figure 1.13), the MR can reach 60%–70% ($V_{AP} - V_P \approx 60$ mV for $V_P \approx 100$ mV for the sample from Figure 1.13d), well above the output electrical signals that can be obtained with semiconductor lateral channels. These results can be explained not only by the long spin lifetime in CNT but also by the short dwell time resulting from the high velocity of the electrons inside CNTs [54]. To update the above text written in 2010 for the Handbook version of 2016, I can say that the works on carbon-based materials described above have been extended by an intense research on graphene, which turns out to be the best material for propagation of spin currents to long distance without spin relaxation [98]. Other 2-dimensional (2D) materials, such as the transition metal dichalco-genides, also present an undoubted interest for spintronics [99]. However, the most fascinating direction on this axis of 2D electronic systems could be the exploitation of the topological properties of the 2D electrons at the Rashba interfaces [100] and surfaces or interfaces of the materials called topological insulators [101]. Shortly speaking, we can say that the conversion from charge to spin current [102] or from spin to charge current [103] by Rashba or topological insulator interfaces will probably be the most efficient way to create spin currents from charge currents or to detect electrical spin currents in spintronic devices. The exploitation of the topological properties of magnetic skyrmions also opens great perspectives [104].

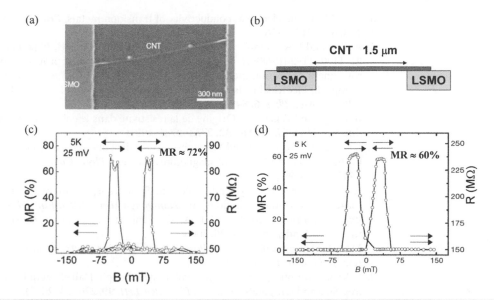

FIGURE 1.13   Spin transport through carbon nanotubes (CNTs) [54]. (a) Electron microscope image of a CNT between LSM electrodes. (b) Schematic side view of the device. (c), (d) Examples of MR experimental results. The resistance, R, of the device is predominantly due to the tunnel resistance at the LSMO/CNT interfaces, and its change between the parallel and antiparallel magnetic configuration is about 72% (c) or 60% (d).

Finally, with qbits based on the spin state of a quantum box, nanospintronics is one of the ways to explore quantum computing. This is certainly an exciting perspective but beyond the scope of this chapter, and, on this point, we refer to a recent review by Trauzettel et al. [105].

1.10 CONCLUSION

In about 20 years, we have seen spintronics considerably increasing the capacity of our hard disks, extending the hard disk technology to mobile appliances such as cameras or portable multimedia players, entering the automotive industry and the biomedical technology and, with TMR and spin transfer, getting ready to enter the RAM of our computers or the microwave emitters of our cell phones. The research of today on the spin-transfer phenomena, on multiferroic materials, on spintronics with semiconductors and molecular spintronics, open fascinating new fields and are also very promising in the "beyond CMOS" perspective. Spintronics is still a young baby in the science family, but this baby is growing remarkably well.

REFERENCES

1. M. N. Baibich, J. M. Broto, A. Fert et al., Giant magnetoresistance of (001) Fe/(001) Cr magnetic superlattices, *Phys. Rev. Lett.* **61**, 2472–2475 (1988).
2. G. Binash, P. Grünberg, F. Saurenbach et al., Enhanced magnetoresistance in layered magnetic structures with antiferromagnetic interlayer exchange, *Phys. Rev. B* **39**, 4828–4830 (1989).

3. N. F. Mott, The electrical conductivity of transition metals, *Proc. Roy. Soc. A* **153**, 699–718 (1936).

4. A. Fert and I. A. Campbell, Two current conduction in nickel, *Phys. Rev. Lett.* **21**, 1190–1192 (1968); A. Fert and I. A. Campbell, Transport properties of ferromagnetic transition metals, *J. Phys.* **32**, C1–46–50 (1971); A. Fert and I. A. Campbell, Electrical resistivity of ferromagnetic nickel and iron based alloys, *J. Phys. F Met. Phys.* **6**, 849–871 (1976).

5. B. Loegel and F. Gautier, Origine de la résistivité dans le cobalt et ses alliages dilués, *J. Phys. Chem. Sol.* **32**, 2723–2735 (1971).

6. J. Barnas, O. Baksalary, and Y. Bruynseraede, Effect of interchannel transitions in the current-in-plane giant magnetoresistance, *J. Phys.: Condens. Matter* **7**, 6437–6448 (1995).

7. A. Fert, A. Barthélémy, and F. Petroff, Spin transport in magnetic multilayers and tunnel junctions, in *Nanomagnetism: Ultrathin Films and nanostructures*, F. Mills and J. A. C. Bland (Eds.), Elsevier, Amsterdam, the Netherlands, pp. 153–226, 2006.

8. M. I. D'yakonov and V. I. Perel', Possibility of orienting electron spins with current, *JETP Lett.* **13**, 467–469 (1971).

9. L. Vila, T. Kjimura, and Y. Otani, Evolution of the spin Hall effect in Pt nanowires: Size and temperature effects, *Phys. Rev. Lett.* **99**, 226604 (2007).

10. S. O. Valenzuela and M. Tinkham, Direct measurement of the spin Hall effect, *Nature* **442**, 176–179 (2006).

11. T. Seki, Y. Hasegawa, S. Mitani et al., Giant spin Hall effect in perpendicularly spin-polarized FePt/Au devices, *Nat. Mater.* **7**, 125–128 (2008).

12. A. Fert, A. Friederich, and A. Hamzic, Hall effect in dilute magnetic alloys, *J. Magn. Magn. Mater.*, **24**, 231–257 (1981).

13. Y. Niimi, M. Morota, D. H. Wei et al., Extrinsic spin Hall effect induced by iridium impurities in copper, *Phys. Rev. Lett.* **106**, 126601 (2011).

14. I. Dzyaloshinskii, A thermodynamic theory of 'weak' ferromagnetism of antiferromagnetics, *J. Phys. Chem. Solids* **4**, 241–255 (1958).

15. T. Moriya, Anisotropic superexchange interaction and weak ferromagnetism, *Phys. Rev.* **120**, 91–98 (1960).

16. A. Fert and P. M. Levy, Role of anisotropic exchange interactions in determining the properties of spin-glasses, *Phys. Rev. Lett.* **44**, 1538–1541 (1980).

17. A. Fert, *Mater. Sci. Forum* **59–60**, 439–480 (1990).

18. S. Mühlbauer, B. Binz, F. Jonietz et al., Skyrmion lattice in a chiral magnet, *Science* **323**, 915–919 (2009).

19. S. Heinze, K. Von Bergmann, M. Menzel et al., Spontaneous atomic-scale magnetic skyrmion lattice in two dimensions, *Nat. Phys.* **7**, 713–718 (2011).

20. S. Emori, U. Bauer, S.-M. Ahn, E. Martinez, and G. S. D. Beach, Current-driven dynamics of chiral ferromagnetic domain walls, *Nat. Mater.* **12**, 611–616 (2013).

21. L. Berger, Low field magnetoresitance and domain drag in ferromagnets, *J. Appl. Phys.* **49**, 2656 (1978).

22. P. Grünberg, R. Schreiber, Y. Young et al., Layered magnetic structures: Evidence for antiferromagnetic coupling of Fe layers across Cr interlayers, *Phys. Rev. Lett.* **57**, 2442–2445 (1986).

23. R. E. Camley and J. Barnas, Theory of giant magnetoresistance effects in magnetic layered structures with antiferromagnetic coupling, *Phys. Rev. Lett.* **63**, 664–667 (1989).

24. P. M. Levy, S. Zhang, and A. Fert, Electrical conductivity of magnetic multilayered structures, *Phys. Rev. Lett.* **65**, 1643–1646 (1990).

25. S. S. P. Parkin, N. More, and K. P. Roche, Oscillations in exchange coupling and magnetoresistance in metallic superlattice structure: Co/Ru, Co/Cr, and Fe/Cr, *Phys. Rev. Lett.* **64**, 2304–2307 (1990).

26. T. Shinjo and H. Yamamoto, Large magnetoresistance of field-induced giant ferromagnetic multilayers, *J. Phys. Soc. Jpn.* **59**, 3061–3064 (1990).

27. C. Dupas, P. Beauvillain, C. Chappert et al., Very large magnetoresistance effects induced by antiparallel magnetization in two ultrathin cobalt films, *J. Appl. Phys.* **67**, 5680–5682 (1990).

28. D. H. Mosca, F. Petroff, A. Fert et al., Oscillatory interlayer coupling and giant magnetoresistance in Co/Cu multilayers, *J. Magn. Magn. Mater.* **94**, L1–L5 (1991).

29. S. S. P. Parkin, R. Bhadra, and K. P. Roche, Oscillatory magnetic exchange coupling through thin copper layers, *Phys. Rev. Lett.* **66**, 2152–2155 (1991).

30. B. Dieny, V. S. Speriosu, S. S. P. Parkin et al., Giant magnetoresistance in soft ferromagnetic multilayers, *Phys. Rev. B* **43**, 1297–1300 (1991).

31. S. S. P. Parkin, Applications of magnetic nanostructures, in *Spin Dependent Transport in Magnetic Nanostructures*, S. Maekawa and T. Shinjo (Eds.), Taylor & Francis, London, pp. 237–279, 2002.

32. C. Chappert, A. Fert, and F. Nguyen Van Dau, The emergence of spin electronics in data storage, *Nat. Mater.* **6**, 813–823 (2007).

33. P. P. Freitas, H. Ferreira, D. Graham et al., Magneto-resistive biochips, *Europhys. News* **34/6**, 224–226 (2003).

34. M. Jullière, Tunneling between ferromagnetic films, *Phys. Lett. A* **54**, 225–226 (1975).

35. S. Maekawa and O. Gäfvert, Electron tunneling between ferromagnetic films, *IEEE Trans. Magn. MAG* **18**, 707–708 (1982).

36. J. S. Moodera, L. R. Kinder, T. M. Wong et al., Large magnetoresistance at room temperature in ferromagnetic thin film tunnel junctions, *Phys. Rev. Lett.* **74**, 3273–3276 (1995); T. Miyazaki and N. Tezuka, Giant magnetic tunneling effect in $Fe/Al_2O_3/Fe$ junction, *J. Magn. Magn. Mater.* **139**, L231–L234 (1995).

37. M. Bowen, V. Cros, F. Petroff et al., Large magnetoresistance in Fe/MgO/FeCo(001) epitaxial tunnel junctions on GaAs(001), *Appl. Phys. Lett.* **79**, 1655–1657 (2001).

38. S. Yuasa, T. Nagahama, K. Fukushima et al., Giant room-temperature magnetoresistance in single-crystal Fe/MgO/Fe magnetic tunnel junctions, *Nat. Mater.* **3**, 868–871 (2004).

39. S. S. P. Parkin, C. Kaiser, A. Panchula et al., Giant tunneling magnetoresistance at room temperature with MgO(100) tunnel barriers, *Nat. Mater.* **3**, 862–867 (2004).

40. Y. M. Lee, J. Hayakawa, S. Ikeda et al., Effect of electrode composition on the tunnel magnetoresistance of pseudo-spin-valve magnetic tunnel junction with a MgO tunnel barrier, *Appl. Phys. Lett.* **90**, 212507 (2007).

41. L. Jiang, H. Naganuma, M. Oogane et al., Large tunnel magnetoresistance of 1056% at room temperature in MgO based double barrier magnetic tunnel junctions, *Appl. Phys. Express* **2**, 083002 (2009).

42. I. I. Oleinik, E. Y. Tsymbal, and D. G. Pettifor, Structural and electronic properties of $CoAl_2O_3/Co$ magnetic tunnel junction from first principles, *Phys. Rev. B* **62**, 3952–3959 (2000); I. I. Oleinik, E. Y. Tsymbal, and D. G. Pettifor, Atomic and electronic structure of $Co/SrTiO_3/Co$ magnetic tunnel junctions, *Phys. Rev. B* **65**, 020401 (2002).

43. J. Mathon and A. Umerski, Theory of tunneling magnetoresistance in a junction with a nonmagnetic metallic interlayer, *Phys. Rev. B* **60**, 1117–1121 (1999).

44. Ph. Mavropoulos, N. Papanikolaou, and P. H. Dederichs, Complex band structure and tunneling through ferromagnet/insulator/ferromagnet junctions, *Phys. Rev. Lett.* **85**, 1088–1091 (2000).

45. X.-G. Zhang and W. H. Butler, Large magnetoresistance in bcc Co/MgO/Co and FeCo/MgO/FeCo tunnel junctions, *Phys. Rev. B* **70**, 173407 (2004).

46. J. P. Velev, K. D. Belashchenko, D. A. Stewart et al., Negative spin polarization and large tunneling magnetoresistance in epitaxial $Co/SrTiO_3/Co$ magnetic tunnel junctions, *Phys. Rev. Lett.* **95**, 216601 (2005).

47. J. M. De Teresa, A. Barthélémy, A. Fert et al., Role of metal-oxide interface in determining the spin polarization of magnetic tunnel junctions, *Science* **286**, 507–509 (1999).

48. M. Bowen, M. Bibes, A. Barthélémy et al., Nearly total spin polarization in La$_{2/3}$Sr$_{1/3}$MnO$_3$ from tunneling experiments, *Appl. Phys. Lett.* **82**, 233–235 (2003).

49. T. Ishikawa, T. Marukame, H. Kijima et al., Spin-dependent tunneling characteristics of fully epitaxial magnetic tunneling junctions with a full-Heusler alloy Co$_2$MnSi thin film and a MgO tunnel barrier, *Appl. Phys. Lett.* **89**, 192505 (2006).

50. P. LeClair, J. K. Ha, J. M. Swagten et al., Large magnetoresistance using hybrid spin filter devices, *Appl. Phys. Lett.* **80**, 625–627 (2002).

51. W. P. Pratt, S. F. Lee, J. M. Slaughter et al., Perpendicular giant magnetoresistances of Ag/Co multilayers, *Phys. Rev. Lett.* **66**, 3060–3063 (1991); J. Bass and W. P. Pratt Jr., Current-perpendicular (CPP) magnetoresistance in magnetic metallic multilayers, *J. Magn. Magn. Mater.* **200**, 274–289 (1999).

52. L. Piraux, J.-M. George, C. Leroy et al., Giant magnetoresistance in magnetic multilayered nanowires, *Appl. Phys. Lett.* **65**, 2484–2486 (1994); A. Fert and L. Piraux, Magnetic nanowires, *J. Magn. Magn. Mater.* **200**, 338–358 (1999).

53. T. Valet and A. Fert, Theory of the perpendicular magnetoresistance in magnetic multilayers, *Phys. Rev. B* **48**, 7099–7013 (1993).

54. L. E. Hueso, J. M. Pruneda, V. Ferrari et al., Transformation of spin information into large electrical signals using carbon nanotubes, *Nature* **445**, 410–413 (2007).

55. G. Schmidt, D. Ferrand, L. W. Molenkamp et al., Fundamental obstacle for electrical spin injection from a ferromagnetic metal into a diffusive semiconductor, *Phys. Rev. B* **62**, 4790–4793 (2000).

56. E. I. Rashba, Theory of electrical spin injection: Tunnel contacts as a solution of the conductivity mismatch problem, *Phys. Rev. B* **62**, R16267–R1670 (2000).

57. A. Fert and H. Jaffrès, Conditions for efficient spin injection from a ferromagnetic metal into a semiconductor, *Phys. Rev. B* **64**, 184420 (2001); A. Fert, J.-M. George, H. Jaffrès, and R. Mattana, Semiconductors between spin-polarized sources and drains, *IEEE Trans. Electron Dev.* **54**, 921–932 (2007); H. Jaffrès, A. Fert, and J.-M. George, *Phys. Rev. B.* **67**, 140408(R) (2010).

58. M. Johnson and R. H. Silsbee, Thermodynamic analysis of interfacial transport and of the thermomagnetoelectric system, *Phys. Rev. B* **35**, 4959–4972 (1987).

59. Z. G. Yu and M. E. Flatté, Electric-field dependent spin diffusion and spin injection into semiconductors, *Phys. Rev. B* **66**, 201202 (2002).

60. T. Kimura, J. Hamrle, and Y. Otani, Estimation of spin-diffusion length from the magnitude of spin-current absorption: Multi-terminal ferromagnetic/non-ferromagnetic hybrid structures, *Phys. Rev. B* **72**, 014461 (2005).

61. S. Takahashi and S. Maekawa, Spin currents in metals and superconductors, *J. Phys. Soc. Jpn.* **77**, 031009 (2008).

62. F. J. Jedema, M. S. Nijboer, A. T. Filip et al., Spin injection and spin accumulation in all-metal mesoscopic spin valves, *Phys. Rev. B* **67**, 085319 (2003).

63. T. Kimura and Y. Otani, Large spin accumulation in a permalloy-silver lateral spin valve, *Phys. Rev. Lett.* **99**, 196604 (2007).

64. B. T. Jonker and M. E. Flatté, Electrical spin injection and transport in semiconductors, in *Nanomagnetism: Ultrathin Films, Multilayers and Nanostructures*, D. L. Mills and J. A. C. Bland (Eds.), Elsevier, Amsterdam, the Netherlands, pp. 227–272, 2006.

65. A. Cottet, T. Kontos, S. Sahooo et al., Nanospintronics with carbon nanotubes, *Semicond. Sci. Technol.* **21**, S78–S95 (2006).

66. N. Tombros, C. Jozsa, M. Popinciuc et al., Electronic transport and spin precession in single graphene layers at room temperature, *Nature* **448**, 571–574 (2006); M. Ohishi, M. Shiraishi, R. Nouchi et al., Spin injection into graphene at room temperature, *Jpn. J. Appl. Phys.* **46**, 25–28 (2007); W. Han, K. Pi, K. McCreary et al., Tunneling spin injection into single layer graphene, *Phys. Rev. Lett.* **105**, 16702–16704 (2010).

67. J. C. Slonczewski, Current-driven excitation of magnetic multilayers, *J. Magn. Magn. Mater.* **159**, L1–L7 (1996).
68. L. Berger, Emission of spin waves by a magnetic multilayer traversed by a current, *Phys. Rev. B* **54**, 9353–9358 (1996).
69. M. Tsoi, A. G. M. Jansen, J. Bass et al., Excitation of a magnetic multilayer by an electric current, *Phys. Rev. Lett.* **80**, 4281–4284 (1998).
70. F. J. Albert, J. A. Katine, F. J. Albert et al., Spin-polarized current switching of a Co thin film nanomagnet, *Appl. Phys. Lett.* **77**, 3809–3811 (2000).
71. J. Grollier, V. Cros, A. Hamzic et al., Spin-polarized current induced switching in Co/Cu/Co pillars, *Appl. Phys. Lett.* **78**, 3663–3665 (2001).
72. W. H. Rippart, M. R. Pufall, S. Kaka et al., Direct-current induced dynamics in $Co_{90}Fe_{10}/Ni_{80}Fe_{20}$ point contacts, *Phys. Rev. Lett.* **92**, 027201 (2004).
73. A. Dussaux, B. Georges, J. Grollier et al., Large microwave generation from current-driven magnetic vortex oscillators in magnetic tunnel junctions, *Nat. Commun.* **1**, 8 (2010).
74. A. Ruotolo, V. Cros, B. Georges et al., Spin transfer induced dynamics of an array of interacting vortices in a periodic potential, *Nat. Nanotechnol.* **4**, 528–532 (2009).
75. I. Žutić, J. Fabian, and S. Das Sarma, Spintronics: Fundamentals and applications, *Rev. Mod. Phys.* **76**, 323–410 (2004).
76. H. Ohno, A. Shen, and F. Matsukura, (Ga,Mn)As: A new diluted magnetic semiconductor based on GaAs, *Appl. Phys. Lett.* **69**, 363–365 (1996).
77. V. F. Motsnyi, P. Van Dorpe, W. Van Roy et al., Optical investigation of spin injection into semiconductors, *Phys. Rev. B* **68**, 245319 (2003).
78. A. T. Hanbicki, M. J. van't Erve, and R. Magno, Analysis of the transport process providing spin injection through an Fe/AlGaAS Schottky barrier, *Appl. Phys. Lett.* **82**, 4092–4094 (2003).
79. Y. Lu, V. G. Truong, P. Renucci et al., MgO thickness dependence of spin injection efficiency in spin-light emitting diodes, *Appl. Phys. Lett.* **93**, 152102 (2008).
80. X. Lou, C. Adelmann, S. A. Crooker et al., Electrical detection of spin transport in lateral ferromagnet–semiconductor devices, *Nat. Phys.* **3**, 197–202 (2007).
81. O. Van't Erve, A. T. Hanbicki, M. Holub et al., Electrical injection and detection of spin-polarized carriers in silicon in a lateral transport geometry, *Appl. Phys. Lett.* **91**, 212109 (2007).
82. M. Tran, H. Jaffrès, C. Deranlot et al., Enhancement of the spin accumulation at the interface between a spin-polarized tunnel junction and a semiconductor, *Phys. Rev. Lett.* **102**, 036601 (2009).
83. S. P. Dash, S. Sharma, R. S. Pald et al., Electrical creation of spin polarization in silicon at room temperature, *Nature* **462**, 491–494 (2009).
84. S. Datta and B. Das, Electronic analog of the electro-optic modulator, *Appl. Phys. Lett.* **56**, 665–667 (1990).
85. H.-C. Koo, J.-H. Kwon, J. Eom et al., Control of spin precession in a spin-injected field effect transistor, *Science* **325**, 1515–1518 (2009).
86. M. König, S. Wiedmann, and C. Brüne, Quantum spin Hall insulator state in HgTe quantum wells, *Science* **318**, 766–770 (2007).
87. D. Hsieh, Y. Xia, D. Qian et al., Observation of time-reversal-protected single-Dirac-Cone topological-insulator states in Bi_2Te_3 and Sb_2Te_3, *Phys. Rev. Lett.* **103**, 146401 (2009).
88. T. Zhao, A. Scholl, F. Zavaliche et al., Electrical control of antiferromagnetic domains in multiferroic $BiFeO_3$ films at room temperature, *Nat. Mater.* **5**, 823–829 (2006); H. Béa, M. Bibes, F. Ott et al., Mechanisms of exchange bias with multiferroic $BiFeO_3$ epitaxial films, *Phys. Rev. Lett.* **100**, 017204 (2008); J. D. Burton and E. Y. Tsymbal, Prediction of electrically induced magnetic reconstruction at the manganite/ferroelectric interface, *Phys. Rev. B* **80**, 1744061–1744065 (2009).
89. M. Bibes and A. Barthélémy, Oxide spintronics, *IEEE Trans. Electron Dev.* **54**, 1003–1023 (2007).

90. Y.-H. Chu, L. W. Martin, M. B. Holcomb et al., Electric-field control of local ferromagnetism using a magnetoelectric multiferroic, *Nat. Mater.* **7**, 478–482 (2008).

91. D. Chiba, M. Sawicki, Y. Nishitami et al., Magnetization vector manipulation by electric field, *Nature* **455**, 515–518 (2008).

92. M. Weisheit, S. Fähler, A. Marty et al., Electric field-induced modification of magnetization in thin-film ferromagnets, *Science* **315**, 349–351 (2007).

93. A. Ohtomo and H. Y. Hwang, A high-mobility electron gas at the $LaAlO_3/SrTiO_3$ heterointerface, *Nature* **427**, 423–426 (2004).

94. M. Basletic, J.-L. Maurice, C. Carrétéro et al., Mapping the spatial distribution of charge carriers in $LaAlO_3/SrTiO_3$ heterostructures, *Nat. Mater.* **7**, 621–625 (2008); C. Cen, S. Thiel, J. Mannhart et al., Oxide nanoelectronics on demand, *Science* **323**, 1026–1028 (2009).

95. M. K. Niranjan, Y. Wang, S. S. Jaswal et al., Prediction of a switchable two-dimensional electron gas at ferroelectric oxide interfaces, *Phys. Rev. Lett.* **103**, 016804 (2009).

96. V. Garcia, S. Fusil, K. Bouzehouane et al., Giant tunnel electroresistance for non-destructive readout of ferroelectric states, *Nature* **460**, 81–84 (2009).

97. Z. H. Xiong, D. Wu, V. Vardeny et al., Giant magnetoresistance in organic spin-valves, *Nature* **427**, 821–824 (2004).

98. B. Dlubak, M. B. Martin, C. Deranlot et al., Highly efficient spin transport in epitaxial graphene on SiC. *Nat. Phys.* **8**, 557–561 (2012).

99. Z. Wang, D. K. Ki, H. Chen, H. Berger, A. H. MacDonald, and A. F. Morpurgo, Strong interface-induced spin–orbit interaction in graphene on WS2. *Nat. Commun.* **6**, 8339 (2015).

100. E. I. Rashba, Properties of semiconductors with an extremum loop: 1. Cyclotron and combinational resonance in a magnetic field perpendicular to the plane of the loop. *Sov. Physics, Solid State* **2**, 1109–1122 (1960).

101. M. Z. Hasan and C. L. Kane, Colloquium: Topological insulators, *Rev. Mod. Phys.* **82**, 3045–3067 (2010).

102. K. Kondou, R. Yoshimi, A. Tsukazaki et al., Fermi level dependent charge-to-spin current conversion by Dirac surface state of topological insulators, *Nat. Phys.* **12**, 1027–1031 (2016).

103. J. C. Rojas Sanchez, L. Vila, G. Desfonds et al., Spin-to-charge conversion using Rashba coupling at the interface between non-magnetic materials, *Nat. Commun.* **4**, 2944 (2013).

104. R. Wiesendanger, Nanoscale magnetic skyrmions in metallic films and multilayer: a new twist for spintronics, *Nat. Rev. Mater.* **1**, 16044 (2016).

105. B. Trauzettel, M. Borhani, M. Trif et al., Theory of spin qubits in nanostructures, *J. Phys. Soc. Jpn.* **77**, 031012 (2008).

Section II
Magnetic Metallic Multilayers

2

Basics of Nanothin Film Magnetism

Bretislav Heinrich, Pavlo Omelchenko, and Erol Girt

2.1 INTRODUCTION

Ultrathin magnetic films have been studied and used in low-dimensional systems and materials employed in spintronics. In these systems, the interfaces play a crucial role. The magnetic properties of interface atoms are different from those in the bulk. First-principle calculations indicate that the bulk properties are almost fully acquired just a few atomic layers from the interface. Ultrathin films behave like giant magnetic molecules having their magnetic properties given by an admixture of the interface and bulk properties. The ability to admix the interface and bulk magnetic properties and combine the magnetic films together by nonmagnetic interlayers allows one to engineer unique magnetic materials. In the following three sections, the emphases will be put on the basic magnetic properties: magnetic anisotropy, exchange coupling, and exchange bias (EB) coupling. Each section starts with a general theoretical introduction, followed by most common experimental techniques, and a brief summary of important experimental studies, which played a crucial role in the development of magnetic nanostructures. We apologize to those who feel that their work and contributions were either insufficiently covered or perhaps even omitted. This is certainly not intentional. The goal of these sections is not to provide a comprehensive account of this field. The page limitation allows one to provide only basic ideas that are needed for specialists and graduate students entering the field of magnetic nanostructures and spintronics applications. There are many references in this chapter to important books and review articles where further details and references can be found.

Since a large range of problems is covered, the following provides a brief guided tour through these sections. In Section 2.2, the concept of free energy of magnetic anisotropy is described, starting from a phenomenological description satisfying the lattice symmetry requirement. It is shown that the magnetic anisotropy exerts a torque on a magnetic moment by an effective field given by Equation 2.7. It is also shown that the interfaces, due to their decreased lattice symmetry, play a crucial role in magnetic anisotropies. In ultrathin films, the overall magnetic properties are then given by adding the bulk and interface properties using the scaling with the film thickness shown in Equation 2.11 allowing one to use a concept of giant magnetic molecules. The theory of magnetic anisotropy in 3D transition elements is described in Section 2.2.1. It is shown that it is caused by the spin-orbit contribution to the energy of electron states occupying the valence bands. Néel's model of magnetic anisotropy based on a magnetic atom pair interaction is described in Section 2.2.2. This allows one to include the magnetoelastic contributions, which, in a strained lattice, results in perpendicular and in-plane uniaxial fields. MacLaren's and Victora's extension of Néel's model also includes, in addition to a pair interaction between two ferromagnetic atoms, a pair interaction between a ferromagnetic atom and nonmagnetic nearest neighbor substrate atoms, allowing one to include the dependence of interface anisotropy on the type of nonmagnetic substrate. Dzyaloshinski-Moriya exchange anisotropy energy and noncollinear magnetism in ultrathin films are discussed in Section 2.2.3. The measurements of magnetic anisotropies by ferromagnetic resonance (FMR) are described in Section 2.2.4. The interpretation of the data in this section is based on a detailed solution of the Landau-Lifshitz-Gilbert (LLG) equations of motion. It requires evaluating the effective fields arising from the magnetic anisotropies including the demagnetizing energy and magnetoelastic contributions. Expressions for the effective fields and corresponding torque equations are shown in detail. The results of magnetic anisotropies in seminal ultrathin film systems are presented in Section 2.2.5.

Interlayer exchange coupling is covered in Section 2.3. The phenomenology of the bilinear and biquadratic interface exchange coupling is discussed in Section 2.3.1. The theory of bilinear exchange coupling using quantum well states (QWS) in a normal metal (NM) spacer is discussed in Section 2.3.2. The mechanism of biquadratic exchange coupling based on interface roughness is discussed in Section 2.3.3. It is also shown that in any system that exhibits a lateral inhomogeneity, one must expect additional energy terms. They originate from intrinsic magnetic energy terms that fluctuate in strength across the sample interface. These additional terms have the next higher angular power compared with that of the intrinsic term, and they have only one sign. The static and dynamic techniques employed in measurements of the interlayer exchange coupling are discussed in Section 2.3.4. The seminal studies of interlayer exchange coupling are summarized in Section 2.3.5. It covers the Cr(001) spacers with the long-range spin-density waves (SDWs) and the Cu(001) spacer where the interlayer exchange coupling is generated by QWS passing through the Fermi surface.

Interlayer exchange coupling in ferromagnetic films with a nonuniform orientation of magnetic moments is described in Section 2.4. The bulk exchange coupling parameter, A, and effective fields are covered in Section 2.4.1. Selecting intrinsic magnetic parameters plays a crucial role in designing magnetization reversal of perpendicular magnetic recording media and memory nanopillars. This is discussed in Section 2.4.2. Results of measurements of exchange the coupling parameter A and saturation magnetization Ms in Co alloyed films by using synthetic antiferromagnet (SAF) film structures is shown in Section 2.4.3.

The EB coupling is covered in Section 2.5. The EB mechanism is an important part of spintronic devices. This mechanism originates in the very thin region between the antiferromagnetic (AF) and ferromagnetic materials (F), where the magnetism, structure, and the role of interface morphology abruptly change and acquire unique features, which are strongly dependent on sample preparation. In this respect, there is no unified theory applicable to these systems. However, some basic principles governing the EB mechanism have been found and are useful in designing and fabricating spin valve and spin torque devices where it provides a pinning mechanism for hard magnetic layers. In this section, a brief review of basic models for EB is presented, starting from the Meiklejohn–Beam model and then followed by Malozemoff and Mauri models, respectively.

2.2 MAGNETIC ANISOTROPY

The energy of a ferromagnet (F) depends on the orientation of the magnetic moment with respect to the crystalline axes and the shape of the ferromagnet. This dependence is called magnetocrystalline anisotropy. The free energy contribution due to magnetocrystalline in a cubic crystal has to satisfy the lattice symmetry and can be written to the lowest order in the magnetization components,

$$\mathcal{F}_a = -\frac{K_1}{2}\left[\left(\frac{M_x}{M_s}\right)^4 + \left(\frac{M_y}{M_s}\right)^4 + \left(\frac{M_z}{M_s}\right)^4\right], \tag{2.1}$$

where:

K_1 is the cubic anisotropy parameter (erg/cm^3)

$M_x, M_y,$ and M_z are the components of $\vec{M}_s$ along the three cube axes

$\mathcal{F}_a$ can also be rewritten in a more common form using the directional cosines of the magnetization components ($\alpha_{x,y,z} = M_{x,y,z} / M_s$),

$$\mathcal{F}_a = K_1 \left[\alpha_x^2 \alpha_y^2 + \alpha_x^2 \alpha_z^2 + \alpha_y^2 \alpha_z^2 \right]. \tag{2.2}$$

The equivalence of Equations 2.1 and 2.2 follows from the relation $\alpha_x^2 + \alpha_y^2 + \alpha_z^2 = 1$.

For hexagonal crystal symmetry, the anisotropic free energy density can be expressed in lowest orders by [1]

$$\mathcal{F}_a = K_2 \sin^2(\theta) + K_4 \sin^4(\theta), \tag{2.3}$$

where:

$$\sin^2(\theta) = 1 - \left(\frac{M_z}{M_s} \right)^2, \tag{2.4}$$

where θ is the angle of magnetization with respect to the hexagonal c axis. K_2 and K_4 are hexagonal anisotropy parameters. The first term in Equation 2.3 represents an example of the magnetic energy density in systems having a single axis of symmetry. One can introduce in general a uniaxial anisotropy

$$\mathcal{F}_u = -\frac{K_u}{M_s^2} \left(\vec{M}_s \cdot \vec{u} \right)^2, \tag{2.5}$$

where:

K_u is a uniaxial anisotropy parameter
$\vec{u}$ is a unit vector that specifies the orientation of the twofold axis which can be oriented in the film plane and along the film surface normal.

The variation of the free energy density with magnetization direction in the crystal results in a torque density that acts so as to align the magnetization along a direction that minimizes the free energy density. This torque density can be written as

$$\vec{L} = \vec{M}_s \times \vec{H}_a, \tag{2.6}$$

where

$$\vec{H}_a = -\left(\frac{\partial \mathcal{F}_a}{\partial M_x} \right) \hat{u}_x - \left(\frac{\partial \mathcal{F}_a}{\partial M_y} \right) \hat{u}_y - \left(\frac{\partial \mathcal{F}_a}{\partial M_z} \right) \hat{u}_z, \tag{2.7}$$

an effective magnetic field that exerts the same torque on the magnetization as does a real magnetic field [2]. $\hat{u}_{x,y,z}$ are unit vectors along the $x-$, $y-$, and $z-$axes.

The description of magnetic coupling will be mostly restricted to ultrathin magnetic films. In ultrathin films, magnetic variations across the thickness of the film are mainly suppressed. The magnetization in such thin films is uniform for internal exchange fields that are significantly larger than typical anisotropy fields. The direct exchange interaction fields between spins lead to a ferromagnetic state and, at the same time, make any spatial variation of the magnetization density relatively costly in free energy [2] in ultrathin films. This means that the magnetic moments on lattice sites across the film thickness are nearly parallel to each other. This is not exactly correct, but greatly simplifies the treatment of magnetic properties. The limits of this concept are described in [3]. The film can be usefully considered to be ultra-thin when its thickness does not exceed the exchange length δ, see [3],

$$\delta = \left(\frac{A}{2\pi M_s^2} \right)^{0.5},$$ (2.8)

where:

    A is the exchange stiffness parameter, see Section 2.4.1, (for Fe $A \simeq 2 \times 10^{-6}$ erg/cm)

    M_s is the magnitude of the saturation magnetization

It is clear that the free energy density of a thin film is likely to be quite different when the magnetization is directed along the film normal as compared with the case in which the magnetization lies in the film plane. The source of this free energy difference is the magnetic fields generated by the shape of the film. When the magnetization lies in the plane of a film, a few millimeters in lateral dimensions but a few nanometers thick, the magnetic field generated by the magnetization density discontinuity at the film edges can be ignored (the approximation of infinite lateral dimensions). However, when this same uniform magnetization has a component, M_z, directed along the film normal, the discontinuity in magnetization at the film surfaces generates an internal demagnetizing field $H_d = -4\pi M_z$. The interaction between this demagnetizing field and the magnetization density contributes a term to the free energy density having the form of a uniaxial anisotropy

$$\mathcal{F}_d = 2\pi M_z^2.$$ (2.9)

This uniform demagnetizing energy is true only in the magnetic continuum. The treatment of demagnetizing energy in atomic lattices with magnetic moments localized around the lattice sites is described in Section 2.2.4, see Equation 2.34.

    Another important source of difference between a bulk magnetic crystal and a magnetic thin film comes about because the thin film is grown on a substrate. If the crystal structure of the thin film is not exactly matched to the crystal structure of the substrate, the bonding between film and substrate must result in the deformation of the film. In general, the film structure becomes distorted both in-plane and out-of-plane. The simplest case is that in which a film retains fourfold symmetry in plane, but the lattice spacing in the perpendicular direction changes to keep a constant atomic volume, and the thin film adopts a tetragonal symmetry. For tetragonal symmetry, the magnetocrystalline free energy density can be written as

$$\mathcal{F}_a = -\frac{K_1^{\parallel}}{2} \left[\left(\frac{M_x}{M_s} \right)^4 + \left(\frac{M_y}{M_s} \right)^4 \right]$$

$$-\frac{K_1^{\perp}}{2} \left(\frac{M_z}{M_s} \right)^4 - K_{\perp} \left(\frac{M_z}{M_s} \right)^2,$$ (2.10)

where $K_1^{\parallel}$, $K_1^{\perp}$, and $K_{\perp}$ are anisotropy parameters for magnetization components parallel and perpendicular to the film surface, respectively. An example of this case is the growth of an Fe(001) film on the (001) surface of a silver substrate [3]. The atoms in the fcc Ag(001) surface net form a square array 2.889 Å on a side. The atoms on the bcc Fe(001) surface planes form a square array 2.866 Å on a side. Thus, the fourfold hollows on the silver surface can accommodate the Fe(001) surface atoms if the iron net is expanded in-plane by 0.8%. At the same time, the spacing between iron atomic layers is reduced by 1.5%. This lattice mismatch then leads to a tetragonal distortion, which results in the uniaxial anisotropy energy described by the last term in Equation 2.11.

The atomic layers in the vicinity of an interface generally have different magnetic properties from those that are farther away from the interface, i.e., in the bulk, Ref. [4]. The total free energy functional can be written as the sum of a surface term plus a volume term [3]

$$\mathcal{F}_{\text{tot}} = \frac{\mathcal{F}_{\text{int}}\left(\vec{M}\right)}{d} + \mathcal{F}_b\left(\vec{M}\right), \tag{2.11}$$

where:

$\mathcal{F}_{\text{int}}\left(\vec{M}\right)$ includes interface energies in erg/cm^2
d is the film thickness

It has been observed that each of the parameters in Equations 2.11 and 2.5 usually exhibits a thickness dependence, such that

$$K_i = K_{i,b} + \left(\frac{K_{i,\text{int}}}{d}\right), \tag{2.12}$$

where:

$K_{i,b}$ is a thickness independent volume anisotropy with the units of erg/cm^3
$K_{i,\text{int}}$ is an interface energy in erg/cm^2

In fact, the interface anisotropy $K_{i,\text{int}}$ is found to be sensitive to the structure and chemical composition of the film surfaces and their interfaces. An example is shown in Figure 2.1 of the perpendicular, in-plane uniaxial $2K_u^{\parallel}/M_s$, fourfold (cubic) $2K_1^{\parallel}/M_s$, and $4\pi M_{s,\text{eff}} - 2K_u^{\perp}/M_s$ anisotropy fields in Au/Fe/GaAs(001) films as a function of $1/d$, where d is the film thickness. A detailed discussion of $4\pi M_{s,\text{eff}}$ can be found in the text after Equation 2.20 in Section 2.2.2.

2.2.1 Magnetic Anisotropies: Theory

2.2.1.1 First Principles Electronic Band Calculations

First-principle calculations of magnetic anisotropies are based upon the solution of Schrödinger's equation for interacting electrons with spin. The

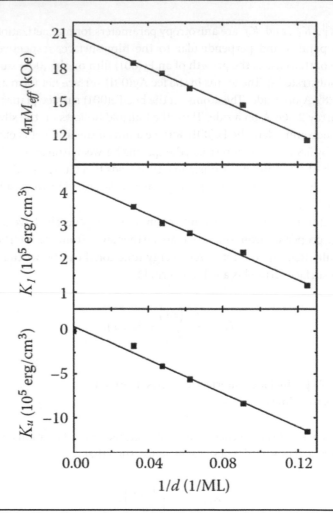

FIGURE 2.1 The effective demagnetizing field perpendicular to the film surface $4\pi M_{eff} = 4\pi M_{s,eff} - 2K_{u,int}^{\perp}/dM_s$, the in-plane fourfold (cubic) anisotropy parameter, K_1, and the uniaxial anisotropy parameter, K_u, are plotted as functions of $1/d$, where d is the Fe film thickness. The Fe films were grown on GaAs(001) and covered by 20 ML of Au(001). The in-plane uniaxial axis is directed along the $[1\,\bar{1}\,0]$ crystallographic direction. The in-plane anisotropies are comprised of bulk and interface contributions: $K_{1,b}^{\parallel} = 4.3 \times 10^5\,\text{erg/cm}^3$, $K_{1,int}^{\parallel} = -0.036\,\text{erg/cm}^2$, $K_{u,int}^{\parallel} = -0.13\,\text{erg/cm}^2$, and $4\pi M_{s,eff} = 20.3\,\text{kOe}$, $K_{u,int}^{\perp} = 0.88\,\text{erg/cm}^2$, where b and int stand for the bulk and interface parts of the corresponding anisotropies. (After Urban, R. et al., *Phys. Rev. Lett.* 87, 217201, 2001. Copyright 2001 by the American Physical Society. With permission.)

form of the equations to be solved is discussed by Gay and Richter [5]. The magnetic anisotropies are caused by the spin-orbit interaction. A simple argument can be used to describe the spin-orbit interaction term. In the rest coordinate system of moving electrons, the electric field $\vec{E}$ appears as an effective field $\vec{B} \sim \vec{p} \times \vec{E}$ due to the Lorentz transformation. This leads to a relativistic contribution to the electron Hamiltonian of the form $\mathcal{H}_{so} \sim \vec{B} \cdot \vec{S}$, where $\vec{S}$ is the spin momentum. For an electron passing an atomic potential having radial symmetry $\vec{E} = -\dfrac{dV}{dr}\dfrac{\vec{r}}{r}$, this results in $\mathcal{H}_{so} \sim \dfrac{1}{r}\dfrac{dV(r)}{dr}\vec{S} \cdot (\vec{r} \times \vec{p})$, which can be rewritten as $\left(\dfrac{1}{r}\dfrac{dV}{dr}\right)\vec{S} \cdot \vec{L}$, where $\vec{L}$ is the orbital momentum. Therefore, for a radially symmetric potential, the Hamiltonian for the spin-orbit interaction is given by

$$\mathcal{H}_{so} = \xi(r)\vec{S} \cdot \vec{L}, \tag{2.13}$$

where ξ is the spin-interaction parameter, which is proportional to the radial derivative of the atomic potential. In lattices exhibiting a high degree of symmetry, the crystal field hybridizes the d-orbitals and results in zero-orbital momentum. The spin-orbit interaction introduces orbital momentum.

When the majority and minority spin bands are well separated, it was shown by Bruno [6, 7] that the magnetic anisotropy in a second-order perturbation calculation for a uniaxial system can be expressed as

$$\Delta E_{so} \simeq -\frac{\xi}{4\mu_B}\left[m_{orb}^{\perp} - m_{orb}^{\parallel} \right], \tag{2.14}$$

where $m_{orb}^{\perp, \parallel}$ is the orbital moment for directions perpendicular and parallel to the film. This means that the uniaxial magnetic anisotropy is proportional to the difference in the components of the orbital momentum along the uni-axial axis and the plane perpendicular to that axis. This is a particularly useful concept for the systems with reduced dimensionality and symmetry. At the surface, the atom dimensionality is effectively reduced from 3 to 2, and the lattice symmetry is also reduced. This reduced lattice symmetry results in an appreciable interface anisotropy. Weller et al. [7], using x-ray magnetic circular dichroism (XMCD) on Au/Co/Au(111) textured wedge samples, measured an anisotropy in the orbital moment components parallel and perpendicular to the film surface. This anisotropy in the orbital moment components resulted in $m_{orb}^{\perp} - m_{orb}^{\parallel} > 0$, and, therefore, the easy axis was oriented perpendicular to the surface. Equation 2.14 leads to a uniaxial anisotropy energy, which is only a factor 2 larger than that obtained from the magnetic measurements on Au/Co/Ag(111) textured samples. This was in good agreement considering the simplicity of Bruno's model.

In order to calculate anisotropies accurately, it is necessary to calculate the ground state energy of the system under investigation using relativistic local density approximation (LDA) band calculations for various orientations of the magnetization vector with respect to the crystalline axes. This is a formidable task since the energy differences in question are of the order of meV per atom and are very small compared with the total ground state energy, which is of the order of eV per atom.

The precision with which the ground state energies could be calculated improved with time until it is now claimed that anisotropy energies for Fe, Ni, and Co monolayers (ML), which are of the order of 1 meV per atom, can be calculated with an uncertainty of the order of 0.001 meV per atom (1 meV/atom corresponds to a surface energy of 2.45 erg/cm² for atoms arranged on a Cu(001) surface net-a square net 2.56 Å on a side). See the review article by Wu and Freeman [8].

Magnetic anisotropy calculations are usually carried out for smooth and perfect surface interface planes whereas in real specimens these interface planes are usually rough on a scale of at least ±1 atomic layer. Therefore, one cannot expect perfect agreement between calculated and observed anisotropy coefficients no matter how exactly one evaluates the theoretical model.

2.2.2 MAGNETO-ELASTICITY: NÉEL MODEL

In 1954, Néel introduced a phenomenological model for the ferromagnetic state based upon the sum of pairwise interactions between ferromagnetic atoms [9]. This model was meant to help understand the origin of surface anisotropies as well as to elucidate the physics of magnetoelastic phenomena. The pair interaction was written as the sum of Legendre polynomials:

$$w(r,\theta) = g(r) + L(r)P_2(\cos\theta) + g(r)P_4(\cos\theta) + \dots \qquad (2.15)$$

where:

   r is the separation of the atom pair
   $L(r)$ and $g(r)$ are phenomenological parameters
   θ is the angle between the magnetization direction and the line that joins the atom pair

Usually, only the first two terms of the above series are used to discuss magnetic anisotropies and magnetoelastic effects. The first term $g(r)$ is isotropic and results in a volume magnetostriction [10]. Thus, ignoring a constant term, the pair interaction can be written as

$$w(r,\theta) = L_f(r)(\vec{u}_r \cdot \vec{m})^2, \qquad (2.16)$$

where:

   $L_f(r)$ is an atom pair interaction parameter between the magnetic sites
   $\vec{u}_r$ is a unit vector directed along the line joining the atom pair
   $\vec{m}$ is a unit vector in the direction of the magnetization vector

This simple model can be used to calculate the free energy density for an ultrathin film in terms of the parameter L_f. Such a free energy function will consist of the sum of a term proportional to the film thickness, a volume term, plus a term independent of film thickness, a surface term. It is clear that the free energy density must depend upon any strains introduced by film growth on a lattice mismatched substrate because it depends upon the angles between the magnetization direction and the orientation of the lines joining nearest neighbor atoms. In Chapter 8 of Ref. [10], Chikazumi demonstrates the use of the Néel model to deduce the form of the magnetoelastic coupling energy for cubic crystals. The presence of a magnetic moment will change the lattice. Using the strain tensor $e_{i,j}(i,j=x,y,z)$, one can write the bulk contribution to the magnetoelastic energy as

$$\mathcal{E}_{\text{mag}-el} = B_1\left[e_{xx}\left(\alpha_1^2 - \frac{1}{3}\right) + e_{yy}\left(\alpha_2^2 - \frac{1}{3}\right) + e_{zz}\left(\alpha_3^2 - \frac{1}{3}\right)\right]$$
$$+ B_2\left(e_{xy}\alpha_1\alpha_2 + e_{yz}\alpha_2\alpha_3 + e_{zx}\alpha_3\alpha_1\right),$$

(2.17)

where:

α_1, α_2, and α_3 are the directional cosines of the magnetic moment with respect to x, y, and z (usually major crystallographic axes)

B_1 and B_2 are the basic magnetoelastic parameters

The elastic energy is given by

$$\mathcal{E}_{el} = \frac{1}{2}C_{11}\left(e_{xx}^2 + e_{yy}^2 + e_{zz}^2\right) + \frac{1}{2}C_{44}\left(e_{xy}^2 + e_{yz}^2 + e_{zx}^2\right)$$
$$+ C_{12}\left(e_{xx}e_{yy} + e_{xx}e_{zz} + e_{yy}e_{zz}\right),$$

(2.18)

where C_{11}, C_{12}, and C_{44} are the elastic moduli.

For a (001) oriented cubic lattice whose in-plane lattice parameter has been altered by $e_{\parallel} = \Delta a / a$, and whose lattice parameter perpendicular to the plane has been altered by $e_{\perp} = \Delta c / a$, then the condition that the film must be stress-free along the film normal leads to

$$\frac{e_{\perp}}{e_{\parallel}} = -\frac{2C_{12}}{C_{11}},$$

(2.19)

see also Sander [11]. From Equations 2.2 and 2.19 and the condition $M_x^2 + M_y^2 + M_z^2 = M_s^2$, the strain contribution to the free energy density can be written as

$$\mathcal{F}_{me} = \text{const.} + B_1\left[e_{\perp} - e_{\parallel}\right]\left(\frac{M_z}{M_s}\right)^2$$
$$= \text{const.} - B_1\left(1 + \frac{2C_{12}}{C_{11}}\right)e_{\parallel}\left(\frac{M_z}{M_s}\right)^2.$$

(2.20)

This is a typical expression for a uniaxial anisotropy perpendicular to the film surface with the coefficient of uniaxial anisotropy equal to $K_{\perp} = B_1(1 + 2C_{12}/C_{11})$. For iron grown on GaAs(001), the in-plane strain $e_{\parallel} = -0.0136$. $C_{11} = 2.29 \times 10^{12}\,\text{dyn/cm}^2$, $C_{12} = 1.34 \times 10^{12}\,\text{dyn/cm}^2$ [11], and Equation 2.19 leads to $e_{\perp}/e_{\parallel} = -1.17$; and $B_1 = -34.3 \times 10^6\,\text{erg/cc}$ [11]. Therefore, the magnetoelastic field perpendicular to the surface using Equation 2.20 and Equation 2.7 is 1.18 kOe. Since for the bulk Fe at RT

$4\pi M_s = 21.5$ kG, the perpendicular uniaxial magnetoelastic field leads to $4\pi M_{s,\text{eff}} = 4\pi M_s - 1.18 = 20.3$ kG, in good agreement with measurements, see Figure 2.1.

The above Néel model results in an anisotropic free energy density that is independent of the substrate or overlayer materials. In order to correct this defect, Victora and MacLaren [12] have proposed an extension of the Néel theory that includes in addition to a pair interaction between two ferromagnetic atoms also a pair interaction between a ferromagnetic atom and nonmagnetic nearest neighbor atoms having the form

$$W(r,\theta) = L_m(r)(\vec{u}_r \cdot \vec{m})^2, \tag{2.21}$$

where:

$\vec{u}_r$ is a unit vector in the direction of the line joining a ferromagnetic atom with a nonmagnetic atom

$\vec{m}$ is a unit vector parallel with the magnetization vector

In this version of the theory, at least two interaction parameters are required, L_f and L_m. MacLaren and Victora have determined these two parameters for the Co(001)/Pd system by comparison of the resulting anisotropic free energy expression with the free energy obtained from first principles calculations using the experimentally observed strained lattice spacings as measured by Engel et al. [13]. The results of their first principles calculations were in good agreement with experiment. They used the values of L_f, L_m so determined to calculate the anisotropy coefficients for (111) and (110) oriented Co/Pd surfaces. The results were in reasonable agreement with the experimental data of Engel et al. [13].

2.2.3 DZYALOSHINSKI-MORIYA INTERACTION AND NONCOLLINEAR MAGNETIC CONFIGURATIONS

The low dimensionality of magnetic nanostructures can lead to additional potential gradients, which then result in additional electric fields. A typical example occurs at the interface between magnetic and nonmagnetic materials. The lack of inversion symmetry leads to a gradient of the lattice potential, which will, just as for the spin-orbit interaction, result in an additional energy term due to the spins at the sites i and j. It was shown by Dzyaloshinski [14] and Moriya [15] that this results in the energy interaction term

$$\mathcal{H}_{DM} = \sum_{i,j} \vec{D}_{i,j} \cdot (\vec{S}_i \times \vec{S}_j), \tag{2.22}$$

where $\vec{D}$ is a vector dependent on the lattice symmetry and is proportional to the spin-interaction parameter ξ. Notice that this term acts to tilt the magnetic moments away from collinearity. Since this term arises in the first-order perturbation expansion of the spin interaction, its contribution to the total

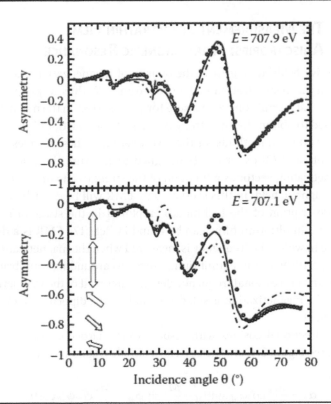

FIGURE 2.2 fcc Fe films six atomic layers thick were grown on a Cu(001) substrate using pulsed laser deposition (PLD). The measured x-ray reflectivity θ–2θ asymmetry parameter (using the right and left hand polarized x-ray beams) is shown by the open circles for two different x-ray energies, which straddle the Fe-L_3 absorption edge. The data were fit using the two magnetic configurations. The noncollinear configuration in six atomic layers is shown in the inset. Computer fits using the noncollinear configuration is shown by the solid lines. The dashed lines are the best fits one can get for a fully collinear configuration. See details in Ref. [16](Courtesy of Jo Stoehr; After Meyerheim, H.L. et al., *Phys. Rev. Lett.* 103, 267202, 2009. Copyright 2009 by the American Physical Society. With permission.)

Hamiltonian can be important and result in noncollinear magnetic moments in magnetic nanostructures. $\mathcal{H}_{DM}$ is well known in weak magnetic materials where a weak magnetic moment arises from an AF configuration in systems, which lack inversion symmetry. Recently, noncollinear configurations were found in ultrathin metastable metallic films. Meyerheim et al. [16] used soft x-ray resonant magnetic scattering in fcc Fe on Cu(001) to demonstrate the existence of a noncollinear magnetic structure in 6Fe(fcc)/Cu(001); the integer represents the number of atomic layers. Resonant magnetic scattering of a circularly polarized light beam using high scattering angles combined with first principles density functional theory (DFT) allows one to determine the magnetic moments in the individual atomic layers. The solid lines in Figure 2.2 show that the magnetic moments at the Fe/Cu interface are clearly noncollinear, and their orientation requires the presence of both the collinear exchange coupling and the Dzyaloahinski-Moriya interaction.

2.2.4 THE MEASUREMENT OF ULTRATHIN FILM ANISOTROPIES: FERROMAGNETIC RESONANCE

The effective fields arising from the density of magnetization energy exert torques on the magnetization Equations 2.6 and 2.7. Thus, any experimental technique that permits one to measure torques exerted on the magnetization can be used to obtain the anisotropy coefficients of Equations 2.11 and 2.26. There are two main methods for the investigation of these torques: (1) FMR experiments and (2) experiments designed to measure the orientation of the magnetization vector as a function of the strength and orientation of an applied magnetic field. In this section, only FMR will be described. A more general description of the techniques employed in the study of magnetic anisotropies in ultrathin films can be found in Ref. [17]. FMR is a dynamic technique in which the frequency is measured when the magnetization, having been perturbed from equilibrium, precesses around its equilibrium orientation. The precessional frequency depends upon all of the magnetic fields to which the magnetization is subject, including the dipolar and anisotropy effective fields.

The directional cosines with respect to the lattice coordinate system (X, Y, Z), see Figure 2.3, are then given by simple trigonometry:

$$\alpha_X = \frac{M_x}{M_s}\cos\varphi_M \sin\theta_M - \frac{M_y}{M_s}\sin\varphi_M - \frac{M_z}{M_s}\cos\varphi_M \cos\theta_M \tag{2.23}$$

$$\alpha_Y = \frac{M_x}{M_s}\sin\varphi_M \sin\theta_M + \frac{M_y}{M_s}\cos\varphi_M - \frac{M_z}{M_s}\sin\varphi_M \cos\theta_M \tag{2.24}$$

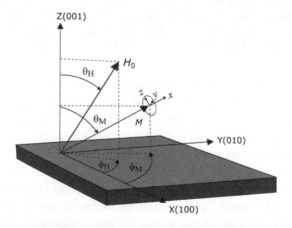

FIGURE 2.3 The laboratory system of coordinates is described by X, Y, Z. $\vec{M}$ and $\vec{H}$ are the magnetization and applied magnetic field, respectively. The x, y, z coordinates are used to describe the magnetization dynamics. The x-axis lies parallel with the static magnetization vector $\vec{M}_s$. y-axis is in the X–Y plane and perpendicular to the x–Z plane. (With kind permission from Springer Science+Business Media: *Magnetic Heterostructures: Advances and Perspectives in Spinstructures and Spintransport*, Exchange coupling in magnetic multilayers, Vol. 227 (2008), 183, Heinrich, B.)

$$\alpha_Z = \frac{M_x}{M_s}\cos\theta_M + \frac{M_z}{M_s}\sin\theta_M \qquad (2.25)$$

In this section, a simple example will be worked out at some length. The configuration of the magnetic field and the applied dc field is shown in Figure 2.3. For simplicity, let us assume that the contributions of the anisotropy to the free energy is given by the in-plane and perpendicular uniaxial terms and that the applied field and the static magnetization lie in the plane of the film, $\theta_{M,H} = \pi/2$. The Gibbs energy density is then given by

$$\mathcal{G} = \mathcal{F}_c - \frac{K_u^{\perp}}{2}\alpha_Z^2 - K_u^{\|}\frac{(\vec{u}\cdot\vec{M})^2}{M_s^2} - \vec{M}\cdot\vec{H}, \qquad (2.26)$$

where $\mathcal{F}_c$ is the cubic anisotropy shown in Equation 2.1. Substituting Equations 2.23 through 2.25 in Equation 2.26, one obtains the effective fields given by Equation 2.7. In FMR one usually uses a small transverse field $\vec{h}$ in addition to the dc applied field $\vec{H}_{dc}$. In this case, the torque on the magnetic moment leads to a precessional motion around the effective field having the transverse rf components $m_y, m_z \ll M_s$. The response of the magnetization is usually described by the LLG equations of motions

$$\frac{1}{\gamma}\frac{\partial\vec{M}}{\partial t} = -\left[\vec{M}\times\vec{H}_{\text{eff}}\right] + \frac{G}{\gamma^2 M_S}\left[\vec{M}\times\frac{\partial\hat{n}}{\partial t}\right] \qquad (2.27)$$

$$= -\left[\vec{M}\times\left(\vec{H}_{\text{eff}} - \frac{\alpha}{\gamma}\frac{\partial\hat{n}}{\partial t}\right)\right], \qquad (2.28)$$

where the first term on the right represents the torque due to the effective field, and the second term is the Gilbert damping torque. G is the Gilbert damping parameter, $\alpha = G/\gamma M_s$, $\vec{m}$ is the unit vector in the direction of the instantaneous magnetization $\vec{M}$, and γ is the absolute value of the gyromagnetic ratio, see further details in Ref. [3]. We will show solutions for the in-plane FMR where the static magnetic field and saturation magnetization are in the plane ($\theta_M = \pi/2$) of a laterally extended film (considering only demagnetizing fields perpendicular to the film surface). In this case, we will use a coordinate system with the in-plane and out-of-plane rf magnetization described by $m_{\|}$ and $m_{\perp}$, respectively. Solutions for $\vec{m}$ were found using the ansatz $\vec{m}$, $\vec{h} \sim \exp^{i\omega t}$, where ω is the angular microwave frequency. rf driving field h is assumed to be in-plane and perpendicular to the external field H_{dc}. Inserting the effective field $\vec{H}_{\text{eff}}$ in Equation 2.28 and keeping only terms linear in $m_{\|}$, $m_{\perp}$ one gets

$$i\frac{\omega}{\gamma}m_{\|} + \left(\mathbf{B}_{\text{eff}} + i\alpha\frac{\omega}{\gamma}\right)m_{\perp} = 0, \qquad (2.29)$$

$$\left(\mathbf{H}_{\text{eff}} + i\alpha\frac{\omega}{\gamma}\right)m_{\|} - i\frac{\omega}{\gamma}m_{\perp} = M_s h\,\cos(\varphi_M - \varphi_H). \qquad (2.30)$$

The LLG equation of motion also includes a large-term proportional only to the large effective field components. This term is zero after minimizing the total Gibbs potential in Equation 2.26. It states that the dc torque is zero in equilibrium, see Ref. [3]. The effective fields $\mathbf{B_{eff}}$ and $\mathbf{H_{eff}}$ are

$$\mathbf{B_{eff}} = H_{dc}\cos(\varphi_M - \varphi_H) + 4\pi M_{eff}$$

$$+ \frac{K_1^{\|}}{2M_s}(3 + \cos 4\varphi_M) + \frac{K_u^{\|}}{M_s}\left(1 + \cos 2(\varphi_M - \varphi_u)\right), \quad (2.31)$$

$$\mathbf{H_{eff}} = H_{dc}\cos(\varphi_M - \varphi_H) + \frac{2K_1^{\|}}{M_s}\cos 4\varphi_M$$

$$+ \frac{2K_u^{\|}}{M_s}\cos 2(\varphi_M - \varphi_u). \quad (2.32)$$

Notice that the cubic anisotropy for the in-plane FMR contributes only with the in-plane $K_1^{\|}$, the out-of-plane cubic anisotropy can be neglected because its effective field depends as the 3th power in small rf magnetization component $m_{\|}$. Here, the demagnetizing factor $4\pi M_s$ is grouped together with the perpendicular uniaxial anisotropy contribution:

$$4\pi M_{eff} = 4\pi M_s - \frac{2K_u^{\perp}}{M_s}. \quad (2.33)$$

The demagnetizing field is usually described by a magnetic continuum in which the demagnetizing field is $4\pi M_s$. This approach is incorrect in ultra-thin films where the atomic magnetic moments are localized around the atomic sites. The discreteness of atomic moments results in a demagnetizing field that changes across the film thickness. The demagnetizing field from a given atomic layer decreases exponentially away from its surface with a decay length corresponding to the in-plane lattice spacing. Consequently, inside the film, the demagnetizing field decreases when approaching the film surface. This results in a perpendicular surface dipole – dipole uniaxial energy term

$$E_{d-d,s} = -c_{str}\left(2\pi M_s^2\right)a_0\alpha_z^2, \quad (2.34)$$

where a_0 is the interplanar spacing (in cm). The structural factor $c_{str} = 0.425$ and 0.234 for the bcc and fcc lattice, respectively. For bcc Fe(001), this dipole-dipole interface term is $K_{u,int}^{\perp} = 0.06\,\mathrm{erg/cm^2}$. See further details in Ref. [3].

Notice that only the in-plane fourfold effective field affects FMR. The effective fourfold field perpendicular to the surface includes only a third-order term in the perpendicular component m_z, see Equation 2.1. The transverse rf susceptibility χ_y can be found from the system of Equations 2.29 and 2.30:

$$\chi_y \equiv \chi_y' - i\chi_y'' \equiv \frac{m_y}{h}$$

$$= \frac{M_s\left(\mathbf{B_{eff}} + i\alpha\dfrac{\omega}{\gamma}\right)}{\left(\mathbf{B_{eff}} + i\alpha\dfrac{\omega}{\gamma}\right)\left(\mathbf{H_{eff}} + i\alpha\dfrac{\omega}{\gamma}\right) - \left(\dfrac{\omega}{\gamma}\right)^2}\cos(\varphi_M - \varphi_H), \quad (2.35)$$

where χ'_y and χ''_y are dispersive and absorptive parts of the rf susceptibility, respectively. The term $\cos(\varphi_M - \varphi_H)$ represents a decreasing sensitivity by having noncollinear configuration due to dragging the saturation magnetization behind the applied field. FMR occurs when the denominator of the susceptibility function χ_y is at a minimum. Neglecting a small damping contribution to the resonance condition,

$$\left(\frac{\omega}{\gamma}\right)^2 = \mathbf{B}_{\mathrm{eff}}\mathbf{H}_{\mathrm{eff}}\,|_{H_{FMR}}. \tag{2.36}$$

Equation 2.35 shows that the dependence of x''_y on the externally applied field can be mathematically represented for a narrow FMR linewidth ($\Delta H \ll H_{FMR}, B_{FMR}$) by a Lorentzian function centered around the FMR field H_{FMR} with a half width at half maximum given by $\Delta H = \alpha\omega/\gamma$, neglecting terms quadratic in $\alpha\dfrac{\omega}{\gamma}$ and the magnetization dragging behind the field H.

$$\chi''_y \approx \left(\frac{M_s \mathbf{B}_{\mathrm{eff}}}{\mathbf{B}_{\mathrm{eff}} + \mathbf{H}_{\mathrm{eff}}}\right)_{H = H_{FMR}} \times \frac{\Delta H}{(\Delta H)^2 + (H_{dc} - H_{FMR})^2} \tag{2.37}$$

A simple calculation can be carried out for perpendicular FMR, where $\vec{M}_s$ and $\vec{H}$ are perpendicular to the film surface. In this case, the effective in-plane fourfold anisotropy field is proportional to the third power of the rf magnetization components and can be neglected. It is convenient to use a coordinate system with the x-axis parallel to the in-plane uniaxial anisotropy and the z-axis perpendicular to the film surface. The perpendicular anisotropy $K_1^\perp$ leads to an effective field $2K_1^\perp / M_s$ in the perpendicular direction and results in an effective perpendicular dc field of $4\pi M_{\mathrm{eff}} - 2K_1^\perp / M_s$. The effective field due to the in-plane uniaxial anisotropy can be easily obtained using the second term in Equation 2.26 and partial derivatives, Equation 2.7, with respect to the magnetization components $m_{x,y}$; $h_{x,u}^{\|} = \left(2K_u^{\|} / M_s^2\right)m_x$. Using the LLG equation, one obtains

$$\left(\frac{\omega}{\gamma}\right)^2 = \left(H_{dc}^\perp - 4\pi M_{\mathrm{eff}} + \frac{2K_1^\perp}{M_s}\right)$$
$$\times \left(H_{dc}^\perp - 4\pi M_{\mathrm{eff}} + \frac{2K_1^\perp}{M_s} - \frac{2K_u^{\|}}{M_s}\right). \tag{2.38}$$

Notice that the in-plane uniaxial anisotropy leads to an elliptically polarized rf magnetization precession even in the perpendicular configuration.

The in-plane FMR measurements allow one to determine the in-plane anisotropy fields. The perpendicular FMR allows one to determine the perpendicular fourfold anisotropy field $2K_1^\perp / M_s$. In ultrathin films, the in-plane and perpendicular fourfold anisotropy fields are often dramatically different (even in sign) for a film thickness of less than 10 ML. This is caused by an additional anisotropy arising from an inhomogeneous lateral distribution of the interface perpendicular uniaxial anisotropy, see Section 2.3.3. The gyromagnetic ratio ($\gamma \sim g$ factor) can be determined either by using out-of-plane FMR measurements in which the external field is rotated from the in-plane to the perpendicular FMR configuration or by using FMR measurements at two or more different microwave frequencies. However, the g factor can be dependent for on the in-plane and out-of-plane FMR. In this case, g factor has to be determined by using FMR at different microwave frequencies for the in-plane and out-of-plane FMR.

In FMR experiments using microwave standard waveguides, one usually maintains a constant microwave frequency and sweeps the applied field. In microstrip and coplanar waveguides, FMR measurements can be also carried out by sweeping the microwave frequency at the constant applied magnetic field. The resonance field corresponds to a maximum in $\chi^{\parallel}$; see the review articles by Heinrich [18], Farle [19], and Poulopoulos and Baberschke [20]. An example of the use of FMR to determine magnetic anisotropies is shown in Figure 2.4.

Note that FMR measurements yield effective fields. It is, therefore, necessary to know M_s in order to obtain the anisotropy coefficients K_i. Usually, the saturation magnetization must be measured in a separate experiment using a sensitive magnetometer such as a superconducting quantum interference device (SQUID) or a vibrating sample magnetometer (VSM). However, the strength of the FMR absorption signal is proportional to M_s, and it proves to be possible to obtain M_s with an accuracy of a few percent by means of very careful absorption measurements [21].

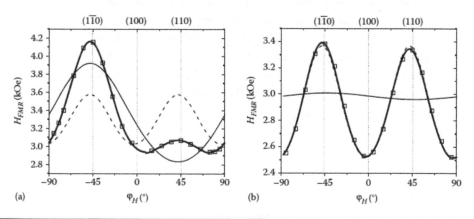

FIGURE 2.4   The angular dependence of the resonant magnetic field H_{res} at 24 GHz. **(a)** 20Au/16Fe/GaAs(001). **(b)** 20Au/70Fe/GaAs(001). The integers represent the number of atomic layers. Both specimens were grown using molecular beam epitaxy (MBE). The angle of the applied magnetic field, φ_H, was rotated in the plane of the film and measured with respect to the $[100]_{Fe}$ crystallographic direction. The data are plotted using open squares. The contributions of the fourfold $K_1^{\parallel}$ and uniaxial $K_u^{\parallel}$ magnetic anisotropies are shown using the dashed and thin solid lines, respectively. The thick solid line is a computer fit to the data using the following magnetic parameters: **(a)** 20Au/16Fe/GaAs – $K_1^{\parallel} = 2.61 \times 10^5$ erg/cm^3, $K_u^{\parallel} = -5.01 \times 10^5$ erg/cm^3, $4\pi M_{eff} = 16.8$ kOe; **(b)** 20Au/70Fe/GaAs – $K_1^{\parallel} = 3.77 \times 10^5$ erg/cm^3, $K_u^{\parallel} = -0.2 \times 10^5$ erg/cm^3, $4\pi M_{eff} = 19.6$ kOe. The g factor was fixed at $g = 2.09$ and the hard axis of the in-plane uniaxial anisotropy lies along the $[1\,\bar{1}\,0]$ direction. Clearly, the 70 ML thick Fe film and the 16 ML thick Fe film exhibit different magnetic anisotropies. The significantly smaller in-plane uniaxial anisotropy in the 70Fe film compared to the 16Fe film reveals a complex interface origin. (After doctoral thesis of Dr. Bartek Kardasz. With permission.)

2.2.5 ANISOTROPY DATA

More detailed data and discussion are provided in the review article by Heinrich and Cochran [17]; in particular, refer to their table 2.1 for anisotropy parameters measured for interfaces between Fe(001) and vacuum, Ag, Cu, Au, and Pd. The review article by Farle [19] contains an extended discussion of the Ni/Cu(001) and the Gd/W(001) systems. The case of Fe grown on GaAs(001) has become very interesting because the small mismatch between the Fe(001) and the GaAs(001) interface makes this combination an obvious choice for devices based on a hybrid ferromagnetic/semiconductor system. The Fe/GaAs(001) system is exhaustively discussed in the monumental review article by Wastlbauer and Bland [22]; this article includes a compilation of measured anisotropy coefficients plus relevant references. Bayreuther and coworkers extensively studied the in-plane uniaxial anisotropy [23, 24].

Seki et al. [25] have shown that the thin films of the alloy $Fe_{38}Pt_{62}$, having a uniaxial perpendicular anisotropy parameter as large as $K_u^{\perp} = 1.8 \times 10^7$ erg/cc, can be grown on an MgO(001) substrate at substrate temperatures as low as 300°C. These films, 18 nm thick, were grown by means of co-deposition of Fe and Pt by sputtering on a (1 nm Fe + 40 nm Pt) buffer layer deposited on the MgO(001) crystal at room temperature. The films exhibited a long-range intermetallic order having the $L1_0$ structure. Such films may be useful for ultrahigh-density recording media.

2.2.5.1 Vicinal Templates

A special case of decreased symmetry is provided by film growth on a template having a self-assembled network of oriented atomic steps. If a crystal is cut so that the surface plane makes a small angle with respect to a principal crystallographic plane, the result is a surface containing many monatomic steps. Such a surface is called a vicinal surface. In a series of experiments in which Fe films and Fe films covered by Ni films were grown on vicinal Ag substrates, it was discovered that the specimens exhibited an in-plane two-fold magnetic anisotropy in which the direction of the uniaxial anisotropy axis was correlated with the orientation of the vicinal surface steps [26]. A few years later, systematic studies by Krams et al. [27] have shown that the in-plane uniaxial anisotropy due to a stepped surface is a consequence of the lattice mismatch between the magnetic thin film and the substrate upon which it has been grown. See further discussion of magnetic anisotropy on vicinal surfaces in Ref. [17]. Recently, Li et al. [28] studied the in-plane uniaxial anisotropy for the structure Fe/Ag(1,1,10). The Ag substrate has step edges along the $[\bar{1}10]$ direction with the average terrace width of 2 nm. The Fe films were grown by molecular beam epitaxy (MBE) as a wedged shape along the $[\bar{1}10]$ direction of the Ag substrate, i.e., along the [100] direction of the Fe film. For temperatures less than 200 K, they observed strong oscillations of the uniaxial magnetic anisotropy as a function of the Fe thickness with a period of 5.7 atomic layers, see Figure 2.5. For Au-covered samples, the uniaxial anisotropy axis oscillated between the perpendicular and parallel (P) directions to the atomic steps. This behavior is attributed to QWS in

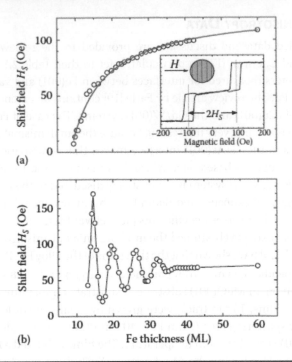

(a)

(b)

FIGURE 2.5   The shift field (saturation field) measured as a function of the Fe film thickness for a Fe film grown on a vicinal surface [Fe/Ag(1,1,10)] at (a) RT and (b) 5K. The inset shows a typical MOKE signal. (After Li, J. et al., *Phys. Rev. Lett.* 102, 207206, 2009. Copyright 2009 by the American Physical Society. With permission.)

the Fe film. At room temperature, the magnetization of Fe is parallel with the atomic steps, independent of the Fe layer thickness, see Figure 2.5a.

2.3 INTERLAYER EXCHANGE COUPLING

2.3.1 PHENOMENOLOGY OF MAGNETIC COUPLING

The simplest form of magnetic coupling per unit area of an ultrathin film trilayer structure consisting of two ferromagnetic films (F) separated by an NM spacer, F1/NM/F2, can be described by an interface bilinear form

$$\mathcal{E}_1 = -J_1 \vec{m}_1 \cdot \vec{m}_2, \tag{2.39}$$

J_1 is the exchange coupling coefficient in erg/cm^2
$\vec{m}_1$ and $\vec{m}_2$ are unit vectors along the magnetic moments in layers 1 and
    2, respectively

Another common coupling equation has a biquadratic form

$$\mathcal{E}_2 = J_2(\vec{m}_1 \cdot \vec{m}_2)^2, \tag{2.40}$$

where J_2 describes the strength of biquadratic coupling. Cases of the bilinear coupling J_1 are found having either a positive sign (+) or a negative sign (−). For a +sign, the minimum of the bilinear energy term is reached for a parallel

(P) orientation of the magnetic moments; for a −sign, the minimum energy corresponds to antiparallel (AP) magnetic moments. J_2 is almost exclusively found to be positive, and the minimum of the biquadratic energy term is reached when the magnetic moments are oriented perpendicularly to each other. A detailed description of bilinear and biquadratic coupling terms will be carried out in Section 2.3.3.

2.3.2 QUANTUM WELL STATES AND INTERLAYER EXCHANGE COUPLING

The first successful model of interlayer exchange coupling was introduced by Mathon et al. [29]. They correctly pointed out that exchange coupling is primarily a property of the NM spacer and is related to the confinement of Fermi surface electrons in the NM. This model was quickly extended to include the spin-dependent electron reflectivity at the F/NM interfaces [30, 31]. One has to include the itinerant nature of the 3d, 4sp electrons in the ferromagnetic (F) layers. The interlayer bilinear coupling, J_1, is given by the difference in energy between the AP and P alignment of the magnetic moments in the F/NM/F structure [32],

$$J_1 = \frac{1}{2S}\left(E_{\uparrow\downarrow} - E_{\uparrow\uparrow}\right), \tag{2.41}$$

where S is the area of the film. Calculations of energy differences are simplified by using the *force theorem*. The main problem is how to treat electron correlations self-consistently. The force theorem says that the energy difference between the two configurations is well accounted for by taking the difference in single particle energies. It is adequate to take an approximate spin-dependent potential and to calculate the single particle energies in the P and AP configurations. This difference in energy is very close to that obtained from self-consistent calculations, see the further discussion in Ref. [32]. In fact, this section closely follows Stiles's Section 4.3 in [32]. This procedure based on the force theorem significantly simplifies the calculation of exchange coupling and interface magnetic anisotropies. In calculations of the interlayer exchange coupling energies, one does not create a large error by neglecting spin-orbit interactions, while in calculations of the interface anisotropies spin-orbit coupling is the crucial ingredient. Single particle energy calculations require one to evaluate the electron energy for four QWS, see Figure 2.6. For thick F layers, one finds large energy contributions. However, these large contributions cancel out in the difference Equation 2.41. In order to avoid mistakes in this procedure, it is better to calculate the cohesive energy of the QWS by subtracting the bulk contributions,

$$\Delta E_{QWS} = E_{\text{tot}} - E_F V_F - E_{NM} V_{NM}, \tag{2.42}$$

where $V_{F,NM}$ and $E_{F,NM}$ are the total volumes and bulk energies for F and NM layers, respectively.

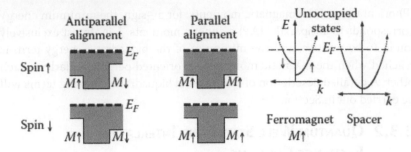

FIGURE 2.6   Quantum wells employed in the calculation of the interlayer exchange coupling energy. These spin-dependent potentials correspond reasonably well to a Co/Cu/Co(001) system. On the left side, the four panels show quantum wells for spin-up and spin-down electrons and for parallel and antiparallel alignment of the magnetic moments. The gray regions show the occupied states. (With kind permission from Springer Science+Business Media: *Ultrathin Magnetic Structures*, Interlayer Exchange Coupling, Vol. III, 99 (2005) Stiles, M.D.)

2.3.2.1 Quantum Interference

Let us consider a simple 1D model in which an electron having a wave vector $k_\perp$ travels inside the NM spacer, of width d, and is partially reflected at the interfaces, F/NM (interface A) and NM/F (interface B). The reflection coefficients are $R_{A,B} = r_{A,B}\exp(i\phi_{A,B})$. After multiple interference, the electron density of states (EDS) changes. The phase of the wavefunction after a round trip changes by

$$\Delta\phi = 2k_\perp d + \phi_A + \phi_B, \tag{2.43}$$

The amplitude after multiple reflections is given by a sum of round trips

$$\sum_{1}^{\infty} r_A r_B e^{i\Delta\phi} = \frac{r_A r_B e^{i\Delta\phi}}{1 - r_A r_B e^{i\Delta\phi}}. \tag{2.44}$$

The denominator becomes small when one obtains a constructive interference, $\Delta\phi = 2n\pi$. For energies less than the potential barrier at the interface $r_A = r_B = 1$ and one gets perfect QWS. For energies greater than the barrier energy, the QWS become broader resonances due to an extension of the wave amplitude into the surrounding F layers. By changing the NM spacer thickness, these states can be made to pass through the Fermi energy, see Figure 2.7, which leads to an oscillatory behavior of the cohesive energy and consequently to an oscillatory interlayer exchange coupling energy. The first clear experimental observation of QWS was presented by Himpsel and Ortega [33, 34] using photoemission and inverse photoemission using a nonmagnetic layer on top of a magnetic layer.

In the first approximation, the change in the density of states (DOS) due to interference, $\Delta n(\varepsilon)$, should be proportional to $r_A r_B \cos(\Delta\phi)$, the spacer

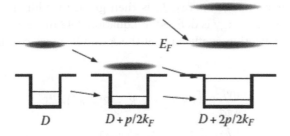

FIGURE 2.7   Evolution of quantum well (QW) states as a function of the film thickness. The solid lines represent bound states (localized in the QW) and resonance states are shown by the "fuzzy ellipses". E_F is the Fermi energy. (With kind permission from Springer Science+Business Media: *Ultrathin Magnetic Structures, Interlayer Exchange Coupling*, Vol. III, 99, (2005) Stiles, M.D.)

width d and the DOS per unit length, $(2/\pi)(dk_\perp/d\varepsilon)$ [31]. Therefore, $\Delta n(\varepsilon)$ per spin can be written as

$$\Delta n(\varepsilon) \simeq \frac{2d}{\pi}\frac{dk_\perp}{d\varepsilon}r_A r_B \cos(\Delta\phi). \qquad (2.45)$$

For multiple scattering, one has to use the expression in Equation 2.44. It is relatively easy to show that Equation 2.45 can be generalized to [35]

$$\Delta n(\varepsilon) = -\frac{1}{\pi}\operatorname{Im}\frac{d}{d\varepsilon}\Big[\ln\big(1 - r_A r_B e^{i\Delta\phi}\big)\Big]. \qquad (2.46)$$

Note that Equation 2.46 equals Equation 2.45 for small reflection coefficients. The cohesive energy is then given by

$$E_{\text{coh}} = -\int_{-\infty}^{E_F} d\varepsilon\,(\varepsilon - E_F)\Delta n(\varepsilon). \qquad (2.47)$$

Using integration by parts, one gets

$$E_{\text{coh}} = \frac{1}{\pi}\operatorname{Im}\int_{-\infty}^{E_F} d\varepsilon\,\ln\big(1 - r_A r_B e^{i\Delta\phi}\big). \qquad (2.48)$$

For fixed thickness d, the integral oscillates rapidly as a function of $k_\perp$. Only energies close to the Fermi level will generate nonzero contributions. It can be shown that in these regions for large d [32]

$$E_{\text{coh}} = \frac{\hbar v_F}{2\pi d}\sum_n \frac{1}{n}\operatorname{Re}\Big(\big(r_A r_B\big)^n e^{in\Delta\phi(k_F)}\Big). \qquad (2.49)$$

For small reflection coefficients

$$E_{\text{coh}} \simeq \frac{\hbar v_F}{2\pi d}r_A r_B \cos\big(2k_F d + \phi_A + \phi_B\big). \qquad (2.50)$$

The interlayer exchange energy, J_1, is then given by adding all cohesive energies (cohesive energy is defined in Equation 2.42 in Figure 2.7, assuming the same reflection coefficients at the A and B interfaces. In the limit of large d

$$
\begin{aligned}
J_1 &\simeq \frac{\hbar v_F}{4\pi d}\operatorname{Re}\Big[\big(R_-R_\downarrow + R_\downarrow R_- - R_-^2 - R_\downarrow^2\big)e^{i2k_Fd}\Big] \\
&= -\frac{\hbar v_F}{4\pi d}\operatorname{Re}\Big[(R_- - R_\downarrow)^2 e^{i2k_Fd}\Big],
\end{aligned}
\tag{2.51}
$$

where $R_\uparrow$ and $R_\downarrow$ are the corresponding reflectivities, for the majority and minority electrons spins, assuming symmetric interfaces $(R_A = R_B)$. The exchange coupling in this simple 1D limit is inversely proportional to the film thickness, d, and its associated oscillatory spatial period as a function of d is given by the Fermi spanning vector $2k_F$. In 3D space, one has to take into account all k-vectors parallel to the surface. These k-vectors, resulting from the lattice periodicity, are conserved in going from F to NM regions. The total cohesive energy per unit area involves the integration of the QWS over the interface Brillouin zone. The integrand for E_{coh} oscillates rapidly with the $\vec{k}$-wavevectors except on areas of the Fermi surface where opposite sheets of the Fermi surface are nearly parallel. The vectors connecting these parts of the Fermi surface are called *critical spanning vectors*. The spanning k-vectors for (001) interfaces for simple metals such as Cu are shown in Figure 2.8.

The exchange coupling involves the difference in cohesive energies (see Equation 2.42) for P and AP configurations of the magnetic moments. In its asymptotic form, this coupling can be written as

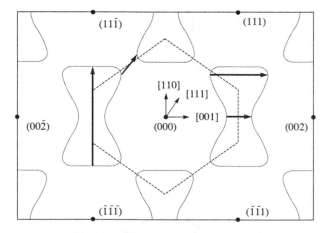

FIGURE 2.8 The [1$\bar{1}$0] cross-section of the fcc Cu Fermi surface. The hexagon of dashed lines outlines the first Brillouin zone. The solid dots represent reciprocal k-vectors. All three important orientations are present. The critical spanning vectors in the extended Brillouin zone are denoted by the solid arrows. Along the [001] direction the two critical spanning vectors are located at the belly and neck of the Fermi surface. (After Bruno, P., *J. Physics: Condensed Matter* 11, 9403, 1999. With permission.)

$$J \simeq \sum_{\alpha} \frac{\hbar v_{\perp}^{\alpha} \kappa^{\alpha}}{16\pi^2 d^2} \mathrm{Re}\left((R_{\uparrow}^{\alpha} - R_{\downarrow}^{\alpha})^2 e^{iq_{\perp}^{\alpha}d} e^{i\chi^{\alpha}} \right), \qquad (2.52)$$

where the summation is over all critical points of the Fermi surface corresponding to the film crystalline orientation perpendicular to the film surface

$v_{\perp}^{\alpha}$ are Fermi velocities at the spanning vectors

$q_{\perp}^{\alpha}$ is the length of a critical spanning wavevector

χ^{α} is the phase associated with the type of critical point

κ^{α} is the average radius of curvature of the Fermi surface at the critical point, see Ref. [32]

$R_{\uparrow}^{\alpha}$ and $R_{\downarrow}^{\alpha}$ are corresponding reflectivities

The periods of the observed exchange coupling oscillations as the film thickness is varied are in good agreement with those obtained in de Haas–van Alphen measurements, see Table 4.1 in Ref. [32]. A detailed discussion of the calculations of exchange coupling for Co/Cu/Co(001), Fe/Au/Fe(001), and Fe/Ag/Fe(001) systems can be found in Ref. [32]. The quantitative agreement for the exchange coupling between theory and experiment is far from being good. The main reason is that the interfaces in real samples are far from being ideal, and measurements are often not carried out in the asymptotic thickness limit.

Comprehensive studies of exchange coupling and its relationship to QWS were carried out by the Qiu group at the University of California at Berkeley [36] (see references within the topical review from Qiu) using a wedged Cu spacer in Co/Cu/Co(001) structures grown on Cu(001) single crystal substrates. This system was particularly convenient for such studies because Cu has a simple Fermi surface whose sp bands can be easily separated from the other energy bands, see the Fermi surface of Cu in Figure 2.8. Cu and Co can be grown in the (001) orientation with atomically flat interfaces. Angular-resolved photoemission spectroscopy (ARPES) of QWS was carried out at the advanced light source (ALS) of the Lawrence Berkeley National Lab. The DOS is significantly increased at energies corresponding to the QWS. This allows one to follow the QWS as a function of energy for different Cu thicknesses.

In Figure 2.9, ARPES measurements show the formation of QWS corresponding to the belly direction of the fcc Cu Fermi surface, see Figure 2.8. The study was carried out for 20 ML thick Co grown on a Cu(001) substrate and with a Cu wedge grown on top of the Co layer. The ARPES oscillations have clearly shown the QWS corresponding to the sp electrons in the Cu layer. The periodicity of the oscillations corresponding the belly direction (photoemission spectra were obtained along the surface normal) was found to be 5.88 atomic layers at the Fermi level, and this is exactly the periodicity of the long period exchange coupling oscillations observed for Co/Cu/Co(001) systems.

The interlayer exchange coupling rarely behaves as outlined in the above theory. The interface roughness often smears out the presence of

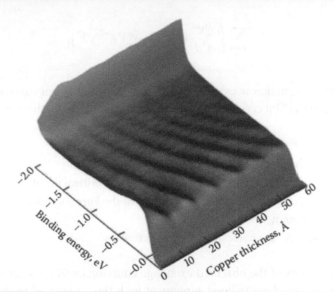

FIGURE 2.9 Photoemission spectra obtained along the surface normal corresponding to the belly direction of the fcc Cu Fermi surface [36]. Oscillations in intensity as a function of the Cu layer thickness and electron energy demonstrate the formation of quantum well states (QW). (After Qiu, Z.Q. et al., *J. Phys.: Cond. Matter* 14, R169, 2002. With permission.)

short-wavelength oscillations and even significantly decreases the strength of long-wavelength interlayer coupling oscillations. Various contributions to magnetic interlayer coupling and the role of interface roughness and interface chemistry are extensively described in Ref. [37]. Here, the discussion will be limited to the biquadratic exchange coupling.

2.3.2.2 Quantum Well States in an Irreversible Process of Spin Pumping

It is worth pointing out that recently, Montoya et al. [38] have made experimental observations of the quantum well states in the irreversible process of spin pumping in a GaAs/Fe/Au/Pd structure. Spin pumping is a form of magnetic interface damping at the F/NM interface, resulting in a spin accumulation in the NM (Au). The increase in damping in F due to spin pumping is given by [39],

$$\alpha_{sp} = \frac{g\mu_B}{4\pi M_s}\frac{g_{\text{eff}}}{d_F}$$

(2.53)

where:

g is the Lande factor

μ_B is the Bohr magneton

M_s is the saturation magnetization of the F layer

d_F is the thickness of the F layer

g_{eff} is an effective spin mixing conductance which will depend on the F/NM$_1$ interface, the material properties of the NM$_i$ and thickness of the NM$_i$ for all layers i

This result is in good agreement for most F/NM or $F/NM_1/NM_2$ spin pumping experiments [40]. However, in the $17Fe/d_{Au}Au/50Pd$ structure, where the numbers represent the thicknesses of layers in monolayers, an oscillatory behavior dependent on the thickness of the Au layer was observed in Ref. [38]. This result cannot be described by the current theory of spin pumping. It was found that the oscillation has two length scales: 8–10 and 2–3 atomic layers, which are very close to those found for the quantum well oscillations of interlayer exchange coupling in Au(001) associated with the spanning k-vectors along the belly and neck of the Au Fermi surface. Spin pumping involves electron transitions from $-\vec{k}_F$ to $\vec{k}_F$ with the spin-flip, which is similar to that in interlayer exchange coupling. It looks like the quantum confinement in the accumulated spin-density, created by spin pumping in Au, enhances these transitions along the belly and neck areas of the Fermi surface of Au.

2.3.3 BIQUADRATIC EXCHANGE COUPLING

The presence of biquadratic exchange coupling was observed at the same time by Heinrich et al. [41] for Co/Cu/Co(001) trilayers and by Ruehrig et al. [42] for Fe/Cr/Fe(001) trilayers. The evidence for biquadratic exchange coupling in Ref. [41] was obtained using magnetization loops. In order to properly explain the observed critical fields, one needed to use an angular-dependent exchange coupling parameter in the form of

$$J(\theta) = J_1 - J_2 \cos(\theta), \qquad (2.54)$$

where θ is the angle between the magnetic moments. Consequently, the corresponding interlayer exchange energy was given by

$$E(\theta) = -J(\theta)\cos(\theta) = -J_1\cos(\theta) + J_2\cos^2(\theta). \qquad (2.55)$$

Ruehrig et al. observed a perpendicular orientation of the magnetic moments in an Fe/wedged Cr spacer/Fe(001) whisker sample in which the Cr interlayer was grown with a linearly variable thickness. They explained their observed domain patterns on Fe film by adding a term B_{12} allowing one to reach a noncollinear orientation between the magnetic moments in the Fe film ad Fe-whisker. The total interlayer exchange coupling $E(\theta)$ was given by

$$E(\theta) = -A_{12}\cos(\theta) - B_{12}\cos^2(\theta). \qquad (2.56)$$

Clearly, these two concepts are identical. Slonczewski soon after that proposed a theoretical interpretation [43]. He realized that fluctuations in the NM interlayer thickness could result in an additional coupling term. The nonmagnetic layers at different parts of the sample have different thicknesses and consequently generate different strengths of magnetic coupling between the two ferromagnetic layers. Short-wavelength oscillations can even result in changing the coupling from ferromagnetic to AF. His model is applicable when lateral variations in the bilinear interlayer exchange field are suppressed by corresponding lateral variations of the bulk exchange field. This means that local angular variations from the average direction of the magnetic moments are small so that, in this case, the problem can be treated by perturbation theory. The magnetic moments are frustrated across the film surface by a variable interlayer coupling. Consequently, there is an additional energy term that prefers to orient the magnetic moments in the F layers perpendicularly to each other. This additional coupling has then an angular dependence given by $\cos^2(\theta)$, for which the name biquadratic exchange coupling was coined. Its strength is given by the competition between variations in the interlayer exchange coupling field, $\Delta J_1 / M_s d$, and the in-plane intralayer exchange field, $2Ak^2 / M_s$. Where d is the thickness of the F layers and the length of the k-vector, k, is given by the average size of atomic terraces, L; $k = \pi / L$. Slonczewski has shown that the interlayer exchange coupling fluctuations are decreased due to exchange averaging. Physical meaning of biquadratic exchange coupling can be expressed for ultrathin films thinner than the lateral variations in the atomic terrace size L by

$$J_2 \cong \Delta J_1 \frac{\Delta J_1 / M_s d}{2Ak^2 / M_s}. \tag{2.57}$$

Notice that the large fraction describes the exchange averaging effect. A more general description can be found in Solnczewski Ref. [43] and Stiles section 4.4.1 in Ref. [32]. The above expression shows that biquadratic coupling has only a positive sign, and therefore it always encourages a perpendicular orientation of the magnetic moments. The angle between the magnetic moments is given by a competition between the bilinear, biquadratic magnetic couplings, and the magnetic anisotropies, see Section 2.3.4.1. In zero field, this angle can range continuously from 0 to π.

It is often believed that biquadratic exchange coupling occurs only from short-wavelength exchange coupling oscillations where the exchange coupling changes its sign between two subsequent atomic terraces. This is not correct. Any lateral variations in magnetic coupling strength (including ferromagnetic coupling) will result in biquadratic exchange coupling. Once the magnetic moments are arranged in a noncollinear state, the magnetic frustration due to an inhomogeneous magnetic coupling strength results in biquadratic magnetic coupling.

The Slonczewski idea of additional energy terms due to imperfect interfaces is more general. It has been shown [44] that:

In any system that exhibits a lateral inhomogeneity, one must expect additional energy terms. They originate from intrinsic magnetic energy terms that fluctuate in strength across the sample interface. These additional terms have the next higher angular power compared with that of the intrinsic term, and they have only one sign.

The power of the higher order angular term has to satisfy the requirements of the sample symmetry including time inversion symmetry. Here are several examples: (a) Variations of the interlayer exchange coupling results in a $\cos^2(\theta)$ angular term; (b) variations in a uniaxial interface perpendicular anisotropy results in an angular-dependent term having the form $\sin^2(\theta)\cos^2(\theta)$, where θ is the angle between the magnetic moment and the film normal: this term has an easy axis at 45° from the surface normal; (c) variation in the in-plane uniaxial anisotropy results in an in-plane fourfold anisotropy with the easy axis rotated by 45° with respect to the uniaxial anisotropy axis; (d) variation in the EB field results in a uniaxial anisotropy with the easy axis perpendicular to the EB field direction.

2.3.4 MEASURING TECHNIQUES

2.3.4.1 Static Measurements

In SQUID and VSMs, the total magnetic moment is measured, usually along the direction of the applied field. In the magneto-optical Kerr effect (MOKE), the magnetic moment is commonly measured in the polar and longitudinal configurations. In the longitudinal configuration, the dc field is applied parallel to the film surface, and it is usually applied in the plane of incidence of the laser beam. In the polar configuration, the incident laser beam impinges nearly perpendicularly on the film surface, and the MOKE signal is mainly sensitive to the magnetic moment component that is perpendicular to the film surface.

The discussion in this section will be limited to the micromagnetics of trilayer structures consisting of a crystalline ultrathin film F1/NM/F2 structure. For simplicity, these calculations will be limited to cubic materials with the film surface oriented in the (001) plane. Films of 3d transition group elements are often, but not exclusively, grown on (001) templates. This is a particular case but it involves all the ingredients needed to formulate calculations for any other configurations involving an arbitrary direction of the applied field and crystalline orientation. The discussion below should be viewed as an example taken from a user's manual.

The total free energy per unit area for the in-plane configuration (the magnetic moments $\vec{M}_{s1}$ and $\vec{M}_{s2}$, and the field $\vec{H}$ applied parallel to the film surface) is given by

$$\mathcal{F} = \sum_{i=1,2} \left[-\frac{K_{1,i}^{\parallel}}{2}\left(\alpha_{X,i}^4 + \alpha_{Y,i}^4\right) - K_{u,i}^{\parallel}\frac{\left(\vec{n}_{u,i} \cdot \vec{M}_i\right)^2}{M_{s,i}^2} - K_{u,i}^{\perp}\alpha_{Z,i}^2 - \vec{M}_i \cdot \vec{H} \right] d_i$$

$$-J_1\left(\vec{m}_1 \cdot \vec{m}_2\right) + J_2\left(\vec{m}_1 \cdot \vec{m}_2\right)^2, \tag{2.58}$$

where:

$\alpha_{X,i}$ $\alpha_{Y,i}$ and $\alpha_{Z,i}$ are the directional cosines between the magnetization vector $\vec{M}_i$ and the crystallographic axes [100], [010], and [001], respectively

$K_{1,i}^{\parallel}$ $K_{u,i}^{\parallel}$, and $K_{u,i}^{\perp}$ are the parameters that describe the in-plane effective fourfold (cubic) anisotropy, the in-plane uniaxial anisotropy, and the perpendicular uniaxial anisotropy

$\vec{n}_{u,i}$ are the directions of the in-plane uniaxial axes

$\vec{m}_i$ are unit vectors directed along the magnetizations of the coupled films

J_1 and J_2 are the bilinear and biquadratic coupling coefficients

The indices i = 1 and 2 describe the properties of the layers F1 and F2, respectively. In ultrathin films, the magnetic moments across the film are locked together by intralayer exchange coupling, and they can be considered to be giant magnetic molecules [3]. For ultrathin films, the role of the interface anisotropies associated with a uniformly magnetized sample can be included in the effective anisotropy parameter, see Equation 2.12. The energy expression in Equation 2.58 is valid for a wide range of magnetic ultrathin films such as Fe on Ag(001) [3] and GaAs(001) [45, 46] templates. One can easily generalize it by using the appropriate film symmetry.

For the in-plane configuration, $\theta_{M_{1,2}} = \theta_H = \pi / 2$. The static equilibrium is found by minimizing the total Gibbs energy with respect to the angles φ_{M_1} and φ_{M_2} for a given angle φ_H, see Figure 2.3.

There are a number of minimization procedures available, and they are usually implemented by individual groups according to their liking. One should realize that minimum energy solutions can exhibit metastable states. Usually, one looks for the lowest energy state. This means that no hysteresis is present in the magnetization measurements. Experimentally, this is not often the case, especially when using films that were grown on GaAs(001) substrates [45]. Therefore, it is imperative to carry out MOKE measurements with the specimen in the lowest energy state. The lowest energy state for a given magnetic field can be achieved by cycling the magnetic state at the given applied field using a transverse ac magnetic field, which increases in amplitude to some preselected maximum, and then the amplitude is gradually decreased to zero. This has to be repeated for each applied dc field. An example of such a procedure can be found in Ref. [45] using exchange coupled GaAs/Fe/Au/Fe/Au(001) structures.

Several examples of simulated hysteresis loops using exchange coupled magnetic bilayers F1/NM/F2 are shown in Figure 2.10. The external field was applied along the magnetic easy axis of the thicker film. In that case, the thicker film remains oriented close to the easy axis. It is the thinner film that undergoes a full angular dependence due to the presence of interlayer AF coupling, see the insets in Figure 2.10. Since the two magnetic moments are different it is easy to identify the contributions from each individual layer.

The family of curves in Figure 2.10 shows the role of biquadratic magnetic coupling. When the biquadratic magnetic coupling becomes comparable to the bilinear coupling, one is not able to achieve an AP configuration of magnetic moments in small magnetic fields. In zero field, the biquadratic coupling can lead to an angle between the two magnetic moments ranging

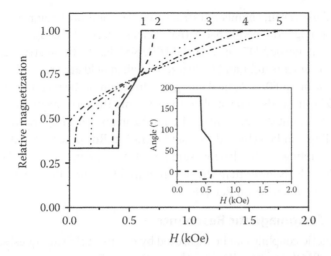

FIGURE 2.10 Simulations of magnetization curves using a set of exchange coupling parameters. The lines 1, 2, 3, 4 and 5 correspond to $J_2 = 0.0, 0.01, 0.04, 0.06$, and 0.08 erg/cm², respectively. The bilinear coupling parameter was kept constant, $J_1 = -0.1$ erg/cm². The calculations were carried out for the magnetic parameters obtained using GaAs/8Fe/Au/16Fe/Au(001) structures [46], where the integers represent the number of atomic layers. 16Fe: $K_1^\parallel = 3.1 \times 10^5$ erg/cm³; 8Fe: $K_1^\parallel = 1.33 \times 10^5$ erg/cm³. $4\pi M_s = 21.5$ kOe; in-plane uniaxial anisotropies were omitted in order to keep the easy magnetic axis in both films directed along the [100] crystallographic axis. The applied field was oriented along the magnetic easy axis [100]. The inset shows the field dependence of the magnetization angle of 16Fe (dashed line) and 8Fe (solid line) for $J_1 = -0.1$ erg/cm² and $J_2 = 0$. Note that the first jump brings the magnetic moment of the thinner 8Fe film from the parallel orientation over the first hard axis, [110], close to the second easy axis, [010]. The second jump corresponds to pulling the magnetic moment over the second hard axis, [1 1̄ 0], resulting in an antiparallel configuration of the magnetic moments. (With kind permission from Springer Science+Business Media: *Magnetic Heterostructures: Advances and Perspectives in Spinstructures and Spintransport*, Exchange coupling in magnetic multilayers, *Springer Tracts in Modern Physics*, Vol. 227, 185, (2008), Heinrich, B.)

from 180° to 90°. The biquadratic magnetic coupling can also affect the approach to saturation. For even small J_2, the saturation field is reached without a jump in the magnetic moment, see Figure 2.10. The saturation field can be easily determined when there is no discontinuity in the magnetic moment. The deviation from saturation can be treated using a small angle expansion. It is easy to show that the saturation field along the easy direction can be described by the simple expression

$$H_{\text{sat}} + \frac{2K_{\text{eff}}}{M_s} = -\frac{J_1 - 2J_2}{M_s}\left(\frac{1}{d_1} + \frac{1}{d_2}\right), \tag{2.59}$$

where it has been assumed that the saturation magnetization M_s is the same for both films. K_{eff} is an effective anisotropy field obtained from the minimization of Equation 2.58.

In F1/NM/F2 structures, a negative bilinear J_1 (AF coupling) can only be measured using MOKE. A positive (ferromagnetic) bilinear coupling can be measured using spin engineered structures. An additional magnetic F0 layer plus a nonmagnetic metal spacer that creates a large AF coupling between F0 and F1 is needed [47], e.g., F0/NM/F1/NM/F2. In these structures, the magnetic moments in F1 and F2 in a zero applied field are oriented AP to the magnetization in F0. A dc field along the direction of the magnetic moment of the F0 layer needs to be applied to overcome the ferromagnetic coupling and thus to orient the moments in F1 and F2 in an AP configuration. Strong AF coupling can be achieved by using ultrathin Ru spacers [47]. A positive bilinear coupling can be also measured in F1/NM/F2 structures if the nucleation field of F2 is lower than the field required to switch F1 from parallel to antiparallel configuration Ref. [48].

2.3.4.2 Ferromagnetic Resonance

The magnetic coupling can be measured by means of rf techniques, see Refs. [18, 3]. In FMR, one usually fixes the microwave frequency and sweeps the field. However, with network analyzers (NAs), this is not a limitation; one can set the field and sweep the frequency. When the field is held constant, the angle between the magnetic moments is fixed. This is a simpler situation compared to a regular FMR measurement (holding the frequency constant and sweeping the field) where the angle between the magnetic moments changes in noncollinear configurations for samples, which are not fully saturated (usual in measurements at low microwave frequencies). For a saturated sample, the difference between constant field and constant frequency sweeps is minimal.

Interpretation of FMR results can be carried out using rf solutions of Equation 2.60.

$$\frac{1}{\gamma_{1,2}}\frac{\partial \vec{M}_{1,2}}{\partial t} = -\left[\vec{M}_{1,2}\times\vec{H}_{\mathrm{eff},1,2}\right] + \frac{G_{1,2}}{\gamma_{1,2}^2 M_{s,1,2}}\left[\vec{M}_{1,2}\times\frac{\partial \vec{m}_{1,2}}{\partial t}\right], \qquad (2.60)$$

where $\gamma_{1,2}$, $\vec{M}_{1,2}$, and $G_{1,2}$ are the absolute values of the electron gyromagnetic ratios, magnetic moments, and Gilbert damping parameters of layers 1 and 2, respectively. The damping parameter is often expressed in the form of a dimensionless parameter $\alpha = G/\gamma M_s$. The first term on the right-hand side of Equation 2.60 represents the precessional torque, and the second term represents the well-known Gilbert damping torque. The effective fields $\vec{H}_{\mathrm{eff},1,2}$ are given by the derivatives of the magnetic Gibbs energy density, $\mathcal{G}$, and Equation 2.7.

$$\mathcal{G}_i = -\frac{K_{1,i}^{\parallel}}{2}\left(\alpha_{X,i}^4 + \alpha_{Y,i}^4\right) - K_{u,i}^{\parallel}\frac{\left(\vec{n}_{u,i}\cdot\vec{M}_i\right)^2}{M_{s,i}^2} - K_{u,i}^{\perp}\alpha_{Z,i}^2 - \vec{M}_i\cdot\vec{H}$$

$$+ \frac{\left[-J_1\left(\vec{m}_1\cdot\vec{m}_2\right) + J_2\left(\vec{m}_1\cdot\vec{m}_2\right)^2\right]}{d_i}, \qquad (2.61)$$

Where $\vec{M}_i$ is the magnetization densities. The partial derivatives for $H_{i,\text{eff}}$ are taken with respect to $M_{x,i}$, $M_{y,i}$, and $M_{z,i}$. In FMR, the rf transversal components are usually negligible compared to the magnetization component parallel to the x-axis (static effective fields). In ultrathin films with a symmetric rf driving field, any variations of the magnetization across the film thickness can be neglected. The total rf magnetic moment along the rf driving field $\vec{h}$ is then obtained by adding the rf magnetization components, along the rf field h, multiplied by the film thickness and lateral area for the individual layers.

The interlayer coupling described by the bilinear J_1 and biquadratic J_2 interlayer exchange coupling parameters lead to a coupled system of the LLG equation of motion. Coupled equations for the rf magnetization components lead to two solutions. The precessional motions are coupled and result in an acoustic mode in which the magnetic moments in the two layers precess in phase and in an optical mode in which the magnetic moments precess in antiphase. The simplest interpretation of the coupling can be obtained in the saturated case when the dc magnetic moments are parallel to the applied field. The isolated films F1 and F2 must have different resonant frequencies (fields) in order to be able to observe acoustic and optical modes. For a homogeneous rf driving across the magnetic structure if the resonance frequencies (fields) of the two films are exactly the same, then the strength of the optical mode is zero. For co-planar transmission line the rf driving is not homogeneous and consequently the optical peak can be observed even for the same resonance fields. Different resonant fields are easy to establish by choosing different film thicknesses for the films F1 and F2. The interface anisotropies scale with $1/d$ and in consequence the isolated films exhibit two separate resonance frequencies (fields), and the optical mode becomes observable. The sign of the coupling can be determined from the relative positions of the acoustic and optical modes. Calculated spectra for a magnetic double layer F1/NM/F2 for P and AP coupling are shown in Figure 2.11. For AP coupling, the acoustic and optical peaks move to higher magnetic fields at a fixed FMR frequency. The acoustic peaks keep increasing their intensity with increasing AP coupling whereas the optic peaks get weaker with increasing coupling. The acoustic peaks gradually approach a fixed magnetic field point, which is located between the resonance peaks for the uncoupled films. For P coupling, this trend is exactly the opposite. The resonance fields decrease with an increasing coupling. It is the film with a higher resonance field, which approaches the fixed point.

In the saturated state (collinear magnetic moments), the overall strength of the interlayer coupling, J_{eff}, is given by the superposition of bilinear and biquadratic interlayer couplings,

$$J_{\text{eff}} = J_1 - 2J_2. \tag{2.62}$$

Further details of studying interlayer exchange coupling including a noncollinear configuration of the magnetic moments can be found in Ref. [37].

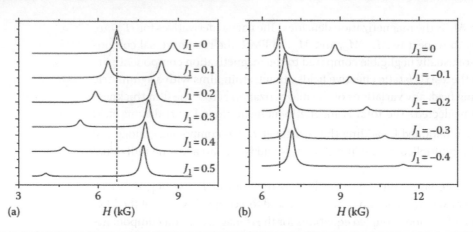

FIGURE 2.11   Simulations of acoustic and optical resonance absorption peaks at $f = 36$ GHz as a function of the bilayer exchange coupling strength in an F1/NM/F2 structure for parameters taken from GaAs/8Fe/Au/16Fe/Au where the integers are in a number of atomic layers. In panel (a), $J_1 = 0.0, 0.1, 0.2, 0.3, 0.4,$ and 0.5 erg/cm². In panel (b), $J_1 = 0.0, -0.1, -0.2, -0.3,$ and 0.4 erg/cm². The following magnetic parameters were used: 16Fe: $K_1^{\parallel} = 3.1 \times 10^5$ erg/cm³, $K_{u,s}^{\perp} = 0.88$ erg/cm², and $K_u^{\parallel} = 3.3 \times 10^4$ erg/cm³; 8Fe: $K_1^{\parallel} = 1.33 \times 10^5$ erg/cm³, $K_{\perp,s} = 0.82$ erg/cm², and $K_{u,\text{eff}}^{\parallel} = -1.14 \times 10^6$ erg/cm³. $4\pi M_s = 21.5$ kG, $g = 2.09$, and $\alpha = 0.009$. The in-plane uniaxial easy axes for the 16Fe and 8Fe films were oriented along the [110] and [1 $\bar{1}$ 0] directions, respectively. The applied field was oriented along the [1 $\bar{1}$ 0] crystallographic axis. (With kind permission from Springer Science+Business Media: *Magnetic Heterostructures: Advances and Perspectives in Spinstructures and Spintransport*, Exchange coupling in magnetic multilayers, Springer Tracts in Modern Physics, Vol. 227, 185 (2008), Heinrich, B.)

2.3.5 Experimental Results

2.3.5.1 Interlayer Exchange Coupling through Cu(001) and Cr(001)

Several examples will be shown to demonstrate the basic behavior and role of interfaces in the interlayer exchange coupling in magnetic ultrathin film structures. No attempt is made to account for the complete work done in this field. A more detailed review containing a large number of references can be found in Ref. [37].

2.3.5.2 Simple Normal Metal Spacers: Cu

It was pointed out in Section 2.3.2 that comprehensive studies of exchange coupling and its relationship to QWS were carried out by the Qiu group at the University of California, Berkeley [36] (see references within the topical review from Qiu). They used a wedged Cu spacer in Co/Cu/Co(001) structures grown on Cu(001) single crystal substrates. The copper spacer was oriented along the [001] crystallographic direction and provided two critical spanning vectors at the belly and at the neck of the Fermi surface, see Figure 2.8. The belly, 5.88, and neck, 2.67, atomic layer periodicities can be understood by means of the extended Brillouin zone picture, see the solid arrows in Figure 2.8. In this case, one subtracts from the regular spanning vector inside the first Brillouin zone (from belly to belly, from neck to neck) a k-vector having the atomic layer periodicity ($4\pi/a$). The oscillatory period for the exchange coupling through Cu is given by

$$2k^e d_{Cu} - \phi_A - \phi_B = 2\pi n, \tag{2.63}$$

where:

$k^e = k_{BZ} - k$, $k_{BZ} = 2\pi / a$ is a Brillouin zone vector

n is an integer

$\phi_{A,B}$ are the phase shifts of the electron wavefunctions upon reflection at the two boundaries of the potential well formed by the Cu layer surrounded by Co and vacuum

a is the lattice spacing of Cu

Equation 2.63 explains the long and short-wavelength coupling oscillation periods as due to the belly and neck spanning k-vectors, respectively. They are caused because the strength of the exchange coupling is evaluated at discrete atomic layer separations. This is often called the *aliasing effect*. Oscillations in ARPES intensity as a function of the Cu layer thickness clearly demonstrated the formation of QWS along the belly and neck of the Fermi surface, e.g., see Figure 2.9, showing the presence of long-wavelength oscillations across the belly of the Fermi surface.

The QWS form the underlying basis for the presence of the interlayer exchange coupling. To ensure the direct comparison of the exchange coupling periodicity with the QWS variation as a function of the Cu spacer thickness, half of the wedged sample was covered by a Co film 3 ML thick. XMCD measurements are only surface sensitive, and consequently, F and AF coupling can be determined by monitoring the XMCD signal coming from the 3 ML thick Co. Images of the DOS (using photoemission measurements) at the belly and neck of the Fermi surface were obtained by scanning the photon beam across the Cu wedge on the Co/Cu side of the wedged Cu film. Figure 2.12c shows the observed XMCD signal with maxima and minima intensities corresponding to AP and P couplings, respectively. Clearly long and short-wavelength oscillations are easily visible, see Figure 2.12. The coupling between the ferromagnetic layers is determined by the energy difference between the P and AP alignment of the magnetic moments.

The coupling between the ferromagnetic layers is determined by the energy difference between the P and AP alignment of the magnetic moments.

$$2J \sim E_{AP} - E_P = \int_{-\infty}^{E_F} E\Delta D dE, \qquad (2.64)$$

where $\Delta D = D_{AP} - D_P$ is the difference of the DOS between AP and P alignment of the magnetic moments. For the P configuration of the magnetic moments, the minority spins are confined and form well-defined QWS. At the neck of the Fermi surface, the minority spins are completely confined by the spin potential of the Co. At the belly of the Fermi surface, they are only partially confined. Whenever the energy of a QWS crosses the Fermi level, it adds energy to E_P making the P configuration of the magnetic moments unfavorable. Fitting the XMCD data using

$$J = -\frac{A_1}{d^2} \sin\left(\frac{2\pi}{\Lambda_1} + \Phi_1 \right) - \frac{A_2}{d^2} \sin\left(\frac{2\pi}{\Lambda_2} + \Phi_2 \right), \qquad (2.65)$$

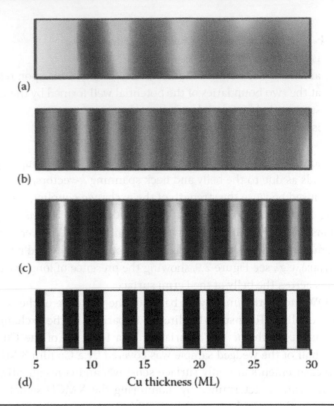

FIGURE 2.12 (a) QW states at the belly of the Cu Fermi surface. (b) QW states at the neck of the Cu Fermi surface. (c) XMCD from the top 3 atomic layers of Co evaporated over the Cu wedged spacer. See further details in [36]. The dark and light regions correspond to ferromagnetic and antiferromagnetic coupling, respectively. (d) Calculated interlayer coupling. Notice the remarkable agreement between the theoretical predictions and experiment for the sign of the exchange coupling (After Qiu, Z.Q. et al., *J. Phys.: Cond. Matter* 14, R169, 2002. With permission.)

resulted in $\Lambda_1 = 5.88$ ML, $\Lambda_2 = 2.67$ ML. XMCD is not able to determine the strength of the coupling.

2.3.5.3 Spin-Density Wave (SDW) in Cr

Fe-whisker/Cr/Fe(001) wedge structures, see Figure 2.14, have played a crucial role in the study of interlayer exchange coupling. The Cr spacers are expected to contain SDWs with short and long-wavelength periods, see the spanning k-wavevectors in Figure 2.13.

Unguris et al. [49] used scanning electron microscopy with polarization analysis (SEMPA) to study Fe-whisker/Cr/Fe(001) samples. They showed that the exchange coupling oscillates with a short-wavelength period of $\simeq 2\text{ML}$. The SEMPA images revealed in a very explicit way the presence of both short-wavelength and long-wavelength oscillations in the thickness range from 5 to 80 ML, see Figure 2.15.

The period of the short-wavelength oscillations, $\lambda = 2.11$ ML, was found to be slightly incommensurate with the Cr lattice spacing; the period of the long-wavelength oscillations was found to be 12 ML. These two basic

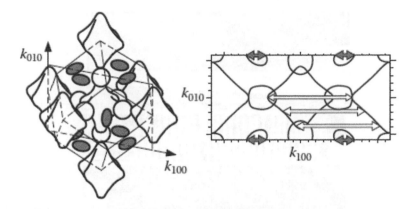

FIGURE 2.13   The left panel shows a representation of the Cr Fermi surface in the paramagnetic state. The gray shaded areas are ellipsoids centered at the N points in the bcc reciprocal lattice. The right panel shows a slice through the Fermi surface as indicated in the left panel. The gray shaded arrows are the critical spanning vectors at the N-centered ellipsoids and the white arrows indicate the nested parts of the Fermi surface that give rise to spin-density wave antiferromagnetism [32] (With kind permission from Springer Science+Business Media: *Ultrathin Magnetic Structures*, Interlayer Exchange Coupling, Vol. III, 99 (2005), Stiles, M.D.).

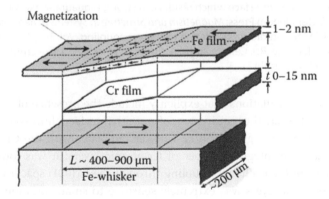

FIGURE 2.14   A schematic view of a Fe-whisker/Cr/Fe(001) sample containing a wedged Cr spacer as employed by the NIST group for their SEMPA measurements. The arrows show the directions of the magnetization. The Fe-whisker is demagnetized by a single 180° domain wall. The direction of the magnetization in the Fe film follows the sign of the local interlayer exchange coupling. (After Unguris, J. et al., *Phys. Rev. Lett.* 67, 140, 1991. Copyright 1991 by the American Physical Society. With permission.)

periods were expected from the complex Cr Fermi surface, see Figure 2.13. The incommensurate nature of the short-wavelength oscillations in Cr, 1 = 2.11 ML, results in phase slips of exchange coupling at 24 and 44 atomic layers of Cr (see Figure 2.15 and further details in Ref. [50]).

Heinrich et al. carried out quantitative exchange coupling measurements on Fe-whisker/Cr/Fe(001) samples [3, 51]. The objective was to grow samples with the best available interfaces to measure quantitatively the strength of the exchange coupling and to compare these coupling strengths

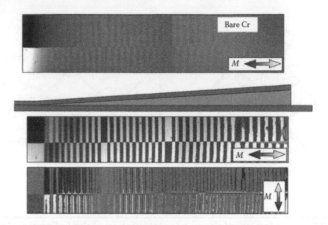

FIGURE 2.15   The direction of the magnetic moment in a Fe film grown over a Cr wedge is shown in the second SEMPA image. The top SEMPA image shows the presence of a magnetic moment in the Cr wedge grown on a Fe-whisker substrate. White and black contrast indicate parallel and antiparallel orientations of the magnetic moment with respect to the Fe-whisker magnetization. Note in the bottom SEMPA image that at the boundaries between the parallel and antiparallel orientation of the magnetic moments a perpendicular component of the magnetic moment is present. This is caused by magnetic frustration due to a partial completion of the top Cr/Fe interface which results in a strong biquadratic exchange coupling. (After Plenum Press: *Magnetism and Structure in Systems of Reduced Dimension*, SEMPA studies of oscillating exchange coupling, Vol. 309, 107 (1993), J. Unguris, D.T. Pierce, R.J. Celotta, and J.A. Stroscio. Work of U.S. Government, not subject to U.S. copyright.)

with *ab initio* calculations that explicitly include the presence of spin-density waves in the Cr. The requirement of smooth interfaces limited our study to samples that were grown on Fe-whisker templates with the Cr spacers terminated at an integral number of Cr atomic layers. It was found that the strength of the exchange coupling through the Cr(001) spacer even for these smooth Cr layers was extremely sensitive to small variations in the growth conditions. Cr ultrathin films do not possess a robust spin-density wave; the spatial variation of the spin moments is extremely sensitive to the interface structures. In our studies, we concentrated on samples for which the Cr thickness ranged from 4 to 13 atomic layers where the role of interfaces was most pronounced. The measured exchange coupling was found to be reproducible only in those samples that exhibited true layer by layer growth, see Ref. [51]. Quantitative Brillouin Light Scattering (BLS) studies of the exchange coupling in Fe-whisker/Cr/Fe(001) have been discussed in Refs. [3, 51, 52]: the main results are shown in Figure 2.16. These studies showed that the exchange coupling through Cr(001) contains both oscillatory bilinear J_1 and positive biquadratic J_2 exchange coupling terms. The exchange coupling first becomes AF at 4 ML of Cr. For Cr spacer thicknesses $d_{Cr} < 8$ ML, the strength of the short-wavelength oscillations is quite weak, ~0.1 erg/cm^2. The exchange coupling in this range is AF. This is due to the presence of an AF long-wavelength bias. This AF bias is peaked in the range between 6 and 7 ML. It is interesting to note that the strength of the long-wavelength AF

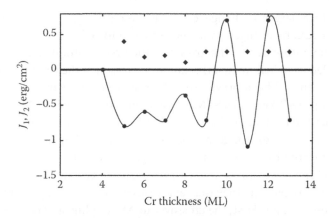

FIGURE 2.16   The thickness dependence of the bilinear $J_1(\cdot)$ and biquadratic J_2 (solid diamonds) exchange coupling. The biquadratic coupling can be measured only for AF coupled samples. The values of J_2 for F coupled samples (10 and 12 ML) were assumed to be the same as those for the AF coupled samples whose spacer thicknesses were 9, 11, and 13 ML of Cr. Note that the coupling becomes AF for thicknesses greater than 4 ML and the thickness dependence of J_1 has a broad maximum around 7 ML of Cr. This behavior is caused by long-wavelength oscillations in the coupling strength. Strong short-wavelength oscillations, from antiferromagnetic to ferromagnetic coupling, appear for Cr thicknesses greater than 9 ML. (After Heinrich, B. et al., *Phys. Rev. B* 59, 14520, 1999. With permission.)

bias is very nearly the same as that observed in Fe/Cr/Fe(001) epitaxial multilayers prepared by sputtering where a relatively large interface roughness annihilates the presence of the short-wavelength oscillations [53]. Exchange coupling in the sputtered films exhibited long-wavelength oscillations with a rapidly decreasing coupling strength for thicknesses greater than 10 ML of Cr. For Cr spacers thicker than 8 ML, $d_{Cr} > 8$ ML, the exchange coupling in films grown on a whisker substrate is dominated by the short-wavelength oscillations. In this thickness range, the samples are AF coupled for an odd number of Cr atomic layers and ferromagnetically (F) coupled for an even number of Cr atomic layers.

The coupling between the Fe and Cr atoms at the Fe/Cr interface is expected to be strongly AF [54] and, consequently, the spin-density wave in Cr is locked to the orientation of the Fe magnetic moments. Since the period of the short-wavelength oscillations is close to 2 ML, one would expect antiferromagnet (AF) coupling for an even number of Cr atomic layers and F coupling for an odd number of Cr atomic layers. For the period 2.11 ML, the first-phase slip in the short-wavelength coupling is predicted to occur at 20 ML, see Figure 2.16. Surprisingly, the BLS and SEMPA measurements showed clearly that the phase of the short-wavelength oscillations is exactly opposite to that expected. It is also important to note that the strength of the exchange coupling was found to be much less than that obtained by Stoeffler and Gautier using first-principle calculations, $J_1 \sim 30$ erg/cm^2 [55]. Our studies showed that the strength of the bilinear exchange coupling J_1 is very sensitive to the initial growth conditions: a lower initial substrate temperature resulted in a larger exchange coupling strength. The bilinear exchange coupling could be changed by as much as a factor of 5 by varying the substrate temperature during the growth of the first Cr atomic layer [56]. This behavior led us to believe that the atomic formation of the Cr layer was more complex than had been previously acknowledged. Angular-resolved Auger spectroscopy (ARAES) [57], STM [58], and proton-induced ARAES [59] have shown that the formation of the Fe/Cr(001) interface is strongly affected by an interface atom exchange mechanism (interface alloying). Freyss et al. [60] investigated the phase of the exchange coupling for intermixed Fe/Cr interfaces. The calculations were carried out using two mixed atomic layers $Fe/Fe_{1-x}Cr_x/Cr_{1-x}Fe_x/Cr$. They were able to account for two important experimental observations. First, the crossover to AF coupling and onset of short-wavelength oscillations was predicted to occur at 4–5 ML of Cr, in good agreement with our observations, see Figure 2.16, and in agreement with NIST studies using the SEMPA imaging technique. Second, the phase of the coupling was reversed for a concentration of $x \parallel 0.2$. AF and F coupling was obtained at an odd and an even number of Cr atomic layers, respectively, in perfect agreement with experiment.

2.4 BEYOND THE MACROSPIN MODEL

In all of the above discussions, it was assumed that the magnetic moments in F films are locked together by very strong intralayer exchange coupling and that the reversal of the magnetic moment behaves like a single macrospin. This is well satisfied for weak interlayer coupling and thin F films requiring a small external field which are not able to compete with large fields generated by exchange coupling. Good examples of MOKE measurements of the magnetic moment as a function of applied field (hysteresis loops) for this behavior are shown in the Fe/Cu/Fe(001) bcc structures, see Ref. [61]. However, with strong interlayer coupling one requires a large magnetic field and can't ignore the competition between the coupling and Zeeman energies in the applied magnetic field. This is particularly true for high-density memory media. In principle, one has to minimize the total energy involving the exchange couplings, applied magnetic field, and magnetic anisotropies. Let us briefly discuss the main concepts describing the bulk exchange coupling; the density of exchange coupling energy, exchange coupling field, and exchange stiffness parameter A.

2.4.1 BULK EXCHANGE COUPLING: EXCHANGE STIFFNESS PARAMETER

Let us first briefly summarize the basic concepts of intralayer exchange coupling (bulk exchange coupling) for the case when saturation magnetization is constant, $M_x^2 + M_y^2 + M_z^2 = M_s^2$; for the rest of this section, we will refer to intralayer exchange coupling as simply exchange coupling. Minimum exchange energy for an F film is reached when the magnetization vector is a constant, independent on its position in the film. For an inhomogeneously distributed magnetization the exchange energy increases. An inhomogeneous component of the density of exchange coupling energy for slowly varying orientation of magnetization $\vec{M}(\vec{r})$ in an isotropic (cubic) F is given by (for details see Section 3.3.3 of Chapter 3 by A.S. Arrott and Refs. [62, 63, 64, 65])

$$E_{\text{exch}} = A \sum_{m_x, m_y, m_z} \nabla \vec{m} \cdot \nabla \vec{m} = -A\vec{m} \cdot \Delta\vec{m}, \qquad (2.66)$$

where A is the exchange stiffness constant and $\vec{m}$ is a unit vector along the local saturation magnetization $\vec{M}(\vec{r})$; ∇ and Δ are Nabla and Laplace ($\nabla \cdot \nabla$) operators, respectively. A variational principle is employed to minimize the total energy [62]. This converts energy into a torque equation [62] allowing one to introduce effective fields for the individual contributions to the total magnetic energy. For cubic F the effective field from the exchange coupling is given by [64]

$$\vec{H}_{\text{exch}} = \frac{2A}{M_s^2} \Delta\vec{M}, \qquad (2.67)$$

Notice that the effective field can be directly concluded from the right side of Equation 2.66 considering that the self-energy involves a factor two in the denominator.

In order to demonstrate the concept of the intralayer effective exchange field, let us use a hcp film with (A/B/A) hexagonal atomic layers stacking with the film normal along the (0001) axis. One can start from a simple Heisenberg model for nearest neighbor interaction in F

$$E_{\text{Heis}} = -J \sum_{i,j} \vec{S}_i \cdot \vec{S}_j, \qquad (2.68)$$

where the spin on the site i, j is in units of $\hbar$, and J is the exchange energy in erg. Using a simple relationship between the magnetic moment on site i; $\vec{\mu}_i = -g\mu_B \vec{S}_i$ one can write for an effective field from site j on site i

$$\vec{H}_{\text{eff}}(j) = -J \frac{1}{g\mu_B} \vec{S}_j. \qquad (2.69)$$

For a slowly varying spin along the film normal (z direction) one can expand $\vec{S}_j$ around $\vec{S}_i$ in a Taylor series using nearest neighbor ($n.n$) exchange interaction, one obtains

$$\vec{H}_{\text{eff}}(j) = -J \frac{1}{g\mu_B} N_{nn} d^2 \frac{\partial^2 \vec{S}_i}{\partial z^2}, \qquad (2.70)$$

where N_{nn} is the number of n.n above (and below) the atomic site i and a_{nn} is the distance between the adjacent atomic planes i and $j = i \pm 1$. Converting the atomic spin $\vec{S}_i$ into the saturation magnetization by using the atomic volume $V_{atom} = 2\Delta d$, where 6Δ is the base area and $2d$ is the vertical size of the hcp unit chemical cell, respectively. The effective exchange field due to slowly varying magnetization $\vec{M}(z)$ after simple algebraical steps is given by

$$\vec{H}(z)_{exch} = JS^2 N_{nn} \frac{d}{2\Delta} \frac{1}{M_s^2} \frac{\partial^2 \vec{M}(z)}{\partial z^2}. \tag{2.71}$$

Comparing Equations 2.67 with 2.71 one gets

$$2A = JS^2 N_{nn} d \frac{1}{2\Delta}. \tag{2.72}$$

In SAF experiments the total energy is expressed using an areal energy density. The area for an atom in the (0001) plane is equal to 2Δ. Therefore in hcp (0001) structures the exchange energy per unit surface area (areal density) for two subsequent atomic planes, E_{exch} is given by

$$E_{exch} = -JS^2 N_{nn} \frac{1}{2\Delta} \cos\big(\theta(i) - \theta(i+1)\big), \tag{2.73}$$

where for SAF structures, see Figure 2.20, $\theta_{i,i+1}$ are the in-plane magnetization angles with respect to the in-plane applied field $\vec{H}$. Comparing Equations 2.72 with 2.73 leads to the total exchange energy areal density for N atomic layers

$$E_{exch} = -2A \frac{1}{d} \sum_{i=1}^{i=N-1} \cos\big(\theta(i) - \theta(i+1)\big). \tag{2.74}$$

The lowest exchange energy areal density, $-2A \frac{1}{d} N$, is reached for a macrospin when $\theta_i = \theta_{i+1}$. Assuming that a ground state of F with fully aligned magnetic moments corresponds to zero exchange energy then the contributions with $\theta_i \neq \theta_{i+1}$ corresponds to a slowly varying magnetization across the film thickness which increases the exchange energy areal density and one can write,

$$E_{exch} = 2A \frac{1}{d} \sum_{i=1}^{i=N-1} \cos\big(\theta(i) - \theta(i+1)\big) \tag{2.75}$$

2.4.2 Nonuniform Magnetization Reversal in Magnetic Recording Media

The design of magnetization reversal is critical for spintronic, magneto-electronic, and magnetic recording applications. In this section, we will discuss how the intrinsic magnetic material parameters: the saturation magnetization, M_s, magnetic anisotropy, K and the exchange stiffness, A, are used to control magnetization reversal with focus on magnetic recording media. We will also discuss how M_s and A can be varied by alloying magnetic materials with 3d, 4d, and 5d elements (Figure 2.17).

In the case of uniform magnetization reversal, M_s and K have to be known to calculate the switching field, H_s. The analytical model used to calculate magnetization reversal, in this case, is called the Stoner-Wohlfarth model [66]. In the more general case, the magnetization reversal is nonuniform and an additional parameter A has to be also known to calculate H_s. In this section, we will discuss a simple case of nonuniform magnetization reversal in nanometer-size magnetic grains. Utilizing nonuniform magnetization reversal in magnetic recording media in hard drives was in part responsible for the increase of recording density during the last ten years. Nonuniform magnetization reversal may also play a key role in the development of the future magnetic solid-state memory. In the following section, we will discuss how the formation of a spin spiral in the F/NM/F trilayer structure can be used to measure A. This technique was employed to measure A in Co alloys, the magnetic materials also used in the current magnetic recording media [67, 68, 69].

The magnetic recording media consist of thin film layers with an ensemble of nanometer-size, single-domain magnetic grains. In currently manufactured magnetic recording media, the grains have hexagonal close-packed, hcp, crystal structure and are made of Co alloys with uniaxial crystal anisotropy. To ensure that all grains grow along the easy axis of magnetocrystalline anisotropy, along the [0001] crystal direction, magnetic layers are grown on top of [0001] textured hcp Ru seed layers. The magnetic grains are separated

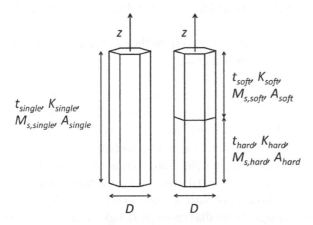

FIGURE 2.17   Magnetic grains with the single and bilayer structures. Grains grow along the z-axis that is also the easy magnetocrystalline anisotropy direction. t_{single}, K_{single}, $M_{s,single}$, and A_{single} are the thickness, anisotropy, saturation magnetization and stiffness of the single layer, respectively. The bilayer magnetic grain consists of low anisotropy *soft* layer on the top with thickness t_{soft}, and intrinsic magnetic parameters K_{soft}, $M_{s,soft}$ and A_{soft}, and a large anisotropy *hard* layer on the bottom with thickness t_{hard}, and intrinsic magnetic parameters K_{hard}, $M_{s,hard}$ and A_{hard}. The diameter (in the case of cylindrical grains) of the single and bilayer grains are the same and equal D.

by weakly magnetic or nonmagnetic oxide grain boundaries to achieve a weak direct exchange interaction between the grains, for more details on the microstructure see Nolan et al. [70]. In hard drives, magnetic media is deposited over an aluminum or glass disk that during the read/write process rotates to up to 15,000 rotations per minute. Information is written by a recording head that is an electromagnet and is located near the media surface. Details of recording physics are discussed in a review article by Richter [71].

To increase the recording density of magnetic media, the size of magnetic grains has to be reduced. This ensures that there is a large enough number of grains per bit to achieve the desired signal-to-noise ratio (SNR). From statistical arguments, the SNR is directly related to the number of grains/bit. The stored bits must be thermally stable. Thus, a decrease of grain volume, V, requires a proportional increase in K of grains in order to keep $KV / k_B T > 40$ [71], where T is the temperature at which hard drives operate, and k_B is the Boltzmann constant. If K of magnetic grains increases, the recording head field has to also increase to reverse magnetic grains in a bit. Since the maximum field that can be generated by recording head is limited, it would follow that the increase of magnetic recording areal density is also limited. This *trilemma* correlates three recording media parameters, SNR, thermal stability and writability.

The head field required to reverse the last grain of the written bit is given as [72],

$$H_{\text{sat}} = H_{s,\text{grain}} + H_{\text{stray}} - H_{\text{ex}}, \tag{2.76}$$

where:

$H_{s,\text{grain}}$ is the field required to reverse a single magnetic grain

H_{stray} is the stray field from surrounding grains that stabilizes the last grain

H_{ex} is the exchange field from the neighboring grains that helps to reverse the last grain

The change from a uniform to nonuniform reversal in a magnetic grain will primarily affect $H_{s,\text{grain}}$. Thus, for simplicity, we will neglect H_{stray} and H_{ex} fields. Furthermore, we will assume that the external magnetic field is perpendicular to the magnetic media film surface, i.e., parallel with the easy magnetocrystalline anisotropy direction of a magnetic grain. This is also a simplification since the data are usually recorded at the trailing edge of a recording head where the fields are not normal to the recording medium surface [71].

If grains in magnetic media consist of a single layer (material) the switching field is $H_{s,\text{grain}} = H_{s,\text{single}} = 2K_{\text{single}} / M_{s,\text{single}}$, where K_{single} and $M_{s,\text{single}}$ are the anisotropy and saturation magnetization of the single layer, respectively. Here we assumed a uniform magnetization reversal of the single layer grain.

Now consider grains in magnetic media with a bilayer structure, so-called exchange spring media, see Figure 2.17. The top layer is called *soft* due to lower magnetic anisotropy and the bottom layer is called *hard* due to a comparatively larger magnetic anisotropy. Initial experimental results on the reversal of recording media with a bilayer grain structure and perpendicular to the film surface magnetic anisotropy showed that ferromagnetic interaction between magnetic layers could be used to facilitate the writing process in magnetic recording [73]. Theoretical work on magnetization reversal of a bilayer grain as a function of intrinsic properties of each layer is discussed in Refs. [74, 75, 76]. Experimental evidence of domain wall assisted switching in nanometer-size magnetic bilayer grains is discussed in Ref. [77].

The magnetization reversal of this bilayer grain will start with magnetization reversal (nucleation) at the top of the *soft* layer as shown in Figure 2.18a. The nucleation field, H_n in the *soft* layer of the bilayer grain can be calculated as Ref. [78],

$$H_n = \frac{2K_{soft}}{M_{s,soft}} + \frac{2\pi^2 A_{soft}}{4t_{soft}^2 M_{s,soft}} \tag{2.77}$$

where:

K_{soft} is the magnetic anisotropy of the *soft* layer
$M_{s,soft}$ is the saturation magnetization of the *soft* layer
A_{soft} is the exchange constant of the *soft* layer
t_{soft} is the thickness of the *soft* layer

The last term in the above equation accounts for the exchange field in the *soft* layer caused by having the magnetic moments of the *soft* layer pinned at the *soft/hard* layer interface by the exchange coupling from the *hard* layer. As the external magnetic field is increased (in opposite direction to the magnetized grain) the formed nucleation propagates to the *soft/hard* interface where it becomes pinned, Figure 2.18b. Further increase in the field

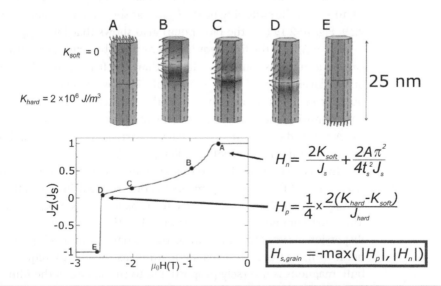

FIGURE 2.18   Compression of domain wall at the *soft-hard* interface after it was nucleated in the *soft* end. Calculations are performed assuming $A_{soft} = A_{hard} = A = 1 \times 10^{-6}$ erg/cm $= 1 \times 10^{-11}$ J/m , $J_{s,soft} = \mu_0 M_{s,soft} = J_{s,hard} = \mu_0 M_{s,hard} = 0.5$ T, and μ_0 is the vacuum permeability. (After Suess, D. et al., *J. Magn. Magn. Mater.* 321, 545, 2009. With permission.)

first compresses the domain wall, Figure 2.18c and 2.18d, and finally, magnetization in the whole grain reverses, Figure 2.18e when the external field is large enough to overcome the pinning field of the bilayer structure given by

$$H_p = 2\left(\frac{K_{hard}}{M_{s,hard}} - \gamma \frac{K_{soft}}{M_{s,soft}} \right) / \left(1 + \gamma^{0.5} \right)^2 \tag{2.78}$$

where:

K_{hard} is the magnetic anisotropy of the *hard* layer
$M_{s,hard}$ is the saturation magnetization of the *hard* layer
A_{hard} is the exchange constant of the *hard* layer

and

$$\gamma = \frac{M_{s,soft} A_{soft}}{M_{s,hard} A_{hard}} \tag{2.79}$$

Combining results from Equations 2.77 and 2.78, the switching field of entire grain is given by $H_{s,grain} = H_{s,bilayer} = -\max\left(|H_p|, |H_n| \right)$. This is shown in Figure 2.19 where the dotted line represents H_n calculated from Equations 2.77, and the dashed line represents H_p calculated from Equations 2.78. Here we assumed that $K_{hard} = 2 \times 10^7 \, erg/cm^3$, $K_{soft} = 0$, $M_{s,single} = M_{s,soft} = M_{s,hard} = 400 \, emu/cm^3$, and $A_{soft} = A_{hard} = 1 \times 10^{-6} \, erg/cm$. If $|H_p| > |H_n|$ the domain first nucleates at the top of the *soft* layer, then as the external field increases, compresses at the *soft/hard* layer interface, and finally sweeps through the hard layer after the external field exceeds the H_p as shown in the left inset Figure 2.19 (and in Figure 2.18). If $|H_p| \le |H_n|$ the domain nucleates at the top of the *soft* layer and immediately sweeps through both *soft* and *hard* layers resulting in a sharp drop of the magnetization from M_s to $-M_s$ at the switching field $H_{s,bilayer}$ as shown in the right inset of Figure 2.19. Figure 2.19 (see the blue point) also shows that for $K_{single} = K_{hard}$ and $M_{s,single} = M_{s,hard}$, the field required to reverse the single layer grain, $H_{s,single}$, can be a much as four times larger than the field required to reverse the bilayer layer grain, $H_{s,bilayer}$. Suess et al. [79] have shown that increasing the number of layers in a grain can further reduce the reversal field without compromising the total magnetic anisotropy and therefore thermal stability. A detailed discussion of the energy barrier in the single and multilayer magnetic grains are presented in Ref. [79, 72].

In this section, we showed that knowing intrinsic parameters M_s, K, and A are crucial for predicting magnetization reversal in magnetic materials. While M_s and K are usually easy to measure, measuring A is not trivial. The most common way to determine A is through a measurement of the dispersion curve of the spin waves. The exchange stiffness is particularly difficult to measure in thin magnetic films, as the energy required to excite bulk magnons is inversely proportional to the square of the film thickness. To illustrate how serious a problem this is, the first-order magnon mode in a 10-nm-thick cobalt film has a frequency of approximately 500 GHz. This frequency is not accessible with FMR. In thin magnetic films, neutron

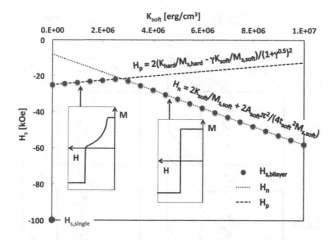

FIGURE 2.19 Switching, nucleation and pinning fields of the bilayer magnetic grain as a function of K_{soft}. H_n is calculated from Equation 2.77, H_p from Equation 2.78, and $H_{s,bilayer}$ is $-\max\left(|H_p|,|H_n|\right)$. The second and third quadrant of $M(H)$ loops show the difference in the magnetization reversal in the case of $|H_p|>|H_n|$ (left inset) and $|H_p|\leq|H_n|$ (right inset). The switching field of the single layer grain is $H_{single}=2K_{single}/M_{s,single}=100$ kOe. Calculations are performed assuming $t_{soft}=t_{hard}=12.5nm$, $K_{single}=K_{hard}=2\times10^7$ erg/cm³, $M_{s,soft}=M_{s,hard}=400$ emu/cm³ and $A_{soft}=A_{hard}=1\times10^{-6}$ erg/cm.

scattering measurements can only be carried out in reflectivity mode due to the large penetration depth of neutrons. For these reasons, the spin-wave-dispersion methods of determining exchange stiffness are typically limited to films exceeding 30 nm in thickness. In thin films, the presence of interfaces and the strain induced by the mismatch between the measured magnetic films and seed layers can affect values of A. In the next section we will present a simple method that is employed to measure A of hcp-Co alloys, the magnetic materials currently used in the recording media. This method can be used to measure A in ultrathin films and thus determine the effects of interfaces on A.

2.4.3 EXCHANGE STIFFNESS PARAMETER FOR CO AND CO ALLOY SPUTTER DEPOSITED FILMS

Magnetic films with a thickness less than the exchange length, δ see Equation 2.8, can be treated as single-spin objects. Such an approximation allows for signification simplification in the treatment of magnetic structures. However, for some applications, having a nonuniform distribution of magnetization can be advantageous, as was discussed in Section 2.4.1. If properly taken into account, the spatially nonuniform magnetization can be used to probe the magnetic properties of materials. An example of such a case can be found in Ref. [69]. In this work, the authors investigate the magnetization reversal of an F/NM/F (Co/Ru/Co) SAF structure where the two ferromagnetic layers of Co are antiferromagnetically coupled via the RKKY interaction, across the NM layer of Ru. An applied magnetic field will induce

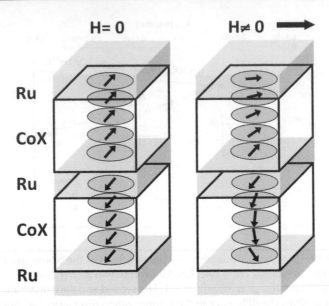

FIGURE 2.20 Schematic of synthetic antiferromagnets of Cobal layers antiferromagnetically coupled by Ruthenium. (After Eyrich, C. et al., *Phys. Rev. B* 90, 235408, 2014. With permission.)

opposing, uniform torques on the two ferromagnets, attempting to bring them into parallel alignment. This will compete with the antiferromagnetic coupling, which prefers antiparallel alignment. Since the RKKY coupling is an interface effect, the net torque will grow in favor of the applied field as we probe further away from the coupled interfaces. This will result in a nonuniform rotation of the magnetization throughout the film thickness and the formation of an *exchange spring*, see Figure 2.20.

In Ref. [69], the authors measured the magnetization of antiferromagnetically coupled Co(10 nm)/Ru/Co(10 nm) structures. The thickness of the Ru interlayer was kept between 0.38 and 0.45 nm in order to maximize antiferromagnetic coupling. The structure was prepared by sputter deposition having hcp texture. The normal to the film was oriented along the [0001] Co crystallographic axis. The samples were polycrystalline and therefore had no in-plane anisotropy and large demagnetizing field (significantly larger then uniaxial anisotropy) held the magnetic moment in the film plane. M_s was determined by VSM and SQUID.

The micromagnetics model used to interpret the magnetic behavior of the structure assumed that magnetic moments in each atomic plane are parallel and rotate with respect to an applied in-plane field. Antiferromagnetic coupling between the Co films rotates the antiferromagnetic axis perpendicular to the applied field H in order to maximize the Zeeman energy. This is justified since the samples have in-plane magnetization and no in-plane anisotropy. Nearest neighboring vertical atomic planes interact through the direct exchange interaction and since their rotation is not uniform, will contribute to the total exchange energy of the structure as described in Section 2.4.1. The total magnetic energy can be written as follows,

$$E_{mag} = E_{RKKY} + E_{ex} + E_Z \tag{2.80}$$

$$E_{RKKY} = J \; \cos(\theta_N - \theta_{N+1})$$

$$E_{ex} = \frac{2A}{d}\left[\sum_{i=1}^{N-1}\cos(\theta_i - \theta_{i+1}) + \sum_{i=N+1}^{2N-1}\cos(\theta_i - \theta_{i+1})\right] \tag{2.81}$$

$$E_Z = -M_s H d \sum_{i=1}^{2N}\cos(\theta_i).$$

where:

E_{RKKY} is the energy due to the interlayer exchange coupling and depends on the difference in the orientation of magnetization of the RKKY coupled interfaces

E_{ex} is the exchange energy between neighboring atomic planes, see Equation 2.75

E_Z is the Zeeman energy

θ_i is the angle of the magnetization in an atomic plane i, with respect to the applied H field

N is the number of atomic planes in each ferromagnetic layer

d is the spacing between atomic planes

A is the exchange stiffness

J is the interlayer coupling constant. For SAF the sign of interlayer exchange coupling was changed compared to Equation 2.39 in order to have an antiferromagnetic coupling parameter J positive, $J > 0$

By minimizing the total magnetic energy, Equation 2.80, the authors were able to fit a measured $M(H)$ curve of Co/Ru/Co and to extract values for A of Co and J, for details on the minimization procedure see Ref. [69]. Furthermore, the authors investigated the effects of alloying Co with X = Cr, Fe, Ni, Ru, Pd, Pd, and Pt on M_s, A, and J, see Table 2.1 and Figure 2.21. Both A and M_s were found to have a linear dependence on the amount of dopant concentration, η. However, interlayer coupling constant was found to be $J \propto M_s^2$, see Figure 2.22, in agreement with predictions from mean field theory, Ref. [35].

The effect that alloying has on M_s and A is qualitatively consistent with the density functional theory (DFT) calculations [69]. From the calculations, the authors find that the magnetic moment of a substituted Cr atom is anti-ferromagnetically coupled to the Co neighbors and has a very large magnetic moment of -2.2 μ_B. The magnetic moment of substituted Pt and Ru atoms coupled ferromagnetically to the Co atoms. The induced magnetic moments in these atoms are 0.39 μ_B for Pt and 0.48 μ_B for Ru. The DFT results also show that the addition of Cr and Ru atoms reduces the magnetic moment of the surrounding Co atoms while the addition of Pt slightly increases the neighboring Co moments, see Table 2.2. The large antiferromagnetically coupled Cr moment combined with the reduction in the magnetic moments

TABLE 2.1

M_s for bulk Co is 1440 emu/cm³ [80]. Ms, A and J are measured in CoX(10 nm)/Ru(0.38 nm)/CoX(10 nm) with X = Cr, Fe, Ni, Ru, Pd and Pt.

	η	Ms (CoX) (emu/cm³)	A (CoX) (10⁻⁶erg/cm)	J (CoX) (erg/cm²)
Co	0	1247 ± 37	1.55 ± 0.09	4.20 ± 0.13
$Co_{100-\eta}\,Cr_\eta$	3	1110 ± 37	1.30 ± 0.09	3.19 ± 0.13
	6	960 ± 37	1.16 ± 0.09	2.10 ± 0.13
	12	715 ± 37	0.90 ± 0.09	0.90 ± 0.13
$Co_{100-\eta}\,Fe_\eta$	10	1347 ± 37	1.55 ± 0.09	4.20 ± 0.13
	20	1447 ± 37	1.55 ± 0.09	4.96 ± 0.13
$Co_{100-\eta}\,Ni_\eta$	8	1222 ± 37	1.29 ± 0.09	3.61 ± 0.13
	16	1160 ± 37	1.12 ± 0.09	2.94 ± 0.13
	2	1235 ± 37	1.36 ± 0.09	3.61 ± 0.13
	4	1160 ± 37	1.21 ± 0.09	3.19 ± 0.13
	6	1097 ± 37	1.05 ± 0.09	2.52 ± 0.13
	8	985 ± 37	0.85 ± 0.09	2.10 ± 0.13
	10	927 ± 37	0.74 ± 0.09	1.60 ± 0.13
$Co_{100-\eta}\,Ru_\eta$	10	973 ± 37 [•]	0.78 ± 0.09 [•]	1.68 ± 0.13 [•]
	12	860 ± 37	0.60 ± 0.09	1.22 ± 0.13
	15	723 ± 62	0.48 ± 0.12	3.30 ± 0.13 [*]
	20	536 ± 62	0.26 ± 0.12	3.01 ± 0.13 [*]
	20	673 ± 37 [•]	0.33 ± 0.09 [•]	3.40 ± 0.13 [*•]
$Co_{100-\eta}\,Pd_\eta$	9	1097 ± 37	1.26 ± 0.09	3.32 ± 0.13
	17	1060 ± 37	0.99 ± 0.09	2.56 ± 0.13
$Co_{100-\eta}\,Pt_\eta$	3	1210 ± 37	1.41 ± 0.09	3.70 ± 0.13
	6	1210 ± 37	1.41 ± 0.09	3.49 ± 0.13
	9	1085 ± 37	1.41 ± 0.09	2.98 ± 0.13
	12	1097 ± 37	1.30 ± 0.09	2.52 ± 0.13

[•]   indicates that **M(H)** measurements were performed at 10 K, and

[*]   indicates the presence of 1 nm Co interface layer between CoRu and Ru spacer layers

of Co atoms in CoCr alloys explains why the addition of Cr reduces M_s more than the other elements in Figure 2.21.

The trends in A can be qualitatively understood from the change in the magnetic moments as a result of the alloying. The small moment attributed to the Ru atom combined with the reduction in the moment of its Co neighbors amounts to a significant decrease in the overall exchange coupling constant. While Cr reduces the Co neighboring moments more than Ru, it turns out from the DFT calculations that the very large moment on the Cr atom compensate in part for this reduction in the exchange stiffness. Pt in CoPt has a very small moment, however, Pt increases the moment of the Co neighbors and therefore affects A the least. The measured trends match this qualitative conclusion. The results of DFT calculations of A in Co and CoX (X = Cr, Ru, Pt) are summarized in Table 2.2. These results show that alloying can be used to

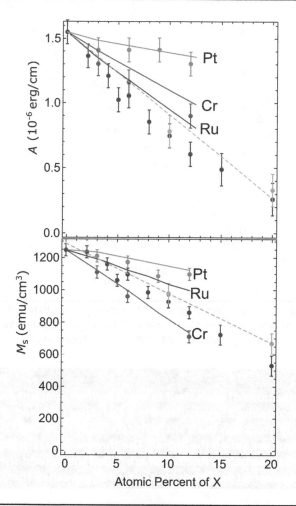

FIGURE 2.21   Experimental (circles) and DFT calculated (lines) values of a) M_s, and b) A for CoRu, CoCr, and CoPt alloys. All M_s and A data are measured at 298 K. Since the calculations did not account for the temperature effects, M_s and A of $Co_{90}Ru_{10}$ and $Co_{80}Ru_{20}$ were also measured at 10 K, these data are shown by the squares and dashed lines.

separately vary M_s and A, and allow for the tailoring magnetic materials for nonuniform magnetic reversal.

2.5 EXCHANGE BIAS

A metallic film in contact with an AF can display a hysteresis loop shifted along the field axis after cooling in an applied field. A typical example is shown in Figure 2.23. In this measurement, Jungblut et al. [81] used a ferromagnetic (F) permalloy (py) film, $Ni_{79}Fe_{21}$, and an AF $Fe_{43}Mn_{57}$ film in a crystalline sandwich structure grown on a crystalline Cu(111) substrate.

The hysteresis loop shows two clear features. The loop is shifted by $H_{eb} \sim 275\,Oe$, and the hysteresis loop is 180 Oe wide. The unidirectional shift of the magnetization reversal loop is called EB.

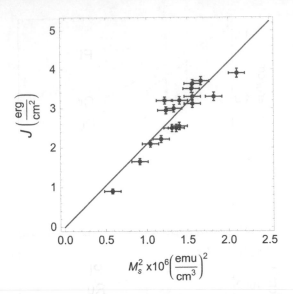

FIGURE 2.22 J as a function of M_s^2 for all CoX alloys in CoX(10 nm)/Ru(0.38 nm)/CoX(10 nm) structures. The solid line is a linear fit through the experimental data. (After Eyrich, C. et al., *Phys. Rev. B* 90, 235408, 2014. With permission.)

TABLE 2.2
The contributions of each shell to the magnetic moment and the exchange coupling are calculated in a Co lattice with and without an impurity atom X in the 0 shell. The six in-plane nearest neighbors to the 0 shell are called the 1 shell. The three nearest neighbors above and three below the 0 shell are called the second shell, and the second nearest neighbors to the 0 shell are called the 3 shells

X	$\mu_i(\mu B)$				$\Delta A_0^{(s)}$ (mRy)				ΔA_{av}
	0	1	2	3	0	1	2	3	
Co	1.63	1.63	1.63	1.63	0	0	0	0	0
Pt	0.39	1.67	1.66	1.67	−13	−0.6	−0.8	0.3	−0.4
Ru	0.48	1.58	1.59	1.67	−14	−3.5	−3	−0.1	−1.6
Cr	−2.2	1.45	1.47	1.6	−0.7	−2.8	−2.5	−1	−1.4

μ_0 magnetic moment of X atom in the shell 0, and μ_1, μ_2, and μ_3 magnetic moments of Co atoms in the shells 1, 2, and 3, respectively. The $\Delta A_0^{(s)}$ is the change in the molecular field exchange constant in individual shells due to the presence of an X atom in shell 0, and $\Delta A_{av} = -\dfrac{1}{N_s}\sum_{s=0}^{N_s}\Delta A_0^{(s)}$. (From Eyrich, C. *Phys. Rev. B* 90, 235408 (2014). With permissions.)

2.5.1 THE MEIKLEJOHN–BEAN MODEL

EB was discovered by Meiklejohn and Bean [82] in 1956 in their study of oxidized Co particles. They explained EB by means of the simplified picture shown in Figure 2.24.

The Meiklejohn–Bean model (MB) can be easily quantified using the total energy for the F/AF structure

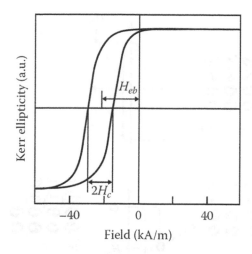

FIGURE 2.23   The magnetization reversal loop for a py film exchange coupled to an antiferromagnetic (AF) $Fe_{43}Mn_{57}$ film in a crystalline structure grown on Cu(111). Note that the magnetization reversal loop is shifted by 275 Oe with respect to zero field. This is called the EB effect. (Courtesy of Coehoorn, R., Phillips Research Laboratory [81]; After Jungblut, R. et al., *J. Appl. Phys.* 75, 6659, 1994. With permission.)

$$\mathcal{E} = -M_F H t_F \cos(\theta - \beta) + K_F t_F \sin^2\beta$$
$$+ K_{AF} t_A \sin^2\alpha - J_{eb}\cos(\beta - \alpha), \tag{2.82}$$

M_F is the saturation magnetization of F

$K_{F,AF}$ are the anisotropy energy densities for F and AF, respectively

J_{eb} is the interface F/AF exchange energy (erg/cm²)

t_F and t_{AF} are the thicknesses of F and AF layers, respectively

$\alpha, \beta,$ and θ are the angles between the axis of the AF moments, the magnetic moment of F, and the external field H and the reference direction given by the AF uniaxial anisotropy axis, respectively. The uniaxial axis for AF and F are along the same direction, see Figure 3.8 in Ref. [37]

The minimum of energy is given by $\partial\mathcal{E}/\partial\alpha = \partial\mathcal{E}/\partial\beta = 0$. Solutions can be found numerically by introducing two parameters: $H_{eb}^\infty = -J_{eb}/(M_F t_F)$ and $R = K_{AF} t_{AF}/J_{eb}$. Solutions are shown in Figures 2.25 and 2.26.

The MB model is obviously a gross simplification. In fact, a completely uncompensated moment in AF at the interface would lead to a bias field several orders of magnitude larger than is observed. For a fully compensated AF interface plane, no EB develops: a uniaxial anisotropy is induced in F with the magnetic axis oriented perpendicular to the spin moment in AF [83]. This implies that a realistic F/AF interface is far from being ideal. It is affected by surface roughness and a complex interface chemistry. This creates an interface with a complex magnetic structure, which cannot, in

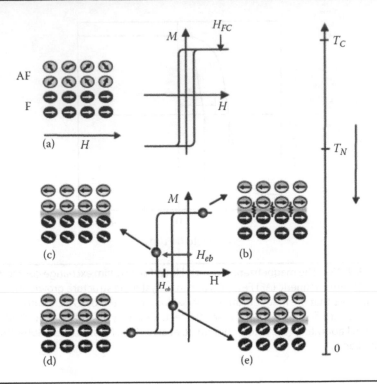

FIGURE 2.24 A simplified view of the EB mechanism. (a) For temperatures greater than the ordering Néel temperature (T_N) of AF, the exchange field between AF and F is zero. (a) After cooling in a positive magnetic field $T < T_N$ the AF moment becomes ordered, and, consequently, an exchange coupling develops between the magnetic moments of AF and F at the AF/F interface. An EB is established when the magnetic moments of AF are held sufficiently firmly along the easy axis of the AF internal anisotropy: usually the sample must be cooled to a temperature less than T_N. This required temperature is called the blocking temperature (T_B). (c) and (e) In ultrathin magnetic films, the reversal of the magnetization proceeds by a rotation of the magnetic moment. (d) Reversal of the F moment also reverses the sign of the coupling at the AF/F interface. This means that the exchange field due to AF acts as an effective field (unidirectional anisotropy) resulting in a shift of the hysteresis loop. In this figure, the coupling between the AF and F moments at the interface was assumed to be positive, and, consequently, the centre of the hysteresis loop moved to a negative field. This is a case of positive EB. (With kind permission from Springer Science+Business Media: *Magnetic Heterostructures*, EB effect of ferro-/antiferromagnetic structures, Vol. 227 (2008), 95, Zabel, D. and Bader, S.D.)

general, be described. However, there are some common features that are shared by a number of EB systems: (a) In the saturated state, the F and AF spins are collinear at the F/AF interface; (b) A small fraction of the uncompensated magnetic moment is pinned to AF and is exchange coupled to F; (c) The sign of the EB is determined by the sign of the exchange coupling between the pinned interfacial and F spins. A schematic picture of an F/AF interface is shown in Figure 2.27.

Direct imaging of interfacial magnetic spins in a Co/NiO EB structure was demonstrated by Ohldag et al. [84]. Using XMCD (determining the magnetic moment in F) and x-ray magnetic linear dichroism (XMLD)

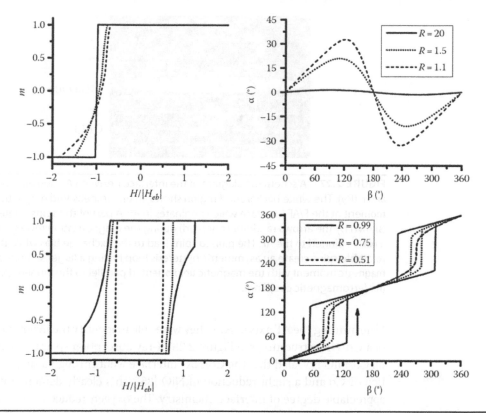

FIGURE 2.25   Simulations of several hysteresis loops and the AF spin orientation during the magnetization reversal. (With kind permission from Springer Science+Business Media: *Magnetic Heterostructures*, EB effect of ferro-/antiferromagnetic structures, Vol. 227, 95 (2008), Zabel, D. and Bader, S.D.)

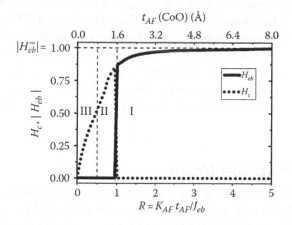

FIGURE 2.26   Meiklejohn and Bean model: The phase diagram of the EB field $|H_{EB}|$ (solid line) and coercive field H_c (dotted line) as a function of the parameter R. Notice that for $R < 1$, there is no EB. (With kind permission from Springer Science+Business Media: *Magnetic Heterostructures*, EB effect of ferro-/antiferromagnetic structures, Vol. 227, 95 (2008), Zabel, D. and Bader, S.D.)

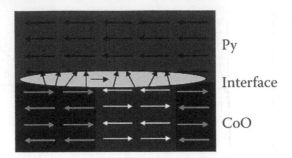

FIGURE 2.27 A schematic diagram of the interface between AF domains in CoO and F (Py). The white background region shows an uncompensated magnetic moment at the F/AF interface which originates in AF. A part of the spins rotate (shown by the arrows randomly oriented) during the magnetization reversal and a part is pinned to the AF. The pinned spins lead to the exchange bias while the rotatable spins create an asymmetric hysteresis loop having a larger saturated magnetic moment with the magnetic field oriented parallel to the pinned spins (for ferromagnetic coupling).

(determining the AF axis in AF), they were able to identify the magnetic state of a Co/NiO structure, see Figure 2.23. X-ray absorption spectra measured on Co/NiO revealed that the Co/NiO interface went through a slight oxidation of Co and a slight reduction of NiO [85]. This clearly demonstrates an appreciable degree of interface chemistry. The oxygen-reduced NiO showed an XMCD signal indicating that the Ni at the interface is magnetic. The second and third layers show that the AF axis of the NiO was collinear with the magnetic moment of the interfacial Ni. This clearly indicates that the interfacial Ni provides uncompensated spins at the Co/NiO interface. The EB in their samples was very weak indicating that most of the uncompensated spins followed the magnetic moment of Co, and only a small fraction of the uncompensated spins was pinned and therefore contributed to the EB. This is strongly supported by the XMCD images in the first and second layers of Figure 2.28: the magnetic moment of the interface Ni is mainly parallel to Co magnetization. The third layer in Figure 2.28 shows another important feature of AF in exchange-biased systems; NiO does not support a single AF domain.

2.5.2 MALAZEMOFF MODEL

Malozemoff [86] recognized that the randomness of the interface exchange interaction between F and AF leads to the breaking of AF into domains. Malozemoff has shown that if it is favorable to keep a single magnetic domain in F, then the interface roughness in AF prefers to break AF into magnetic domains with the domain walls perpendicular to the interface. Malozemoff found that H_{eb} is proportional to the domain wall energy in AF and inversely proportional to the total magnetic moment in F,

$$H_{eb} \simeq \frac{2(A_{AF}K_{AF})^{1/2}}{\pi^2 M_F t_F},$$

(2.83)

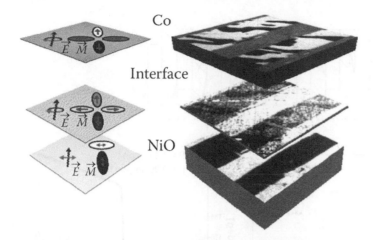

Co

Interface

NiO

FIGURE 2.28 XMCD and XMLD images of a cleaved NiO(100) sample covered by eight atomic layers of Co. The two top layers show ferromagnetic XMCD images recorded at the top Co and middle Ni L absorption edges. The bottom layer shows an XMLD image of NiO. The icons on the left side show the contrast for different domain orientations and the electric polarization of the x-ray beam. (After Ohldag, H. et al., *Phys. Rev. Lett.* 87, 247201, 2001. Copyright 2001 by the American Physical Society. With permission.)

where:

$A_{AF} = J_{AF}/a_{AF}$ is the exchange stiffness coefficient for AF

J_{AF} is the exchange coupling energy for AF

a_{AF} is the lattice constant of AF

2.5.3 Mauri Model

A simple model that includes a domain wall in AF was introduced by Mauri et al. [87]. A sketch describing this geometry is shown in Figure 2.29. The total magnetic energy is

$$\mathcal{E} = -M_F H t_F \cos(\theta - \beta) + K_F t_F \sin^2 \beta$$

$$-J_{eb}\cos(\beta - \alpha) - 2(A_{AF}K_{AF})^{\frac{1}{2}}(1 - \cos\alpha), \tag{2.84}$$

where the last term represents the energy of an AF domain. The shift H_{eb} is given by

$$H_{eb} = -H_{eb}^{\infty} \frac{P}{(1+P^2)^{1/2}}, \tag{2.85}$$

where the parameter $P = (A_{AF}K_{AF})^{1/2}/J_{eb}$ describes the stiffness of the AF EB structure. For large P, the EB shift H_{eb} approaches a maxim value H_{eb}^{∞} asymptotically, see Figure 2.29. In both models, H_{eb} is always inversely proportional to the F layer thickness. This is expected because the F layer was treated in the ultrathin film limit.

Experimentally, it is found that the onset of H_{eb} occurs at a temperature lower than the Néel temperature T_N of AF. This lower temperature is called

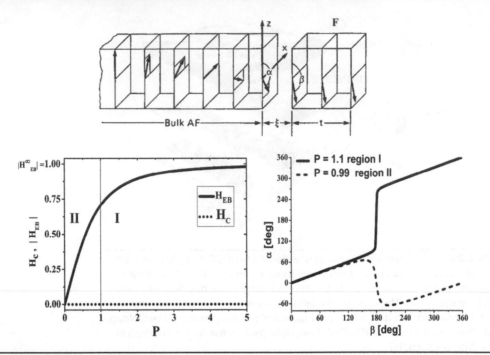

FIGURE 2.29   The Mauri model is shown in the upper diagram. In the Mauri model, the AF domain wall is parallel to the interface. Notice that the spin directions of only one AF sublattice are shown, the other AF sublattice having antiparallel spin moments is omitted for simplicity. The left figure in the lower set shows the dependence of H_{eb} on the parameter $P = (A_{AF}K_{AF})^{1/2} / J_{eb}$ describing the magnetic stiffness of the AF; the right figure shows the dependence of the angle α on the angle β. (With kind permission from Springer Science+Business Media: *Magnetic Heterostructures*, EB effect of ferro-/antiferromagnetic structures, Vol. 227, 95 (2008), Zabel, D. and Bader, S.D.)

the blocking temperature T_B. For thick AF films, $T_B \lesssim T_N$ whereas for thin films $T_B \ll T_N$ [88]. This is obviously associated with the onset of a sufficiently large wall energy for AF, see Equations 2.83 and 2.85.

AF materials are available both as metallic oxides and as metallic alloys. Among the oxides, the monoxides NiO, CoO, and $Ni_xCo_{1-x}O$ are the most commonly used. NiO and CoO have Néel temperatures (R_N) of 525 and 293 K, respectively. In spintronic device applications, the AF materials are mainly metallic alloys of Fe-Mn, Ni-Mn, Ir-Mn, Pt-Mn, and Rh-Mn. For the AF γ phase of the Fe-Mn alloy (with 30 and 35 atomic % of Mn), T_N lies in the range of 425–525 K; and $J_{eb} \sim 0.1 \, erg/cm^2$ at the $Fe_{46}Mn_{54}$/py interface. In the Ni-Mn/py interface, $J_{eb} \sim 0.27 \, erg/cm^2$. A detailed discussion and further references to work on AF materials used in EB systems can be found in the article by Berkowitz and Takano [89], in Section 13.4.3 in *Magnetism* by Stoehr and Siegmann [90], and in Chapter 3 in *Magnetic Heterostructures* by Radu and Zabel [91]. These chapters greatly motivated the authors in writing this section on exchange bias. Figures 2.24 through 2.27, and 2.29 were provided through the courtesy of [91].

2.6 ACKNOWLEDGMENTS

We thank my colleagues Dr. B. Kardasz, Dr. O. Mosendz, Dr. G. Wolterdorf, Dr. Suess, and Prof. J.C. Cochran and Prof. A.S. Arrott for stimulating discussions and invaluable help in the preparation of this manuscript. We have also held extensive discussions with Prof. H. Zabel, Prof. J. Stöhr, Z.Q. Qiu and Dr. M. D. Stiles, Dr. J. Unguris. M. Przybylski, Y.Z. Wu, and Dr. H. Meyerheim. Several figures presented in these sections were obtained through their courtesy. Figures 2.9 and 2.12 were provided by courtesy of Z.Q. Qiu. Figures 2.6, 2.7, 2.13, 2.14, and 2.15 were provided by courtesy by Dr. M. Sties and Dr. J. Unguris. Figure 2.2 was provided by courtesy of Dr. H. Meyerheim. Figure 2.5 by courtesy of Prof. Przybylski and Yizheng Wu. Figure 2.28 was provided by courtesy of Prof. Jo Stoehr. We thank especially the Natural Science Engineering Research Council (NSERC) Canada for continued research funding, which makes this work possible. B. Heinrich also expresses his thanks to Prof. J. Kirschner, Max Planck Institute in Halle, for providing him with a generous support during him summer research semesters in Germany where the first edition of this manuscript was partly prepared.

REFERENCES

1. D. Sanders, The magnetic anisotropy and spin reorientation of nanostructures and nanoscale films, *J. Phys.: Condens Matter* **16**, R603–R636 (2004).
2. W. Brown, *Micromagnetics*, Robert E. Krieger Publishing Co., Huntington, New York, 1978.
3. B. Heinrich and J. F. Cochran, Ultrathin metallic magnetic films: Magnetic anisotropies and exchange interaction, *Adv. Phys.* **42**, 523–639 (1993).
4. B. Heinrich, Magnetic nanostructures. From physical principles to spintronics, *Can. J. Phys.* **78**, 161–199 (2000).
5. J. G. Gay and R. Richter, Electronic structure of magnetic thin films, in *Ultrathin Magnetic Structures*, J. A. C. Bland and B. Heinrich (Eds.), Vol. I, Chapter 2.1, Springer Verlag, Berlin, Heidelberg, p. 22, 2004.
6. P. Bruno, Tight-binding approach to the orbital magnetic moment and magnetocrystalline anisotropy of transition monolayers, *Phys. Rev. B* **39**, 865 (1989).
7. D. Weller, J. Stöhr, R. Nakajima, et al., Macroscopic origin of magnetic anisotropy in Au/Co/Au probed by x-ray magnetic circular dichroism, *Phys. Rev. Lett.* **75**, 3752 (1995).
8. R. Wu and A. J. Freeman, Spin-orbit induced magnetic phenomena in bulk metals and their surfaces and interfaces, *J. Magn. Magn. Mater.* **200**, 498 (1999).
9. L. Néel, Anisotropie magnetique superficielle et surstructures d'orientation, *J. Phys. Radium* **15**, 225 (1954).
10. S. Chikazumi and S. H. Charap, *Physics of Magnetism*, John Wiley & Sons Inc., New York, 1964.
11. D. Sander, The correlation between the mechanical stress and magnetic anisotropy in ultrathin films, *Rep. Prog. Phys.* **62**, 809 (1999).
12. R. H. Victora and J. M. MacLaren, Theory of magnetic interface anisotropy, *Phys. Rev. B* **47**, 11583 (1993).
13. B. N. Engel, C. D. England, R. A. van Leeuwen, M. H. Wiedman, and C. Falco, Magnetocrystalline and magnetoelastic anisotropy in epitaxial co/pd superlattices. *J. Appl. Phys.* **70**, 5873 (1991).

14. I. Dzyaloshinski, A thermodynamic theory of "weak" ferromagnetism of anti-ferromagnet, *J. Phys. Chem. Solids* **4**, 241 (1958).

15. T. Moriya, Anisotropic superexchange interaction and weak ferromagnetism, *Phys. Rev.* **120**, 91 (1960).

16. H. L. Meyerheim, J.-M. Tonnerre, L. Sandratskii, et al., A new model for magnetism in ultrathin fcc Fe on Cu(001), *Phys. Rev. Lett.* **103**, 267202 (2009).

17. B. Heinrich and J. F. Cochran, Magnetic ultrathin films, in *The Handbook of Magnetism and Advanced Magnetic Materials*, H. Kroenmueller and S. Parkin (Eds.), Vol. 4, Wiley, New York, pp. 2285–2305, 2007.

18. B. Heinrich, Ferromagnetic resonance in ultrathin film structures, in *Ultrathin Magnetic Structures*, B. Heinrich and J. A. C. Bland (Eds.), Vol. II, Springer Verlag, Berlin, Heidelberg, pp. 195–222, 1994.

19. M. Farle, Ferromagnetic resonance of ultrathin metallic layers, *Rep. Prog. Phys.* **61**, 755 (1998).

20. P. Poulopoulos and K. Baberschke, Magnetism in thin films, *J. Phys.: Condens. Matter* **11**, 9495 (1999).

21. Z. Celinski, K. B. Urquhart, and B. Heinrich, Using ferromagnetic resonance to measure magnetic moments of ultrathin films, *J. Magn. Magn. Mater.* **166**, 6–26 (1997).

22. G. Wastlbauer and J. A. C. Bland, Structural and magnetic properties of ultra-thin epitaxial Fe films on GaAs(001) and related semiconductor substrates, *Adv. Phys.* **54**, 137–219 (2005).

23. R. Moosbuhler, F. Bensch, M. Dumm, and G. Bayreuther, Epitaxial Fe films on GaAs(001): Does the substrate surface reconstruction affect the uniaxial magnetic anisotropy? *J. Appl. Phys.* **91**, 8757 (2002).

24. M. Brockmann, M. Zolfl, S. Miethaner, and G. Bayreuther, In-plane volume and interface magnetic anisotropies in epitaxial Fe films on GaAs(001), *J. Magn. Magn. Mater.* **198**, 384–386 (1999).

25. T. Seki, T. Shima, K. Takanashi, Y. Takahashi, E. Matsubara, and K. Hono, L1-0 ordering off-stochiometric FePt(001) thin films at reduced temperature, *Appl. Phys. Lett.* **82**, 2461 (2003).

26. B. Heinrich, S. T. Purcell, J. R. Dutcher, K. B. Urquhart, J. F. Cochran, and A. S. Arrott, Structural and magnetic properties of ultrathin Ni/Fe bilayers grown epitaxially on Ag(001), *Phys. Rev. B* **64**, 5334–5336 (1988).

27. P. Krams, F. Lauks, R. L. Stamps, B. Hillebrands, G. Güntherodt, and H. P. Oepen, Magnetic anisotropies of ultrathin Co films on Cu(001) and Cu(1113) substrates, *J. Magn. Magn. Mater.* **121**, 479F (1993).

28. J. Li, M. Przybylski, F. Yildiz, X. D. Ma, and Y. Z. Wu, Oscillatory magnetic anisotropy originating from quantum well states in fe films, *Phys. Rev. Lett.* **102**, 207206 (2009).

29. J. Mathon, M. Villeret, and D. M. Edwards, Exchange coupling in magnetic multilayers: Effect of partial confinement of carriers, *J. Phys. Condens. Matter* **4**, 9873 (1992).

30. M. D. Stiles, Exchange coupling in magnetic heterostructures, *Phys. Rev. B* **48**, 7238 (1993).

31. P. Bruno, Theory of interlayer magnetic coupling, *Phys. Rev. B* **52**, 411 (1995).

32. M. D. Stiles, Interlayer exchange coupling, in *Ultrathin Magnetic Structures*, J. A. C. Bland and B. Heinrich (Eds.), Vol. III, Springer Verlag, Berlin, Heidelberg, pp. 99–142, 2005.

33. F. J. Himpsel, Fe on Au(001): Quantum-well states down to a monolayer, *Phys. Rev. B* **44**, 5966 (1991).

34. J. E. Ortega and F. J. Himpsel, Quantum well states as mediators of magnetic coupling in superlattices, *Phys. Rev. Lett.* **69**, 844 (1992).

35. P. Bruno, Theory of interlayer exchange interactions in magnetic multilayers, *J. Phys. Condens. Matter* **11**, 9403 (1999).

36. Z. Q. Qiu and N. V. Smith, Topical review: Quantum well states and oscillatory magnetic interlayer coupling, *J.Phys.: Condens. Matter* **14**, R169 (2002).

37. B. Heinrich, Exchange coupling in magnetic multilayers, in *Magnetic Heterostructures: Advances and Perspectives in Spinstructures and Spintransport*, H. Zabel and S. D. Bader (Eds.), Vol. 227, Springer Tracts in Modern Physics, pp. 183–250, 2008.

38. E. Montoya, B. Heinrich, and E. Girt, Quantum well state induced oscillation of pure spin currents in Fe/Au/Pd(001) systems, *Phys. Rev. Lett.* **113**, 136601 (2014).

39. Y. Tserkovnyak, A. Brataas, G. E. W. Bauer, and B. I. Halperin, Nonlocal magnetization dynamics in ferromagnetic heterostructures, *Rev. Mod. Phys.* **77**, 1375 (2005).

40. B. Heinrich, Y. Tserkovnyak, G. Woltersdorf, A. Brataas, R. Urban, and G. E. W. Bauer, Dynamic exchange coupling in magnetic bilayers, *Phys. Rev. Lett.* **90**, 187601 (2003).

41. B. Heinrich, J. F. Cochran, M. Kowalewski et al., Magnetic anisotropies and exchange coupling in ultrathin fcc Co(001) structures, *Phys. Rev. B* **44**, 9348 (1991).

42. M. Ruehrig, R. Schaefer, A. Hubert et al., Domain observations on Fe/Cr/Fe layered structures: Evidence for a biquadratic coupling effect, *Phys. Stat. Sol. A* **125**, 635 (1991).

43. J. C. Slonczewski, Fluctuation mechanism for biquadratic exchange coupling in magnetic multilayers, *Phys. Rev. Lett.* **67**, 3127 (1991).

44. B. Heinrich, T. Monchesky, and R. Urban, Role of interfaces in higher order angular terms of magnetic anisotropies: Ultrathin film structures, *J. Magn. Magn. Mater.* **236**, 339 (2001).

45. T. L. Monchesky, B. Heinrich, R. Urban, K. Myrtle, M. Klaua, and J. Kirschner, Magnetoresistance and magnetic properties of Fe/Cu/Fe/GaAs(100), *Phys. Rev. B* **60**, 10242–10251 (1999).

46. R. Urban, G. Woltersdorf, and B. Heinrich, Gilbert damping in single and multilayer ultrathin films: Role of interfaces in nonlocal spin dynamics, *Phys. Rev. Lett.* **87**, 217204 (2001).

47. S. S. P. Parkin and D. Mauri, Spin engineering: Direct determination pf the Ruderman–Kittel–Kasuya–Yosida far-field range function in ruthenium, *Phys. Rev. B* **44**, 7131 (1991).

48. E. Girt and H. J. Richter, Antiferromagnetically coupled perpendicular recording media, *IEEE Trans. Magn.* **39**, 2306 (2003).

49. J. Unguris, R. J. Cellota, and D. T. Pierce, Observation of two different oscillation periods in the exchange coupling of Fe/Cr/Fe(001), *Phys. Rev. Lett.* **67**, 140 (1991).

50. D. T. Pierce, J. Unguris, and R. J. Celotta, Investigation of exchange coupled magnetic layers by SEMPA, in *Ultrathin Magnetic Structures*, B. Heinrich and J. A. C. Bland (Eds.), Vol. II, Chapter 2.3, Springer Verlag, Berlin, Heidelberg, pp. 117–147, 2005.

51. B. Heinrich, J. F. Cochran, T. Monchesky, and R. Urban, Exchange coupling through spin-density waves in Cr(001) structures: Fe-whisker/Cr/Fe(001) studies, *Phys. Rev. B* **59**, 14520 (1999).

52. J. F. Cochran, Brillouin light scattering intensities for a film exchange coupled to a bulk substrate, *J. Magn. Magn. Mater.* **169**, 1, (1997).

53. E. Fullerton, M. J. Conover, J. E. Mattson, C. H. Sowers, and S. D. Bader, Oscillatory interlayer coupling and giant magnetoresistance in epitaxial Fe/Cr(211) and (100) superlattices, *Phys. Rev. B* **48**, 15755 (1993).

54. C. Carbone and S. F. Alvarado, Antiparallel coupling between Fe layers separated by a Cr interlayer: Dependence of the magnetization on the film thickness, *Phys. Rev. B* **36**, 2433 (1987).

55. D. Stoeffler and F. Gautier, Magnetic properties of ferro/antiferromagnetic-based sandwiches, in *Magnetism and Structure in Systems of Reduced Dimensions*, R. F. C. Farrow, B. Dienny, M. Donath, A. Fert, and B. D. Hersmeier (Eds.), NATO-ASI B, Vol. 309, Plenum, New York, p. 411, 1993.

56. B. Heinrich, J. F. Cochran, D. Venus, K. Totland, C. Schneider, and K. Myrtle, Role of interface alloying in Fe-whisker/Cr/Fe(001) structures, angular resolved auger electron and MOKE studies, *J. Magn. Magn. Mater.* **156**, 215 (1996).

57. D. Venus and B. Heinrich, Interfacial mixing of ultrathin Cr films grown on an Fe whisker, *Phys. Rev. B* **53**, R1733 (1996).

58. A. Davis, J. A. Strocio, D. T. Pierce, and R. J. Celotta, Atomic-scale observations of alloying at the Cr-Fe(001) interface, *Phys. Rev. Lett.* **76**, 4175 (1996).

59. R. Pfandzelter, T. Igel, and H. Winter, Intermixing during growth of Cr on Fe(001) studied by proton- and electron-induced Auger-electron spectroscopy, *Phys. Rev. B* **54**, 4496 (1996).

60. M. Freyss, D. Stoeffler, and H. Dreysee, Interfacial alloying and interfacial coupling in Cr/Fe(001), *Phys. Rev. B* **56**, 6047 (1997).

61. B. Heinrich, Z. Celinski, J. F. Cochran, A. S. Arrott, K. Myrtle, and S. T. Purcell, Bilinear and biquadratic exchange coupling in bcc Fe/Cu/Fe trilayers: Ferromagnetic-resonance and surface magneto-optical Kerr-effect studies, *Phys. Rev. B* **47**, 5077 (1993).

62. J. R. Macdonald, Ferromagnetic resonance and internal field in ferromagnetic materials, *Proc. Phy. Soc. London Sec. A* **64**, 968–983 (1951).

63. J. R. Macdonald, Spin exchange effects in ferromagnetic resonance, *Phys. Rev.* **103**, 280 (1956).

64. B. Heinrich, J. F. Cochran, and M. Kowalewski, Effective fields in magnetic thin films: Application to the Co/Cu and Fe/Cr systems, in *Frontiers in Magnetism of Reduced Dimension Systems*, V. G. Bar'yakhtar, P. E. Wigen, and N. A. Lesnik (Eds.), NATO-ASI, Vol. 49, pp. 161–210, 1998.

65. W. S Ament and G. T. Rado, Electromagnetic effects of spin wave resonance in ferromagnetic metals, *Phys. Rev.* **97**, 1558 (1955).

66. E. C. Stoner and E. P. Wohlfarth, A mechanism of magnetic hysteresis in heterogeneous alloys, *IEEE Trans. Magn.* **26**, 3475 (1991).

67. E. Girt, W. Huttema, O. N. Mryasov et al., A method for measuring exchange stiffness in ferromagnetic films, *J. Appl. Phys.* **109**, 07B765 (2011).

68. C. Eyrich, W. Huttema, M. Arora et al., Exchange stiffness in thin film Co alloys, *J. Appl. Phys.* **111**, 07C919 (2012).

69. C. Eyrich, A. Zamani, W. Huttema et al., Effects of substitution on the exchange stiffness and magnetization of Co films, *Phys. Rev. B* **90**, 235408 (2014).

70. T. P. Nolan, J. D. Risner, S. D. Harkness IV et al., Microstructure and exchange coupling of segregated oxide perpendicular recording media, *IEEE Trans. Magn.* **43**, 639 (2007).

71. H. J. Richter, Recent advances in the recording physics of thin-film media, *IEEE Trans. Magn.* **32**, R147 (1999).

72. D. Suess, J. Lee, J. Fidler, and T. Schrefl, Exchange-coupled perpendicular media, *J. Magn. Magn. Mater.* **321**, 545 (2009).

73. E. Girt and H. J. Richter, Antiferromagnetically coupled perpendicular recording media, *IEEE Trans. Magn.* **39**, 2306 (2003).

74. R. H. Victora and X. Shen, Exchange coupled composite media for perpendicular magnetic recording, *IEEE Trans. Magn.* **41**, 2828 (2005).

75. D. Suess, T. Schrefl, S. Fhler et al., Exchange spring media for perpendicular recording, *Appl. Phys. Lett.* **87**, 012504 (2005).

76. A. Yu. Dobin and H. J. Richter, Domain wall assisted magnetic recording, *Appl. Phys. Lett.* **89**, 062512 (2006).

77. E. Girt, A. Yu. Dobin, B. Valcu, H. J. Richter, X. Wu, and T. P. Nolan, Experimental evidence of domain wall assisted switching in composite media, *IEEE Trans. Magn.* **43**, 2166 (2007).

78. H. Kronmüller and H. R. Hilzinger, Incoherent nucleation of reversed domains in Co_5Sm permanent magnets, *J. Magn. Magn. Mater.* **2**, 3 (1976).

79. D. Suess, Multilayer exchange spring media for magnetic recording, *Appl. Phys. Lett.* **89**, 113105 (2006).

80. J. M. D. Coey, *Magnetism and Magnetic Materials*, Cambridge University Press, New York, 2010.

81. R. Jungblut, R. Coehoorn, M. T. Johnson, J. Aan de Stege, and A. Reinders, Orientational dependence of the exchange biasing in MBE grown Ni80Fe20/Fe50Mn50 bilayers, *J. Appl. Phys.* **75**, 6659 (1994).

82. R. H. Meiklejohn and C. P. Bean, New magnetic anisotropy, *Phys. Rev.* **102**, 1413 (1956).

83. T. C. Schulthess and W. H. Butler, Consequence of spin-flop coupling in exchange biased films, *Phys. Rev. Lett.* **81**, 4516 (1998).

84. H. Ohldag, T. J. Regan, J. Stoehr et al., Spectroscopic identification and direct imaging of interfacial magnetic spins, *Phys. Rev. Lett.* **87**, 247201 (2001).

85. T. J. Regan, H. Ohldag, C. Stamm et al., Chemical effects at metal/oxide interfaces studied by x-ray-absorption spectroscopy, *Phys. Rev. B* **64**, 214422 (2001).

86. A. P. Malozemoff, Random-field model of exchange anisotropy at rough ferromagnetic-antiferromagnetic interfaces, *Phys. Rev. B* **35**, 3679 (1987).

87. D. Mauri, H. C. Siegmann, P. S. Bagus, and E. Kay, Simple model for thin ferromagnetic film exchange coupled to an antiferromagnetic substrate, *J. Appl. Phys.* **62**, 3047 (1987).

88. J. Nogues and I. Schuller, Exchange bias, *J. Magn. Magn. Mater.* **192**, 203 (1999).

89. A. E. Berkowitz and K. Takano, Exchange anisotropy—A review, *J. Magn. Magn. Mater.* **200**, 552 (1999).

90. J. Stoehr and H. Ch. Siegmann, Surfaces and interfaces of ferromagnetic metals, in *Magnetism: From Fundamentals to Nanoscale Dynamics*, Vol. 152, Chapter 13.4.3, Springer, Berlin, Heidelberg, pp. 617–629, 2006.

91. H. Zabel and S. D. Bader, Exchange bias effect of ferro-/antiferromagnetic structures, in *Magnetic Heterostructures*, Springer Tracks in Modern Physics, Vol. 227, Springer, Berlin, Heidelberg, pp. 95–184, 2008.

3

Micromagnetism as a Prototype for Complexity

Anthony S. Arrott

3.1 INTRODUCTION

Magnetostatic dipole-dipole interactions play an important role in determining the properties of nanomagnetic systems along with the exchange coupling and magnetocrystalline anisotropy, which were discussed in Chapter 2. The integral-differential equations of micromagnetics lead to complexity. These equations are non-linear because the magnetization is a vector of constant magnitude, $M = m\,M_s$, where m is a unit vector in the direction of the magnetization. The nature of the dipole-dipole interactions favor patterns that are divergence free in volume and put magnetic charges on the surfaces that create demagnetizing fields as opposed to applied fields. In spintronics with its spin-polarized currents in semiconductors, the magnetic moment of the electron is just the tail of the dog. Much of the work on Giant Magnetoresistance (see Chapter 4, Volume 1), Tunneling Magnetoresistance (see Chapters 11, 12, and 13, Volume 1), and Magnetic Random Access Memories (see Chapter 13, Volume 3) comes from clever ways of avoiding the complexities of dipole-dipole interactions. The use of ultrathin films reduces the role of dipole-dipole interactions in the switching of the direction of magnetization, while it eliminates the variation of magnetization in the z-direction, perpendicular to the plane of the ultrathin film. By 1890, J. A. Ewing already understood the importance of the dipole-dipole interaction in ferromagnetism. Ewing coined the term "hysteresis".

The present author addressed the "Past, present and future of soft magnetic materials" at the turn of the century [1]. The "past" reintroduced Ewing

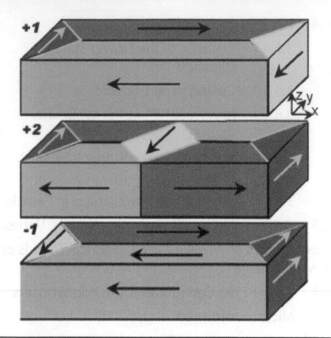

FIGURE 3.1   The Landau structure as presented in domain theory is shown in the top panel with the four domains separated by four 90° walls and one 180° Bloch wall. This is a vortex structure with winding number +1. It does not have inversion symmetry. The structure with a diamond in the center has inversion symmetry and a winding number of +2. The structure in the bottom panel is an antivortex with winding number –1, where on following the circumference clockwise the magnetization rotates counterclockwise by one cycle.

to audiences that had never known the importance of his work. 100 years later, Ewing would have understood much of what was presented at present-day conferences. The "present" was a tribute to Alex Hubert who was the major figure in the theory of *Magnetic Domains*. The book of that name [2], which he coauthored with Rudolf Schäfer, is the most comprehensive work examining the many aspects of domains in magnetic materials. Hubert was a collector of seashells. The book is organized as a biological treatise. It remains a challenge for micromagnetism to account for the over 1000 beautiful illustrations of magnetic configurations. The study of three-dimensional magnetization configurations was envisioned as the "future." It is the focus of this chapter.

In 1935, Landau provided the first three-dimensional model of a structure that minimized the effects of the dipole-dipole interaction; see Figure 3.1. In the 1970s a beautiful new technology was created using magnetic bubbles for which understanding of the dipole-dipole interactions was critical. This chapter is about three-dimensional magnetic structures where some of the simplifications from ultrathin films no longer apply. In this second edition, note is taken of the understanding of these structures that results from advances in computing power. What is certainly more prevalent than at the time of the first edition is the wide use of parallel processing to treat bigger problems faster. Part of this comes from gamers who have motivated the development of the GPU with thousands of CPUs for the cost of, at the most, a few dollars per CPU, using CUDA programming. Random Access Memory has become cheaper and cheaper. This and the use of the 64-bit word facilitate attacks on problems too big for CUDA.

The work of Abert, in his thesis and as the author of programs for both finite element and finite grid using CUDA, is recommended as an up-to-date starting source for potential users of micromagnetics [3]. Commercially available micromagnetic codes are listed there. Perhaps the most used code is OOMMF. This a public domain finite grid code from NIST with more than 20000 hits on Google. It is described as a "code with a steep learning curve."

This chapter is about the three-dimensional patterns of magnetization in (generalized) cylinders of iron of circular, hexagonal, and rectangular cross-section. The *nanobrick* presented here in detail is a parallelepiped with typical dimensions $X = 130$ nm, $Y = 80$ nm, and $Z = 50$ nm, typically using $2 \times 2 \times 2$ nm^3 voxels. Currently, problems with 1000 times more voxels are carried out on a desktop computer. The iron nanobrick is compared with an ellipsoid of comparable dimensions to show that the main actors in this drama of complexity are *tube structures* capable of playing many roles. Tube structures terminate at surfaces in patterns like the center-back of a head where the hair tries to lie in the surface yet twists as it goes away from the center. Hubert has named these as *swirls* [4] in a remarkable paper that remains of interest after almost 30 years.

The swirls are a three-dimensional effect requiring a $\partial m_z/\partial z$ to account for their structure. The center of the swirl need not correspond to the geometric center of a cylinder. Consider a cylinder of length L and radius R with swirls on the end surfaces, $z = \pm L/2$. If the swirls have the opposite handedness, this is *inversion symmetry*, favored by the Landau-Lifshitz formulation of micromagnetics. In three-dimensional magnetic patterns, inversion symmetry is broken, generally. This is already the case when the swirls have the same handedness. There is a contour line, $m_z = 1$, that goes from one swirl to the other. In most cases that line moves from off the z-axis to on the z-axis as one proceeds in from an end surface. In some cases that line forms a helix from the top to the bottom surface.

To describe such patterns, one can look at surfaces of constant m_z. Such *isosurfaces* surround the path of the line $m_z = 1$, but the curl of the magnetization need not be centered on that line. An isosurface of the maximum of $|\boldsymbol{curl\ m}|$ would be more appropriate, but not easily programmed. Expediency leads to the use of isosurfaces of constant component of m_z or a constant value of a scalar such as an energy density or magnetic charge. In the first edition of this chapter, those isosurfaces were referred to as *vortices*, sometimes correctly and sometimes not. In this edition, they are referred to as *tubes*.

In the detailed example of the nanobrick, the tube starts out as a vortex at high, but not too high, magnetic field H, and then in lower applied fields morphs into the famous *Landau structure* of Figure 3.1. Each tube has a life of its own. It is shown here that a tube can take a helical pattern as a means of reducing the magnetostatic energy, which comes from the magnetic charge that results from $-M_s$ ***div m***. Because ***div B*** = 0, ***div H*** = $-f$ ***div M***. It is seen that $-$***div M*** is a source term for the field ***H***. The f is a factor (4π or 1) that depends on the units that may be chosen to discuss magnetism (units are treated in detail in Section 3.2.). A field that arises from $-$***div M*** is called a demagnetizing field. The self-energy of a local magnetic charge density is positive. In magnetically soft materials, that self-energy is minimized by creating divergence-free patterns with magnetic charge appearing only on surfaces, as like charges repel.

In a soft magnetic material, the surface charges create fields that are approximately equal and opposite to any externally applied fields. The boundary conditions created by the magnetostatics require non-uniform magnetization patterns that decrease the parallel alignment of the electron spins, which are favored by the *exchange interaction* responsible for ferro-magnetism. That decrease in the degree of local alignment of the spins is called an increase in the *exchange energy* as discussed below. The boundary conditions increase the exchange energy and lead to ***div M*** in the bulk.

These magnetic charges are paid for by lowering the increase in the exchange of energy with respect to the divergence-free patterns that satisfy the same boundary conditions. Once those magnetic charges exist in the bulk, they interact with unlike charges attracting, but not annihilating, one another because that would increase the local exchange energy. The attractive magnetostatic interaction accounts for the movement of the swirls off center at the ends of the cylinder, and also for what happens when the conditions of geometry and applied magnetic fields are right for the onset of helical magnetic patterns.

When a swirl moves away from a symmetry axis, it develops the beginning of the helical tube pattern to spread out the magnetic charge and at the same time to put the positive magnetic charge closer to the negative magnetic charge. Under recently determined conditions, the helix can propagate from one swirl to the other, providing, at last, an explanation for the striped domain patterns on hexagonal iron whiskers with [111] orientation along the long axis, discovered by Hanham 40 years ago [5]. The basic mechanism is with the arrangement of patches of magnetic charge into the formation of magnetic *extended quadrupoles* when the charge density is mostly confined to a plane, as is the case on a surface. There are four helical tubes of magnetic charge density that form one tube of quadrupole-like magnetic charge density.

If the diameter of the tube of a quadrupole-like magnetic charge measured in the z-direction is h, then the pitch of the helix makes the spacing between the centers of adjacent turns of the helix ~2.5 h for a helix with a period of 750 nm in a 320 nm diameter cylinder of iron; see Section 3.7.

The tube configurations discussed here have minimal problems from surfaces and can be manipulated by sufficiently small fields and currents to be attractive for nanoscale devices.

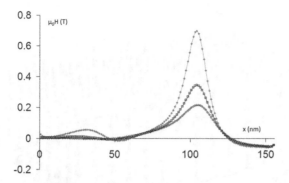

FIGURE 3.2   External fields for a $156 \times 96 \times 60$ nm³ nanobrick using 1.25-nm cubes for calculation. The scan is on a line of constant y through the center of the swirl at positions 1, 10, and 20 nm above the centers of the uppermost calculation cells. The large peak in the H_z field is from one swirl on the top surface. The swirl on the bottom surface is 60 nm away and creates a negative contribution to H_z that is negligible on the top surface. There are also large H_z fields at the four vertical corners of the nanobrick.

The external magnetic fields from a vortex in an ultrathin film are small because the magnetic charge on one surface cancels the effects of the charge on the other surface. In a nanobrick, the swirls on the two ends of the tube can be moved far apart so that the external fields can be greater than one-quarter of the saturation induction of iron; see Figure 3.2.

The swirls can be moved easily and rapidly over large distances. They become highly non-linear oscillators that serve as prototypes for the complexity of bifurcations and chaos.

The calculations for the nanobrick serve also as a primer on the effective use of modern codes for micromagnetism using fields from magnets and fields from electrical currents to provide symmetry breaking.

3.2 SYMMETRY BREAKING IN A CYCLE OF MAGNETIZATION

For an ellipsoid in a mathematical model with continuous variation of magnetization, the application of a large applied magnetic field H along the i-direction parallel to the long x-axis can result in a uniform magnetization. For an iron nanobrick in a high field, this is not the case. If the magnetization is along i at the x-axis it will deviate from i for finite y and z. At high fields, that deviation is outward at the positive x-surface and inward at the $-x$-surface. This is called the *flower state*. It applies to fields along any of the three principal axes of the nanobrick. In this discussion of magnetization processes, the first example is field H_z applied perpendicular to the largest face of the nanobrick. The flower state is discussed in the caption of Figure 3.3.

Discussion of the visual patterns and details of the calculations are found in the figure captions throughout this chapter.

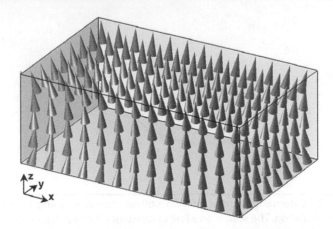

FIGURE 3.3   An iron nanobrick with dimensions of $156 \times 96 \times 60$ nm^3 in the flower state at a field of $\mu_0 H_z = 2.2$ T, showing the directions of the magnetization as hollow cones on three sides of the brick. Each cone represents the average components of the magnetization in a volume of 72 nm^3 combining 36 computational cells in a plane. The computational cells are cubes with 2-nm edges. The magnetic surface charge density on the top surface is $\sigma_m \approx M_s$, where M_s is the saturation magnetization of iron at room temperature, taken here to be 1714 emu or $\mu_0 M_s = 2.154$ T. This charge produces a magnetic field that is in the plane of the top surface and points outward. The charge density on the bottom surface is $\sigma_m \approx -M_s$. The field from this is inward in the plane of the bottom surface. There is also magnetic charge density on the side surfaces that varies almost proportional to z, measured from the center of the brick. The flower state appears in high fields for nanobricks of all dimensions, including a cluster of nine iron atoms at $T = 0$ K in a cube with one of the atoms in the body-centered position. The flower state has inversion symmetry at equilibrium. Changing the uniformly applied field will cause the magnetic moments to precess, yet the inversion symmetry is maintained. Changing the field non-uniformly by passing a current along the z-axis will break the inversion symmetry.

As the field is lowered, the magnetization pattern changes continuously, starting at a critical field, from the *flower state* to the *curling state* in which the magnetization has components that circulate around the central axis; see Figure 3.4. The sense of the circulation, *cw* or *ccw*, is determined while breaking the *reflection symmetry* of the flower state. This is as a result of the dipole-dipole interaction using the concept of *splay saving*, which reduces the magnetic charge density by having opposite signs for $\partial m_z / \partial z$ and $(1/\rho) \partial(\rho m_\rho)/\partial \rho$.

The energy contours present a *three-tined fork* [6]. The paths to both the right and the left become lower in energy than the path along the central tine. It is useful to provide a bias field so that the magnetization does not stay on the higher energy path. If one relies on the round-off error of a computation to provide the bias, the computation time can be much too long. An electric current I_z is passed, briefly, up the z-axis to choose the path and save computation time.

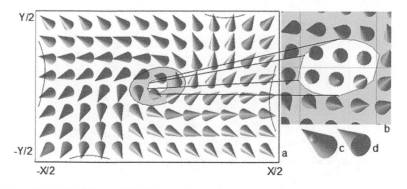

FIGURE 3.4   An iron nanobrick with dimensions of $156 \times 96 \times 60$ nm³ in the curling state at a field of $\mu_0 H_z = 0.8$ T. The contour lines are for direction cosines $m_z = 0.95$ and $m_z = 0.8$. The area between these is shaded gray. The cones on the left, **a**, are the average of the magnetization for nine cells of 4-nm cubes in the topmost plane centered at $z = 28$ nm. In the blow up on the right, **b**, the cones are the magnetization in each cell. As the field is lowered below $\mu_0 H_z = 0.9$ T, each cone on the top surface rotates counterclockwise from its position in the flower state as it would if a current were passed up the z-axis. This configuration at and near the surface is a *surface soliton* called a *swirl*. The position where the component along each edge changes sign is the center of an *edge soliton*. Each cone is treated as if it were centered on a position one-third of the distance from the apex, denoted, in detail at **c** in the lower right, as a sphere superimposed on the cone that points out of the plane of the page. The cones are hollow, so that when viewed from below, light is usually not reflected. The hollow cone at **d** points into the page. Using cones to indicate the direction of magnetization creates artificial waves in a visual pattern when the base of the cone moves from one side of a grid line to the other, for example, the line of cones at $x = 0$.

3.2.1 Breaking Inversion Symmetry and Choosing a Handedness

The inversion symmetry of the flower state may or may not be maintained in the curling state. Note that inversion symmetry for a magnetic dipole is the opposite to that of an electric dipole. Consider the moments at two points equidistant from a center of inversion symmetry on opposite ends of a line through that center. For electric dipoles, the moments point in opposite directions. For magnetic dipoles, they point in the same direction. If the inversion symmetry were maintained on the transition to the curling state, the circulation at the bottom surface would be opposite to that on the top surface; see Figure 3.5, where the "top" and "bottom" are along the x-axis. This leads to complex magnetization patterns as the field is lowered [1]. Away from the axis, the in-plane magnetization reverses on going from top to bottom, but the component along the axis remains $m_z = 1$.

There are many ways to break the inversion symmetry to arrive at the *Landau configuration*. Here the discussion is focused on achieving that configuration by passing a small current along the z-axis while lowering H_z from saturation and then passing a small current along the y-axis to further break

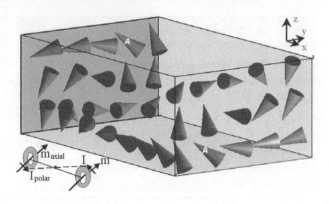

FIGURE 3.5   Curling with inversion symmetry for an iron nanobrick with dimensions of $112 \times 70 \times 42$ nm³. The moments near the front surface at $x = X/2$ show a *cw* circulation. The moments near the back surface at $x = -X/2$ show a *ccw* circulation. The line of moments joining the front and back surfaces shows a Néel wall in the form of an end-over-end helix. Inversion symmetry is maintained. Note that the two cones labeled *A* are connected by a line through the center of the brick and have the same orientation. The rendering of the magnetization as cones shows an object that does not have inversion symmetry (a line joining the tips of the cones does not go through the center of the brick). It is the magnetization at the grid points that has the inversion symmetry. The field along the *x*-axis where the two surfaces are separated by 112 nm was used for this illustration of inversion symmetry. When the field is along the *z*-axis, the two surfaces are separated by 42 nm. Then the exchange energy presents a sufficiently large barrier that the *cw/ccw* structure is suppressed; see Figure 3.6. The distinction in the meaning of inversion symmetry between a polar vector and an axial vector is illustrated in the lower left corner using the polar vector *i* for the current in a ring to produce an axial vector *m*. The two current vectors, equidistant from the center of inversion symmetry, point in opposite directions, while the magnetizations they produce point in the same direction.

the symmetry. That is just one of many paths from saturation to the Landau configuration. A more complicated path is discussed in Section 3. 7

3.2.2 Response of Isosurfaces to Changing Fields

In the presence of the field from a small current I_z, the flower pattern develops a slight swirl even at the highest fields. This is analogous to the magnetization of a ferromagnetic material in a small field above its Curie temperature. When the temperature is lowered below T_c, the response to the small field increases rapidly as the spontaneous magnetization develops. In the nanobrick, below a critical field H_{zc}, there is a spontaneous contribution to the angle of rotation of the moments away from their direction in the flower state. The swirls are centered on the z-axis; see Figure 3.6. As the swirls develop with decreasing H_z, the action is first concentrated at the core of the swirl. As the field is decreased to zero, the moments in the corners are the last to fully participate in the curling pattern, turning from pointing out, along a line at 45°, to pointing perpendicular to that line. The places where the magnetization with respect to an edge changes sign is the center of an

FIGURE 3.6 The curling pattern that develops on reducing $\mu_0 H_z$ from 0.9 T for the nanobrick in Figure 3.4 is called the *l*-vortex state. The circulation is cw (or ccw) in all planes of constant *z*. The *l*-vortex state persists in zero field for the dimensions of $102 \times 70 \times 42$ nm³, but, if Z > 42 nm, the *l*-vortex state becomes unstable at a low field and moves off center. When there is a computational cell centered at the origin, the cells along the z-axis all have $m_z = 1$. Here the center of the vortex is midway between two cells on the y-axis displaced in x by 1 nm on either side of the vortex center. When the cells are grouped in bundles of 49 cells in a plane to produce the cones shown here, the center of the bundles nearest the core of the swirl are displaced by 7 nm from the axis. This is far enough from the center of the swirl that the magnetization lies close to the surface when *H* goes to zero. The development of the *l*-vortex state is followed here with the 42-nm-high brick in the presence of a small current I_z. Note the progress of the corner moments. These change little at high fields. At lower fields, they rotate so they are parallel to one of the two corner edges. In the lowest fields, they lose their radial component and at the same time turn down in response to the dipole field from the central core of the vortex structure. Note that the hollow cone in the lower left corner of the large panel for 0.00 T appears darker at the bottom of the cone where the light enters. This cone points into the iron nanobrick as is the case for all corners. It is more evident in the lower left.

edge soliton. In Figure 3.6, the solitons move from the center of each of the top edges to the corners of the nanobrick.

In general, reversal of the magnetization of the edges takes place by the propagation of edge solitons, either from one end or the other or from both at the same time. It can also happen through the formation of a pair of solitons somewhere along an edge, with the two solitons moving to opposite ends to reverse the edge. The propagation of edge solitons was important in the switching of the original thin film MRAM bits that had lateral dimensions in the order of a micron. In 1998, most of the work in micromagnetics concentrated on calculating hysteresis loops, without much attention paid to the visualization of the reversal processes. Solitons were overlooked even when the visualizations were presented that showed them. They had to be pointed out to the presenters. (Solitons along the axis of the nanobrick are seen in Figure 3.13 and discussed in Section 3.7.) The swirls are surface solitons.

Figure 3.6 shows the curling pattern that develops on reducing H_z. It is seen that the sense of the circulating component is forced to be the same at all z by the small current along z. The radial component is inward for $z < 0$ and outward for $z > 0$ as a result of the dipole-dipole interaction. For the exchange interaction, the energy does not depend on the angle of the in-plane magnetization in cylindrical symmetry. The m_z-isosurfaces define the tube structure. The role of these tubes in magnetism has often been overlooked.

The direction cosines of the magnetization M are denoted by m_x, m_y, and m_z. Starting from 1, m_z decreases with distance from the z-axis. Tubes of constant m_z, called m_z-isosurfaces, connect the top and bottom surfaces; see Figure 3.7. The cross-sections of the tubes in planes of constant z are elliptical (reflecting the geometry of the brick) and centered on the z-axis. The ellipses are largest in the midplane $z = 0$. The magnetization does not lie in the m_z-isosurfaces except in the midplane. Elsewhere it has a radial component as well as a circulating component.

3.2.3 DISPLACING THE SWIRLS

Bias fields perpendicular to the z-axis can move the vortex structure off center. The vortex moves perpendicular to the bias field with the swirls remaining above one another, but the ellipsoidal cross-sections distort and need not be centered with respect to the swirls. That is for equilibrium configurations. In dynamic responses, the swirls move with respect to each other. The m_z-isosurface takes on a life of its own. The tubes can bulge, twist, bend, and even take helical patterns. If the bias fields are derived from a current i_x along the x-axis in the +x-direction, the upper swirl moves in the +x-direction and the lower swirl moves in the −x-direction as a result of fields in the +y-direction at the top and in the −y-direction at the bottom. The H_y fields from i_x are largest at the top and bottom surfaces. When a uniform H_y field is superimposed on the field from i_x, one can independently manipulate the two swirls in any H_z.

3.2.4 ONSET OF THE LANDAU STRUCTURE

If i_x is maintained while H_z is lowered, there is a critical H_z, below which the displacements of the swirls in opposite directions increase rapidly with decreasing H_z. The development of the *S-tube state* from the *I-vortex state* is shown in Figure 3.7. If the dimensions are at the threshold for the instability of the *I*-vortex state, the correspondence of the *S*-tube state with the well-known Landau configuration is not obvious. When viewed in the central cross-section, $z = 0$, the pattern more closely reflects the Landau configuration; see Figure 3.8. When the dimensions X and Y are much larger than

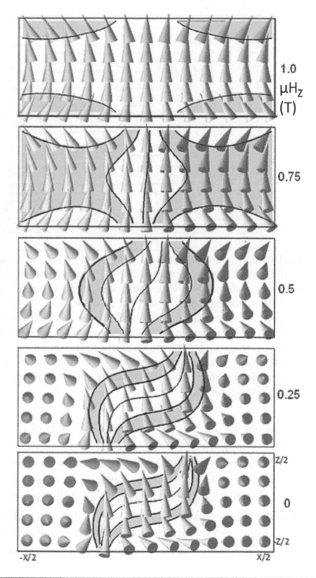

FIGURE 3.7 The S-tube state develops from the *I*-vortex state with decreasing H_z for the iron brick with dimensions of $156 \times 100 \times 60$ nm³. Each cone represents 144 cubes with 4-nm edges in a plane of constant *y*. The coarse grid is used in this calculation for visual purposes, even though it is too crude for accurate assessment of critical fields. The width of the iron nanobrick was increased by one computational cell compared to Figures 3.3 and 3.4 in order to show the central cross-section, $y = 0$, of the *I*-vortex state. The formation of the Stube state is followed using m_z-isosurfaces that connect the top and bottom swirls of the Ivortex state. The intersections with the midplane at $y = 0$ for $\mu_0 H_z > 0.4$ T or at $y = 4$ nm for $\mu_0 H_z < 0.4$ T are shown for the contour lines in that plane for $m_z = 0.95$ and $m_z = 0.80$ with the central white regions corresponding to $m_z > 0.95$ and the gray regions to $0.8 < m_z < 0.95$. A small bias current in the *x*-direction creates a field, H_y, that is positive at the top and negative at the bottom, tilting the magnetization in the +*x*-direction at the top and in the −*x*-direction at the bottom. To make this clearer, a contour for $m_y = 0$ is shown in the middle of the central white region. The tilt of the $m_y = 0$ contour in the second panel at 0.75 T is the result of the bias current. The larger displacement of the swirls in the panel at 0.50 T is almost all the result of a spontaneous displacement of the swirls in opposite directions to form the S-tube state. In zero field, the swirls reach their maximum displacement and sit near $x = (X–Y)/2, y = y_1, z = Z/2$ and $x = -(X–Y)/2, y = y_1, z = -Z/2$, where y_1 is a small displacement; see Figure 3.9. At $Z = 60$ nm, the *I*-state vortex can persist as a structure in an unstable equilibrium all the way to $H_z = 0$. The width of the contours at $z = 0$ is essentially the same for the unstable I-state vortex and the S-state tube.

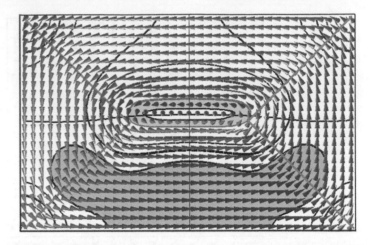

FIGURE 3.8 Central cross-section for $z = 0$ for an iron nanobrick with dimensions of $156 \times 100 \times 60$ nm^3. The core of the S-tube and center of the 180° Bloch wall is the central white region where $mz > 0.95$. In the surrounding gray region, $0.95 > m_x > 0.8$. In the dark gray region, $y < 0, -0.4 < m_z < -0.2$. The open contour around the central region marks $m_z = 0$. In the corners, $m_z \approx 0.8$. A negative H_z is required to turn the corners down for the $Z = 60$ nm structure, while for $Z = 42$ nm in Figure 3.6, the corners are already down at $H_z = 0$. The slightly wavy horizontal line is the contour for $m_x = 0$. The vertical thin line is the contour for $m_y = 0$. The gray lines from the four corners show the positions of walls in the divergence-free van den Berg construction [2].

the values used here, the swirls sit at the ends of the *180° Bloch wall* of the Landau configuration at $x = (X-Y)/2, y = y_1, z = Z/2$ and $x = -(X-Y)/2, y = y_1, z = -Z/2$, where y_1 is a small displacement; see Figure 3.9 for a portrait of the S-tube that is the heart of the Landau configuration. If the current i_x were in the $-x$-direction while H_z was lowered through the critical H_z, the swirls would sit at $x = -(X-Y)/2, y = -y_1, z = Z/2$, and $x = (X-Y)/2, y = -y_1, z = -Z/2$. The statics and dynamics of switching between these two configurations, called the S^+ tube and the S^- tube, are emphasized in this chapter.

3.2.5 Landau's Structure

Landau's structure, as shown in Figure 3.1, was postulated for a large iron brick where the anisotropy causes a clearer distinction between the *domains* and the *domain walls*. The Landau structure does not have inversion symmetry, but the structure in Figure 3.1 with a diamond in the center, often observed in the 1950s by researchers at General Electric and General Motors, does have inversion symmetry. Landau put the walls at 90° to avoid a discontinuity in the component of m perpendicular to the walls of the *end-closure domains*. The magnetization was assumed to be in the z-plane everywhere except in the 180° Bloch wall. The Bloch wall creates *surface charges*, $\sigma = n \cdot m$, which Néel eliminated by having the magnetization lie on the surface as it turned through 180°. These are the *Néel caps* on the Bloch wall first calculated by LaBonte in the 1960s in his treatment of a never-ending Bloch wall; see Figure 3.10. Consequently, it was pointed out that the Néel cap

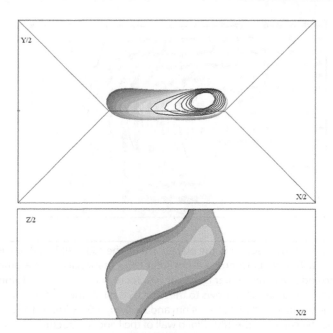

FIGURE 3.9   Portrait of an S-tube state core showing contours of $m_z = 0.95$ in successive planes in y or z spaced by 1.5 nm. The S-tube state is a 3D structure that is not fully characterized by a single slice through the m_z-isosurfaces. The upper panel shows that the swirl is offset from the core of the 180° wall, which itself is offset from the plane $y = 0$. The silhouette of the structure in the lower panel has contributions from several slices in y. The lines in the upper panel show the positions of the 180° Bloch wall and the four 90° walls delineating the two end-closure domains in the classic structure first postulated by Landau in 1935. These lines are also the positions of the walls obtained by van den Berg in his solution to the problem of ideally soft magnetic materials in the limit of ultrathin films.

was an extension of one or the other of the end-closure domains with that extension terminating in a swirl at the opposite end [7]. The swirls satisfy the topological necessity of the magnetization pointing out of the surface in at least two places. One of the first SEMPA (scanning electron microscopy with polarization analysis) experiments was to show that this is the case in iron whiskers [8]. Structures, where both swirls are on the same surface, are possible [9] but not considered here. Which end of the Bloch wall claims the swirl depends on the sense of rotation in the Néel cap, which itself depends on the sense of rotation of the magnetization around the y-axis on transversing the Bloch wall in the y-direction. The diamond structure, also shown in Figure 3.1, requires additional tubes as there are now two swirls and two half anti-swirls on the top and bottom surfaces. But that discussion will be left for another time.

The Bloch wall with the Néel caps is seen in Figure 3.10 for an S-tube in the iron nanobrick for a cross-section in the plane $x = 0$. This structure is essentially the same as that calculated by LaBonte in the 1960s for a never-ending 180° Bloch wall [10]. That there is a regular vortex structure around a line in the x-direction has long been noted without recognizing that the vertical section of that circulation is actually the core of an S-tube connecting

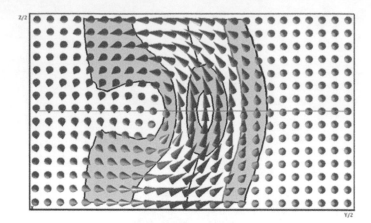

FIGURE 3.10   Cross-section in the plane $x = 0$ for the $156 \times 100 \times 60$ nm³ iron nanobrick. Each cone represents a 4-nm cubic computational cell. The almost vertical bowed line in the center is the contour for $m_x = 0$. It is centered along the top and bottom surfaces but bows to the right by 4 nm at the center of the nanobrick. In the white region about $y = 4$ nm and $z = 0$, $m_z > 0.95$. This is the core of the S-tube and the center of the 180° Bloch wall of the Landau structure. The light line at $z = 0$ is also the contour for $m_y = 0$. Along that line, the cones rotate through an angle greater than 180° from $y = -Y/2$ to $y = Y/2$. This is the Bloch wall separating the two principal domains of the Landau structure, where in the white region on the far left, $m_x < -0.95$ and in the white region on the far right, $m_x > 0.95$. In the gray region on the left, $-0.95 < m_x < -0.8$. In the gray region on the right, $0.95 > m_x > 0.8$. In the gray region in the center, $0.95 > m_z > 0.8$. This figure is in one to one correspondence with LaBonte's calculation of the cross-section of a neverending Bloch wall from the 1960s, which clearly showed the existence of the Néel caps to the left of the center on both surfaces. The magnetization in the Néel caps and the core of the S-tube circulate around the white bulge on the left, which forms the core of a partial vortex in the x-direction. This partial vortex accounts for the component $<m_x>$ that accompanies the transformation of the I-vortex to the S-tube as shown in Figure 3.14.

the upper and lower surfaces. The swirls do not appear in LaBonte's calculation for they are displaced to infinity in the never-ending Bloch wall. They do not appear in Figure 3.10 because there the cross-section is midway between the two swirls. But the displacement of the core of the S-tube does appear in Figures 3.8 and 3.9 and in LaBonte's calculation. The reason for the displacement has long been understood. It is to make room for the Néel caps that not only remove magnetic charge from the surfaces but also minimize volume charge by curling about the line parallel to the x-axis to form part of the x-axis vortex structure.

3.2.6 The Nano-Ellipsoid

The S-tube state for a nano-ellipsoid is shown in Figure 3.11. The nano-ellipsoid avoids the discussion of what happens along all 12 edges in the nanobrick. The central core of the S-tube in the nano-ellipsoid is the Bloch wall terminating in the displaced swirls. The S-tube is illustrated by an

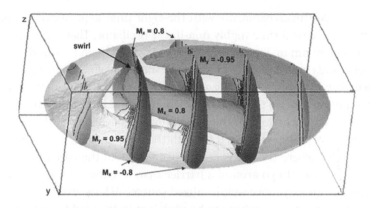

FIGURE 3.11 The Landau structure in an ellipsoid ($260 \times 161 \times 100$ nm³) calculated using R. Hertel's finite element program *TetraMag*. This program results in "slices" which allow viewing of the internal structure. The isosurfaces at $m_y = +0.95$ and $m_y = -0.95$ delineate the combined closure domains and Néel walls at each end of the core of the Bloch wall, outlined by the isosurface $m_z = 0.8$. The core of the Bloch wall terminates at the surfaces in swirls. The Bloch and Néel walls separate the regions of high magnetization in the $+x$ and $-x$ directions, indicated by the darker regions of the five planes perpendicular to the x-axis. The same sequence of fields used in Figure 3.7 to produce the Landau structure starting from the high field flower state. But, the flower state exists only to the extent that the finite elements are not sufficiently effective in producing an exact ellipsoid. The ellipsoid is simpler than the brick because there are no corners or edges. For the ellipsoid, this configuration is one of the eight ground states which differ in the choice of polarization with respect to the z-axis, the handedness of the circulation about the z-axis and whether the upper swirl is on the left in an S^- state as shown here or on the right in an S^+ state. The two swirls and the core of the Bloch wall are displaced from the $y = Y/2$ plane just as shown for the brick in Figure 3.9. The swirls can be displaced in opposite directions by the field from a current in the x-direction as shown in Figure 3.7 for the brick. A uniform H_y will move both the swirls toward one end or the other of the ellipsoid. A large H_y will drive an S^+ pattern into an I^* pattern (see Figure 3.16) near the end of the ellipsoid. Then, on decreasing Hy, the I^* pattern becomes an S^+ pattern again if biased by a field in $//_x$. If the bias field is in $-H_x$ the I^* pattern becomes an S^- pattern. In a high enough H_y field, the swirls move together out one end of the ellipsoid.

m_z-isosurface for $m_z = 0.8$ which corresponds to the 3-4-5 triangle with acos $(m_z) \approx 37°$. (The m_z-isosurfaces in Figure 3.7 are also for $m_z = 0.8$. In Figure 3.9, for the portrait of the S-tube, the m_z-isosurfaces are for $m_z = 0.95$.) The Néel caps and the closure domains are represented by the m_y-isosurfaces for $m_y = \pm0.95$ in Figure 3.11. The contours lines on the circular slices at various values of x, show the m_x components. The principal domains of the Landau structure in the ellipsoid are suggested by contours with $m_x > 0.8$ and $m_x < -0.8$.

3.2.7 MANIPULATING THE LANDAU STRUCTURE

Near the critical field, it is easy to drive the swirls back and forth between their small offset positions. At $H_z = 0$ it is harder. But it can still be done. It is easier to do this dynamically with the right time sequence of the fields. The moving swirls form a highly non-linear oscillator. They can move over distances of 50 nm in 200 ps. The swirls carry with them localized external magnetic fields with $\mu_0 H_z \approx 0.5$ T from the surface magnetic charges. At resonance, they will oscillate as long as the energy is supplied to compensate for the damping losses. As the swirls oscillate back and forth in the x-direction they also make excursions in the y-direction as they follow paths of almost constant energy. This is in contrast with the motion of the swirls in slowly varying fields where the paths are perpendicular to the paths of constant energy; it is easier to go around a barrier than to climb over it. The ability to easily and quickly move well-localized sources of large external magnetic fields is a phenomenon waiting to be exploited in the world of nanoscience and nanotechnology.

3.2.8 CORE REVERSAL BY PAIR CREATION

When H_z is increased in the $-k$-direction, the swirls move back toward the z-axis reaching the axis at a critical field that is lower in magnitude than the critical field for forming the displaced swirls on the decreasing field in the k-direction. During all these changes the $m_z = 1$ line still goes from the bottom surface to the top surface. But at a high enough field in the $-k$-direction, a pair of solitons are created that propagate from near the midplane outward toward the surfaces, reversing the core of the vortex as they propagate; see Figure 3.12.

Each soliton contains a singularity in a continuum model. On a finite grid, the core of the vortex centers itself on a position between the grid points, so that the singularity itself is a mathematical point between the grid points. This should be the case for a real lattice where the center of the vortex would lie between the atoms. But that is a classical description for which there is no quantum mechanical calculation to support the concept of the atoms maintaining a rigid magnetic moment in the core of the soliton. The magnetic moment density can vary in direction as well as magnitude across an atom [11].

3.3 THE LLG EQUATIONS AND THEIR NUMERICAL SOLUTIONS

The above description is for a vortex that intersects the top and bottom surfaces. Starting with fields in the x- or y-directions, vortices can form with swirls on the end and/or side surfaces. In either case, the removal or reversal of these fields returns the system to the state where the swirls are on the top and bottom surfaces. This process can be quite complex with the swirls moving from one surface to another or by the vortices leaving the brick and then re-entering. Starting with the Landau structure in zero field, there is a rich landscape of responses, steady and dynamic, to fields and field gradients applied in the x-y plane.

Solving the Landau-Lifshitz-Gilbert (LLG) equations of motion explains most of the above. A modern review of micromagnetics is given in Volume 2

FIGURE 3.12 Pair creation in a reversed applied magnetic field $\mu_0 H_z = -0.8$ T. The centers of the propagating singularities are in the middle of the four central cones in planes of constant z in each of the panels A, B, C, D on the right. The panel on the left is a cut in a plane of constant y through the two cones just above that center. By symmetry, the two central columns of cones on the left are two views of each cone of the four central cones. In panels A, B, C, D the four central cones all rotate in unison as one views this as a sequence of layers one above the other. It is also the time sequence for any given layer as the singularity passes up the axis. For the upper singularity, $\partial m_z/\partial z$ is positive and in the lower singularity, it is negative. For splay canceling, a negative m_r is required for the first case and a positive m_r for the lower case, as $\nabla \cdot m \approx m_r/r + \partial m_z/\partial z$. As m_r is positive in both regions, there is splay saving in the case of the lower singularity. The lower one propagates more quickly as the splay saving lowers the energy barrier.

of *The Handbook of Magnetism* where 33 authors over 500 pages provide a modern review of micromagnetics. The chapters on numerical methods are quite detailed with hundreds of equations [12].

There are only a few cases where these equations can be treated by analytical techniques. The LLG equations are simultaneous integral-differential equations for the direction cosines of the components of the magnetization. The differential part comes from the exchange energy responsible in the first place for ferromagnetism. Any variation of the moments from parallel alignment locally changes the negative exchange energy in the positive direction. The integral part comes from the dipole-dipole interactions in which, if there are N moments, there are $N(N1)/2$ pairs to consider.

3.3.1 ZEEMAN ENERGY

The independent variable is the applied magnetic field H_{ext}. The *Zeeman energy* is $E_{ext} = -\mathbf{\mu} \cdot \mathbf{B}$ for a magnetic moment $\mathbf{\mu}$ in an externally applied magnetic induction, $\mathbf{B}$, defined by the force on a moving charge. In free space, $\mathbf{B} = \mu_0 \mathbf{H}_{ext}$, where $\mu_0 \equiv 4\pi\ 10^{-7}$ H/m, is an arbitrary constant called the permeability of free space in SI units, which were enacted by a one-vote margin at a conference in 1931 dominated by electrical engineers with little appreciation of magnetism. Magneticians have resisted the adoption of these units in favor of Gaussian units for the reasons of good physical insight. In Gaussian units $\mu_0 = 1$. The magnetic moment of a nanobrick is $\mathbf{\mu} = <M>V$, where $<M>$ is the average magnetization in the volume V. In SI units, the magnetization $\mathbf{M}$ is replaced by the magnetic polarization $\mathbf{J}$, where $\mathbf{\mu} = <J>V/\mu_0$. In terms of the unit vector $\mathbf{m}$ representing the direction of magnetization or magnetic polarization, the Zeeman energy in SI units is $E_{ext} = -J_s\,\mathbf{m} \cdot \mathbf{H}_{ext}\ V = -M_s\,\mathbf{m} \cdot (\mu_0 \mathbf{H}_{ext})\ V$, where J_s and M_s are the spontaneous polarization and the spontaneous magnetization, respectively, of the iron nanobrick. For iron, $J_s = 4\ \pi\ 1714/10^4$ T = 2.154 T, where 1714 comes from M_s for iron in Gaussian units and the 10^4 comes from the conversion of Gauss to Tesla. In Gaussian units the polarization has the same units as the magnetization but differs by the factor 4π; that is, $J_s = 4\pi M_s$. In SI units $J_s = \mu_0 M_s$. This means that one cannot convert from SI to Gaussian just by setting $\mu_0 = 1$. The magnetic susceptibility, which is dimensionless in both systems, differs by that factor of 4π.

The engineers removed the 4π's in Maxwell's equations from where they properly belong, in front of the source terms. In Maxwell's equations, M appears as a current source given by $\mathbf{j}_m = \nabla \times \mathbf{M}$. That 4π will come up again when the dipole-dipole interaction is considered. The anisotropy and the exchange energies depend only on the directions of magnetization, so the question of units should not affect their energy expressions.

3.3.2 ANISOTROPY

When a magnetic moment is rotated with respect to the atomic lattice, there is a change in energy because of spin-orbit coupling to the lattice. The vector derivatives of the *anisotropy energy density* with respect to the magnetization components produce effective fields acting on the magnetization. These are not real fields and do not necessarily obey the Rabi-Schwinger theorem [13], that a magnetic moment in a real field can be treated classically, but they are treated as real fields in the LLG equations. As the anisotropy fields play very minor roles in the behavior of an iron nanobrick, quantum mechanical subtleties will be ignored here.

The magnetization patterns for an iron nanobrick and those for a permalloy nanobrick are indistinguishable in a zero field at the level of visually comparing graphs of lines of constant components of the reduced magnetization. The anisotropy becomes a player in the magnetization patterns when the dimensions are in microns rather than nanometers. Even then the anisotropy in single crystal iron is much more noticeable than in polycrystalline iron when the grain size is less than the dimensions of the element being considered.

Magnetostatics act as a constraint on the magnetization of adjacent grains forcing them to have close to the same normal component to their mutual boundary. The anisotropy rotates the magnetization away by a small amount from the pattern that would be there in the absence of anisotropy. That small amount accounts for the phenomena called buckling patterns or ripple structures where there are Néel-like walls perpendicular to the direction of the magnetization. The helical pattern in hexagonal iron whiskers occurs whether the anisotropy is considered or not for dimensions less than 1 μm. Anisotropy requires large distances to be effective in iron. To see where anisotropy is important in spintronics visit Chapter 2 in this volume.

3.3.3 Exchange Interaction

The effective field from the exchange interaction can be written using the Laplacian of the direction cosines of the magnetization Δm, leading to nine terms in the torque equations representing curvatures in each of the components in each of the three directions. By definition, the exchange energy density e_{ex} is A times the sum of the squares of the nine first derivatives of the three components of the magnetization direction, with respect to the three axes. The coefficient A is called the exchange stiffness constant. For the unit vector m, $|m|^2 = 1$ and hence $m \cdot \partial m/\partial x = 0$; e_{ex} can be written as

$$e_{ex} = A(\nabla m)^2 = -\ Am \cdot \nabla^2 m, \tag{3.1}$$

where the Laplacian Δ can be written also as

$$\nabla^2 m = \nabla(\nabla \cdot m) - \nabla \times (\nabla \times m). \tag{3.2}$$

That an expression with only first derivatives can be replaced by an expression with only second derivatives is the property of a unit-vector field. To see this, put the unit-vector field along the y-direction and then rotate one moment into the x-direction, with respect to its neighbors, by a small angle. Then $(dm_x/dx)^2 = -d^2m_y/dx^2$. The minus sign appears in front of the Laplacian.

Equation 3.2 can be considered as the definition of the vector Laplacian. When expressed in Cartesian coordinates there are nine terms in $\nabla\ (\nabla \cdot m)$ and 12 terms in $\nabla \times (\nabla \times m)$, but the six terms in the first cancel the six terms in the second. The vector Laplacian in Cartesian coordinates has nine second derivatives, whereas the starting expression has the squares of the nine first derivatives. The vector Laplacian has 15 components in cylindrical coordinates and 18 in spherical coordinates.

The appeal of Equation 3.2 is that the *curl* can be envisioned by the Stokes theorem for the line integral around an area and the divergence by the Gauss law for a pillbox. Despite all its terms, a simple insight can be seen by looking at the vector Laplacian in spherical coordinates. For the *spherical hedgehog*, the magnetization is along the radial vector everywhere, $m_r = \hat{r}$ (The spherical hedgehog has no swirl, a real hedgehog does.) The curvature is 1/r in each of the directions orthogonal to the radial vector $\hat{r}$. One of the 18 terms is $2m_r/r^2\hat{r}$, with no derivatives. The exchange energy density is $2A/r^2$. If this is integrated for a sphere of radius R, the total energy is $8\pi AR$. The energy density is all from one term in $\nabla\,(\nabla\cdot m)$. There is no curl in the spherical hedgehog. This hedgehog has a point singularity at its center. Next, take the hedgehog pattern and rotate every spin by π around the z-axis. Now in spherical coordinates, the directions of the spins are given by $m = -\cos(2\theta)$ $R + \sin(2\theta)\,\Theta$. This is a splay-saving pattern that decreases $\nabla\,(\nabla\cdot m)$. It will increase $|\nabla\times(\nabla\times m)|$.

An easy way to visualize the component of the vector Laplacian in the direction of the magnetization is to put the unit-vector magnetization in the x-direction for one voxel, add the x-components of the six nearest neighbor voxels, and subtract the number six, then divide by the square of the voxel side. This measure of curvature will be a negative number and that is why there is a minus sign in the expression for the exchange energy with the vector Laplacian.

As an exercise in vector analysis, the exchange energy was evaluated for that splay-saving pattern using: (1) the liquid crystal expression $A\cdot\{(\nabla\cdot m)^2 + (\nabla\times m)^2\}$, (2) the first derivative expression $A\,(\nabla m)^2$, and (3) the second derivative vector Laplacian $-A\,m\cdot\{\nabla\,(\nabla\cdot m) - \nabla\times(\nabla\times m)\}$ for each of the three choices of the coordinate system. There are so many terms that it is easy to make a mistake, so it helped to know the answer before starting. If one looks at $A\,(\nabla m)^2$, the exchange energy as originally defined, there are nine derivatives each of which is squared. Rotating the spin by 180° about the z-axis changes the signs of the derivatives but does not change the squares of the derivatives. The energy density must be the same as for the spherical hedgehog. The splay energy is reduced, but the total energy remains the same. The energy for the sphere with the splay-saving pattern is also $8\pi AR$.

Splay saving is very important as it reduces the magnetic charge density, but not because it lowers the exchange energy as it appears to do in the liquid crystal formulation of the strain energy.

The hedgehog shows the scale of the exchange energy. It increases in proportion to a length. For a 100 nm sphere in the splay-saving configuration, $8\pi AR$ is ~500 picoergs for iron. The numerical solution using the LLG equations for a 100 nm iron sphere removes most of the enormous magnetostatic energy that a hedgehog would have. The pattern is a vortex with much of the exchange energy, ~400 picoergs, in the two swirls at the north and south poles. When the exchange energy of a sphere of iron is calculated using the full power of micromagnetics, the result shows that, as the sphere gets larger, the total exchange energy approaches that of the hedgehog, within one part in 100 by a radius of 150 nm.

Another configuration where it is simple to determine the exchange energy is when the magnetization circulates around the axis making an angle χ with the radial vector from the z-axis. It doesn't matter for the exchange whether that angle is 90° for which the **div m** = 0 (the pattern of a z-axis vortex) or 0° for which the curl m = 0 (the pattern of a *cylindrical hedgehog*). If m_ρ = 1 in cylindrical coordinates, ρ, ϕ, and z, the vector Laplacian is $-1/\rho^2$ in the ρ-direction. If m_ϕ = 1, the Laplacian is unchanged, and the energy density is also A/ρ^2. The first case is pure splay and the second case is pure curl. The total energy for a cylinder of length 2 L and radius R is $4\pi A \log (R/R_0)L$, where R_0 is the radius of the hole that Néel first put down the center to avoid the logarithmic singularity in estimating the critical radius for the onset of curling in an infinitely long cylinder.

The toroid is a simple problem. There is no divergence, so it is only necessary to use Stokes theorem twice to find **curl(curl (m))** for a toroidal shell magnetized with components circulating about either the minor radius or the major radius.

The energy for producing a vortex-antivortex pair in an ultrathin film (no dependence on z) is $8\pi Ah$ where h is the height of the film.

In a divergence-free pattern of magnetization, the only contributions to the exchange field come from $\nabla \times m$. In cylindrical coordinates, there are 15 terms in the Laplacian. Even so, it is easier to think about divergence-free patterns using the Laplacian rather than $(\nabla m)^2$. In the simplest magnetization pattern for a vortex: $m_\rho = 0$, $m_\phi = \tanh (a\rho)$ and $m_z = \text{sech} (a\rho)$, where $1/a$ is a constant called the effective exchange length, as it measures the region where the exchange energy is more important than the dipole-dipole energy that would come from the z-component of the magnetization at a surface. There are no derivatives with respect to ϕ or z, there is no m_r component, and $\nabla \cdot m = 0$. In this case, there are still five terms in the Laplacian. When the dot product of the Laplacian and the magnetization is considered, two of the terms cancel and two of the terms combine, leaving the local exchange energy density $e_{ex}(r)$ as

$$e_{ex}(\rho) = A\left(\frac{\text{sech}^2(a\rho)}{a^2} + \frac{\tanh^2(a\rho)}{\rho^2} \right).$$

(3.3)

Even after these simplifications, the integration of Equation 3.3 requires approximations [14]. It would be helpful to have the total exchange energy in terms of $<m_z>$, but that requires the evaluation of $\int \rho \text{sech}(a\rho) d\rho$, which does not have an analytic expression. Progress in micromagnetics is difficult without numerical methods.

The exchange fields calculated analytically, in the simple case with m_z = sech($a\rho$) and m_ρ= tanh($a\rho$), do not quite point in the same direction as the magnetization, indicating that these are not self-consistent solutions of the torque equations. They are, however, quite good approximations to the computed magnetization pattern around a vortex in an ultra-thin circular disk (for which there is no significant dependence on z) using the full micromagnetic equations including all the dipole-dipole interactions [14].

3.3.4 DIPOLE-DIPOLE INTERACTIONS

The treatment of the dipole-dipole interaction is different in the finite element program *TetraMag* by Hertel [15] and the finite grid program *LLG Micromagnetic Simulator* by Scheinfein [16]. In *TetraMag* the moments in each element are used to solve Poisson's equation for the magnetic potential from which the fields are derived. In the LLG Micromagnetic Simulator, the dipoles on a uniform grid are summed using fast Fourier transforms, which require that every grid point is treated the same, so that the interactions depend only on the vector connecting any two grid points. Using fast Fourier transforms reduces the calculation from the order of N^2 to the order of N $\ln(N)$. Fortunately, the equations of micromagnetics have attracted mathematicians who have brought sophisticated methods to bear on the problem of treating the dipole-dipole problem, sufficiently sophisticated to be beyond the intent of this chapter.

There is a problem with the finite grid approach in treating boundaries that are not aligned with the grid. This is avoided here by choosing the parallelepiped as the object of interest. One solution to the jagged-boundary problem is to make *boundary corrections* by treating local regions at the order of $n(n-1)/2$ and farther regions at the level of $N \ln(N)$ for all N points, adding $Nn(n-1)/2$ calculations, where n is the number of points in the local region. It takes but a small n for $n(n-1)/2$ to be bigger than $\ln(N)$.

In Section 3.7, for circular geometry, the jagged boundaries appear in the graphics for the charge density. The LLG equations are solved with boundary corrections. but the graphics using the correct magnetization do not correct for the jagged edges in plotting *div m*.

The dipole-dipole energy is written using the demagnetizing field. This follows from fact that the sources of $\nabla \times B$ are currents and there are no sources for $\nabla \cdot B$. The vector H is a mixed vector combining the source vector M with the field vector B. In Gaussian units the combination that defines H is $H \equiv B - 4\pi M$, where the 4π belongs in front of the source vector. In SI units H is defined as $\mu_0 H \equiv B - J$. The sources of $\nabla \times H$ are real currents. The sources of $\nabla \cdot H$ are magnetic charges given by $-\nabla \cdot M$. This part of H that derives from those charges is called the demagnetizing field H_D. In SI units, $\nabla \cdot H_D = -(J_s/\mu_0) \nabla \cdot m$. In Gaussian units, $\nabla \cdot H_D = -4\pi M_s \nabla \cdot m$. The demagnetizing field energy is the integral over the nanobrick to get

$E_{dem} = -(1/2)\int H_D \cdot J dV$ in SI units and $E_{dem} = -(1/2)\int H_D \cdot M dV$ in Gaussian units. The factor (1/2) comes from these being self-energies.

For a sphere uniformly magnetized in the z-direction, in Gaussian units $H_D = 4\pi(1/3) M_s \mathbf{k}$ and $M = M_s \mathbf{k}$ for which $E_{dem} = -(1/2) 4\pi(1/3)M_s^2$. In SI units, $J = J_s \mathbf{k}$ and $H_D = (1/3)J_s/\mu_0 \mathbf{k}$ for which $E_{dem} = -(1/2)(1/3)J_s^2 V / \mu_0$

or $E_{dem} = -(1/2)(1/3)\mu_0 M_s^2 V$. In all cases, the $(1/3)$ is the demagnetizing coefficient N for a sphere. In the general ellipse, $N_x + N_y + N_z = 1$. In Gaussian units, $4\pi N$ is called the demagnetizing factor.

3.3.5 THE TORQUE EQUATION

The basic equation of micromagnetics is the torque equation for the precession with time t of an electron in a magnetic induction B. The electron has a magnetic moment $\mu = -g\mu_B S$ and an angular momentum $S\hbar$, where $\hbar$ is the reduced Planck's constant, g is very close to 2 and $S = 1/2$. The minus sign appears because the spin and the moment are in opposite directions for a negatively charged electron. The ratio of the angular momentum to the magnetic moment of the electron is $\gamma_e = -g\mu_B/\hbar$. These are used in a classical equation of motion, where the angular momentum of the magnetic moment is $L\mu = \mu/\gamma_e$ and the torque acting on that moment is $\mu \times B$; that is,

$$\left(\frac{1}{\gamma_e}\right)\frac{d\mu}{dt} = \mu \times B.$$

(3.4)

A mystery of this equation lies in questions about nutation. Nevertheless, it works. One can replace B by $\mu_0 H$ in the torque. Dividing both sides by a small volume and letting μ stand for the moment in that volume, μ can be replaced by M to give

$$\left(\frac{1}{\gamma_e}\right)\frac{dM}{dt} = \mu_0 M \times H,$$

(3.5)

which is in SI units, but produces the torque equation in Gaussian units by replacing $\mu_0 = 4\pi\ 10^7$ H/m by a dimensionless 1. It is common practice to absorb the μ_0 into the gyromagnetic ratio to write

$$\left(\frac{1}{\gamma_0}\right)\frac{dM}{dt} = M \times H,$$

(3.6)

where $\gamma_0 \equiv \mu_0 \gamma_e$. As M, the magnetization, appears on both sides of the equation, it can be replaced by J, the magnetic polarization ($J = \mu_0 M$), to obtain

$$\left(\frac{1}{\gamma_0}\right)\frac{dJ}{dt} = J \times H\left\{-\nabla_j(e_{ext})\right\}.$$

(3.7)

The H in Equations 3.5 to 3.7 is treated as an effective field by taking minus the vector gradient of the energy density terms with respect to the magnetization or the polarization as in Equation 3.8,

$$H_{eff} = -\nabla_J(e_{tot}) = \frac{-\nabla_m(e_{tot})}{J_s} = \frac{-\nabla_m(e_{tot})}{(\mu_0 M_s)}$$

(3.8)

The price for using the magnetic polarization as the variable in the torque equations is to put some μ_0's into expressions for the anisotropy and exchange energies where there is no physical reason for them being there. The *effective field* includes the applied field, the demagnetizing field from all the other magnetic moments, the exchange field from the variation of the moment direction with position, the anisotropy field, and the *damping field*.

The *damping field* was formulated by Gilbert to be proportional to the rate of change of the components of the magnetization [17]. The damping in iron comes from the repopulation of the Fermi surfaces of spin-up and spin-down electrons as the magnetization direction is rotated locally. These spin currents dissipate energy. Spin currents can be created externally and used to cause the magnetization to rotate by forced repopulation of the Fermi surfaces. The two processes differ in the sign of the coefficient of the contribution of $\partial m/\partial t$ to the effective field. There is much more about the subject of damping in Chapter 2 in this book. Spin currents are not discussed in this chapter because the author has not yet applied them to the nanobrick.

3.4 APPLYING THE MICROMAGNETIC EQUATIONS OF MOTION

This chapter describes in some detail the results of calculations for the iron nanobrick. The parameters are those of iron except for the calculation of the equilibrium configurations where the *damping coefficient* α is greatly increased from the low value of iron to the value that gives critical damping. The dynamic calculations use $\alpha = 0.02$ and the equilibrium calculations use $\alpha = 1$. For $\alpha = 0.02$, the time constant for the approach to equilibrium is typically 10's of ns. For $\alpha = 1$, the time constants can be less than 1 ns, unless the system is near a critical point for which the torques vanish. Then the time can be too long to compute and the usual approach to equilibrium as exponential changes to a 1/t approach. As critical points are of interest in describing magnetic configurations, it is necessary to have techniques to obtain the answers more quickly.

The time steps used in micromagnetic calculations are small compared to the time resolution adequate to describe the fastest of dynamic responses. The torque equations have mathematical instabilities that often require that the time steps be as small as a fraction of a femtosecond. The time step must be smaller when the grid size is smaller.

The grid size itself should be smaller, by at least a factor of two, than the exchange length λ given by $\lambda^2 = A/K$, where A is exchange stiffness constant of Equation 3.1 and K is a magnetostatic energy density; $K = \mu_0 M_s^2/2$.

In classical micromagnetics [18], the language is that of Gaussian units, where M, H, and B all have the same dimensions. Then, μ_0 is replaced by 4π in the magnetostatic energy used in the definition of the exchange length. The use of a grid spacing that is too large can lead to results that are completely misleading. In a treatment of magnetization processes in a nanobox with square sides, $X = Y > Z$, it was shown that the moments in the I_z vortex have $m_z = 0$, even at the axis, when the grid is greater than 3λ [19]. Critical fields are sensitive to the grid size even for grid size $< \lambda/10$.

3.4.1 BIAS FIELDS

The fields from currents used to break the symmetry at critical points also provide a means of avoiding prohibitively long computation times. Such bias fields are called anticipatory fields [20] as they select among three prongs of a three-tined fork. Without a proper bias field, one can stay on the central tine even though the other tines lead to lower energies. If an anticipatory field is not used, the middle tine will be abandoned after numerical round-off errors propagate exponentially with time for more than ten-time constants.

Away from critical points, the total energy comes to equilibrium exponentially, while the components of the magnetization and individual terms in the energy come to equilibrium as damped oscillators. The convergence of a calculation is achieved when the extrapolation of the exponential or the damped oscillators to infinite time no longer changes significantly with time. From the magnetization at any three points equally spaced in time, it is simple to extrapolate to infinite time to get the apparent equilibrium magnetization for a simple exponential approach. For example, if the points are m_a, m_b, and m_c, the extrapolated $m_f = (m_b{}^2 - m_a\,m_c)/(2m_b - m_a - m_c)$ provides criteria for terminating the calculation.

At a critical field, it is convenient to use a bias field that takes the magnetization from one configuration to another quickly and then analyze the results of that calculation to obtain what would have happened in the absence of that bias field. This is called the *path method* [21].

3.4.2 INTERNAL ENERGY

The path method relies on the insensitivity of the *internal energy* E_{int} to the values of the external fields needed to reach configurations that have the same values of the average components of the magnetization $<m_x>$, $<m_y>$, $<m_z>$. E_{int} is the sum of all the energy terms excluding the Zeeman term. The energy, divided by the volume V, along an equilibrium path can be written as

$$\frac{E}{V} = -\left\langle m^* \right\rangle \cdot (J_s H - \nabla_{\langle \mathbf{m} \rangle}(E_{int})\,|_{\langle \mathbf{m} \rangle = \langle \mathbf{m}^* \rangle}), \tag{3.9}$$

where $<m^*>$ is $<m>$ at some point along the path. Knowing the gradient of the internal energy E_{int}, with respect to the components of the average magnetization as a function of $<m>$ along an equilibrium path, one can calculate the field necessary to reach equilibrium (not necessarily stable) for a given $<m^*>$ along that path. The path method assumes that knowledge of E_{int} along a path that is close to a given equilibrium path will give the same result; that is, $\nabla<m>(E_{int})|<m>=<m^*>$, which is assumed to be insensitive to small bias fields.

In the path method the calculations of the magnetic response to applied fields are carried out for a sufficient number of fields to determine the functional form of E_{int}, but no more than necessary. Once one has an analytic expression for $E_{int}(<m>)$ of a particular type of pattern, one can produce the entire dependence of the magnetization on the field. When this works, it can

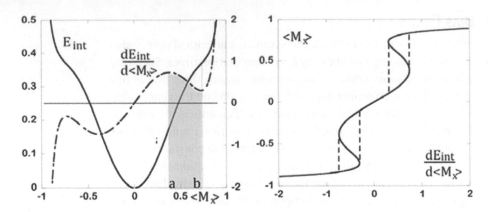

FIGURE 3.13 Illustration of the path method. The dependence of the internal energy E_{int} upon $<M_x>$ is shown in the left panel as the solid curve for an arbitrarily constructed $E_{int} = E_2<M_x>^2 + E_4<M_x>^4 + E_6<M_x>^6)/(1-<M_x>^{12})$ to produce four inflection points and also mimic the approach to saturation. The system is in an unstable equilibrium for the region between the inflection points at **a** and **b**. The dashed line shows $dE_{int}/d<M_x>$, which is shown again in the right panel as the independent variable ($\sim H_x$) to produce a magnetization curve with hysteresis.

save many orders of magnitude in computation time. To determine whether it works can take time, but that need only be done once for a given type of system. The path method does not work when there are two or more independent responses to the applied field.

3.4.3 USING ANTICIPATORY FIELDS

An example of the use of the path method and anticipatory fields is the calculation of the field for a vortex parallel to the z-axis to re-enter an iron nanobox [19] after being driven out through the $y = Y/2$ surface by a field H_x. After the vortex leaves, the magnetization is in a *C-state* which can be viewed as a virtual I_z-vortex just outside the brick. When H_x is reduced sufficiently, the vortex should re-enter the surface through which it exited. The three-tined fork, in this case, is the fact that the virtual vortex must choose the direction of magnetization for the core in order to re-enter. The virtual vortex does not have polarization before it enters unless the *C-state* itself has a bias in the direction of the vortex that exited, which can happen for particular geometries. If a bias field along the z-axis is not applied, the system stays on the central tine and the vortex does not re-enter until long after a round-off error provides an initial bias. If a bias field is applied along the z-axis, it slightly changes the H_x at which the vortex enters, but it changes by many orders of magnitude how long one would have to wait for that to happen.

3.4.4 USING THE PATH METHOD

A hypothetical example of the path method is given in Figure 3.13. The left panel in this figure shows the dependence of the internal energy E_{int} upon $<M_x>$ as the solid curve for an arbitrarily constructed E_{int} given by

$$E_{\text{int}} = \frac{\left(E_2 \langle m_x \rangle^2 + E_4 \langle m_x \rangle^4 + E_6 \langle m_x \rangle^6 \right)}{\left(1 - \langle m_x \rangle^{12} \right)} \tag{3.10}$$

to produce four inflection points and also mimic the approach to saturation. The system is in unstable equilibrium for the region between the inflection points at **a** and **b**. The dashed line shows $dE_{\text{int}}/d<m_x>$, which is shown again in the right panel as the independent variable for producing a magnetization curve with hysteresis. For equilibrium, the applied field must be equal and opposite to the internal field given by $-dE_{\text{int}}/d<m_x>$, which explains why there are no minus signs in this illustration of the path method. E_{int} can be constructed by analysis of the calculations in regions of stable equilibrium to interpolate the regions of unstable equilibrium. E_{int} can also be calculated using the regions of unstable equilibrium if the timescale of changes is appropriate for obtaining close to equilibrium configurations while the configuration is moving in time as a whole.

The path method works only for large damping. For small damping, the magnetization moves on a path of almost constant energy, while the path method has the magnetization moving in the direction of the maximum gradient of the energy.

3.5 AROUND THE $<M_z>$–H_z HYSTERESIS LOOPS

The computations are designed to obtain E_{int} as a function of $<M>$ and $<M>$ as a function of H_z for each configuration using appropriate bias fields when necessary. Hysteresis loops are shown for $<m_z>$ and $<m_x>$ in Figure 3.14, with and without a small bias field from currents $i_x = 0.1$ mA. The iron brick has dimensions of $50 \times 80 \times 130$ nm^3. The sequence of configurations starting from high H_z includes:

3.5.1 THE FLOWER STATE

The magnetization splays out from the center on the top surface and inward toward the axis on the bottom surface, with most of the magnetization along the $+z$-axis; see Figure 3.3 and the section of Figure 3.14 labeled **c**, $\mu_0 H_z \sim 1$.

3.5.2 THE CURLING STATE WITH INVERSION SYMMETRY

The circulation is cw on one-half of the nanobrick and ccw on the other; see Figure 3.5. The magnetization dependence on H is not shown in Figure 3.14. (The reader will be spared the complexities of magnetization processes proceeding from this state.)

3.5.3 THE I_z-VORTEX STATE

The same handedness throughout the nanobrick is achieved by applying a bias field from a current along the z-axis, which can be removed once the handedness is chosen.

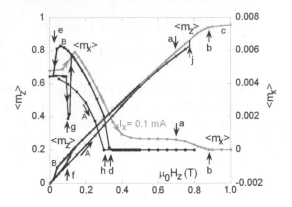

FIGURE 3.14 Hysteresis loops for an iron nanobrick with dimensions of $130 \times 80 \times 50$ nm^3. The first quadrant of the major hysteresis loop is shown for both $<m_z>$ and $<m_x>$, the latter with much magnification (its ordinate is on the right). The gray curves for $<m_z>$ and $<m_x>$ were calculated with a bias current $i_x = 0.1$ mA to anticipate the transitions to and from the *I* state. The bias current has little effect on $<m_z>$ but facilitates following the transitions from the effects it has on $<m_x>$. The gray curve for $<m_x>$ measures the degree of curling between points **a** and **b** and measures the susceptibility of the *I* state for forming the *S* state in the field region between 0.3 and 0.4 T. The two black curves for $<m_x>$ show the sharp transition between the *S* and *I* state near 0.3 T that occur in the absence of the bias current. The curve for $<m_x>$ labeled **A** is for the vortex core m_z in the $-z$-direction after coming from saturation at a high negative field. The curve for $<m_x>$ labeled **B** is for the vortex core m_z in the $+z$-direction with the transition occurring at a slightly higher field, **d** compared to **h**. All the hysteresis at low fields is from the switching of the magnetization directions in the four corners of the nanobrick. For $<m_z>$ the black curves, one of which hides the gray curve, are for no bias current. The curve for $<m_z>$ labeled **A** is for the core of the vortex in the $-z$-direction with the transition for the state with the core m_z in the $+z$-direction taking place by pair creation and propagation at **j**. There are two different states within the $+z$-direction and the other has the core in the $-z$-direction. This would give different values for the magnetization in zero field if it were not for the almost complete compensation of the net magnetization by the moments in the four corners, which are opposed, to the core magnetization in both cases.

The gray curves in Figure 3.14 were obtained in the presences of a bias field from a current along the x-axis, $i_x = 0.1$ mA, which anticipates the transformation from the I_z-vortex state to the S_z-tube state as H_z is lowered.

The degree of the displacement of the swirls is tracked by $<m_x>$. The inflection point on the gray curve at $\mu_0 H_z = 0.32$ T corresponds to the field at which the transition takes place in the absence of the bias field, labeled **d**. The bias current, i_x, does not produce any $<m_x>$ in the flower state but does in the curling state, so that $<m_x>$ tracks the onset of curling at **b** and the approach to saturation of that effect at **a** on the gray curves.

3.5.4 THE LANDAU-TYPE CURLING STATE

The I_z vortex distorts spontaneously into either the S_z^+-tube or the S_z^--tube depending on the bias field applied from a current along the x-axis, see Figures 3.6 to 3.10. The presence of the S_z-tube is signaled by the presence of

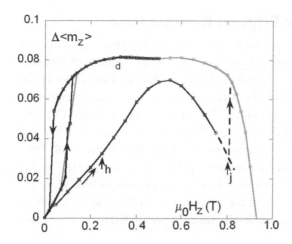

FIGURE 3.15   The hysteresis loops for the difference $\Delta<m_z> = <m_z> - <m_z>_D$ found by subtracting $<m_z>_D = H_z/H_s$ from each $<m_z>$ in Figure 3.14. The hysteresis in low fields accompanies the reversal of the magnetization in the four vertical edges. All four vertical edges flipped on decreasing the field in a single large step in the field on decreasing the field, but on increasing the field in smaller steps, two flip back first and then after a second step in the field the other two flip back. When resolved in time, the four edges flip independently from one another. The onset of the S-tube is signaled by $\Delta<m_z>$ beginning to decrease below **d**. The lower curve is for the magnetization of the core opposite to the applied field. The S-tube goes back to the I-vortex at **h**. The higher slope for the reversed field reflects the smaller demagnetizing field for the trapped vortex. Point **j** is where the core reverses by pair creation and propagation.

$<m_x>$ in the absence of bias fields. The presence of the S_z-tube has a minor effect on $<m_z>$ that can be noticed after subtracting the demagnetizing field, the dominant effect of the magnetostatic energy.

3.5.5 States of the Vertical Edges

The Landau-type state has four choices of polarization, plus (*p*) or minus (*m*) for the virtual vortices along the four vertical edges. Without additional bias fields, the four vertical edges for the S_z^+-tube state are either *pppp*, *ppmm*, or *mmmm*, labeled *cw* from the corner (−*X*/2, *Y*/2). For the S_z^--tube state, the sequence is *pppp*, *mmpp*, *mmmm*. Each of these states has its range of stability with hysteresis in the minor loops for the switching of these edges by soliton propagation; see 3.7, where *edge solitons* play a major role in achieving the Landau structure. The minor hysteresis loops at low fields are shown in Figure 3.14 and in more detail in Figure 3.15, where the mean effect of the demagnetizing field has been subtracted. All of this is avoided in the ellipsoid, which has no edges.

3.5.6 The *S*-Tube in Reversed Field

The extension of the *S*-tube in the *x*-direction is maximum for $H_z = 0$. For $H_z < 0$, the field is opposite to the magnetization of the core of the *S*-tube, but most of the magnetization outside the tube follows H_z. When H_z is sufficiently negative, the S-tube returns to the z-axis becoming a very compact version of the I_z vortex. The $<m_x>$ component goes to zero for a smaller magnitude of H_z when the fields are in the direction opposite to the core magnetization; compare points **d** and **h** in Figure 3.14. The curves labeled **A** are calculated for negative fields and then replotted in the first quadrant for the direct comparison with the positive fields. Those curves would be obtained directly in the first quadrant if the magnetization process started with large negative fields and then proceeded to positive fields.

3.5.7 The Trapped I_z-Vortex State

When the field is large in the reversed direction, the magnetization is negative everywhere except in the core of the vortex. The demagnetizing field becomes smaller for the trapped state leading to a higher slope of m_z vs. H_z in A compared to B.

3.5.8 The Transient Core Reversal State

Starting with the trapped ^+I_z vortex, as the field becomes more negative, near $-\mu_0 H_z = 0.8$ T, a pair of point singularities is created near the midplane. These propagate up and down the z-axis reversing the magnetization of the core to produce the $-I_z$ vortex with the same circulation as maintained in all these processes; see Figure 3.12 and point **j** in Figure 3.14. The derivative of m_z with respect to z changes sign at each singularity but the derivatives of m_ρ with respect to ρ do not, providing a clear picture of the role of splay saving in Figure 3.12.

3.5.9 Saturation

At a high negative H_z, the $-I_z$-vortex state goes to the flower state with the moments splaying outward from the center of the bottom surface and inward toward the center of the top surface, with most of the magnetization along the $-z$-axis

3.6 DISCUSSION OF MAGNETIZATION PROCESSES

The programs for micromagnetic calculations keep track of the components of the magnetization for each grid point for every-so-many iterations as well as the net magnetizations and each term in the energy of the nanobrick as a whole. These are analyzed to gain insight into the competition among the energy terms and how they lead to the various magnetization patterns. The leading energy term is the Zeeman energy. Even when the external fields are all zero, the configuration in a real nanobrick is one that reflects the past history of the magnetization. On the computer one can arbitrarily assign a configuration A and calculate the configuration B that minimizes the energy starting from A. Then, if $H_z = 0$, the dominant term in the energy of the iron nanobrick is the dipole-dipole energy. At all other fields, the competition is between the Zeeman energy and the dipole-dipole energy with the exchange playing a supporting role and the anisotropy almost no role at all.

3.6.1 IDEALLY SOFT MAGNETIC MATERIALS

For a magnetic system to act as a conductor, the magnetization pattern has to adjust itself to produce the required surface charges while remaining divergence free within the volume. It can do this if there is no anisotropy and the system is large enough that the increase in exchange energy required by the divergence-free pattern is very small. A large enough ferromagnetic body without anisotropy behaves as a magnetic conductor with no net field inside the body, except that in a singly connected body, the magnetic conductor cannot topologically escape the need for two swirls. For a large enough body the magnetization pattern is divergence free everywhere except in the vicinity of the swirls. This is called the ideally soft magnetic material. This is realized experimentally in iron whiskers with $X = Y \sim 0.1$ mm and $Z \sim 10$ mm just below the Curie temperature where the magnetic anisotropy goes to zero much faster than the spontaneous magnetization [22]. Even though this work was inspired by measurements at high temperature, the calculations are all for low temperature where thermal agitation is completely neglected, except for its effect on the material constants that are those of ambient temperature.

The electrical charge on the surface of an electrical conductor is a very small fraction of the charge on a surface atom. The magnetic charge on the surface of a magnetic conductor is limited by the finite moment of the surface atom. As the charge necessary to cancel an applied field at a corner of a brick goes to infinity, the corners become saturated (in the direction of the net field) as the external field penetrates the surface.

In high fields in the curling state, the iron nanobrick also mimics a magnetically soft material as long as the high field is not so high as to force the flower state.

3.6.2 DEMAGNETIZING FACTOR

To a good approximation, the dipole-dipole energy is quadratic in $<m_z>$ for all the states of the nanobrick described above. The Zeeman energy is, of course, linear in $<m_z>$. The magnetization curve is then approximately linear in H_z until the flower state is approached at high fields. The magnetization is given to a good approximation by $H_z -<m_z> H_s =0$, where $<m_z>H_s$ is called the effective demagnetizing field and H_s is a fitting parameter. To emphasize the role of the supporting actors in this drama and to show the non-quadratic terms in E_{dem}, a magnetization $<m_z>_D \equiv H_z/H_s$ is subtracted from each point in Figure 3.14 to produce Figure 3.15, where the value of H_s

has been chosen to make the difference $\Delta <m_z> = <m_z> - <m_z>_D$ independent of H_z over much of the range of H_z.

3.6.3 VORTEX STATICS AND DYNAMICS

To some extent, the iron nanobrick behaves like an ideally soft magnetic material, but the S-tube has its own life. The I-vortex builds up during the transition from the flower state to the curling state. It costs exchange energy to do this. During the buildup, the exchange energy is proportional to the deviation of $<m_z>$ from unity. The derivative of the exchange energy with respect to $<m_z>$ is constant during this process, giving rise to a constant effective exchange field that aids the Zeeman field in maintaining the magnetization.

Once the I state breaks free of its confinement to the symmetry axis and becomes the S-tube, it is like a moving domain wall in which changes in exchange energy are slight and compensated by changes in magnetostatic self-energy. In the dynamic response in the constant applied field, the tube can move back and forth between the S^+- and the S^--tube states along paths of constant internal energy, along which there are oscillations in the exchange energy and the magnetostatic energy that are equal and opposite to one another. When the S-tube is free to move, the exchange energy has little effect on the magnetization loop.

3.6.3.1 Vortex Contributions to the Magnetization

The core of the tube does have an effect on the magnetization loop because the core is magnetized. It is a separate permanent magnet that has its own magnetization loop, reversing only in fields of the order of $\mu_0 H_z = 1$ T. The volume of that permanent magnet changes somewhat with the field, shrinking as the field is lowered from saturation and continuing to shrink as the field is increased in the reverse direction.

3.6.3.2 The I_z-Vortex State

In the nanobrick, the curling pattern with the swirls centered on the z-axis is called the I_z-vortex state. The m_z-isosurfaces are elliptical in cross-section. The central bulge along the x-axis corresponds to the 180° wall of the Landau structure. As field H_z is lowered, the bulge extends toward the positions $x = \pm(X/2-Y/2)$. When the I_z-vortex state is maintained down to $H_z = 0$, for $|x| > (X/2-Y/2)$ and $y = 0$, magnetization lies almost in the midplane with $|m_y| \sim 1$ corresponding to the closure-domain pattern of the Landau structure.

For Z above a critical thickness that depends somewhat on X and Y, the I_z-vortex is not stable for $H_z = 0$, but it is always an equilibrium state that persists as long as there is no symmetry-breaking field or the inevitable effect of computational round-off error has not yet developed. Once one sees the correspondence between the Landau structure and the S-tube state, one can view the I_z-vortex state as the Landau structure with its two swirls centered on the z-axis. Or one can view the Landau structure as an S-tube with its two swirls moved off the z-axis. The two swirls can be manipulated to move along the top and bottom surfaces distorting the 180° Bloch wall as they move.

3.6.3.3 The S-tube States

The swirls of the I_z-vortex state can be manipulated. In response to a current i_x in the +x-direction, the I_z-vortex state takes the S_z^+-tube configuration. For the dimensions of the iron nanobricks chosen for this chapter, the S_z^+-tube configuration appears spontaneously below a critical magnitude of H_z. In the absence of bias fields in the x- or y-directions, the two swirls of the spontaneous S_z^+- and S_z^--tube states have coordinates $(x_s, y_s, Z/2)$ and $(-x_s, y_s, -Z/2)$, respectively, where x_s and y_s increase with decreasing H_z, reaching a maximum at $H_z = 0$. As x_s increases it is accompanied by an increase in $<m_x>$ as shown in Figure 3.14. This occurs because there is a displacement of the core in the $-y$-direction increasing the volume in which the magnetization in the +x-direction is dominant. In the midplane where $x = 0$, the core of the vortex with $m_z = 1$ and the two Néel caps with $m_y = 1$ at the top and $m_y = -1$ at the bottom form a circulating magnetization pattern on one side of the S-tube. The circulation is around an x-axis displaced from the midplane in the $-y$ direction, as originally calculated by LaBonte for a never-ending Bloch wall; see Figure 3.10.

The lowest Z for the spontaneous appearance of an S_z-tube structure is $Z_{crit} = 25$ nm for $Y = 35$ nm with X varying from 120 to 126 nm. For $X = 119$ nm the S_z-tube configuration goes to the I_z-vortex state. For $X = 127$ nm, an S_z-tube structure is unstable with respect to the formation of an I_x-vortex along the x-axis. For $X = 130$ nm and $Y = 80$ nm, the spontaneous S_z-tube structure occurs for $Z > 42$ nm.

There are also limits on the sizes of the nanobrick for which the I_z-vortex state is stable in the absence of a magnetic field. If the nanobrick is too small, the I_z-vortex state moves away from the axis and disappears out the nearby Y face as H_z is reduced. The range of dimensions (X, Y, Z) and of applied fields $(H_x, H_y, $ or $H_z)$ for which the S_z^+- and S_z^--tube states and the I_z-vortex states are stable has been studied by T. L. Templeton [23] who includes the effects of the configurations in the four vertical edges in his elaborate phase diagrams.

3.6.3.4 The I_z^*-Vortex State

The spontaneous S_z^+-tube is distorted by applying a uniform field H_y. The swirl on the left moves toward the swirl on the right, which moves only slightly to the right. There is a critical field H_y ($\mu_0 H_y \sim 0.1$ T) at which the left swirl catches up to the right swirl. The I_z^*-vortex state is the Landau structure with both swirls on the same end of the 180° Bloch wall. The centers of the two swirls are at the same x-position, but the small displacements of the swirls in the y-direction are in opposite y-directions. The m_z-isosurfaces are far from symmetric. The 180° Bloch wall is attached to one side of the I_z^*-vortex. The central cross-section of the m_z-isosurfaces bulges to include the Bloch wall, see Figure 3.16. On lowering H_y there is a critical field for

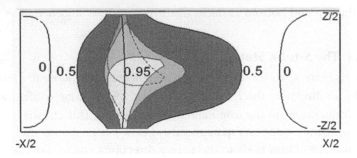

FIGURE 3.16 The *I**-vortex formed by displacing both swirls to the same end of the Bloch wall, using a field H_y. The two swirls have the same x-coordinate but are displaced in y by −4 nm on the bottom surface and +4 nm on the top. The regions in light gray ($m_z > 0.8$) and dark gray ($0.8 > m_z > 0.5$) are for m_z-isosurfaces intersecting the $y = 0$ plane. The slightly curved, almost vertical line is the contour for $m_y = 0$ in that plane. The *I**-vortex maintains a memory of the Bloch wall as it bulges to the right. The area in white labeled 0.95 is for the plane $y = −4$ nm and the dotted contour is in the plane $y = 4$ nm. The *I**-vortex has two-fold symmetry on rotation about the x-axis.

re-nucleating the S_z-tube that restores the Landau state with the swirls on opposite ends as H_y goes to zero. The transitions from S_z to I_z^* in H_y are not quite continuous and hysteresis occurs.

The transitions from I_z^* to S_z depends on a bias to select between S_z^+ and S_z^-. From the view of the swirls, one of the swirls traveled further to form the I_z^*-vortex state in the large H_y. The choice of which swirl travels back along the nanobrick when H_y is reduced is, again, a three-tined fork. If no decision is made, the I_z^*-vortex state persists for a long time in a region where an S_z-tube would be more stable. A small bias field in H_x would make the selection. For a finite step in H_y, the dynamics make the selection. In experimental studies at Grenoble of faceted nanogems of iron, it was the facets on the top surface that favored the upper swirl as the one that determined the handedness on reducing H_y [24].

The transition to the I_z^*-vortex state from the spontaneous S_z-tube state can also be made by applying a large current, i_y (~20 mA). The latter produces a large gradient field in the z-direction ($\mu_0 H_z = 0.06$ T at the end surface) which stabilizes the I_z^*-vortex state in a first-order jump, with the complication of turning the magnetization in the corners into the direction of the gradient field, that is *mmmm* goes to *mppm* for the four corners ordered clockwise from ($−X/2,Y/2$). Removing the large i_y does not restore the corners to their original configurations.

3.6.3.5 Switching

A dramatic result of the attack on the micromagnetics of the nanobrick was the discovery by computation that the two swirls of the Landau structure, when viewed as the ends of the S-tube, could be switched back and forth, using modest driving fields, over long distances in short times carrying with them large external fields [25]. The switching of the S-tube had already been observed experimentally in 2004 but not specifically identified with the reversal of the positions of the two swirls [26]. The switching between two stable states is discussed here in terms of the energy landscape correlated with the positions of the two swirls. This is a gross simplification, but in equilibrium and for heavily damped dynamics there is some usefulness in thinking about the internal energy along and near the equilibrium path.

The combination of the field from a current i_x, producing a field that is $+H_y$ at the top surface and $-H_y$ at the bottom, and a uniform H_y permits the independent manipulation of the positions of the two swirls in any given H_z. The internal energy at equilibrium in the combined fields changes with the positions of the two swirls. At constant H_z, one has an energy landscape with minima at the symmetric positions that the two swirls for S^+ or S^- take in the absence of bias fields.

If the motion of the two swirls between S^+ and S^- is determined by a slowly varying current oscillating between $+i_x$ and $-i_x$, the internal energy along the path goes through the minima when the current goes through zero. The displacement in x from that equilibrium has the separation between the two swirls first increase reaching a maximum for the highest current, returning to equilibrium as the current goes through zero, and then having the displacement in x go toward zero as the swirls approach each other. But before they reach each other, the swirls have reached an energy position where it is all downhill toward a stable position that lies beyond the far equilibrium position in a zero current. That position will be reached if the current is maintained at the critical current for switching, $i_x = 1.2$ mA, which produces a maximum field $\mu_0 H_z = 0.06$ T at the surfaces for the nanobrick $156 \times 96 \times 60$ nm^3.

3.6.3.6 Double-Well Potential

Along the path to the critical current for switching from the S^+ to S^- configuration, the internal energy contour is one-half of a double-well potential. That potential is completely determined at each current up to the critical current. The inflection point on that curve is reached at the critical current. If the motion is calculated using large damping, a path beyond the inflection point can be determined from the damped dynamic response and the full double-well potential can be determined. This path leads through the position where the separation in x of the swirls goes through zero, but when the separation in x is zero, the separation in y is not zero. The paths of the two swirls on the opposite faces of the nanobrick are narrow ellipses tilted in the x-y planes. The double-well potential is defined along the ellipses. The swirls do not pass over the maximum in the potential where the displacements in x and y are both zero.

3.6.3.7 Forced Oscillations

An example of the non-linear oscillations of the S-tube with large amplitude is found using a 1.28 GHz driving current $i_x = 0.6$ mA with a period of 780 ps. This current is one-half of that necessary to reverse the S-tube with slowly varying currents. The swirls move in elliptical paths on the top and bottom

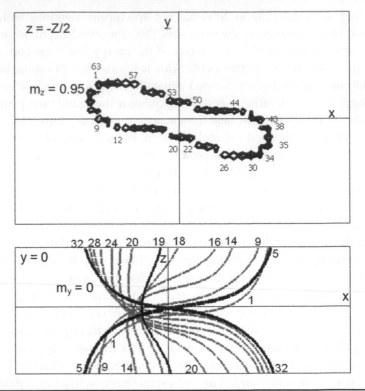

FIGURE 3.17 The dynamic response of a nanobrick with dimensions of 156 × 96 × 60 nm³ to an ac driving current $i_x = 0.6$ mA with a period $\tau = 780$ ps. The path of the swirl on the bottom surface is followed in the upper panel by tracing the contours $m_z = 0.95$ in steps of 12.8 ps. The calculation was carried out with a grid of 4-nm cubes, which exaggerates the interaction of the swirl with the grid resulting in the steps of 4 nm in both the x- and y-directions. A much smoother ellipse is obtained for a grid of 1.25 nm At no time does the central core of the S-tube lie in a plane, let alone in the $y = 0$ plane, but the contours with $m_y = 0$ in the plane $y = 0$ are used in the bottom panel to reflect the distortions of the m_z-isosurfaces, which to be fully appreciated require 3D movies that someday should be on youtube.

surfaces that avoid the region of the local maximum in the internal energy at the center of the faces, as they follow contours of almost constant energy near the saddle point on the $+y$ and $-y$ sides of the origin. The path on the top surface is shown in Figure 3.17 for one period of oscillation. The path on the bottom surface is the mirror image in either the $x = 0$ or $y = 0$ plane with the two ends of the m_z-isosurface moving counterclockwise along the paths. When the swirls pass one another at $x = 0$, the two ends are displaced in $y = \pm 8$ nm. At their greatest separation $x = \pm 36$ nm at $y = 16$ nm.

3.6.3.8 Mired at the Central Maximum

In the dynamic response with low damping, the path of either swirl can go through the origin but generally not at the same time. But in one dynamical calculation, the two swirls came through the origin at the same time during one cycle of a damped oscillation. The two swirls then stayed there for a time equal to the period of the non-linear oscillator. During this time a higher harmonic of the dynamic response, corresponding to a wave propagating up and down the z-axis, provided the driving force to allow the two vortices to move away from the central energy maximum.

3.6.3.9 Vertical Edges as Partial Antivortices

The vertical edges of the nanobrick are each one-quarter of a virtual vortex that is just outside the corners. These partial-virtual vortices are antivortices with a winding number of –1. An example of an antivortex is given in Figure 3.1, where the Bloch wall is the core of the antivortex. Each of the corner partial-virtual antivortices can be described using Preisach diagrams (with sloping sides). When a corner reverses there is a change in the exchange energy in the region between the corners and the core of the S or I-vortex in the center. The change in exchange energy with $<m_z>$ during the reversal of a corner is an exchange field that adds to or subtracts from the applied field at the same time that the magnetization of the antivortex is changing its contribution to the magnetization. The flipping of an antivortex shifts the sloping line of m_z vs. H_z both sideways from the exchange field effect and up or down from the change in a magnetic moment.

The composite picture is then of five Preisach diagrams added to the ideal soft magnet plus some exchange energy to be provided for the buildup of the five Preisach regions.

3.6.4 ELLIPSOIDS

From the beginning with Ewing, the ellipsoid has been useful in understanding magnetism. With the work of Stoner and Wohlfarth, it became the model for understanding the origin of permanent magnets and the applications to magnetic recording. The earliest micromagnetic calculations addressed the particles in magnetic recording media. It was concluded that magnetic recording depended so much on particle-particle interactions that understanding the micromagnetics of a single ellipsoid was not sufficient for understanding magnetic tapes. Yet the properties of a single ellipsoid led to a fuller appreciation of the Landau structure.

3.6.4.1 Absence of the Flower State in Ellipsoids

When the calculations are carried out for an ellipsoid, the flower state does not occur. Splay saving works right up to saturation. The swirl in the ellipsoid is not at a flat surface. It is the curvature that forestalls the breakdown of splay saving. The ellipsoid goes directly from saturation to the swirl with the handedness chosen by a symmetry-breaking field. As the swirl forms, the exchange energy increases in proportion to $(M_s - <M_z>)$. The linear increase in exchange energy with decreasing $<M_z>$ is a constant exchange field that adds to the applied field. When H_z is increasing below saturation, the exchange field brings the ellipsoid to saturation at a lower H_z than one would obtain for a paramagnet with infinite susceptibility to reach M_s. The magnetization is linear in the applied field with the slope determined completely by the demagnetizing field. The demagnetizing field line is offset by the constant exchange field.

For the ellipsoid, this is true for a field along any of the three principal axes. The slope is different for each axis because of the change in the demagnetizing factor. The constant offset is different because the curvatures of the surface change the contribution of the exchange energy to the energy of formation of the swirl.

3.6.4.2 Curling in an Ellipsoid

To predict the straight line of $<M_z>$ versus H_z for an ellipsoid at high fields, all one needs is Osborn's formulas [27] for the demagnetizing factors of the ellipsoid and a single number for $\partial E_{ex}/\partial <M_z>$ for the chosen axis. The latter can be obtained from a micromagnetic calculation of $<M_z>$ at a single field below saturation. Precise agreement with the analytic formulas has been found using *TetraMag* [9] to calculate the properties of the mathematical ellipsoid with a triangular mesh on the boundaries; see Figure 3.11. A full micromagnetic calculation of $<M_z>$ versus H_z for the approach to saturation for an ellipsoid would require very long computational times because the torques become very small as the swirl saturates.

Modeling the results of the calculations can lead to insights that greatly shorten the computational time for a given problem. The ellipsoid at high fields is an extreme example in which one calculation in a single field, where the torques are large and the relaxation time short, produces the entire magnetization "curve" in the region where swirl remains along the axis and $\partial E_{ex}/\partial <M_z>$ remains constant with change in H_z. In this case, the field for saturation is determined precisely. The instability field at which the I_z vortex moves off axis cannot be determined without calculating the pattern changes when the swirls move off axis.

3.6.4.3 Reversal of a "Stoner-Wohlfarth Particle"

Below a second critical field, the central position of the swirls on the ends of the principal axis of the ellipsoid becomes an energy maximum and the swirls move off center if the dimensions of the ellipsoid are sufficiently large. At small enough dimensions, the ellipsoid remains "uniformly magnetized" in all fields. The magnetization process is limited to rotations in the Stoner-Wohlfarth model. It is assumed that the exchange energy does not change in the process. That model has served as the starting point for 60 years for understanding magnetization processes as a competition between the Zeeman energy and the anisotropy energy, where the anisotropy energy includes the dipole-dipole interactions and the crystalline anisotropy. Variations of the exchange energy in "uniform" rotation would also appear as an addition to the anisotropy. Even an ellipsoid in the Stoner-Wohlfarth model requires bias fields for the uniform rotation of the magnetization away from a principal axis.

3.6.4.4 Bias fields in the Stoner-Wolhfarth Model

The importance of bias fields for the reversal of the magnetization was first pointed out by D.O. Smith at the 2nd MMM conference in Boston in 1956. The subject of the session was the failure of experiments to show switching with the time constant predicted by the Landau-Lifshitz equations. The experimentalists were asking what was wrong with the Landau-Lifshitz equations. Smith showed that if the experiments and the theory are done using bias fields they agree.

3.6.4.5 Propagating Singularities in Particles

In a micromagnetic calculation, one can eliminate all the geometrical biases and cause the small ellipsoid to reverse its magnetization through a non-uniform distortion in which a singularity propagates along the principal axis starting at the swirl. The field must be applied fast enough that the round-off errors in the numerical calculation do not have sufficient time to nucleate the uniform rotation by the displacement of the swirls. Even then one needs a symmetry-breaking field to choose the handedness of the swirls. Here again, the numerical round-off error can provide the handedness. The field must be larger than necessary and applied fast enough that the round-off error favors the reversal through singularity propagation rather than through uniform rotation.

The reversal can also take place with the creation of a vortex-antivortex pair away from the ends where the swirls reside.

3.6.5 PAIR CREATION AND PROPAGATION IN THE I_z VORTEX

A rule for the formation of a pair of singularities in micromagnetics has been given by Sebastian Gliga who worked in high-energy physics before coming to magnetism [28]. If enough energy is provided to create each of the singularities in a given region, the program for solving the micromagnetic equations will find the solution in which the pair is created. The energy for the creation of a vortex-antivortex pair is of the order of $8\pi A\lambda$, where λ is the exchange length and A is the exchange constant of Equation 3.1.

As the field becomes more negative for a ^+I_z-vortex state, the magnetization turns to the $-z$ direction everywhere except in the immediate vicinity of the axis of the vortex. There is a wall in which exchange energy becomes higher as the field in the $-z$-direction becomes more negative. In a bcc lattice, the singularity appears at the center of a tetrahedron of iron atoms. The four moments can no longer sustain the wall when their m_z decreases to a critical value. A pair of solitons is created which then propagate to reverse the core of the vortex.

The creation and annihilation of vortices was envisioned in the original discussion of point singularities in the micromagnetics of a cylinder [7] in the 1970s but the terminology of skyrmion was not in the literature of magnetism until Tretiakov and Tchernyshyov [29] recognized "that a peculiar annihilation of a vortex-antivortex pair observed numerically by Hertel and Schneider [97] represents the formation and a subsequent decay of a skyrmion." These had been studied for many years, but this gave them a new name and popularity.

3.7 RECENT STUDIES OF WHISKERS AND CYLINDRICAL TOROIDS ON THE MICRON SCALE

In the original manuscript for this chapter, Section 3.7 was to obtain the Landau structure starting at high fields along the long axis and arriving by a path that did not require fields from currents. This would match the experimental approach, which generally required cycling between positive and negative fields. With 40 years of experiments on the magnetization processes in measuring the susceptibility of iron whiskers, it seemed that there were too many effects to be explained with the computational tools available at that time, particularly because dimensions of the whiskers were so many orders of magnitude larger (10^{16}) than an exchange-length voxel.

The new age of computing, with desktop computers that would have been called super computers not too long ago, has shown calculations with dimensions still very small compared to the experimental whiskers are able to provide insights into their magnetization processes. Already, from computations with the smallest dimension near 300 nm, the behavior of systems with smallest dimensions of 100 µm is understandable. These systems include:

a. The formation of the Landau structure starting from saturation along the long axis in real iron whiskers with the [100] orientation as studied by Heinrich and coworkers in the 1970's using ac susceptibility [22].

b. The formation of striped domain patterns on hexagonal iron whiskers with the hard [111] axis of magnetization along the long axis as seen by Hanham [5].

c. The role of exchange springs in the interaction of domain walls with surfaces.

d. The formation of single domain walls in toroidal geometry.

e. The suppression by magnetostatics of the effects of anisotropy on the approach to saturation in polycrystalline materials permitting the use of the full saturation magnetization of iron in a transformer.

3.7.1 BARBER-POLE CONFIGURATIONS IN IRON WHISKERS

The whisker is considered as having one central tube of magnetization around the line with $m_x = 1$ that goes from the center of the swirl on one end to the center of the swirl on the other end for the case that the two end swirls have the same handedness. It is now seen how a single handedness can occur without using a current along the whisker. Once a single handedness is established, it takes a field of the order of $\mu_0 H = 1$ T to return the whisker to the inversion symmetry state preferred by the inversion symmetry of the LLG equations. The whiskers were not subjected to such high fields. The whiskers are below the Curie temperature when they form.

The are several possible states that can occur when starting at moderate fields with the same handedness at each end. To achieve the Landau state in a [100] whisker, it is necessary to reverse the magnetization on two adjacent side edges. It is less complicated if the cross-section of the whisker is rectangular, breaking the degeneracy of the square cross-section.

A Landau-like structure can appear on adjacent faces at opposite ends of the square cross-section whisker. This effect was seen by Graham in the 1950s. The author published a cut out drawing of the three-dimensional transition domain between the two Landau structures [30] in the 1990s after Graham told him that the original wooden model was lost.

The reversal of the magnetization of the edges takes place through the propagation of edge solitons, either from one end or the other or from both at the same time. It can also happen via the creation of a pair of solitons somewhere along an edge, with the two solitons moving to opposite ends to reverse the edge.

As there are four side edges, it matters whether one, two, three, or four edges reverse. There are stable states in which edges are partially reversed with one, two, or more solitons. The Landau structure in a rectangular cross-section whisker has two closely adjacent edges magnetized opposite to the central core. The two reversed edges can also be on two nonadjacent edges, accounting for a state seen in the ac susceptibility studies where the losses are strongly field dependent.

When the swirls are on the two ends and the four sides are magnetized in the same direction, there are two states, one of which has the central I_x-vortex with its core magnetized in the same direction as the four edges, while the other state has the four edges opposite to the core. In either case, in addition to the central vortex, there are four edge vortices, which are quarter-vortices with their core trapped at the edge or they can be virtual vortices with their cores lying just outside of the whisker.

When the field is reduced, the end swirls move off center and the central vortex becomes a tube of magnetization around the line $m_x = 1$ that goes from one end swirl to the other. When the swirl approaches one of the four end edges, it is repulsed from that edge. That the swirl moves off center when the field is lowered is discussed below. The moving off center was an important unanswered question in the original manuscript of this chapter.

The swirl cannot cross an end edge without first having the direction of the magnetization on that edge reverse by soliton motion. Again, that can take place when a soliton breaks away from a corner or by the creation of a pair of solitons along that edge, instigated by the approaching swirl.

Before the end edge reverses by soliton motion, a second swirl appears on the side surface. Both swirls are connected to the central vortex in complicated patterns that are more readily grasped by three-dimensional plots of tubes of the energy density. The energy-density isosurfaces have arms going from the central vortex to each of the swirls (two on each end). Sometimes even a third swirl appears near the end. Once the edge reverses, the end swirl merges with the side swirl to produce the Landau structure via the propagation of solitons along adjacent edges on the smaller of the side surfaces. This is where it is better to have a rectangular rather than a square cross-section to avoid the Graham configuration. If the solitons propagate from both ends but not on the same two edges, the structure is complicated. This is one reason why it was often necessary to cycle the magnetization to achieve the field dependence of the ac susceptibility assigned to the presence of the Landau structure.

The side surface becomes magnetized in the direction opposite to the central vortex, which now should be called a tube because the core of the x-vortex does not come to the surface at the center of a swirl. The tube in m_z, described in Section 3.4, coexists with the vortex in m_x. The tube around m_x interacts with the two long edge quarter-vortices that are magnetized in the same x-direction. The combined structure of the central tube and the two like-directed quarter-vortices is called the Bloch wall with Néel caps as first calculated by LaBonte. The contour of that tube, where $m_x = 0$, is the domain wall separating the regions of reversed magnetization as shown in Figure 3.10.

From this picture, it is clear that it should make a difference whether the Néel caps of LaBonte's wall are pointing toward or away from the surface through which the domain wall moves with fields that are large enough in either direction to saturate the central cross-section of the whisker. The Néel caps act as springs as the wall approaches either side of the whisker. That spring is stronger if the Néel caps are displaced from the center of the Bloch wall in the direction of the approaching side surface. That this effect was never observed in the measurements of the ac susceptibility reflects the structure of a real wall. The real wall is not translationally invariant as calculated by LaBonte. It has reversals of the direction of the Néel caps from place to place along the wall as seen in all studies of domain walls intersecting surfaces and thoroughly discussed in *Magnetic Domains*. The net strength of the exchange spring is then the same approaching either side surface.

That the tube of m_x has a life of its own is seen in the calculations related to the barber-pole effect, first seen experimentally by Hanham [5] and recently in the calculations of Templeton [23] for the hexagonal [111] whisker. The barber-pole effect is the tube of m_x taking on a helical pattern as one proceeds from the swirl on one end to the swirl on the other. The beginning of the helix has been present in calculations of swirls for the last two decades. It usually goes smoothly into the central vortex near the end surfaces.

The propagation all the way from one end of the swirl to the other first appeared in a hysteresis loop in a calculation for a 150 nm-diameter hexagonal iron whisker 1500 nm high. It gave an account of the reversal of the six edges by both the creation of soliton pairs along the edges and the breaking away of a soliton from either of the corners where the side edge ends. The helical pattern of the m_x-tube formed sequentially in a critical negative field. It was stable for more negative fields until all six edges reversed relative to the core. At an even more negative field, the core reversed and aligned with the edges. If the field was increased (less negative) after the barber pole formed, it was stable until, at a still negative field, it switched back by soliton propagations to the state where the six edges and the central vortex were aligned in

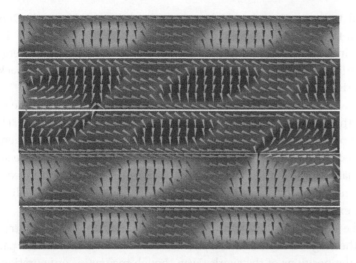

FIGURE 3.18 The appearance of a 0.6 μm periodic structure in a [100] oriented iron whisker with dimensions of $1.54 \times 0.20 \times 0.24$ μm³. The four long sides are shown as if they were unwrapped, with a repetition of the front surface to see the configuration on each of the four long edges. The color code is red to the right, green to the left. Both blue and yellow are downward on the page. The core of the helical structure comes close enough to the side surfaces every 0.6 μm to turn the magnetization perpendicular to the x-axis and slightly reversed. On two of the edges on the backside, two edge solitons have reversed part of the edges and part of the top, bottom, and backside surface, showing green to the left.

the original direction of the core. The formation of a stable helix requires a diameter greater than 99 nm.

The barber-pole pattern observed on the surface of the hexagonal whisker is an outward manifestation of the helix when the radius of the helix brings the center of the m_z-tube close enough to the surface to change the magnetization direction in patches along the six edges.

It was then calculated that the barber-pole effect still occurred on the removal of the anisotropy of Fe, which is a small effect at 150 nm. It was next calculated that the replacement of the hexagonal cross-section by a cylindrical cross-section lowered the range of stability somewhat, but the barber-pole effect was still there on the cylindrical surface. After searching, conditions were found for the helical periodic structure to appear in a whisker with close to square cross-section; see Figure 3.18.

Unstable helices are found by looking at the patterns of magnetization reversal calculated previously, once one knows where to look.

During the course of these discoveries, it was paramount to find the driving forces for the formation of the swirl and its preference for moving away from the center of symmetry while connecting to the vortex down the centerline by a path that traced out a helix of diminishing radius starting at the surface. And then to explain why sometimes it propagates all the way to the far swirl.

All of these searches fixed upon the idea of using splay saving as the mechanism.

The discussion of the hedgehog in Section 3.3.3 explains why this was the wrong approach. Changing the hedgehog into splay saving configurations does not change the exchange energy. It does in liquid crystal theory applied to magnetism, but Hubert in his treatise on *Magnetic Domains* showed two decades ago why one cannot use (div m)^2 + (curl m)^2 for the energy in micromagnetic calculations. Yet, thinking in terms of divergences and curls was sufficiently useful in that the author brought them back into magnetism in Equation 3.2.

In writing the revision of this chapter, the author took the time to calculate the splay and curl terms in Equation 3.2 and found that the exchange energy of the splay-saving configuration was the same as the hedgehog. It appeared to be perfectly obvious for the energy to be written as the sum of the squares of the first derivatives. All the energy to be that was saved in the divergence term appears in the curl term using Equation 3.2. The helix could not be explained from splay saving.

The answer is that the splay is created by the exchange energy in trying to meet the magnetostatic boundary conditions at the surfaces and, in particular, at the positions of the swirls. Even though magnetic configurations can be created that are divergence free except at the surface, these are not torque-free solutions of the LLG equations. Those torques create divergences in the bulk. In the swirl, they are near to the surface. In the helical pattern, they are throughout the bulk. The divergences are both positive and negative. The self-magnetostatic energy of the regions of positive or negative divergence is paid for by the minimizing of the exchange energy taking into account the magnetostatics of the boundaries and the applied field. Once these regions of charge density exist, it pays to move them to lower the magnetostatic energy.

In ultrathin film memory elements, a rectangle can take either the *S* or the *C* configuration. In the *S* configuration, the magnetic charges of opposite sign are on the diagonally opposite far corners. In the *C* configuration, they are on corners along the same longer edge but not on the same shorter edge because that would cost too much exchange energy. The *C*-state is lower energy for a rectangle that is not too close to a square and not too small.

Given four patches of charge, two of each sign, the lowest energy configuration is a square with charges arranged with the four nearest neighbors of opposite sign and two pairs of diagonal neighbors of the same sign. This is an extended quadrupole. The extended quadrupole appears in the barber pole. It also has been noted in the chemistry and biology of the arrangement of charges on surfaces. The helical configuration is analyzed in Figures 3.19 and 3.20 to make the case that the helix takes advantage of the energy to be saved from the magnetostatic interactions of unlike charges. The usefulness of thinking about magnetic charges has been presented in an editorial titled "Visualization and Interpretation of Magnetic Configurations Using Magnetic Charge" printed in IEEE Magnetic Letters [31] and illustrated in "Using magnetic charge to understand soft-magnetic materials" [32].

3.7.2 MICROTRANSFORMERS

The simplest of all magnetic configurations is the cylindrical toroid. A single LaBonte wall would divide the cylindrical toroid into two domains, one clockwise and one counterclockwise. The energy of the LaBonte wall would

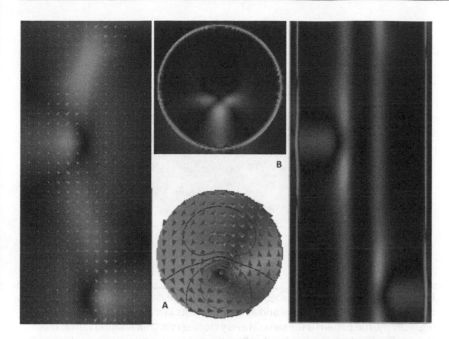

FIGURE 3.19   Magnetization **A** and demagnetizing fields **B** for one pitch of 750 nm of the helical magnetization patterns in an iron cylinder of radius 160 nm. The central figures are crosssections at constant z. The patterns rotate as z is changed. The panels on the sides are meridian cuts. The positions of the colors move up and down as the angle of the meridian cut is changed. The demagnetizing fields in a slice at constant z point along the sides of a rectangle with green, blue, red, and yellow in the middle of the sides. The green region points left, the blue region points up, the red region points right, and the yellow region points down indicating the presence of four regions of charge density on the four corners of the rectangle. The charge is positive in the corner between the yellow and the green and in the opposite corner between the blue and the red. The charge is negative at the other two corners. There are no charges along the radial line from the center as $\partial m_z/\partial z$ $(1/\rho)$ cancels $\partial(\rho m_\rho)/\partial\rho$. This is seen by the direct calculation of **div m** as shown in Figure 3.20. To reconcile the two figures, it is necessary to remember that the magnetic charge is $-\textbf{\textit{div m}}$.

increase linearly with the radius of the wall making the wall unstable. The wall can be stabilized by a magnetic field from a current in a wire through the hole in the doughnut. It could be further stabilized by passing a second current anti-parallel to the first through the doughnut itself, producing a field that changes sign between the inner and outer radius. Yet it is not necessary to use currents to produce a stable LaBonte wall in the toroid.

If the Néel caps are directed toward the inner radius, they provide strong exchange springs that keep the wall from leaving. The direction of the Néel caps will be directed inward if the domain wall is nucleated at the outer radius. This is accomplished using a current through the doughnut itself, as the field will then be at maximum at the outer radius. When this is modeled using micromagnetics there are several interesting results.

The first result relates to the process of magnetization before the current is sufficient to nucleate a domain wall. The others are about what happens

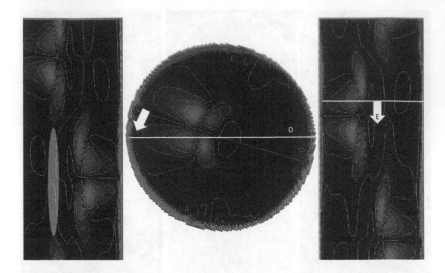

FIGURE 3.20   Four helical tubes of magnetic charge density shown as a cut in the z-plane in the central panel and as meridian cuts in x and y on the left and right, respectively. The pattern is the same in every cross-section but rotating with z. The charge density along the line labeled **E** on the right is the same as along the line labeled **D** in the center. As **E** is moved down the page, the pattern of the cut in z rotates counterclockwise changing the charge along line **D**. The gray ellipse in the left side meridian cut represents the charges that are not there directly across from all the charges that are there. The charge density in red sees the charge density in blue across and downward as well as directly above. The length of the gray ellipse is 3/2 of the length of the red or blue regions of the charge distribution. This accounts for the observation that the outer red moves to where the outer blue is presently on a rotation of 2π/7. The outer charges are at the same distance from the inner charges as the inner charges are from the cylinder axis.

after nucleation. Three different metallurgical preparations were considered. A single grain of iron, in which the direction of the anisotropy axes could be chosen in any orientation, was considered first. The other two preparations were polycrystalline iron. For one of these the grain size was smaller than the smallest dimension of the toroidal cylinder. The other had a grain size so small (less than 8 nm) that the exchange energy averaged out the effects of crystalline anisotropy. The latter is the ideally soft magnetic material.

Ultrafine grains have been the subject of research in magnetism for over 25 years. No one has yet to produce a commercial magnetic material with higher saturation than grain-oriented Fe(Si). Material with nanocrystalline grains produced by mechanical milling proved too mechanically hard to compact into useful material because mechanical strength is inversely proportional to grain size.

Cases, where grain size was smaller than the smallest dimension, showed that it was not necessary to produce nanocrystalline grains. With grain sizes in the range of 100 nm, the material still averages over the effects of the anisotropy, but because of magnetostatic interactions rather than exchange. When the magnetization in the toroid tries to rotate in the direction preferred in each of the grains, the magnetic charge develops at the

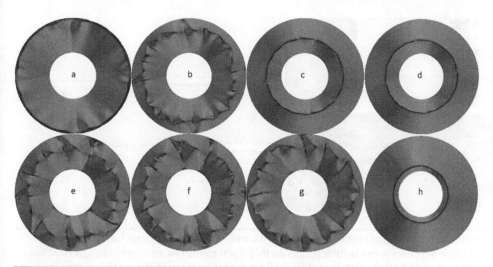

FIGURE 3.21 Effect of anisotropy on the nucleation of a domain wall at the outer radius in a cylindrical toroid with radii $R_o = 1000$ nm and $R_i = 400$ nm with height $L = 80$ nm for a differing microstructure. For *a* to *d*, the configuration in a single crystal is changing with time. In *a*, nucleation is just starting. In *e* there are 80 grains just after nucleation as is the case for **b**. In *f* there are 800 and in *g* and *h* there are 8000 grains. From *b* to *d*, the vortices and antivortices are annihilating in pairs. The last four pairs shown in *d* are about to annihilate. The wall is stable in *h*, which also shows the ripple structure in the almost completely magnetized domains. The grain size does not significantly change the structure of the nucleation process.

grain boundaries. The fields from those charges keep the magnetization directions very close to the mean direction of the magnetization determined by the circular boundaries of the toroid. For this to happen, it is necessary that the height of the toroid be less than 300 nm.

If the height is greater than that, the magnetization in adjacent grains can each rotate toward the z-axis, one with $+m_z$ and one with $-m_z$. When the height is less than 300 nm each grain produces enough magnetic charge on the top and bottom surface to suppress those rotations. Toroids 300 nm high can be stacked to any height if the separation between the toroids is 4 nm and the magnetization still does not rotate into the z-axis; see Figure 3.20.

The magnetization in the polycrystalline toroid is close to saturation in the direction of the circulating magnetization, but it is not actually saturated because a ripple structure develops wherein the magnetization alternately rotates inward and outward in the plane of the toroidal cylinder.

When a current is used to try to reverse the magnetization, the ripple structure increases as a means to lower the Zeeman energy without producing much charge on the surfaces. It is not sensitive to the crystalline microstructure.

At a critical current, surface charges appear as full reversal nucleates at many places on the outer or inner cylindrical surface depending on whether the current is in a wire through the hole in the doughnut or through the doughnut itself. Despite multiple nucleations, the magnetization in the doughnut

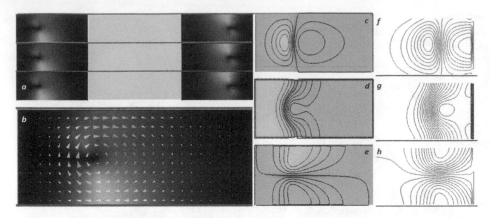

FIGURE 3.22 Continuous stacking of cylindrical toroids with an 8-nm gap to suppress rotation of the magnetization into the z-direction (4-nm is sufficient to do this). Each toroid has an outer radius of 1000 nm, an inner radius of 400 nm and a height of 240 nm. A vertical cross-section through the stack is shown in all panels. In *a*, three consecutive toroids are shown to illustrate that the magnetization lies close to the planes where the toroids are separated. It takes very little magnetic charge to suppress the rotations into the z-direction. In *b*, an enlargement of the cross-sections on the left show the partial vortex that circulates around the z-axis. The Néel caps are red, pointing toward the z-axis and green pointing outward. The blue is the center of the Bloch wall. Black is either into the paper as it is on the far left or out of the paper as it is in the vortex and on the right. The panels labeled *c* through *h* are contours for the three components of the magnetization, m_x in the lower panels, m_y in the middle panels, and m_z in the upper panels. More contours are shown in the panels to the right, to illustrate the compressing of the exchange springs as the wall approaches the inner radius. Panel *d* shows the two domains. Panel *c* shows the Bloch wall. Panel *e* shows the Néel caps displaced to the right of the Bloch wall.

re-organizes itself to form one LaBonte wall that separates the volume into two countercirculating domains; see Figure 3.21.

As nucleated, the Néel caps are not all in the same direction and the wall is decorated by vortices and antivortices. With time, these annihilate in pairs until there is a single direction for the caps all the way around; see Figure 3.21. But if any of the vortices have their cores in opposite directions, they do not annihilate and the direction of the Néel caps changes from place to place along the circular wall. This effect is eliminated by carrying out the nucleation of the wall in a vertical field of $\mu_0 H_z = 0.4$ T. The nucleation of the single domain wall is insensitive to the grain size.

The calculations have been carried out for infinitely high stacks using periodic boundary conditions for toroids with outer radii up to 5 μm, which is almost big enough to use to produce transformers with close to the full saturation magnetization of iron.

Calculations of single crystal picture frames at the level of five micron dimensions should be applicable to large sizes if magnetostriction is taken into account. In going to large sizes for polycrystalline toroids, it is not obvious at what dimensions the averaging over anisotropy by magnetic charges at the grain boundary will no longer be effective. Fortunately, the sizes of the calculated toroids have reached the range where they can be used to interpret the surface observations of magneto-optics on bigger systems.

In over 1,000 pictures in their treatise *Magnetic Domains*, Rudi Schäfer and the late Alex Hubert have shown how complex and beautiful the patterns of magnetization can be. It now seems possible that many of those 1,000 pictures could be reproduced using micromagnetics on smaller systems (Figure 3.22) The studies of toroidal geometry have now been adapted to propose the use of patterned polycrystalline iron in motors, generators and transformers [33]. Substantial gains in performance are to be achieved by operating close to the full saturation inductance of iron.

ACKNOWLEDGMENTS

Bretislav Heinrich and the author started their studies of the magnetic response of iron whiskers in 1969. Dan S. Bloomberg, Murray J. Press, Scott D. Hanham, Amikam Aharoni, T. L. Templeton, and J.-G. Lee have all contributed to the partial understanding of a long list of phenomena observed in whiskers. The micromagnetics of a nanobrick is a continuation of that work made possible by Riccardo Hertel, Attila Kakay, and Mike Scheinfein, who have provided tools for, and participated in, this attack on complexity and now have added parallel processing using graphic processing units to speed up the calculations. This work has been supported by the Natural Sciences and Engineering Research Council of Canada from 1968 to 1998. Since then Dr. Harold Weinstock of the Air Force Office of Scientific Research sponsored the work on ultrafine grain sizes that led to the study of doughnuts at Morgan State University with Professor Conrad Williams and Dr. Ezana Negusse and at Simon Fraser University with Professor Karen Kavanagh. Templeton, Hanham, and Scheinfein each contributed to the development of this manuscript.

REFERENCES

1. A. S. Arrott, The past, present, and future of soft magnetic materials, *J. Magn. Magn. Mater.* **215–216**, 6–10 (2000).
2. A. Hubert and R. Schäfer, *Magnetic Domains. The Analysis of Magnetic Microstructures*, Springer, Berlin, 1998. Chapter 3 of *Magnetic Domains* remains, after almost two decades, the key reference in the magnetism of soft magnetic materials.
3. C. W. Abert, Dissertation, discrete mathematical concepts in micromagnetic computations 2013. This is summarized at http://micromagnetics.org/micromagnetism/. The full thesis is at: www.physnet.uni-hamburg.de/services/biblio/dissertation/dissfbPhysik/___Volltexte/Claas_Willem___Abert/Claas%20Willem___Abert.pdf. Abert appears on youtube comparing finite grid and finite element micromagnetics.
4. A. Hubert. The role of "magnetization swirls" in soft magnetic materials, *J. Phys. Colloq.* **49**, C8, C8-1859–C8-1864 (1988). This superb article can be downloaded from: https://hal.archives-ouvertes.fr/jpa-00229109/document
5. S. D. Hanham, Thesis, The magnetic behaviour of the (111)-oriented iron whisker, This is available at: summit.sfu.ca/system/files/iritems1/3926/b12516338.pdf.
6. A. J. Newell and R. T. Merrill, The curling nucleation mode in a ferromagnetic cube, *J. Appl. Phys.* **84**, 4394 (1998).

7. A. S. Arrott, B. Heinrich, and A. Aharoni, Point singularities and magnetization reversal in ideally soft ferromagnetic cylinders, *IEEE Trans. Magn.* **15**, 128 (1979). See also: A. Hubert and W. Rave, Arrott's ideal soft magnetic cylinder, revisited. *J. Magn. Magn. Mater.* **184**, 67 (1998).

8. M. R. Scheinfein, J. Unguris, M. H. Kelley, D. T. Pierce, and R. J. Celotta, Scanning electron microscopy with polarization analysis (SEMPA)— Studies of domains, domain walls and magnetic singularities at surfaces and in thin films, *J. Magn. Magn. Mater.* **93**, 109–115 (1991); *Phys. Rev. B* **43**, 3395 (1991).

9. A. S. Arrott and R. Hertel, Formation and transformation of vortex structures in soft ferromagnetic ellipsoids, *J. Appl. Phys.* **103**, 07E39 (2008).

10. A. E. LaBonte, Two dimensional Bloch-type domain walls in ferromagnetic thin films, *J. Appl. Phys.* **40**, 2450–2458 (1969).

11. A. Arrott, Antiferromagnetism in metals, in *Magnetism*, H. Suhl and G. T. Rado (Eds.), Vol. III B, Academic Press, New York, 1966.

12. H. Kronmüller and S. Parkin, *Handbook of Magnetism and Advanced Magnetic Materials*, Vol. 2, Micromagnetics, John Wiley & Sons, Chichester, 2007. See in particular: J. E. Miltat and J. Donahoe, *Numerical Micromagnetics: Finite Difference Methods*, John Wiley & Sons, New York, 2007; T. Schrefl et al., *Numerical Methods in Micromagnetics (Finite Element Method)*, John Wiley & Sons, New York, 2007.

13. I. Rabi, N. E. Ramsey, and J. Schwinger, Use of rotating coordinates in magnetic resonance problems, *Rev. Mod. Phys.* **26**, 167–171 (1954).

14. A. S. Arrott. Introduction to micromagnetics, in *Ultrathin Magnetic Structures IV*, B. Heinrich and J. A. C. Bland (Eds.), Springer-Verlag, Berlin, 2005. See p. 131 for the approximation to the integral of $\tanh^2 (\rho/\lambda)/\rho$. This reference follows micromagnetics from a single iron atom imbedded in a lattice to a 4-nm thick circular disk of 96 nm diameter. The present chapter continues that development of thicknesses where the patterns of magnetization are fully 3D.

15. A concise summary of *TetraMag* is found in *TetraMag* - A general-purpose finite element micromagnetic simulation package and high resolution large-scale micromagnetic simulations with hierarchical matrices, both by Riccardo Hertel and Attila Kakay. These are available by inserting *TetraMag* Hertel into the Google search box.

16. A full description of Michael R. Scheinfein's LLG Micromagnetic Simulator is available by inserting "Scheinfein micromagnetic simulator" into the Google search box. The integrated graphics are native to the Microsoft operating system, but that is one of the strengths of the LLG Micromagnetic Simulator. Visualization is very important in micromagnetics.

17. The story of Gilbert damping starts with an abstract by T. Gilbert in 1955. Gilbert's idea was formalized by R. Kikuchi. On the minimum of magnetization reversal time. *J. Appl. Phys.* **27**, 1352–1357 (1956). Fifty years later, Gilbert commented on his work in 2004. *IEEE Trans. Magn.* **40**, 3443. The basis for Gilbert's phenomenology is firmly established in experiments by R. Urban, G. Woltersdorf, and B. Heinrich, *Phys. Rev. Lett.* **87**, 2173 (2001). The theoretical foundation is of current interest as described in Chapter 2 by Bret Heinrich and in the interesting exchange in *Physical Review Letters* that can be accessed by entering Hickey Replies in the Google search box.

18. Classical micromagnetics refers to two monographs by William Fuller Brown, Jr. published in 1962 and 1963 by Interscience, New York: *Magnetostatic Principles in Ferromagnetism* and *Micromagnetism*. These are still worth reading today, not only for the physics but also for the style. Brown was also an English teacher. A detailed discussion of the conversion between Gaussian units and SI units is given in the introduction of *Ultrathin Magnetic Structures I*, B. Heinrich and J. A. C. Bland (Eds.), Springer-Verlag, Berlin, 1994.

19. D. Dotze and A. S. Arrott, Micromagnetic studies of vortices leaving and entering square nanoboxes, *J. Appl. Phys.* **97**, 10E307 (2005).

20. A. S. Arrott and R. Hertel, Mode anticipation fields for symmetry breaking, *IEEE Trans. Magn.* **43**, 2911 (2007).
21. M. R. Scheinfein and A. S. Arrott, Increased efficiency and accuracy in micro-magnetic calculations of switching asteroids, *J. Appl. Phys.* **93**, 6802 (2003).
22. For the years 1971–1975, the *Proceedings of the MMM Conference* were published in books by the American Institute of Physics. These are known in magnetism as the "lost years" because the work was so infrequently cited and are still not referenced in the Web of Science (The AIP now has its conference proceedings online at $28 per paper.) In those years Heinrich and Arrott studied iron whiskers at the Curie temperature. The conjectured curling pattern just below Tc appeared in the seldom-cited and hard-to-find paper by B. Heinrich and A. S. Arrott, in *Proceedings of the International Conference of Magnetism ICM-73*, Vol. IV, Publishing House NAUKA, Moscow, Russia, pp. 556–561, 1974.
23. T. L. Templeton, S. D. Hanham, A. S. Arrott, Helical patterns of magnetization and magnetic charge density in iron whiskers, *AIP ADVANCES* 8, 056022 (2018).
24. F. Cheynis, A. Masseboeuf, O. Fruchart et al., Controlled switching of Néel caps in flux-closure dots, *Phys. Rev. Lett.* **102**, 107201 (2009).
25. A. S. Arrott and R. Hertel, Large amplitude oscillations (switching) of bi-stable vortex structures in zero field, *J. Magn. Magn. Mater.* **322**, 1389–1391 (2010).
26. S. B. Choe, Y. Acremann, A. Scholl et al., Vortex core driven magnetic dynamics, *Science* **304**, 420–422 (2004). Their Figure 1 panel IV shows the oscillation of a swirl in the Landau structure. See also the discussion by Xiaowei Yu. 2009. in *Time-resolved x-ray imaging of spin-torque-induced magnetic vortex oscillation*. Ph.D. Thesis, Applied Physics, Stanford.
27. J. A. Osborn, Demagnetizing factors of the general ellipsoid, *Phys. Rev.* **67**, 351 (1945).
28. S. Gliga, Energy thresholds in the magnetic vortex core reversal, *J. Phys.: Conf. Ser.* **303**, 012005 (2011).
29. O. A. Tretiakov and O. Tchernyshyov, Vortices in thin ferromagnetic films and the skyrmion number, *Phys. Rev. B* **75**, 012408 (2007).
30. J.-G. Lee, S. A. Govorkov, and A.S. Arrott, Iron whisker domain patterns imaged by garnet films *J. Appl. Phys.* **79**, 6051 (1996).
31. A. S. Arrott, Visualization and interpretation of magnetic configurations using magnetic charge, *IEEE Magnetic Letters* 7, 1108505 (2016).
32. A. S. Arrott and T. L. Templeton, Using magnetic charge to understand soft-magnetic materials, *AIP ADVANCES* **8**, 047301 (2018).
33. A. S. Arrott, C. M. Williams, E. Negusse, Magnetic charges suppress effects of anisotropy in polycrystalline soft ferromagnetic materials, *AIP ADVANCES* **8**, 056114 (2018).

4

Giant Magnetoresistance

Jack Bass

4.1 INTRODUCTION AND OVERVIEW

This chapter reviews experimental data on giant magnetoresistance (GMR) in magnetic multilayers composed of ferromagnetic (F) and nonmagnetic (N) metals. Since Chapter 5 reviews the theory underlying GMR [1], theoretical issues in this chapter are treated only as needed to motivate and understand experimental data. As background, the review begins with a brief overview of the MR for a single magnetic film—called anisotropic magnetoresistance (AMR). It then covers, in more detail, the traditional current-in-plane (CIP) MR and the current-perpendicular-to-plane (CPP) MR in multilayers. Results on current-at-an-angle-to-the-plane (CAP) MR in multilayers [2, 3] and MR in granular films (GMR) [4–6] are treated only briefly. Figure 4.1a shows the different current directions for CIP and CPP. In a real CIP film, the aspect ratio in Figure 4.1a is inverted—the length of the CIP film in the direction of the current flow is much greater than the total thickness of its layers.

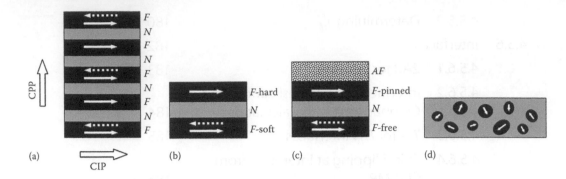

(a)   CIP   (b)   (c)   (d)

FIGURE 4.1 Sample geometries: The magnetic field is applied horizontally (in-plane). (a) A multilayer composed of alternating ferromagnetic (*F*) and nonmagnetic (*N*) metals, and *F*-layer moment orientations for the *P*-state (all solid arrows pointing to the right) and *AP*-state (all dotted arrows pointing to the left). The arrows outside the multilayer show the current directions for CIP and CPP geometries. (b) A trilayer composed of a soft *F*-layer, the moment of which reverses at low field, and a hard *F*-layer, the moment of which stays fixed until a much higher field. (c) An EBSV, where an antiferromagnet (*AF*) pins the moment of an adjacent *F*-layer, leaving the other "free" *F*-layer to reverse its magnetization at low fields. (d) A granular *N* alloy, containing *F*-inclusions with initially randomly aligned moments that align in a large enough magnetic field.

Since the discovery and early history of GMR are covered in Chapter 1 [7], this chapter touches only briefly upon those topics. This review is also not encyclopedic but rather focuses on a number of questions of special physical interest that include the following: (1) How strong is scattering at *F*/*N* interfaces and how important is it in GMR? Such scattering is a new transport phenomenon arising with multilayers. When GMR was first discovered, almost nothing was known about such scattering. In the CIP geometry on thin films, earlier analyses of electron scattering at the metal/air surfaces had assumed a single parameter that characterized the probability of diffuse (as opposed to specular) reflection at the interface [8]. Understanding CIP- and CPP-GMRs, in contrast, requires ascertaining both the strength of the scattering of electrons at an *F*/*N* interface and the asymmetry of such scattering for electrons with magnetic moments oriented along ($\uparrow$) or opposite to ($\downarrow$) the moment of the *F*-metal. In CIP-MR, current flow is parallel to the layers and interfaces, whereas, in CPP-MR, it is perpendicular to both. Given this difference, are interface parameters the same in the two cases? Are the relative weightings of scattering in the bulk *F*-metals and at the *F*/*N* interfaces similar? Especially for the CPP-MR, there is an interest also in the scattering at the N_1/N_2 and *F*/*S* (*S* = superconductor) interfaces. (2) Are the bulk and interfacial contributions to CIP- and CPP-MRs dominated by band structure effects, or by scattering from impurities? (3) Can GMR be adequately understood with semiclassical models, or are quantum-coherence effects needed?

This chapter is organized as follows: Section 4.2 contains definitions of AMR, GMR, and various parameters that are needed later. Section 4.3 compares CIP- and CPP-MRs and gives examples of CAP and granular MR data. Section 4.4 more fully covers CIP-MR data and what we have learned from them. Section 4.5 covers CPP-MR data and what we have learned from them. Section 4.6 contains a summary and conclusions.

For more information about GMR, and additional references, we recommend several earlier reviews [9–15].

4.2 DEFINITIONS AND BACKGROUND

4.2.1 ANISOTROPIC MAGNETORESISTANCE

The resistivity ρ of a thin film of an F-metal to which a magnetic field H is applied in the plane of the film was found long ago [16] to vary with the angle θ between the current I and the magnetization M as [1, 17]

$$\rho(\theta) = \rho_\perp + (\rho_\parallel - \rho_\perp)\cos^2\theta \qquad (4.1)$$

Here, the normally larger $\rho_\parallel$ and smaller $\rho_\perp$ are the limiting resistivities when the current is parallel or perpendicular to the F-layer magnetization. AMR arises from the asymmetry of scattering by the spin–orbit interaction, which is standardly stronger for electrons moving parallel to the magnetization [1]. AMR in the alloy permalloy, $Ni_{1-x}Fe_x$ with $x \sim 0.2$, was the source of the magnetoresistance in the magnetoresistive read heads in computers prior to the use of GMR [7]. When the CIP-MR is small, account must be taken of the MR contribution from AMR.

4.2.2 GIANT MAGNETORESISTANCE

GMR is the change in the electrical resistance, R, of a magnetic multilayer composed of alternating F and N-layers, upon application of H, usually in the layer plane. Figure 4.1 shows examples of the different types of multilayers that give GMRs: (a) an $[F/N]_n$ multilayer repeating n identical F and N-layers, (b) an $[F/N/F]$ multilayer containing a magnetically soft layer that reverses in a small field and a magnetically hard one that requires a larger field, (c) an $[AF/F/N/F]$ exchange-biased spin valve (EBSV) containing a free F-layer that reverses in a small field and an F-layer that is "pinned" by an adjacent antiferromagnet (AF) to reverse only at a larger field, and (d) a granular alloy containing F-inclusions in an N-matrix. In all four cases, so long as the F-layers are not coupled ferromagnetically, the sample will show a magnetoresistance as H is swept from a large positive value to a large negative one. As described in Chapter 1 [7], GMR was discovered in samples in which Fe-layers were coupled antiferromagnetically across Cr layers [18, 19], so that at $H = 0$, the magnetizations of the F-layers were oriented antiparallel (AP) to each other. This state gave the highest resistance, R_{AP}. Application of H large enough to align all of the Fe layer moments parallel (P) to each other gave a lower resistance, R_P. The resulting MR is shown in Chapter 1.

It was quickly discovered that large MRs did not require such AF coupling. As an example, Figure 4.2 shows the MR for a magnetically uncoupled, simple $[Co/Ag]_n$ multilayer. The largest value of R in Figure 4.2, R_0, occurs in the as-prepared (virgin) state where $M = 0$. A later study [20] showed that

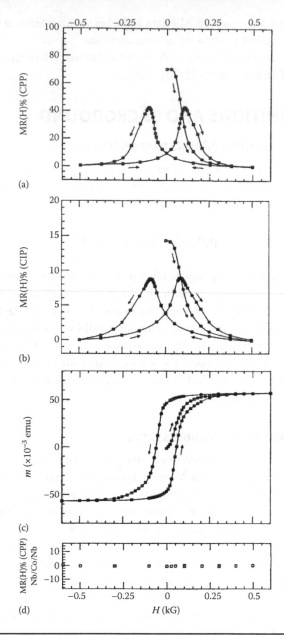

(a)

(b)

(c)

(d)

H (kG)

FIGURE 4.2   (a) CPP-MR, (b) CIP-MR, and (c) magnetization, $M(H)$, for [Co(6 nm)/ Ag(6 nm)]^{60} multilayers, and (d) negligible CPP-MR for a single Co(9 nm) film. In the text, the maximum CPP-MR in the "virgin" state is labeled R_0, the maximum after saturation at large H is labeled R_{peak}, and the minimum is labeled R_p. (After Pratt, W.P. et al., *Phys. Rev. Lett.* 66, 3060, 1991. With permission.)

this state most closely approaches the AP-state (but probably does not quite reach it—see Figure 4.3) because the micron-sized domains in the adjacent F-layers are initially oriented close to AP. In contrast, the peaks in R in Figure 4.2 after H has been taken to saturation do not occur at $H = 0$, but rather at the coercive field, H_c. Although the total sample magnetization at H_c is again $M = 0$, now the relative orientations between moments of domains in the

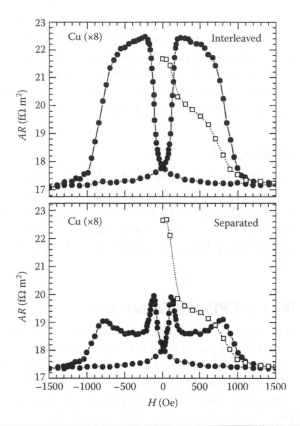

FIGURE 4.3   Comparison of CPP data for $A\Delta R$ versus H(Oe) for (top) "interleaved" [Co(6)/Cu(20)/Co(1)/Cu(20)]8 and (bottom) "separated" [Co(6)/Cu(20)]8 [Co(1)/Cu(20)]8 samples. The use of two different Co thicknesses leads to an AP-state of maximum $A\Delta R$ after saturation that is larger than the virgin state value of $A\Delta R$ at $H = 0$ (contrast with Figure 4.2). The behavior of the "separated" sample will be discussed in Section 4.5.9. (After Eid, K. et al., *Phys. Rev. B* 65, 054424, 2002. With permission.)

adjacent layers are close to random. Thus, these peaks are not AP-states, and R_{peak} is much less than the resistance of the virgin state, R_0.

For F-inclusions in N-hosts, GMR gives qualitatively similar curves to Figure 4.2, except that R_0 is barely larger than R_{peak} at the "peaks" at H_c after saturation, and R_{peak} occurs at much higher magnetic fields ($H_c \sim 1$ kG for GMR vs $H_c \sim 0.1$ kG for GMR) [4]. Again, at these peaks, the moments of the individual inclusions are presumably aligned randomly, rather than AP. The large values of H_c for granular samples have limited their use in devices.

As just noted, the largest MR (GMR) occurs when the order of the magnetic moments of the two F-layers changes from P (usually at high H) to AP (usually at low H). Assuming that these two states can be achieved, we formally define the GMR as

$$\text{MR} = \frac{R_{AP} - R_P}{R_P} = \frac{\Delta R}{R_P} \tag{4.2}$$

As explained in Chapter 1, for multilayers containing only a single pair of metals N and F, R_{AP} is always greater than R_P, i.e. $R_{AP} > R_P$, giving a positive

MR. But we will see below, that multilayers composed of certain combinations of different F- and/or N-metals—$F1$, $F2$, and/or $N1$, $N2$—can give a negative MR ($R_{AP} < R_P$). Such an "inversion" occurs when the scattering anisotropy of $F2$ in Figure 1.4c is inverted from that of $F1$—that is, electrons in $F2$ are now weakly scattered when the F-layer and electron moments are antiparallel, giving a "shorting" in the AP-state [1].

4.2.3 MR VERSUS M

When $R_{AP} > R_P$, one expects a maximum $R = R_{AP}$ to occur when $M = 0$, and a minimum $R = R_P$, to occur when M is maximum. Both MR(H) and M(H) then vary smoothly between these limits, as shown for a simple $[F/N]_n$ multilayer in Figure 4.2. But, as explained above, and shown in Figure 4.2, for magnetically uncoupled F-layers, the value of R at $M = 0$ can depend upon the sample history.

4.2.4 CIP- AND CPP-MR GEOMETRIES

GMR can be measured in two limiting geometries as shown in Figure 4.1. Current flows in the layer planes (CIP) MR and current flows perpendicular-to-the-layer-planes (CPP) MR [21]. As already noted, GMR has also been measured with a current flow at an angle to the layer planes (CAP)-MR [2, 3]. CAP-MR interpolates between CIP- and CPP-MRs and is usually more complex to analyze. We, thus, focus on the CIP- and CPP-MRs.

4.2.5 ACHIEVING AP-STATES

From the definition in Equation 4.2, to determine GMR, it is necessary to produce both AP and P-states. The P-state can always be achieved by applying H large enough to align all the F-layer magnetic moments in the multilayer. The problem is to achieve the AP-state. GMR was discovered in Fe/Cr multilayers [18, 19] using AF exchange coupling between the adjacent Fe-layers to achieve the AP-state at field $H = 0$. But, for devices such coupling is disadvantageous since it requires a large H to switch from AP to P. Fortunately, it was soon shown that the AP-state could be obtained without such coupling (see, e.g., Refs. [20, 22]), as already illustrated in Figure 4.2. In the following, we describe different techniques used to reliably and controllably produce AP-states, mostly at low H. In most of the multilayers of interest, the layers are much wider than they are thick. Due to demagnetization effects, the magnetization usually prefers to lie in the layer plane. If the layer diameters are less than microns, and the N-layer thicknesses are much smaller than microns, dipolar coupling will cause the adjacent F-layers to couple antiferromagnetically. If instead, the layer diameters are microns or larger, the dipoles at the ends of the layers will be so far out that such coupling is negligible.

1. *An mm diameter multilayer with uncoupled, identical F-layers*: If dipolar coupling in an $[F/N]_n$ multilayer is small, then making the N-layer thick enough to minimize exchange coupling (see Chapter 2

[23]) between the adjacent F-layers, will still give a GMR as shown in Figure 4.2. Figure 4.2 illustrates that the field dependences of the CIP- and CPP-MRs are usually similar.

2. *Nanowires with coupled identical layers*: In nanowires, where the F-layers are much wider than thick (so that their magnetizations tend to lie "in-plane"), dipolar coupling will usually produce AF ordering of the adjacent F-layers. By alternating two very different thicknesses of N, such coupling can be used to give AF coupling of each pair of close F-layers but little or no coupling of widely separated pairs [24].

3. *Multilayers with uncoupled, nonidentical F-layers (extension of Figure 4.1b)*: A well-defined AP-state can be obtained by using two F-layers with different coercive fields. These can either be F-layers of different thicknesses, or grown on different underlayers [22], or two different F-layers [25]. The top panel in Figure 4.3 shows a CPP-MR example for a $[Co(6)/Cu(20)/Co(1)/Cu(20)]_8$ multilayer, with layer thicknesses in nanometers. Now, because of the very different values of H_c for the 6 nm and 1 nm thick Co layers, the multilayer reaches (and maintains for a short range of values of H) R_{AP} after saturation. In contrast to Figure 4.2, R_0 for the "initial" state is now slightly below R_{AP}. This difference suggests that R_0 in Figure 4.2 might also be slightly below the actual R_{AP}. The bottom panel in Figure 4.3 shows a CPP-MR example for a "separated" multilayer. These data will be examined in Section 4.5.9. An example of CPP-MR with two different F-metals, Co and Py = permalloy = $Ni_{1-x}Fe_x$ with $x \sim 0.2$, is given in Ref. [26]. The different values of H_c—$H_c(Co) \sim 150$ Oe and $H_c(Py) <$ 20 Oe—yield $R_{peak} \approx R_{AP}$ as in Figure 4.3.

4. *EBSV (Figure 4.1c)*: A sharp transition and good control over the AP-state can be achieved with an EBSV [27] of the form $AF/F/N/F$, where AF is an antiferromagnet (AF), such as FeMn, IrMn, or CoO. Here, first heating the sample to above the blocking temperature of the AF-layer, and then cooling to room temperature in the presence of an applied magnetic field, produces in the AF-layer an internal order that pins, via the exchange interaction, the magnetization of the adjacent F-layer to a much higher field than the coercive field of the other "free" F-layer. For a sample with a macroscopic cross-sectional area, the "free" F-layer can be exchange decoupled from the pinned F-layer by making the N-layer thick enough (typically >5 nm). An example of CIP-MR data for such multilayers is shown in Figure 4.4. Similar results for CPP-MR are given in Ref. [28]. In both cases, the equality of the values of R for the $-H$- and $+H$ P-states shows that the AF-layer contributes only a constant term to R that is independent of whether the adjacent F-layer is aligned along or opposite to its "pinned" direction. If, however, the sample area is at the nanoscale, as needed for modern read heads, then two adjacent F-layers with magnetizations in the layer planes will couple antiferromagnetically via dipolar coupling. Real read heads now minimize this problem by replacing the AF/F of the "pinned layer" by a synthetic AF of the

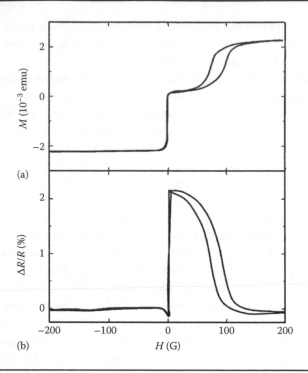

FIGURE 4.4   Magnetization curve (a) and CIP-MR (b) for an EBSV multilayer Si/Ni Fe(15)/Cu(2.6)/NiFe(15)/FeMn(10)/Ag(2) (layer thicknesses in nm). The in-plane field *H* is parallel to the exchange anisotropy (EA) field created by the FeMn. The current flows perpendicular to this direction. (After Dieny, B. et al., *Phys. Rev. B* 43, 1297, 1991. With permission.)

form $AF/F2/Ru/F2$, where a Ru thickness ~0.6 nm gives AF coupling between the two F-layers as illustrated in Figure 4.5. Here, exchange coupling between the AF and adjacent $F2$ sets the orientation of the synthetic AF and the presence of the two, oppositely magnetized $F2$ layers greatly reduces the magnetic coupling with the separate "free" layer that reverses under the action of a small applied magnetic field.

4.2.6 ASYMMETRY PARAMETERS

As explained in Chapters 1 and 5 [1, 7], the models standardly used by experimentalists assume that electrons passing through an F/N multilayer can be separated into two-current streams, one with its moments "up" and the other with its moments "down," relative to a chosen fixed axis (e.g., the direction of the initial applied field H). As one stream passes through a given F-metal (assumed monodomain), its moments can be along ($\uparrow$) or opposite to ($\downarrow$) the magnetization of the F-metal. If, as the electrons flow, their moments do not "flip" so as to mix up and down currents, this assumption gives a "two-current" (2C) model [18, 29–32]. In the limit of diffusive transport, within the bulk of the F-metal, this model has two parameters: $\rho_F^\uparrow$ and $\rho_F^\downarrow$. These parameters can be combined to give two alternative dimensionless parameters: $\alpha_F = \rho_F^\downarrow/\rho_F^\uparrow$ or $\beta_F = (\rho_F^\downarrow - \rho_F^\uparrow)/(\rho_F^\downarrow + \rho_F^\uparrow)$. Values of α_F for F-based

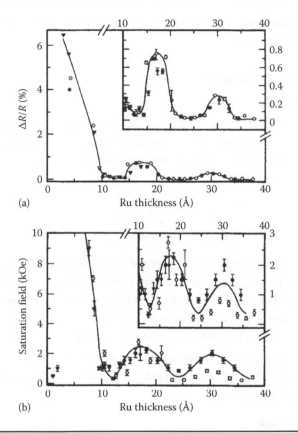

(a)

(b)

FIGURE 4.5 (a) CIP-MR and (b) saturation field, H_{sat}, versus Ru thickness, t_{Ru}, for Si(111)/Ru(10)/[Co(2)/Ru(t_{Ru})]$_{20}$/Ru(5) multilayers (thicknesses in nanometer) deposited at temperatures of 313 K (filled circles), 398 K (open circles), or 473 K (crosses). (After Parkin, S. et al., *Phys. Rev. Lett.* 64, 2304, 1990. With permission.)

alloys were derived a few decades ago [33] from measurements of deviations from Matthiessen's rule (DMR) in three-component F-based alloys. For GMR analysis, it is often more convenient to use β_F, along with an enhanced resistivity $\rho_F^* = \left(\rho_F^{\downarrow} + \rho_F^{\uparrow}\right)/4 = \rho_F/\left(1 - \beta_F^2\right)$. Here, ρ_F is the standard resistivity of the F-metal, which can be measured separately on films deposited in the same way as the GMR multilayers of interest. For the CPP-MR, one also defines similar quantities for the F/N interface: the asymmetry parameter $\gamma_{F/N} = \left(AR_{F/N}^{\downarrow} - AR_{F/N}^{\uparrow}\right)/\left(AR_{F/N}^{\downarrow} + AR_{F/N}^{\uparrow}\right)$ and the enhanced specific resistance $2AR_{F/N}^* = \left(AR_{F/N}^{\downarrow} + AR_{F/N}^{\uparrow}\right)/2$, where A is the area through which the CPP-current flows, and $R_{F/N}$ is the CPP F/N interface resistance. Because the CIP-MR can rarely be written in closed form with bulk and interfacial scattering contributions separated, it is usually difficult to derive quantitative values of bulk or interfacial scattering asymmetries. In the CPP-MR, in contrast, the applicability of simple two-current series-resistor (2CSR) [30–32], or more complex Valet–Fert (VF) [32] models, often allow bulk and interfacial contributions and parameters to be separated. For the CPP-MR, we will list tables of values, and in several cases, compare them with values found by completely independent techniques, or else with calculated values involving no adjustable parameters.

4.2.7 LENGTH SCALES

4.2.7.1 CIP-MR

The dominant length scales in the CIP-MR are the mean-free-paths (λ) for the scattering of electrons in the F-metal ($\lambda_F^\uparrow$ and $\lambda_F^\downarrow$) and the N-metal (λ_N) [1, 34]. For an N-metal, λ_N is generally -estimated from the metal's resistivity, ρ_N, using equation [35]:

$$\lambda_N = \frac{(\rho_B l_B)}{\rho_N} \tag{4.3}$$

Here, the parameter ($\rho_B l_B$) is a property of the N-metal of interest. Values for a number of metals, determined mostly from size-effect studies, have been collected together in Ref. [8]. For an F-metal, the situation is complicated by the presence of the two mean-free-paths, $\lambda_F^\uparrow$ and $\lambda_F^\downarrow$. If for a given F-metal or alloy one is much longer than the other, then the longer one can be estimated from Equation 4.3. In the absence of knowledge of the ratio, $\lambda_F^\uparrow$ and $\lambda_F^\downarrow$ must be determined independently. In Section 4.4.3.7, we will describe an experiment that did so.

4.2.7.2 CPP-MR

The dominant length scales in the CPP-MR are the lengths that electrons diffuse before flipping their spins, the spin-diffusion lengths l_{sf}^F in the F-metal or alloy, and l_{sf}^N in the N-metal or alloy [1, 32]. In the 2CSR [30–32] and VF [32] models used to analyze most CPP-MR data, the mean-free-paths do not appear at all. In Section 4.5.5, we will explain how l_{sf}^F and l_{sf}^N are measured and list some values of each. More complete listings, with values also determined by other techniques, are contained in Ref. [36].

At 4.2 K, electrons are scattered by impurities and fixed defects. Such scattering is mostly large angle, in which case any spin-flipping during the scattering does not mix spin currents [1, 32]. At higher temperatures, however, electron–magnon (and perhaps also electron–phonon scattering) can mix spin currents. Then the spin-current mixing length can become important in both CIP- and CPP-MRs.

4.3 COMPARE CIP-MR, CPP-MR, AND CAP-MR

In this section, we compare data on CIP-MR, CPP-MR, and CAP-MR.

Figure 4.2 compares values at 4.2 K for (a) CPP-MR; (b) CIP-MR; and (c) magnetization $M(H)$, for a $[Co/Ag]_{60}$ multilayer with $t_{Co} = t_{Ag} = 6$ nm; and also (d) the CPP-MR for a Co(9 nm) film between the Nb leads. The multilayer CPP-MR is several times larger than the multilayer CIP-MR, and both are much larger than the CPP-MR for a single Co film, which is <1%.

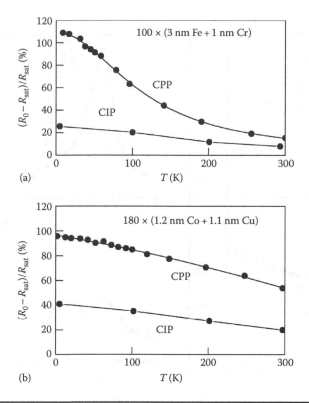

FIGURE 4.6 CPP-MR and CIP-MR versus temperature T (K) for (a) [Fe(3)/Cr(1)]$_{100}$ and (b) [Co(1.2)/Cu(1.1)]$_{180}$ multilayers (thicknesses in nanometer). (After Bass, J. et al., *J. Magn. Magn. Mater.* 200, 274, 1999; Bass, J. et al., Erratum: *J. Magn. Magn. Mater.* 296, 65, 2006; Gijs, M.A.M. et al., *Phys. Rev. Lett.* 70, 3343, 1993; Gijs, M.A.M. et al., *J. Appl. Phys.* 75, 6709, 1994. With permission.)

Similar data for uncoupled magnetic granules requires as standard a much larger H to reach the P-state than do the CIP- or CPP-MRs for uncoupled F-layers [3, 4].

Figure 4.6 compares the CPP- and CIP-MRs for (a) [Fe(3)/Cr(1)]$_{100}$ and (b) [Co(1.2)Cu(1.1)]$_{180}$ as functions of temperature T from 4.2 to 300 K. For both metal pairs, the CPP-MR is larger than the CIP-MR at all temperatures. Both MRs decrease with the increasing T, due to the increase in the resistivity (R_P) as a result of enhanced phonon scattering, as we now explain (MR decrease can also be due to some mixing of spin currents, see Section 4.2.7.2). The resistivities of both the F and N-layers increase substantially as T increases from cryogenic (4.2 K) to room (293 K) temperature, as the scattering of electrons by phonons becomes increasingly important. As a rough guide, for sputtered or evaporated polycrystalline samples of nominally "pure" elemental metals (e.g., Cu, Ag, Co, Fe), the resistivity ratio ($RRR = \rho(293\ \text{K})/\rho(4.2\ \text{K})$) is typically ~2. However, RRRs as large as 4 have been seen for Cu, and RRRs closer to 1 can be found for concentrated alloys, such as Py [8, 12]. If such bulk resistivities dominate R, $R(T)$ will increase by roughly a factor of 2 from 4.2 to 293 K. If, in contrast, interface scattering dominates, then the increase

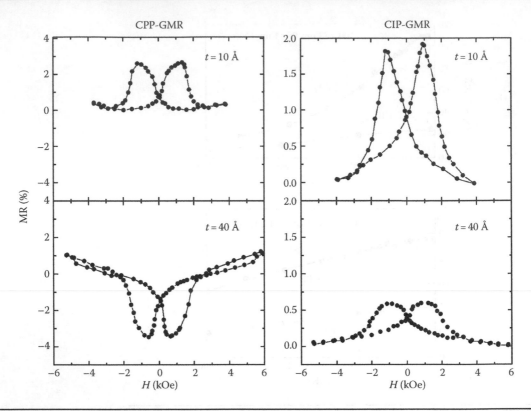

FIGURE 4.7 Comparisons of CPP-MR (left) and CIP-MR (right) for [Ni$_{95}$Cr$_5$(t)/Cu(4 nm)/Co(0.4 nm)/Cu(4 nm)]20 multilayers with t = 1 and 4 nm. The CPP-MR inverts, but the CIP-MR does not. (After Vouille, C. et al., *Phys. Rev. B* 1999; Vouille, C. et al., *J. Appl. Phys.* 81, 4573, 1997. With permission.)

of R, and the variation of $\Delta R/R$, will depend upon the temperature dependence of interface scattering, which must be derived from the experiment. In general, interface scattering seems to vary much less rapidly than bulk—see Refs. [37–39], discussion of the CIP-MR(T) in Section 4.4.3.11, and of the CPP-MR(T) in Section 4.5.8.

Figure 4.7 compares CPP-MR (left) and CIP-MR (right) [40] for hybrid spin valves with different thicknesses of a Ni$_{95}$Cr$_5$ alloy that has negative β_{NiCr} but positive $\gamma_{NiCr/Cu}$. For thin NiCr, the positive interface dominates and both CPP- and CIP-MRs are "positive." For thick enough NiCr, the negative bulk dominates the CPP-MR, making it negative. But the increasing bulk contribution only reduces the magnitude of the CIP-MR. These data illustrate that the relative importance of bulk and interface contributions usually differ for CIP- and CPP-MRs. For more discussion see Ref. [40, 41].

Using V-grooves cut into an insulating substrate and then depositing layers normal to the substrate produced CAP [2] data. Levy et al. [42] derived a relationship between the smaller CIP, intermediate CAP, and larger CPP resistances that allow the prediction of the CPP-MR from measurements of the CIP- and CAP-MRs. Alternatively, by depositing the metals perpendicular to one side of the V-grooves, starting with a highly conductive, thick Cu- or Au-layer before the multilayer, stopping the multilayer before the end of the V-side, and then depositing a thick Cu or Au-layer on top to connect with the thick Cu- or Au-layer on the bottom of the adjacent V-groove, Gijs et al. [3, 37] converted the V-groove geometry [3] into an almost CPP one. For an overview of these latter studies see Ref. [11].

4.4 CIP-MR

4.4.1 MEASURING CIP-MR

The advantage of the CIP-MR is that a typical long (~mm), thin (<1 μm) multilayer has a resistance R large enough to be easily measured with standard laboratory equipment. However, sample and lead geometries must be handled with some care to measure ΔR and R separately. In practice, most studies simply provide the GMR ratio, MR = $\Delta R/R$, for which knowledge of the detailed geometry is not essential. Said another way, if the current distribution is independent of the applied magnetic field H, the MR ratio will be independent of the applied current and will be a property of the sample at a given H. However, interpreting the MR is often not trivial, because it is affected by changes in both ΔR and R.

There are two standard ways to measure CIP-MR. The first involves a long, narrow sample with side contacts long and narrow enough to minimize the current flow into them. If the end contacts have low resistance compared to the sample, this geometry should give a uniform current and allow direct determinations of ΔR and R separately. The widely used second case involves a rectangular sample with four leads in a straight line. Unless the sample is much longer than it is wide, and the leads are well separated, the current density will be nonuniform, and separate determinations of ΔR and R require calculations to correct this nonuniform density. As noted above, to avoid the effects of current nonuniformity, most published studies focus on MR = $\Delta R/R$.

4.4.2 SOME CIP-MR EQUATIONS

In general, a quantitative analysis of CIP-MR is difficult because the current flow through the sample is not uniform, and the CIP-MR depends upon the ratios of the thicknesses of the F- and N-layers to the mean-free-paths in those layers. If the F- and N-layer thicknesses are similar, the current is larger in the metal (usually the N-metal) with the lower resistivity. Including contributions both within the bulk F- and N-layers and at the F/N interfaces generally involves a large number of parameters [29, 34], most of which cannot easily be independently determined. Thus, analyses are mostly qualitative and highly simplified. As two examples, Ref. [29] fit data on Fe/Cr trilayers assuming a single mean-free-path in both the Fe and Cr and asymmetric scattering only at the interfaces, whereas [43] fit data on Py/Cu EBSVs completely neglecting scattering at the Py/Cu interfaces, but assuming five mean-free-paths, two in the Cu (parallel (∥) and perpendicular (⊥) to the planes), and three in the Py ((∥) and (⊥) for spin-up electrons and isotropic for spin down). As we will see next, data have also been fitted using phenomenological equations that do not formally distinguish between bulk or interfacial scattering. Further below we will see arguments in favor of interfacial scattering often dominating the CIP-MR. For these reasons, analyses of CIP-MR data rarely yield quantitative parameters that can be compared with measurements by techniques other than CIP-MR.

Two phenomenological equations have been used to fit CIP-MR data as only t_N or t_F is varied. The physics underlying these equations is discussed in Chapter 5 [1] and Ref. [44].

For varying t_N [10, 14, 45]:

$$\frac{\Delta R}{R} = (\Delta R / R)_o \exp\left(\frac{-t_N/\lambda_N}{(1+t_N/t_0)}\right) \tag{4.4}$$

In Equation 4.4, the CIP-MR is a monotonically decreasing function of t_N. The t_N/t_0 term in the denominator gives a simple "dilution" effect [45]; the thicker t_N, the larger the fraction of current that flows through it, and the smaller the fraction scattered asymmetrically in the F-metal or at the F/N interfaces. The $\exp(t_N/\lambda_N)$ term is a mean-free-path effect; the thicker t_N, the harder for electrons scattered in one F-metal, or at its F/N interface, to get to the other F-metal before being scattered.

For varying t_F [10, 14, 46]:

$$\frac{\Delta R}{R} = \frac{A\left[1-\exp(-t_F/\lambda_F)\right]}{(1+t_F/t_0)} \tag{4.5}$$

To qualitatively understand Equation 4.5, assume that the numerator corresponds to a normalized value of ΔR and the denominator to a normalized value of R. As t_F increases, the numerator grows on scale λ_F, first linearly with t_F, but then gradually saturating, as t_F becomes much larger than λ_F. The denominator continuously grows linearly with increasing t_F, but on a scale t_0, where the ratio t_F/t_0 represents the fraction of R due to F. Coupling the growth and saturation of ΔR, with the slower but continuing growth of R, predicts an initial increase in $\Delta R/R$, a maximum, and then a slower decrease.

We will compare these equations with data in Section 4.4.3.4.

4.4.3 Some Basic CIP-MR Questions and Answers

4.4.3.1 How Does Coupling Vary with t_N?

As described in Chapter 1 [7], GMR was discovered [18, 19] in Fe/Cr samples with the Cr thickness chosen to give a strong AF coupling between the adjacent F-layers. The maximum MR then occurred for $H = 0$, and the minimum MR for H above a saturation field, H_s. In Ref. [18], modest deviations of the Cr thickness from that which gave best AF coupling left the form of the MR data unchanged but reduced its magnitude. The question naturally arose, "How does the coupling between the F-layers (e.g., Fe) across a given N-layer (e.g., Cr) vary with the thickness, t_N, of N?" Parkin et al. [47] showed, for carefully prepared samples, that the coupling and MRs oscillated together with decreasing coupling strength as t_N increased. Figure 4.5 shows an example of such behavior for F = Co and N = Ru.

4.4.3.2 Is AF Coupling Necessary for GMR?

The answer is "No." As noted above, all that is needed is the ability to change the magnetic order from P at high field to AP at lower field [22, 48]. In Figure 4.5 we see that AF coupling (large H_s), which gives AP order at $H = 0$, gives the largest CIP-MR, and that ferromagnetic coupling (small H_s), which forces the retention of approximate P order at all Hs, gives the smallest CIP-MR. From the continuous curves, we infer that zero coupling should give at least partial AP order and an intermediate CIP-MR. Figure 4.2 shows examples of data for no coupling. For devices, no coupling is preferred, so that the device can respond to small H. See the discussion of how to achieve the AP-state in Section 4.2.5.

4.4.3.3 Can Other F/N Pairs Give MRs Competitive with Fe/Cr?

Not long after the discovery of GMR, the pure-metal pairs Co/Cu and Co/Ag were shown to give comparably large MRs. Subsequently, pairs with alloy F-metals, such as (Py = $Ni_{80}Fe_{20}$)/Cu and $Co_{91}Fe_9$/Cu were also found to give significant MRs, for Py, with much smaller H_c. For references to these studies see Ref. [10]. For devices, inserting a thin layer of Co at the Py/Cu interface combined the advantages of small H_c (for switching at low H) with high MR [49]. In contrast, a variety of other combinations of F- and N-metals gave smaller CIP-MRs (see Ref. [14, pp. 136–137]). For a discussion of the band structure matching and other effects that contribute to the physics underlying these differences, see Chapter 5 [1].

4.4.3.4 How Do CIP-MRs Vary with t_F and t_N?

As the analysis of CIP-MR can involve many parameters, early investigators developed phenomenological equations to describe their data. Figures 4.8 and 4.9 show examples of multilayer data with only t_N or t_F varied. Equations 4.4 and 4.5 used for the fits are given in Section 4.4.2. The physics underlying these behaviors is discussed in Chapter 5 [1] and Ref. [44].

4.4.3.5 Is CIP-MR Dominated by Spin-Dependent Scattering in Bulk F or at F/N Interfaces?

Early interpretations favoring both possibilities were published [29, 43]. To investigate the alternatives in more detail, investigators began to change the interfaces in different ways, of which we give some examples. In Fe/Cr multilayers, the authors of Refs. [49–51] inserted at the Fe/Cr interfaces impurities like Cr that give $\beta_F < 1$ (V, Mn) and impurities unlike Cr that give $\beta_F > 1$ (Ge, Al, Ir). For V and Mn, the CIP-MRs decreased relatively slowly with increasing impurity layer thickness, similar to the behavior for Cr. In contrast, for Ge, Al, and Ir, the CIP-MRs decreased rapidly with impurity layer thickness. The authors attributed these differences to the qualitative differences in β_F, as expected for spin-dependent interfacial scattering. Speriosu et al. [52] heated

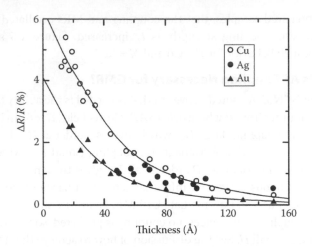

FIGURE 4.8 CIP-MR = ($\Delta R/R$) (%) versus N-layer thickness for samples of the form Si/Co(7)/N/Py(4.7)/FeMn(7.8)/N(1.5) with N = Cu, Ag, or Au and thicknesses in nanometer. The solid curves are fits to Equation 4.4. (After Dieny, B. et al., *J. Appl. Phys.* 69, 4774, 1991; Dieny, B. et al., *Phys. Rev. B* 45, 806, 1992. With permission.)

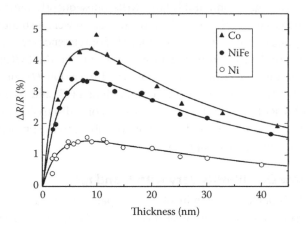

FIGURE 4.9 CIP-MR = ($\Delta R/R$) (%) versus F-layer thickness for multilayers of the form Si/F/Cu(2.2)/Py(4.7)/FeMn(7.8)/Cu(1.5), with F = Ni, Py = $Ni_{0.8}Fe_{0.2}$, or Co and thicknesses in nanometer. The curves are fits to Equation 4.5. (After Dieny, B. et al., *J. Appl. Phys.* 69, 4774, 1991; Dieny, B. et al., *Phys. Rev. B* 45, 806, 1992. With permission.)

Py/Cu and Co/Cu samples to extend the intermixed regions of the interfaces. For Py/Cu, they found decreases in MR with increasing intermixing, which they attributed to the growth of nonmagnetic alloys at the interfaces, which increased spin-independent scattering there. In contrast, in Fe/Cr multilayers, increased intermixing increased the CIP-MR [53, 54].

Parkin [49] used EBSVs to study the effects at 293 K of inserting thin Co layers at Py/Cu interfaces or within the bulk Py-layers, and of inserting thin Py-layers at Co/Cu interfaces. Figure 4.10 shows that the CIP-MR grew with more Co at the Py/Cu interface until the Co contribution dominated,

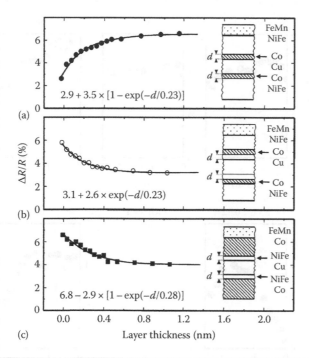

FIGURE 4.10 Dependence of room temperature CIP-MR = ($\Delta R/R$)(%) on (a) Co interface layer thickness, d, in EBSVs of the form Si/Py(5.3−d)/Co(d)/Cu(3.2)/Co(d)/Py(2.2−d)/FeMn(9)/Cu(1); (b) distance d of a 0.5 nm thick Co layer from the Py/Cu interfaces in EBSVs of the form Si/Py(4.9−d)/Co(0.5)/Py(d)/Cu(3)/Py(d)/Co(5)/Py(1.8−d)/FeMn(9)/Cu(1); and (c) Py interface layer thickness, d, in EBSVs of the form Si/Co(5.7 − d)/Py(d)Cu(2.4)/Py(d)/Co(2.9−d)/FeMn(10)/Cu(1). Py = Ni0.8Fe0.2s is designated as NiFe, and layer thicknesses are in nanometers. (After Parkin, S.S.P., *Phys. Rev. Lett.* 71, 1641, 1993. With permission.)

whereas moving the Co into the Py, or inserting Py at Co/Cu interfaces, decreased the CIP-MR. In all three cases, the scale over which the effect occurred was only 0.15−0.3 nm. He argued that these behaviors showed that interfacial scattering was dominant.

In contrast, the data for Co/Cu-based EBSVs in Figure 4.11 show that inserting fractional monolayers (MLs) of various impurities at the Co/Cu interfaces always decreased the MR, whereas inserting the same ML within the bulk of the Co, a distance x from the interface, could decrease, increase, or leave the MR unchanged, depending upon both the impurity and x. The authors attributed these different behaviors to deviations of the β_F for these impurities in Co from the scattering anisotropy of the undoped Co. They concluded that surface scattering was important, but that bulk scattering by impurities could also be important much further into the Co than the fraction of a nanometer, as seen in Ref. [49].

As illustrated in Figure 4.7, for thin F-layers, interface effects predominate, but bulk effects grow in thicker layers. The relative importance of bulk and interfacial scattering depends upon both the particular host F-metal or the alloy and the particular N-metal. If impurities are selectively inserted

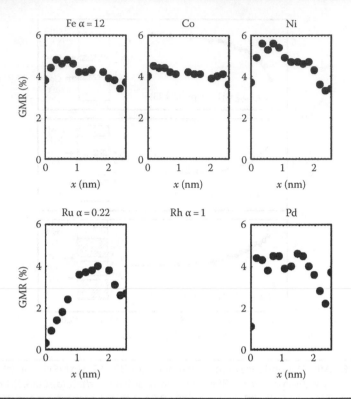

FIGURE 4.11 GMR (%) = CIP-MR versus distance x of the specified impurity from a Co/Cu interface. (After Marrows, C.H. et al., *Phys. Rev. B* 63, 220405R, 2001. With permission.)

into a "pure" host F-metal, these effects depend on just where the impurities are inserted into the host F-metal or at the interface.

4.4.3.6 What Is the Fundamental Quantity for CIP-MR?

Dieny et al. [55] argue that the fundamental quantity for the CIP-MR is the change in sheet conductance, $\Delta G = \Delta R/(R_P R_{AP})$. Figure 4.12 shows the different behaviors of MR = $\Delta R/R$ and ΔG for three samples that differ only in the capping layer: (a) cap = Ta(5), (b) cap = Cu(10)/Ta(5), and (c) cap = Cu(30)/Ta(5), where the layer thicknesses are in nanometer. The data collapse from three different curves for MR, to a single curve for ΔG, consistent with the prediction in Ref. [55] of $\Delta G = n(\Delta G_{per}) - \Delta G_{corr}$, where "$\Delta G_{per}$ is independent of the number of periods n in the multilayer, and ΔG_{corr} corrects for boundary effects and is independent of n." Despite the advantage argued for ΔG, most researchers have stuck with the dimensionless CIP-MR, where knowing the precise sample dimensions is not crucial.

4.4.3.7 How Long Are Mean-Free-Paths in F-Metals?

As explained in Chapter 5 [1] and Section 4.2.7.1 above, the CIP-MR depends upon three mean-free-paths: $\lambda_F^\uparrow$ (majority = long) and $\lambda_F^\downarrow$ (minority = short), and one in the N-metal, λ_N. λ_N can be found from Equation 4.3. Gurney et al. [56] described a technique for estimating $\lambda_F^\uparrow$ and $\lambda_F^\downarrow$. For three standard

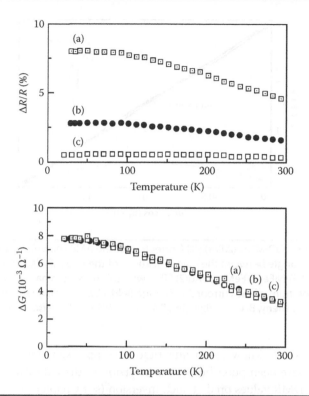

FIGURE 4.12 Top: CIP-MR, and bottom: ΔG, versus temperature for multilayers (a), (b), and (c) with differing capping layers—see text. (After Dieny, B. et al., *Appl. Phys. Lett.* 61, 2111, 1992. With permission.)

F-metals or alloys, they reported room temperature values of (in nanometer): $\lambda_{Py}^{\uparrow} = 4.6 \pm 0.3$, $\lambda_{Co}^{\uparrow} = 5.5 \pm 0.4$, and $\lambda_{Fe}^{\downarrow} = 1.5 \pm 0.2$, and, $\lambda_{Py}^{\uparrow} \leq 0.6$, $\lambda_{Co}^{\uparrow} \leq 0.6$, and $\lambda_{Fe}^{\downarrow} = 2.1 \pm 0.5$.

4.4.3.8 How Does Interface Roughness Affect the CIP-MR?

The published evidence suggests that the effects of interface roughness on the CIP-MR are not simple but depend upon both the particular pair of F- and N-metals studied, and the structure of the roughness. The first model of CIP-MR [29] assumed that the MR was dominated by the scattering of electrons at interfacial roughness. Attempts to deduce the effects of such roughness soon followed. Early studies gave what appeared to be conflicting results: greater roughness was claimed to increase the CIP-MR in Fe/Cr [53, 54, 57], but to decrease it in Py/Cu [52]. A detailed study of Fe/Cr [58] concluded that the specific structure of the roughness was important, with the CIP-MR magnitude increasing with increasing vertical roughness amplitude, or with decreasing lateral correlation length. For a more detailed discussion, see Ref. [14].

4.4.3.9 How Do Values of β_F from CIP-MR and DMR Compare?

In the CIP-MR, the difficulty of separating bulk and interface scattering contributions makes it hard to isolate the bulk (β_F) and interface ($\gamma_{F/N}$) scattering

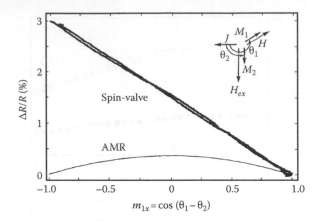

FIGURE 4.13 CIP-MR = ($\Delta R/R$) (%) (corrected for the AMR shown) versus the cosine of the angle between the magnetizations of the two NiFe layers in a Si/NiFe(6)/Cu(2.6)/NiFe(3)/FeMn(6)/Ag(2) EBSV with thicknesses in nanometer. Inset shows orientations of the current J, exchange field, Hex, and magnetizations Mi and $M2$. (After Dieny, B. et al., *Phys. Rev. B* 43, 1297, 1991. With permission.)

anisotropies. We know of no quantitative comparisons for the CIP-MR. Examples have been published of the inversion of the CIP-MR [44, 59, 60], where the DMR values predict such inversion (see Chapters 1 and 5 [1, 7]). Such examples show a qualitative understanding of the physics underlying the CIP-MR.

4.4.3.10 How Does the CIP-MR Vary with Angle?

For a sample in which the magnetization of one F-layer is held fixed, and the other allowed to rotate, the CIP-MR is expected to grow closer as $[1 - \cos(\theta_1 - \theta_2)]$ [1, 14], where $\theta_1 - \theta_2$ is the angle between the magnetizations of the two F-layers. Figure 4.13 shows such a variation for a Py-based EBSV with one Py layer pinned and the other rotated by a 10 Oe field. The CIP-MR data in Figure 4.13 are corrected for a modest AMR effect (shown by the curve labeled AMR). For more details, see Ref. [14].

4.4.3.11 How Does the CIP-MR Vary with Temperature?

As illustrated in Figure 4.6, for Fe/Cr and Co/Cu multilayers, the CIP-MR generally decreases monotonically with increasing temperature. Part of the decrease is due to the phonon-scattering driven increase in the bulk resistivity components of the sample resistance, R_P, which forms the denominator of the GMR. As explained in Section 4.3, the bulk contributions to R_P generally increase by roughly a factor of two between 4.2 K and room temperature (293 K). For Co/Cu, doubling R_P, and holding ΔR nearly constant, would account for most of the decrease in CIP-MR in Figure 4.6. For Fe/Cr, in contrast, the decrease is more rapid, suggesting that ΔR also decreases strongly with increasing T. Contributors to this latter decrease could include interband transitions [1] and scattering-driven intermixing of the two spin currents [61].

4.4.3.12 Can CIP-MR Be Increased by Modifying Top and Bottom Interfaces?

In the moderate thickness EBSVs used for devices, such as read heads, scattering from the interfaces at the bottom and top of the sample can affect the CIP-MR. Specifically, if such scattering can be made specular, holding other factors constant, one would expect to reduce the total R, and thus increase the CIP-MR. Several papers have reported enhanced CIP-MRs in samples with NiO and other top and bottom layers, and in samples subject to oxidation of the top layer. Most attributed these enhancements to increased specular reflection at the interfaces with such layers. A number of such studies are examined in Ref. [14], which concludes that the sources of the observed enhancements are not yet fully understood.

4.5 CPP-MR

In contrast to the CIP-MR, the CPP resistance R of a thin film with widths ~mm (giving $A \sim mm^2$) and thickness ~100 nm (giving "length" ~100 nm) is ~10^{-8} Ω, difficult to measure and much too small for devices. Area miniaturization, with sample widths <100 nm, is now raising CPP resistances to ~Ω, leading to device-potential and device-related research. Tunneling MR, which also uses the CPP geometry, is now the device preference for read heads and magnetic random access memory (MRAM) [62]. However, because of the large AR product of tunneling samples, there is a likely need for alternatives as A decreases still further. The CPP-MR with just metals is one candidate for the next generation of read heads. We begin this section with a discussion of the different techniques used to measure the CPP-MR, and their relative advantages and disadvantages.

4.5.1 MEASURING CPP-MR

Three techniques have been used to measure the CPP-MR. They are (1) short, wide samples between superconducting cross-strips [12], (2) micron- or nanosized pillars, and (3) electrodeposited nanowires [13].

1. The first technique [21, 63–65] involves a short-($\leq\mu$m) sample, sandwiched between wide (~1 mm) crossed superconducting (S) strips. Such S leads give lead resistances due only to the F/S interfaces, which can be measured separately [64]. The "capacitor-like" geometry gives uniform current flow [65]. Slater et al. [66] developed a way to measure CPP-MR using micron-sized top superconducting contacts on an otherwise mm² area sample. The disadvantages of superconducting leads are (1) the limitation to cryogenic temperatures (typically 4.2 K using the usual superconductor, Nb, which stays superconducting to only about 9 K [35]); and (2) the need for a high sensitivity, high precision, bridge to measure the typically very small sample resistances ($R \sim 10$ nΩ) [63, 67]. Cyrille et al. [68] and Highmore et al. [69] avoided the need for a sensitive bridge by using one hundred to five

hundred ~10 μm size samples in series. The advantages of superconducting leads are their ability to (a) achieve uniform current distributions and contacts with known and reproducible lead resistances and (b) combine a wide variety of metals with control of AP and P-states via the full range of available techniques.

2. The second technique involves patterning a multilayer into micronsized [70] or more recently, nanosized pillars (see, e.g., Ref. [71]). The advantage is the ability (a) to combine a wide variety of F- and N-metals and (b) to allow measurements of CPP-MR from cryogenic temperatures to room temperature. The disadvantages are difficulties in (a) achieving uniform current distributions [72], (b) reproducibly preparing nanosize samples, and (c) controlling, determining, and including contact resistances in equations of interest (simplified equations that neglect them can potentially give large uncertainties).

3. The third technique involves nanowires electrodeposited into nanopores in either plastics [73–75] or more ordered structures, such as Al_2O_3 [76]. The nanopore diameters can range from ~20 to 100 nm. The long, thin wire geometry gives a uniform current flow and large enough resistances for convenient measurement. Most published studies have the disadvantage of not knowing the number of wires. Contact resistance can be a problem, but long samples can make them unimportant. Recently, procedures have been developed to contact to just single nanowires [77, 78]. As yet, only a modest number of different F/N combinations (mostly Co/Cu, Fe/Cu, and Py/Cu) have been studied with this technique [13]. Its advantages are its simplicity and the ability to measure MRs from cryogenic to high temperatures. Its limitations include the difficulty in making very thin layers—interfaces are usually "thicker" than with other deposition techniques; difficulty in making EBSVs; and the need for care to minimize coupling to achieve P and AP-states at relatively small fields.

4.5.2 Some 2CSR and VF Equations for CPP-MR Analysis

For details of the theoretical analysis of CPP-MR, see Chapter 5 [1]. We briefly describe here, the simplest picture of the underlying physics and give some 2CSR [30–32] and VF [32] equations used by experimentalists to analyze CPP-MR data. The 2CSR equations hold when all layers are thinner than their spin-diffusion lengths. The more general VF equations apply to thicker layers.

4.5.2.1 2CSR Model and Equations

As noted above, if scattering of electrons transiting a multilayer does not mix the moment up and down currents, then a 2C model results. If, in addition, the spin-diffusion lengths in the F- and N-layers are much longer than the layer thicknesses, and spin-flipping at interfaces is negligible, then for free-electron Fermi surfaces and diffusive transport, it was quickly shown [30, 32] that the CPP-MR should be given by a 2CSR model in which the total specific resistance, AR(up) or AR(down) for each electron-moment direction is just the series sum of the specific resistance of each layer ($\rho_N^{\uparrow} t_N$, $\rho_N^{\downarrow} t_N$, $\rho_F^{\uparrow} t_F$, and $\rho_F^{\downarrow} t_F$) and the specific resistance of each interface ($AR_{F/N}^{\uparrow}$ and $AR_{F/N}^{\downarrow}$). As in Section 4.1, $\uparrow$ means that the electron moment is along the local F-moment, and $\downarrow$ means it is opposite to the local F-moment. For a simple $[F/N]_n$ multilayer, where each F- and N-layer is just repeated, in the AP-state, AR(up) and AR(down) are the same, and AR_{AP} takes a simple form, just half of the common value for each. Here, we give some general 2CSR model equations that we employ in later analyses. These use as unknowns, the three parameters: β_F, $\gamma_{F/N}$, and $2AR_{F/N}^*$.

We start first with simple $[F/N]_n$ or $[F/N]_n F$ multilayers and superconducting (S) leads. Such leads have the advantage that the contact resistance, $2AR_{S/F}$, can be measured independently [12, 64, 79, 80]. However, they give the following complications: For simple $[F/N]_n$ multilayers, the contact of the top N-layer with the top S-lead causes the proximity effect to turn the top N-layer superconducting, and there are only $(2n - 2)$ F/N interfaces. Conversely, for $[F/N]_n F$ multilayers, there are $n + 1$ F-layers. For simplicity, we neglect these complications in the following three equations used with S-contacts [12, 31]:

$$AR_{AP} = 2AR_{S/F} + n\left(\rho_N t_N + \rho_F^* t_F + 2AR_{F/N}^*\right) \qquad (4.6)$$

$$A\Delta R = \frac{n^2\left(\beta_F \rho_F^* t_F + \gamma_{F/N} 2AR_{F/N}^*\right)^2}{AR_{AP}} \qquad (4.7)$$

$$\sqrt{AR_{AP} A\Delta R} = n\left[\beta_F \rho_F^* t_F + \gamma_{F/N} 2AR_{F/N}^*\right] \qquad (4.8)$$

We turn next to nanowires. For long nanowires, the lead resistance can be neglected, giving the following 2CSR equations useful for determining the desired parameters [38, 81].

$$R_{AP} = n\left[\rho_F^* t_F + 2AR_{F/N}^* + \rho_N t_N\right] \qquad (4.9)$$

$$\left[\frac{\Delta R}{R_{AP}}\right]^{-1/2} = \left[\frac{\left(\rho_F^* t_F + 2AR_{F/N}^*\right)}{\left(\beta_F \rho_F^* t_F + \gamma_{F/N} 2AR_{F/N}^*\right)}\right] + \left[\frac{\rho_N t_N}{\left(\beta_F \rho_F^* t_F + \gamma_{F/N} 2AR_{F/N}^*\right)}\right] \qquad (4.10)$$

If the square root in Equation 4.10 is plotted versus t_N for the two different fixed thicknesses t_F of the F-metal, then, for each thickness of t_F, the data should give a straight line, and the ordinate value where the two lines cross should be just $(\beta_F)^{-1}$ [38].

Finally, for micropillars, the sample resistance is often not large compared to the lead resistances, which then cannot be neglected, and in many cases are not known. Determining their contribution is usually non-trivial. However, simply neglecting them might lead to significant uncertainties in derived parameters. Nanopillars reduce this problem, which decreases as the nanopillar diameters decrease.

4.5.2.2 Valet–Fert (VF) Model and Some Equations with Finite l_{sf}

VF [32] showed that the same assumptions (free-electron Fermi surfaces and dominant diffusive scattering) that give the 2CSR model when $l_{sf}^F \gg t_F$, and $l_{sf}^N \gg t_N$, give more general equations that depend upon l_{sf}^F and l_{sf}^N when $l_{sf}^F \leq t_F$ and $l_{sf}^N \leq t_N$. The most general VF equations are so complex that they require numerical analysis. But the equations sometimes simplify enough for a quantitative solution. Here, we give examples that have been used in determining l_{sf}^F.

We start with superconducting leads and a symmetric EBSV of the form Nb/FeMn/F(t_F)/N/F(t_F)/Nb, where FeMn is an AF used to pin the magnetization of the adjacent F-layer. In the limits $t_F \gg l_{sf}^F$ (e.g., $l_{sf}^{Py} \sim 5.5$ nm [36, 82]) and $t_N \ll l_{sf}^N$ (e.g., $l_{sf}^{Cu} > 200$nm [36]), VF theory gives $A\Delta R$ as a variant of Equation 4.6 where l_{sf}^F replaces t_F in both the numerator and denominator, and the denominator reduces to just the magnetically active central region between the l_{sf}^F thicknesses of the two F-layers [12]:

$$A\Delta R = \frac{\left(2\beta_F \rho_F^* l_{sf}^F + \gamma_{F/N} 2AR_{F/N}^*\right)^2}{\left[2\rho_F^* l_{sf}^F + \rho_N t_N + 2AR_{F/N}^*\right]} \tag{4.11}$$

Equation 4.11 is independent of t_F, thus predicting that $A\Delta R$ should become constant for $t_F \gg l_{sf}^F$.

We end with an equation used for nanowires. For a multilayer nanowire with a long $t_F \gg l_{sf}^F$, but $t_N \ll l_{sf}^N$, VF theory gives the equation [38, 81]:

$$\left[\frac{R_P}{\Delta R}\right] = \frac{[(1 - (\beta_F)^2]t_F}{2p(\beta_F)^2 l_{sf}^F} \tag{4.12}$$

Here, p is an assumed fraction of AP alignment in the nominal AP-state. As illustrated by the data in Figure 4.2, a multilayer with equal thickness F-layers usually does not reach the AP-state after saturation. From Equation 4.12, a plot of $[R_P/\Delta R]$ versus t_F should give a straight line with slope inversely proportional to l_{sf}^F.

4.5.3 CPP-MR Test of the 2CSR Model at 4.2 K

In Section 4.5.2, we listed some equations used to determine the CPP-MR parameters. Before describing what we know about those parameters, we present a set of confirmatory tests of the 2CSR that do not require explicit knowledge of specific parameters. Such tests have been made at 4.2 K using multilayers of Co with both Ag [31, 83] and Cu [84]. We focus here on Co/Ag. As we will see in Section 4.5.5.2, Co was a fortunate choice, as l_{sf}^{Co} is unusually long.

The tests involve comparing Equations 4.6 through 4.8, for multilayer data with the N-metal Ag and a dilute $Ag_{96}Sn_4$ alloy. Alloying with 4 at%

Sn gave $\rho_{AgSn} \sim 180$ nΩ m $\gg \rho_{Ag} \sim 10$ nΩ m [31], but AgSn gives only weak spin–orbit scattering [85] so that the alloy spin-diffusion length stays long. Equations 4.6 through 4.8 then make very explicit, directly testable, 2CSR model predictions about the different behaviors of the data for Co/Ag and Co/AgSn multilayers.

We rewrite Equation 4.6 for $[F/N]_n$ multilayers with the Co-thickness held fixed at $t_{Co} = 6$ nm and the total sample thickness fixed at $t_T = nt_F + nt_N$. We assume superconducting crossed leads and neglect differences between n and $n \pm 1$.

Equation 4.6 then becomes [31]:

$$AR_{AP} = 2AR_{Nb/Co} + \rho_N t_T + n\left[\left(\rho^*_{Co} - \rho_N\right)(6) + 2AR^*_{Co/N}\right] \quad (4.6')$$

Equation 4.6 predicts that plots of AR_{AP} versus n should give straight lines, with a much larger intercept for AgSn, but only about a third the slope (due to the negative contribution from $(\rho^*_F - \rho_N)t_F$). Figure 4.14 shows $AR_0 \approx AR_{AP}$ and AR_P -versus n for $[Co(6)/Ag]_n$ and $[Co(6)/AgSn]_n Co(6)$ multilayers. As predicted, the data for Co/Ag fall along straight lines, and the -intercept agrees with an independent prediction from measurements of $2AR_{Nb/Co}$ and ρ_{Ag} [31, 64]. The downward curvatures of the broken lines at small n for the Co/Ag data are due to the presence of only $2n - 2$ interfaces. The data for Co/AgSn have, the predicted, much larger intercept and smaller slope. Because

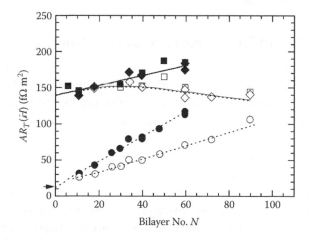

FIGURE 4.14 $AR_T(H)$ versus bilayer number N for $[Co(6)/Ag(t_{Ag})]_N$ (circles and broken lines) and $[Co(6)/AgSn(t_{AgSn})]_N Co(6)$ (squares or diamonds and solid or dashed curves) multilayers with fixed $t_{Co} = 6$ nm and fixed $t_T = 720$ nm (as closely as possible with integer N). The filled symbols are for AR_0 (approximating AR_{AP}) and the open symbols are for AR_P. The arrow to the lower left indicates the independently predicted value of the Co/Ag ordinate intercept. The dotted curves for the Co/Ag data are best fits using the 2CSR model. The solid and dotted curves for the Co/AgSn data are similar fits with two slightly different choices of parameters. The downturns of the dashed curves for Co/Ag at small N are an "end effect" due to the presence of only $2N - 1$ interfaces. (After Pratt Jr., W.P. et al., *J. Appl. Phys.* 73, 5326, 1993. With permission.)

of the extra Co layer, all of the n AgSn layers and $2n$ Co/AgSn interfaces contribute to AR_T.

Turning next to Equation 4.7, it stays unchanged, independent of whether t_T stays fixed, except for inserting Co for F. From Figure 4.14, due to ρ_{Ag} being small, for Co/Ag, AR_{AP} is approximately linear in n, and Equation 4.7 predicts that $A\Delta R$ should grow closely linearly with n. Thus, we expect AR_P to also be linear in n, which it is. In contrast, because ρ_{AgSn} is large, for Co/AgSn and small n, AR_{AP} is almost constant. Equation 4.7 then predicts that $A\Delta R$ should grow initially as n^2. This means that AR_P should be curved, falling further below AR_{AP} as n grows. Such behavior is illustrated by the dashed curve in Figure 4.14 and is shown explicitly in Refs. [31, 83].

Lastly, the most important test involves revisiting Equation 4.8 for fixed t_{Co} = 6 nm to give a square root quantity predicted to grow strictly linearly with n, with zero intercept and slope that is independent of ρ_N [31]. That is, for a dilute enough AgSn alloy, so that $\gamma_{Co/AgSn} = \gamma_{Co/Ag}$ and $2AR_{Co/AgSn} = 2AR_{Co/Ag}$, the 2CSR model predicts that the data for Ag and AgSn should fall along exactly the same straight line passing through the origin. Requiring a straight line to pass through zero is a much more stringent test of a model prediction than a straight line with unknown intercept. Despite the very different data for Co/Ag and Co/AgSn in Figures 4.14 and Figure 4.15 shows that the prediction is obeyed not only for the data of Figure 4.14 but also for independent data on Co/Ag samples with fixed $t_{Co} = t_{Ag}$ = 6 nm and variable t_T [84]. The combination of data in Figures 4.15 and 4.16 is strong evidence for the 2CSR model for Co and Ag.

4.5.4 How to Determine β_F, $\gamma_{F/N}$, AND $2AR*_{F/N}$

The tests described in Section 4.5.3 support the applicability of the 2CSR model to CPP data for multilayers with F- and N-metals for which l_{sf}^F and l_{sf}^N are long. For simple $[F/N]_n$ multilayers with layer thicknesses $t_F \ll l_{sf}^F$ and $t_N < l_{sf}^N$, and where ρ_F, ρ_N, and (with superconducting leads) $2AR_{S/F}$ have been separately measured, there are only three unknown parameters: β_F, $\gamma_{F/N}$, and $2AR_{F/N}^*$. Determining these parameters requires -measuring AR_{AP} and AR_P on multilayers with systematic variations of t_F, t_N, and n, and analyzing the data using equations like Equations 4.6 through 4.10. A detailed study of Co/Ag with superconducting leads, using several sets of multilayers, is given in Ref. [65]. Comparable studies using nanowires are described in Refs. [38, 81]. If the value of β_F for a given F-metal or alloy has already been determined in the same laboratory, then, in principle, $2AR_{F/N}^*$ and $\gamma_{F/N}$ could be determined simply from either measurements of AR_{AP} and AR_P versus n for $[F/N]_n$ multilayers with fixed t_F, or measurements of AR_{AP} and $A\Delta R$ on EBSVs. In practice, uncertainty in AR_{AP} for the multilayers and fluctuations in AR_{AP} for the EBSVs (due to the usually large resistivities of thick F-metal and AF layers), limit the reliability of each procedure alone. For better results, recent determinations of $2AR_{F/N}^*$ and $\gamma_{F/N}$ combined measurements on $[F/N]_n$ multilayers and EBSVs—for example, Refs. [86, 87].

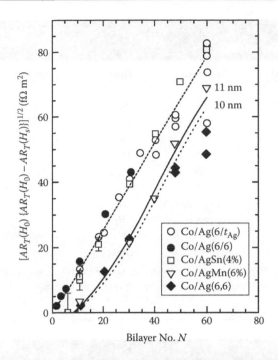

FIGURE 4.15 $\sqrt{AR_o(AR_o - AR_p)}$ – versus bilayer number N for $[Co(6\ nm)/N(t_x)]_N$ multilayers with fixed total thickness $t = 720$ nm, except for the samples indicated by filled circles, which had the form $[Co(6\ nm)/Ag(6\ nm)]_N$. The spin-diffusion lengths, l_{sf}^N, of Ag and AgSn are long enough so that the data stayed (to within uncertainties) on the straight line passing through the origin predicted by Equation 4.12. In contrast, the spin-diffusion lengths of AgMn and AgPt are short enough to cause deviations from this line. The solid and broken curves passing through the AgPt and AgMn data are fits using VF theory. The deduced values of l_{sf}^N are given in Table 4.3. (After Bass, J. et al., *Mater. Sci. Eng. B* 31, 77, 1995. With permission.)

4.5.4.1 β_F from CPP-MR for Elemental *F*-Metals: Co, Fe, Ni

The values of β_F for elemental *F*-metals are not necessarily intrinsic. While the band structure must play a role, an important role may also be played by the scatterers in the given *F*-metal. At 4.2 K, the dominant scatterers in sputtered, -evaporated, or electrodeposited nominally elemental *F*-metals are not phonons or other electrons, but a usually unknown collection of "dirt," composed of defects (grain boundaries and dislocations) and residual impurities. Even at room temperature, 293 K, the scattering from such "dirt" is typically comparable to that from phonons. Thus, one cannot count on β_F being intrinsic. To analyze new CPP-MR data involving a given *F*-metal, it is thus safest to have a value of β_F derived from multilayers prepared in the same way as those to be analyzed. Despite this caveat, Table 4.1 shows that the values of β_{Co}, the only *F*-metal measured by several different investigators, span a surprisingly narrow range, 0.38–0.48 (except for V-grooves), at low T (4.2–20 K) whether determined from the more reliable R_0 (see Figure 4.2) or from R_{peak}. Moreover, values of β_{Co} obtained by the same

TABLE 4.1
β_F for the Elemental F-Metals: Co, Fe, and Ni

F-Metal	β_F	T (K)	Technique	Field	ρ_F (nΩ m)	References
Co	0.48 ± 0.06	4.2	S	H_0	68	[65]
Co	0.46 ± 0.05	4.2	S	H_0	58	[12, 88]
Co	0.38 ± 0.06	4.2	S	H_{peak}	58	[12, 88]
Co	0.48 ± 0.04	4.2	S	H_0	24	[12, 89]
Co	0.49 ± 0.08	20	NW	H_{peak}	400	[39]
Co	0.43 ± 0.08	300	NW	H_{peak}	520	[39]
Co	0.36 ± 0.04	77,300	NW	H_{peak}	160	[39, 81, 90]
Co	0.27 ± 0.05	4.2	V-groove	H_0	53	[37]
Co	0.26 ± 0.05	300	V-groove	H_0		[37]
Fe	0.78 ± 0.05	4.2	S	H_0	38	[91]
Ni	0.14 ± 0.02	4.2	S	Other	33	[92]

T_0 his table lists the F-metal, β_F, the temperature T (K) of the measurement, the technique used (S, superconducting leads; NW, nanowires), the estimated AP-state (R_0 or R_{peak}—see Figure 4.2), the measured or estimated F residual resistivity, ρ_F, and the reference.

TABLE 4.2
β_F at 4.2 K for F-Based Alloys from DMR [34] and CPP-MR

F-Host	Imp	β_F (DMR)	References	β_F (CPP-MR)	References
Ni	**Fe**	**+0.83 ± 0.1**	**C&F**	**+0.65 ± 0.1**	**[93]**
				$+0.76 \pm 0.07$	[94]
Ni	FeCo			$\sim$+0.82	[95]
Ni	**Cu**	**+0.53 ± 0.05**	**C&F**	**+0.19 ± 0.04**	**[40]**
Ni	Cr	-0.43 ± 0.1	C&F	-0.13 ± 0.01	[40]
Ni	Cr	-0.43 ± 0.1	C&F	-0.35 ± 0.1	[95]
Co	**Cr**	**−0.5**	**C&F**	**−0.10 ± 0.02**	**[40]**
Co	Mn	-0.1	C&F	-0.03 ± 0.03	[40]
Co	**Fe**	**+0.85 ± 0.1**	**C&F**	**+0.65 ± 0.05**	**[96]**
Fe	Cr	-0.6 ± 0.1	C&F	-0.22 ± 0.06	[40]
Fe	V	-0.78 ± 0.05	C&F	-0.11 ± 0.03	[40, 97]

The four cases β_F and l_{sf}^N where derived simultaneously are given in bold. Listed are the host F-metal, the impurity, β_F from DMR and its reference, and β_F from CPP-MR and its reference.

investigator decrease only slightly from 4.2 to 300 K [37–39]. One interpretation is that the defect, impurity, and phonon scatterers happen to give anisotropies similar to that of the band structure.

4.5.4.2 Compare β_F for F-Alloys from CPP-MR and DMR

If the concentration of a given impurity in a given host F-metal is large enough to dominate scattering, then β_F for that alloy should be intrinsic. As noted in Chapter 1 [7], in the 1970s–1980s, studies of deviations from Matthiessen's rule (DMR) in three-component F-based alloys gave estimates of β_F for a range of F-alloys [33]. We examine some of these alloys.

Table 4.2 compares values of β_F derived from CPP-MR experiments with those found from studies of impurity- and temperature-driven DMRs [33]. Unfortunately, most of the CPP-MR values listed were derived from the 2CSR model, which assumes $l_{sf}^F = \infty$ in the F-alloy. Since values of l_{sf}^F in the host F-metals (see Figure 4.18) are finite, and because adding impurities should reduce l_{sf}^F (see Figure 4.18), such values are unlikely to be reliable in magnitude. We, thus, list them without bolding. All of these values agree with DMR results in sign, but disagree in -magnitude. The four cases where values of β_F and l_{sf}^F were derived simultaneously are given in bold. The three that can be directly compared with DMR values agree with them reasonably well.

4.5.5 Spin-Flipping Within N- and F-Metals

In this section, we first briefly review how values of l_{sf}^N in "pure" N-metals and alloys, and of l_{sf}^F in "pure" F-metals and alloys, are found from CPP-MR measurements. We then provide examples of the results obtained. More complete information about how l_{sf}^N and l_{sf}^F are determined from both CPP-MR and alternative techniques, and extensive lists of values for each, are given in a recent review [36].

Just as β_F is "intrinsic" in an F-metal only when the main scatterer is known, l_{sf}^N and l_{sf}^F are "intrinsic" only when the main scatterer in the N- or F-metal (or alloys) is known. For this reason, N-based alloys were first used to test the CPP-MR techniques for finding l_{sf}^N. We begin with such tests, which allow quantitative comparisons with determinations from completely independent information.

4.5.5.1 Determining l_{sf}^N

Figures 4.14 and 4.15 showed that data for $[Co/Ag]_n$ and $[Co/AgSn]_n$ multilayers with fixed t_{Co} and long l_{sf}^N agreed with the 2CSR model prediction that a plot of the square root in Equation 4.8 versus n should give a straight line passing through the origin. If, instead, l_{sf}^N is shortened, by adding an impurity that flips electron spins by spin–orbit or spin–spin interactions, the data should fall below this line, by a larger fraction the smaller is n. Yang et al. [98] showed how such deviations can be used to find l_{sf}^N for dilute alloys. Figure 4.15 shows such data for AgPt (strong spin–orbit interaction [85]) and AgMn (strong spin–spin interaction [61]). Table 4.3 compares selected values from CPP-MR with electron spin-resonance (ESR) measurements of spin–orbit scattering [85] for nonmagnetic impurities, or separate calculations for the magnetic impurity Mn [61]. More comparisons are given in Ref. [36]. The good agreements between CPP and ESR result in further support for using VF theory for such data.

With the quantitative support of Table 4.3 for the VF theory in hand, we turn to another technique developed in Refs. [99, 102] to measure l_{sf}^N in nominally "pure" N-metals or dilute N-based alloys. In this case, a thickness t_N of the N-metal (or alloy) of interest is sputtered into the middle of an EBSV of the form: FeMn(8)/Py(24)/Cu(10 or 20)/N(t_N)/Cu(10 or 20)/Py(24), and ln $(A\Delta R)$ is plotted against t_N. If no spin-flipping occurs within N, $A\Delta R$ should

TABLE 4.3
Comparing Alloy Values of l_{sf}^N from CPP-MR and ESR

Alloy	Tech.	l_{sf}^N (CPP)	l_{sf}^N (ESR)	References
Ag(4% Sn)	ML	≈7		[98]
Ag(6% Pt)	ML	≈10	≈7	[98]
Ag(6% Mn)	ML	≈11	≈12*	[98]
Cu(6% Pt)	ML	≈8	≈7	[98]
Cu(6% Pt)	SV	11 ± 3	≈7	[99]
Cu(22.7% Ni)	ML	7.5	6.9	[100]
Cu(22.7% Ni)	SV	8.2 ± 0.6	7.4	[101]

This table specifies the alloy, the technique used (multilayer [ML] or EBSV [SV]), and the values of l_{sf}^N in nm from CPP-MR and ESR [36]. The superscript "*" indicates that the listed value of l_{sf}^N was calculated for spin–spin scattering in Ref. [61].

be given by Equation 4.11, except that the denominator would now also contain the additional terms $\rho_N t_N + 2AR_{N/Cu}$. As t_N grows, $A\Delta R$ should decrease as a linear function of t_N in the denominator. If, however, spin-flipping occurs in N, on a length scale l_{sf}^N, then $A\Delta R$ should decrease exponentially as $\exp(-t_N/l_{sf}^N)$. If this decrease dominates, a plot of $\ln(A\Delta R)$ versus t_N should be close to a straight line with a slope determined by l_{sf}^N. In practice, data such as those in Figure 4.16 are fit numerically with VF theory, including the contributions from both phenomena just described. The values of l_{sf}^N for the metals shown in Figure 4.16 are $l_{sf}^{Ag} \geq 40\,\text{nm}$, $l_{sf}^V \geq 40\,\text{nm}$, $l_{sf}^{Nb} = 25_{-5}^{+\infty}\,\text{nm}$, $l_{sf}^W = 4.8 \pm 1\,\text{nm}$, and $l_{sf}^{CuPt} = 11 \pm 3\,\text{nm}$. A complete listing of values of l_{sf}^N for a variety of metals and alloys, also measured by other techniques, is given in Ref. [36].

4.5.5.2 Determining l_{sf}^F

Values of l_{sf}^F have also been determined by two techniques, one using superconducting leads and one using nanowires.

4.5.5.2.1 S Leads, EBSVs

With superconducting leads, the sample is a symmetric EBSV, with two equal thicknesses t_F of the F-layers, one of which is pinned by the AF FeMn. If $l_{sf}^F = \infty$, $A\Delta R$ grows approximately linearly with t_F, as illustrated by the dotted line in Figure 4.17 for CoFe. If, in contrast, l_{sf}^F is finite, $A\Delta R$ breaks away from the $l_{sf}^F = \infty$ line approximately at $t_F = l_{sf}^F$, and then saturates at a constant value for $t_F \gg l_{sf}^F$, as predicted in Equation 4.11. Examples of data are shown in Figure 4.17 for CoFe and Co [80]. Analysis of the behavior of Co in Figure 4.17 might be complicated by spin-flipping at the Co/Cu interfaces (see Sections 4.5.6.4 and 4.5.9).

4.5.5.2.2 Nanowires

For nanowires, the samples are $[F/Cu]_n$ multilayers with F = Co or Py. For F = Co, all layer pairs were identical, and the equation of interest was

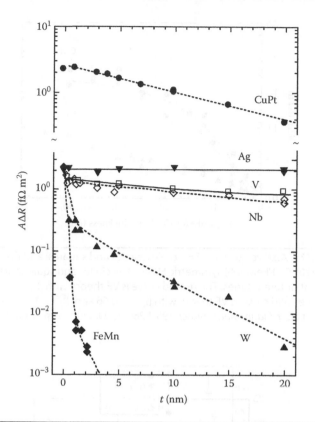

FIGURE 4.16 $A\Delta R$ (log scale) versus t_X (in nm) for single layer inserts X of CuPt, FeMn, Ag, V, Nb, and W into EBSVs of the form [FeMn/(8)/Py(24)/Cu(10)/X/Cu(10)/ Py(24)]. The solid curves are fits assuming $l_{sf}^N = \infty$. The dashed curves are fits to the finite values of l_{sf}^N as given in the text. (After Park, W. et al., *Phys. Rev. B* 62, 1178, 2000. With permission.)

Equation 4.12. In Ref. [81], plots of $(R_P/\Delta R)$ versus t_{Co} gave straight lines, the slopes of which were used to infer values of $l_{sf}^{Co} \sim 59 \pm 18\,\mathrm{nm}$ at 77 K and $l_{sf}^{Co} \sim 38 \pm 12\,\mathrm{nm}$ at 300 K [38]. The data for Co at 4.2 K in Figure 4.17 are consistent with this relatively long value at 77 K.

For F = Py [13, 24], the nanowires had two different N-layer thicknesses to better set AP-states. Here, using Equation 4.11 with p = 1 gave $l_{sf}^{Py} = 4.3 \pm 1\,\mathrm{nm}$ at 77 K, consistent with the value of 5.5 ± 1 at 4.2 K from Ref. [82] found with superconducting contacts.

Figure 4.18 collects together values of l_{sf}^F for a set of alloys plus Co, Ni, and Fe, all plotted versus $1/\rho_F$ [38, 80, 82, 91, 92, 95, 96]. This graph approximates a plot of l_{sf}^F versus λ_F, where λ_F is the effective mean-free-path. Note that l_{sf}^{Co} is much longer than would have been extrapolated from the alloy data; the residual resistivity of Co is likely dominated by stacking faults, which may be weak spin-flippers.

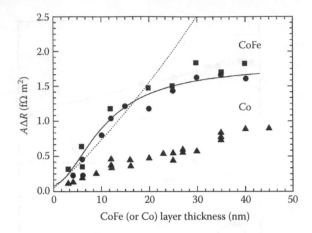

FIGURE 4.17 $A\Delta R$ versus t_F for $F = Co_{91}Fe_9$ (circles and squares) and Co (triangles) in $[F(t_F)/Cu(20)/F(t_F)/FeMn(8)]$ symmetric EBSVs. The circles and squares are CoFe runs made at different times. The dotted curve is VF theory with $\beta_{CoFe} = 0.66$ and $l_{sf}^{CoFe} = \infty$; the solid curve is VF theory with $\beta_{CoFe} = 0.66$ and $l_{sf}^{CoFe} = 12$ nm. (After Reilly, A.C. et al., *J. Magn. Magn. Mater.* 195, L269, 1999. With permission.)

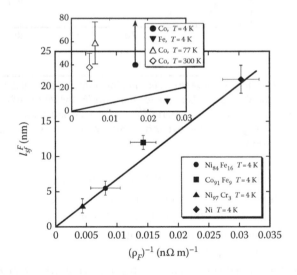

FIGURE 4.18 I_{sf}^F versus $1/\rho_F$ for several F-metals and F-based alloys. The solid line (repeated in the inset) is a straight line fit to the data in the main figure, constrained to go through zero. Note that the Ni symbol is in the main figure, but the Fe and Co symbols are in the insert. References: Co [38, 96]; Fe [91]; Ni [92]; Py [82]; CoFe [80]; NiCr [95]. (After Bass, J. et al., *J. Phys. Condens. Matter* 19, 183201, 2007. With permission.)

4.5.6 INTERFACES

In this section, we consider first $2AR$ for interfaces, then spin-scattering anisotropy, $\gamma_{F/N}$, for F/N interfaces, then spin-flipping at interfaces, and finally potential effects of interface roughness.

TABLE 4.4
$2AR_{N1/N2}$ or $2AR^*_{F/N}$ at 4.2 K

Metals	2AR(exp)	2AR (perf.)	2AR (50–50)	References
Ag/Au	0.1	0.09	0.13	[103, 104]
Co/Cu*	1.0	0.9	1.1	[12, 104]
Fe/Cr*	1.6	1.7	1.5	[104, 105]
Pd/Pt	0.3	0.4	0.4	[104, 106]
Pd/Ir	1.0	1.1	1.1	[104]
Au/Cu	0.3	0.45	0.7	[103, 107, 108]
Ag/Cu	0.1	0.45	0.6	[103, 108]
Pd/Cu	0.9	1.5	1.6	[108, 109]
Pd/Ag	0.7	1.6	2.0	[108]
Pd/Au	0.5	1.7	1.9	[108]
Pt/Cu	0.9			[109]
Co/Ag*	1.2			[65]
Ni/Cu*	0.36			[92]
W/Cu	3.1			[99]
Nb/Cu	2.2			[99]
V/Cu	2.3			[99]
Co/Ru*	~1			[110]
Co/Al*	11			[87]
Fe/Al*	8			[87]
Py/Al*	9			[87]
Py/Cu*	1			[26]
Fe/Cu*	1.5			[91]
Co/Pt*	1.7			[86]
Py/Pd*	0.4			[86]
Fe/Nb*	2.7			[86]
Fe/V*	1.4.			[86]
Co/Cr*	1.0			[40]
Py/Cr	1.9			[40]

Values of $2AR_{N1/N2}$ or $2AR_{FN/^*}$ are in fΩ m^2 and are rounded to significant figures. The first five pairs are lattice-matched ($\Delta a/a \leq 1\%$–2%); the other pairs are not.
* Indicates $2AR$. Orientations are (111) for fcc and (011) for bcc.

4.5.6.1 2AR for Interfaces

Procedures for determining $2AR^*$ for F/N interfaces are usually complex, as outlined in Section 4.5.4. In contrast, procedures for determining $2AR$ for $N1/N2$ interfaces are simple enough to be described in more detail below. We will see in Table 4.4 that values of both $2AR^*$ and $2AR$ for lattice-matched ($\Delta a/a \leq 1\%$–2%) pairs with the same crystal structure agree surprisingly well with calculations that involve no adjustable parameters.

An issue raised at the start of this review was whether the CPP-MR is dominated by scattering at F/N interfaces or in the bulk of the F-metal. Equations 4.7 and 4.10 show that their -relative importance is determined by comparing the products $\beta_F \rho_F^* t_F$ for bulk and $\gamma_{F/N} 2AR_{F/N}^*$ at interfaces.

Tables 4.1, 4.2, 4.4, and 4.5 give the information needed to evaluate these quantities for given layer thicknesses. Usually, for thin t_F layers, interface scattering dominates, but sometimes thick F-layers can make bulk scattering predominate.

4.5.6.2 Interface Specific Resistances and Comparisons with Calculations

We start with $2AR_{N1/N2}$, values of which have been determined using two techniques.

1. In the first, multilayers of a fixed total thickness, t_T, are sequentially subdivided into more and more subunits with equal thicknesses $t_{N1} = t_{N2} = t_T/2n$ of $N1$ and $N2$. Such a procedure leaves the total thicknesses of $N1$ and $N2$ fixed at $t_T/2$, but adds more interfaces as t_n decreases. So long as t_n is larger than the interface thickness, t_I, the total specific resistance, AR_T, should grow linearly with n as [103]:

$$AR_T = 2AR_{S/Co} + 2\rho_{Co}(10) + AR_{Co/N1} + AR_{Co/N2}$$
$$+ \rho_{N1}\left(\frac{t_T}{2}\right) + \rho_{N2}\left(\frac{t_T}{2}\right) - AR_{N1/N2} + 2nAR_{N1/N2} \quad (4.13)$$

Here, we have assumed superconducting (S) leads, and 10 nm thick Co layers outside the nonmagnetic multilayer to eliminate proximity

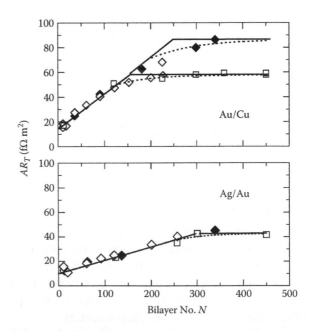

FIGURE 4.19 AR versus N for $[N1/N2]_N = [Au/Cu]_N$ and $[Ag/Au]_N$ multilayers with equal thickness ($t_{N1} = t_{N2} = t_T/2N$) layers of $N1$ and $N2$, sputtered at standard (diamonds) or half (squares) rates. Open symbols are for fixed total sample thickness t_T = 360 nm; filled symbols are for t_T = 540 nm. (After Henry, L.L. et al., *Phys. Rev. B* 54, 12336, 1996. With permission.)

effects. Figure 4.19 [103] shows examples of such data for Au/Cu and Ag/Au. The data initially grow linearly with n, but eventually, saturate to an approximation to a uniform alloy when t_n is less than the interface thickness [103].

2. In the second, a multilayer of N-layers of fixed thicknesses $t_{N1} = t_{N2}$ is inserted into the center of an EBSV [99]. Again, AR_T should grow linearly with n. But now this growth also contains a term $n(\rho_{N1}t_{N1} + \rho_{N2}t_{N2})$ that must be corrected for. This correction makes this technique more uncertain than the other. On the other hand, as described in Ref. [99] and Section 4.5.6.4 below, this technique can also be used to derive a spin-flip probability at the $N1/N2$ interface.

Values of $2AR$ (or $2AR^*$) found with these techniques are collected together in Table 4.4 for elemental F- and N-metals (but including also F = Py), along with calculations for several of the interfaces. Additional values for some F-alloys are given in Ref. [40]. Atomic planes within layers are closest-packed, corresponding to (111) for face-centered-cubic (fcc) and (011) for body-centered-cubic (bcc) layers. Of special interest are lattice-matched metal pairs, with the same crystal structure and lattice parameters, a, almost the same (i.e., $\Delta a/a \leq 1\%$–2%). Then $2AR$ or $2AR^*$ can be calculated with no adjustable parameters for a given interfacial structure [104, 111–114] as described in Chapter 5 [1]. The simplest structure is a perfectly flat interface. The next simplest are N-layers of a 50%–50% random alloy. Table 4.4 contains experimental and calculated values of $2AR$ (or $2AR^*$ for F/N interfaces) for both closely matched pairs (first five) and unmatched pairs (all the others). For the closely matched pairs, the experimental values fall close to (often between) the values calculated for perfect interfaces and for two MLs of a 50%–50% alloy. For the nonlattice-matched pairs, the calculations are in the right ballpark, but 50% to a factor of 4 too large. A separate test comparing experimental values and calculations of residual resistivities for dilute alloys [115] found the results to be sensitive to lattice strains, which provides a plausible explanation for much of the discrepancies between data and calculations for nonlattice-matched pairs.

4.5.6.3 $\gamma_{F/N}$ for F/N Interfaces

Table 4.5 lists values of $\gamma_{F/N}$ for F/N interfaces with F and N -elemental metals, except for the well-studied alloy Py. Inferred values for other F-alloys with N = Cu or Cr are given in Ref. [40]. The same calculations that gave the values of $2AR^*$ in Table 4.4, gave the following values for $\gamma_{Co/Cu}$ and $\gamma_{Fe/Cr}$: $\gamma_{Co/Cu}(\text{perf}) = +0.60$; $\gamma_{Co/Cu}(50–50) = +0.68$ and $\gamma_{Fe/Cr}(\text{perf}) = -0.49$; $\gamma_{Fe/Cr}(50–50) = -0.35$. Our choices for "best" experimental values from Table 4.5 are $\gamma_{Co/Cu} = +0.77 \pm 0.04$ and $\gamma_{Fe/Cr} = -(0.7 \pm 0.15)$. For Co/Cu, the calculated values are 80%–90% of the experimental one, with the disordered value closer to experiment. In contrast, the Fe/Cr values are only 50%–70% of the experimental one, with the ordered value closer to experiment.

TABLE 4.5
Values of $\gamma^{F/N}$

F/N	$\gamma^{F/N}$	T (K)	Tech.	R	References
Co/Cu	+0.77 ± 0.04	4.2	S	R_0	[12]
Co/Cu	+0.85 ± 0.15	77	NW		[81, 90]
Co/Cu	+0.53 ± 0.08	4.2	V-gr.	R_0	[37]
Co/Cu	+0.46 ± 008	300	V-gr.	R_0	[37]
Co/Cu	+0.55 ± 0.07*	20	NW	R_{peak}	[39]
Co/Cu	+0.40 ± 0.10*	300	NW	R_{peak}	[39]
Co/Ag	+0.84 ± 0.03	4.2	S	R_0	[65]
Fe/Cr	−(0.7 ± 0.15)	4.2	S	R_0	[105]
Fe/Cu	+0.55 ± 0.07	4.2	S	R_0	[91]
Ni/Cu	+0.29 ± 0.15	4.2	S	Other	[92]
Py/Cu	+0.7 ± 0.1	4.2	S	Other	[79]
Co/Ru	−0.2	4.2	S	Other	[110]
Co/Al	+0.1 ± 0.08	4.2	S	R_0	[87]
Fe/Al	+0.05 ± 0.02	4.2	S	R_0	[87]
Py/Al	+0.025 ± 0.01	4.2	S	R_0	[87]
$Co_{91}Fe_9$/Al	+0.1 ± 0.01	4.2	S	R_0	[87]
Co/Pt	+0.38 ± 0.06	4.2	S	R_0	[86]
Py/Pd	+0.41 ± 0.14	4.2	S	R_0	[86]
Fe/V	−(0.27 ± 0.08)	4.2	S	R_0	[86]
Fe/Nb	−(~0.14)	4.2	S	R_0	[86]
Co/Cr	−(0.24 ± 0.17)	4.2	S	Other	[40]
Py/Cr	−(0.03 ± 0.03)	4.2	S	Other	[40]

This table lists the *F/N* pair, $\gamma^{F/N}$, the measuring temperature, the technique used, the *AP*-state (or a note that another method was used), and the reference. The * by Co/Cu values from Ref. [39] means that we chose the larger of two alternatives.

4.5.6.4 Spin-Flipping at Interfaces from CPP-MR

4.5.6.4.1 N1/N2 Interfaces

In Section 4.5.6.2, we described how to determine $2AR_{N1/N2}$ by measuring AR after inserting into the middle of the Cu layer of Py/Cu/Py EBSVs *n*-layers of a multilayer of two *N*-metals $[N1/N2]_n$, each of fixed thickness (say $t_{N1} = t_{N2} = 3$ nm). Measurements of $A\Delta R$ on these same EBSVs give quite different information. They allow determination of the probability of spin-flipping at the *N1/N2* interface, characterized by a parameter $\delta_{N1/N2}$ as described in Refs. [99, 102]. Just as spin-flipping in an inserted bulk metal of thickness t_N led to the exponential decrease in $A\Delta R$ with increasing t_N in Figure 4.16, spin-flipping at the *F/N* interfaces in an inserted $[F(3)/N(3)]_n$ multilayer should lead to an exponential decrease in $A\Delta R$ with increasing *n*. Just as the data in Figure 4.16 with ML inserts allowed derivation of l_{sf}^N, similar data [99] for multilayer inserts allow the derivation of the spin-flipping probability $\delta_{N1/N2}$. Published data are given in Ref. [36]. We list only a few representative examples: $\delta_{N1/N2}$: ~ 0 for Cu/Ag; 0.07 ± 0.04 for Cu/V; 0.19 ± 0.05 for Cu/Nb, and 0.96 ± 0.1 for Cu/W [99].

The physics underlying the values of $\delta_{N1/N2}$ is not yet established. The values obtained are associated with interfaces that are intermixed, typically to 3–4 ML [103]. Whether similar values would be obtained for "perfect" interfaces (flat and with $N1$ and $N2$ completely separated) is yet to be determined. Theoretical guidance is needed.

4.5.6.4.2 F/N Interfaces

A general procedure for determining $\delta_{F/N}$ is described in Ref. [116] and has been applied to Co/Cu [116], Co/Ni [117], $Co_{90}Fe_{10}$/Cu [118], Co/Ru [119], Co/Ag [120], and Co/Pt [121]. It consists of inserting into the center of a symmetric Py-based double EBSV, a symmetric $[F(3)/N(t)]_n F(3)$ multilayer, where t is chosen to give *ferromagnetic coupling* between adjacent F-layers. A common coercive field of the two pinned Py-layers that is much larger than the coercive field of the ferromagnetically coupled F-layers should then give a simple hysteresis curve with well-defined AP- and P-states. If $\delta_{F/N}$ is nonzero, a plot of AΔR vs n should give data that tends toward saturation as n grows. Fitting the data with separately determined parameters allows isolation of the only unknown, $\delta_{F/N}$. For most pairs studied so far, values of $\delta_{F/N}$ range up to ~0.35 [116–120], in general agreement with ones estimated earlier by Manchon et al. [122] by comparing data on CPP-MR and spin-torques. Nguyen et al. [121] reported a larger value, $\delta_{Co/Pt} = 0.9^{+0.5}_{-0.2}$, for Co/Pt. Such a large value for $\delta_{Co/Pt}$ is supported by two other studies [123, 124].

4.5.6.5 Interface Roughness

Most studies of CPP-MR simply take the interfaces as they are formed. As described in Section 4.5.6.2, and Table 4.4, the values of $2AR$ for several intermixed interfaces of lattice-matched metals agree surprisingly well with no-free-parameter calculations. In several cases, the calculated values are not sensitive to interfacial intermixing at the interface. However, the calculations do not take account of the physical roughness of the interfaces—that is both the perfect interfaces and the boundaries of the intermixed interfaces are assumed to be flat.

Three attempts have been made to explicitly determine the effects of increasing interfacial roughness, with conflicting results. Increasing roughness was produced by increasing the sputtering pressure, and inferred from observations of broadening and decreasing of the heights of low angle x-ray peaks. In Co/Ag [125], and one study of Fe/Cr [105], the CPP-MR was reported to decrease with increasing interfacial roughness. In another study of Fe/Cr [126], the CPP-MR was reported to increase. The value of $2AR^*_{Fe/Cr}$ found in Ref. [105] falls between those calculated for perfect or two ML of intermixed 50%–50% Fe(Cr) alloy interfaces, and the calculation predicts a decrease in $2AR^*_{Fe/Cr}$. with increasing intermixing.

4.5.7 ATTEMPTS TO ENHANCE CPP-AR AND CPP-MR

For standard multilayer nanopillars with "length" <100 nm and area $A \sim 10^{-2}$ μm^2, the total resistance is typically ~1 Ω, still too small for impedance matching in devices, such as read heads. Moreover, the CPP-MRs of standard samples subject to the dimensional constraints of read heads are smaller than desired. Several techniques have been tried to address these problems. We discuss three: (1) lamination of the F-layer to increase the CPP-MR, (2) current confinement via nano-oxide layers to increase AR without decreasing the CPP-MR, and (3) search for half-metallic F-metals.

4.5.7.1 Laminated F-Layers

If the enhanced interface specific resistance, $2AR^*_{F/N}$ for an F/N pair is large, and the scattering asymmetry, $\gamma_{F/N}$, of the F/N interface is larger than the asymmetry, β_F, of the F-layer, then a potential way to increase both $A\Delta R$ and the CPP-MR is to laminate the F-layers by inserting very thin N-layers between very thin F-layers. Very thin N-layers ensure that the thin F-layers are ferromagnetically exchange-coupled so that they behave like a single F-layer, but with enhanced properties. Several groups found lamination to increase AΔR and/or the CPP-MR [127–130]. By themselves, the increases were not large enough to justify the extra processing required for lamination. Spin-flipping at interfaces could also weaken the effect.

4.5.7.2 Current Confinement via Nano-Oxide Layers

Fujiwara and Mankey [131] first proposed to increase the resistance, R (and perhaps also ΔR), of a micropillar EBSV by inserting into the central N-layer a thin nano-oxide layer (NOL), composed of an insulator containing a set of small conducting channels (pinholes). This insertion greatly reduces the area through which the current passes through the NOL (Current-Confined Paths (CCP)), thus increasing R. Subsequent studies showed that NOLs can increase not only R but also ΔR and the CPP-MR (see, e.g., [132–134]). Issues still being studied include producing channels with uniform areas and uniform distributions, both needed for best performance [see, e.g., Ref. 15]. NOLs also become relatively less effective as EBSV diameters go below ~40 nm [135].

4.5.7.3 Search for Half-Metallic F-Layers

In principle, inclusion of a true half-metallic F-layer, in which current is carried by only one electron spin, should give an infinite CPP-MR. In practice, F-alloys that are predicted to be half-metallic when they contain no defects, have yet to give the hoped-for CPP-MRs, perhaps because of defects leading to incomplete half-metallic behavior, along with changes in properties of such alloys at their interfaces. The first study of a NiMnSb-based Heusler alloy [136] gave CPP-MRs larger than the equivalent CIP-MRs but even at 4.2 K, they were only ~10%. Subsequently, studies of a variety of Heusler alloys and different N-metals have given larger CPP-MRs at 293 K than this first study did at 4.2 K (for a comprehensive listing see Ref. [15]). Here we mention just a few recent studies with especially large CPP-MRs and AΔR at 293 K. Unfortunately, to maximize crystalline perfection, the largest values required annealing to temperatures well above the ~573 K, which is likely an upper limit for minimizing the degrading of magnetic properties of the samples and shields. In 2011, an EBSV annealed to only 523 K [137] gave CPP-MRs up to 12% with AΔR ~4 f Ωm^2. Annealing a pseudo-SV to 673 K gave

CPP-MR ~42% and AΔR ~9.5 f Ωm^2. [138]. In 2014, a pseudo-SV annealed to 903 K gave CPP-MR ~200% and AΔR ~21.5 f Ωm^2 [139]. In contrast, annealing to only 623 K reduced the CPP-MR to ~26% and AΔR to ~11 f Ωm^2 [139].

4.5.8 TEMPERATURE DEPENDENCES OF CPP-MR PARAMETERS

Three studies of how CPP-MR parameters for Co/Cu multilayers vary with T reached similar conclusions. One used a V-groove geometry incorporating Cu connections to more closely approach a CPP geometry [37], and two others used nanowires [38, 39]. Combining the results listed in those papers suggests that the changes with temperature of CPP-MR are due mainly to phonon-scattering driven increases of the resistivities, ρ_{Co} and ρ_{Cu}, and decreases of the spin-diffusion length, l_{sf}^{Co}, of Co. In contrast, the asymmetry parameters, β_{Co} and $\gamma_{Co/Cu}$ (see Tables 4.2 and 4.5), and the interface resistance, $2AR_{Co/Cu}$ (see Refs. [37–39]), all seem to vary only weakly with T.

4.5.9 INTERLEAVED VERSUS SEPARATED MULTILAYERS

In Section 4.5.3, we showed that the CPP-MR of multilayers with Co and various N-layers much thinner than their spin-diffusion lengths can be well described by a 2CSR model. An -interesting property of this model is that the CPP-MR does not depend upon the order of the F-moments. That is, the AR should be the same for a sample with F-moments oriented ↑↓↑↓↑↓ (interleaved) or with F-moments oriented ↑↑↑↓↓↓ (separated). In contrast to this prediction, the first such comparison, involving [Co/Cu/Py/Cu]$_n$ versus [Co/Cu]$_n$[Py/Cu]$_n$ multilayers [140], showed clear differences, similar in form to those in Figure 4.3. These differences could be explained by the presence of a short spin-diffusion length $l_{sf}^{Py} = 5.5$nm in Py [82], which made the 2CSR model inappropriate. A similar study involving Fe and Co showed similar differences that can also be plausibly explained by spin-flipping within the Fe-layers [141]. However, Bozec et al. [142] then reported new results like those in Figure 4.3 for samples of the form [Co(1)/Cu/Co(8)/Cu]$_n$ and [Co(1)/Cu]$_n$[Co(8)/Cu]$_n$. Since l_{sf}^{Co} is long (see, e.g., Figure 4.17), these differences cannot be so easily explained. They have been attributed to (1) mean-free-path effects that require ballistic transport and specular reflection at interfaces (see Ref. [1]); (2) magnetic moment orientation effects of the adjacent F-layers (also called mean-free-path effects, but with different physics from that just noted [142]); and (3) standard VF theory, but adding spin-flipping at Co/Cu interfaces [143]. For further references, and a detailed examination of this topic, see Appendix C of Ref. [36] and Ref. [15].

4.6 SUMMARY AND CONCLUSIONS

We described results of studies of CIP- and CPP-MRs with a wide range of F/N-metals, and with several different kinds of samples: simple F/N multilayers; hybrid spin valves; and EBSVs. In the following, we first give our best answers to the questions that we raised in Section 4.1 and then conclude with some still unanswered questions.

4.6.1 CIP-MR

1. Interface scattering almost always dominates with small layer thicknesses. Bulk scattering can sometimes be made to dominate with larger thicknesses.

2. Data can be qualitatively understood with a semiclassical, two-current model with mean-free-paths as characteristic lengths. Qualitative agreements of parameters with those from deviations from Matthiessen's rule studies are affirmed by observations of the expected inversion of CIP-MR. However, we know of no fully quantitative comparisons with other measurements or calculations.

3. Interface structure can affect the CIP-MR, but available studies suggest that details of the effects depend upon the particular F/N pair, and the physical nature of the structure or roughness.

4. The presence of large CIP-MRs with certain F/N pairs, but not with others, argues that band structures are important. But we know of no successful quantitative agreements between data and calculations.

4.6.2 CPP-MR

1. Interface scattering almost always dominates for small layer thicknesses. Bulk scattering can sometimes be made to dominate with larger thicknesses.

2. Data can often be quantitatively understood with a semiclassical two-current model with spin-diffusion lengths, l_{sf}^F and l_{sf}^N, as characteristic lengths. When $t_F \ll l_{sf}^F$ and $t_N \ll l_{sf}^N$, a 2CSR model often works well. When $t_F \geq l_{sf}^F$ and/or $t_N \geq l_{sf}^N$, the more general VF model works. There is no compelling evidence for "quantum-coherence effects."

3. For dilute $F(N)$ alloys, measured values of $\beta_{F(N)}$ agree semiquantitatively with values determined from measurements of deviations from Matthiessen's rule in similar alloys.

4. For the two lattice-matched F/N pairs, Co/Cu and Fe/Cr, measured values of $\gamma_{F/N}$ agree semiquantitatively with the calculated ones.

5. Measured values of l_{sf}^F and l_{sf}^F generally agree well with values obtained by other techniques.

6. For lattice-matched metal pairs, measured values of $2AR_{N1/N2}$ and $2AR_{F/N}^*$ agree quantitatively with no-free-parameter calculations. These agreements suggest that band structures usually dominate $2AR$, with interfacial intermixing usually playing a secondary role.

7. There is now good evidence for spin-flipping at both N1/N2 and F/N interfaces, with values of δ occasionally ranging up to $\delta \sim 1$.

4.6.3 CIP-MR versus CPP-MR

Studies of inverse CIP- and CPP-MRs strongly suggest that the relative weightings of bulk and interface contributions differ for the CIP- and CPP-MRs.

4.6.4 Some Unanswered Questions

1. Is the source of observed spin-flipping at N1/N2 and F/N interfaces just spin–orbit effects on band structures, or is scattering within interfacial alloys important? And, is it necessary to include spin–spin interactions for F/N interfaces?

2. What physics underlies the measured values of $2AR_{S/F}$ for (superconducting Nb)/F interfaces?

3. Do changes in the physical roughness of interfaces affect $2AR_{N1/N2}$, $2AR^*_{F/N}$, and/or $\gamma_{F/N}$? If so, how are variations affected by the amplitude and length scales of such roughness?

4. The analyses of CPP-MR in this review assume that scattering within F- and N-layers is diffusive, even when the thicknesses of individual layers are less than the layer mean-free-paths. The ability of this assumption to fit data is attributed to interfaces that are rough enough to eliminate quantum coherent scattering. So an unanswered question is: will more nearly perfect interfaces give deviations from the simple 2CSR and VF models used in this review?

5. Can a combination of new F- and N-metals and alloys give CPP-MRs and ARs large enough to beat the competition for higher density read heads, without needing annealing to above 573 K?

REFERENCES

1. E. Y. Tsymbal, D. G. Pettifor, and S. Maekawa, Giant magnetoresistance: Theory, in *Handbook of Spin Transport and Magnetism*, E. Y. Tsymbal and I. Žutić (Eds.), Taylor & Francis, Boca Raton, FL, Chapter 5, pp. 95–114, 2011.
2. T. Ono and T. Shinjo, Magnetoresistance of multilayers prepared on microstructured substrates, *J. Phys. Soc. Jpn.* **64**, 363 (1995).
3. M. A. M. Gijs, M. T. Johnson, A. Reinders et al., Perpendicular giant magnetoresistance of Co/Cu multilayers deposited under an angle on grooved substrates, *Appl. Phys. Lett.* **66**, 1839 (1995); M. A. M. Gijs, S. K. J. Lenczowski, J. B. Giesbers et al., Perpendicular giant magnetoresistance using microlithography and substrate patterning techniques, *J. Magn. Magn. Mater.* **151**, 333 (1995)
4. J. Q. Xiao, J. S. Jiang, and C. L. Chien, Giant magnetoresistance in nonmultilayer magnetic systems, *Phys. Rev. Lett.* **68**, 3749 (1992).
5. A. Berkowitz, J. R. Mitchell, M. J. Carey et al., Giant -magnetoresistance in heterogeneous Cu–Co alloys, *Phys. Rev. Lett.* **68**, 3745 (1992).
6. T. L. Hylton, Limitations of magnetoresistive sensors based upon the giant magnetoresistive effect in granular magnetic composites, *Appl. Phys. Lett.* **62**, 2431 (1993).
7. A. Fert, Historical overview: From electron transport in magnetic materials to spintronics, in *Handbook of Spin Transport and Magnetism*, E. Y. Tsymbal and I. Žutić (Eds.), Taylor & Francis, Boca Raton, FL, Chapter 1, pp. 1–20, 2011.
8. J. Bass, Metals, electronic transport phenomena, in *Landolt-Bornstein Numerical Data and Functional Relationships in Science and Technology*, K. H. Hellwege and J. L. Olsen (Eds.), Group III, V.15a, Springer-Verlag, Berlin, Germany, p. 1, 1982.
9. P. M. Levy, Giant magnetoresistance in magnetic layered and granular materials, in *Solid State Physics Series*, Vol. 47, H. Ehrenreich and D. Turnbull (Ed.), Academic Press, New York, NY, p. 367, 1994.

10. B. Dieny, Giant magnetoresistance in spin-valve multilayers, *J. Magn. Magn. Mater.* **136**, 335 (1994).

11. M. A. M. Gijs and G. E. W. Bauer, Perpendicular giant magnetoresistance of magnetic multilayers, *Adv. Phys.* **46**, 285 (1997).

12. J. Bass and W. P. Pratt, Current-perpendicular (CPP) magnetoresistance in magnetic metallic multilayers, *J. Magn. Magn. Mater.* **200**, 274 (1999); J. Bass and W. P. Pratt, *Erratum: J. Magn. Magn. Mater.* **296**, 65 (2006).

13. A. Fert and L. Piraux, Magnetic nanowires, *J. Magn. Magn. Mater.* **200**, 338 (1999).

14. E. Y. Tsymbal and D. G. Pettifor, Perspectives of giant magnetoresistance, in *Solid State Physics*, Vol. 56, H. Ehrenreich and F. Spaepen (Eds.), Academic Press, San Diego, CA, pp. 113–239, 2001.

15. J. Bass, CPP magnetoresistance of magnetic multilayers: A critical review, *J. Magn. Magn. Mater.* **408**, 244 (2016).

16. W. Thomson, On the electro-dynamic qualities of metals: Effects of magnetization on the electrical conductivity of nickel and of iron, *Proc. Roy. Soc.* **8**, 546 (1857).

17. T. R. McGuire and R. I. Potter, Anisotropic magnetoresistance in ferromagnetic 3d alloys, *IEEE Trans. Magn. Magn.* **11**(4), 1018 (1975).

18. M. N. Baibich, J. M. Broto, A. Fert et al., Giant magnetoresistance of (001)Fe/(001)Cr superlattices, *Phys. Rev. Lett.* **61**, 2472 (1988).

19. G. Binasch, P. Grunberg, F. Saurenbach, and W. Zinn, Enhanced magnetoresistance in layered magnetic structures with antiferromagnetic interlayer exchange, *Phys. Rev. B* **39**, 4828 (1989).

20. J. A. Borchers, J. A. Dura, J. Unguris et al., Observation of antiparallel order in weakly coupled Co/Cu multilayers, *Phys. Rev. Lett.* **82**, 2796 (1999).

21. W. P. Pratt, S. F. Lee, J. M. Slaughter, R. Loloee, P. A. Schroeder, and J. Bass, Perpendicular giant magnetoresistances of Ag/Co multilayers, *Phys. Rev. Lett.* **66**, 3060 (1991).

22. J. Barnas, A. Fuss, R. E. Camley, P. Grunberg, and W. Zinn, Novel magnetoresistance effect in layered magnetic structures: Theory and experiment, *Phys. Rev. B* **42**, 8110 (1990).

23. B. Heinrich, Basics of nano-thin film magnetism, in *Handbook of Spin Transport and Magnetism*, E. Y. Tsymbal and I. Žutić (Eds.), Taylor & Francis, Boca Raton, FL, Chapter 2, pp. 21–46, 2011.

24. S. Dubois, L. Piraux, J. M. George, K. Ounadjela, J. L. Duvail, and A. Fert, Evidence for a short spin diffusion length in permalloy from the giant magnetoresistance of multilayered nanowires, *Phys. Rev. B* **60**, 477 (1999).

25. A. Chaiken, P. Lubitz, J. J. Krebs, G. A. Prinz, and M. Z. Harford, Low-field spin-valve magnetoresistance in Fe/Cu/Co sandwiches, *Appl. Phys. Lett.* **59**, 240 (1991).

26. Q. Yang, P. Holody, R. Loloee et al., Prediction and measurement of perpendicular (CPP) giant magnetoresistance of Co/Cu/Ni$_{84}$Fe$_{16}$/Cu multilayers, *Phys. Rev. B* **51**, 3226 (1995).

27. B. Dieny, V. S. Speriosu, S. S. P. Parkin, B. A. Gurney, D. R. Wilhoit, and D. Mauri, Giant magnetoresistance in soft ferromagnetic multilayers, *Phys. Rev. B* **43**, 1297 (1991).

28. S. D. Steenwyk, S. Y. Hsu, R. Loloee, J. Bass, and W. P. Pratt Jr., A comparison of hysteresis loops from giant magnetoresistance and magnetometry of perpendicular-current exchange-biased spin-valves, *J. Appl. Phys.* **81**, 4011 (1997).

29. R. E. Camley and J. Barnas, Theory of giant magnetoresistance effects in magnetic layered structures with antiferromagnetic coupling, *Phys. Rev. Lett.* **63**, 664 (1989).

30. S. Zhang and P. M. Levy, Conductivity perpendicular to the plane of multilayered structures, *J. Appl. Phys.* **69**, 4786 (1991).

31. S. F. Lee, W. P. Pratt, Q. Yang et al., Two-channel analysis of CPP-MR data for Ag/Co and AgSn/Co multilayers, *J. Magn. Magn. Mater.* **118**, L1 (1993).

32. T. Valet and A. Fert, Theory of perpendicular magnetoresistance in magnetic multilayers, *Phys. Rev. B* **48**, 7099 (1993).
33. I. A. Campbell and A. Fert, Transport properties of ferromagnets, in *Ferromagnetic Materials*, Vol. 3, E. P. Wolforth (Ed.), North-Holland, Amsterdam, the Netherlands, p. 747, 1982.
34. R. Q. Hood and L. M. Falicov, Theory of the negative magnetoresistance of ferromagnetic-normal metallic multilayers, *Phys. Rev. B.* **46**, 8287 (1992); R. Q. Hood and L. M. Falicov, Effects of interfacial roughness on the magnetoresistance of magnetic metallic multilayers, *Phys. Rev. B* **49**, 368 (1994).
35. N. W. Ashcroft and N. D. Mermin, *Solid State Physics*, W. B. Saunders, Philadelphia, PA, 1976.
36. J. Bass and W. P. Pratt, Spin-diffusion lengths in metals and alloys, and spin-flipping at metal/metal interfaces: An experimentalist's critical review, *J. Phys. Condens. Matter* **19**, 183201 (2007).
37. W. Oepts, M. A. M. Gijs, A. Reinders, R. M. Jungblut, R. M. J. van Gansewinkel, and W. J. M. de Jonge, Perpendicular giant magnetoresistance of Co/Cu multilayers on grooved -substrates: Systematic analysis of the temperature dependence of spin-dependent scattering, *Phys. Rev. B* **53**, 14024 (1996).
38. L. Piraux, S. Dubois, A. Fert, and L. Belliard, The temperature dependence of the perpendicular giant magnetoresistance in Co/Cu multilayered nanowires, *Europhys. J. B* **4**, 413 (1998).
39. B. Doudin, A. Blondel, and J. P. Ansermet, Arrays of multilayered nanowires, *J. Appl. Phys.* **79**, 6090 (1996).
40. C. Vouille, A. Barthelemy, F. Elokan Mpondo et al., Microscopic mechanisms of giant magnetoresistance, *Phys. Rev. B* **60**, 6710 (1999).
41. C. Vouille, A. Fert, A. Barthelemy, S. Y. Hsu, R. Loloee, and P. A. Schroeder, Inverse CPP-GMR in (A/Cu/Co/Cu) multilayers (A = NiCr, FeCr, FeV) and discussion of the spin asymmetry induced by impurities, *J. Appl. Phys.* **81**, 4573 (1997).
42. P. M. Levy, S. Zhang, T. Ono, and T. Shinjo, Electrical transport in corrugated multilayered structures, *Phys. Rev. B* **52**, 16049 (1995).
43. B. Dieny, Quantitative interpretation of giant magnetoresistance properties of permalloy-based spin-valve structures, *Europhys. Lett.* **17**, 261 (1992).
44. A. Barthelemy and A. Fert, Theory of the magnetoresistance in magnetic multilayers: Analytical expressions from a semiclassical approach, *Phys. Rev. B* **43**, 13124 (1991).
45. S. S. P. Parkin, Dependence of giant magnetoresistance on Cu-layer thickness in Co/Cu multilayers: A simple dilution effect, *Phys. Rev. B* **47**, 9136 (1993).
46. B. Dieny, V. S. Speriosu, S. Metin et al., Magnetotransport properties of magnetically soft spin-valve structures, *J. Appl. Phys.* **69**, 4774 (1991); B. Dieny, P. Humbert, V. S. Speriosu et al., Giant magnetoresistance of magnetically soft sandwiches: Dependence on temperature and on layer thickness, *Phys. Rev. B* **45**, 806 (1992).
47. S. S. P. Parkin, N. More, and K. P. Roche, Oscillations in exchange coupling and magnetoresistance in metallic superlattice structures: Co/Ru, Co/Cr, and Fe/Cr, *Phys. Rev. Lett.* **64**, 2304 (1990).
48. E. Velu, C. Dupas, D. Renard, J. P. Ranard, and J. Seiden, Enhanced magnetoresistance of ultra-thin $(Au/Co)_n$ multilayers with perpendicular anisotropy, *Phys. Rev. B* **37**, 668 (1988).
49. S. S. P. Parkin, Origin of enhanced magnetoresistance of magnetic multilayers: Spin-dependent scattering from magnetic interface states, *Phys. Rev. Lett.* **71**, 1641 (1993).
50. P. Baumgart, B. A. Gurney, D. R. Wilhoit, T. Nguyen, B. Dieny, and V. S. Speriosu, The role of spin-dependent impurity scattering in Fe/Cr giant magnetoresistance multilayers, *J. Appl. Phys.* **69**, 4792 (1991).
51. B. A. Gurney, P. Baumgart, D. R. Wilhoit, D. Dieny, and V. S. Speriosu, Giant magnetoresistance of Fe/Cr multilayers: Impurity scattering model of the influence of third elements deposited at the interface, *J. Appl. Phys.* **70**, 5867 (1991).

52. V. Speriosu, J. P. Nozieres, B. A. Gurney, B. Dieny, T. C. Huang, and H. Lefakis, Role of interfacial mixing in giant magnetoresistance, *Phys. Rev. B* **47**, 11579 (1993).

53. F. Petroff, A. Barthelemy, A. Hamzic et al., Magnetoresistance of Fe/Cr superlattices, *J. Magn. Magn. Mater.* **93**, 95 (1991).

54. E. E. Fullerton, D. M. Kelly, J. Guimpel, I. K. Schuller, and Y. Bruynseraede, Roughness and giant magnetoresistance in Fe/Cr superlattices, *Phys. Rev. Lett.* **68**, 859 (1992).

55. B. Dieny, J. P. Nozieres, V. S. Speriosu, B. A. Gurney, and D. R. Wilhoit, Change in conductance is the fundamental measure of spin-valve magnetoresistance, *Appl. Phys. Lett.* **61**, 2111 (1992).

56. B. A. Gurney, V. S. Speriosu, J.-P. Nozieres, H. Lefakis, D. R. Wilhoit, and O. U. Need, Direct measurement of spin-dependent conduction-electron mean-free-paths in ferromagnetic metals, *Phys. Rev. Lett.* **71**, 4023 (1993).

57. D. M. Kelly, I. K. Schuller, V. Korenivski et al., Increases in giant magnetoresistance by ion irradiation, *Phys. Rev. B* **50**, 3481 (1994).

58. R. Schad, P. Belien, G. Verbanck et al., Giant magnetoresistance dependence on the lateral correlation length of the interface roughness in magnetic superlattices. *Phys. Rev. B* **59**, 1242 (1999).

59. J. M. George, L. G. Pereira, A. Barthelemy et al., Inverse spin-valve-type magnetoresistance in spin engineered multilayered structures, *Phys. Rev. Lett.* **72**, 408 (1994).

60. J.-P. Renard, P. Bruno, R. Megy et al., Inverse magnetoresistance in the simple spin-valve system $Fe_{1-x}V_x/Au/Co$, *Phys. Rev. B* **51**, 12821 (1995).

61. A. Fert, J. L. Duvail, and T. Valet, Spin relaxation effects in the perpendicular magnetoresistance of magnetic multilayers, *Phys. Rev. B* **52**, 6513 (1995).

62. J. Åkerman, Magnetoresistive random access memory, in *Handbook of Spin Transport and Magnetism*, E. Y. Tsymbal and I. Žutić (Eds.), Taylor & Francis, Boca Raton, FL, Chapter 35, pp. 685–698, 2012.

63. J. M. Slaughter, W. P. Pratt Jr., and P. A. Schroeder, Fabrication of layered metallic systems for perpendicular resistance measurements, *Rev. Sci. Instrum.* **60**, 127 (1989).

64. C. Fierz, S. F. Lee, J. Bass, W. P. Pratt Jr., and P. A. Schroeder, Perpendicular resistance of thin Co films in contact with superconducting Nb, *J. Phys.: Condens. Mater.* **2**, 9701 (1990).

65. S. F. Lee, Q. Yang, P. Holody et al., Current perpendicular and parallel giant magnetoresistances in Co/Ag multilayers, *Phys. Rev. B* **52**, 15426 (1995).

66. R. D. Slater, J. A. Caballero, R. Loloee, and W. P. Pratt Jr., Perpendicular-current exchange-biased spin valve structures with micron-size superconducting top contacts, *J. Appl. Phys.* **90**, 5242 (2001).

67. D. L. Edmunds, W. P. Pratt Jr., and J. R. Rowlands, 0.1 ppm four-terminal resistance bridge for use with a dilution refrigerator, *Rev. Sci. Instrum.* **51**, 1516 (1980).

68. M. C. Cyrille, S. Kim, M. E. Gomez, J. Santamaria, K. M. Krishnan, and I. K. Schuller, Enhancement of perpendicular and parallel giant magnetoresistance with the number of bilayers in Fe/Cr superlattices, *Phys. Rev. B* **62**, 3361 (2000).

69. R. J. Highmore, M. G. Blamire, R. E. Somekh, and J. E. Evetts, Magnetoresistance of Cu/Ni multilayers, *J. Magn. Magn. Mater.* **104–107**, 1777 (1992).

70. M. A. M. Gijs, S. K. J. Lenczowski, and J. B. Giesbers, Perpendicular giant magnetoresistance of microstructured Fe/Cr magnetic multilayers from 4.2 to 300 K, *Phys. Rev. Lett.* **70**, 3343 (1993).

71. M. AlHajDarwish, H. Kurt, S. Urazhdin et al., Controlled normal and inverse current-induced magnetization switching and magnetoresistance in magnetic nanopillars, *Phys. Rev. Lett.* **93**, 157203 (2004).

72. S. K. J. Lenczowski, R. J. M. van de Veerdonk, M. A. M. Gijs, J. B. Giesbers, and H. H. J. M. Janssen, Current-distribution effects in microstructures for perpendicular magnetoresistance measurements, *J. Appl. Phys.* **75**, 5154 (1994).

73. A. Blondel, J. P. Meir, B. Doudin, and J.-Ph. Ansermet, Giant magnetoresistance of nanowires of multilayers, *Appl. Phys. Lett.* **65**, 3019 (1994).

74. L. Piraux, J. M. George, J. F. Despres et al., Giant magnetoresistance in magnetic multilayered nanowires, *Appl. Phys. Lett.* **65**, 2484 (1994).

75. K. Liu, K. Nagodawithana, P. C. Searson, and C. L. Chien, Perpendicular giant magnetoresistance of multilayered Co/Cu nanowires, *Phys. Rev. B* **51**, 7381 (1995).

76. P. R. Evans, G. Yi, and W. Schwarzacher, Current perpendicular to plane giant magnetoresistance of multilayered nanowires electrodeposited in anodic aluminum oxide membranes, *Appl. Phys. Lett.* **76**, 481 (2000).

77. B. Doudin, J. F. Wegrowe, D. Kelly, et al., Magnetic and transport properties of electrodeposited nanostructured nanowires, *IEEE Trans. Magn.* **34**, 968 (1998).

78. X.-T. Tang, G.-C. Wang, and M. Shima, Layer thickness dependence of CPP giant magnetoresistance in individual Co/Ni/Cu multilayer nanowires grown by electrodeposition, *Phys. Rev. B* **75**, 134404 (2007).

79. W. P. Pratt Jr., S. D. Steenwyk, S. Y. Hsu et al., Perpendicular-current transport in exchange-biased spin-valves, *IEEE Trans. Magn.* **33**, 3505 (1997).

80. A. C. Reilly, W. Park, R. Slater et al., Perpendicular giant magnetoresistance of Co$_{91}$Fe$_9$/Cu exchange-biased spin-valves: Further evidence for a unified picture, *J. Magn. Magn. Mater.* **195**, L269 (1999).

81. L. Piraux, S. Dubois, and A. Fert, Perpendicular giant magnetoresistance in magnetic multilayered nanowires, *J. Magn. Magn. Mater.* **159**, L287 (1996).

82. S. Steenwyk, S. Y. Hsu, R. Loloee, J. Bass, and W. P. Pratt Jr., Perpendicular current exchange biased spin-valve evidence for a short spin diffusion length in permalloy, *J. Magn. Magn. Mater.* **170**, L1 (1997).

83. W. P. Pratt Jr., S. F. Lee, Q. Yang et al., Giant magnetoresistance with current perpendicular to the layer planes of Ag/Co and AgSn/Co, *J. Appl. Phys.* **73**, 5326 (1993).

84. J. Bass, P. A. Schroeder, W. P. Pratt et al., Studying spin-dependent scattering in magnetic multilayers by means of perpendicular (CPP) magnetoresistance measurements, *Mater. Sci. Eng. B* **31**, 77 (1995).

85. P. Monod and S. Schultz, Conduction electron spin-flip scattering by impurities in copper, *J. Phys. Paris* **43**, 393 (1982).

86. A. Sharma, J. A. Romero, N. Theodoropoulou, R. Loloee, W. P. Pratt Jr., and J. Bass, Specific resistance and scattering asymmetry of Py/Pd, Fe/V, Fe/Nb, and Co/Pt interfaces, *J. Appl. Phys.* **102**, 113916 (2007).

87. A. Sharma, N. Theodoropoulou, T. Haillard et al., Current perpendicular to plane (CPP) magnetoresistance of ferromagnetic (F)/Al interfaces (F = Py, Co, Fe, and Co$_{91}$Fe$_9$) and structural studies of Co/Al and Py/Al, *Phys. Rev. B* **77**, 224438 (2008).

88. P. Holody, Perpendicular giant magnetoresistance: Study and application of spin dependent scattering in magnetic multilayers of cobalt/copper and nickel(84)iron(16)/copper, PhD thesis, Michigan State University, East Lansing, MI, 1996.

89. N. J. List, W. P. Pratt Jr., M. A. Howson, J. Xu, M. J. Walker, and D. Greig, Perpendicular resistance of Co/Cu multilayers prepared by molecular beam epitaxy, *J. Magn. Magn. Mater.* **148**, 342 (1995).

90. L. Piraux, S. Dubois, C. Marchal et al., Perpendicular magnetoresistance in Co/Cu multilayered nanowires, *J. Magn. Magn. Mater.* **156**, 317 (1996).

91. D. Bozec, Current perpendicular to the plane magnetoresistance of magnetic multilayers. Physics and astronomy, PhD thesis, Leeds University, West Yorkshire, UK, 2000.

92. C. E. Moreau, I. C. Moraru, N. O. Birge, and W. P. Pratt Jr., Measurement of spin diffusion length in sputtered Ni films using a special exchange-biased spin valve geometry, *Appl. Phys. Lett.* **90**, 012101 (2007).

93. P. Holody, W. C. Chiang, R. Loloee, J. Bass, W. P. Pratt Jr., and P. A. Schroeder, Giant magnetoresistance in copper/permalloy multilayers, *Phys. Rev. B* **58**, 12230 (1998).

94. L. Vila, W. Park, J. A. Caballero et al., Current perpendicular magnetoresistances of NiCoFe and NiFe permalloys, *J. Appl. Phys.* **87**, 8610 (2000).

95. W. Park, R. Loloee, J. A. Caballero et al., Test of unified picture of spin dependent transport in perpendicular (CPP) giant magnetoresistance and bulk alloys, *J. Appl. Phys.* **85**, 4542 (1999).

96. D. K. Kim, Y. S. Lee, H. Y. T. Nguyen, et al., *IEEE Trans. On Magn.* **46**, 1374 (2010).

97. S. Y. Hsu, A. Barthelemy, P. Holody, R. Loloee, P. A. Schroeder, and A. Fert, Towards a unified picture of spin dependent transport in perpendicular giant magnetoresistance and bulk alloys, *Phys. Rev. Lett.* **78**, 2652 (1997).

98. Q. Yang, P. Holody, S. F. Lee et al., Spin diffusion length and giant magnetoresistance at low temperatures, *Phys. Rev. Lett.* **72**, 3274 (1994).

99. W. Park, D. V. Baxter, S. Steenwyk, I. Moraru, W. P. Pratt, and J. Bass, Measurement of resistance and spin-memory loss (spin relaxation) at interfaces using sputtered current perpendicular-to-plane exchange-biased spin valves, *Phys. Rev. B* **62**, 1178 (2000).

100. S. Y. Hsu, P. Holody, R. Loloee, J. M. Rittner, W. P. Pratt, and P. A. Schroeder, Spin-diffusion Lengths of $Cu_{1-x}Ni_x$ using current perpendicular to plane magnetoresistance measurements of magnetic multilayers, *Phys. Rev. B* **54**, 9027 (1996).

101. R. Loloee, B. Baker, and W. P. Pratt Jr., Unpublished, 2006.

102. D. V. Baxter, S. D. Steenwyk, J. Bass, and W. P. Pratt Jr., Resistance and spin-direction memory loss at Nb/Cu interfaces, *J. Appl. Phys.* **85**, 4545 (1999).

103. L. L. Henry, Q. Yang, W. C. Chiang et al., Perpendicular interface resistances in sputtered Ag/Cu, Ag/Au, and Au/Cu multilayers, *Phys. Rev. B* **54**, 12336 (1996).

104. R. Acharyya, H. Y. T. Nguyen, R. Loloee et al., Specific resistance of Pd/Ir interfaces, *Appl. Phys. Lett.* **94**, 022112 (2009).

105. A. Zambano, K. Eid, R. Loloee, W. P. Pratt, and J. Bass, Interfacial properties of Fe/Cr multilayers in the current-perpendicular-to-plane geometry, *J. Magn. Magn. Mater.* **253**, 51 (2002).

106. S. K. Olson, R. Loloee, N. Theodoropoulou et al., Comparison of measured and calculated specific resistances of Pd/Pt interfaces, *Appl. Phys. Lett.* **87**, 252508 (2005).

107. H. Kurt, W. C. Chiang, C. Ritz, K. Eid, W. P. Pratt, and J. Bass, Spin-memory loss and CPP-magnetoresistance in sputtered multilayers with Au, *J. Appl. Phys.* **93**, 7918 (2003).

108. C. Galinon, K. Tewolde, R. Loloee et al., Pd/Ag and Pd/Au interface specific resistances and interfacial spin-flipping, *Appl. Phys. Lett.* **86**, 182502 (2005).

109. H. Kurt, R. Loloee, K. Eid, W. P. Pratt, and J. Bass, Spin-memory loss at 4.2 K in sputtered Pd, Pt, and at Pd/Cu and Pt/Cu interfaces, *Appl. Phys. Lett.* **81**, 4787 (2002).

110. K. Eid, R. Fonck, M. A. Darwish, W. P. Pratt, and J. Bass, Current-perpendicular-to-plane magnetoresistance properties of Ru and Co/Ru interfaces, *J. Appl. Phys.* **91**, 8102 (2002).

111. K. M. Schep, P. J. Kelly, and G. E. W. Bauer, Ballistic transport and electronic structure, *Phys. Rev. B* **57**, 8907 (1998).

112. M. S. Stiles and D. R. Penn, Calculation of spin-dependent interface resistance, *Phys. Rev. B* **61**, 3200 (2000); D. R. Penn and M. D. Stiles, Spin transport for spin diffusion lengths comparable to the mean-free-path, *Phys. Rev. B* **72**, 212410 (2005).

113. K. Xia, P. J. Kelly, G. E. W. Bauer, I. Turek, J. Kudrnovsky, and V. Drchal, Interface resistance of disordered magnetic multilayers, *Phys. Rev. B* **63**, 064407 (2001).

114. P. X. Xu, K. Xia, M. Zwierzycki, M. Talanana, and P. J. Kelly, Orientation-dependent transparency of metallic interfaces, *Phys. Rev. Lett.* **96**, 176602 (2006).

115. P. X. Xu and K. Xia, Ab-initio calculations of alloy resistivities, *Phys. Rev. B* **74**, 184206 (2006).

116. B. Dassonneville, R. Acharyya, H. Y. T. Nguyen, R. Loloee, W. P. Pratt Jr., and J. Bass, A way to measure electron spin-flipping at F/N interfaces and application to Co/Cu, *Appl. Phys. Lett.* **96**, 022509 (2010).

117. H. Y. T. Nguyen, R. Acharyya, E. Huey et al., Conduction electron scattering and spin-flipping at sputtered Co/Ni Interfaces, *Phys. Rev.* **82,** 220401(R) (2010).

118. H. Y. T. Nguyen, R. Acharyya, W. P. Pratt Jr., and J. Bass, Spin-flipping at sputtered Co90Fe10/Cu interfaces, *J. Appl. Phys.* **109**, 07C903 (2011).

119. M. A. Khasawneh, C. Klose, W. P. Pratt Jr., and N. O. Birge, Spin-memory loss at Co/Ru interfaces, *Phys. Rev. B* **84**, 014425 (2011).

120. H. Y. T. Nguyen, R. Loloee, W. P. Pratt Jr., and J. Bass, Spin-flipping at sputtered Co/Ag interfaces, *Phys. Rev. B* **86**, 064413 (2012).

121. H. Y. T. Nguyen, W. P. Pratt Jr., and J. Bass, Spin-flipping in Pt and at Co/Pt interfaces, *J. Magn. Magn. Mater.* **361**, 30 (2014).

122. A. Manchon, N. Strelkov, A. Deac, A. Vedyayev, and B. Dieny, Interpretation of the relationship between current perpendicular to plane magnetoresistance and spin torque amplitude, *Phys. Rev. B* **73**, 184418 (2006).

123. J.-C. Rojas-Sanchez, N. Reyren, P. Lacszkowski et al., Spin pumping and inverse spin Hall effect in platinum, the essential role of spin-memory loss at metallic interfaces, *Phys. Rev. Lett.* **112**, 106602 (2014).

124. Y. Liu, Z. Yuan, R. J. H. Wesselink, A. A. Starikov, and P. J. Kelly, Interface enhancement of Gilbert damping from first principles, *Phys. Rev. Lett.* **113**, 207202 (2014).

125. W. C. Chiang, W. P. Pratt Jr., M. Herrold, and D. V. Baxter, Effect of sputtering pressure on the structure and current-perpendicular-to-the-plane magneto-transport of Co/Ag multilayered films, *Phys. Rev. B* **58**, 5602 (1998).

126. M. C. Cyrille, S. Kim, M. E. Gomez, J. Santamaria, K. M. Krishnan, and I. K. Schuller, Effect of sputtering pressure-induced roughness on the microstructure and the perpendicular giant magnetoresistance of Fe/Cr superlattices, *Phys. Rev. B* **62**, 15079 (2000).

127. K. Eid, W. P. Pratt, and J. Bass, Enhancing current-perpendicular-to-plane magnetoresistance (CPP-MR) by adding interfaces within ferromagnetic layers, *J. Appl. Phys.* **93**, 3445 (2003).

128. H. Oshima K. Nagasaka, Y. Seyama, et al., Spin filtering effect at inserted interfaces in perpendicular spin valves, *Phys. Rev. B* **66**, 140404(R) (2002).

129. M. Saito, CPP mode magnetic sensing element including a multilayer free layer biased by an antiferromagnetic layer, US Patent 6947263 (2003).

130. F. Delille, A. Manchon, N. Strelkov et al., Thermal variation of current perpendicular-to-plane giant magnetoresistance in laminated and nonlaminated spin valves, *J. Appl. Phys.* **100**, 013912 (2006).

131. H. Fujiwara and G. J. Mankey, CPP spin-valve device, US patent 6560077B2 (2000).

132. K. Nagasaka, Y. Seyama, L. Varga, Y. Shimizu, and A. Tanaka, Giant magnetoresistance properties of specular spin valve films in a current perpendicular to Plane structure, *J. Appl. Phys.* **89**, 6943 (2001).

133. M. Takagishi, K. Koi, M. Yoshikawa, T. Funayama, H. Iwasaki, and M. Sahashi, The applicability of CPP-GMR heads for magnetic recording, *IEEE Trans. Magn.* **38**, 2277 (2002).

134. H. Fukuzawa, H. Yuasa, and H. Iwasaki, CPP-GMR films with a current-confined-path nano-oxide layer (CPP-NOL), *J. Phys. D: Appl. Phys.* **40**, 1213 (2007).

135. K. Nakamoto et al., CPP-GMR read heads with current screen layer for 300 Gbit/in2 recording, *IEEE Trans. Magn.* **44**, 95 (2008).

136. J. A. Caballero, Y. D. Park, J. R. Childress et al., Magnetoresistance of NiMnSb-based multilayers and spin-valves, *J. Vac. Sci. Tech. A* **16**, 1801 (1998).

137. M. J. Carey, S. Maat, S. Chandrashekariaih et al., Co2MnGe-based current-perpendicular-to-the-plane giant magnetoresistance spin-valve sensors for recording head applications, *J. Appl. Phys.* **109**, 093912 (2011).

138. Y. K. Takahashi, A. Srinivasan, B. S. Varaprasad et al., Large magneto-resistance in current-perpendicular-to-plane pseudospin valves using a Co2Fe(Ge0.5Ga0.5) magnetic layer, *Appl. Phys. Lett.* **98**, 152501 (2011); *Erratum: Appl. Phys. Lett.* **98**, 18990 (2011).

139. Y. Du, T. Furubayashi, T. T. Sasaki et al., Large magnetoresistance in current-perpendicular-to-plane pseudo-spin-valves using Co2Fe(Ga0.5Ge0.5) Heusler alloy and AgZn spacer, *Appl. Phys. Lett.* **107**, 112405 (2015).

140. W. C. Chiang, Q. Yang, W. P. Pratt Jr., R. Loloee, and J. Bass, Variation of mul-tilayer magnetoresistance with ferromagnetic layer sequence: Spin-memory loss, *J. Appl. Phys.* **81**, 4570 (1997).

141. D. Bozec, M. J. Walker, B. J. Hickey, M. A. Howson, and N. Wiser, Comparative study of the magnetoresistance of MBE-grown multilayers: $[Fe/Cu/Co/Cu]_N$ and $[Fe/Cu]_N[Co/Cu]_N$, *Phys. Rev. B* **60**, 3037 (1999).

142. D. Bozec, M. A. Howson, B. J. Hickey et al., Mean free path effects on the cur-rent perpendicular to the plane magnetoresistance of magnetic multilayers, *Phys. Rev. Lett.* **85**, 1314 (2000).

143. K. Eid, D. Portner, J. A. Borchers et al., Absence of mean-free-path effects in CPP-magnetoresistance of magnetic multilayers, *Phys. Rev. B* **65**, 054424 (2002).

5

Spin Injection, Accumulation, and Relaxation in Metals

Mark Johnson

5.1 INTRODUCTION

The main focus of this chapter is the early development of *spin injection* in metals and related phenomena prior to the year 2000. This includes a review of the origination of the lateral spin valve. More recent related developments are discussed in Chapters 7–9 of Volume 3 of this book. A second focus is the application of the spin-injection technique, using the lateral spin valve geometry, to observe the Datta–Das conductance oscillation in a gated two-dimensional electron system (2DES) structure.

There are many topics that relate to the effects of carrier spin on transport in electronic systems. Among the most fundamental, spin-dependent tunneling (SDT) [1] (see also Chapter 11, Volume 1) has the longest history and offers a technique for measuring the spin polarization of electric current in a ferromagnet. Magnetic tunnel junctions (MTJs) [2–4] form a device family with the largest magnetoresistance ratio (MR) and with the highest promise for application (see also Chapter 12, Volume 3). Giant magnetoresistance (GMR) [5, 6] (see also Chapter 4, Volume 1) provides a descriptive understanding of spin-dependent scattering in ferromagnets and in thin-film ferromagnet/nonmagnetic (*F/N*) metal multilayers. The spin injection phenomenology [7] has a history that is second to SDT. Developed for the study of *F/N* metal interfacial spin transport and the study of spin-relaxation processes, spin injection introduced a number of ideas and techniques that have been commonly adopted. Researchers today invoke fundamental concepts such as spin injection, spin accumulation, spin-polarized currents in ferromagnet/nonmagnet structures, and pure spin currents. All these concepts were described, defined, and demonstrated in a series of experimental and theoretical papers in the mid-1980s [7–10].

These spin-injection experiments were the first to use a ferromagnet/nonmagnetic metal/ferromagnet ($F1/N/F2$) device structure, using a geometry that has become known as the lateral spin valve (LSV), and the first to demonstrate a spin-mediated magnetoresistance (at temperatures above 4 K) that was controlled by using a magnetic field to manipulate the magnetization orientations $\vec{M}_1$ and $\vec{M}_2$ of $F1$ and $F2$ between parallel and antiparallel configurations [7]. The $F1/N/F2$ device family was later expanded by the introduction of the current-in-plane (CIP) spin valve [5, 11] and the current-perpendicular-to-the-plane (CPP) spin valve [12]. The spin-injection theory was the first to use unique conductances, $g_\uparrow$ and $g_\downarrow$, for up and down-spin electrons in both ferromagnetic and nonmagnetic materials in a $F/N/F$ structure. Finally, the spin-injection technique has been successful in studies of spin transport in novel categories of materials, enabling measurements of spin relaxation times, τ_s, in a variety of materials systems. While transmission electron spin resonance (TESR) had been used to measure transverse spin relaxation times, T_2, of order ns or longer in bulk metals at low temperature [13], spin injection has been capable of measuring relaxation times of 0.1 ns and shorter and has been applied to metal, semiconductor, semimetal, and superconducting films. The treatment of spin injection in this chapter will cover these themes.

The concept that an electric current in a metal could be spin-polarized is nearly 70 years old. Mott calculated scattering rates for up-spin and down-spin itinerant electrons in the exchange-split d-band of transition metal ferromagnets [14] and deduced that the electrons carrying an electrical current in a ferromagnetic metal had a net spin polarization P,

$$P = \frac{(J_\uparrow - J_\downarrow)}{(J_\uparrow + J_\downarrow)} \neq 0, \tag{5.1}$$

where $J_\uparrow$ and $J_\downarrow$ are spin-dependent partial currents that sum to form the total electric current, $J = J_\uparrow + J_\downarrow$.

This concept can be described with the aid of Figure 5.1a, which shows the density of states as a function of energy, $N(E)$, for the 3d band of a transition metal ferromagnet F such as Ni, Fe, Co, or an alloy of these or similar constituents. Because of exchange splitting U_{ex}, the majority, down-spin subband is filled with more electrons than the minority, up-spin subband, and this difference accounts for the equilibrium, spontaneous magnetization of F. Another consequence of U_{ex} is that the density of states at the Fermi level, $N(E_F)$, is different for the down- and up-spin subbands, $N_\downarrow(E_F) \neq N_\uparrow(E_F)$. The density of states at E_F of the minority spin subband is larger than that of the majority spin subband in the simplified schematic representation of Figure 5.1a.

The conductivity of a metal, $g = 1/\rho$ with ρ the resistivity (note that notations g and σ are both used for conductivity, and σ is used elsewhere in this book), can be related to $N(E_F)$ by the Einstein relation,

$$g = e^2 D N(E_F),$$

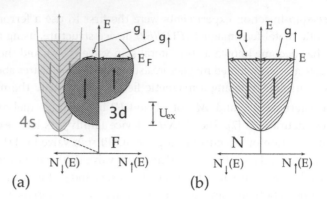

FIGURE 5.1   Density of states as a function of energy for (a) the exchange split 3*d* band of a transition metal ferromagnet and (b) a free electron nonmagnetic metal, $T = 0$. (After Johnson, M., *Magnetoelectronics*, Copyright 2004, figure 1.8, with permission from Oxford University Press, Elsevier, Oxford, UK.)

where

   D is the diffusion constant

   e is the electron charge

Mott showed that the scattering rates for up- and down-spin subbands were different, that scattering between subbands was relatively rare, and, therefore, that electrons in the two subbands do not intermix. It follows that a conductance, g_s, can be defined individually for each spin subband,

$$g_{\uparrow,\downarrow} = e^2 DN_{\uparrow,\downarrow}(E_F),\qquad(5.2)$$

where D is assumed to be the same (in metals) for up- and down-spin carriers. It is clear from Figure 5.1a that $g_\uparrow \neq g_\downarrow$ and Equation 5.1 can be rewritten as

$$P = \frac{(g_\uparrow - g_\downarrow)}{(g_\uparrow + g_\downarrow)} \neq 0,\qquad(5.3)$$

by noting that $J \propto g$. The conceptually important point is that the electric current in a ferromagnet is the sum of spin-up and spin-down partial currents, and these partial currents can be modeled as independent. This idea is often called the "two current model" of charge and spin transport.

   A transition metal ferromagnet also has a free electron 4s band that intersects the Fermi surface, depicted in Figure 5.1a with light shading. Since an electric current in F involves all electrons within $k_B T$ of the Fermi surface, where k_B is Boltzmann's constant and T is temperature, an accurate model of spin-polarized currents must take account of contributions from the unpolarized 4s band. It is common practice for experimental results to be discussed in a way that neglects the 4s band, and this practice will be used within the pedagogical context of this chapter. More formally, an acceptable model can be based on combining the 4s and 3d bands into a single, hybridized spin-polarized band [15]. Figure 5.1b shows a free electron density of states diagram characteristic of a nonmagnetic metal N. For a nonmagnetic metal in equilibrium, the Pauli exclusion principle demands that the number of up spins equals that of the down spins. An early contribution of spin-injection theory [9] is the observation that spin-relaxation scattering rates in nonmagnetic materials are very low, and the subsequent assertion that a conductance, g_s, can be defined individually for each spin subband s in a nonmagnetic material.

Many of the advances in the field of spin-polarized transport in the solid state have involved tunneling, and experimental studies began with superconducting tunneling spectroscopy about 30 years ago. As reviewed in Chapter 11, Volume 1, planar tunnel junctions were fabricated using thin, superconducting aluminum films, aluminum oxide tunnel barriers, and a ferromagnetic metal counterelectrode. By deconvolving the measured tunneling conductance, Tedrow and Meservey [1] could deduce the net polarization P of the tunnel current crossing the F/S interface, directly proportional to the asymmetry of the density of states of F at the Fermi level,

$$P = \frac{[N_\uparrow(E_F) - N_\downarrow(E_F)]}{[N_\uparrow(E_F) + N_\downarrow(E_F)]}. \tag{5.4}$$

These experiments provided both an experimental confirmation of Mott's assertion and a quantitative estimate of the polarization, $P \approx 10\% - 45\%$ for polycrystalline films of nickel, iron, or cobalt. They also demonstrated that a polarized current could tunnel across a barrier at a boundary of F and maintain its polarization. A few years later, the SDT experiments of Julliere [2] and Maekawa and Gafvert [3] introduced the sandwich MTJ device structure, $F1/I/F2$ where I is an insulating tunnel barrier that is most commonly used in technological applications. Although the first planar MTJ measurements were performed at cryogenic temperatures, spin-polarized scanning tunneling microscopy (SP STM) experiments gave very good SDT results at room temperature [16, 17] and encouraged the development of planar MTJs that could be operated at room temperature [4]. These MTJs have been the dominant device for magnetic random access memory (MRAM), which is the most important integrated electronics application of the physics of spin-dependent transport (refer to Chapter 13, Volume 3). However, the broad approach of spin injection/detection and geometries in the LSV family continue to be important for novel applications (refer to Chapters 16 and 17, Volume 3) and for spin-transport and spin-orbit interaction (SOI) studies of novel materials.

5.1.1 TRANSMISSION ELECTRON SPIN RESONANCE

The tunneling experiments of Tedrow and Meservey led to a fundamental question: Could a spin-polarized current, equivalently denoted as a magnetization current J_M of oriented dipoles, enter a nonmagnetic material and maintain its polarization over a penetration depth of significant length? The prevailing view at the time was based on the observation that the RKKY interaction, a coupling between dilute magnetic impurities in a nonmagnetic metal host that is mediated by conduction electrons, has a length scale of

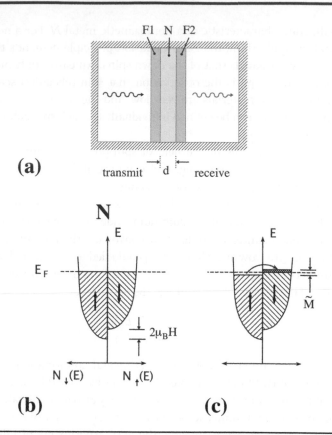

FIGURE 5.2 (a) Schematic cross-section of a TESR experiment. Resonantly tuned radio-frequency (rf) transmit and receive cavities are separated by a nonmagnetic *N* metal foil sample. The transmission was enhanced when the foil was coated by thin ferromagnetic films, *F*1 and *F*2, on either side. Note that the sketch is not to scale. (b) The density of states representation of Pauli paramagnetic splitting in a *N* foil in the presence of external magnetic field *H*. (c) Resonant absorption of rf photons creates a nonequilibrium population of spin-polarized electrons in *N*.

order 1 nm. An analogous argument would suggest that a spin-polarized current injected into a nonmagnetic metal *N* would quickly lose its polarization. The contrary view was presented by Aronov, who predicted that the penetration depth was relatively long and proposed that current J_M could be injected successfully into nonmagnetic metals [18], semiconductors [19], and superconductors [20].

At about the same time, Silsbee was studying spin diffusion in nonmagnetic metals using TESR [21]. Stimulated by Meservey's observation of spin-polarized tunnel currents, thin ferromagnetic films *F*1 and *F*2 were deposited on either side of a high-purity metal foil *N*. The trilayer sample was mounted in a spectrometer where the foil separated the "transmit" and "receive" cavities (Figure 5.2a). As discussed below, the transmission amplitude for samples coated with ferromagnetic films was substantially enhanced when compared to samples with no ferromagnetic coatings. This suggested that spin-polarized electrons were transported across the *F*/*N* interface and diffused into *N*, and these results motivated the first spin-injection experiment.

5.1.1.1 Nonequilibrium Spin Populations

The physics of spin accumulation and spin diffusion are closely related to the principles of electron spin resonance, and there are similarities in the techniques for measuring spin-diffusion lengths and spin-relaxation times. Beginning with a reminder of basic definitions, the magnetic dipole moment of a conduction electron is related to its spin-angular momentum, $\vec{S}$, by

$$\vec{\mu} = \frac{-ge}{mc} \vec{S}, \tag{5.5}$$

where:

- $-e$ is the electron charge
- m is the electron mass
- c is the speed of light
- $g = 2.0$ for free electrons

Choosing a coordinate system that allows for the measurement of spin to be quantized along a convenient axis, the electrons are described as having an up or down spin; the angular momentum is $s_z = s\hbar$, where $s = \pm 1/2$ is the spin quantum number, and it follows that

$$\mu = \pm \frac{e}{2mc} \hbar = \pm \mu_B, \tag{5.6}$$

with μ_B being the Bohr magneton.

If a magnetic field H is applied along the quantization axis, each electron acquires a Zeeman energy associated with its magnetic moment, $\pm \mu_B H$. Figure 5.2b depicts the energy band of free electrons in a nonmagnetic metal in the presence of a magnetic field. The relative shift of the two spin subbands is called Pauli paramagnetism. The subband with moments oriented parallel (antiparallel) with the field H has energy lower (higher) by $\mu_B H$. The net splitting between the subbands, $\Delta E = 2 \mu_B H$, is equivalent to the energy required for a transition from one spin subband to the other. The transition energy can be supplied by absorption of a photon of energy

$$\hbar\omega = \frac{e}{mc} \hbar H \quad \text{or} \quad \omega = -\gamma H = \frac{e}{mc} H, \tag{5.7}$$

where γ is the gyromagnetic ratio. As depicted in Figure 5.2c, spin-up electrons of relatively low energy absorb photons and move to higher energy spin-down states. The result is a nonequilibrium distribution of spin-polarized electrons, $\tilde{M}$, near the Fermi level. The electrochemical potential of down-spin electrons is raised by the addition of carriers from the up-spin subband, and the electrochemical potential of up-spin electrons is depleted by the loss of carriers. One can say that the down-spin subband is pumped to a nonequilibrium level by the resonant absorption of photons. This is a nonequilibrium process because the spin-subband electrochemical potentials relax to equilibrium in the absence of a flux of appropriate photons. In

Figure 5.2c, the energy height of $\tilde{M}$ is less than the Pauli paramagnetic splitting because of spin relaxation. Spin-relaxation processes, described later in this chapter, are reverse transitions and act to limit and deplete the population of spin-down electrons. The resulting nonequilibrium magnetization is a balance between the spin pumping of resonant absorption and the spin depletion caused by spin relaxation. The nonequilibrium population is often called "spin accumulation," and similar populations appear in optical orientation experiments [22].

In TESR, microwave photons are supplied to a "transmit" cavity (Figure 5.2a) and absorbed within a skin depth on one side of a foil sample. The nonequilibrium population of spin-polarized electrons diffuses across the foil and, within a skin depth of the other side, spin relaxation results in reverse transitions that emit microwave photons into the "detect" cavity. An experimentally convenient frequency range for a microwave source and detector is called X-band, $\nu \approx 10$ GHz, $\lambda \approx 3$ cm. Using Equation 5.7, the resonant condition for free electrons is $\nu = 2.8 \times 10^6$ Hz/Oe, and X-band TESR experiments require a resonant field of about $H = 3600$ Oe.

5.1.1.2 Spin Relaxation and Diffusion

The transmitted spin resonance "signal" detected in the receive cavity is described as an amplitude of microwave radiation that varies as a function of the magnetic field, $A(H)$, because typically the frequency is fixed, and the magnetic field is varied. The signal has maximum amplitude centered at the resonant frequency "line," and the characteristic line shape of $A(H)$ can be related to the spin-transport processes that occur in the metal sample. Examples are shown in the schematic drawings of Figure 5.3. The panels on the left depict a cross-section of a sample N of thickness d. The experiment is designed such that a population of nonequilibrium spin-up electrons is introduced within a skin depth on the left side of N, the external field H is perpendicular to the plane of the cross-section, and the "receive" cavity is on the right side of N. The panels on the right show the line shapes $A(H)$ of resonant signals for several different conditions.

Experiments require low spin-scattering rates, $1/T_2$ where T_2 is the spin-relaxation time for N, and the bulk metal foil samples at low temperature have a very high conductivity. In Figure 5.3a, the conductivity is so high that the foil thickness is less than the electron mean free path, $d < \lambda$. The spin-polarized electrons introduced on one side travel with ballistic trajectories, at the Fermi velocity υ_F, across the foil. Throughout the foil, and in particular on the other side, there is a nonequilibrium population of "hot" electrons, near the Fermi surface, having electron spin with the same phase. As the ballistic electrons scatter off the surface on the other side, there is some probability that the scattering event will flip the electron spin by π radians, diminishing the nonequilibrium population and emitting a microwave photon in the process. If the field H were held at a constant value, the ballistic electrons reaching the other side would not change their phase and the spin would be up. As the magnetic field is monotonically varied, the spin phase changes at a constant rate along the ballistic trajectory. The nonequilibrium

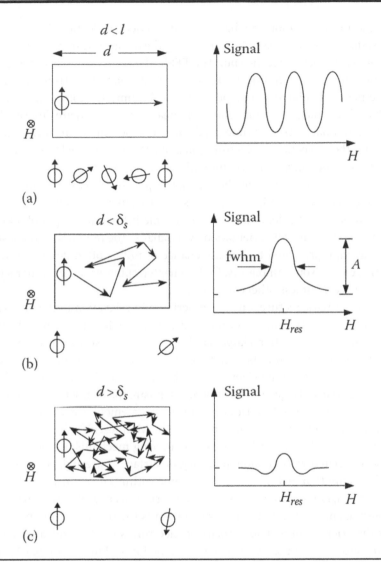

FIGURE 5.3   Resonant emission of photons detected by the receiving cavity as a function of field H for several different conduction electron transport conditions across a foil with thickness d. (a) Ballistic transport results in Larmor waves. (b) Diffusive transport, $d < \delta_s$. Thin limit. (c) Diffusive transport, $d > \delta_s$. Thick limit. Left panels: in cross-section of foil, sketch shows a schematic description of the trajectory of a spin-polarized electron. External field H is perpendicular to the plane of the drawing (parallel to the plane of the foil). Right panels: the shape of the signal detected at the receiver.

population at the right edge will have spin up (down) when the magnetic field, Fermi velocity, and trajectory length correspond to phase rotations of $n\pi$, with n even (odd). The "receive" signal amplitude $A(H)$ takes the shape of repeating oscillations with undiminished amplitude, called Larmor waves [23] because the spin precession occurs at the Larmor frequency.

Larmor waves are not commonly observed for several reasons. A fundamental reason relates to the effects of spin relaxation. A variety of physical mechanisms can result in changes of the spin orientation of an electron, and the most common mechanism is discussed in Section 5.1.1.3. Spin relaxation can be discussed by introducing the mean time between the scattering events that affect the spin. For TESR, the transverse relaxation time, T_2, measures the mean time it takes for the projection of electron spin in the plane perpendicular to the applied field to become random. Experiments such as direct magnetization relaxation measurements [24] are described by the longitudinal relaxation time, T_1, defined as a measure of the mean time it takes for the projection of electron spin along the axis parallel to the applied field to become random. It is commonly assumed that $T_1 = T_2$ in metals [25] and the distinction between them is not important. It is also common to describe spin relaxation in metals by using a parameter τ_s, called a "spin-flip time," which can be defined as the mean time it takes for a spin of a given orientation to change its orientation by π radians. The randomization of spin orientation, represented by a phase change of $\pi/2$, is often described by τ_s, but this carries some ambiguity. In this chapter, spin relaxation is characterized by time T_2 unless otherwise noted.

In a heuristic picture, one can describe spin relaxation in the following way. In every scattering event, the collision results in a random and small change of phase, $\delta\theta$. After a large number of N_s of such scattering events, the phase changes $\delta\theta$ accumulate, and the spin phase has no relation to the initial orientation; the spin orientation has been completely randomized. For a given material and a given relaxation mechanism, the characteristic number of scattering events is $N_s = 1/\alpha$. An equivalent statement is $\alpha \equiv \tau/T_2$, where τ is the mean electron scattering time given by Mathiesen's rule. Parameter α can be expected to differ for different scattering mechanisms, for example, those associated with phonon, impurity, or surface scattering. However, the crude assumption that α is roughly the same within a given material, for any mechanism, is approximately correct and permits a crude description of spin scattering and diffusion. For many metals, α is the order of 0.001, meaning that the orientation of a spin-polarized electron is randomized after about 1000 scattering events. Described in a detailed model in Section 5.3.2, each scattering event marks a "step" in the path of a conduction electron in its diffusive, random walk motion. The spin-diffusion length, δ_s, is defined as the average distance the electron diffuses before its spin orientation is randomized, $\delta_s = \sqrt{DT_2}$, where D is the electron diffusion constant. Using $D \sim v_F^2 \tau$, δ_s can be estimated as $\delta_s \sim \sqrt{v_F^2 \tau T_2} \sim \sqrt{\alpha} v_F \tau \sim 30\ell$.

In Figure 5.3b, the foil thickness is less than the spin-diffusion length, $d < \delta_s$. The nonequilibrium population $\tilde{M}$ is induced near the left surface of N. Since there is no "spin pumping" inside N, the electron distribution is in equilibrium, and, therefore, a gradient of electrochemical potential exists, which drives self-diffusion of the spin-polarized electrons across the foil. All conduction electrons leave the left surface with the same spin phase and acquire a precessional phase angle at the same rate. At any given time, some of the electrons reaching the right surface have moved across the foil width in only a few steps, but others have taken many steps to reach the other side. In the limit of negligible field H, all the electrons reach the other side with nearly the original phase (e.g., spin up), and the signal is maximum. As the magnitude of H is increased, the electrons that have spent a longer time in transit have acquired a phase angle large enough that the spin orientation is random. Spin relaxation imposes a limit on the contributions of the electrons with the most phase accumulation. The result is the Lorentzian line shape depicted in the right panel of Figure 5.3b. The half width at half maximum is $H_{hw} = 1/(\gamma T_2)$.

When the foil thickness is larger than a spin-diffusion length (Figure 5.3c), the amplitude of microwave emission detected at the receive cavity is diminished. In this case, very few electrons have trajectories that go straight across the foil. By contrast, the common trajectory has many steps, and many of the electrons precess by π radians or more if they have reached the right side before spin relaxation. The resulting line shape shows a small dip that represents a net negative spin polarization, which then asymptotically approaches zero with further increases of H. These symmetric features are called side lobes of the resonant line.

The rigorous derivation of fitting functions for TESR line shapes involves solving the Bloch equations in a reference frame that is rotating at the precessional Larmor frequency, ω_L. The identical physical processes occur in spin-injection experiments, but they occur at zero frequency. The spin-injection experiment is, therefore, more easily described, both conceptually and rigorously, and formal solutions will be presented in Section 5.2. However, it should be clear from Figure 5.3b that the relaxation time, T_2, can be deduced from analysis of TESR data. The number of results is limited because only a few metals could be prepared with appropriate thickness and conductivity.

5.1.1.3 Elliott–Yafet Spin-Relaxation Mechanism

Whereas any scattering event can perturb an orbital state, tipping the spin-angular momentum requires a magnetic torque. In the absence of impurities with local magnetic moments, there are few interactions capable of providing such a torque. In metals, the dominant spin-relaxation mechanism is the weakly relativistic spin–orbit interaction derived independently by Yafet [26] and Elliott [27]. An electron with spin, $\vec{s}$, moves in a periodic potential, $U(r)$, with a weakly relativistic Fermi velocity, v_F (Figure 5.4). In the rest frame of the electron, any scattering event that changes the electron trajectory is a perturbation of the periodic electric field, which transforms as a magnetic

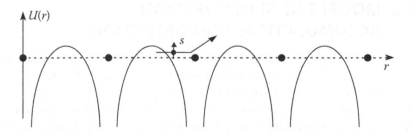

FIGURE 5.4   An electron with spin $\vec{s}$ moves in the periodic potential $U(r)$ of a metal with Fermi velocity v_F. A weakly relativistic spin–orbit interaction can tip the spin orientation. (After Johnson, M., *Magnetoelectronics*, Copyright (2004), figure 1.8, with permission from Elsevier, Oxford, UK.)

field. This effective magnetic field "pulse" can exert a torque on the spin. Formally, the effect derives from a term in the Dirac equation [26].

$$\frac{h}{4m^2c^2}(\nabla U \times \vec{p}) \cdot \vec{s},\tag{5.8}$$

where:

　　m　is the electron mass
　　c　is the speed of light

Inspecting Equation 5.8, any scattering event that changes the momentum, $\vec{p}$, can couple weakly to the spin. Scattering events with impurities, phonons, grain boundaries, and surfaces can affect $\vec{s}$ with different strength. As mentioned above, a phenomenological parameter, $\alpha = \tau/T_2$, can be used to describe the spin-scattering probability per scattering event, for any given material. In Elliott–Yafet theory, the parameter $c_\gamma^2 \propto \tau/T_1$ is a property of a given material and is relatively large (small) for high (low) Z elements (recall $T_1 = T_2$ for metals).

The utility of introducing parameter α to describe the probability of flipping an electron spin, per scattering event and averaged over all scattering events, has been discussed. However, a counter example can be used to illustrate the limits of such simple pictures. The ratio $\alpha = \tau/T_2$ has been measured experimentally for bulk aluminum by TESR [28] and by spin injection [7, 10], and the value $\alpha = 0.001$ was found by both techniques. Since Elliott–Yafet theory predicts that α should be small for Al, an element with a low value of Z, the measurement of such a relatively large value was a mystery. The contradiction was solved by detailed theoretical research [29]. The Fermi surface of polyvalent aluminum is not spherical, and most of the volume is characterized by a low spin–orbit interaction and a small spin-scattering rate, $\alpha \approx c_\gamma^2 \approx 10^{-4}$. However, a few small pockets exist in which the spin–orbit coupling is quite large and the spin-scattering rate is very rapid. Although these pockets are a small portion of the Fermi surface volume, they dominate spin scattering to the extent that the mean spin-scattering rate measured experimentally is indeed an order of magnitude smaller than the rate over most of the volume, a rate consistent with Elliott–Yafet theory.

5.2　MODELS OF SPIN INJECTION, ACCUMULATION, AND DETECTION

The spin-injection phenomenology is unusual in that detailed theoretical modeling preceded the experimental results. This section begins with a review of a microscopic transport model that gave quantitatively accurate predictions of the first experimental measurements, and which continues to be used for the analysis of recent spin-injection experiments in semiconductors and mesoscopic devices. The Johnson–Silsbee thermodynamic theory is also reviewed. This treatment permitted a detailed analysis of charge and spin transport across a F/N interface and is also recognized as the earliest contribution to a new field called "spin caloritronics" [30], discussed in detail

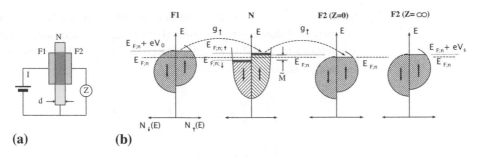

FIGURE 5.5 (a) Cross-section sketch of the pedagogical geometry of spin-injection experiment. The nonmagnetic metal sample N has thickness d. Single domain ferromagnetic films serve as a spin injector ($F1$) and detector ($F2$). A bias current can be driven through the injector and into N. The detector circuit has an arbitrary impedance Z. (b) Density of states drawings for microscopic transport model of spin injection, accumulation, and detection. (After Johnson, M. et al., *Phys. Rev. B* 37, 5312, 1988. With permission.)

in Chapter 8, Volume 3. A derivation of the line shape of the Hanle effect will be presented in Section 5.3.2, in the same section with experimental results.

5.2.1 MICROSCOPIC TRANSPORT MODEL

Pictures of nonequilibrium populations of spin-polarized electrons have been introduced in the context of TESR experiments. When Silsbee et al. [21] coated a metal foil sample with a ferromagnetic film on each surface, a large enhancement of the amplitude of the TESR signal was observed. Silsbee explained these results as an enhanced transfer of spin-polarized electrons across the $F1/N$ interface, along with enhanced transmission at the second, $N/F2$ interface. He proposed that similar transport phenomena would occur in a dc, electrically biased experiment.

Silsbee's idea [31] can be explained by a simple pedagogical geometry (Figure 5.5a) and the microscopic transport model described in Figure 5.5b. A bias current J_e driven through a single domain ferromagnetic film $F1$ and into a nonmagnetic metal sample N carries magnetization across the interface (with unit area A) and into N at the rate J_M. Because current J_M may have arbitrary direction and carry spin magnetization with arbitrary polarization orientation, in general, J_M is a second-rank tensor [8]. A convenient spin quantization axis may be chosen, and then spin polarization has a positive or negative value and J_M is a vector. For one-dimensional flow (e.g., Figure 5.5a), positive current I_M can be reduced to a scalar.

Referring to Figure 5.5b, there is a finite interface resistance between $F1$ and N, so that $F1$ is biased up by a small voltage V_0, $E_{F;f1,\uparrow} = E_{F;f1,\downarrow} = E_{F;n} + eV_0$. The electric current from $F1$ to N is

$$J_{e,1} = \left(\frac{1}{e}\right)\left[g_\uparrow(E_{F;f1,\uparrow} - E_{F;n}) + g_\downarrow(E_{F;f1,\downarrow} - E_{F;n})\right] = (g_\uparrow + g_\downarrow)V_0. \quad (5.9)$$

In Figure 5.5b, the up-spin conductance $g_\uparrow$ dominates. In the steady state, current and charge conservation are strict, the total charge in N must be conserved, and the increase of nonequilibrium up spins must be compensated by an equal decrease in the population of down spins. There is also a small partial current of down-spin electrons that creates a nonequilibrium population of down spins, and this is compensated by an equal decrease of up spins. The net magnetization current is

$$J_M = \left(\frac{\mu_B}{e}\right)\left[g_\uparrow(E_{F;f1,\uparrow} - E_{F;n}) - g_\downarrow(E_{F;f1,\downarrow} - E_{F;n})\right]$$

$$= \left(\frac{\mu_B}{e}\right)(g_\uparrow - g_\downarrow)V_0. \tag{5.10}$$

The ratio of J_M to J_e is

$$\frac{J_M}{J_e} = \frac{g_\uparrow - g_\downarrow}{g_\uparrow + g_\downarrow}\frac{\mu_B}{e} = \eta_1\frac{\mu_B}{e} \quad \text{or} \quad J_M = \eta_1\frac{\mu_B}{e}J_e. \tag{5.11}$$

This defines the interfacial spin-polarization coefficient η_1. The interfacial spin current is not necessarily the same as the fractional polarization p_f in the bulk of F, and the interfacial parameter, η, is similar, but not identical, to P (refer to Equation 5.1).

Silsbee identified the penetration length of polarized carriers injected into N to be identically the same as the TESR spin depth $\delta_s = \sqrt{DT_2}$. Consider a sample with thickness d (refer to Figure 5.5a) larger than an electron mean free path but smaller than a spin depth, $d < \delta_s$. In the steady state, J_M is the rate that spin magnetization is added to the sample region, and relaxation at the rate $1/T_2$ is steadily removing magnetization by spin relaxation. The nonequilibrium magnetization,

$$\tilde{M} = \frac{I_M T_2}{Ad}, \tag{5.12}$$

that results in a balance between the source and sink rates is called *spin accumulation*. It represents a difference in spin-subband electrochemical potential in N, $\tilde{M} \propto N_{n,\uparrow}(E_{F\uparrow}) - N_{n,\downarrow}(E_{F\downarrow})$ and is depicted by dark gray shading in the N layer in Figure 5.5a.

A second ferromagnetic film, $F2$, can be added, fabricated to be in interfacial contact with the sample region, and connected to ground through a low-impedance current meter detector ($Z = 0$ in Figure 5.5b). A positive current, $I_d \propto N_n(E_{Fn,\downarrow}) - N_n(E_{Fn})$, where E_{Fn} is the average chemical potential of the two spin subbands, is driven across the $N/F2$ interface and through the current detector when the magnetizations $M1$ and $M2$ are parallel. When $M1$ and $M2$ are antiparallel, the current $I_d \propto N_n(E_{Fn,\downarrow}) - N_n(E_{Fn})$ is negative. Conceptually, this induced electric current is the converse of the injection process and is an interface effect: A gradient of spin-subband electrochemical potential across the $N/F2$ interface causes an interfacial electric field that drives an electrical current, either positive or negative depending on the sign of the gradient, across the interface. This is an emf source and current conservation demands that a clockwise (counterclockwise) current must be driven in the detecting loop. The current induced across the $N/F2$ interface is [9].

$$J_{e,2} = \left(\frac{1}{e}\right)\left[g_\uparrow(E_{F;n,\uparrow} - E_{F;f2}) + g_\downarrow(E_{F;n,\downarrow} - E_{F;f2})\right]$$

$$= \frac{1}{e}\left[(E_{F;n} - E_{F;f2})(g_\uparrow + g_\downarrow) + \frac{\mu_B \tilde{M}}{\chi}(g_\uparrow - g_\downarrow)\right].$$

(5.13)

Electrode $F2$ may be connected to the ground through a high-impedance voltmeter detector ($Z = \infty$ in Figure 5.5a, b). It is convenient to think of $F2$ as a "floating electrode," giving no concern to experimental issues of how to configure an appropriate ground for the voltage measurement. In this case, a positive (negative) voltage, V_d, is developed at the $N/F2$ interface [8] when $M1$ and $M2$ are parallel (antiparallel). It will be convenient to denote detected voltage V_d as a spin-dependent voltage V_s. The voltage V_s is directly related to the interfacial, spin-subband electrochemical potential gradient described above. The high-impedance ($Z \to \infty$) forces the interfacial current flow to vanish, and the rise (or fall) of the electrochemical potential of $F2$ is the mechanism that stops the current flow. Setting $J_{e,2} = 0$, the voltage V_s is found to be

$$V_s = \frac{\eta_2 \mu_B}{e} \frac{\tilde{M}}{\chi},$$

(5.14)

where η_2 is defined in analogy with η_1 as the ratio of the difference and sum of spin-up and spin-down conductances, and χ is the Pauli susceptibility.

The magnitude of the spin accumulation is described by Equation 5.14. A physical interpretation is easily made after multiplying both sides of the equation by e. On the left-hand side, eV_s represents the energy of a spin-polarized electron at the top of the nonequilibrium distribution. On the right-hand side, $\tilde{M}/\chi$ has units of magnetic field and represents the effective magnetic field that is associated with the spin-polarized electrons. The product $\mu_B \tilde{M}/\chi$ represents the effective Zeeman energy of the spin-polarized electron at the top of the distribution. The polarization η_2 is the fractional efficiency with which the effective Zeeman energy is measured. Equation 5.14 thus describes the effective Zeeman energy of a spin-polarized electron at the top of the nonequilibrium spin population, arising from the effective magnetic field associated with all of the spin-polarized electrons and measured with efficiency η_2.

Equations 5.12 and 5.14 can be combined to give the spin-coupled transresistance R_s that is observed in a spin-injection/detection experiment. In the pedagogical geometry of Figure 5.5a, spin-polarized electrons are injected across the $F1/N$ interface having area A and are confined to a volume Ad, defined by the thickness d of the sample. The case of $d < \delta_s$ is analogous with the TESR thin limit. Using an Einstein relation to relate the electron diffusion constant D to the measured resistivity, the transresistance for the thin limit is [32]

$$R_s = \frac{\eta_1 \eta_2}{\chi} \frac{\mu_B^2}{e^2} \frac{T_2}{\text{Vol}}, \tag{5.15}$$

$$R_s = \eta_1 \eta_2 \frac{\rho \delta_s^2}{\text{Vol}} = \eta_1 \eta_2 \frac{\rho \delta_s^2}{Ad}, \tag{5.16}$$

$$\Delta R = 2R_s = \frac{\eta_1 \eta_2 T_2}{\chi \text{Vol}} \cdot \text{constant}. \tag{5.17}$$

Equation 5.16 is valid for a confined geometry, such as the pedagogical geometry of Figure 5.5a. Equation 5.16 is also valid for the unconfined geometry of the LSV, discussed in Section 5.3.1, when volume Vol is properly defined.

5.2.2 ATTRIBUTES OF SPIN-ACCUMULATION DEVICES

Equation 5.17 is a simple form that illustrates one of the remarkable features of devices that operate with spin injection, detection, and accumulation. The resistance modulation ΔR is inversely proportional to the sample volume occupied by the spin accumulation $\tilde{M}$. At first, this is a result of the simple observation that the units of $\tilde{M}$ are magnetic moments divided by volume. It immediately follows that given a constant number of magnetic moments, shrinking the volume results in a large value of $\tilde{M}$. This simple scaling rule, however, is uniquely different from the more common kinds of devices. It is highly common that the performance of a device fails to scale with shrinking dimensions. In other words, the output level tends to diminish, rather than remain constant, as the dimensions of the device structure are reduced. It is a remarkable contrast that the output level of a spin-accumulation device can be expected to increase as its dimensions are reduced. The important caveats are (1) that the spin-relaxation time is not dramatically shortened and (2) the interfacial injection and detection efficiencies are not dramatically reduced. It will be shown in Section 5.3.5.2 that the simple inverse scaling rule of Equation 5.17 has been confirmed for sample volumes that vary by ten decades.

Inverse scaling is also a characteristic that is uniquely different from GMR. In GMR, interfacial spin-dependent scattering is approximately constant and independent of sample size. The magnetoresistance ratio, MR = $\Delta R/R$, is expected to remain constant. A second important difference is that the electrical output of spin-accumulation devices is bipolar; the output voltage V_s is positive (negative) when magnetizations $M1$ and $M2$ are parallel (antiparallel). Bipolar output has been demonstrated in several experiments [7, 33]. If V_s is in superposition with a nonzero baseline voltage, $|V_B|$, then V_s is relatively high (low) when $M1$ and $M2$ are parallel (antiparallel). This is opposite to the convention of GMR devices, where the resistance is relatively low (high) when $M1$ and $M2$ are parallel (antiparallel). A third difference is that spin accumulation can be measured in a nonlocal geometry, and a fourth difference is that spin accumulation can be reduced to zero by the Hanle effect and the application of a perpendicular magnetic field.

5.2.3 Johnson–Silsbee Thermodynamic Theory

By applying the methods of nonequilibrium thermodynamics to study the transport of charge, heat, and nonequilibrium magnetization in discrete and continuous systems, Johnson and Silsbee extended the linear dynamic equations of thermoelectricity to describe interface thermoelectric and thermomagnetoelectric (spin-charge) effects [8]. These fully general equations can be used to study issues of charge and spin transport in F/N systems and give a deeper understanding of the phenomena.

5.2.3.1 Linear Equations of Motion

Beginning with a review of the thermoelectric system, flows of charge and heat are driven by an electric field and a gradient of temperature. Classically, the dynamic laws are derived for continuous systems and thermoelectric effects are described in terms of bulk properties of conductors, such as the absolute thermopower ε, the thermal conductivity k, and the electrical conductivity σ (recall that g and σ are synonymous and σ is often used in thermodynamic derivations). In the classical treatment of thermoelectric effects, the junctions between two conductors are considered to be in complete thermal equilibrium; no temperature or voltage difference appears across junctions that are assumed to have infinite conductance. This need not be the case, however, and Johnson–Silsbee theory developed equations describing a number of transport effects specific to the behavior of a junction, between two conductors, that has nonzero resistance. Differences of nonequilibrium magnetization $\tilde{M}$ across an interface can drive currents of spin and charge, and the classical thermoelectric system was generalized to include thermomagnetoelectric effects across metal–metal interfaces. The same generalization was then applied to bulk systems. In order to include currents of nonequilibrium spin magnetization, J_M, the magnetization potential H^* was identified as the generalized force associated with the flux J_M, $-H^* = \left(\tilde{M}/\chi\right) - H$. This expression allows for the possible presence of an external magnetic field H, but $H = 0$ is assumed for the discussion below.

The formal approach uses an entropy production calculation, where a *flux* of interfacial current I_N (or bulk current J_N) of a thermodynamic parameter N (charge, heat, and nonequilibrium spin magnetization) is associated with a generalized force, or affinity, F_N (differences or gradients of voltage, temperature, and magnetization potential). Each flux can, in general, be driven by each of the generalized forces, so that each flux can be expanded in powers of F_N. Only the first-order terms are kept for linear response theory, and the coefficients are known as the kinetic coefficients $L_{m,n}$.

For the discrete case of two metals separated by an interface with intrinsic electrical conductance G, electronic transport across the interface is given by the following linear dynamic transport equations [8]:

$$
\begin{pmatrix} I_q \\ I_Q \\ I_M \end{pmatrix} = -G
\begin{pmatrix}
1 & \dfrac{k_B^2 T}{e\varepsilon} & \dfrac{\eta\mu_B}{e} \\[2ex]
\dfrac{k_B^2 T^2}{e\varepsilon} & \dfrac{ak_B^2 T}{e^2} & \eta'\dfrac{\mu_B}{\varepsilon}\left[\dfrac{k_B T}{e}\right]^2 \\[2ex]
\dfrac{\eta\mu_B}{e} & \eta'\dfrac{\mu_B T}{\varepsilon}\left[\dfrac{k_B}{e}\right]^2 & \xi\dfrac{\mu_B^2}{e^2}
\end{pmatrix}
\begin{pmatrix} \Delta V \\ \Delta T \\ \Delta - H^* \end{pmatrix}.
\quad (5.18)
$$

The kinetic coefficients $L'_{m,n}$ can be provided phenomenologically or estimated within a specific transport model. Coefficients $L'_{1,3} = L'_{3,1}$ have been derived from the microscopic transport model (refer to Equation 5.14). The phenomenological parameter η' is comparable with η but would be measured under conditions associated with heat flow, and ε is an energy-dependent transport parameter [8]. Finally, the $L'_{3,3}$ term describes the self-diffusion of nonequilibrium spin populations, and $\xi \approx 1$ is a very good approximation.

The same approach can be followed for a continuous medium, and electronic transport inside a bulk conductor is given by these linear dynamic transport equations [8]:

$$
\begin{pmatrix} J_q \\ J_Q \\ J_M \end{pmatrix} = -\sigma
\begin{pmatrix}
1 & \dfrac{a''k_B^2 T}{eE_F} & \dfrac{p\mu_B}{e} \\[2ex]
\dfrac{a''k_B^2 T^2}{eE_F} & \dfrac{a'k_B^2 T}{e^2} & p'\dfrac{\mu_B}{E_F}\left[\dfrac{k_B T}{e}\right]^2 \\[2ex]
\dfrac{p\mu_B}{e} & p'\dfrac{\mu_B T}{E_F}\left[\dfrac{k_B}{e}\right]^2 & \zeta\dfrac{\mu_B^2}{e^2}
\end{pmatrix}
\begin{pmatrix} \nabla V \\ \nabla T \\ \nabla - H^* \end{pmatrix}.
\quad (5.19)
$$

The kinetic coefficients $L_{m,n}$ are similar to those derived for interfacial transport. For example, $L_{1,3} = L_{3,1} = p_f(\mu_B/e)$ describes the flow of a magnetization current associated with an electric current, with fractional polarization p_f. Values of p_f are $p_f \approx 0.35–0.45$, according to experimental measurements [34]. In a nonmagnetic material, $P_n = 0$. The fractional polarization constant p' would be associated with thermal gradients and has not yet been experimentally measured. Constants a' and a'' can be related to the thermal conductivity k and thermopower ε with a free electron model. Finally, the term $L_{3,3} = \zeta(\mu_B/e)^2$, with $\zeta = 1$, describes self-diffusion of nonequilibrium spins. In most cases, gradients and differences of temperature are small, heat flow is minimal, and all terms except $L_{1,1}$, $L_{3,1}$, $L_{1,3}$, and $L_{3,3}$ are negligible.

5.2.3.2 Charge and Spin Transport Across an *F/N* Interface

Measurements of the average fractional spin polarization in a ferromagnet have provided values for transition metal ferromagnets of $p_f = 0.35–0.6$. In spin-injection experiments, however, the fractional interfacial polarization, η, typically has values of $0.04–0.25$. Because of this discrepancy, details of charge and spin transport across the *F/N* interface may play an important role. Referring to Figure 5.6a, a ferromagnetic metal *F* and nonmagnetic material *N* are in interfacial contact, and, for simplicity, the cross-sectional

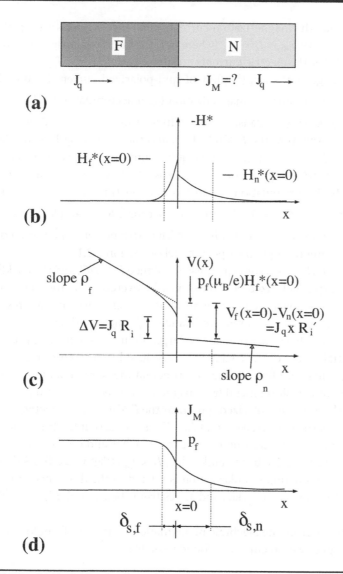

FIGURE 5.6   (a) Model for the flow of charge and spin currents, J_q and J_M, at the interface between a ferromagnetic metal and nonmagnetic material, $x = 0$ at the interface; (b) magnetization potential. The nonequilibrium spin population decays in F and N with characteristic lengths $\delta_{s,f}$ and $\delta_{s,n}$, respectively; (c) voltage; (d) current of spin magnetization, J_M. (After Johnson, M. et al., *Phys. Rev. B* 35, 4959, 1987. With permission.)

area of the interface is taken to be unity and flows are isothermal. A constant electric current J_q is imposed (notation J_q is used for the convenience of a reader who seeks further details in [8, 35]), and the solution for the resultant magnetization current J_M is calculated. There are three distinct regions of electron transport, the bulk of F and of N, and the F/N interface. Equation 5.19 is used to relate currents to potential gradients and to describe steady-state flows in each of the materials F and N. Equation 5.18 is used to relate interfacial currents with differences of potential across the interface. Under the assumption of no interfacial spin relaxation, boundary conditions demand that the magnetization currents of all three regions are equal at the interface ($x = 0$), $J_{M,f} = J_M = J_{M,n}$ (where J_M is the interfacial magnetization current) and that the electric currents are also equal, $J_{q,f} = J_q = J_{q,n}$.

In the general case, the flow of spin-polarized current J_M into N generates a nonequilibrium magnetization (spin accumulation) $-H_n^* = \tilde{M}_n/\chi_n$ in N (Figure 5.6b). Because of spin diffusion in N, $-H_n^*(x)$ decreases as x increases away from the F/N interface, and the decay length is the classical spin-diffusion length, $\delta_{s,n}$. The nonequilibrium spin population can also diffuse *backward*, along $-x$, going back across the F/N interface and into F. The nonequilibrium population $-H^*$ in F is not constrained to match that in N, $-H_f^*(x=0) \neq -H_n^*(x=0)$ (Figure 5.6b) because, for example, the susceptibilities χ_f and χ_n can be quite different. The existence of $-H_f^*$ was ignored in the simple microscopic transport model of Section 5.2.1.

The interface has some intrinsic resistance $R_i = 1/G$, and the backflow of diffusing, spin-polarized electrons must be overcome by the imposed current. The backflow thereby acts as an additional, effective interface resistance (Figure 5.6c). The apparent resistance R_i' measured at the interface, $J_q R_i' = V_f(x=0) - V_n(x=0)$, may be larger than the intrinsic resistance, R_i. The spatial extent of the backflow is described by the spin-diffusion length in F, $\delta_{s,f}$. There are few reliable experimental measurements for $\delta_{s,f}$, but an estimate for transition metal ferromagnetic films is $\delta_{s,f} \approx 10$ nm [36].

The backflow of polarized spins near the F/N interface effectively cancels a portion of the polarized current J_{Mf}. The result is that the fractional polarization of the magnetization current that reaches and crosses the interface, J_M, is reduced relative to the bulk value, $J_M < J_{M,f}$ (Figure 5.6d). As described below, this reduction of polarization is related to the difference in resistivities of the F and N materials and arises from the L_{33} and L_{33}' self-diffusion terms.

After appropriate algebraic manipulations, a general form for the interfacial magnetization current is found to be [8]:

$$J_M = \frac{\eta\mu_B}{e} J_q \left[\frac{1 + G(p_f/\eta)r_f(\xi - \eta^2)/(\zeta_f - p_f^2)}{1 + G(\xi - \eta^2)\left[(r_n/\zeta_n) + r_f/(\zeta_f - p_f^2)\right]} \right], \tag{5.20}$$

where:

$r_f = \delta_f\rho_f = \delta_f/\sigma_f$
$r_n = \delta_n\rho_n = \delta_n/\sigma_n$
$G = 1/R_i$

Noting that $\xi, \zeta_f, \zeta_n \approx 1$, this can be rewritten as follows [8, 37]:

$$J_M = \frac{\eta\mu_B}{e} J_q \left[\frac{1 + G(p_f/\eta)r_f(1 - \eta^2)/(1 - p_f^2)}{1 + G(1 - \eta^2)\left[r_n + r_f/(1 - p_f^2)\right]} \right] \tag{5.21}$$

It is important to note that spin transport is governed by the relative values of the intrinsic interface resistance $1/G$, the resistance of a length of normal material equal to a spin depth, r_n, and the resistance of a length of ferromagnetic material equal to a spin depth, r_f. Typical values of these resistances are easily estimated. A thin transition metal ferromagnetic film has resistivity and spin depth of roughly 20 $\mu\Omega$ cm and 15 nm [36], respectively, giving a value $r_f \sim 10^{-11}$ Ω cm^2 that is nearly temperature independent. Similarly, a nonmagnetic metal film has resistivity and spin depth of roughly 2 $\mu\Omega$ cm and 0.1–1.0 μm, respectively, at low temperature, giving a value in the range $r_n \sim 2 \times 10^{-11}$ to 2×10^{-10} Ω cm^2. A typical value for R_i can be found from the contact resistance R_c measured between a narrow Permalloy electrode and a mesoscopic Ag wire in a lithographically patterned structure, $R_c \approx 10^{-11}$ Ω cm^2 [36]. Since all of the typical characteristic values fall within a range of a factor of 10, all of the terms in Equation 5.21 are expected to be important for the general case.

It is instructive to examine several limiting cases of Equation 5.21. One case is that of high interfacial conductance, $G \to \infty$ (equivalently, low interfacial resistance, $R_i \to 0$). An appropriate experimental system is a multilayer, CPP GMR sample grown in an ultra-high vacuum (UHV) [12]. In this case, $R_i \approx 3 \times 10^{-12}$ Ω cm$^2 \ll r_f$ may justify the high conductance approximation. Equation 5.17 reduces to the simpler form [37]:

$$J_M = p_f \frac{\mu_B}{e} J_q \left[\frac{1}{1 + (r_n/r_f)(1 - p_f^2)} \right], \tag{5.22}$$

and we see that the polarization of the injected current is reduced from that in the bulk ferromagnet by the *resistance mismatch* factor $(1 + M')^{-1} = [1 + (r_n/r_f)(1 - p_f^2)]^{-1}$. Using the above estimates for r_f and r_n, the mismatch factor can be expected to be as large as $M' \sim 20$. This result is readily described. With no interface resistance, the spin accumulation in N flows back across the interface because of self-diffusion. The nonequilibrium spin populations in N and F are closely coupled, and there are two consequences. First, spin accumulation acts as a "bottleneck" for spin *and charge* flow, and there is an apparent interface resistance, even though the intrinsic resistance is assumed to be zero. Second, the backflow of polarized spins partially cancels the forward biased polarized current and reduces the fractional polarization of current crossing the interface. These effects depend directly on the transport properties of carriers in N. If spin relaxation and carrier diffusion are rapid, $r_n \leq r_f$, then spin accumulation (and the spin "bottleneck") is negligible. All of the polarized current carried in F is dumped into N, where it rapidly relaxes [8].

The case of high interfacial resistance R_i is even more important. Spin accumulation in N can be large, but the resistive barrier prevents back diffusion [38]. The nonequilibrium spin population in F remains small, and the voltage drop across the interface is almost entirely due to R_i. The interfacial magnetization current is now given by

$$J_M = \eta \frac{\mu_B}{e} J_q, \tag{5.23}$$

and the fractional polarization is dominated by the interface parameter η. The resistive barrier may be asymmetric with respect to spin. For example, spin-up (or spin-down) electrons may have preferential transmission. The reflection of polarized carriers may diminish the polarization of current in F as it approaches the interface and, in general, the limit $\eta \leq p_f$ is imposed.

A claim that "resistance mismatch" was a fundamental obstacle [39] that would prevent spin injection across a ferromagnetic metal/semiconductor interface was based on a calculation using the infinite interface conductance limit, Equation 5.22. However, a ferromagnetic metal/semiconductor interface is always characterized by a Schottky barrier, tunnel barrier, or low conductance Ohmic contact. In each case, the intrinsic interface resistance is high. As one example, for Ohmic contact to an InAs heterostructure [40], one has $R_i > 10^{-4}\ \Omega\ \text{cm}^2 \gg r_f, r_n$, and electrical spin injection and detection [41] have been observed. For a second example, spin injection from a ferromagnetic metal across a Schottky barrier contact and into GaAs has been demonstrated, using optical detection [42]. As a third example, an MgO tunnel barrier has been shown to provide highly polarized injection into GaAs at room temperature [43]. Contrary to the erroneous assertion of [39], efficient electrical spin injection and detection using transition metal ferromagnetic films and low transmission barriers with an InAs heterostructure was crucial to the development of the spin-injected FET (Spin FET) and the observation of the Datta–Das conductance oscillation (Section 5.4).

Rashba has treated the case of high interface resistance, $R_i \gg r_f, r_n$, using a tunneling calculation to show that "resistance mismatch" effects disappear when transport across a ferromagnet/semiconductor interface is mediated by a resistive barrier [44]. The resulting form of the fractional polarization of the injected current agrees with Equation 5.20 within factors of order unity.* Thus, calculations based on nonequilibrium thermodynamics and those based on tunneling formalism give the same result.

More generally, the result described by Equation 5.23 is valid for many experimental conditions independent of the detailed nature of the interface resistance. Specifically, the junction does not need to be a perfect tunnel barrier. Any resistance that creates a voltage drop that blocks the backflow of polarized spins will permit efficient spin injection across the interface.

Equations 5.18 and 5.19 for charge and spin transport can be used to derive the diffusion equations for charge and spin. These equations are more commonly recognized and are useful for solving specific problems. The derivation is straightforward [38], but the boundary conditions for spin diffusion deserve some comment. For any geometry used to study spin injection and diffusion, charge current is conserved: The charge current that enters a sample at the injector is equal to that which exits at the ground. However, spin orientation is *not conserved*. If a carrier with an oriented spin is injected into the sample, there is no constraint on the orientation of the carrier removed

---

* Specifically, Equation A11 of Ref. [8] is seen to be identically the same as Equation 5.18 of Ref. [44] by using the following identifications between variables in the former and latter: $\eta^* = \gamma$, $p = \Delta\sigma/\sigma$, $n = \Delta\Sigma/\Sigma$, $G(\xi - \eta^2) = 1/r_c$, $\delta_n/(\sigma_n \zeta_n) = r_N$, and $\delta_f/[\sigma_f(\zeta_f - p_f^2)] = r_N$.

at the ground. For LSVs that have F injectors and detectors and are characterized by spin diffusion in N, the constraint $I_M = \eta(\mu_B/e)I_q$ is imposed at the $F1/N$ injector interface, but J_M is unrelated to J_q in N because $P_n = 0$ in a nonmagnetic material.

5.3 TECHNIQUES AND EXPERIMENTAL RESULTS

The spin-injection technique enables the basic research of spin transport in a variety of materials systems. Spin injection can study issues such as spin relaxation and interfacial spin polarization directly, whereas GMR techniques provide such information indirectly, at best. This section begins with a brief overview of experimental techniques and then reviews the experimental results from the original spin-injection experiments on bulk aluminum samples.

5.3.1 IN-PLANE MAGNETORESISTANCE MEASUREMENTS

The techniques of spin injection and accumulation have been applied to several different geometries, but the nonlocal geometry used in the LSV is the most common. The microscopic transport model (Section 5.2.1) introduced a "floating electrode," $F2$, to make a voltage measurement of the spin accumulation (Figure 5.5a). In any real experiment, the voltage must be measured between $F2$ and a voltage ground that is chosen to minimize the effects of any background voltages. Figure 5.7a shows "floating electrode" $F2$ along with an appropriate choice for ground for the measurement of voltage $V_{F2} \equiv V_s$. The ground is provided at a portion of the sample that is remote from the charge flow associated with the bias current.

The *nonlocal geometry* introduced by Johnson and Silsbee [7] provides a physical embodiment of the schematic solution shown in Figure 5.7a. A quasi-one-dimensional metal wire is chosen to lie along the $\hat{x}$ axis and to have a cross-sectional area A. Ferromagnetic electrodes $F1$ and $F2$ cross the N wire near the middle (Figure 5.7b). This structure was introduced by Johnson and Silsbee [7] and is often called an LSV. The electric bias current, I, enters the sample wire at $x = 0$ and is grounded at the left end, $x = -b$. Electric current in the region $x < 0$ is uniform and is characterized by equally spaced lines of constant potential (dash/dot lines in Figure 5.7b). There is zero current flow to the right of $x = 0$, and the portion of the wire $x > 0$ is equipotential. In the absence of nonequilibrium effects, a voltage measurement V from the electrode $F2$ to a ground at $x = b$ would give $V = 0$. The ideal geometry described here is achieved when the electrodes have a negligible width, $W_{F1}, W_{F2} \ll W$ (as depicted in panel (b) where F1 and F2 are shown as lines), and have uniform contact across the width of the wire. Departures from ideality result in a small, spin-independent baseline voltage $V \neq 0$ [45].

The electric bias current also is a current of spin-polarized electrons that are injected at the $F1/N$ interface, at $x = 0$, and they diffuse into N along the $\hat{x}$ axis. Diffusion is equal for $x > 0$ and $x < 0$, and the approximate volume occupied by the nonequilibrium spin population in this unconfined

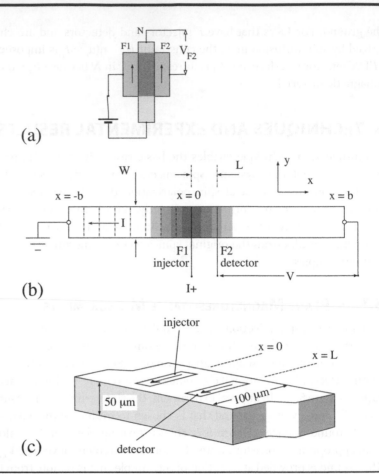

FIGURE 5.7 (a) Schematic cross-sectional sketch of the geometry of a spin-injection experiment, showing the floating detector $F2$. The measurement can be made by choosing the nonlocal ground shown. (b) Schematic top view of nonlocal, quasi-one-dimensional geometry used in the original spin-injection experiment as well as in recent mesoscopic metal wire samples. Dotted lines represent equipotentials characterizing electrical current flow. Gray shading represents the diffusing population of nonequilibrium spin-polarized electrons injected at $x = 0$, with darker shades corresponding to the higher density of polarized electrons. (c) Perspective sketch of bulk Al sample used in the original spin-injection experiment. (After Johnson, M. et al., *Phys. Rev. Lett.* 55, 1790, 1985. With permission.)

geometry is $2A\delta_s$, where the factor of 2 comes from a length of a spin depth on both the positive and negative sides of $x = 0$. The density of diffusing spin-polarized electrons is depicted in Figure 5.7b by the shaded region, with darker shades representing higher density.

When the electrode at $x = L$ is also a ferromagnetic film, $F2$, a measurement of V records a spin-dependent voltage that is relatively high (low) when the magnetization orientation $\vec{M}_2$ is parallel (antiparallel) with $\vec{M}_1$. The difference between high and low values is $\Delta R = 2R_s$ (Equation 5.17). Ferromagnetic films $F1$ and $F2$ are fabricated to have different coercivities, and a small, external magnetic field H applied along the $\hat{y}$ axis is used to manipulate the magnetizations $\vec{M}_1$ and $\vec{M}_2$ between parallel and antiparallel. These measurements with in-plane magnetic fields are similar to those that were later used for studies of GMR and are often called the "magnetoresistance" or "in-plane magnetoresistance" technique.

Since $V = 0$ in the absence of nonequilibrium spin effects, this measurement uniquely discriminates against any background voltages. Furthermore, the electric current density is zero to the right of $x = 0$, and the current of spin-polarized electrons, detected as a spin accumulation at $x = L$, is, therefore, a *pure spin current*, driven by self-diffusion. Such a pure spin current was first demonstrated in the original spin-injection experiment [7].

A detector $F2$ located at a separation L from the injector, $L < \delta_s$, is in the "thin limit." Using Equation 5.17 and the volume Vol $= 2A\delta_s$, the transresistance in this geometry is [32]

$$R_s = \eta_1 \eta_2 \frac{\rho \delta_s}{2A}. \tag{5.24}$$

A detector $F2$ located at a separation from injector $L < \delta_s$ is in "the thick limit," and the transresistance is [7]

$$R_s = \eta_1 \eta_2 \frac{\rho \delta_s}{2A} e^{-L/\delta_s}. \tag{5.25}$$

Because of the explicit exponential dependence $e^{-L/\delta s}$ in Equation 5.25, in-plane magnetoresistance measurements can be performed on a number of samples that are identically prepared while having variable separation L between injector and detector. A plot of the spin transresistance, $R_s(L)$, directly gives a measure of the spin-diffusion length, δ_s. A separate measurement of the diffusion constant from the resistivity ρ then yields a measurement of the spin-relaxation time τ_s.

5.3.2 Nonresonant TESR: Hanle Effect

Precessional dephasing of spin-polarized electrons was described in Section 5.1.1.2 for TESR. The same dynamics occur for the nonequilibrium population of spin-polarized electrons in spin-injection experiments. The experimental technique employs an external magnetic field, applied perpendicular to the plane of the spins. It is called the *Hanle* effect after Walter Hanle performed optical pumping studies and provided an explanation for the depolarization of mercury vapor luminescence in a transverse magnetic field [46].

To derive the line shape experimentally observed in the Hanle effect, we begin with a one-dimensional random walk model that gives a description of the physical principles that is both intuitive and rigorous. The parameters of the model are chosen for the original experiment performed on a "wire" of bulk, high-purity aluminum characterized residual resistivity ratio ($T = 4$ K) of $RRR = 10^4$. The aluminum was rolled into a foil with thickness 50 μm, narrow wires were cut and fastened to a substrate, and then annealed. Each

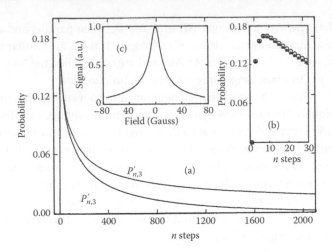

FIGURE 5.8 (a) The unnormalized probability $P_{n,3}$ that a spin-polarized electron took n steps to arrive at the detector located at $L = 3$ in the absence of relaxation. $P'_{n,3}$ is the reduced probability in the presence of relaxation. (b) A detail of (a) showing the distribution of arrival times for a small number of steps. (c) Signal calculated numerically using this model. (After Johnson, M. et al., *Phys. Rev. Lett.* 55, 1790, 1985. With permission.)

aluminum wire sample was about 100 μm wide and 50 μm thick [7, 10] and is sketched in a perspective view in Figure 5.7c. After the process, the sample was measured to have $RRR = 1100$.

The discrete random walk model [7] is similar to the Lewis and Carver [47] model of TESR. Referring to Figure 5.7b, c, spin-polarized electrons are injected in steady state at $x = 0$ and eventually relax by a T_2 process. Transverse motion is neglected, and we assume that each electron undergoes a random walk along the x-axis. Every τ seconds it moves a discrete step of length $\lambda = v_F\tau$, where v_F is the Fermi velocity, either forward or backward. Whenever an electron is at $x = L = p$, it has a small probability of exiting from the bar through the detector. For this sample of electrons, we plot the distribution of times spent in the bar in Figure 5.8. Parameters for the experimental results presented below are $\tau = 9$ ps, $\lambda = 17$ μm, and $p \sim 3$. In the absence of relaxation, the probability $P_{n,p}$ is just the number of ways to get to $x = p$, in n steps, which is the binomial coefficient ${}_nC_{(n+p)/2}$ divided by 2^n:

$$P_{n,p} = \binom{n}{(n+p)/2} 2^{-n}, \quad n + p \text{ even}, \tag{5.26}$$

$$= 0, \quad n + p \text{ odd}. \tag{5.27}$$

The peak probability occurs at around nine steps, as expected for a random walk. There follows a long tail that eventually falls off as $n^{-1/2}$; it takes a long time for electrons to diffuse away from the origin in a one-dimensional bar. In the presence of relaxation, there is a probability per step of $\alpha \equiv \tau/T_2$ ($\alpha = 0.001$ for the results below) that the orientation of a spin is randomized by its collision, and the spin is lost to the system. The probability that a spin-polarized electron arrives at the detector after n steps without relaxation is reduced from $P_{n,p}$ by the factor $\exp(-n\tau/T_2)$ and is plotted in Figure 5.8 as $P'_{n,3}$.

To determine the effect of an imposed magnetic field H parallel to $\hat{z}$, we suppose that the injected electrons are polarized along $\hat{y}$, and the detector is a spin analyzer with orientation along $\hat{y}$ as well. A moment that points at angle θ to $\hat{y}$ will register a magnetization proportional to $\cos\theta$. An external field along $\hat{z}$ will cause every spin to precess through a phase angle of $\delta\theta = \gamma H\tau$ in each step. The signal at the analyzer in a static field H is the sum of contributions of all the electrons, weighted by the probability $P'_{n,p}$ that they started from the injector n steps earlier and multiplied by the precession factor $\cos(\gamma H n\tau)$:

$$\text{Signal} \propto \sum e^{-n\tau/T_2} \, {}_nC_{(n+p)/2}2^{-n}\cos(\gamma H n\tau). \tag{5.28}$$

The signal as a function of field, calculated numerically from Equation 5.28 with $p = 3$, is plotted in Figure 5.8c. In this example, the detector is well within a spin-diffusion length of the injector. Recall from the discussion of TESR that this is the "thin limit." The moments dephase before the net magnetization has precessed, and the feature width is characterized by T_2, which determines the tail in Figure 5.8a. For larger probe separations and/or shorter mean free paths (achieved at higher temperature), the regime changes to $\delta_s > L$. This is known as the "thick limit," in which the detector preferentially samples electrons with relatively long travel times, and the Hanle signal develops side lobes. The width and shape become influenced by the arrival-time distributions as well as the decay time, T_2. In this discrete model, there is a clear physical interpretation of the signal: It is the Fourier transform of the probability $P'_{n,p}$ of finding an unrelaxed moment at the detector at time t.

Although the random walk model is rigorously correct and gives good physical insight into the physics of precessional dephasing, the same physics is described by solving the Bloch equations with a diffusion term [9]. The advantage of the latter is that the solutions are readily adapted as fitting functions to experimental data. The picture described by the Bloch equations, the same description used in Section 5.1.1 for TESR, is the following. Spin-polarized electrons diffuse across a distance, L, from the injector to the detector. Under the influence of a perpendicular magnetic field, each spin precesses by a phase angle that is proportional to the time it takes to reach the detector. Since the electrons are moving diffusively, there is a distribution of arrival times. In the limit of a zero external magnetic field, all the diffusing, polarized electrons that reach the detector have the same phase, as long as they have spent a time less than T_2 in the sample. For sufficiently large field, the spin-phase angles of the electrons reaching the detector at any one time are completely random. As T_2 is increased, there is a larger distribution of arrival times at the detector, and a very small field is needed to randomize the distribution of phases. The characteristic field, H_{hw}, is given by the condition that the product of the precessional frequency and T_2 is a complete precessional rotation angle of 2π, $H_{hw} = 1/\gamma T_2$. It is explicit, in this inverse relationship, that long values of T_2 result in narrow lineshapes of the Hanle effect.

The dependence of the spin-coupled voltage V_s on the transverse field, detected at ferromagnetic electrode $F2$ and presented for the quasi-one-dimensional sample geometry of Figure 5.7, has the form [9]

$$V_s = \frac{\eta_1\eta_2\hbar^2\pi^2 IT_2}{e^2 m^* k_F\, 2A\delta_s}\exp\left(\frac{-L_x f(H_\perp)}{\sqrt{DT_2}}\right)[F1(H_\perp), F2(H_\perp)], \qquad (5.29)$$

where

$$F1(H_\perp) = \frac{1}{f(H_\perp)}\left[f(H_\perp)\cos[g(H_\perp)] - \frac{\gamma H_\perp T_2}{f(H_\perp)}\sin[g(H_\perp)]\right] \qquad (5.30)$$

$$F2(H_\perp) = \frac{1}{f(H_\perp)}\left[\frac{\gamma H_\perp T_2}{f(H_\perp)}\cos[g(H_\perp)] + f(H_\perp)\sin[g(H_\perp)]\right] \qquad (5.31)$$

$$g(H_\perp) = \frac{(d/\sqrt{DT_2})\gamma H_\perp T_2}{f(H_\perp)}, \quad f(H_\perp) = \sqrt{1 + (\gamma H_\perp T_2)^2}.$$

m^* is the electron effective mass
k_F is the Fermi wavevector

Functions $F1(H_\perp)$ and $F2(H_\perp)$ determine the shape of $V_s(H_\perp)$, either "absorptive" or "dispersive" for parallel or crossed injector and detector polarizations, respectively, and examples are shown below. Experimentally, the advantage of using the Hanle effect is that a single measurement gives relaxation time T_2 from the width of the line shape, and polarization η is then the sole parameter to be fit to the amplitude.

5.3.3 EXPERIMENTS: IN-PLANE MAGNETORESISTANCE MEASUREMENTS

Figure 5.9 shows data for a sample with $L = 50$ μm using in-plane fields, $H = H_y$ (refer to Figure 5.7) [7, 10]. As the field is swept along the easy axis of both $F1$ and $F2$ from negative to positive values, the positive voltage for $H < H_{C,1}$ is a measure of spin accumulation, R_s, with M_1 and M_2 parallel. The region $H_{C,1} < H < H_{C,2}$ represents the region where magnetizations M_1 and M_2 are reorienting between parallel and antiparallel, and the detected voltage V_s drops from positive to negative. In the region $H > H_{C,2}$, the orientations M_1 and M_2 return to parallel, and the original, positive voltage is regained. A field sweep from positive to negative values shows the same feature in the field range -100 Oe $< H < -80$ Oe, as expected for the hysteresis of the ferromagnetic films. Note that the voltage V_s is bipolar, positive for parallel magnetizations and negative for antiparallel magnetizations.

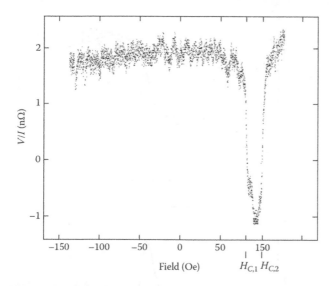

FIGURE 5.9 Magnetoresistance data presented in units of resistance as $R = V/I$. The external field is applied along the $\hat{y}$ axis, in the plane of $F1$ and $F2$, starting at $H_y = -130$ Oe, sweeping up and stopping at $H_y = -130$ Oe. The dip occurs when $\bar{M}_1$ and $\bar{M}_2$ change their relative orientation from parallel to antiparallel. A similar dip occurs for the symmetric negative field when H_y is swept from positive to negative. $L = 50$ μm, $T = 27$ K. (After Johnson, M. et al., *Phys. Rev. Lett.* 55, 1790, 1985. With permission.)

An important ramification of this first *spin-injection* experiment was the first demonstration of magnetization-dependent resistance modulation in a $F1/N/F2$ device structure. The ratio of the difference of resistance $\Delta R = 2R_s$ to the resistance of a segment of Al wire of length L was $\Delta R/R \approx 4\%$ for thin limit samples, somewhat larger than the early spin valve results that were reported a few years later [5].

5.3.4 EXPERIMENTS: HANLE MEASUREMENTS

Examples of the Hanle effect from Al wire samples are shown in Figure 5.10, presented in units of resistance, $R = V/I$. In Figure 5.10a, the injector and detector magnetizations are oriented along perpendicular axes and the resulting lineshape, dispersive in appearance, is fit (solid line) using Equations 5.29 and 5.31. For these data, $L = 300$ μm, $T = 4.3$ K, and the deduced fitting parameters are $\eta = 5.0\%$ and $T_2 = 10$ Ns. Note that the voltage, V_s, is bipolar, positive when the injected spins have precessed to be parallel with the detector and negative when precession has caused the spins to be antiparallel with the detector magnetization. Also note that when the injector and detector are fabricated from the same material, we have $\eta_1 = \eta_2 = \eta$.

In general, the Hanle data are fit to a mixture of absorptive and dispersive contributions, using a linear combination of Equations 5.30 and 5.31 to represent the relative orientation between injector and detector. Figure 5.10b presents more typical data, primarily absorptive in appearance, for a sample with $L = 50$ μm and $T = 21$ K. The deduced fitting parameters are $\eta = 7.5\%$,

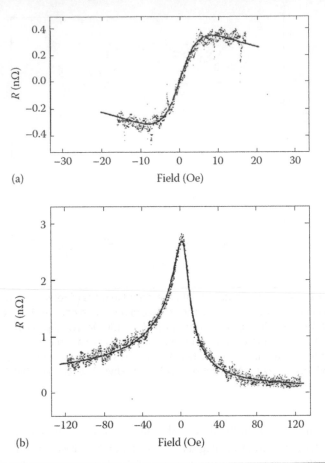

FIGURE 5.10   Examples of Hanle data from bulk Al wire sample presented in units of resistance as $R = V/I$. (a) Example of dispersive lineshape, $L = 300$ µm, T = 4.3 K. Solid line is fit to equations in the text: $T_2 = 10$ Ns, $\eta = 0.050$. (b) Example of Hanle data having absorptive lineshape, with a small admixture of dispersive character. $L_x = 50$ µm, $T = 21$ K. Solid line is fit to equations in the text: $T_2 = 7.0$ Ns, $\eta = 0.075$. (After Johnson, M., *Magnetoelectronics*, Copyright 2004, figure 1.8, with permission from Elsevier, Oxford, UK.)

$T_2 = 7.0$ Ns, $C_1^2 = 0.84$, and $C_2 = -0.39$, where C_1 and C_2 are the fractional contributions of absorptive and dispersive components [10].

The absorptive Hanle effect line shape appears when the magnetization orientations of the injector and detector are aligned parallel and a dispersive shape results when the two orientations are orthogonal. When the orientations have antiparallel alignment, the Hanle effect should have an absorptive shape but with opposite sign, appearing as a dip rather than a peak. This prediction was confirmed by sweeping magnetic field along the $\hat{y}$ axis and stopping the sweep at a field $H_{C,1} < H_y < H_{C,2}$ (refer to Figure 5.7), such that the voltage is negative, and the magnetization orientations are antiparallel. The field H_y is reduced to zero, and the antiparallel alignment is maintained because of hysteresis, assuming the loops of the injector and detector magnetizations to be square. The field is then swept along the $\hat{z}$ axis and the

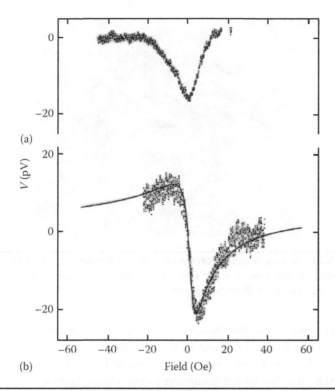

FIGURE 5.11   (a) A field sweep along $\hat{y}$ is halted between $H_{C,1}$ and $H_{C,2}$ (refer to Figure 5.9). The magnetizations $\vec{M}_1$ and $\vec{M}_2$ are now antiparallel. The field along $\hat{y}$ is reduced to zero, and a field sweep along $\hat{z}$ is performed. The resulting Hanle signal now appears with opposite sign, as a dip rather than a peak. (b) An example of the Hanle effect in a sample where the magnetizations have unusual orientations. $L = 200\ \mu m$. (After Johnson, M. et al., *J. Appl. Phys.* 63, 3934, 1988. With permission.)

voltage is recorded. The resulting data, shown in Figure 5.11a, show a Hanle effect with a dip rather than a peak. The Hanle data shown in Figure 5.11b are unusual because the dispersive component is stronger than the absorptive contribution. The angle between magnetization axes of injector and detector was determined to be 124°, an angle beyond perpendicular. These data give an example of the range of injector and detector parameters that were observed. The deduced values of η for these first experiments on bulk samples ranged from 5% to 8%. Values of T_2 were in the range 1–10 Ns for temperatures 4 K < T < 45 K. These parameters will be discussed further in Section 5.3.5.1.

These experiments demonstrated that spin injection is a highly sensitive technique for the detection of spin-polarized electrons. The noise floor of the data represents the threshold of detection, and it corresponds to about 1 nonequilibrium spin out of 10^{12} equilibrium spins. Given the sample volume, typical data with V_s/I the order of 1 nΩ represent a population of roughly one million nonequilibrium spins, several orders of magnitude better sensitivity than TESR. Recently, the Hanle technique has been used with in-plane

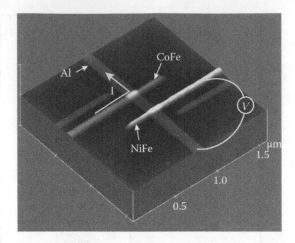

FIGURE 5.12   Scanning atomic force micrograph of the mesoscopic spin-injection device in the nonlocal geometry of Figure 5.7. (After Valenzuela, S.O. et al., *Appl. Phys. Lett.* 85, 5914, 2004. With permission.)

magnetoresistance measurements to study spin diffusion and relaxation in silicon [48] (see Chapters 17 and 18, Volume 2).

A structure comprising one or more thin ferromagnetic films, and having a resistance that varied according to the magnetization orientation of the ferromagnet was first realized using anisotropic magnetoresistance [49]. Anisotropic magnetoresistance sensors were used in read heads from 1991 to 2001 and represented a true paradigm shift for the magnetic recording industry. The first MTJ experiments [2, 3] also represented devices with magnetoresistive states. The early spin-injection experiments gave a new and clear demonstration that an electron spin state could be converted to a resistance or a voltage and helped open the possibilities for the devices and applications that followed.

5.3.5 Experiments: Mesoscopic Metal Structures

One result of the broad interest in spintronics is a renewed interest in the topic of spin injection in thin-film metal wires. The experiments typically involve mesoscopic LSVs in which the sample dimensions, including the separation L between injector and detector, are the order of the electron mean free path, ~100 nm in N. The dimensions of mesoscopic samples have the size scale required for device applications and this generates further interest. These experiments use the Johnson–Silsbee nonlocal geometry that is sketched in Figure 5.7. An atomic force microscope (AFM) image of a typical mesoscopic device is shown in Figure 5.12. The aluminum wire (N) has width $w = 100$–150 nm and thickness 6–10 nm. Ferromagnetic electrodes $F1$ and $F2$ are fabricated using different materials, CoFe and NiFe, respectively, which have intrinsically different coercivities. The F electrodes are fabricated using a three-angle shadow-evaporation technique. The gray wires that do not touch N result from off-angle deposition and can be ignored when understanding the device. The Al is oxidized prior to deposition of $F1$ and

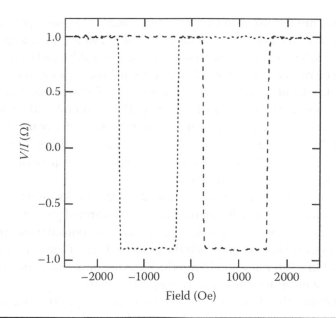

FIGURE 5.13 Example of magnetoresistance data from one of the devices of Ref. [33]. Note the width of the dips, the nearly symmetric bipolar output, and the large magnitude, $\Delta R = 1.9\ \Omega$. (After Valenzuela, S.O. et al., *Appl. Phys. Lett.* 85, 5914, 2004. With permission.)

F2, forming tunnel barriers at the electrode interfaces. In Figure 5.12, *L* is about 200 nm.

The spin-relaxation time in narrow, thin wires is typically so short that the Hanle method would require perpendicular magnetic fields sufficiently large so that the anisotropy fields of *F*1 and *F*2 are exceeded, $\vec{M}_1$ and $\vec{M}_2$ tip to orientations perpendicular to the film plane, and injected spins no longer precess because the external field is now parallel with the axis of spin polarization. Therefore, most experiments use in-plane fields to manipulate the relative magnetizations M_1 and M_2 between parallel and antiparallel orientations to measure resistance dips, $\Delta R = 2R_s$. An example of such data is shown in Figure 5.13 for the LSV shown in Figure 5.12 ($T = 4$ K). The slope of a semilog plot of values $\Delta R(L)$ gives the inverse of the spin-diffusion length. At low temperature, the values of δ_s varied with the thickness of the *N* wire, $\delta_s = 200$ nm (thickness of 6 nm) and $\delta_s = 420$ nm (thickness of 10 nm). Using Equation 5.25, all parameters were measured with the exception of η_1 and η_2. Since the product $\eta_1\eta_2$ is the sole-fitting parameter, there is high confidence in the deduced value of the fractional polarization of the injected current. The authors find values of the geometric mean $\bar{\eta} = \sqrt{\eta_1\eta_2}$ as high as $\bar{\eta} = 25\%$ (4 K) and 16.5% (295 K) [33], for tunnel barriers with impedance values of order 1–10 KΩ. These values corresponded to very large amplitudes of $\Delta R = 2R_s$, larger than 2 Ω at 4 K and larger than 1 Ω at 295 K. Such a large value of ΔR is larger than values of ΔR observed in CPP spin valves and may have ramifications for magnetoelectronic device applications.

Without discussing all of the recent work, a few interesting experiments can be reviewed. Additional studies of nonlocal effects in metallic lateral spin valves are discussed in Chapters 7–9, Volume 3. Mesoscopic Cu wires were studied over a temperature range from 4 to 295 K [50]. The ferromagnetic electrodes were cobalt with slightly different widths, and tunnel barriers were fabricated between the F electrodes and the Cu wire. The Cu wires had a resistivity of 3.4 $\mu\Omega$ cm at 4 K. The Hanle effect was used to measure a spin-relaxation time, $(T = 4$ K$)$ $T_2 = 22$ ps $(\delta_s = 550$ nm$)$, and a fractional polarization with the average value for the two electrodes of $\eta = 5.5\%$ was deduced. The ratio $\tau/T_2 = \alpha_{Cu} = 0.00066$ can be determined. These spin-injection experiments in Cu also studied the temperature dependence of spin transmission across the $F/I/N$ junctions.

A second experiment has reproduced the large value of ΔR observed in thin Al films [51]. Furthermore, it directly confirmed predictions of the Johnson–Silsbee model: Spin currents in N are not coupled to electric currents but instead are driven by self-diffusion, and spin diffusion is isotropic [7, 9, 38]. On a 100 nm wide, 15 nm thick Al wire, ferromagnetic detectors were placed on either side of a ferromagnetic injector, at equal distances of $|L_x| = 300$ nm. The aluminum was oxidized before deposition of the cobalt ferromagnetic electrodes so that a -tunnel barrier, having an impedance of about 20 KΩ, separated the F and N layers. The injected current was grounded at one end of the wire. The voltages of the two Co detecting films were compared and found to be equal. Since one of the detectors was along the current path and the other was in the nonlocal portion of the wire, the equivalence of the detected spin-dependent voltages demonstrated that the spin accumulation diffused equally in both directions. Quantitatively, the magnitude of ΔR was about 0.15 Ω. By analyzing $\Delta R(L)$, a long spin-diffusion length of $\delta_s = 850$ nm was measured $(T = 4$ K$)$. The Al resistivity was low, $\rho = 4$ $\mu\Omega$ cm, and the authors deduced values $\eta = 12\%$ for their tunnel junctions, $T_2 = 110$ ps, and $\alpha = \tau/T_2 = 10^{-4}$. Allowing for the different thickness of the Al wire and differing values of η, the magnitude of spin accumulation is quite comparable with that observed by Valenzuela and Tinkham [33].

As a final example of a mesoscopic system, an LSV using a novel, nonplanar geometry has been introduced [52]. Called a film edge nonlocal (FENL) spin valve, the device is fabricated across the edge of a multilayered film stack so that the critical dimension L is determined by the thickness of one of the films in the stack. Using this geometry, length L can be an order of magnitude smaller than the lengths that are typical of planar LSVs [33, 50, 51]. The multilayer film consists of two F layers, $F1$ (Ni$_{81}$Fe$_{19}$) and $F2$ (Co$_{90}$Fe$_{10}$), separated and capped by insulating layers. After deposition, the multilayer stack was masked with photoresist, leaving a portion near one edge exposed. This portion was completely removed by ion milling, creating a new edge. The removal process allowed the individual layers to terminate cleanly against the new edge. A narrow copper wire (N) was deposited across the edge of the multilayer film, making electrical contact separately to each F layer. The center-to-center spacing was $L = 42 \pm 2$ nm in several prototype devices.

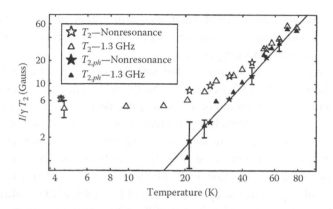

FIGURE 5.14   Comparison of relaxation times T_2 between spin–injection measurements [7] and TESR performed at 1.3 GHz [28], for bulk Al samples of comparable purity. (After Johnson, M. et al., *Phys. Rev. Lett.* 55, 1790, 1985. With permission.)

Magnetoresistance measurements were made with an in-plane field, using both local and nonlocal configurations. The spin-diffusion length could not be measured because L could not be varied over a large range of values. Using a value $\delta_s \approx 450$ nm, found in LSVs made with copper with similar resistivity [53], the FENL spin valves are characterized by $L \approx \delta_s/10 \ll \delta_s$. The observed values of $\Delta R = 2R_s$, about 2.6 mΩ at room temperature, are relatively large. The temperature dependence ΔR (295 K)/ΔR (77 K) was relatively small, which can be expected when ΔR (T) is weakly dependent on the exponential because its argument is very small, $L/\delta_s \ll 1$.

The greater importance of this geometry is the demonstration that spin transport is efficient through the edge of a ferromagnetic film, an observation that enables new kinds of spin valves and MTJs that have dimensions that are not limited by a lithographic feature size but rather are determined by the film thickness. Thin films can be made with thicknesses of order 1 nm, and the FENL spin valve demonstrates a technique for fabricating spintronic device structures with truly nanometer dimensions.

5.3.5.1 EXPERIMENT: SPIN-RELAXATION TIMES

As discussed in Section 5.1.1.2, the simplest model of spin relaxation assigns a scattering probability, $\alpha = \tau/T_2$, per scattering event and averaged for all kinds of events. The mean scattering time, τ, is found from Matthiesen's rule and represents a mean time that is independent of the nature of the scattering. A more careful interpretation of the Yafet–Elliott theory of spin–orbit interactions assumes that electron scattering with impurities, grain boundaries, phonons, and sample surfaces are all different, and each is associated with a different spin-scattering probability.

The first spin-injection experiments, performed on bulk aluminum wires, included measurements of T_2 over a temperature range 4 K $< T <$ 55 K. The values $T_2(T)$ are shown in Figure 5.14, plotted as line width $1/\gamma T_2$ (open stars). Spin relaxation from scattering with surfaces and impurities dominate the residual width at low temperature. This width is subtracted from the data in order to isolate the temperature dependence of spin relaxation from phonon scattering (solid stars).

Spin relaxation in aluminum foil samples of comparable purity has been studied by TESR [28]. In order to minimize the effect of g anisotropies in aluminum, measurements of T_2 from spin injection are compared with data measured at the relatively low frequency of 1.3 GHz. Values of T_2 from resonance measurements are also plotted in Figure 5.14 (open triangles). The residual width is subtracted from the data to isolate the temperature dependence of spin relaxation from phonon scattering (closed triangles). The values of $T_{2,ph}$, adjusted by subtraction of the residual line width, are in extremely good agreement for the two different techniques. This agreement is a graphic demonstration that the physics of the Hanle effect is exactly the same as the physics of TESR, with a simple adjustment for the rotating frame of reference in the latter. The low-temperature spin-relaxation time in Figure 5.14 corresponds to a spin-diffusion length of about 0.4 mm.

TESR is an effective technique for measuring T_2, but it suffers from some limitations. The deduced value T_2 is inversely proportional to the line width (half width at half maximum), $\Delta H_{hw} = 1/\gamma T_2$, and long values of T_2 result in narrow line widths that are easily detected and fit. As T_2 becomes short, on the order of 1 Ns or less, ΔH_{hw} becomes so large that the resonance shape can be difficult to distinguish from other variations in the baseline produced by the spectrometer. Using the spin-injection technique, the separation, L, between injector and detector can be varied, measurements of $V_s(L)$ can be used to determine δ_s, and a determination of the spin-relaxation time T_2 then follows (τ_s and T_2 are considered to be equal, noting that T_2 is commonly used in TESR and τ_s is commonly used for describing films). Spin-injection experiments in mesoscopic samples, where L can vary in the range 50 nm $< L <$ μm, have resulted in measurements of τ_s on the order of 1 ps, thereby extending the range of measured values of τ_s (or T_2) by three orders of magnitude.

TABLE 5.1
Examples of Spin-Relaxation Times for a Variety of Metals, Measured by a Variety of Techniques

Material	T (K)	Technique	$T2$ (s)	α
Cu, bulk [47]	10	TESR	1.8×10^{-8}	7.5×10^{-3}
Al, bulk [54]	4	TESR	1.7×10^{-8}	1.6×10^{-3}
Al, bulk [28]	4	Spin injection	1.4×10^{-6}	1.0×10^{-3}
Al, film [7]	4	Spin injection	1.1×10^{-10}	1.0×10^{-4}
Ag, film [36]	77	Spin injection	3.5×10^{-12}	4.5×10^{-3}
Ag, film [36]	295	Spin injection	2.6×10^{-12}	4.5×10^{-3}
Au, film [55]	295	Spin injection	4.5×10^{-11}	5.0×10^{-4}

Contributions to spin relaxation may come from scattering with impurities, phonon, or surfaces. TESR and spin-injection measurements on high conductivity, bulk samples at low temperature are dominated by surface scattering. The value $\alpha = \tau/T_2$ is an average that uses a simple Drude time for τ.

Table 5.1 shows a variety of values of T_2 for metal samples, measured by several different techniques. The primary purpose of Table 5.1 is to give a sampling of values that are representative of the different materials classes, bulk foil, and thin film. Some remarks can also be made. Consistent with the simple model of spin relaxation described by $\alpha = \tau/T_2$, relaxation times for high resistivity films are much shorter than times for low resistivity foils. The aluminum film studied by Urech et al. [51] has surprisingly large values of δ_s and τ_s. The discrepancy of a factor of 10 in the values of α for this thin film, compared with bulk aluminum, is not understood. With this as an exception, values of α vary by one order of magnitude while values of the relaxation time vary by four orders of magnitude. Since α is expected to differ for different materials, the data in Table 5.1 give support to the simple picture of spin relaxation.

5.3.5.2 Experiment: Inverse Scaling in Lateral Spin Valves

Among the characteristics of spin injection and accumulation device structures discussed in Section 5.2.2, the idea of "inverse scaling" is perhaps the most unique. Equation 5.17 can be used for analyzing and comparing results for a variety of spin-injection experiments performed on mesoscopic samples in the limit $L < \delta_s$. Noting that the susceptibility χ may vary for different materials, Equation 5.17 can be rewritten as

$$\frac{\Delta R \cdot \text{Vol}}{\eta_1 \eta_2 T_2} \approx \text{constant.} \tag{5.32}$$

Equation 5.32 is independent of temperature, although T_2 may have intrinsic temperature dependence. Values of $\Delta R \cdot \text{Vol}/\eta_1\eta_2 T_2$ are shown in Table 5.2 for several recent experiments, along with experimentally derived parameters. All experiments were performed using LSVs, with the third volume dimension given by twice the spin-diffusion length. Furthermore, the ΔR measurements were achieved using an injector/detector spacing, L, that is comparable with δ_s. In order to make a thin limit ($L \ll \delta_s$) comparison that is independent of injector/detector separation L, the values of ΔR are extrapolated to a value $L \approx 0$, resulting typically in an increase by a factor e^1. These values for experiments on three different N materials are in good agreement, with a variation that is less than a factor of 2.

This inverse scaling analysis can be extended to include results from the original LSV [7, 10]. Comparing values in Table 5.2 for bulk and thin-film Al samples, the volume differs by ten orders of magnitude, but the inverse scaling parameter $\Delta R \cdot \text{Vol}/\eta_{av}^2 T_2$ agrees within about 10%. It is remarkable that inverse scaling is upheld so well over ten decades of volume change.

A few remarks can be made about the experimentally measured values of interfacial spin polarization η. First, the values of η observed in spin-injection experiments are much smaller than the limit of $p_f = 0.4$–0.6 assumed for transition metal ferromagnetic films, and the reasons are not yet fully understood. Second, the temperature dependence of η is consistent among

TABLE 5.2
Verification of Inverse Scaling as Predicted by Equation 6.3 and Described in the Text

Material	Vol (cm$_3$)	$\eta_{a\upsilon}$	T2 (ps)	δ_s (µm)	L (µm)	$\Delta R'$ (Ω)	$\Delta R'$Vol/ (Ω; cm$_3$/s)
Al, film	3×10^{-16}	0.17	50	0.20	0.20	2.7	5.9×10^{-4}
Cu, film	5.9×10^{-15}	0.055	22	0.55	0.43	1.2×10^{-2}	1.0×10^{-3}
Ag, film	4.5×10^{-15}	0.22	3.3	0.19	0.30	3.9×10^{-2}	1.1×10^{-3}
Al, bulk	4×10^{-6}	0.06	9000	400	50	2.0×10^{-9}	5.4×10^{-4}

Values of ΔR measured in the thick limit, L $\sim \delta_s$, are extrapolated to L $\rightarrow$ 0 in order to make a thin limit comparison. Thin-film data from Al, Cu, and Ag are from Refs. [33,36,50], respectively, and bulk Al data are from Ref. [7].

most experiments. The value decreases as temperature increases up to 295 K. This may involve spin asymmetries at the injecting interface [50], but a more complete understanding is needed. Third, the condition $R_i \ll r_f\, r_n$ is rarely realized, and, in those few cases, the effects of resistance mismatch are correctly treated using Equation 5.22.

To give some examples of η, for $F/I/N$ tunnel junctions using aluminum oxide barriers, at low temperature, η = 5.5% was measured for Co/Cu samples [50], and η = 15%–25% was measured for $CoFe/Al$ samples [33]. In recent work, MgO barriers were used in Permalloy/Ag samples and η = 11% was measured [55]. The agreement is better at room temperature, η = 12% for Co/Al samples [51] and η = 16% for $CoFe/Al$ [33]. For low-resistance F/N junctions, η = 25% was measured for Co/Au junctions on mesoscopic Au wires at 4 K [56], and η = 24% was measured for $NiFe/Ag$ junctions on mesoscopic Ag wires [36]. The results showing high values of η for LSVs with low interface resistance are important for another reason: They demonstrate that the polarization efficiency η and the output modulation ΔR are independent of the interface resistance and, therefore, independent of the device output impedance, Z_{out}.

5.4 SPIN-INJECTION IN SEMICONDUCTORS: OBSERVATION OF THE DATTA–DAS CONDUCTANCE OSCILLATION

Measurement of the spin lifetimes of carriers in semiconductors has been studied optically and with ESR for many years [57]. Although Aronov had proposed the possibility of spin injection into semiconductors in the 1970s [19], interest in the topic was kindled by the proposal to fabricate a spin-injected field effect transistor (Spin FET) [58] (refer also to Chapter 6, Volume 2). In the Datta–Das structure, a ferromagnetic source and drain are connected by a two-dimensional electron system (2DES) channel, with the source–drain distance L on the order of an electron ballistic mean free path. The magnetizations of both source and drain were oriented along the axis of the channel, the $\hat{x}$ axis. Injected electrons travel along $\hat{x}$ with a weakly relativistic Fermi velocity, $v_F/c \sim$ 1% (where c is the speed of light). An intrinsic electric field E_z, perpendicular to the 2DEG plane, transforms as an effective magnetic field H_y^* in the rest frame of the carriers [59]. Carriers are injected at the source with their spin axes oriented along $\hat{x}$, precess under the influence of H_y^* and arrive at the drain with a spin phase θ that depends on their transit time, $θ \propto L/v_F$. The source–drain conductance was predicted to be proportional to the projection of the carrier spin on the magnetization orientation of the drain and therefore to be a function of θ. By applying a monotonically increasing gate voltage V_G to the channel, the electric field E_z, the effective field H_y^*, and θ would be varied, and an oscillatory modulation of source–drain conductance was predicted to result.

The Datta–Das Spin FET has become the paradigm device of spintronics. The geometry is similar to the LSV but two conceptual differences are important. First, the original Johnson–Silsbee phenomenology, applied to metals, involves diffusive charge and spin transport and the development of spin accumulation. By contrast, the Datta–Das device was designed to employ ballistic charge and spin transport, and no spin accumulation is expected. Second, the transport of spin-polarized carriers in the Datta–Das device is determined by internal electric fields related to Rashba spin–orbit coupling (SOC), and these fields can be modulated by a gate voltage. This has no analogue in metal-based devices. However, the Datta–Das conductance oscillation does have an analogue in classical spin physics: Larmor waves [23], described in Section 5.1.1.2. and Figure 5.3. The former derives from the coherent spin precession of ballistic electrons in a 2DES and the axis of precession is the effective magnetic field $\mathbf{H}^*$, which is perpendicular to the applied electric field $\mathbf{E}$. The latter derives from the coherent precession of ballistic conduction electrons in a high-purity metal foil, and the axis of precession is the externally applied magnetic field $\mathbf{H}$.

Experimental observation of the Datta–Das conductance oscillation, using a Spin FET with both electrical injection and detection, was achieved in 2009 [60]. An independent experimental technique confirmed these results in 2015 [61]. Although these experiments demonstrated the remarkable spin-transport effects predicted by Datta and Das 20 years earlier, the magnitude of the oscillation is very small and has only been observed at cryogenic temperatures. Furthermore, the effect has been observed only in compound semiconductor heterostructures with large SOC. An indirect observation of Rashba SOC spin precession in a gated device also used a compound semiconductor heterostructure at low temperature [62].

A single quantum well (SQW) in an epitaxial indium arsenide heterostructure has characteristics that are ideal for observing the Datta–Das effect. This two-dimensional electron system (2DES) has carriers with a ballistic mean free path (l) of one micron or more, and there is an extremely large Rashba SOC. The large SOC can cause spin precession of a ballistic conduction electron with a spatial wavelength $λ \ll l$. Using high-resolution electron beam lithography, a device structure can be fabricated to have a narrow channel with source and drain having a small separation, $L < l$.

In a heuristic description of the SOC mechanism, the starting point is to note that the confining potential of the SQW has structural asymmetry.

A doping layer inserted on one side of the SQW or band bending that results from boundary conditions imposed by a surface are examples of such an asymmetry. The gradient of the confining potential represents intrinsic electric fields. For an asymmetric potential, fields on opposite sides of the SQW have opposite sign but do not cancel. It follows that there is a residual intrinsic electric field at the position of the 2DES. The field is perpendicular to the plane of the substrate (and of the 2DES) and has substantial magnitude, E_z, even in the absence of gate voltage.

Figures 5.15a, b show a schematic perspective view of a gated LSV structure used in the experiments of Ref. [60]. A current source is connected to a ferromagnetic film injector that has in-plane magnetization with orientation along the axis of a narrow 2DES channel. The current is grounded at the left end of the channel, using the nonlocal LSV geometry. Spin-polarized carriers are injected into the channel with spin axis aligned along $\hat{x}$. A fraction of the injected carriers have ballistic trajectories along the + x-axis, moving with a weakly relativistic Fermi velocity, $v_{F,x} \sim c/300$. If the intrinsic electric field E_z is negligible and no gate voltage is applied ($V_G = 0$), the carrier spin orientation is unchanged during the lifetime of the trajectory (Figure 5.15a).

The dynamics are different when $E_z \neq 0$. In the rest frame of the carrier, field E_z is transformed into an effective magnetic field, H_y^* (Figure 5.15b). This field, equivalently called the Rashba field, $B_{R;y}$ [59], because it formally derives from the Rashba spin–orbit parameter α, is perpendicular to the electron velocity and the electric field vectors.

In this case, electrons injected with spin orientation along the + x-axis precess under the torque of the effective field. This is the basic concept of Datta and Das, who further predicted that adding a gate to the channel and applying a variable gate voltage V_G would modulate the magnitude of E_z. The effective field $\mathbf{H}^*$ would change proportionately and the rate of spin precession is therefore modulated as a function of V_G.

In Figure 5.15, a ferromagnetic electrode detector has a magnetization orientation also along the x-axis. The center-to-center distance from the injector to the detector is L. For a 2DES with a long ballistic mean free path, $l \gg L$, most of the carriers injected with momentum $\hbar k_x$ reach the detector without a scattering event that would randomize spin phase. The detector voltage V will be relatively high (low) when the spin orientation of the ballistic carriers in the vicinity of the detector is parallel (antiparallel) with the magnetization orientation of the detector. The specific prediction of Datta and Das was that the channel conductance of the Spin FET would oscillate periodically as a function of monotonically increasing gate voltage. The simple schematic description in Figure 5.15 is for a case in which the intrinsic field, $E_z(V_G = 0)$, is negligible but internal field E_z is strongly modulated by gate voltage V_G. In Figure 5.15a, the gate voltage is zero, there is no torque on the spin of ballistic electrons in the channel, the spin orientation of the carriers is parallel with that of the detector at $x = L$, and the detected voltage V is relatively high (Figure 5.15c). Figure 5.15b depicts the case with the value of V_G chosen such that $\mathbf{H}^*$ causes spin precession by π radians during the transit from injector to detector. The carrier spin is antiparallel with the detector

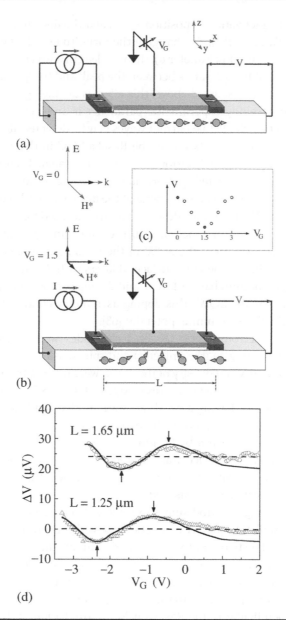

FIGURE 5.15 Schematic representation of the Datta–Das experiment, adapted to the NLSV geometry. Ferromagnetic injector and detector electrodes are fabricated on a gated 2DES heterostructure patterned as a narrow channel. The distance L between electrodes is fixed and is less than a ballistic mean free path. The electrons are injected with spin orientation along the $+x$-axis, and both injector and detector have parallel magnetization orientations. A portion of the injected electrons (shown as solid circles) have ballistic trajectories along the x-axis. (a) $V_G = 0$. Fields **E** and **H*** are negligibly small. No torque is applied to the spin of the electron, and the spin orientation is not changed along the trajectory. The detector magnetization orientation and the spin orientation of proximal carriers are parallel, and detector voltage V is high (solid symbol in panel **c**). (b) $V_G = 1.5$ V This voltage is chosen to correspond with spin precession by π for a typical device. Associated with electric field **E** is effective magnetic field **H***. The spin-polarized electrons precess in response to the torque applied by **H***. For the chosen gate voltage, the detector magnetization orientation and the spin orientation of proximal carriers are antiparallel, and detector voltage V is low [solid symbol in panel (c)]. (c) As V_G is varied, the wavelength λ changes and the detector records a periodic conductance oscillation $\Delta V(V_G)$. (d). Experimental data that demonstrate the Datta–Das conductance oscillation for two values of L. Symbols: data. Solid lines: fits. Details of the fits are found in [60] and [64]. (After Koo, H.C. et al., *Science* 325, 1515, 2009. With permission.)

magnetization at position x = L, and the detected voltage V is relatively low (Figure 5.15c). By changing the magnitude of $\mathbf{H}^*$ (and therefore the Rashba spin–orbit parameter α), the wavelength λ (α) changes and the detector voltage at fixed L varies between the peak and trough values.

As discussed above, this effect is an analogue of Larmor waves, but the controlling experimental parameter is an electric field rather than a magnetic field. While Larmor waves may be detected along with a diffusive TESR feature in a bulk metal, the Rashba SOC in the 2DES of Spin FET is large and the spin-relaxation rate $1/\tau_s$ is very rapid. There is no spin accumulation because injected spin orientation becomes random after only a few scattering events and an analogue of the diffusive Hanle effect is not observed.

The technique described above was used for the first experimental demonstration of coherent spin precession in the channel of a Spin FET [60]. The location of the SQW in the InAs heterostructure is shallow, about 36 nm below the chip surface. Bilayers of InAlAs/InGaAs form both the top and bottom barrier layers, and doping is provided by an n + layer of InGaAs below the SQW. This doping asymmetry contributes to the asymmetry in the SQW confining potential [63].

Ferromagnetic ($N_{81}Fe_{19}$) electrodes were fabricated on top as a spin polarizing source (injector) and spin sensitive drain (detector). The carrier mobility and density of the 2DEG were μ = 50,000–60,000 cm^2 V^{-1}s^{-1} and N_s = 1.8–2.8 $\times$ 10^{12} cm^{-2}, respectively, at T = 1.8 K. A dry mesa etch defined the channel, having a width w_c = 8 μm. Electron beam lithography and lift-off were used to pattern the two ferromagnetic electrodes with lateral dimensions of 0.4 μm by 80 μm and 0.5 μm by 40 μm. After fabrication of the NLSV, the device was covered by a dielectric and a metal gate electrode. Samples with two values L of channel length, L = 1.25 and 1.65 μm, were fabricated and measured.

Figure 5.15d shows examples of the observed Datta–Das conductance oscillation [60]. A magnetic field, with magnitude $B_{a,x}$ = 0.5 T, is externally applied to orient the source and drain magnetizations to be along the x-axis. The gate voltage is varied monotonically over the range $-3 \leq V_G \leq 3$ V while the nonlocal channel conductance is measured. Describing the top trace (L = 1.65 μm) first, the injected spin orientation (along the + x-axis) is perpendicular to the Rashba field. An oscillation of the voltage detected at the drain, as a function of gate voltage V_G, is observed. The range of V_G is sufficiently large that Figure 5.15d shows more than a full cycle of conductance oscillation. Data for a Spin FET with L = 1.25 μm also are shown in Figure 5.15d as the bottom trace.

The conductance oscillations shown in Figure 5.15d were quantitatively analyzed using only a single free fitting parameter, a small arbitrary phase shift. The analysis required direct measurements of the oscillation amplitude A and the Rashba SOC, α (V_G), for the range of gate voltages applied. The amplitude was determined using standard nonlocal lateral spin valve magnetoresistance measurements and an in-plane magnetic field externally applied along the y-axis. Hysteretic dips were observed in the small range of field between the values of switching field for the injecting and detecting films, where magnetizations $M1$ and $M2$ are antiparallel. The magnitude of the dips, $A = \Delta V = 6 \pm 0.2 \ \mu V (\Delta R = 6 \pm 0.2 \ m\Omega)$, is used as the amplitude of the fitting functions in Figure 5.15d. To determine the Rashba parameter α (V_G), Shubnikov de Haas oscillations were measured, and the observed beat patterns were analyzed [60]. As examples of several parameters of interest, the Rashba spin–orbit parameter with zero gate voltage is α $(V_G = 0) = 9.0 \times 10^{-12}$ eV-m and the magnitude of the effective magnetic field (also at zero gate voltage) is $B_{R,y} = 8.5$ T.

The fits shown in Figure 5.15d were generated using the theory of Datta and Das [58] for the voltage recorded at the detector:

$$V = A \cos\left(2m^*\alpha L / \hbar^2 + \phi\right). \tag{5.17}$$

In the above equation, m^* is the effective mass, $\hbar$ is Planck's constant divided by 2π, and ϕ is a small phase shift that may be related to shielding effects from the ferromagnetic films. The quality of both fits (solid lines in Figure 5.15d) is very good. The two data sets, for $L = 1.65$ and $L = 1.25$ μm, can be discussed by comparing their half-wavelengths. As the channel length L decreases from 1.65 μm to 1.25 μm the half-wavelength increases from ΔV_G = 1.24 V to 1.53 V. Thus, the Datta–Das theory successfully describes the dependence of the oscillation with channel length. The detailed shape of the fit has been reproduced through numerical simulations [64] that are accompanied by a physical explanation.

Inspection of Figure 5.15b suggests a second experimental technique for demonstrating the Datta–Das conductance oscillation. If spin–orbit parameter α is held constant, ballistic spin-polarized electrons will precess with a constant wavelength:

$$\lambda = \pi\hbar^2 / \alpha m^* \tag{5.18}$$

The detector voltage measured as a function of displacement x from the injector, $V(x)$, would oscillate with this same wavelength. An experiment using this approach [61] succeeded in measuring two full wavelengths of the oscillation. The authors used the same InAs SQW heterostructure that was used for the Datta–Das technique, and the geometry of the device structure was very similar. A narrow ferromagnetic film was fabricated as a spin injector, using the nonlocal geometry to prevent an Ohmic resistance contribution in the device region. Instead of using a ferromagnetic film as a detector, the authors employed the Spin Hall Effect [65] and confirmed the measurements using the Inverse Spin Hall Effect [66]. The Spin Hall force **F** is a cross product of carrier velocity **v** and spin **S** and is perpendicular to both. Along the ballistic trajectory, **F** is maximum when **S** is along the z-axis and here Spin Hall voltage V_H has a maximum value, either positive or negative. When **S** is along the x-axis, **F** is zero and $V_H = 0$. The detector took the form of a Hall cross with narrow transverse arms. For a constant value of spin-injected current, the voltage between the arms was measured as an external magnetic field changed the magnetization orientation of the ferromagnetic injector from parallel to antiparallel with the x-axis. Since this change corresponds to changing the initial spin orientation by π radians, the Spin Hall force $\mathbf{F}(x)$ and the Spin Hall voltage $V(x)$ change signs. The difference $\Delta V_H(x)$ is therefore proportional to the spin orientation $\mathbf{S}$ of the ballistic electrons at the detector position, x.

For the experiments of Ref. [61], it was not feasible to fabricate a single channel with multiple Hall detectors. Instead, the authors fabricated several identical nonlocal intrinsic Hall effect devices and the discrete channel length L was systematically varied. The width of the 2DES channel was 750 nm. The following sample characteristics were determined (at 1.8 K): carrier density $n = 2.0 \times 10^{12}$ cm^2, mobility $\mu = 60,000$ cm^{2}V^{-1}s^{-1}, and mean free path l = 1.61 μm. From the beat frequency in Shubnikov de Haas measurements, the Rashba spin–orbit parameter was found to be $\alpha = 8.93 \times 10^{-12}$ eV-m, at $V_G = 0$. Using these numbers and m* = 0.05 m$_0$ (m$_0$ = 9.1 $\times$ 10^{-31} kg), the Datta–Das wavelength can be calculated, $\lambda = 0.54$ μm (<<l = 1.61 μm). Plotting values of $\Delta R_H(L)$ showed two complete wavelengths of the conductance oscillation predicted by Datta and Das. It is important to note that there is no arbitrary phase shift in the data of Ref. [61]. The data values are determined only by the injector and detector distance L.

ACKNOWLEDGMENTS

The author is grateful for many important collaborations, beginning with Prof. R. H. Silsbee and extending to numerous colleagues. Specifically, the colleagues whose work is mentioned in this chapter which includes J. Clarke, P. R. Hammar, B. R. Bennett, M. J. Yang, J. Eom, R. Godfrey, A. McCallum, H. C. Koo, S. H. Han, W. Y. Choi, and J. Chang. The author gratefully acknowledges the support of the Office of Naval Research.

REFERENCES

1. P. M. Tedrow and R. Meservey, Spin-dependent tunneling into ferromagnetic nickel, *Phys. Rev. Lett.* **26**, 192 (1971).
2. M. Julliere, Tunneling between ferromagnetic films, *Phys. Lett. A* **54A**, 225 (1975).
3. S. Maekawa and U. Gafvert, Electron-tunneling between ferromagnetic-films, *IEEE Trans. Magn.* **MAG-18**, 707 (1982).
4. J. S. Moodera, L. R. Kinder, T. M. Wong, and R. Meservey, Large magnetoresistance at room temperature in ferromagnetic thin film tunnel junctions, *Phys. Rev. Lett.* **74**, 3273 (1995).
5. G. Binasch, P. Grnberg, F. Saurenbach, and W. Zinn, Enhanced magnetoresistance in layered magnetic structures with antiferromagnetic interlayer exchange, *Phys. Rev. B* **39**, 4828 (1989).
6. M. N. Baibich, J. M. Broto, A. F. Fert et al., Giant magnetoresistance of (001)Fe/(001)Cr magnetic superlattices, *Phys. Rev. Lett.* **61**, 2472 (1988).
7. M. Johnson and R. H. Silsbee, Interfacial charge-spin coupling: Injection and detection of spin magnetization in metals, *Phys. Rev. Lett.* **55**, 1790 (1985).
8. M. Johnson and R. H. Silsbee, A thermodynamic analysis of interfacial transport and of the thermomagnetoelectric system, *Phys. Rev. B* **35**, 4959 (1987).
9. M. Johnson and R. H. Silsbee, Coupling of electronic charge and spin at a ferromagnetic–paramagnetic interface, *Phys. Rev. B* **37**, 5312 (1988).

10. M. Johnson and R. H. Silsbee, The spin injection experiment, *Phys. Rev. B* **37**, 5326 (1988).
11. B. Dieny, V. S. Speriosu, B. A. Gurney, S. S. P. Parkin, D. R. Whilhoit, and D. Mauri, Giant magnetoresistance in soft ferromagnetic multilayers, *Phys. Rev. B* **43**, 1297 (1991).
12. Q. Yang, P. Holody, S.-F. Lee et al., Spin flip diffusion length and giant magnetoresistance at low temperatures, *Phys. Rev. Lett.* **72**, 3274 (1994).
13. S. Schultz, M. R. Shanabarger, and P. M. Platzman, Transmission spin resonance of coupled local-moment and conduction-electron systems, *Phys. Rev. Lett.* **19**, 749 (1967).
14. N. F. Mott, The electrical conductivity of transition metals, *Proc. Roy. Soc. A* **153**, 699 (1936).
15. M. B. Stearns, Simple explanation of tunneling spin-polarization of Fe, Co, Ni and its alloys, *J. Magn. Magn. Mater.* **5**, 167 (1977).
16. M. Johnson and J. Clarke, Spin-polarized scanning tunneling microscope: Concept, design, and preliminary results from a prototype operated in air, *J. Appl. Phys.* **67**, 6141 (1990).
17. R. Weisendanger, H. J. Guntherodt, G. Guntherodt, R. J. Gambino, and R. Ruf, Observation of vacuum tunneling of spin-polarized electrons with the scanning tunneling microscope, *Phys. Rev. Lett.* **65**, 247 (1990).
18. A. G. Aronov, Spin injection in metals and polarization of nuclei, *JETP Lett.* **24**, 32 (1976).
19. A. G. Aronov and G. E. Pikus, Spin injection into semiconductors, *Sov. Phys. Semicond.* **10**, 698 (1976).
20. A. G. Aronov, Spin injection in a superconductor, *Sov. Phys. JETP* **44**, 193 (1976).
21. R. H. Silsbee, A. Janossy, P. Monod, Coupling between ferromagnetic and conduction-spin-resonance modes at a ferromagnetic normal-metal interface, *Phys. Rev. B* **19**, 4382 (1979).
22. F. Meier and B. P. Zakharchenya (Eds.), *Optical Orientation*, North-Holland, New York, 1984.
23. A. Janossy P. Monod, Spin waves for single electrons in paramagnetic metals, *Phys. Rev. Lett.* **37**, 612 (1976).
24. A. Elezabi, M. R. Freeman, and M. Johnson, Direct measurement of the conduction electron spin-lattice relaxation time T_1 in gold, *Phys. Rev. Lett.* **77**, 3220 (1996).
25. C. P. Slichter, *Principles of Magnetic Resonance, Springer Series in Solid-State Sciences*, Vol. 1, 3rd edn., Springer-Verlag, New York, 1990.
26. Y. Yafet, Calculation of the g factor of metallic sodium, *Phys. Rev.* **85**, 478 (1952).
27. R. J. Elliott, Theory of the effect of spin-orbit coupling on magnetic resonance in some semiconductors, *Phys. Rev.* **96**, 266 (1954).
28. D. Lubzens and S. Schultz, Observation of an anomalous frequency dependence of conduction electron spin resonance in Al, *Phys. Rev. Lett.* **36**, 1104 (1976).
29. J. Fabian and S. Das Sarma, Spin relaxation of conduction electrons in polyvalent metals: Theory and a realistic calculation, *Phys. Rev. Lett.* **81**, 5624 (1998).
30. M. Johnson, Spin caloritronics and the thermomagnetoelectric system, *Solid State Commun.* **150**, 543 (2010).
31. R. H. Silsbee, Novel method for the study of spin transport in conductors, *Bull. Magn. Reson.* **2**, 284 (1980).
32. M. Johnson, Spin polarization of gold films via transport, *J. Appl. Phys.* **75**, 6714 (1994).
33. S. O. Valenzuela and M. Tinkham, Spin-polarized tunneling in room-temperature mesoscopic spin valves, *Appl. Phys. Lett.* **85**, 5914 (2004).
34. B. Nadgorny, R. J. Soulen, Jr., M. S. Osofsky et al., Transport spin polarization of Ni_xFe_{1-x}: Electronic kinematics and band structure, *Phys. Rev. B* **61**, R3788 (2000).

35. M. Johnson, Charge–spin coupling at a ferromagnet–nonmagnet interface, *J. Supercond.* **16**, 679 (2003).
36. R. Godfrey and M. Johnson, Spin injection in mesoscopic silver wires: Experimental test of resistance mismatch, *Phys. Rev. Lett.* **96**, 136601 (2006).
37. M. Johnson and R. H. Silsbee, Ferromagnet–nonferromagnet interface resistance, *Phys. Rev. Lett.* **60**, 377 (1988).
38. M. Johnson and J. Byers, Charge and spin diffusion in mesoscopic metal wires and at ferromagnet/nonmagnet interfaces, *Phys. Rev. B* **67**, 125112 (2003).
39. G. Schmidt, D. Ferrand, L. W. Molenkamp, A. T. Filip, and B. J. van Wees, Fundamental obstacle for electrical spin injection from a ferromagnetic metal into a diffusive semiconductor, *Phys. Rev. B* **62**, R4790 (2000).
40. P. R. Hammar, B. R. Bennett, M. J. Yang, and M. Johnson, Observation of spin polarized transport across a ferromagnet/two dimensional electron gas interface, *J. Appl. Phys.* **87**, 4665 (2000).
41. P. R. Hammar and M. Johnson, Detection of spin-polarized electrons injected into a two-dimensional electron gas, *Phys. Rev. Lett.* **88**, 066806 (2002).
42. H. J. Zhu, M. Ramsteiner, H. Kostial, M. Wassermeier, H.-P. Schonherr, and K. H. Ploog, Room temperature spin injection from Fe into GaAs, *Phys. Rev. Lett.* **87**, 016601 (2001).
43. X. Jiang, R. Wang, R. M. Shelby et al., Highly spin-polarized room-temperature tunnel injector for semiconductor spintronics using MgO (100), *Phys. Rev. Lett.* **94**, 056601 (2005).
44. E. I. Rashba, Theory of electrical spin injection: Tunnel contacts as a solution of the conductivity mismatch problem, *Phys. Rev. B* **62**, R16267 (2000).
45. M. Johnson and R. H. Silsbee, Calculation of nonlocal baseline resistance in a quasi-one dimensional wire, *Phys. Rev. B* **76**, 153107 (2007).
46. W. Hanle, Uber magnetische beeinflussung der polarisation der resonanzfluoreszenz, *Z. Phys.* **30**, 93 (1924).
47. R. B. Lewis and T. R. Carver, Spin transmission resonance: Theory and experimental results in lithium metal, *Phys. Rev.* **155**, 309 (1967).
48. I. Appelbaum, B. Huang, and D. J. Monsma, Electronic measurement and control of spin transport in silicon, *Nature* **447**, 295 (2007).
49. R. P. Hunt, Magnetoresistive head, U.S. Patent no. 3,493,694 (1970).
50. S. Garzon, I. Žutić, and R. A. Webb, Temperature dependent asymmetry of the nonlocal spin-injection resistance: Evidence for spin non-conserving interface scattering, *Phys. Rev. Lett.* **94**, 176601 (2005).
51. M. Urech, V. Korenivski, N. Poli, and D. B. Haviland, Direct demonstration of decoupling of spin and charge currents in nanostructures, *Nano Lett.* **6**, 871 (2006).
52. A. McCallum and M. Johnson, Film edge nonlocal spin valves, *Nano Lett.* **9**, 2350 (2009).
53. T. Kimura, T. Sato, and Y. Otani, Temperature evolution of spin relaxation in a NiFe/Cu lateral spin valve, *Phys. Rev. Lett.* **100**, 066602 (2008).
54. S. Schultz and C. Lathan, Observation of electron spin resonance in copper, *Phys. Rev. Lett.* **15**, 148 (1965).
55. J.-H. Ku, J. Chang, H. Kim, and J. Eom, Effective spin injection in Au film from Permalloy, *Appl. Phys. Lett.* **88**, 172510 (2006).
56. Y. Fukuma, L. Wang, H. Idzuchi and Y. Otani, Enhanced spin accumulation obtained by inserting low-resistance MgO interface in metallic lateral spin values, *App. Phys. Lett.* **97**, 012507 (2010).
57. D. Stein, K. V. Klitzing, and G. Weimann, Electron-spin resonance on GaAs-$Al_xGa_{1-x}As$ heterostructures, *Phys. Rev. Lett.* **51**, 130 (1983).
58. S. Datta and B. Das, Electronic analog of the electro-optic modulator, *Appl. Phys. Lett.* **56**, 665 (1990).
59. Y. A. Bychkov and E. I. Rashba, Properties of a 2D electron-gas with lifted spectral degeneracy, *JETP Lett.* **39**, 78 (1984).

60. H. C. Koo, J. H. Kwon, J. Eom, J. Chang, S. H. Han, and M. Johnson, Gate voltage control of spin precession in a spin injected field effect transistor, *Science* **325**, 1515 (2009).

61. W. Y. Choi, H. J. Kim, J. Chang, S. H. Han, H. C Koo, and M. Johnson, Electrical detection of coherent spin precession using the ballistic intrinsic spin Hall effect, *Nat. Nanotechnol.* **10**, 666 (2015).

62. J. Wunderlich, B. G. Park, A. C. Irvine et al., Spin Hall effect transistor, *Science* **330**, 1801 (2010).

63. Y. H. Park, H. C. Koo, J. Chang, S.-H. Han, and M. Johnson, High mobility in a two dimensional electron system with a thinned barrier, *Solid State Commun.* **151**, 1599 (2011).

64. A. N. M. Zainuddin, S. Hong, L. Siddiqui, S. Srinivasan and S. Datta, Voltage-controlled spin precession, *Phys. Rev.* B **84**, 165306 (2011).

65 M. I. Dyakonov and V. I. Perel, Possibility of orienting electron spins with current. *JETP Lett.* **13**, 467 (1971).

66. S.O. Valenzuela and M. Tinkham, Direct electronic measurement of the spin Hall effect, *Nature* **442**, 176 (2006).

6

Magnon Spintronics
Fundamentals of Magnon-Based Computing

Andrii V. Chumak

6.1 INTRODUCTION

A spin wave is a collective excitation of the electron spin system in a magnetic solid [1]. Spin-wave characteristics can be varied by a wide range of parameters including the choice of the magnetic material, the shape of the sample as well as the orientation and size of the applied biasing magnetic field [2, 3]. This, in combination with a rich choice of linear and nonlinear spin-wave properties [4], renders spin waves excellent objects for the studies of general wave physics. One- and two-dimensional soliton formation [5, 6], nondiffractive spin-wave caustic beams [7–10], wave-front reversals [11, 12], and room-temperature Bose–Einstein condensation of magnons [13–16] is just a small selection of examples.

On the other hand, spin waves in the GHz frequency range are of large interest for applications in telecommunication systems and radars. Since the spin-wave wavelengths are orders of magnitude smaller compared to electromagnetic waves of the same frequency, they allow for the design of micro- and nanosized devices for analog data processing (e.g., filters, delay lines, phase shifters, isolators – see, e.g., special issue on circuits, *System and Signal Processing*, Volume 4, No. 1–2, 1985). Nowadays, spin waves and their quanta, magnons, are attracting much attention also due to another very ambitious perspective. They are being considered as data carriers in novel computing devices instead of electrons in electronics. The main advantages offered by magnons for data processing are [17–21]:

+ *Wave-based computing.* If data is carried by a wave rather than by particles, such as electrons, the phase of the wave allows for operations with vector variables rather than scalar variables and, thus, provides an additional degree of freedom in data processing [20–29]. This opens access to a decrease of the number of processing elements, a decrease in footprint [20–26], parallel data processing [27], non-Boolean computing algorithms [28, 29], etc.

+ *Metal- and insulator-based spintronics.* Magnons transfer spin not only in magnetic metals/semiconductors but also in magnetic dielectrics like the low-damping ferrimagnetic insulator

Yttrium-Iron-Garnet (YIG) [30–33]. In YIG, magnons can propagate over centimeter distances, while an electron-carried spin current is limited by the spin diffusing length and does not exceed one micrometer in metals and semiconductors. Moreover, a magnon current does not involve the motion of electrons and thus, it is free of Joule heat dissipation.

✦ ***Wide frequency range from GHz to THz.*** The wave frequency limits the maximum clock rate of a computing device. The magnon spectrum covers the GHz frequency range used nowadays in communication systems [4, 33–35], and it reaches into the very promising THz range [36–38]. For example, the edge of the first magnonic Brillouin zone in YIG lies at about 7 THz [32, 39, 40].

✦ ***Nanosized structural elements.*** The minimum sizes of wave-based computing elements are defined by the wavelength λ of the wave used. Among a wide choice of waves in nature, spin waves seem to be one of the most promising candidates since the wavelength of the spin wave is only limited by the lattice constant of a magnetic material and allows for operations with wavelengths down to the few nm regime (probably the first steps in this direction for magnetic insulators were taken by [41, 42]). Moreover, the frequency of short-wavelength exchange magnons increases with increasing wavenumber as well as the group velocity [43–45].

✦ ***Pronounced nonlinear phenomena.*** In order to process information, nonlinear elements are required in order for one signal to be manipulated by another (like semiconductor transistors in electronics). Spin waves have a variety of pronounced nonlinear effects that can be used for the control of one magnon current by another, for suppression or amplification [2–4, 46, 47]. Such magnon-magnon interactions were used for the realization of a magnon transistor and they open access to all-magnon integrated magnon circuits [24].

The field of science that refers to information transport and processing by spin waves is known as *magnonics* [4, 17, 48, 49]. The utilization of magnonic approaches in the field of spintronics, hitherto addressing electron-based spin currents, gave birth to the field of *magnon spintronics* [18, 19, 50]. Magnon spintronics comprises magnon-based elements operating with analog and digital data as well as converters between the magnon subsystem and electron-based spin and charge currents.

The most recent and advanced realizations of spin-wave logic devices including Boolean and non-Boolean logic gates (e.g., magnonic holographic memory, pattern recognition, prime factorization problem) are addressed in Chapter 18 of this book (Volume 3) written by Alexander Khitun and Ilya Krivorotov.

The current chapter addresses a selection of fundamental topics that form the basis of the magnon-based computing and are of primary importance for the further development of this concept. First, the transport of spin-wave-carried information in one and two dimensions that is required for the realization of logic elements and integrated magnon circuits is covered. Second, the converters between spin waves and electron (charge and spin) currents are discussed. These converters are necessary for the compatibility of magnonic devices with modern CMOS technology. The chapter starts with basics on spin waves and the related methodology. In addition, the general ideas behind magnon-based computing are presented. The chapter finishes with conclusions and an outlook on the prospective use of spin waves.

6.2 BASICS OF MAGNON SPINTRONICS

In this section, a basic knowledge of spin waves in the most commonly used structure, a spin-wave waveguide in the form of a narrow strip, is given. Formulas that can be used for the calculation of spin-wave dispersions as well for spin-wave lifetimes are presented along with the discussions of the main factors that should be considered at the micro and nanoscale. In addition, an estimation of the properties of spin waves with wavelength down to a few nanometers is performed in order to give an understanding of the future potential of the field of magnonics. Finally, magnetic materials used in magnonics are discussed along with the methodologies for the fabrication of spin-wave structures as well as for spin-wave excitation and detection.

6.2.1 SPIN-WAVE DISPERSION RELATIONS

The main spin-wave characteristics can be obtained from the analysis of its dispersion relation, i.e., the dependence of the wave frequency ω on its wavenumber k. In the simplest case, there are two main contributors to the spin-wave energy: long-range dipole-dipole and short-range exchange interactions [2, 3]. As a result, the dispersions of spin waves are complex and significantly different from the well-known dispersion of light or sound in uniform media. Moreover, the dispersion relations in in-plane magnetized films are strongly anisotropic due to the dipolar interaction [51]. In most practical situations, spin waves are studied in spatially localized samples such as thin films or strips which are in-plane magnetized by an external magnetic field. The geometry of a spin-wave waveguide, namely its thickness d and its width w, is a key parameter defining the spin-wave dispersion along with the effective saturation magnetization M_s of the magnetic material, exchange constant A_{ex}, and the applied magnetic field $\mu_0 H$ ($\mu_0 = 1.257 \cdot 10^{-6}$ (T m)/A is the magnetic permeability of vacuum).

In order to calculate spin-wave dispersion relations $\omega(k)$, a theoretical model developed in Refs. [52, 53] can be used. This model takes into account both dipolar and exchange interactions as well as spin-pinning conditions at the film surfaces. In order to adopt the model to the case of a spin-wave waveguide of a finite width w, a total wavenumber $k_{total} = \sqrt{k_\perp^2 + k_\parallel^2}$, and $\theta_k = \arctan[k_\perp / k_\parallel]$ should be considered [54, 55]. Here, $k_\parallel$ is the spin-wave

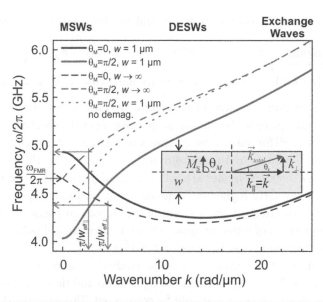

FIGURE 6.1   Spin-wave dispersion characteristics for an infinite in-plane magnetized YIG film (dashed lines) and spin-wave waveguide $w = 1$ μm (solid bold lines) for BVMSW ($\theta_M = 0$) and MSSW ($\theta_M = \pi/2$) geometries calculated using Equation 6.1. The dotted line shows the spin-wave dispersion in the waveguide in the absence of demagnetization. First width mode $n = 1$ is considered, YIG film/waveguide thickness $d = 100$ nm, saturation magnetization $M_s = 140$ kA/m, exchange constant $A_{ex} = 3.5$ pJ/m, a magnetic field of $\mu_0 H_{ext} = 100$ mT is applied in-plane. The inset shows the schematics of the spin-wave wavevector components and corresponding angles for the case of MSSW. Dashed lines show the long and short axis of a spin-wave waveguide.

wavenumber along the waveguide (see inset in Figure 6.1), $k_\perp = n\dfrac{\pi}{w}$ is the perpendicular quantized wavenumber with the number of spin-wave width mode n (please note that in some cases an effective width of the waveguide rather than the real width should be used). In the following $k \equiv k_\parallel$ is used underlying that the wave propagating along the waveguide is of importance. The circular frequency of the spin wave can then be presented:

$$\omega(k) = \sqrt{\left(\omega_H + \omega_M \lambda_{ex}\left[k^2 + (n\pi/w)^2\right]\right)\left(\omega_H + \omega_M \lambda_{ex}\left[k^2 + (n\pi/w)^2\right] + \omega_M F\right)}, \quad (6.1)$$

where $\omega_H = \gamma\mu_0 H_{eff}$, $\omega_H = \gamma\mu_0 M_s$, $\gamma = 1.76\cdot10^{11}$ rad/(s T) is the gyromagnetic ratio, $\mu_0 H_{eff}$ is the effective internal magnetic field, M_s is the saturation magnetization, $\lambda_{ex} = 2A_{ex}/(\mu_0 M_s)^2$ and A_{ex} are the exchange constants. F is given by

$$F = 1 - g\cos^2\left(\theta_M - \theta_k\right) + \frac{\omega_M\, g(1-g)\sin^2\left(\theta_M - \theta_k\right)}{\omega_H + \omega_M \lambda_{ex}\left[k^2 + \left(n\pi/w\right)^2\right]}, \quad (6.2)$$

where $\theta_k = \arctan\left[n\pi/\left(kw\right)\right]$ is the angle between the spin-wave wave vector and the long axis of the waveguide – see inset in Figure 6.1, θ_M is the angle between the magnetization direction and the long axis of the waveguide (in this model $0 \leq \theta_M$, $\theta_k \leq \pi/2$), $g=1-\left[1-\exp\left(-d\sqrt{k^2+\left(n\pi/w\right)^2}\right)\right]/\left(d\sqrt{k^2+\left(n\pi/w\right)^2}\right)$. It is important to note that strictly speaking the dispersion relation Equation 6.1 is valid for the case when $kd < 1$ (however, deviations take place in the dipolar-exchange part of the spectrum only, the model works well for pure exchange waves with large k), it takes into account fully unpinned spins at the bottom and top surfaces of the waveguide, higher-order thickness modes are not considered (in nm-thick samples they are usually few GHz higher in frequency) and magnetic crystallographic anisotropy is omitted (for the account of the anisotropy please see, e.g., Ref. [56]). Comparison of different approaches for calculating dipolar-exchange spin-wave dispersions can be found in Ref. [57].

Nowadays, spin waves are usually studied in nanometer-thick and micrometer-wide waveguides and, therefore, two additional factors, which define spin-wave properties, should be considered. These are the demagnetization and the spin-pinning conditions at the edges of the waveguide. To address these factors, it makes sense to consider waveguides magnetized longitudinally ($\theta_M = 0 \Rightarrow \vec{k} \parallel \vec{M}$) and transversally in-plane ($\theta_M = \pi/2 \Rightarrow \vec{k} \perp \vec{M}$) separately. In the first case, the demagnetization along the longitudinal direction can be ignored since the length of the waveguide is considered to be much larger in comparison to other waveguide dimensions. Therefore, the static internal magnetic field $\mu_0 H_{\mathrm{eff}}$ is uniform and is equal to the externally applied field $\mu_0 H_{\mathrm{ext}}$. The pinning conditions will be mainly defined by the dipolar alternating magnetic fields [58–60] and typically the spins at the edges are partially unpinned. The pinning can be described in terms of the effective width of the waveguide $w_{\mathrm{eff}\parallel} \geq w$. If the spins are fully pinned then $w_{\mathrm{eff}\parallel} = w$ and if the spins are fully unpinned then $w_{\mathrm{eff}\parallel} \rightarrow \infty$. The effective width (for $\theta_M = 0$) can be found in Ref. [58]:

$$w_{\mathrm{eff}\parallel} = w\frac{D_{\mathrm{Dip}}}{D_{\mathrm{Dip}}-2}, \tag{6.3}$$

where $D_{\mathrm{Dip}} = 2\pi(w/d)/[1+2\ln(w/d)]$.

The case of the transversally magnetized spin-wave waveguide $\theta_M = \pi/2$ is more complex since the external magnetic field has to compete with a shape anisotropy that tends to align the magnetization along the long axis in order to minimize the static stray fields [61, 62]. Therefore, the internal magnetic field is smaller than the external field and is nonuniform showing minima at the edges and a maximum in the center. In the assumption that the magnetization is always aligned along the short axis, the effective internal field can be found at [62]:

$$\mu_0 H_{\mathrm{eff}} = \mu_0 H_{\mathrm{ext}} - M_s\frac{\mu_0}{\pi}\left[\arctan\left(\frac{d}{2z+w}\right)-\arctan\left(\frac{d}{2z-w}\right)\right], \tag{6.4}$$

where z is the coordinate transverse to the waveguide ($-w/2 \leq z \leq w/2$). Usually, spin waves propagate in the center area of the waveguide with relatively uniform internal magnetic field. (However, waves can also propagate in the strongly nonuniform magnetic fields close to the edges of the waveguides that are known as edge modes [41, 54, 55, 63].) In the case of the spin-wave propagation in the center, the internal magnetic field is considered to be uniform with a value $\mu_0 H_{\mathrm{eff}}^{\mathrm{max}}$ equal to the maximal value given by Equation 6.4 at $z = 0$. The effective width of the waveguide $w_{\mathrm{eff}\perp}$, in this case, can be defined in different ways, e.g., as the distance between the points where the effective field is reduced by 10%, i.e., to the value $0.9\,\mu_0 H_{\mathrm{eff}}^{\mathrm{max}}$.

The dispersion relations for an infinitely large plane film (dashed lines) as well as for a spin-wave waveguide (bold solid lines) are shown in Figure 6.1 for the cases of the spin-wave waveguide magnetized along the long axis $\theta_M = 0$ and transversally $\theta_M = \pi/2$. Only the first width mode $n = 1$ is shown for simplicity. The magnetic parameters of the commonly used ferrimagnetic material YIG are considered [41, 42, 64, 65]: Saturation magnetization $M_s = 140$ kA/m, exchange constant $A_{ex} = 3.5$ pJ/m, the thickness of the waveguide is $d = 100$ nm, and its width is $w = 1$ μm. The external magnetic in-plane field is $\mu_0 H_{\mathrm{ext}} = 100$ mT. The spin-wave dispersions comprise three main regions: the region of small wavenumber corresponds to dipolar waves usually termed MagnetoStatic Waves (MSWs) [2, 3], the region of large wavenumbers corresponds to exchange waves, and the region between corresponds to the Dipolar-Exchange Spin Waves (DESWs) – see labels above in Figure 6.1. The dipolar wave that propagates along the magnetization direction is usually called a *Backward Volume MagnetoStatic Wave* (BVMSW), the wave that propagates perpendicularly is named the *Magnetostatic Surface Spin Wave* (MSSW) or *Damon-Eshbach mode* [2, 3, 51]. (The spin wave propagating in the out-of-plane magnetized film is named the *Forward Volume MagnetoStatic Wave* (FVMSW) [2, 3, 66] and is beyond the scope of this section.) Please note that with the account of the exchange interaction, the terminology used for the dipolar MSWs is not strictly speaking applicable. However, even in this case one often uses terms like BVMSW- or MSSW-geometries. A special frequency, which is indicated in Figure 6.1 with a black arrow, is the FerroMagnetic Resonance (FMR) frequency $\omega_{\mathrm{FMR}}/2\pi \approx 4.65$ GHz. The FMR describes the case of a uniform magnetization precession $k_\| = k_\perp = 0$. The dispersions for both BVMSW and MSSW modes start from this frequency for the case of a plane film. However, if one considers a waveguide, the frequency of the $k = 0$ wave differs from the FMR frequency for both magnetization configurations since in the waveguide there is always $k_\perp > 0$. The frequency of the BVMSW mode $\omega_{\|,k=0}/2\pi \approx 4.93$ GHz is defined by the frequency of the first standing mode, which corresponds to an MSSW with $k_\perp = \pi/w_{\mathrm{eff}\|} = 2.58\,\mathrm{rad}/\mathrm{\mu m}$ and is higher than the FMR frequency shown by the black arrow in Figure 6.1. Please note that, here, the effective width $w_{\mathrm{eff}\|} \approx 1.22\,\mathrm{\mu m}$ calculated using Equation 6.3 is considered instead of the real waveguide width w. It can be seen in the figure that with an increase in k, the frequency of the BVMSW mode decreases for both, the plane film as well as the waveguide. This happens due to the dipolar interaction and results in a negative group velocity of the dipolar BVMSWs. This negative velocity allows for phenomena like the reverse Doppler shift [67, 68]. With a further increase in k, the frequency of the spin wave increases due to the exchange interaction according to the k^2 law.

The frequency at the starting point $k = 0$ of the MSSW in the waveguide $\omega_{\perp,k=0}/2\pi = 4.03$ GHz lies below the FMR frequency ω_{FMR}. There are two phenomena responsible for such an occurrence. First of all, again, the quantization over the waveguide width plays a role. This time one has to consider that the first width mode corresponds to a BVMSW mode with $k_{\perp} = \pi / w_{\text{eff}\perp} = 4.62$ rad$/\mu$m, where the effective waveguide width $w_{\text{eff}\perp} = 0.68$ μm calculated according to Equation 6.4 and the description below the equation is used. The second mechanism responsible for the further shift of the $\omega_{\perp,k=0}$ frequency down is the demagnetization. According to Equation 6.4, the field in the center of the waveguide is $\mu_0 H_{\text{eff}} = 88.8$ mT and it is smaller than the applied external field $\mu_0 H_{\text{ext}}$. In order to demonstrate the contributions of the demagnetization and quantization individually, an "artificial" dispersion in the absence of the demagnetizing field (11.2 mT) is shown in Figure 6.1 by the dotted line. In Figure 6.1, the first standing BVMSW mode defines the $\omega_{\perp,k=0\,\text{nodemag}} = 4.38$ GHz frequency. Therefore, the frequency $\omega_{\perp,k=0}$ is well below the corresponding frequency $\omega_{\perp,k=0\,\text{nodemag}}$ in the absence of the demagnetization. Finally, it can be observed that with an increase in k the spin-wave dispersion characteristics for the plane film and waveguides are getting closer to one another. It is a consequence of the fact that the exchange spin waves are not angle dependent and are less sensitive to the geometry of the waveguide than the dipolar MSWs.

6.2.2 MAGNON LIFETIME AND FREE-PATH

If spin waves are to be considered as information carriers in future data processing devices, one can formulate a set of requirements for the spin-wave characteristics [19]. Particularly, that the minimum size of a magnonic device should be larger than the wavelengths $\lambda = 2\pi/k$ in order to keep access to the usage of the phase of the spin waves. The minimization of the delays for data transfer between the different magnonic elements requires the maximization of the spin-wave group velocity v_g. In the simplest case, the clock rate of the devices is limited by the spin-wave frequency ω and, therefore, the frequency should also be as high as possible. Further, the parasitic loss in the magnetic devices will be inversely proportional to the spin-wave free-path (or propagation length) $l_{\text{free}} = v_{\text{gr}}\,\tau$, where τ is the spin-wave lifetime. Finally, in order to exploit the spin-wave phases for data processing, the ratio of the spin-wave free-path to the wavelength l_{free}/λ is of importance.

In general, the minimal wavelength of the spin wave is limited by the lattice constant of a magnetic material and, therefore, magnonics has similar fundamental limitations as electronics. Moreover, the decrease in the wavelength (in the assumption of purely exchange spin waves of wavenumbers staying away from the edge of the Brillouin zone) results in an increase of the frequency $\omega \propto k^2$ of the spin wave and the group velocity $v_{gr} \propto k$. But the spin-wave lifetime τ is, in the simplest case, inversely proportional to the spin-wave frequency $\tau \propto 1/\omega$ [3] and, thus, the dependence of the spin-wave propagation length on its frequency is $I_{\text{free}} \propto 1/\sqrt{\omega}$. However, in the case when both dipolar and exchange interactions are considered, the dependence of the free-path on the spin-wave wavenumber $I_{\text{free}}(k)$ is not trivial. In order to estimate it, let us consider the same YIG waveguide discussed above, which is magnetized along its long axis (this magnetization direction is preferable since it requires minimal external field and ensures uniform internal magnetic field). The lattice constant of YIG is approximately 1.24 nm [30–33, 64] and, in order to stay away from the edge of the first Brillouin zone, the minimal wavelength of 5 nm is considered below. The dispersion relation of the first width mode $n = 1$ calculated using Equation 6.1 is shown in Figure 6.2a in logarithmic scale (left axis). It can be seen that the spin wave with a wavelength of 5 nm has a frequency of about 2 THz in YIG. On the right axis, the corresponding spin-wave wavelengths are shown.

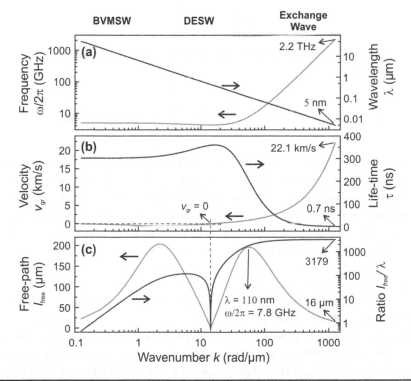

FIGURE 6.2 (a) Spin-wave dispersion characteristics (left axis) and corresponding spin-wave wavelength (right axis). (b) Spin-wave group velocity (left axis) and lifetime (right axis) as functions of spin-wave wavenumber. (c) Spin-wave free-path (left axis) and the ratio of the free-path to the spin-wave wavelength (right axis) as functions of spin-wave wavenumber. The first width mode $n = 1$ is considered in YIG waveguide of width $w = 1$ μm and thickness $d = 100$ nm magnetized along with $\mu_0 H_{\text{ext}} = 100$ mT magnetic field. The Gilbert damping parameter of nano-YIG is assumed to be $\alpha = 1 \cdot 10^{-4}$, which agrees with the latest reports [64].

The group velocity of spin waves is defined as $v_{gr} = \partial\omega / \partial k$ and can be found by taking differentiation of Equation 6.1. The dependence of the velocity on the spin-wave wavenumber is shown in Figure 6.2b on the left axis. It is clear that in this particular case the group velocity is negative for the BVMSW with small wavenumbers, then it passes through the zero value in the DESW region, and increases monotonically in the exchange region reaching values of around 20 km/s.

The main parameters that define the lifetime τ of a spin wave are the spin-wave frequency ω and the Gilbert damping constant of a magnetic material α [2, 3, 69–71] (1/α defines approximately the number of precession periods before it vanishes). In the simplest case of a circular magnetization precession, the lifetime is simply defined as τ = 1/(αω). (Please note that the inhomogeneous broadening of the FMR linewidth ΔH_0 [71] is neglected). However, the magnetization precession in spin-wave waveguides is usually elliptic and the lifetime can be found according to the phenomenological loss theory [3]:

$$\tau = \left(\alpha\omega \frac{\partial\omega}{\partial\omega_H} \right)^{-1}, \tag{6.5}$$

with the aforementioned definition $\omega_H = \gamma\mu_0 H_{\text{eff}}$. Differentiating Equation 6.1 one obtains

$$\tau = \frac{1}{\alpha}\left(\omega_H + \omega_M \lambda_{ex}\left(k^2 + (n\pi/w)^2 \right) + \frac{\omega_M}{2}\left[1 - g\cos^2\left(\theta_M - \theta_k \right) \right] \right)^{-1} \tag{6.6}$$

The dependence of the lifetime on the wavenumber calculated according to Equation 6.6 is shown in Figure 6.2b (right axis) for YIG having a Gilbert damping constant $\alpha = 1 \cdot 10^{-4}$. One sees that in the MSW region the spin-wave lifetime is approximately a few hundred nanoseconds and is practically constant due to the rather flat dispersion and due to the decrease of ellipticity with increasing wavenumber. As opposite, in the exchange region the spin-wave frequency increases rapidly with the wavenumber (Figure 6.2a) and the lifetime drops down to a value around 1 ns.

The spin-wave free-path (the distance which a spin wave propagates before its amplitude decreases to its 1/e value) $I_{\text{free}} = v_{gr}\tau$ is shown in Figure 6.2c on the left axis. It can be clearly observed that the free-path of the long-wavelength BVMSW is rather large and is close to a few hundred micrometers. Moreover, I_{free} is almost proportional to the film thickness t for MSWs and, therefore, this value is much larger in the μm-thick YIG samples that were intensively studied in the recent decades [4]. However, the miniaturization of magnonic elements requires a decrease in the thickness of the structures to the nm scale and the propagation distance of MSWs drastically reduces. Therefore, exchange rather than dipolar waves are primarily of interest for future investigations [43–45] (although nowadays experimental magnonics mainly operate with dipolar MSWs due to the relative simplicity of the methodology [4, 19, 54, 72]). It can be seen in Figure 6.2c that the free-path of DESW is close to zero due to its zero-group velocity [13, 73, 74] (please note that this is not the case for MSSWs – see the dispersion slope in Figure 6.1). With a further increase in k, the spin-wave group velocity increases along with a decrease in the lifetime. Therefore, there is a maximum of the free-path for waves featuring wavelengths of about 100 nm. A further increase in k results in the decrease in I_{free}. Nevertheless, the decrease in the wavelength assumes that long free-paths are not always necessary. The ratio I_{free}/λ, which shows how many wavelengths (i.e., how many unit elements) a wave propagates before it relaxes, is of importance. It can be clearly seen in Figure 6.2c (right axis) that the decrease in the wavelength of the exchange waves results in the increase of the ratio I_{free}/λ. This ratio reaches values above 3000 for waves of a nanometer wavelength. This estimation is very encouraging for the further development of the field of magnonics since it shows that potentially only around $1-\exp(-2\lambda/I_{free})\approx6\cdot10^{-4}$ of the spin-wave energy will be lost for data transport in a unit element of size λ.

6.2.3 MATERIALS FOR MAGNONICS AND METHODOLOGY

As was discussed above, spin waves are usually studied in thin magnetic films or waveguides in the form of narrow strips. The choice of the material plays a crucial role in fundamental as well as in applied magnonics. The main requirements for the magnetic materials are: (*i*) a small Gilbert damping parameter in order to ensure long spin-wave lifetimes; (*ii*) a large saturation magnetization for high spin-wave frequencies and velocities in the dipolar region; (*iii*) high Curie temperatures to provide high thermal stability; and (*iv*) simplicity in the fabrication of the magnetic films and in the patterning processes [75]. The most commonly used materials for magnonics, as well as those with a high potential for magnonic applications, are presented in Table 6.1 (adapted from [75]) together with some selected parameters and estimated spin-wave characteristics: lifetime τ, velocity v_{gr}, free-path I_{free}, and the ratio I_{free}/λ. Only the dipolar MSW is considered since nowadays, for experimental investigations, these are the waves that are mainly used (please note that the values in Ref. [75] differ a bit from the values here since in Table 6.1 the exchange interaction and the ellipticity of magnetization precession is considered). It has to be noted that here MSSW propagating in a plane film perpendicularly to the magnetization orientation rather than BVMSW is considered since it features higher values of the group velocity.

The first material in the table is monocrystalline Yttrium-Iron-Garnet $Y_3Fe_5O_{12}$ (YIG) films grown by high-temperature Liquid Phase Epitaxy (LPE) on Gadolinium Gallium Garnet (GGG) substrates [30–33, 64, 65, 76, 77]. This ferrimagnet was first synthesized in 1956 by Bertaut and Forrat [76] and has the smallest known magnetic loss that results in the lifetime of spin waves being of some hundreds of nanoseconds and, therefore, finds widespread use in academic research [4, 19, 33, 75]. Many of the experimental results presented in this chapter were obtained using LPE YIG. The small magnetic loss is due to the fact that YIG is a magnetic dielectric (ferrite)

TABLE 6.1
Shows a Selection of Magnetic Materials for Magnonic Applications, their Main Parameters, and Estimated Spin-Wave Characteristics

	µm-thick LPE Yttrium-Iron-Garnet (YIG)	nm-thick Yttrium-Iron-Garnet (YIG)	Permalloy (Py)	CoFeB	Heusler CMFS compound
Chemical composition	$Y_3Fe_5O_{12}$	$Y_3Fe_5O_{12}$	$Ni_{81}Fe_{19}$	$Co_{40}Fe_{40}B_{20}$	$Co_2Mn_{0.6}Fe_{0.4}Si$
Structure	single-crystal	single-crystal	poly-crystal	amorphous	single-crystal
Gilbert damping α	$5 \cdot 10^{-5}$	$2 \cdot 10^{-4}$	$7 \cdot 10^{-3}$	$4 \cdot 10^{-3}$	$3 \cdot 10^{-3}$
Sat. magnetization M_0, kA/m	140	140	800	1250	1000
Exchange constant A, pJ/m	3.6	3.6	16	15	13
Curie temperature T_C, K	560	560	550–870	1000	> 985
Typical film thickness t	1–20 µm	5–100 nm	5–100 nm	5–100 nm	5–100 nm
	The following parameters are calculated for dipolar MSSW modes, film magnetized in-plane by the field of 100 mT, for the spin-wave wavenumber $k = 0.1/t$:				
Lifetime for dipolar surface wave τ,	604.9 ns (@ 4.77 GHz)	150.2 ns (@ 4.8 GHz)	1.3 ns (@ 11.1 GHz)	1.6 ns (@ 14.9 GHz)	2.6 ns (@ 12.8 GHz)
Velocity v_{gr}	33.7 km/s (@ $t = 5$ µm)	0.23 km/s (@ $t = 20$ nm)	2.0 km/s (@ $t = 20$ nm)	3.5 km/s (@ $t = 20$ nm)	2.6 km/s (@ $t = 20$ nm)
Free-path l	20.4 mm	35.1 µm	2.7 µm	5.7 µm	6.9 µm
Ratio l/λ,	64.9	27.9	2.1	4.5	5.5
References	[4, 30–33, 64, 65 76, 77]	[41, 42, 64, 78–82]	[54, 71, 72, 83]	[84–86]	[87–90]

The characteristics are calculated using the dipolar approximation for infinite films magnetized in-plane with a 100 mT magnetic field. The lifetime is estimated with ellipticity being considered according to Equation 6.6 and for the case of nonuniform FMR linewidth widening to be zero.

with very little spin-orbit interaction and, consequently, with small magnon-phonon coupling [2, 32]. Moreover, high-quality LPE single-crystal YIG films ensure a small number of inhomogeneities and, thus, suppressed two-magnon scatterings [2, 91]. However, the thickness of these films, which is in the micrometer range, does not allow for the fabrication of YIG structures of nanometer sizes. Therefore, the fabrication of nanostructures became possible only within the last few years with the development of technologies for the growth of high-quality nm-thick YIG films (see the second column in Table 6.1) by means of, e.g., Pulsed-Laser Deposition (PLD) [42, 78–81], sputtering [82], or via modification of the LPE growth technology [41, 64]. Although the quality of these films is still worse when compared to micrometer-thick LPE YIG films, it is already good enough to satisfy the majority of requirements of magnonic applications [19].

The second most commonly used material in magnonics is permalloy (Py) that is a polycrystalline alloy of 80% Ni – 20% Fe (see Table 6.1). This is a soft magnetic material with low coercivity and anisotropies. One of the major advantages of this material is that it has a fairly low spin-wave damping value considering it is a metal, and it can be easily deposited and nanostructured. Therefore, permalloy was intensively used for the investigation of spin-wave physics in micro-structures (see reviews [54, 72]). Nowadays, a large quantity of attention from the community is also focused on CoFeB and half-metallic Heusler compounds. These materials possess smaller Gilbert damping parameters and larger values of the saturation magnetization, and are, therefore, even more suitable for the purposes of magnonics. For example, it was demonstrated that the spin-wave mean free-path in Heusler compounds can reach 16.7 μm [90].

The fabrication of high-quality spin-wave waveguides in the form of magnetic strips is also one of the primary tasks in the field of magnonics. The most commonly used technique for the fabrication of micrometer-thick YIG waveguides is a dicing saw [92] since the width of the waveguide is usually larger than 1 mm. A commonly used method for the patterning of such YIG films is photolithography with subsequent wet etching by means of hot orthophosphoric acid [7, 93]. Different techniques are used to pattern nanometer-thick YIG films: E-beam lithography with subsequent Ar$^+$ dry etching have shown good results [41, 42]. Focused Ion Beam (FIB) milling has also recently shown very promising results and allows for the fabrication of YIG structures with lateral sizes below 100 nm (not published). The same techniques can also be used for the patterning of metallic magnetic films. Frequently, an alternative approach is used; the magnetic material is deposited on a sample covered with a resist mask produced via photo- or electron beam lithography followed by a standard lift-off process. Antennas and the required contact pads are deposited afterwards in subsequent lithography, electron beam evaporation (or sputtering), and lift-off processes.

Modern magnonics consists of a wide range of instrumentation for the excitation and detection of magnons. The main requirements of magnon detection techniques are defined as sensitivity, the range of detectable wavelengths and frequencies, as well as frequency, spatial, and temporal resolution. For the spin-wave excitation techniques, the efficiency of excitation, its coherency as well as the wavenumber range is of primary importance. A broad scope of techniques intensively used nowadays in magnonics, as well as techniques showing much potential, are listed in [75]. Among others, one can define the three main categories of these techniques: microwave approaches, optical technologies, and spintronics approaches. The last one is at the heart of magnon spintronics and is discussed later in more detail. One can attribute the following techniques to the microwave approaches: conventional microstrip (or CoPlanar Waveguide (CPW) or meander-type) antenna-based techniques for spin-wave excitation and detection [4, 94–98], contactless antenna-based approaches to excite coherent spin waves [44, 99, 100], FMR spectroscopy [65, 71, 83, 101], the parametric pumping technique that is usually used for spin-wave amplification [4, 13–16, 43, 102, 103], the Pulsed Inductive Microwave Magnetometer technique (PIMM) [71, 104], and the Inductive Magnetic Probe (IMP) technique [105]. The most commonly used optical techniques are Brillouin Light Scattering (BLS) spectroscopy for the detection of spin waves [54, 72, 106, 107], thermal and nonthermal excitation of spin waves by femtosecond pulsed-laser techniques [49, 108–110], and Magneto-Optical Kerr Effect (MOKE) spectroscopy for spin-wave detection

[99, 100, 110]. With spintronic approaches, one can associate Spin-Transfer Torque (STT) based techniques used for amplification and generation of spin waves [19, 111–120], Spin Pumping (SP) based approaches for spin-wave detection [19, 43, 121–125], and Spin-Polarized Electron Energy Loss Spectroscopy (SPEELSC) [126, 127]. Besides the mentioned techniques, there is a set of other promising approaches: Magneto-Electric (ME) cells [20–23, 128, 129], Magnetic Resonance Force Microscopy (MRFM) [130], detection of magnon-induced heat [131], Nuclear Resonant Scattering (NRS) of synchrotron radiation [132], X-ray detected FMR (XFMR) [133, 134], as well as electron-magnon scattering approaches [135, 136].

The most commonly used technique for spin-wave excitation is inductive excitation with a microwave current sent through a strip-line or CPW antenna. In order to understand the spin-wave signal excitation mechanism, it is useful to consider the waveguide as a reservoir of quasi-classical spins. When the waveguide is magnetically saturated, the mean precessional axis of all the spins is parallel to the bias field. The application of a microwave signal to the strip-line antenna generates an alternating Oersted magnetic field around it. The components of this field which are perpendicular to the bias direction create an alternating torque on the magnetization that results in an increase in the precessional amplitude. The spins precessing under the antenna interact with their nearest neighbors and, if the correct conditions for field and frequency are satisfied, spin-wave propagation is supported. After the propagation, the spin waves might be detected by an identical output microstrip (or CPW or meander-like) antenna [4, 95–98]. The mechanism of spin-wave detection is, by symmetry the inverse of the excitation process.

One of the most powerful techniques in magnonics nowadays is Brillouin Light Scattering (BLS) spectroscopy [4, 54, 72, 106, 107]. The physical basis of BLS spectroscopy is the inelastic scattering of photons by magnons. Analyzing scattered light from a probe beam allows the frequencies and wavenumbers of the scattering magnons to be determined, where the scattered photon intensity is proportional to the spin-wave intensity. The technique is generally used in conjunction with a microwave excitation scheme and, over the last decade, has undergone extensive improvements. BLS spectroscopy now achieves a spatial resolution of 250 nm, and time-, phase-, and wavenumber resolved BLS spectroscopy have been realized.

6.2.4 BASIC IDEAS OF MAGNON-BASED COMPUTING

One of the main strengths of magnonics lies in the benefits provided by the wave nature of magnons for data processing and computation. In the past, the application area of spin waves was mostly related to analog signal processing in the microwave frequency range. For this, applications microwave filters, delay lines, phase conjugators, power limiters, and amplifiers are just a few examples [4, 33–35]. Nowadays, new technologies, allowing, e.g., for the fabrication of nanometer-sized structures or for operation in the THz frequency range, in combination with novel physical phenomena, provide a new momentum to the field and make the advantages discussed earlier accessible for both analog and digital data signal processing. Magnons also possess the potential to be used in the implementation of alternate computing concepts such as non-Boolean computing [137, 138], reversible logic [28, 139], artificial neuromorphic networks [29], and, more generally, the projecting of optical computing concepts [140] onto the nanometer scale. These directions are still at the beginning of their development. The basic ideas behind the standard Boolean logic operations with digital binary data, which are currently the subject of intensive theoretical and experimental studies [19], are addressed here. The more advanced developments in magnon-based computing are presented in Chapter 18 "Spin Wave Logic Devices" (Volume 3) of this book.

The idea of coding binary data into the spin-wave amplitude was first stated by Hertel et al. in 2004 [141]. Micromagnetic simulations revealed that magnetostatic spin waves change their phase as they pass through domain walls. It was suggested to split spin waves into different branches of a ring (spin-wave interferometer). After being merged in the output, their interference depends on the presence of domain walls in the branches. In such a way, a controlled manipulation of phases of spin waves was proposed to be utilized for the realization of spin-wave logical operations. The first experimental studies of spin-wave logic were reported by Kostylev et al. in Ref. [142]. It was proposed that a Mach–Zehnder spin-wave interferometer equipped with current-controlled phase shifters embedded in the interferometer arms can be used to construct logic gates. Following this idea, Schneider et al. realized a proof-of-principle XNOR logic gate shown in Figure 6.3a [143]. The currents applied to Input 1 and Input 2 represent logic inputs: The current-on state, which results in a π-phase shift of the spin wave in the interferometer arm, corresponds to logic "1"; the current-off state corresponds to logic "0". The logic output is defined by the interference of microwave currents induced by the spin waves in the output microstrip antennas: A low amplitude (destructive interference) corresponds to logic "0", and a high amplitude (constructive interference) defines logic "1". The change in the magnitude of the output pulsed signal for different combinations of the logic inputs (see Figure 6.3a) corresponds to the XNOR logic functionality. It was also shown that an electric current can create a magnetic barrier reducing or even stopping the spin-wave transmission. Using this effect also a universal NAND logic gate was demonstrated [143]. Lee et al. [26] have proposed an alternative design of a nanometer-sized logic gate – see Figure 6.3b. In this device, the spin-wave phases are controlled by way of an electric current flowing through a vertical wire placed between the arms of the interferometer. This current defines the logic input, while the amplitude of the output spin wave after interference defines the logic output. The ability to create NOT, NOR, and NAND logic gates was demonstrated using numerical simulations [26].

The drawback of the discussed logic gates is that it is impossible to combine two logic gates without additional magnon-to-current and current-to-magnon converters. This fact stimulates the search for a means to control a

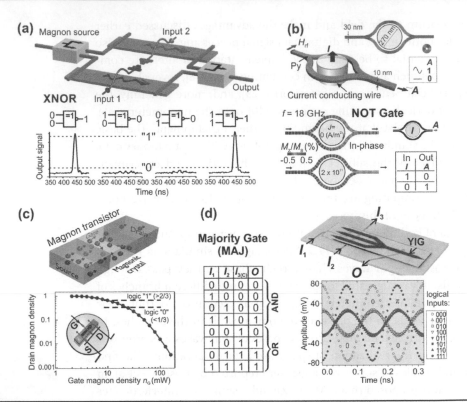

FIGURE 6.3 (a) Spin-wave XNOR logic gate [18, 143]. The gate is based on a Mach–Zehnder interferometer with electric current-controlled phase shifters embedded in the two-magnon conduits. The bottom panel shows the output pulsed spin-wave signals measured for different combinations of the input DC signals applied to the phase shifters. (After Stamps, R.L. et al., *J. Phys. D: Appl. Phys.* 47, 333001, 2014. With permission.) (b) Nanosized Mach–Zehnder spin-wave interferometer designed in a form of a bifurcated Py conduit girdling a vertical conducting wire of 270 nm in diameter (adapted from [26]). Numerical simulation of a NOT logic operation is shown below. (After Lee, K.-S. et al., *J. Appl. Phys.* 104, 053909, 2008. With permission.) (c) Schematic of the operational principle of a magnon transistor [24]: The source-to-drain magnon current (shown with blue spheres) is nonlinearly scattered by gate magnons (red spheres) injected into the gate region. In the bottom panel, the measured drain magnon density is presented as a function of the gate magnon density. The horizontal dashed lines define the drain density levels corresponding to a logic "1" and a logic "0". (After Chumak, A.V. et al., *Nat. Commun.* 5, 4700, 2014. With permission.) (d) Left panel: Truth table of the majority operation. Upper panel: photograph of the majority gate prototype under test in Ref. [25]. Spin waves are excited with microstrip antennas (I_1, I_2, and I_3) in the input YIG waveguides (width 1.5 mm) and propagate through the spin-wave combiner into the output waveguide for the detection with the output antenna (O). Lower panel: Output signal observed with an oscilloscope for different phases of the input signals. The dependence of the output phase on the majority of the input phases is clearly visible. (After Fischer, T. et al., *Appl. Phys. Lett.* 110, 152401, 2017. With permission.)

magnon current by another magnon current. It has been demonstrated that such control is possible due to nonlinear magnon-magnon scattering, and a *magnon transistor* was realized [24] – see Figure 6.3c. In this three-terminal device, the density of the magnon current flowing from the source to the drain (see blue spheres in the figure) is controlled by the magnons injected into the gate of the transistor (red spheres). A magnonic crystal (discussed in the next section) in the form of an array of surface grooves [93] is used to increase the density of the gate magnons and, consequently, to enhance the efficiency of the nonlinear four-magnon scattering process used to suppress the source-to-drain magnon current. It was shown that the source-to-drain current can be decreased by up to three orders of magnitude (see the bottom panel in Figure 6.3c). The potential for the miniaturization of this transistor, for the realization of an integrated magnonic circuit (using the example of an XOR logic gate) as well as for the increase of its operating speed and a decrease in energy consumption are discussed in Ref. [24]. Although the operational characteristics of the presented insulator-based transistor in its proof-of-principle form do not overcome those of semiconductor devices, the presented transistor might play an important role for future *all-magnon technology* in which information will be carried and processed solely by magnons [24]. The main advantage of the all-magnon approach is that it does not require converters between magnons and charge currents in each unit (the converters are needed only in the beginning to code the electric signal into the magnons in a magnonic chip and at the end to read the data out after the processing). The naturally-strong nonlinearity of spin waves is assumed to be used to perform operations with data.

In the approaches discussed above, logic data was coded into spin-wave amplitude (a certain spin-wave amplitude defines logic "1", zero amplitude corresponds to logic "0"). Alternatively, Khitun et al. proposed [21, 128, 144] (see also Chapter 18 in Volume 3 of this book) to use the spin-wave *phase* for digitizing information instead of the *amplitude*. A wave with some chosen phase ϕ_0 corresponds to a logic "0" while a logic "1" is represented by a wave with phase $\phi_0 + \pi$. Such an approach allows for a trivial embedding of a NOT logic element (which requires two transistors in modern CMOS technology) in magnonic circuits by simply changing the position of a readout device by a $\lambda/2$ distance. Moreover, it opens up access to the realization of a *majority gate* in the form of a multi-input spin-wave combiner [21]. The spin-wave majority gate consists of three input waveguides (generally speaking, of any odd number of inputs larger than one) in which spin waves are excited. A spin-wave combiner, which merges the different input waveguides, and an output waveguide in which a spin wave propagates with the same phase as the majority of the input waves. Thus, the majority logic function can be realized due to a simple interference between the three input waves – see the truth table in the left panel of Figure 6.3d. The majority gate can perform not only the majority operation but also AND and OR operations (as well as NAND and NOR operations if to use a half-wavelength long inverter) if one of its inputs is used as a control input [145, 146]. The large potential of the majority gate is also underlined by the fact that a full adder (used in electronics to sum up three bits) can be constructed using only three majority gates while a total of a few tens of transistors are used nowadays in CMOS. This will allow for a drastic decrease in the footprint of magnonic devices when compared to electronics. One more advantage of the majority gate is that it might operate with spin waves of different wavelengths simultaneously, paving the way towards single chip parallel computing [27]. This approach requires the splitting of the signals of the spin waves that have different wavelengths and, therefore, the usage of magnonic crystals or directional coupler (discussed later) is needed.

The first fully-functioning design of the majority gate employing a spin-wave combiner was demonstrated using micromagnetic simulations by Klingler et al. in Ref. [145]. One of the main problems of a spin-wave majority gate based on the spin-wave combiner is the coexistence of different spin-wave modes with different wavelengths at the same frequency. The nonuniformities of the combiner area, through which spin waves propagate, usually act as re-emitters of new spin waves of the same frequency but having different wavenumbers. By choosing a smaller width for the output waveguide (1 μm in this case as opposed to 2 μm-wide input waveguides), it was possible to select the first width mode from the combiner and to ensure the readability of the output signal. The functionality of the majority gate was proven by showing that all excitation combinations with a majority phase of zero are in phase and, simultaneously, in anti-phase to all combinations with a majority phase of π. However, the output signal in [145] was still influenced by exchange spin waves of the same frequency but with shorter wavelengths. To overcome this limitation, isotropic FVMSWs in an out-of-plane magnetized spin-wave majority gate were used in Ref. [146]. A high spin-wave transmission through the new asymmetric design of the majority gate of up to 64%, which is about three times larger than for the in-plane magnetized gate, was achieved. The phases of output spin wave clearly satisfied the majority function (see the truth table in Figure 6.3d), proving the functionality of the gate.

Recently, an experimental prototype of a majority gate utilizing a macroscopic YIG structure was shown by Fischer et al. [25]. The respective device comprises three input lines, as well as one output line and has been structured from a YIG film of 5.4 μm thickness by means of photolithography and wet chemical etching – see top panel in Figure 6.3d. In this geometry, inductively excited spin waves propagate coherently along the three input waveguide and eventually superimpose when they leave the spin-wave combiner, at which point the input lines merge. The resultant spin wave induces an electrical signal in a copper strip-line at the output which is directly mapped by a fast oscilloscope [25]. The logical information encoded in the phase of the spin waves is controlled by upstream adjustable phase shifters. The lower panel of Figure 6.3d shows the measured spin-wave amplitudes for all possible combinations of logic input states. The phase of the output spin wave corresponds to the majority phase of the input lines, according to the truth table of a majority gate. Nevertheless, one can clearly see that although the phase of the output signal is defined according to the majority logic function, the amplitude of the output also depends on the combination of the phases of the input spin-wave signals (it is three times larger for the case 0-0-0, when all waves are in phase, in comparison to, e.g., the 0-0-1 case). Therefore, if to follow the all-magnon computing approach, a nonlinear amplitude normalizer is required in order to be able to send the output spin wave from one majority gate as an input wave into the next majority gate in an integrated magnonic circuit. Another way to go was originally proposed by Khitun [21] and assumes that the data is converted from spin waves into an electric signal using Magneto-Electric cells after each majority gate and is coded back into spin waves in the next element (see the section of Magneto-Electric cells later). In this case, the nonlinearity required for all-magnon computing is replaced by the conversion itself. However, many state-of-the-art approaches to convert AC spin waves into DC voltages, such as the Inverse Spin Hall Effect or the Tunneling Magnetoresistance Effect, are not sensitive to the spin-wave phase. Hence, the electrical readout of the phase information is a challenge. In Ref. [147], the authors have demonstrated the conversion of the spin-wave phase into a spin-wave intensity by local nonadiabatic parallel pumping (see discussion below). The first step for the practical application of phase-to-intensity conversion was realized experimentally for spin waves in a microstructured magnonic waveguide made from permalloy [147]. It was also shown how phase-to-intensity conversion can be used to extract the majority information from an all-magnonic majority gate.

6.3 GUIDING OF SPIN WAVES IN ONE AND TWO DIMENSIONS

The main problems faced as it rates to the transfer of data between information processing elements in two-dimensional spin-wave circuits are discussed in this section. A selection of prospective solutions for the guidance as well as for the processing of data is presented.

6.3.1 MAGNONIC CRYSTALS

Magnonic crystals are artificial magnetic media with periodic variation of their magnetic properties in space. Bragg scattering affects the spin-wave spectrum of such a structure. This leads to the formation of band gaps – regions of the spin-wave spectrum in which spin-wave propagation is prohibited (see right panels in Figures 6.4a and b). Consequently, areas between band gaps allow for selective spin-wave propagation [4, 49, 75, 106, 148]), while the pronounced changes of the spin-wave dispersion near the band-gap edges opens up access to the formation of band-gap solitons [149], the deceleration of spin waves and the appearance of confined spin-wave modes [150]. In spite of the fact that the term "magnonic crystal" is relatively new (it was first introduced by Nikitov et al. in 2001 [151]), this field goes back to the studies of spin-wave propagation in periodical structures that had already been initiated by Sykes, Adam, and Collins in 1976 [152]. However, the early magnonic crystal research activities were mainly devoted to the fabrication of microwave filters and resonators (see reviews [153–155]).

Nowadays, when the coupling of spin-wave modes and demagnetizing effects cannot be neglected on the micro- and nanoscales, many of the magnonic crystal studies are focused on the understanding of the spin-wave physics in the crystals. Besides, recent experimental studies of nonlinear magnonic crystals (see for example [4, 149, 156, 157]) as well as the development of magnonic crystal theories (the model of three-dimensional

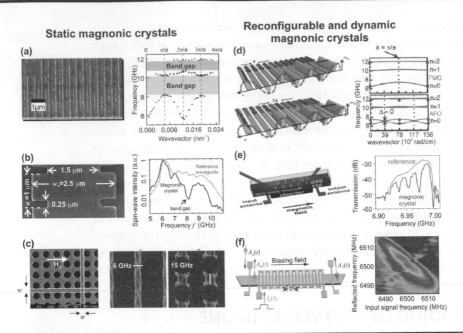

FIGURE 6.4 Realizations of magnonic crystals. (a) One-dimensional magnonic crystal in the form of an array of alternating Py and Co nanostrips and the structure of magnon band gaps measured by BLS spectroscopy [161]. (After Wang, Z.K. et al., *Appl. Phys. Lett.* 94, 083112, 2009. With permission.) (b) Magnonic crystal designed as a Py conduit with a periodically varying width, and spin-wave spectra of this crystal and of a uniform reference conduit [162]. (After Chumak, A.V. et al., *Appl. Phys. Lett.* 95, 262508, 2009. With permission.) (c) Scanning electron microscopy image of a two-dimensional magnonic crystal in the form of an anti-dot array in a Py film and the numerically simulated spatial distribution of spin-wave amplitudes for different magnon modes [164]. (After Tacchi, S. et al., *IEEE Trans. Magn.* 46, 172–178, 2010. With permission.) (d) Reconfigurable magnonic crystal in the form of an array of bi-stable magnetic nanowires and spin-wave spectra obtained for ferromagnetic and antiferromagnetic nanowires orders [148, 171]. (After Topp, J. et al., *PureAppl. Chem.* 83, 2011. With permission.) (e) Schematic of the reconfigurable magnonic crystal consisting of a GGG/YIG/absorber multilayer system (temperature is coded in color) and measured spin-wave transmission characteristics in the thermal landscape and reference data without the projected pattern [167]. (After Vogel, M. et al., *Nat. Phys.* 11, 487, 215. With permission.) (f) Schematics of a dynamic magnonic crystal in the form of a current-carrying meander structure positioned close to the surface of a YIG conduit [165, 174]. A two-dimensional map of reflected signal spectra as a function of the incident signal frequency demonstrating frequency inversion is shown on the right. (After Chumak, A.V. et al., *Nat. Commun.* 1, 141, 2010. With permission.)

magnonic crystals [158], for example) have brought sufficient progress to the general wave physics. Thus, the field of the magnonic crystal is growing rapidly and magnonic crystals have already been given a partial overview in a set of review papers. Such papers include spin-wave dynamics in periodic structures for microwave applications [33, 153–155], magnonics crystals for data processing in general [4, 19], Brillouin Light Scattering studies of planar metallic magnonic crystals [106], micromagnetic computer simulations of width-modulated waveguides [159], photo-magnonic aspects of the anti-dot lattices [49], theoretical studies of one-dimensional monomode waveguides [160], and reconfigurable magnonic crystals [148].

The wide variety of parameters, which define the characteristics of spin waves in a magnetic film, results in a variety of possible designs of magnonic crystals. For example, magnonic crystals can be constructed using two differ-ent magnetic materials (bi-component magnonic crystals [161]) – see Figure 6.4a. Other examples are spin-wave conduits with a periodic variations of their width (Figure 6.4b) [159, 162, 163], periodic dot or anti-dot lattices (Figure 6.4c) [49, 164], arrays of interacting magnetic strips (Figure 6.4d) [106, 148], periodic variations of temperature (Figure 6.4e) [167, biasing magnetic field (Figure 6.4f) [165, 166], thickness [4, 152, 168], saturation magnetization [169], mechanical stress [68, 170], etc. When the classifications of magnonic crystals are spoken of, one could define at least two approaches. First of all, based on their dimensions, magnonic crystals can be divided into *one-*, *two-*, *and three-dimensional magnonic crystals* types. Another way to system-atize the crystals is based on the possibility to vary their parameters with time. The most common are *static magnonic crystals* properties of which are defined by the design of the structure and cannot be changed after their fab-rication. *Reconfigurable magnonic crystals*, whose properties can be changed on demand [148, 167, 171, 172], attract special attention since they allow for the tuning of the functionality of a magnetic element. The same element can be used in applications as a magnon conduit, a logic gate, or a data-buffering element. An example of such a structure is a magnonic crystal in the form of an array of magnetic strips magnetized parallel or anti-parallel to each other [171] – see Figure 6.4d. Another way for the creation of an arbitrary 2D mag-netization pattern in a magnetic film is based on laser-induced heating and was also used to demonstrate a reconfigurable magnonic crystal [167] – see Figure 6.4e. Furthermore, changes in the properties of a magnonic crystal give access to novel physics, if these changes occur on a timescale shorter than the spin-wave propagation time across the crystal. Such magnonic crys-tals are termed *dynamic magnonics crystals* [165, 166, 173]. The first dynamic magnonic crystal was realized in the form of a YIG conduit placed in a time-dependent spatially periodic magnetic field. The field was induced by an elec-tric current sent through a meander-type conducting structure placed close to the surface of the conduit – see Figure 6.4f and could be changed on a tim-escale below 10ns [165]. It has been shown that this dynamic magnonic crys-tal can perform a set of spectral transformations such as frequency inversion and time reversal [174]. Finally, another type of dynamic magnonic crystals, in which a "static" magnonic crystal is moving with respect to the spin waves of a certain velocity, is termed *traveling magnonic crystals* [68, 170]. In this case, one should consider a Doppler effect, in addition to the Bragg scatter-ing. As a result, the band gaps of a traveling magnonic crystal are shifted in frequencies and the velocity of the moving grating defines the frequency shift. One of the realizations of such a crystal is based on the utilization of Surface Acoustic Waves (SAWs) that act as a moving periodic scatterer for spin waves in a YIG film [68, 170].

For magnonic applications, magnonic crystals constitute one of the key elements since they open up access to novel multi-functional magnonic devices [75]. These devices can be used as spin-wave conduits and filters (in fact, any magnonic crystal can serve as a conduit or a filter), sensors, delay lines and phase shifters, components of auto-oscillators, frequency and time

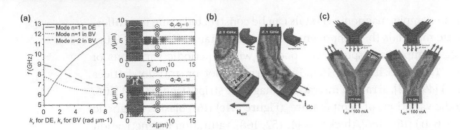

FIGURE 6.5 (a) Left panel: Spin-wave dispersion relations, calculated for T-structure for the $n = 1$ MSSW width mode, and for $n = 1$ and $n = 2$ BVMSW width modes [175]. (After Brächer, T. et al., *Appl. Phys. Lett.* 102, 2132411, 2013. With permission.) Right panels: Snapshots of a simulation showing the basic principle of the investigated mode conversion (the color coding represents the out-of-plane component of the magnetization). The orange bars represent the positions of the simulated microwave antennas (in the top panel antennas are oscillating in phase, in the bottom panel they are in anti-phase). (b) Two-dimensional intensity distributions of spin waves excited with an externally applied magnetic field (left panel) and applied DC pulses with an amplitude of 66.7 mA (right panel) obtained with BLS spectroscopy [176]. (After Vogt, K. et al., *Appl. Phys. Lett.* 101, 042410, 2012. With permission.) (c) Two-dimensional BLS mapping of the spin-wave propagation path illustrating the switching: spin waves are guided through the Y-junction and only propagate in the same direction as the current flow. Red arrows in the insets show the magnetization orientation [177]. (After Vogt, K. et al., *Nat. Commun.* 5, 3727, 2014. With permission.)

inverters, data-buffering elements, power limiters, nonlinear enhancers in a magnon transistor, and components of logic gates (please check Ref. [75] for the corresponding references).

6.3.2 Two-Dimensional Structures

The guiding of information carried by spin waves in two dimensions is required for the realization of magnonic circuits and is one of the important challenges modern magnonics is facing. As was shown in the previous section, spin-wave dispersions are highly anisotropic in in-plane magnetized films and, in the simplest case of dipolar waves, BVMSWs and MSSWs exist in different frequency ranges (see the case of a plane film in Figure 6.1) [2, 3]. Therefore, the realization of simple magnonic circuits of the type of printed circuit boards in electronics is not possible and alternative solutions should be used.

One likely possible way to achieve this was proposed in Ref. [175] with a usage of a T-shaped structure shown in Figure 6.5a. Counter-propagating MSSWs are excited by two microstrip antennas in a permalloy ($Ni_{81}Fe_{19}$) strip (vertical waveguide in the figure), which was saturated along its short axis by an externally applied magnetic field. In the center of this strip, a second perpendicular strip was patterned to support the propagation of BVMSWs along the magnetization. Brillouin Light Scattering spectroscopy was employed to measure the spin-wave intensity in the structure [175]. In an unstructured film, the MSSW-to-BVMSW conversion is forbidden as was discussed above. However, in the waveguides, the quantization of spin waves across the width of the waveguide leads to the modification and to the overlap of spin-wave dispersions (see Figure 6.1). Moreover, the internal field in the perpendicularly magnetized strip is smaller than the internal field in the longitudinally magnetized waveguide due to significant demagnetization. These factors allow for the coexistence of spin waves propagating parallel as well as perpendicular to the magnetization at the same frequency – see the left panel in Figure 6.5a. As a result, the generation of BVMSW by the originally-excited MSSWs was shown experimentally and by means of numerical simulations [175]. In the right panels of Figure 6.5a, the simulated phase fronts of the propagating spin waves are shown with a color code. In the top and bottom panels, the spin waves are excited by the same and opposite phases correspondingly. The MSSWs of the same phases excite the $n = 1$ symmetric BVMSW mode (top panel) while MSSWs of opposite phases excite the asymmetric $n = 2$ BVMSW mode (bottom panel). These results were confirmed by phase-sensitive BLS measurements. The same type of mode conversion was also investigated for larger size YIG T-shaped spin-wave splitter in [176]. It was revealed that the spin wave beams in the outputs of the splitter are generally given by a superposition of both even and odd modes, with the details dictated by the dispersion overlap. By adjusting the frequency of the incident wave, it is possible to alter the character of the output beams or to switch the output off completely.

A similar approach of the dispersion mismatch between the narrow magnonic waveguide and a wide antenna was used even earlier in Ref. [100] to excite spin waves by a global microwave field. This wireless method of spin-wave excitation uses a uniform FMR-based antenna that couples to the microwave field and converts it into finite wavelength spin waves propagating in strip magnonic waveguides. The functionality of this method is demonstrated on micrometer-scale devices using time-resolved scanning Kerr microscopy. This approach is especially important for the field of magnonics since these antennas can be placed at multiple positions on a magnonic chip and can be used to excite mutually coherent multiple spin waves for magnonic logic operations [99, 100].

Another way to guide spin waves in two dimensions was proposed by Vogt et al. in Ref. [177]. Spin-wave propagation in a permalloy waveguide comprising an S-shaped bend was studied using BLS spectroscopy. In opposition to the cases discussed above, this method does not involve any conversion between the different modes. Instead, a nonuniform biasing field was used to magnetize the bent spin-wave waveguide transversally and to ensure the spin-wave propagation conditions for MSSW. For this, a direct current flowing through a gold wire underneath the permalloy waveguide provided a local magnetic field – see Figure 6.5b. The mapped BLS intensity distribution of spin waves in the bent region of the waveguide is shown with color code in the figure (the magnetization of the structure is shown in the insets with red arrows). The left panel shows that spin waves are not able to propagate through the bent structure if it is uniformly magnetized by an external field. As opposite, if a direct current is flowing through the bilayer (see the right panel in Figure 6.5b), the spin wave is guided within the curved waveguide. The advantage of this method is a possibility to control spin-wave propagation (in the simplest case to switch it on or off) and its drawback is the use of an electric current that generates parasitic Joule heat.

The same approach was used in Ref. [178] to guide spin waves in a Y-shaped structure – see Figure 6.5c. Electric-current induced locally generated magnetic fields, rather than uniform external fields, oriented the magnetization into only one arm of the Y-structure perpendicularly to the spin-wave propagation direction ensuring the propagation conditions for MSSWs. The investigated structure is proposed to be used as a spin-wave multiplexer (switch) that can be used, e.g., for the guiding of spin-wave information to one or another magnonic data processing component. An interesting experimental finding in [178] is that the Y-structure is efficient for an angle of 60 degrees between the output arms and is much less efficient for angles of 30 and 90 degrees. One possible explanation is that the formation of caustic beams discussed in the next section takes place in the experiments.

6.3.3 SPIN-WAVE CAUSTICS

Strong spin-wave anisotropy in in-plane magnetized films discussed above is not necessarily a drawback for the transfer of information in two dimensions. For example, it opens access to the guiding of spin waves even in an un-patterned film in the form of *spin-wave caustics* – nondiffractive wave beams with stable sub-wavelength transverse aperture [7–10]. The direction of these beams is controlled by the magnetic field and therefore caustics can selectively transfer information to a spin-wave data processing element [19].

In an anisotropic medium, the direction of the wave group velocity $\vec{v}_g$ indicating the direction of energy propagation generally does not coincide with the direction of the wave vector $\vec{k}$. When the medium's anisotropy is sufficiently strong, the direction of the group velocity of the wave beam may become independent of the wave vectors of the waves forming the beam in the vicinity of a certain carrier wave vector $\vec{k}_c$. In such a case, wave packets excited with a broad angular spectrum of wave vectors (i.e., with a broad spectrum of phase velocities) may be channeled along this direction. The ideal source for such a wave packet is a point source whose size is comparable or smaller than the carrier wavelength of the excited wave packet. A simple excitation source delivering wave packets with wide angular spectra is shown in Figure 6.6a. A microstrip antenna was used to excite BVMSWs into a narrow spin-wave waveguide and the waves were then guided into a continuous area of a film. The transition between the waveguide and the continuous area of the film acts as a point source [7]. Figure 6.6b presents the k_y spectrum of spin waves excited in such a way. The shaded area indicates where wave excitation is effective. The isofrequency curves of BVMSWs for two different magnetic fields are shown in Figure 6.6c. On the linear segment of the iso-frequency curve corresponding to 1840 Oe, the direction of $\vec{v}_g$ is the same for all wave vectors. Thus, a caustic wave beam can be formed, and the energy of this beam will propagate along the caustic directions perpendicular to the linear segments of the isofrequency curve making the angle θ_c with the magnetic field $\vec{H}_0$ – see the two beams measured experimentally using BLS spectroscopy in Figure 6.6d. The condition $d^2 k_z / dk_y^2 = 0$ defines the carrier wave vector $\vec{k}_c$, the group velocity $\vec{v}_g$, and the propagation direction

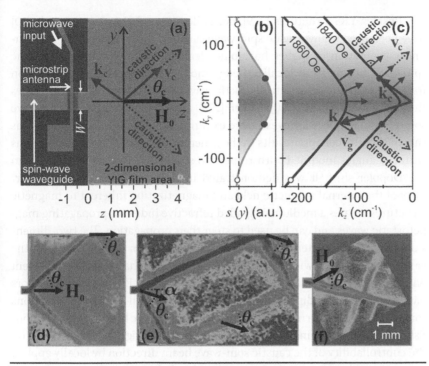

FIGURE 6.6 (a) YIG structure used for investigations of caustic spin-wave beams [7]. (b) Excitation amplitude of the waveguide antenna as a function of the transverse wave vector k_y. The shading marks the area were the spin-wave excitation is efficient. (c) Isofrequency curves for different magnetic fields. The filled circuits show caustic points. (d)–(f) Spin-wave intensity of the propagating caustic beams and their scattering from medium boundaries (d), (e) and a defect (black dot) (f) measured using BLS spectroscopy. (After Schneider, T. et al., *Phys. Rev. Lett.* 104, 197203, 2010. With permission.)

θ_c of the caustic beam [7]. In contrast, for the larger magnetic field 1860 Oe, the caustic points (open red circles in Figures 6.6b, c) are situated outside of the region of the effective excitation of the spin waves by the waveguide opening and therefore no caustics will be formed for this field [7].

Figure 6.6d demonstrates the scattering of caustic spin-wave beams from the YIG film boundaries when the field $\vec{H}_0$ is directed along these boundaries [7]. The boundary region, from which the propagating caustic wave beam scatters, acts as a secondary wave source, whose finite size is of the order of the beam's width. This secondary source also radiates a wave packet with a wide angular spectrum that again forms caustic wave beams propagating at the same angle θ_c to the anisotropy axis defined by $\vec{H}_0$ as the initial wave beam. The rotation of the bias magnetic field clockwise through the angle of 20 degrees with respect to the medium boundary changes the pattern of the beam's reflection. The direction of the secondary, reradiated wave beam is determined by the direction of the inclined anisotropy axis in the medium $\vec{H}_0$ rather than by the rule of reflection in linear optics – see Figure 6.6e. A similar effect is seen in Figure 6.6f, where the bias magnetic field was rotated counterclockwise by 30 degrees, to observe caustic beam scattering at the intentionally made defect in the film (shown as a black dot).

Besides YIG structures [7, 9], permalloy [8] and Heusler materials [10] have been used to investigate spin-wave caustics. In the latter case, the nonlinear generation of higher harmonics leading to the emission of caustic spin-wave beams from localized edge modes was reported. The radiation frequencies of the propagating caustic waves were at twice and three times the excitation frequency [10]. In Ref. [179], a collapse of nonlinear spin-wave solitons and bullets was investigated experimentally and theoretically. It was shown that the collapse results in the generation of spin-wave caustic beams with the angles modified with respect to stationary caustics source due to the Doppler shift. It was demonstrated in Ref. [180] that the nonuniformity of the internal magnetic field and magnetization inherent to magnetic structures creates a medium of graded refractive index for propagating magnetostatic waves and can be used to steer their propagation. The two-dimensional diffraction pattern arising in the far-field region of a ferrite slab in the case of a plane wave with a noncollinear group and phase velocities incident on a slit is investigated theoretically in Ref [181]. In Ref. [182], the authors have combined the dipole-dipole and Dzyaloshinskii–Moriya interactions that resulted in the formation of unidirectional caustic beams in the Damon-Eshbach geometry. Finally, a switchable spin-wave signal splitter based on the controllability of the caustic spin-wave beam direction by locally applied magnetic fields was recently demonstrated in Ref. [183].

6.3.4 DIRECTIONAL COUPLERS

Several different concepts of magnonic logic and signal processing devices will be discussed later, but one of the unsolved problems of magnonic technology is an effective and controllable crossing of magnonic conduits without the interactions that are required for the realization of a functioning magnonic circuit. Unfortunately, a simple X-type crossing structure [146, 184] has a significant drawback, since the crossing point acts as a spin-wave re-emitter into all four connected spin-wave channels. The usage of the third dimension, like it is done in electronics, is also problematic due to the strong anisotropy of spin waves and demagnetizing effects. Thus, an alternative solution for the realization of spin-wave interconnections is necessary. One of the more promising solutions is based on the dipolar interaction between magnetic spin-wave waveguides. Originally, such a spin-wave coupling had been studied theoretically in a "sandwich-like" vertical structure consisting of two infinite films separated by a gap [185, 186]. However, the experimental studies of such structure are rather complicated due to the lack of access to the separate layers that are required for the excitation and detection of propagating spin waves. The configuration of a connector based on two laterally adjacent waveguides, which is well-studied in integrated optics, was recently proposed by Sadovnikov et al. also for magnonics [187] and has been studied experimentally using Brillouin Light Scattering spectroscopy for macroscopic YIG structures [157, 187, 188].

The general idea of the directional coupler is as follows [189]: In the case, where two identical magnetic strip-line spin-wave waveguides are placed

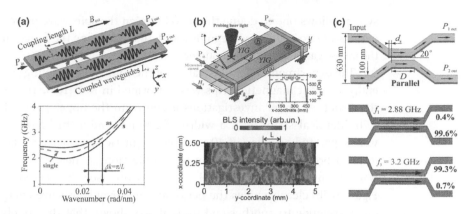

FIGURE 6.7 (a) The operational principle of a directional coupler based on two dipolarly-coupled spin-wave waveguides [189]: Solid black lines illustrate the periodic energy exchange between the two interacting waveguides with a spatial periodicity of 2L. Bottom panel: The red dashed line shows the dispersion characteristic of the lowest spin-wave width mode in an isolated single spin-wave waveguide. Solid black lines show the dispersion curves of the "symmetric" (s) and "anti-symmetric" (as) lowest collective spin-wave modes of a pair of dipolarly-coupled waveguides [189]. (After Wang, Q. et al., *Sci. Adv.* 4, e1701517, 2018. With permission.) (b) Top panel: Schematic of the structure under investigation in Ref. [187]. The inset shows the profile of the static internal magnetic field along the x-axis. Bottom panel: Normalized color-coded BLS spin-wave intensity map. The lower waveguide is excited with a microwave antenna. (After Sadovnikov, A.V. et al., *Appl. Phys. Lett.* 107, 202405, 2015. With permission.) (c) Top panel: Schematic view of the nanoscale directional coupler studied in [189]. The widths of the YIG waveguides are 100 nm, the thickness is 50 nm, the gap 30 nm. Bottom panels: Directional coupler as a crossing of magnonic conduits (middle panel) or as a switchable transmission line (bottom panel). The numbers show the percentage of the spin-wave output energy in each arm. (After Wang, Q. et al., *Sci. Adv.* 4, e1701517, 2018. With permission.)

sufficiently close to one another (see top panel in Figure 6.7a), the dipolar coupling between the waveguides leads to a splitting of the lowest width spin-wave mode of a single waveguide into the symmetric and anti-symmetric collective modes of the coupled waveguides – see the bottom panel in Figure 6.7b. Thus, in a system of two dipolarly-coupled waveguides, two spin-wave modes at different wavenumbers k_s and k_{as} $\left(\Delta k = |k_s - k_{as}|\right)$ will be excited simultaneously in both waveguides. The interference between these two propagating waveguide modes will lead to a periodic transfer of energy from one waveguide to the other as shown in Figure 6.7a, so that the energy of a spin-wave excited in one of the waveguides will be transferred to the other waveguide after propagation over a certain distance that is defined as the coupling length $L = \pi/\Delta k_x$ [185]. The case which is of interest for applications is the one in which a spin wave is originally excited in only one waveguide (see Figures 6.7b and c). The output powers for both waveguides can be expressed as [185]: $P_{1out} = \cos^2[\pi L_w/(2L)]$ for the first waveguide and $P_{2out} = \sin^2[\pi L_w/(2L)]$ for the second one, where L_w is the length of the coupled waveguides, and P_{in} is the input spin-wave power in the first waveguide. The dependence of the normalized output power of the first waveguide can be expressed as $P_{1out}/(P_{1out} + P_{2out}) = \cos^2[\pi L_w/(2L)]$. Thus, the interplay between the length of the coupled waveguides L_w and the coupling length L, which is strongly dependent on several external and internal parameters of the system, allows one to define the ratio between the spin-wave powers at the outputs of two coupled waveguides, and, thus, to define the functionality of the investigated directional coupler.

The experimental demonstration of the spin-wave coupling in two laterally adjacent magnetic strips was performed in Ref. [187]. The sketch of the structure under investigations is shown in the upper panel of Figure 6.7b. Identical strips with a width of 200 μm and a thickness of 10 μm were separated by a gap of 40 μm. The magnetic strips have a trapezoidal form in order to minimize spin-wave reflection at their ends. The spin-wave intensity mapped with BLS spectroscopy (see the bottom panel in Figure 6.7b) clearly shows the periodic transfer of spin-wave energy from one waveguide to another and back. It was shown that the coupling efficiency depends both on the geometry of the spin-wave waveguides and on the characteristics of the spin-wave modes [187]. In the following work, the authors have improved the spin-wave coupling by optimizing the design and have investigated it for different types of MSWs [188]. Also, a nonlinear coupling regime of spin waves in adjacent magnonic crystals has been investigated experimentally in macroscopic waveguides [157]. It was shown experimentally as well as by using a numerical simulation that a nonlinear phase shift of spin waves in the adjacent magnonic crystals leads to a nonlinear switching regime at the frequencies near the forbidden magnonic gap.

The dipolar coupling of nano-rather than macro-scale YIG spin-wave waveguides with parallel and anti-parallel orientations of the static magnetization along the long axis of the waveguides was studied using micromagnetic simulations in Ref. [189]. A general analytic theory describing the spin-wave coupling in the adjacent spin-wave waveguides is also developed in the same work. The coupling length L, over which the energy of the spin waves is transferred from one waveguide to the other, is studied as a function of the spin-wave wavenumber, the geometry of the coupler, the applied magnetic field, and the relative orientation of the static magnetization in the waveguides. Further, the design of the nanoscale directional coupler, in which strong shape anisotropy orients the static magnetization along the direction of the spin-wave propagation (see black arrows in the top panel of Figure 6.7c) and ensures practically reflectionless spin-wave propagation (only a few percent of spin-wave energy was reflected), was proposed [189]. The bottom panels in Figure 6.7c show (simulated spin-wave amplitudes are color-coded) that the change in the spin-wave frequency leads to a change of the coupling length and, as a result, the spin wave can be guided to the second arm of the directional coupler (middle panel) or can be sent back to the same arm (bottom panel). In general, it is demonstrated that the directional coupler can be used as an element to cross waveguides (discussed at the beginning of the section, middle panel in Figure 6.7c), as a controlled multiplexer, a frequency separator, or as a power divider for microwave signals [189]. Moreover, the proposed device has the additional benefit of dynamic reconfigurability of its functionality within a few tens of nanoseconds. The nanometer sizes of the proposed directional coupler make it interesting and useful for the processing of both digital and analog microwave information.

6.4 SPIN-WAVE EXCITATION, AMPLIFICATION, AND DETECTION

Besides the transfer of information carried by spin waves, which is the main topic of the previous section, other important challenges the field of magnon spintronics is facing are the excitation, manipulation, amplification, and, finally, the detection of spin waves. The spin-wave manipulation was already discussed in the part on magnonic crystals and in the section on basic ideas of magnon-based computing. Often spin waves should be amplified in order to compensate spin-wave damping or to restore the intensity of the spin-wave signal after splitting into two channels. The state-of-the-art microwave and spintronics approaches for the amplification of spin waves are discussed here.

The techniques used for the spin-wave excitation and detection in a laboratory have already been overviewed in the section on materials and methodology in magnonics. However, the majority of these approaches (e.g., optical techniques) requires large complex equipment and cannot be implemented into magnonic devices. In this section, the selected techniques for the excitation and detection of spin waves that have the potential to be implemented into a magnonic chip are discussed. As opposite to all-magnon approaches, in which the electron-magnon conversion is not required [24], the majority of computing-oriented spin-wave studies assumes that data should be coded from an electric signal to magnons in each magnonic unit, should be processed and converted back to the electric signal afterwards. This will simplify the clocking of Spin-Wave Devices [23] and will allow for compatibility with existing CMOS technology [20–23]. Therefore, the size of the converters and their efficiency will define the sizes of future devices as well as their power consumption. High-potential microwave and spintronic approaches suitable for this task are discussed. Please note, that one more approach based on the excitation of spin waves by electric field using magneto-eclectic cells [21] will be discussed in the next section.

6.4.1 PARAMETRIC PUMPING

Different methods of spin-wave amplification were discussed, e.g., in Ref. [4] but probably one of the most important is the parametric pumping approach [2]. A parametric process is one in which a temporally periodic variation in some system parameter affects oscillations or waves of another parameter in this system and can lead to their amplification. Perhaps the most well-known of all parametric amplifiers is the child's swing. The phenomenon of parametric instability in spin-wave systems was discovered by Bloembergen and Damon in 1952, through the observation of a nonlinear spin-wave damping effect [190]. An explanation was later proposed by Anderson and Suhl [191] and a corresponding theory was developed. The theory is based on the consideration of interactions between spin-wave eigenmodes in a system, in particular between a uniform FMR mode $k = 0$ and propagating spin-wave modes. When the correct conditions are fulfilled, energy from the driven uniform mode is pumped into the propagating modes, leading to their amplification from a thermal level. Since the alternating magnetization of the pumping mode in such a system is always perpendicular to the static bias field, this mechanism became known as spin-wave instability under *perpendicular pumping* [2]. The order of the parametric instability m is defined simply by the law of conservation of energy $m\omega_p = \omega_1 + \omega_2$, where ω_p is the frequency of the pumping magnon, ω_1 and ω_2 are the frequency of the pumped (or secondary) magnons. Perpendicular parametric pumping is also often described in terms of multi-magnon scatterings. Thus, the instability of the first order $m = 1$ is named three-magnon splitting and the instability of the second order $m = 2$ is named four-magnon scattering.

The related but physically different phenomenon of *parallel pumping* was discovered a few years later by Schlömann, Green, and Milano [192] and occurs when a spin wave receives energy from a double-frequency alternating magnetic field applied along the direction of magnetization. In this case, the magnetic field rather than an alternating magnetization is responsible for the effect and the back influence of the pumped spin waves on the pumping signal is usually ignored. In terms of energy quanta, parallel pumping can be understood as the creation of two magnons from a single pumping photon. The energy and momentum conservation laws for the instability of the spin wave under parallel pumping can be written as [2]:

$$\omega_p = \omega_{m1} + \omega_{m2}$$
$$k_p = k_{m1} + k_{m2}$$
(6.6)

where ω_{m1}, ω_{m2} and k_{m1}, k_{m1} are the frequencies and the wavenumbers of the interacting magnons ω_p and k_p, the frequency and wavenumber of the pumping photons.

From the law of the conservation of energy, in the simplest case, the frequencies of the interacting magnons are both equal to half the pump frequency $\omega_{m1} = \omega_{m2} = \omega_p/2$ – see Figure 6.8a. The photon wavenumber is much smaller than that of the magnons $k_p \ll k_{m1}$, k_{m2}. Therefore, the interacting magnons are usually counter-propagating $k_{m1} = -k_{m2}$ [11, 12]. In order to enable interaction of the double-frequency ω_p pumping field with the magnetization precession of frequency ω_{m1}, an alternative component of the magnetization oriented along the pumping field at frequency $2\omega_{m1}$ is required [103]. In the case of an in-plane magnetized magnetic thin film, this dynamic longitudinal component is nonzero as a consequence of the shape anisotropy, that results in an elliptic magnetization trajectory – see Figure 6.8b. Therefore, parallel pumping is efficient in thin films and the amplification of traveling MSSWs [193] and BVMSWs [194, 195] in YIG waveguides have been demonstrated. The gain of the amplification reached a value of 30

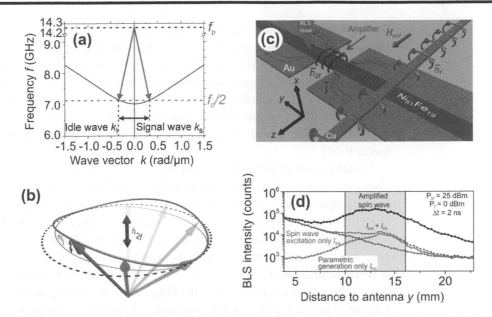

FIGURE 6.8 (a) The parallel pumping process in the particle picture [103]: A photon with a small wave-vector splits into a pair of counter-propagating magnons, leading to the formation of the signal and idler wave at one-half of the photon frequency. The solid dark blue line represents the spin-wave dispersion of the fundamental waveguide mode. (b) Schematic of the elliptical magnetization trajectory in a thin film [103]. This trajectory gives rise to the double-frequency longitudinal dynamic magnetization component m_l that interacts with the microwave pumping field h_{2f} (After Brächer, T. et al., *Physics Report* 699, 1, 2017. With permission.) (c) Sample layout for the localized parallel parametric amplification studied by BLS microscopy [199]. (d) Space-resolved BLS intensity arising from the parametric amplification of the externally excited spin waves (black squares), from the antenna excitation only (red dots) and from parametric generation only (green triangles). The shading marks the position and the length of the amplification region. (After Brächer, T. et al., *Appl. Phys. Lett.* 105, 232409, 2014. With permission.)

dB in some experiments. In all but a few special cases, parametric amplification of magnons in mesoscopic samples must be performed in a pulsed, rather than continuous, pumping regime. The reason for this is that when pumping is applied, the amplitudes of thermal exchange modes, which are degenerate with the magnetostatic magnons of interest, start to grow exponentially [196]. Taking into account that these thermal waves (often named dominant modes) usually exhibit a lower damping, their amplification rate is larger which allows their amplitudes to overcome the amplitude of the signal-carrying spin wave quite fast in spite of their small original amplitudes [196]. In order to avoid undesirable competition between these modes and the amplified spin-wave signal packet, the pumping duration is usually considerably shorter than the characteristic relaxation time of the spin wave. The most comprehensive theory that describes parametric instabilities as well as the interactions between the magnons is named *S-theory* and was developed by L'vov and Zakharov [46, 197].

The amplification of nonlinear eigen-excitations of magnetic media, namely, spin-wave envelope solitons [198] and bullets [12], constitutes a very interesting separate problem since the ratio between the length and the amplitude of a soliton is fixed. The solution was realized through the use of localized pulsed parametric pumping and an amplification of 17 dB of fundamental BVMSW solitons has been demonstrated by Melkov et al. [198]. Moreover, according to the momentum conservation shown in Equation 6.6, the secondary wave of a wave vector $-k_{m2}$ propagates in the opposite direction and is phase-conjugated. Therefore, the Wave-Front Reversal (WFR) of linear [11] as well as nonlinear [12] spin-wave packets was realized.

Finally, parallel parametric pumping is a very efficient mechanism to excite spin waves of arbitrary wavelengths and, therefore, is usually used for the magnon injection (namely amplification of spin waves from thermal level) in the experiments on Bose–Einstein Condensation (BEC) of magnons [13–16]. Using parametric pumping, a pumped magnon density of 10^{18}–10^{19} cm^{-3} can be reached [13]. Although this density is much smaller than that of thermal magnons at room temperature (10^{21}–10^{22} cm^{-3}), it is sufficient to increase the chemical potential of the magnons to the energy of the lowest magnon state even at the room temperature. As a result, the formation of a Bose–Einstein condensate of magnons was reported by Demokritov et al. in Ref. [13]. Later, Serga et al. [15] demonstrated that parametric pumping can create remarkably high effective temperatures in a narrow spectral region of the lowest energy states in a magnon gas that results in strikingly unexpected transitional dynamics of a Bose–Einstein magnon condensate: The density of the condensate increases immediately after the external magnon flow is switched off (evaporative supercooling mechanism [15]). Finally, the first evidence of the formation of a room-temperature magnon supercurrent was reported recently by Bozhko et al. [16]. The appearance of the supercurrent, which is driven by a thermally-induced phase shift in the condensate wavefunction, is evidenced by an analysis of the temporal evolution of the magnon density measured by means of BLS spectroscopy.

The majority of the experiments described above were performed using YIG samples of micrometer thicknesses and macroscopic lateral dimensions. An important breakthrough in the utilization of the parallel parametric pumping at the micrometer-scale was performed in a set of studies by Brächer et al. [102, 199–201] (see also Ref. [103]). The parallel parametric generation of spin waves was studied by means of BLS spectroscopy in a longitudinally magnetized permalloy magnonic waveguide of 2.2 μm width [200]. A 1.2 μm-wide microstrip antenna was placed perpendicularly on top of the spin-wave waveguide to apply the double-frequency pumping magnetic field oriented parallel to the magnetization of the waveguide. First, the parametric instability was investigated in the absence of any coherently-excited spin waves [200]. Considering the quite large wavenumber range of micro-focused BLS spectroscopy (approximately $k_{\max}^{\mathrm{BLS}} \approx 20$ rad/μm), this allows for a direct mapping of the parametrically excited dominant spin-wave mode exhibiting the smallest threshold of the parametric instability. The analysis of the spatial distributions of the generated spin waves has shown that odd, as well as even BVMSW waveguide modes, can be excited parametrically. Furthermore, it was revealed that the generation process takes place underneath the antenna where the pumping field is at its maximum only due to the threshold nature of the parametric instability (the magnetic field away from the antenna is too small to reach the threshold of the parametric instability). Parametric amplification of propagating spin waves was proven by the investigation of the spatial decays of spin waves in Ref. [102]. This time, the pumping field was applied by a microwave current flowing through a copper microstrip line underneath the waveguide. It was shown that amplified spin waves propagate distances of about 30 μm and an amplification factor of approximately 10 dB was achieved. To avoid mode competition with a dominant spin-wave mode and saturation of the amplification, short pulses with pumping powers close to the threshold power of parametric generation were used [102].

The utilization of a localized rather than uniform parametric pumping in a transversally magnetized in-plane magnetized permalloy waveguide was demonstrated in Refs. [199 and 201]. The localization was realized by combining the threshold character of a parametric generation with a spatially confined enhancement of the amplifying microwave field by introducing a narrowed region in a microstrip transmission line – see Figure 6.8c. Figure 6.8d shows the functionality of the localized spin-wave amplifier. Spin waves are excited at the antenna by a 15 ns-long microwave pulse with a carrier frequency of 6 GHz and decay exponentially if no pumping signal is applied (see the red line in the figure). The generation of the dominant spin wave only due to the application of the parametric pumping signal at 12 GHz is shown with the green line. Finally, the amplified spin wave is shown by the black line in Figure 6.8d. An amplification of the propagating wave is clearly visible. Moreover, it was shown by time-resolved measurements that the amplification is efficient as long as the pumping is timed properly with respect to the arrival time of the spin-wave packet. In the case of a strong pumping, this timing is crucial and the spin waves have to arrive prior to the pumping pulse. If the applied pumping is rather weak, the timing becomes less important, a higher gain can be achieved, and the amplification is practically independent of the arrival time of the spin-wave packet [199].

6.4.2 Spin Hall Effect and Spin-Transfer Torque

As was discussed above, the combination of magnonic devices with electronic circuits requires an efficient means of magnon excitation by a charge current. Although magnons can be injected relatively easily by an AC current (e.g., using antenna structures), it is a quite complex problem if a DC current is used. One of the promising solutions is the usage of the *Spin-Transfer Torque* (STT) effect. In 1996, Slonczewski [111] and Berger [112] predicted independently that the injection of a spin-polarized current in a magnetic metallic film can generate a Spin-Transfer Torque strong enough to reorient the magnetization or to excite magnetization precession [202] in this film. In order to generate the spin-polarized current, the charge current is sent through an additional magnetic layer with a fixed magnetization direction.

A device specially designed to excite the magnetization precession is named Spin-Torque Nano-Oscillator (STNO). A first microwave measurement of a spin-torque-driven precession was presented in 1998 by Tsoi et al. [203]. Krivorotov et al. demonstrated experimentally that STT can be used to control the magnetic damping and for the magnetization reversal of a nanomagnet [113].

The excitation of spin waves by STNO was observed by Demidov et al. in 2010 [114]. The authors used BLS spectroscopy in order to perform a two-dimensional mapping of waves emitted by the STNO into an in-plane magnetized permalloy film. It was reported that the emission is directional and depends on the orientation of the applied magnetic field. However, since the propagation length of the emitted waves did not exceed one micrometer, the propagating character of the waves was not proven. Another report on spin-wave generation by STNO was presented by Madami et al. [115]. The authors used a normally magnetized permalloy film, which was probed by BLS spectroscopy – see the left panel in Figure 6.9a. In this case, the radial emission of spin waves propagating over a distance of a few micrometers was

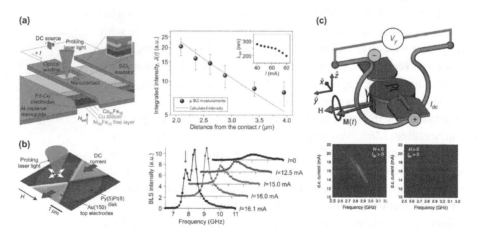

FIGURE 6.9 Magnon excitation by Spin-Transfer Torque. (a) Sample layout for studies of propagating spin waves induced by the Spin-Transfer Torque [115]. An aluminum coplanar waveguide is deposited onto the spin-valve structure, and an optical window for BLS probing is etched into the central conductor of the waveguide close to the STT nano-contact. Right panel: Measured (symbols) and calculated (line) spin-wave intensity as a function of the distance from the center of the point contact. The inset shows the dependence of the spin-wave wavelength on the applied electric current. (After Madami, M. et al., *Nat. Nanotech* 6, 635, 2011. With permission.) (b) The experimental layout of a magnetic nano-oscillator driven by a spin current generated by the SHE [116]. The device comprises a Pt (8 nm)/Py (5 nm) disk of 4 μm diameter with two Au electrodes separated by a 100 nm gap. Right panel: BLS spectra of the thermal spin-wave fluctuations at electric currents below the onset of the auto-oscillation. The spin current induced auto-oscillation peak (marked with a vertical arrow) appears at a current of 16.1 mA. (After Demidov, V. et al., *Nat. Mater.* 11, 1028, 2012. With permission.) (c) Sketch of the measurement configuration a device with two YIG/Pt micro discs (scale bar is 50 μm) [117]. The bias field is oriented transversely to the electric current flowing in Pt. The inductive voltage produced in the antenna by the SHE-STT generated precession of the YIG magnetization is amplified and monitored by a spectrum analyzer. Bottom panels: power spectral density maps measured on a 4 μm YIG/Pt disk at a fixed magnetic field and variable direction of electric current. Magnetization precession amplitude is color-coded. (After Collet, M. et al., *Nat. Commun.* 7, 10377, 2016. With permission.)

observed (see the right panel in Figure 6.9a). The experimentally obtained magnon free-path agrees well with a theoretical estimation.

Another way in which spin-polarized electron current can be generated is based on the *Spin Hall Effect* (SHE) caused by spin-dependent scattering of electrons in a nonmagnetic metal or semiconductor with large spin-orbit interaction [204, 205]. Electrons flowing in a film, magnetized in-plane and transversely to the direction of the current, scatter with spin-asymmetry generating a spin current perpendicular to the film plane. This current, crossing the interface to an attached magnetic layer, generates a STT in this layer. A typical metal used in SHE studies is Pt due to its large conversion factor of charge current to spin current [205, 206]. Its thickness varies between 2 and 10 nanometers; these values are close to the spin-diffusion length in Pt [207, 125].

A great advantage of the SHE as a spin-current source is that a STT can be injected not only into a single local object (the diameter of a typical STNO is less than 100 nm) but also into a large area of a magnetic film. This allows for the realization of spin-wave amplification due to damping compensation. The first experimental observation of SHE-induced damping reduction was reported by Ando et al. [206] in a Py/Pt bilayer. In the studies in Ref. [208], the variation of the damping parameter by a factor of four was demonstrated in Py/Cu/Pt multilayers studied by means of BLS spectroscopy. However, no spin-wave generation was achieved in these experiments, the probable reason being the strong nonlinear redistribution of the injected energy between many magnon modes. In subsequent studies, a modified design of the current-conducting structure containing bowtie-shaped electrodes with a 100 nm gap between them was used [116]. According to the model proposed by the authors, these electrodes allow for an increase in the density of the electric current applied to the Pt layer and for introducing a controlled radiation loss mechanism for the parasitic magnon modes previously disturbing the generation process. Consequently, the injected energy was concentrated onto the small area between the electrodes, and a single bullet-like spatially localized magnon mode was observed. Subsequently, the coherency of these magnons was proven by Liu et al. through microwave measurements [209]. Later, Duan et al. [210] demonstrated microwave oscillations of the magnetization in a ferromagnetic nanowire, where the geometric confinement dilutes the magnon spectrum and, thus, suppresses the parasitic nonlinear energy redistribution.

Another important advantage of the SHE-based STT is that no electric current needs to flow across the magnetic layer and, thus, the usage of a low-damping magnetic dielectric material such as YIG is possible. In the pioneering work of Kajiwara et al. [124], the transmission of a continuous electric signal through a YIG film was demonstrated. For that, the SHE-based STT was used in order to convert an electric current into traveling spin waves. However, the exact conditions required for such a conversion [79, 211, 212], or even for the SHE-STT-based damping compensation [213, 214], were not defined. In this context, special attention has been attracted by the work of Hamadeh et al. where bowtie-like electrodes (see Figure 6.9b) were used for the damping compensation in YIG disks [215]. The decrease of the ferromagnetic resonance linewidth (which is a measure of the magnetic damping) by a factor of three was reported. Another approach was reported by Lauer et al. in Ref. [118] where the threshold of the parametric instability measured by BLS spectroscopy was used to determine the degree of STT-SHE-controlled spin-wave damping. A macroscopically sized YIG(100 nm)/Pt(10 nm) bilayer of 4×2 mm^2 lateral dimensions and pulsed current regime were under investigations rather than microscale structures with a continuous current. A variation in the relaxation frequency of $\pm 7.5\%$ was achieved for an applied current density of $5 \cdot 10^{10}$ A/m^2 depending on current polarity [118]. The SHE-STT amplification of propagating spin waves was studied experimentally in microscopic waveguides based on nanometer-thick YIG/Pt bilayers in Ref. [216]. It was shown that the propagation length of the spin waves in such systems can be increased by nearly a factor of 10. It was also demonstrated that, in the regime where the magnetic damping is completely compensated by the SHE-STT, the amplification of the spin wave was suppressed by the nonlinear scattering of the coherent spin waves from current-induced excitations [216]. An important breakthrough in the field was the demonstration of the SHE-STT excitation and transport of diffusive spin waves by Cornelissen et al. in Ref. [217]. It was shown experimentally that magnons can be excited and fully detected electrically and can transport spin angular momentum in YIG over distances of 40 μm. Gönnenwein et al. reported shortly afterwards [218] on the observation of the same phenomenon due to investigations of a local and nonlocal magnetoresistance of thin Pt strips deposited onto YIG. Later, this phenomenon was used to build a spin-wave majority gate using a multi-terminal YIG/Pt nanostructure [219].

The SHE-STT generation of coherent spin-wave modes in YIG micro-disks by Spin-Orbit Torque (SOT) was reported by Collet et al. [117]. Magnetic micro-disks with diameters of 2 and 4 μm were fabricated based on a hybrid YIG(20 nm)/Pt(8 nm) bilayer and were measured using a microwave antenna around the disks – see the top panel in Figure 6.9c. Color plots of the inductive signal measured as a function of the relative polarities of the magnetic field are presented in the bottom panels of Figure 6.9c. An auto-oscillation signal is clearly visible if the current and field polarities are properly chosen in accordance with the symmetry of SHE [117]. Further investigations of the phenomena using spatially-resolved BLS spectroscopy were reported in Ref. [220]. It was shown that SHE-STT excited spin-wave modes exhibit nonlinear self-broadening preventing the formation of the self-localized magnetic bullet, which plays a crucial role in the stabilization of the single-mode magnetization oscillations. Time-resolved BLS spectroscopy measurements of the SHE-STT excited magnetization precession in YIG/Pt micro-disks was reported in [119]. It was shown that the magnetization precession intensity saturates within a time frame of 20 ns or longer, depending on the current density.

It was also demonstrated that a spin current and consequently, the STT may be induced by a thermal gradient rather than by an electric current and the SHE effect [221–223]. Particularly, recently the generation of coherent spin waves in YIG/Pt nanowires by the application of thermal gradients across the structure was demonstrated in Ref. [224]. The nanowires were investigated at low temperatures by means of electrical measurements. It was shown that the Spin Seebeck Effect (SSE) induced spin current induced by the Ohmic heating of Pt is the dominant drive of auto-oscillations of the YIG magnetization. The magnon generation was also observed in Ref. [225] in a similar structure using time- and space-resolved BLS spectroscopy at room temperature in a pulsed heating regime. In this experiment, it was found that the role of SSE vanished with respect to the heating of the Pt/YIG structure by the DC current pulses. The heating and consequent rapid cooling of the structure resulted in the increase of the chemical potential of magnons and lead to the formation of Bose–Einstein Condensation. All these approaches open the door for the effective conversion of waste heat in future magnonic devices into spin waves or electric currents for further data transfer and processing [131, 226].

6.4.3 SPIN PUMPING AND INVERSE SPIN HALL EFFECT

The field of magnon spintronics assumes that after the information is processed within a magnonic system it needs to be converted back to electronic signals. A conventional way to do this is to use a strip or a coplanar antenna, in which spin waves induce an AC current that is in turn rectified by a semiconductor diode. Another recently discovered way is based on the combination of two physical effects: *Spin Pumping* (SP) and the *Inverse Spin Hall Effect* (ISHE). In 2002, Tserkovnyak, Brataas, and Bauer [121] showed theoretically that magnetization precession in a magnetic film will generate a spin-polarized electric current in an attached nonmagnetic metallic layer. This process manifests itself in the increase in damping of a magnonic system [121, 227, 228]. The electrical detection of the Spin Pumping induced spin current was reported by Costache et al. in 2006 [122]. In the same year, Saitoh et al. [123] reported the observation of Spin Pumping using the ISHE. This effect refers to the generation of a charge current in a nonmagnetic metal by a spin current and is the reciprocal effect of the SHE. Due to this effect, a spin current induced in a Pt film by a precessing magnetization in an adjacent magnetic film is converted into a detectable DC voltage (see, e.g., the review by Hoffman [125]). Since then, the combined SP-ISHE mechanism is used as a convenient detection mechanism of magnons.

The SP-ISHE mechanism allows for measurements of the magnetization precession not only in metallic but also in YIG-based structures. The first report from Kajiwara et al. on the observation of an ISHE voltage in YIG/Pt bilayers [124] was followed by comprehensive studies of this phenomenon. The dependencies on the thicknesses of the nonmagnetic metal [208, 211, 229] and the YIG [230, 231] layer, as well as on the applied microwave power [231] have been reported. The influence of the interface conditions on the spin-pumping efficiency [232, 233] was revealed, and the contributions to the SP effect by different spin-wave modes were studied [234]. An important milestone was the successful implementation of the combined SP-ISHE mechanism for

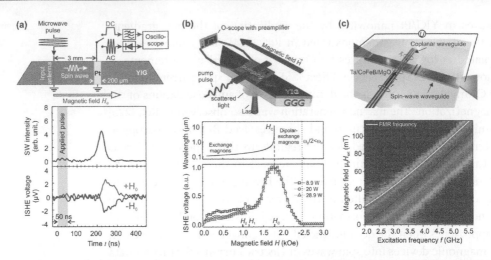

FIGURE 6.10 (a) Schematic illustration of the experimental setup for the Inverse Spin Hall Effect detection of propagating spin waves [235]. A spin-wave packet is excited in the YIG conduit using a microstrip antenna and is detected by the Pt strip placed 3 mm apart as AC and DC signals. Temporal evolutions of the spin-wave intensity (AC signal) and the ISHE voltage (DC signal) generated by a 50 ns-long spin-wave pulse are shown in the bottom panel for different field polarities. (After Chumak, A.V. et al., *Appl. Phys. Lett.* 100, 082405, 2012. With permission.) (b) Sketch of the experimental setup for the investigation of Spin Pumping by parametrically injected exchange magnons [43]. A 10 nm-thick, 3×3 mm² Pt layer is deposited onto the 2.1 μm-thick YIG film. A microwave pumping field is applied using a microstrip antenna. Bottom panel: Normalized dependencies of the ISHE voltage induced by the injected magnons as a function of the bias magnetic field for different pumping powers. The calculated wavelength of the parametrically injected magnons is shown in the middle panel. (After Sandweg, C.W. et al., *Phys. Rev. Lett.* 106, 216610, 2011. With permission.) (c) Schematic of the sample and the geometry of the spin-wave rectification measurements in [45]. A Ta/CoFeB/MgO trilayer is patterned into a spin-wave waveguide with leads to measure the ISHE voltage along the waveguide. A nanoscale coplanar waveguide antenna is placed on top to excite spin waves. Bottom panel: Color-coded measured ISHE voltage as a function of the applied frequency and applied magnetic field. (After Brächer, T. et al., *Nano Lett.* 17, 2017. With permission.)

the detection of propagating spin waves [235]. A typical geometry is shown in Figure 6.10a. A spin-wave detector in form of a 200 μm wide Pt strip was placed 3 mm away from the microstrip antenna. The Pt strip was used for the detection of the time-resolve ISHE voltage, as well as, for reference, as a conventional inductive probe. In the bottom panels of Figure 6.10a, both the AC and the electric signals produced by a 50 ns-long spin-wave pulse are shown. One can clearly see that the ISHE voltage appears with a delay of 200 ns determined by the spin-wave propagation time between the antenna and the detector. The ISHE nature of the electric signal is proven by the fact that the inversion of the direction of the biasing magnetic field results in the switching of the voltage polarity [236] – see Figure 6.10a. In the same experiment, it was also demonstrated that the spin-pumping efficiency does not depend on the spin-wave wavelength. d'Allivy Kelly et al. have used a similar experimental setup to demonstrate the ISHE detection of propagating magnons in a nanometer-thick YIG film [79]. Magnetostatic Surface Spin Waves show nonreciprocal behavior. The propagation direction of these waves can be reversed by a change in the polarity of the bias magnetic field [96]. An electric probe was used to measure the ISHE voltage at different points of a sample in Ref. [237].

It was shown that the ISHE voltage induced by the MSSWs depends on the field orientation. The combined SP-ISHE mechanism opened doors for the access to short-wavelength exchange magnons [43, 45]. The short-wavelength regime is of particular interest for nanosize magnonic applications. In order to excite exchange magnons, the parallel parametric pumping technique discussed above was used [43] – see Figure 6.10b. By the variation of the bias magnetic field, the magnon spectrum is shifted upward or downward to tune the wavelength of the magnon (see the inner panel in Figure 6.10b). The Spin Pumping induced ISHE voltage is shown in the bottom panel of Figure 6.10b. One can see that magnons effectively contribute to the Spin Pumping within a wide range of wavelengths (down to 100 nm in Ref. [43]). Follow-up studies by Kurebayashi et al. [238], where parallel and perpendicular parametric pumping techniques were used for the magnon injection, have evidenced that the spin-pumping efficiency is independent of the magnon wavelength within experimental error.

It was recently shown by Brächer et al. [45] that exchange spin waves coherently excited by a microstrip antenna rather than by parametric pumping can be efficiently detected in a $Ta/Co_8Fe_{72}B_{20}/MgO$ microscaled waveguide – see the top panel in Figure 6.10c. This layer system features large SOTs and a large Perpendicular Magnetic Anisotropy (PMA) constant. The short-wavelength spin waves were excited by nanoscale coplanar waveguides (the smallest feature size was 50 nm) and were detected by way of SP-ISHE voltage as well as using micro-focused BLS spectroscopy [45]. The bottom panel in Figure 6.10c shows the measured SP-ISHE spectra of the excited spin waves. The measured ISHE voltage is shown color-coded as a function of the applied magnetic field and frequency. It is demonstrated that the noise-limited maximum detectable wave vector was about 42 rad μm^{-1} (wavelength is around 150 nm) and is determined by the Fourier spectrum of the excitation source. These short wavelengths are not detectable by, e.g., BLS spectroscopy, which is only sensitive to magnons down to about 300 nm wavelengths. Moreover, it was demonstrated that the spin-wave emission by the CPWs exhibits a strongly preferred emission direction due to the large PMA in the investigated spin-wave waveguide [45].

6.4.4 MAGNETO-ELECTRIC CELLS AND THEIR USAGE FOR SPIN-WAVE LOGIC

For more than four decades the successful scaling of Complementary Metal–Oxide–Semiconductor (CMOS) Field-Effect Transistors (FETs) took place according to Moore's law [239]. The projection of scaling limits and quantum limits on the size of electronic transistors [240] stimulates a search for novel *beyond-CMOS* technologies. The benchmarking of these novel approaches became an important effort and it was nicely reviewed by Nikonov and Young in Ref. [20]. The authors have analyzed 11 new devices operating with three types of new magnetization switching mechanisms: (*i*) the Spin Hall Effect,

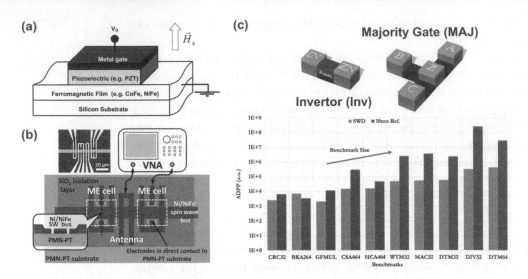

FIGURE 6.11 (a) Schematic view of the Magneto-Electric cell proposed in [144]. (After Khitun, A. et al., *IEEE Trans. Mag.* 44, 2141, 2008. With permission.) (b) Schematic of the experimentally realized ME cell [247]. Spin-wave generation and propagation is measured using a vector network analyzer. The inset shows a cross-section view of the ME cell. (After Cherepov, S. et al., *Appl. Phys. Lett.* 104, 082403, 2014. With permission.) (c) Top panel: Gate primitives used for SWD circuits [22]. Bottom panel: ADP product of all benchmarks for Spin-Wave Devices (green columns) and for the reference 10 nm CMOS (blue columns). (After Zografos, O. et al., *Proc. of the 15th IEEE Int. Conf. on Nanotechn* 686, 2015. With permission.)

(*ii*) ferroelectric switching, and (*iii*) piezoelectric switching. The Spin Hall Effect-based approaches were already described above in detail. Ferroelectric devices rely on electric polarization and are very promising [241–243]. In this section, however, the piezoelectronic devices, which utilize stress and strain mechanisms to switch magnetization, are described. Since excitation, detection, and manipulation of spin waves in these devices is performed using electric fields rather than currents, this approach has a large potential for future spin-wave computing with low-energy consumption.

A feasibility study of logic circuits utilizing spin waves controlled by an electric field for information transmission and processing by Khitun et al. was discussed in Ref. [244]. In the following publications [144], the authors proposed basic elements that include voltage-to-spin-wave and spin-wave-to-voltage converters, spin-wave waveguides, a spin-wave modulator, and a Magneto-Electric (ME) cell. The performance of the basic elements was demonstrated by experimental data as well as with the results of numerical modeling. The combination of the basic elements allowed for the construction of magnetic circuits for NOT and majority logic gates as well as for AND, OR, NAND, and NOR logic gates [21, 144]. The proposed ME cell requires special attention and is shown schematically in Figure 6.11a. It consists of piezoelectric and magnetostrictive films. The ME coupling occurs via strain, or in the case of high frequency, via acoustic waves. As the coupling between voltage and strain in piezoelectric films can be efficient [27], the ME cells are very promising as transducers with a high efficiency as it relates to energy along with a large output signal. An additional advantage of ME-cell-based transducers is their high scalability due to the absence of delocalized magnetic fields inherent to microstrip antennas and current-driven STT elements. A numerical modeling of the ME cells and a concept of magnetic logic circuits engineering is given in Ref. [245]. The utilization of ME cells to control the phase of spin waves was presented in Ref. [246].

Cherepov et al. reported [247] on the spin-wave generation by multiferroic ME cell transducers driven by an alternating voltage. A multiferroic element consisting of a magnetostrictive Ni film and a piezoelectric $[Pb(Mg_{1/3}Nb_{2/3})O_3]_{(1-x)}-[PbTiO_3]_x$ substrate was used for this purpose – see Figure 6.11b. By applying an AC voltage to the piezoelectric, an oscillating electric field was created within the piezoelectric material and resulted in an alternating strain-induced magnetic anisotropy in the magnetostrictive Ni layer. The resulting anisotropy-driven magnetization oscillations propagate in the form of spin waves along a 5 μm wide Ni/NiFe waveguide. The authors noted that the amplitude of the generated spin waves was rather low in this demonstration but can be improved by using materials with lower damping or by geometrical and materials optimization [247].

Dutta et al. [23] proposed a comprehensive scheme for building a clocked nonvolatile Spin-Wave Device by introducing an ME cell that translates information from the electrical domain to the spin domain, Magneto-Electric spin-wave repeaters that operate in three different regimes (spin-wave transmitter, nonvolatile memory and spin wave detector), and a novel clocking scheme that ensures the sequential transmission of information and nonreciprocity. The authors have demonstrated that the proposed device satisfies the five essential requirements for logic application: nonlinearity, amplification, concatenability, feedback prevention and a complete set of Boolean operations [23]. Finally, Zografos et al. presented [22] a design and benchmarking methodology of Spin-Wave Device (SWD) circuits based on micromagnetic modeling. Spin-Wave Device technology is compared against a 10 nm FinFET CMOS technology, considering the key metrics of the area, delay, and power. ME-cell-based spin-wave invertors and majority gates (see the top panel in Figure 6.11c) have been considered as primitives of complex SWD circuits processing up to a few hundreds of bits. Figure 6.11c shows the Area-Delay-Power Product (ADPP) which depicts how much the SWD circuits outperform the 10 nm CMOS reference ones. On average, the area of future SWD circuits is expected to be 3.5 times smaller and the power consumption up to 100 times lower when compared with the 10 nm CMOS reference circuits [22]. Therefore SWD appears as a strong contender for ultra-low power applications.

6.5 CONCLUSIONS AND OUTLOOK

To conclude, magnonics and magnon spintronics are very active and promising fields with a number of breakthrough-developments towards application in data processing that have already been demonstrated or are expected to be demonstrated in the near future. In order to clarify this statement, the concluding section is prepared in the form of brief discussions at a very basic level.

Transport of spin-wave carried data. Data coded into spin-wave amplitude or phase is typically guided by the use of waveguides in the form of strips made of magnetic materials [19, 41, 54, 72]. Besides, spin-wave physics pave the ways for novel approaches to data transport, e.g., in nonpatterned plane films in the form of caustics [7–10] or using dipolarly-coupled wave-guiding structures [187, 189]. A guiding of spin waves in two dimensions is presented in the section on magnonic circuits. The main characteristics of spin waves are given in the section on spin waves in thin-film waveguides.

Types of signals processed using spin waves. Classically, spin waves are investigated with the view on operations with analog information in the GHz frequency range (microwave filters, delay lines, amplifiers, etc.) [33–35, 152–155]. Also, binary digital data can be coded into spin-wave amplitude or phase and conventional logic gates operating with the use of spin waves have been developed – see section 6.2.4 on the magnon-based computing. Alternately, novel computing concepts involving reversible computing [28, 139], neuromorphic networks [29], and quantum computing [248] also have high potential and the field of magnonics is suitable to realize them. Chapter 19 of this book (Volume 3) discusses more magnonics concepts for non-Boolean computing including holographic memory, pattern recognition, and the prime factorization problem.

Advantages for data processing proposed by spin waves: The main advantages proposed by spin waves for data processing were briefly mentioned at the beginning of the chapter and they were discussed in more detail throughout the chapter. Among others one can underline [19]: (*i*) efficient wave-based computing, (*ii*) small loss in insulating materials, (*iii*) wide frequency range up to THz, (*iv*) fundamental size limitations are given by the lattice constant of a magnetic material, (*v*) pronounced nonlinear phenomena, (*vi*) possibility of wireless spin-wave excitation and detection, and (*vii*) rich spin-wave physics toolbox opening new ways for the transport and processing of data.

Typical sizes and operational frequencies of magnonic devices: Nowadays, spin waves are studied experimentally in structures from millimeter down to micrometer sizes [4, 17, 19, 33, 54, 72]. The smallest reported prototypes reported so far have lateral sizes of few hundred nanometers [42, 210, 224, 225]. At the same time, Spin-Wave Devices studied by means of numerical simulation demonstrate promising functionalities at lateral sizes scaled down to at least tens of nanometers [26, 159, 172, 189]. The frequencies of the spin waves are mainly defined by the choice of the magnetic material, by the applied magnetic field and by the wavelengths of the spin wave. Nowadays, spin waves are usually investigated within the frequency range from one GHz to one hundred GHz. Nevertheless, there is intensive work going on to further reduce the size of the devices down to a few tens of nanometers and to increase the operating frequencies to sub-THz and THz frequency ranges.

The long-term perspectives of spin-wave data processing: The perspectives are defined by the potential parameters of future devices, which, consequently, depend on the spin-wave characteristics. The qualitative analysis of the characteristics of exchange spin waves of nanometer wavelengths is performed in the section on spin waves in thin-film waveguides. It was shown, e.g., that a spin wave of 5 nm wavelength has a frequency of about 2 THz in

YIG and can propagate a distance of more than 16 µm, which is more than three thousand times the wavelength. Therefore, spin waves indeed appear to be a promising candidate for the use as an information carrier in future ultrafast low-loss computing systems.

Spin-wave logic devices that have already been demonstrated: At least, the following types of spin-wave logic devices have been realized experimentally at the level of proof-of-concept prototypes: (*i*) spin-wave logic gates in which spin waves are manipulated by DC electric currents [143] (hardly suitable for the realization of an integrated circuit), (*ii*) spin-wave majority gate [25] (suitable for the realization of integrated magnonic circuits after the development and implementation of an energy efficient nonlinear amplitude normalizer), and (*iii*) magnon transistor for all-magnon data processing [24] (the technology is self-consistent but requires, first of all, miniaturization to the nanometer scale), (*iv*) a technique for magnetic microstructure imaging [249], (*v*) a device for prime factorization using spin-wave interference [250], and (*vi*) a micrometer-scale spin-wave interferometer suitable for logic operations [251]. Besides, there are many very promising theoretical investigations in this field that were already discussed in this chapter.

Energy consumption of Spin-Wave Devices: Since the field of spin-wave computing is at an initial stage of its development, the energy consumption can only be estimated with a certain accuracy and this strongly depends on the concrete choice of computing approach. For example, the approach based on the magnon transistor (the estimated energy consumption for a nanoscaled magnon transistor is 5 aJ per bit [24]) will most likely require more energy in comparison to a spin-wave majority gate [21, 25, 145] operating with linear waves. In the latter case, the amplitude of the spin-wave is limited from below by the level of thermal noise rather than by the thresholds of nonlinear processes. This, in combination with the usage of fast exchange waves in low-damping materials such as YIG, should ensure the energy consumption on the aJ level. However, the majority gate approach still requires the realization of an amplitude normalizer in order to combine many gates into a circuit and, currently, it is hard to estimate the total energy consumption of a fully-functioning device. The energy consumption of Spin-Wave Devices based on STT or ME cells converters depends directly on the efficiency of the conversion between spin waves and an electric signal. Nowadays, the energy cost for injecting of information into magnonic elements dominate the energy loss within the magnonic system itself. As an example, in Ref. [27], the authors have estimated that 24 aJ will be needed to excite spin waves per switching of a 100 nm × 100 nm ME cell.

Interfacing of Spin-Wave Devices to CMOS: Any new concept for data processing requires interfacing with existing electronics devices and this is a topic of the section on spin-wave excitation, amplification, and detection. Even in the case of all-magnon computing discussed above (when all information is kept inside of the magnonic system) the data should be transformed from the electric form into magnon-carried signals in the magnonic circuit at the beginning and should be read out at the end. In the case of other approaches, in which data is coded to and from magnonic elements after each operation [23], the interfacing is much more crucial. In particular, the efficiency of the conversion will define the energy consumption of such Spin-Wave Devices. Therefore, the realization of efficient converters by different means represents one of the most challenging tasks of modern magnon spintronics.

Main challenges in the field of magnonics: This chapter, of course, only covers a chosen selection of scientific, engineering, and technological problems. Among them probably the most important challenges are the miniaturization of the magnonic devices down to 10 nm sizes, increase of operating frequency to sub-THz and THz frequency ranges, the development of the approaches for the excitation and the detection of short-wavelength exchange magnons, the development of low-loss and nonreciprocal spin-wave conduits, the reflectionless guiding of spin waves in two dimensions, the realization of highly-efficient spin-wave splitters and combiners, the development of highly-efficient means for spin-wave amplification, the investigations of nonlinear spin-wave phenomena at the nanoscale and, as was discussed above, the development of efficient converters between spin waves and electric signals.

Research directions beyond the scope of this book chapter: The fields of magnonics and magnon spintronics are very versatile and consist of many different research directions. A rough sketch of the selected research subfields is given in Figure 6.12. Some of the directions shown are already well

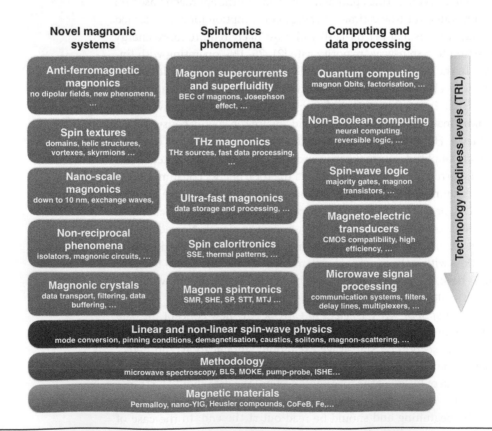

FIGURE 6.12   The variety of problems in modern magnonics and magnon spintronics.

established, others are just at the initial stage of their development but demonstrate much potential. In addition, one can see in Figure 6.12 that the material science, the development of the methodology as well as the investigations of the physical spin-wave phenomena, form the basis for all the research directions.

ACKNOWLEDGMENTS

I am grateful to Thomas Brächer, Qi Wang, Philipp Pirro, Oleksandr Serha, Marjorie Lägel, and Burkard Hillebrands for the inspiring discussions and for the support in the preparation of this chapter. The financial support given by the ERC Starting Grant 678309 MagnonCircuits and DFG within Spin+X SFB/TRR 173 is strongly acknowledged.

REFERENCES

1. F. Bloch, Zur Theorie des Ferromagnetismus, *Z. Phys.* **61**, 206 (1930).
2. A. G. Gurevich and G. A. Melkov, *Magnetization Oscillations and Waves*, CRC, Boca Raton FL, 1996.
3. D. D. Stancil and A. Prabhakar, *Spin Waves: Theory and Applications*, Springer, 2009.
4. A. A. Serga, A. V. Chumak, and B. Hillebrands, YIG magnonics, *J. Phys. D Appl. Phys.* **43**, 264002 (2010).
5. B. A. Kalinikos, N. G. Kovshikov, and A. N. Slavin, Observation of spin wave solitons in ferromagnetic films, *JETP Lett.* **38**, 413 (1983).
6. A. A. Serga, S. O. Demokritov, B. Hillebrands, and A. N. Slavin, Self-generation of two-dimensional spin-wave bullets, *Phys. Rev. Lett.* **92**, 117203 (2004).
7. T. Schneider, A. A. Serga, A. V. Chumak et al., Nondiffractive subwavelength wave beams in a medium with externally controlled anisotropy, *Phys. Rev. Lett.* **104**, 197203 (2010).
8. V. E. Demidov, S. O. Demokritov, D. Birt, B. O'Gorman, M. Tsoi, and X. Li, Radiation of spin waves from the open end of a microscopic magnetic-film waveguide, *Phys. Rev. B* **80**, 014429 (2009).
9. R. Gieniusz, H. Ulrichs, V. D. Bessonov, U. Guzowska, A. I. Stognii, and A. Maziewski, Single antidot as a passive way to create caustic spin-wave beams in yttrium iron garnet films, *Appl. Phys. Lett.* **102**, 102409 (2013).
10. T. Sebastian, T. Brächer, P. Pirro et al., Nonlinear emission of spin-wave caustics from an edge mode of a micro-structured $Co_2Mn_{0.6}Fe_{0.4}Si$ waveguide, *Phys. Rev. Lett.* **110**, 067201 (2013).
11. G. A. Melkov, A. A. Serga, V. S. Tiberkevich, A. N. Oliynyk, and A. N. Slavin, Wave front reversal of a dipolar spin wave pulse in a nonstationary three-wave parametric interaction, *Phys. Rev. Lett.* **84**, 3438 (2000).
12. A. A. Serga, B. Hillebrands, S. O. Demokritov et al., Parametric generation of forward and phase-conjugated spin-wave bullets in magnetic films, *Phys. Rev. Lett.* **94**, 167202 (2005).
13. S. O. Demokritov, V. E. Demidov, O. Dzyapko et al., Bose-Einstein condensation of quasi-equilibrium magnons at room temperature under pumping, *Nature* **443**, 430 (2006).
14. A. V. Chumak, G. A. Melkov, V. E. Demidov, O. Dzyapko, V. L. Safonov, and S. O. Demokritov, Bose–Einstein condensation of magnons under incoherent pumping, *Phys. Rev. Lett.* **102**, 187205 (2009).
15. A. A. Serga, V. S. Tiberkevich, C. W. Sandweg et al., Bose–Einstein condensation in an ultra-hot gas of pumped magnons, *Nat. Commun.* **5**, 3452 (2014).

16. D. A. Bozhko, A. A. Serga, P. Clausen et al., Supercurrent in a room-temperature Bose–Einstein magnon condensate, *Nat. Phys.* **12**, 1057 (2016).

17. V. V. Kruglyak, S. O. Demokritov, and D. Grundler, Magnonics, *J. Phys. D Appl. Phys.* **43**, 264001 (2010).

18. R. L. Stamps, S. Breitkreutz, J. Åkerman et al., Magnon spintronics in the "The 2014 magnetism roadmap", *J. Phys. D: Appl. Phys.* **47**, 333001 (2014).

19. A. V. Chumak, V. I. Vasyuchka, A. A. Serga, and B. Hillebrands, Magnon spintronics, *Nat. Phys.* **11**, 453 (2015).

20. D. I. E. Nikonov and I. A. Young, Benchmarking of beyond-CMOS exploratory devices for logic integrated circuit, *IEEE J. Explor. Solid-State Comput. Devices and Circuits* **1**, 3 (2015).

21. A. Khitun, M. Bao, and K. L. Wang, Magnonic logic circuits, *J. Phys. D Appl. Phys.* **43**, 264005 (2010).

22. O. Zografos, B. Soree, A. Vaysset et al., Design and benchmarking of hybrid CMOS-spin wave device circuits compared to 10nm CMOS, in *Proceedings of the 15th IEEE International Conference on Nanotechnology*, 686 (2015).

23. S. Dutta, S.-C. Chang, N. Kani et al., Non-volatile clocked spin wave interconnect for beyond-CMOS nanomagnet pipelines, *Sci. Rep.* **5**, 9861 (2015).

24. A. V. Chumak, A. A. Serga, and B. Hillebrands, Magnon transistor for all-magnon data processing, *Nat. Commun.* **5**, 4700 (2014).

25. T. Fischer, M. Kewenig, D. A. Bozhko et al., Experimental prototype of a spin-wave majority gate, *Appl. Phys. Lett.* **110**, 152401 (2017).

26. K-S. Lee and S-K. Kim, Conceptual design of spin wave logic gates based on a Mach–Zehnder-type spin wave interferometer for universal logic functions, *J. Appl. Phys.* **104**, 053909 (2008).

27. A. Khitun, Multi-frequency magnonic logic circuits for parallel data processing, *J. Appl. Phys.* **111**, 054307 (2012).

28. A. Bérut, A. Arakelyan, A. Petrosyan, S. Ciliberto, R. Dillenschneider, and E. Lutz, Experimental verification of Landauer's principle linking information and thermodynamics, *Nature* **483**, 187 (2012).

29. T. Brächer and P. Pirro, An analog magnon adder for all-magnonic neurons, *J. Appl. Phys.* **124**, 152119 (2018).

30. S. Geller and M. A. Gilleo, Structure and ferrimagnetism of yttrium and rare-earth-iron garnets, *Acta Crystallogr.* **10**, 239 (1957).

31. H. L. Glass, Ferrite films for microwave and millimeter-wave devices, *Proc. IEEE* **76**, 151 (1988).

32. V. Cherepanov, I. Kolokolov, and V. L'vov, The saga of YIG: Spectra, thermodynamics, interaction and relaxation of magnons in a complex magnet, *Phys. Rep.* **229**, 81 (1993).

33. I. V. Zavislyak and M. A. Popov, Microwave properties and applications of Yttrium Iron Garnet (YIG) films: Current state of art and perspectives, in *Yttrium: Compounds, Production and Applications*, B. D. Volkerts (Ed.), Nova Science Publishers, Inc., pp. 87–125, 2009.

34. J. M. Owens, J. H. Collins, and R. L. Carter, System applications of magnetostatic wave devices, *Circuits Syst. Signal Process.* **4**, 317 (1985).

35. J. D. Adam, Analog signal processing with microwave magnetics, *Proc. IEEE* **76**, 159 (1988).

36. T. Balashov, P. Buczek, L. Sandratskii, A. Ernst, and W. Wulfhekel, Magnon dispersion in thin magnetic films, *J. Phys. Condens. Matter* **26**, 394007 (2014).

37. T.-H. Chuang, Kh. Zakeri, A. Ernst et al., Magnetic properties and magnon excitations in Fe(001) films grown on Ir(001), *Phys. Rev. B* **89**, 174404 (2014).

38. T. Seifert, S. Jaiswal, U. Martens et al., Efficient metallic spintronic emitters of ultrabroadband terahertz radiation, *Nat. Photon* **10**, 483 (2016).

39. J. S. Plant, Spinwave dispersion curves for yttrium iron garnet, *J. Phys. C Solid State Phys.* **10**, 4805 (1977).

40. J. Barker and G. E. W. Bauer, Thermal spin dynamics of yttrium iron garnet, *Phys. Rev. Lett.* **117**, 217201 (2016).

41. P. Pirro, T. Brächer, A. V. Chumak et al., Spin-wave excitation and propagation in microstructured waveguides of yttrium iron garnet/Pt bilayers, *Appl. Phys. Lett.* **104**, 012402 (2014).

42. C. Hahn, V. V. Naletov, G. de Loubens et al., Measurement of the intrinsic damping constant in individual nanodisks of $Y_3Fe_5O_{12}$ and $Y_3Fe_5O_{12}|Pt$, *Appl. Phys. Lett.* **104**, 152410 (2014).

43. C. W. Sandweg, Y. Kajiwara, A. V. Chumak et al., Spin pumping by parametrically excited exchange magnons, *Phys. Rev. Lett.* **106**, 216601 (2011).

44. H. Yu, O. d' Allivy Kelly, V. Cros et al., Approaching soft X-ray wavelengths in nanomagnet-based microwave technology, *Nat. Commun.* **7**, 11255 (2016).

45. T. Brächer, M. Fabre, T. Meyer et al., Detection of short-waved spin waves in individual microscopic spin-wave waveguides using the inverse spin hall effect, *Nano Lett.* **17**, 7234 (2017).

46. V. S. L'vov, *Wave Turbulence under Parametric Excitation*, Springer, 1994.

47. P. E. Wigen, *Nonlinear Phenomena and Chaos Inmagnetic Materials*, World Scientific, Singapore, 1994.

48. V. V. Kruglyak and R. J. Hicken, Magnonics: Experiment to prove the concept, *J. Magn. Magn. Mat.* **306**, 191 (2006).

49. B. Lenk, H. Ulrichs, F. Garbs, and M. Münzenberg, The building blocks of magnonics, *Phys. Rep.* **507**, 107136 (2011).

50. A. D. Karenowska, A. V. Chumak, A. A. Serga, and B. Hillebrands, Magnon spintronics, in *Handbook of Spintronics*, Y. Xu, D. D. Awschalom, and J. Nitta (Eds.), Springer, pp. 1505–1549, 2015.

51. R. W. Damon and J. R. Eshbach, Magnetostatic modes of a ferromagnet slab, *Phys. Chem. Solids* **19**, 308 (1961).

52. B. A. Kalinikos, Excitation of propagating spin waves in ferromagnetic films, *Microw. Opt. Antennas IEE Proc.* **127**, 1 (1980).

53. B. A. Kalinikos and A. N. Slavin, Theory of dipole-exchange spin-wave spectrum for ferromagnetic films with mixed exchange boundary conditions, *J. Phys. C* **19**, 7013 (1986).

54. V. E. Demidov and S. O. Demokritov, Magnonic waveguides studied by microfocus Brillouin Light Scattering, *IEEE Trans. Magn.* **51**, 0800215 (2015).

55. T. Brächer, O. Boulle, G. Gaudin, P. Pirro, Creation of unidirectional spin-wave emitters by utilizing interfacial Dzyaloshinskii-Moriya interaction, *Phys. Rev. B* **95**, 064429 (2017).

56. B. A. Kalinikos, N. V. Kozhus, M. P. Kostylev, and A. N. Slavin, The dipole exchange spin wave spectrum for anisotropic ferromagnetic films with mixed exchange boundary conditions, *J. Phys. Condens. Matter* **2**, 9861 (1990).

57. A. Kreisela, F. Sauli, L. Bartosch, and P. Kopietz, Microscopic spin-wave theory for yttrium-iron garnet films, *Eur. Phys. J. B* **71**, 59 (2009).

58. K. Y. Guslienko, S. O. Demokritov, B. Hillebrands, and A. Slavin, Effective dipolar boundary conditions for dynamic magnetization in thin magnetic stripes, *Phys. Rev. B* **66**, 132402 (2002).

59. R. Verba, G. Melkov, V. Tiberkevich, and A. Slavin, Collective spin-wave excitations in a two-dimensional array of coupled magnetic nanodots, *Phys. Rev. B* **85**, 014427 (2012).

60. R. Verba, Spin waves in arrays of magnetic nanodots with magnetodipolar coupling, *Ukr. J. Phys.* **58**, 758 (2013).

61. C. Bayer, J. Jorzick, S. Demokritov et al., Spin-wave excitations in finite rectangular elements, in *Spin Dynamics in Confined Magnetic Structures III, Band 101 von Topics in Applied Physics*, B. Hillebrands, A. Thiaville (Hg.), Springer, Heidelberg, Berlin, pp. 57–103, 2006, ISBN 978-3- 540-20108-3.

62. R. I. Joseph and E. Schlömann, Demagnetizing field in nonellipsoidal bodies, *J. Appl. Phys.* **36**, 1579 (1965).

63. J. Jorzick, S. O. Demokritov, B. Hillebrands et al., Spin wave wells in nonellipsoidal micrometer size magnetic elements, *Phys. Rev. Lett.* **88**, 47204 (2002).

64. C. Dubs, O. Surzhenko, R. Linke, A. Danilewsky, U. Brückner, and J. Dellith, Sub-micrometer yttrium iron garnet LPE films with low ferromagnetic resonance losses, *J. Phys. D Appl. Phys.* **50**, 204005 (2017).

65. S. Klingler, A. V. Chumak, T. Mewes et al., Measurements of the exchange stiffness of YIG films using broadband ferromagnetic resonance techniques, *J. Phys. D Appl. Phys.* **48**, 015001 (2015).

66. R. W. Damon and H. Van De Vaart, Propagation of magnetostatic spin waves at microwave frequencies in a normally-magnetized disk, *J. Appl. Phys.* **36**, 3453 (1965).

67. D. D. Stancil, B. E. Henty, A. G. Cepni, and J. P. Van't Hof, Observation of an inverse Doppler shift from left-handed dipolar spin waves, *Phys. Rev. B* **74**, 060404(R) (2006).

68. A. V. Chumak, P. Dhagat, A. Jander, A. A. Serga, and B. Hillebrands, Reverse Doppler effect of magnons with negative group velocity scattered from a moving Bragg grating, *Phys. Rev. B* **81**, 140404(R) (2010).

69. V. Kambersky, On the Landau–Lifshitz relaxation in ferromagnetic metals, *Can. J. Phys.* **48**, 2906 (1970).

70. J. Kunes and V. Kambersky, First-principles investigation of the damping of fast magnetization precession in ferromagnetic 3d metals, *Phys. Rev. B* **65**, 212411 (2002).

71. S. S. Kalarickal, P. Krivosik, M. Wu et al., Ferromagnetic resonance linewidth in metallic thin films: Comparison of measurement methods, *J. Appl. Phys.* **99**, 093909 (2006).

72. T. Sebastian, K. Schultheiss, B. Obry, B. Hillebrands, and H. Schultheiss, Micro-focused Brillouin light scattering: imaging spin waves at the nanoscale (Review paper), *Front. Phys.* **3**, 35 (2015).

73. G. A. Melkov, V. I. Vasyuchka, Yu. V. Kobljanskyj, and A. N. Slavin, Wave-front reversal in a medium with inhomogeneities and an anisotropic wave spectrum, *Phys. Rev. B* **70**, 224407 (2004).

74. M. B. Jungfleisch, A. V. Chumak, V. I. Vasyuchka et al., Temporal evolution of inverse spin Hall effect voltage in a magnetic insulator-nonmagnetic metal structure, *Appl. Phys. Lett.* **99**, 182512 (2011).

75. A. V. Chumak, A. A. Serga, and B. Hillebrands, Magnonic crystals for data processing, *J. Phys. D Appl. Phys.* **50**, 244001 (2017).

76. F. Bertaut and F. Forrat, The structure of the ferrimagnetic rare earth ferrites, *Cornpt. Rend. Asad. Sci. (Paris)* **242**, 382 (1956).

77. H. L. Glass, Ferrite films for microwave and millimeter-wave devices, *Proc. IEEE* **76**, 151 (1988).

78. Y. Sun, Y.-Y. Song, H. Chang et al., Growth and ferromagnetic resonance properties of nanometer-thick yttrium iron garnet films, *Appl. Phys. Lett.* **101**, 152405 (2012).

79. O. d'Allivy Kelly, A. Anane, R. Bernard et al., Inverse spin Hall effect in nanometer-thick yttrium iron garnet/Pt system, *Appl. Phys. Lett.* **103**, 082408 (2013).

80. H. Yu, O. d'Allivy Kelly, V. Cros et al., Magnetic thin-film insulator with ultra-low spin wave damping for coherent nanomagnonics, *Sci. Rep.* **4**, 6848 (2014).

81. M. C. Onbasli, A. Kehlberger, D. H. Kim et al., Pulsed laser deposition of epitaxial yttrium iron garnet films with low gilbert damping and bulk-like magnetization, *Appl. Phys. Lett. Mater.* **2**, 106102 (2014).

82. T. Liu, H. Chang, V. Vlaminck et al., Ferromagnetic resonance of sputtered yttrium iron garnet nanometer films, *J. Appl. Phys.* **115**, 17A501 (2014).

83. C. E. Patton, Linewidth and relaxation processes for the main resonance in the spin-wave spectra of Ni–Fe alloy films, *J. Appl. Phys.* **39**, 3060 (1968).

84. A. Brunsch, Magnetic properties and corrosion resistance of $(CoFeB)_{100-x}Cr_x$ thin films, *J. Appl. Phys.* **50**, 7603 (1979).

85. X. Liu, W. Zhang, M. J. Carter, and G. Xiao, Ferromagnetic resonance and damping properties of CoFeB thin films as free layers in MgO-based magnetic tunnel junctions, *J. Appl. Phys.* **110**, 033910 (2011).

86. A. Conca, J. Greser, T. Sebastian et al., Low spin-wave damping in amorphous $Co_{40}Fe_{40}B_{20}$ thin films, *J. Appl. Phys.* **113**, 213909 (2013).

87. C. Liu, C. K. A. Mewes, M. Chshiev, T. Mewes, and W. H. Butler, Origin of low Gilbert damping in half metal, *Appl. Phys. Lett.* **95**, 022509 (2009).

88. S. Trudel, O. Gaier, J. Hamrle, and B. Hillebrands, Magnetic anisotropy, exchange and damping in cobalt-based full-Heusler compounds: An experimental review, *J. Phys. D Appl. Phys.* **43**, 193001 (2010).

89. M. Oogane, T. Kubota, Y. Kota et al., Gilbert magnetic damping constant of epitaxially grown Co-based Heusler alloy thin films, *Appl. Phys. Lett.* **96**, 252501 (2010).

90. T. Sebastian, Y. Ohdaira, T. Kubota et al., Low-damping spin-wave propagation in a micro-structured $Co_2Mn_{0.6}Fe_{0.4}Si$ Heusler waveguide, *Appl. Phys. Lett.* **100**, 112402 (2012).

91. M. Sparks, *Ferromagnetic Relaxation Theory*, McGraw-Hill, 1964.

92. A. Maeda and M. Susaki, Magnetostatic wave propagation in yttrium-iron-garnet with microfabricated surfaces, *IEEE Trans. Magn.* **42**, 3096 (2006).

93. A. V. Chumak, A. A. Serga, B. Hillebrands, and M. P. Kostylev, Scattering of backward spin waves in a one-dimensional magnonic crystal, *Appl. Phys. Lett.* **93**, 022508 (2008).

94. J. D. Adam, Delay of magnetostatic surface waves in y.i.g., *Electr. Lett.* **6**, 718 (1970).

95. A. K. Ganguly and D. C. Webb, Microstrip excitation of magnetostatic surface waves: Theory and experiment, *IEEE Trans. Micr. Theory Techn.* **MTT-23**, 998 (1975).

96. T. Schneider, A. A. Serga, T. Neumann, B. Hillebrands, and M. P. Kostylev, Phase reciprocity of spin-wave excitation by a microstrip antenna, *Phys. Rev. B* **77**, 214411 (2008).

97. V. Vlaminck and M. Bailleul, Current-induced spin-wave Doppler shift, *Science* **322**, 410 (2008).

98. V. E. Demidov, M. P. Kostylev, K. Rott, P. Krzysteczko, G. Reiss, and S. O. Demokritov, Excitation of microwaveguide modes by a stripe antenna, *Appl. Phys. Lett.* **95**, 112509 (2009).

99. Y. Au, E. Ahmad, O. Dmytriiev, M. Dvornik, T. Davison, and V. V. Kruglyak, Resonant microwave-to-spin-wave transducer, *Appl. Phys. Lett.* **100**, 182404 (2012).

100. Y. Au, T. Davison, E. Ahmad, P. S. Keatley, R. J. Hicken, and V. V. Kruglyak, Excitation of propagating spin waves with global uniform microwave fields, *Appl. Phys. Lett.* **98**, 122506 (2011).

101. G. N. Kakazei, T. Mewes, P. E. Wigen et al., Probing arrays of circular magnetic microdots by ferromagnetic resonance, *J. Nanosci. Nanotechn.* **8**, 2811 (2008).

102. T. Brächer, P. Pirro, T. Meyer et al., Parallel parametric amplification of coherently excited propagating spin waves in a microscopic $Ni_{81}Fe_{19}$ waveguide, *Appl. Phys. Lett.* **104**, 202408 (2014).

103. T. Brächer, P. Pirro, and B. Hillebrands, Parallel pumping for magnon spintronics: Amplification and manipulation of magnon spin currents on the micron-scale, *Phys. Rep.* **699**, 1 (2017).

104. K. J. Kennewell, D. C. Crew, M. J. Lwin, R. C. Woodward, S. Prasad, and R. L. Stamps, Interpretation of magnetisation dynamics using inductive magnetometry in thin films, *Surf. Sci.* **601**, 5766 (2007).

105. K. R. Smith, M. J. Kabatek, P. Krivosik, and M. Wu, Spin wave propagation in spatially nonuniform magnetic fields, *J. Appl. Phys.* **104**, 043911 (2008).

106. G. Gubbiotti, S. Tacchi, M. Madami, G. Carlotti, A. O. Adeyeye, and M. P. Kostylev, Brillouin light scattering studies of planar metallic magnonic crystals, *J. Phys. D Appl. Phys.* **43**, 264003 (2010).

107. S. O. Demokritov, B. Hillebrands, and A. N. Slavin, Brillouin light scattering studies of confined spin waves: Linear and nonlinear confinement, *Phys. Rep.* **348**, 441 (2001).

108. A. Kirilyuk, A. V. Kimel, and Th. Rasing, Ultrafast optical manipulation of magnetic order, *Rev. Mod. Phys.* **82**, 2731 (2010).

109. C.-H. Lambert, S. Mangin, B. S. D. Ch. S. Varaprasad et al., All-optical control of ferromagnetic thin films and nanostructures, *Science* **345**, 1337 (2014).

110. H. G. Bauer, J.-Y.Chauleau, G. Woltersdorf, and C. H. Back, Coupling of spin-wave modes in wire structures, *Appl. Phys. Lett.* **104**, 102404 (2014).

111. J. C. Slonczewski, Current-driven excitation of magnetic multilayers, *J. Magn. Magn. Mater.* **159**, L1 (1995).

112. L. Berger, Emission of spin waves by a magnetic multilayer traversed by a current, *Phys. Rev. B* **54**, 9353 (1996).

113. I. N. Krivorotov, N. C. Emley, J. C. Sankey, S. I. Kiselev, D. C. Ralph, and R. A. Buhrman, Time-domain measurements of nanomagnet dynamics driven by spin-transfer torques, *Science* **307**, 228 (2005).

114. V. E. Demidov, S. Urazhdin, and S. O. Demokritov, Direct observation and mapping of spin waves emitted by spin-torque nano-oscillators, *Nat. Mater.* **9**, 984 (2010).

115. M. Madami, S. Bonetti, G. Consolo et al., Direct observation of a propagating spin wave induced by spin-transfer torque, *Nat. Nanotech.* **6**, 635 (2011).

116. V. E. Demidov, S. Urazhdin, H. Ulrichs et al., Magnetic nano-oscillator driven by pure spin current, *Nat. Mater.* **11**, 1028 (2012).

117. M. Collet, X. De Milly, O. D'Allivy-Kelly et al., Generation of coherent spin-wave modes in yttrium iron garnet microdiscs by spin–orbit torque, *Nat. Commun.* **7**, 10377 (2016).

118. V. Lauer, D. A. Bozhko, T. Brächer et al., Spin-transfer torque based damping control of parametrically excited spin waves in a magnetic insulator, *Appl. Phys. Lett.* **108**, 012402 (2016).

119. V. Lauer, M. Schneider, T. Meyer et al., Temporal evolution of auto-oscillations in a YIG/Pt microdisc driven by pulsed spin Hall effect-induced spin-transfer torque, *IEEE Magn. Lett.* **8**, 3104304 (2017).

120. L. J. Cornelissen, J. Liu, R. A. Duine, J. Ben Youssef, and B. J. van Wees, Long-distance transport of magnon spin information in a magnetic insulator at room temperature, *Nat. Phys.* **11**, 1022 (2015).

121. Y. Tserkovnyak, A. Brataas, and G. E. W. Bauer, Enhanced Gilbert damping in thin ferromagnetic films, *Phys. Rev. Lett.* **88**, 117601 (2002).

122. M.V. Costache, M. Sladkov, S.M. Watts, C.H. van der Wal, and B.J. van Wees, Electrical detection of spin pumping due to the precessing magnetization of a single ferromagnet, *Phys. Rev. Lett.* **97**, 216603 (2006).

123. E. Saitoh, M. Ueda, H. Miyajima, and G. Tatara, Conversion of spin current into charge current at room temperature: Inverse spin-Hall effect, *Appl. Phys. Lett.* **88**, 182509 (2006).

124. Y. Kajiwara, K. Harii, S. Takahashi et al., Transmission of electrical signals by spin-wave interconversion in a magnetic insulator, *Nature* **464**, 262 (2010).

125. A. Hoffmann, Spin Hall effects in metals, *IEEE Trans. Magn.* **49**, 5172 (2013).

126. M. Plihal, D. L. Mills, and J. Kirschner, Spin wave signature in the spin polarized electron energy loss spectrum of ultrathin Fe films: Theory and experiment, *Phys. Rev. Lett.* **82**, 2579 (1999).

127. W. X. Tang, Y. Zhang, I. Tudosa, J. Prokop, M. Etzkorn, and J. Kirschner, Large wave vector spin waves and dispersion in two monolayer Fe on W (110), *Phys. Rev. Lett.* **99**, 087202 (2007).

128. A. Khitun and K. L. Wang, Nano scale computational architectures with spin wave bus, *Superlatt. Microstruct.* **38**, 184 (2005).

129. K. L. Wang and P. Khalili Amiri, Nonvolatile spintronics: Perspectives on instant-on nonvolatile nanoelectronic systems, *SPIN* **02**, 1250009 (2012).

130. O. Klein, G. de Loubens, V. V. Naletov et al., Ferromagnetic resonance force spectroscopy of individual submicron-size samples, *Phys. Rev. B* **78**, 144410 (2008).

131. T. An, V. I. Vasyuchka, K. Uchida et al., Unidirectional spin-wave heat conveyer, *Nat. Mater.* **12**, 549 (2013).

132. L. Bocklage, C. Swoboda, K. Schlage et al., Spin precession mapping at ferromagnetic resonance via nuclear resonant scattering of synchrotron radiation, *Phys. Rev. Lett.* **114**, 147601 (2015).

133. J. Goulon, A. Rogalev, F. Wilhelm et al., X-ray detected magnetic resonance at the Fe K-edge in YIG: forced precession of magnetically polarized orbital components, *JETP Lett.* **82**, 696 (2005).

134. G. B. G. Stenning, L. R. Shelford, S. A Cavill et al., Magnetization dynamics in an exchange-coupled NiFe/CoFe bilayer studied by x-ray detected ferromagnetic resonance, *New J. Phys.* **17**, 013019 (2015).

135. A. P. Mihai, J. P. Attané, A. Marty, P. Warin, and Y. Samson, Electron-magnon diffusion and magnetization reversal detection in FePt thin films, *Phys. Rev. B* **77**, 060401(R) (2008).

136. B. Raquet, M. Viret, E. Sondergard, O. Cespedes, and R. Mamy, Electron-magnon scattering and magnetic resistivity in 3d ferromagnets, *Phys. Rev. B* **66**, 024433 (2002).

137. G. Csaba, A. Papp, W. Porod, Spin-wave based realization of optical computing primitives, *J. Appl. Phys.* **115**, 17C741 (2014).

138. S. Khasanvis, M. Rahman, S. N. Rajapandian, and C. A. Moritz, Wave-based multi-valued computation framework, in *2014 IEEE/ACM International Symposium on Nanoscale Architectures (NANOARCH)*, pp. 171–176 (2014).

139. R. Cuykendall and D. R. Andersen, Reversible optical computing circuits, *Opt. Lett.* **12**, 542 (1987).

140. D. C. Feitelson, *Optical Computing: A Survey for Computer Scientists*, MIT Press, 1992.

141. R. Hertel, W. Wulfhekel, and J. Kirschner, Domain-wall induced phase shifts in spin waves, *Phys. Rev. Lett.* **93**, 257202 (2004).

142. M. P. Kostylev, A. A. Serga, T. Schneider, B. Leven, and B. Hillebrands, Spin-wave logical gates, *Appl. Phys. Lett.* **87**, 153501 (2005).

143. T. Schneider, A. A. Serga, B. Leven, B. Hillebrands, R. L. Stamps, and M. P. Kostylev, Realization of spin-wave logic gates, *Appl. Phys. Lett.* **92**, 022505 (2008).

144. A. Khitun, M. Bao, and K. L. Wang, Spin wave magnetic NanoFabric: A new approach to spin-based logic circuitry, *IEEE Trans. Magn.* **44**, 2141 (2008).

145. S. Klingler, P. Pirro, T. Brächer, B. Leven, B. Hillebrands, and A. V. Chumak, Design of a spin-wave majority gate employing mode selection, *Appl. Phys. Lett.* **105**, 152410 (2014).

146. S. Klingler, P. Pirro, T. Brächer, B. Leven, B. Hillebrands, and A. V. Chumak, Spin-wave logic devices based on isotropic forward volume magnetostatic waves. *Appl. Phys. Lett.* **106**, 212406 (2015).

147. T. Brächer, F. Heussner, P. Pirro et al., Phase-to-intensity conversion of magnonic spin currents and application to the design of a majority gate, *Sci. Rep.* **6**, 38235 (2016).

148. M. Krawczyk and D. Grundler, Review and prospects of magnonic crystals and devices with reprogrammable band structure, *J. Phys.-Cond. Matt.* **26**, 123202 (2014).

149. A. B. Ustinov, N. Y. Grigor'eva, and B. A. Kalinikos, Observation of spin-wave envelope solitons in periodic magnetic film structures, *JETP Lett.* **88**, 31 (2008).

150. A. V. Chumak, V. I. Vasyuchka, A. A. Serga, M. P. Kostylev, V. S. Tiberkevich, and B. Hillebrands, Storage-recovery phenomenon in magnonic crystal. *Phys. Rev. Lett.* **108**, 257207 (2012).

151. S. A. Nikitov, P. Tailhades, and C. S. Tsai, Spin waves in periodic magnetic structures - magnonic crystals, *J. Magn. Magn. Mat.* **236**, 320 (2001).

152. C. G. Sykes, J. D. Adam, and J. H. Collins, The waveguides with an array of grooves, *Appl. Phys. Lett.* **29**, 388 (1976)

153. P. Harttemann, Magnetostatic wave planar YIG devices, *IEEE Trans. Magn.* **MAG-20**, 1271 (1984).

154. K. W. Reed, J. M. Owens, and R. L. Carter, Current status of magnetostatic reflective array filters, *Circuits Syst. Signal. Process.* **4**, 157 (1985).

155. W. S. Ishak, Magnetostatic wave technology: A review, *Proc. IEEE* **76**, 171 (1988).

156. A. B. Ustinov, A. V. Drozdovskii, and B. A. Kalinikos, Multifunctional nonlinear magnonic devices for microwave signal processing, *Appl. Phys. Lett.* **96**, 142513 (2010).

157. A. V. Sadovnikov, E. N. Beginin, M. A. Morozova et al., Nonlinear spin wave coupling in adjacent magnonic crystals, *Appl. Phys. Lett.* **109**, 042407 (2016).

158. M. Krawczyk and H. Puszkarski, Plane-wave theory of three-dimensional magnonic crystals, *Phys. Rev. B* **77**, 054437 (2008).

159. S.-K. Kim, Micromagnetic computer simulations of spin waves in nanometre-scale patterned magnetic elements, *J. Phys. D: Appl. Phys.* **43**, 264004 (2010).

160. H. Al-Wahsha, A. Akjouj, B. Djafari-Rouhani, and L. Dobrzynski, Magnonic circuits and crystals, *Surf. Sci. Rep.* **66**, 29 (2011).

161. Z. K. Wang, V. L. Zhang, H. S. Lim et al., Observation of frequency band gaps in a one-dimensional nanostructured magnonic crystal, *Appl. Phys. Lett.* **94**, 083112 (2009).

162. A. V. Chumak, P. Pirro, A. A. Serga et al., Spin-wave propagation in a micro-structured magnonic crystal, *Appl. Phys. Lett.* **95**, 262508 (2009).

163. M. Collet, M. Evelt, V. E. Demidov et al., Nano-patterned magnonic crystals based on ultrathin YIG films, *arXiv:1705.02267* (2017).

164. S. Tacchi, M. Madami, G. Gubbiotti et al., Magnetic normal modes in squared antidot array with circular holes: A combined Brillouin light scattering and broadband ferromagnetic resonance study. *IEEE Trans. Magn.* **46**, 172–178 (2010).

165. A. V. Chumak, T. Neumann, A. A. Serga, B. Hillebrands, and M. P. Kostylev, A current-controlled, dynamic magnonic crystal, *J. Phys. D: Appl. Phys.* **42**, 205005 (2009).

166. A. V. Chumak, A. D. Karenowska, A. A. Serga, and B. Hillebrands, The dynamic magnonic crystal: New horizons in artificial crystal based signal processing, in *Topics in Applied Physics*, Vol. 125: *Magnonics from Fundamentals to Applications*, S. O. Demokritov and A. N. Slavin (Eds.), Springer, Heidelberg, pp. 243–255, 2012.

167. M. Vogel, A. V. Chumak, E. H. Waller et al., Optically-reconfigurable dynamic magnetic materials, *Nat. Phys.* **11**, 487 (2015).

168. A. V. Chumak, A. A. Serga, B. Hillebrands, and M. P. Kostylev, Scattering of backward spin waves in a one-dimensional magnonic crystal, *Appl. Phys. Lett.* **93**, 022508 (2008).

169. B. Obry, P. Pirro, T. Brächer et al., A micro-structured ion-implanted magnonic crystal, *Appl. Phys. Lett.* **102**, 202403 (2013).

170. R. G. Kryshtal and A. V. Medved, Surface acoustic wave in yttrium iron garnet as tunable magnonic crystals for sensors and signal processing applications, *J. Magn. Magn. Mater.* **426,** 666 (2017).

171. J. Topp, D. Heitmann, M. P. Kostylev, and D. Grundler, Making a reconfigurable artificial crystal by ordering bistable magnetic nanowires, *Phys. Rev. Lett.* **104**, 207205 (2010).

172. Q. Wang, A. V. Chumak, L. Jin, H. Zhang, B. Hillebrands, and Z. Zhong, Voltage-controlled nano-scale reconfigurable magnonic crystal, *Phys. Rev. B* **95**, 134433 (2017).

173. A. A. Nikitin, A. B. Ustinov, A. A. Semenov et al., A spin-wave logic gate based on a width-modulated dynamic magnonic crystal, *Appl. Phys. Lett.* **106**, 102405 (2015).

174. A. V. Chumak, V. S. Tiberkevich, A. D. Karenowska et al., All-linear time reversal by a dynamic artificial crystal, *Nat. Commun.* **1**, 141 (2010).

175. T. Brächer, P. Pirro, J. Westermann et al., Generation of propagating backward volume spin waves by phase-sensitive mode conversion in two-dimensional microstructures, *Appl. Phys. Lett.* **102**, 132411 (2013).

176. A. V. Sadovnikov, C. S. Davies, S. V. Grishin et al., Magnonic beam splitter: The building block of parallel magnonic circuitry, *Appl. Phys. Lett.* **106**, 192406 (2015).

177. K. Vogt, H. Schultheiss, S. Jain et al., Spin waves turning a corner, *Appl. Phys. Lett.* **101**, 042410 (2012).

178. K. Vogt, F. Y. Fradin, J. E. Pearson et al., Realization of a spin-wave multiplexer, *Nat. Commun.* **5**, 3727 (2014).

179. M. P. Kostylev, A. A. Serga, and B. Hillebrands, Radiation of caustic beams from a collapsing bullet, *Phys. Rev. Lett.* **106**, 134101 (2011).

180. C. S. Davies, A. Francis, A. V. Sadovnikov et al., Towards graded-index magnonics: Steering spin waves in magnonic networks, *Phys. Rev. B* **92**, 020408(R) (2015).

181. E. H. Lock, Angular beam width of a slit-diffracted wave with noncollinear group and phase velocities, *Phys. Uspekhi* **55**, 1239 (2012).

182. J.-V. Kim, R. L. Stamps, and R. E. Camley, Spin wave power flow and caustics in ultrathin ferromagnets with the Dzyaloshinskii-Moriya interaction, *Phys. Rev. Lett.* **117**, 197204 (2016).

183. F. Heussner, A. A. Serga, T. Brächer, B. Hillebrands, and P. Pirro, A switchable spin-wave signal splitter for magnonic networks, *Appl. Phys. Lett.* **111**, 122401 (2017).

184. M. Balynsky, D. Gutierrez, H. Chiang et al., Magnetometer based on spin wave interferometer, *Sci. Rep.* **7**, 11539 (2017).

185. H. Sasaki and N. Mikoshiba, Directional coupling of magnetostatic surface waves in a layered structure of YIG films, *J. Appl. Phys.* **52**, 3546 (1981).

186. A. V. Krasavin and A. V. Zayats, Active nanophotonic circuitry based on dielectric-loaded plasmonic waveguides, *Adv. Opt. Mater.* **3**, 1662 (2015).

187. A. V. Sadovnikov, E. N. Beginin, S. E. Sheshukova, D. V. Romanenko, Yu. P. Sharaevskii, and S. A. Nikitov, Directional multimode coupler for planar magnonics: Side-coupled magnetic stripes, *Appl. Phys. Lett.* **107**, BQ-06 (2015).

188. A. V. Sadovnikov, S. A. Odintsov, E. N. Beginin, S. E. Sheshukova1, Y. P. Sharaevskii, and S. A. Nikitov, Spin-wave switching in the side-coupled magnonic stripes, *IEEE Trans. Magn.* **55**, 2801804 (2017).

189. Q. Wang, P. Pirro, R. Verba, A. Slavin, B. Hillebrands, and A. V. Chumak, Reconfigurable nano-scale spin-wave directional coupler, *Sci. Adv.* **4**, e1701517 (2018).

190. N. Blombergen and R. W. Damon, Relaxation effects in ferromagnetic resonance, *Phys. Rev.* **85**, 699 (1952).

191. H. Suhl, The theory of ferromagnetic resonance at high signal powers, *J. Phys. Chem. Solids* **1**, 209 (1957).

192. E. Schlömann, J. J. Green, and U. Milano, Recent developments in ferromagnetic resonance at high power levels, *J. Appl. Phys.* **31**, S386 (1960).

193. G. A. Melkov and S. V. Sholom, Amplification of surface magnetostatic waves by parametric pumping, *Sov. Phys.—Tech. Phys.* **35** (1990).

194. A. V. Bagada, G. A. Melkov, A. A. Serga, and A. N. Slavin, Parametric interaction of dipolar spin wave solitons with localized electromagnetic pumping, *Phys. Rev. Lett.* **79**, 2137 (1997).

195. P. A. Kolodin, P. Kabos, C. E. Patton, B. A. Kalinikos, N. G. Kovshikov, and M. P. Kostylev, Amplification of microwave magnetic envelope solitons in thin Yttrium Iron Garnet films by parallel pumping, *Phys. Rev. Lett.* **80**, 1976 (1998).

196. A. V. Chumak, A. A. Serga, G. A. Melkov, V. Tiberkevich, A. N. Slavin, and B. Hillebrands, Parametrically-stimulated recovery of a microwave signal using standing spin-wave modes of a magnetic film, *Phys. Rev. B* **79**, 014405 (2009).

197. V. E. Zakharov and V. S. L'vov, Parametric excitation of spin waves in ferromagnets with magnetic inhomogeneities, *Sov. Phys. Solid State* **14**, 2513 (1973).

198. G. A. Melkov, Yu. V. Kobljanskyj, A. A. Serga, and V. S. Tiberkevich, Nonlinear amplification and compression of envelope solitons by localized nonstationary parametric pumping, *J. Appl. Phys.* **89**, 6689 (2001).

199. T. Brächer, F. Heussner, P. Pirro et al., Time- and power-dependent operation of a parametric spin-wave amplifier, *Appl. Phys. Lett.* **105**, 232409 (2014).

200. T. Brächer, P. Pirro, A. A. Serga, and B. Hillebrands, Localized parametric generation of spin waves in a longitudinally magnetized $Ni_{81}Fe_{19}$ waveguide, *Appl. Phys. Lett.* **103**, 142415 (2013).

201. T. Brächer, P. Pirro, F. Heussner, A. A. Serga, and B. Hillebrands, Localized parallel parametric generation of spin waves in a $Ni_{81}Fe_{19}$ waveguide by spatial variation of the pumping field, *Appl. Phys. Lett.* **104**, 092418 (2014).

202. A. Slavin and V. Tiberkevich, Nonlinear auto-oscillator theory of microwave generation by spin-polarized current, *IEEE Trans. Magn.* **45**, 1875–1918 (2009).

203. M. Tsoi, A. G. M. Jansen, J. Bass et al., Excitation of a magnetic multilayer by an electric current, *Phys. Rev. Lett.* **80**, 4281 (1998).

204. M. I. Dyakonov and V. I. Perel, Current-induced spin orientation of electrons in semiconductors, *Phys. Lett. A* **35**, 459 (1971).

205. J. E. Hirsch, Spin Hall effect, *Phys. Rev. Lett.* **83**, 1834 (1999).

206. K. Ando, S. Takahashi, K. Harii et al., Electric manipulation of spin relaxation using the spin Hall effect, *Phys. Rev. Lett.* **101**, 036601 (2008).

207. V. Castel, N. Vlietstra, J. Ben Youssef, and B. J. van Wees, Platinum thickness dependence of the inverse spin-Hall voltage from spin pumping in a hybrid yttrium iron garnet/platinum system, *Appl. Phys. Lett.* **101**, 132414 (2012).

208. V. E. Demidov, S. Urazhdin, E. R. J. Edwards, and S. O. Demokritov, Wide-range control of ferromagnetic resonance by spin Hall effect, *Appl. Phys. Lett.* **99**, 172501 (2011).

209. R. H. Liu, W. L. Lim, and S. Urazhdin, Spectral characteristics of the microwave emission by the spin Hall nano-oscillator, *Phys. Rev. Lett.* **110**, 147601 (2013).

210. Z. Duan, A. Smith, L. Ynag et al., Nanowire spin torque oscillator driven by spin orbit torques, *Nat. Commun.* **5**, 5616 (2014).

211. C. Hahn, G. de Loubens, O. Klein, M. Kiret, V. V. Naletov, and J. Ben Youssef, Comparative measurements of inverse spin Hall effects and magnetoresistance in YIG/Pt and YIG/Ta, *Phys. Rev. B* **87**, 174417 (2013).

212. J. Xiao and G. E. W. Bauer, Spin-wave excitation in magnetic insulators by spin-transfer torque, *Phys. Rev. Lett.* **108**, 217204 (2012).

213. E. Padron-Hernandez, A. Azevedo, and S. M. Rezende, Amplification of spin waves in yttrium iron garnet films through the spin Hall effect, *Appl. Phys. Lett.* **99**, 192511 (2011).

214. Z. H. Wang, Y. Y. Sun, M. Z. Wu, V. Tiberkevich, and A. Slavin, Control of spin waves in a thin film ferromagnetic insulator through interfacial spin scattering, *Phys. Rev. Lett.* **107**, 146602 (2011).

215. A. Hamadeh, O. d'Allivy Kelly, C. Hahn et al., Full control of the spin-wave damping in a magnetic insulator using spin-orbit torque, *Phys. Rev. Lett.* **113**, 197203 (2014).

216. M. Evelt, V. E. Demidov, V. Bessonov et al., High-efficiency control of spin-wave propagation in ultra-thin yttrium iron garnet by the spin-orbit torque, *Appl. Phys. Lett.* **108**, 172406 (2016).

217. L. J. Cornelissen, J. Liu, R. A. Duine, J. Ben Youssef, and B. J. van Wees, Long-distance transport of magnon spin information in a magnetic insulator at room temperature, *Nat. Phys.* **11**, 1022 (2015).

218. S. T. B. Goennenwein, R. Schlitz, M. Pernpeintner et al., Non-local magnetoresistance in YIG/Pt nanostructures, *Appl. Phys. Lett.* **107**, 172405 (2015).

219. K. Ganzhorn, S. Klingler, T. Wimmer et al., Magnon-based logic in a multi-terminal YIG/Pt nanostructure, *Appl. Phys. Lett.* **109**, 022405 (2016).

220. V. E. Demidov, M. Evelt, V. Bessonov et al., Direct observation of dynamic modes excited in a magnetic insulator by pure spin current, *Sci. Rep.* **6**, 32781 (2016).

221. E. Padron-Hernandez, A. Azevedo, and S. M. Rezende, Amplification of spin waves by thermal spin-transfer torque, *Phys. Rev. Lett.* **107**, 197203 (2011).

222. L. Lu, Y. Y. Sun, M. Jantz, and M. Z. Wu, Control of ferromagnetic relaxation in magnetic thin films through thermally induced interfacial spin transfer, *Phys. Rev. Lett.* **108**, 257202 (2012).

223. M. B. Jungfleisch, T. An, K. Ando et al., Heat-induced damping modification in yttrium iron garnet/platinum hetero-structures, *Appl. Phys. Lett.* **102**, 062417 (2013).

224. C. Safranski, I. Barsukov, H. K. Lee et al., Spin caloritronic nano-oscillator, *Nat. Commun.* **8**, 117 (2017).

225. M. Schneider, T. Brächer, V. Lauer et al., Bose-Einstein condensation of quasi-particles by rapid cooling, arXiv:1612.07305 (2018).

226. G. E. W. Bauer, E. Saitoh, and B. J. van Wees, Spin caloritronics, *Nat. Mater.* **11**, 391 (2012).

227. E. Šimánek and B. Heinrich, Gilbert damping in magnetic multilayers, *Phys. Rev. B* **67**, 144418 (2003).

228. G. Woltersdorf, M. Buess, B. Heinrich, and C. H. Back, Time resolved magnetization dynamics of ultrathin Fe(001) films: Spin-pumping and two-magnon scattering, *Phys. Rev. Lett.* **95**, 037401 (2005).

229. M. Weiler, Experimental test of the spin mixing interface conductivity concept, *Phys. Rev. Lett.* **111**, 176601 (2013).

230. V. Castel, N. Vlietstra, B. J. van Wees, and J. Ben Youssef, Yttrium iron garnet thickness and frequency dependence of the spin-charge current conversion in YIG/Pt systems, *Phys. Rev. B* **90**, 214434 (2014).

231. M. B. Jungfleisch, A. V. Chumak, A. Kehlberger et al., Thickness and power dependence of the spin-pumping effect in $Y_3Fe_5O_{12}$/Pt heterostructures measured by the inverse spin Hall effect, *Phys. Rev. B* **91**, 134407 (2015).

232. C. Burrowes, B. Heinrich, B. Kardasz et al., Enhanced spin pumping at yttrium iron garnet/Au interfaces, *Appl. Phys. Lett.* **100**, 092403 (2012).

233. M. B. Jungfleisch, V. Lauer, R. Neb, A. V. Chumak, and B. Hillebrands, Improvement of the yttrium iron garnet/platinum interface for spin pumping-based applications, *Appl. Phys. Lett.* **103**, 022411 (2013).

234. A. Kapelrud and A. Brataas, Spin pumping and enhanced Gilbert damping in thin magnetic insulator films, *Phys. Rev. Lett.* **111**, 097602 (2013).

235. A. V. Chumak, A. A. Serga, M. B. Jungeisch et al., Direct detection of magnon spin transport by the inverse spin Hall effect, *Appl. Phys. Lett.* **100**, 082405 (2012).

236. M. Schreier, G. E. W. Bauer, V. I. Vasyuchka et al., Sign of inverse spin Hall voltages generated by ferromagnetic resonance and temperature gradients in yttrium iron garnet platinum bilayers, *J. Phys. D: Appl. Phys.* **48**, 025001 (2015).

237. R. Iguchi, K. Ando, Z. Qiu, T. An, E. Saitoh, and T. Sato, Spin pumping by nonreciprocal spin waves under local excitation, *Appl. Phys. Lett.* **102**, 022406 (2013).

238. H. Kurebayashi, O. Dzyapko, V. E. Demidov, D. Fang, A. J. Ferguson, and S. O. Demokritov, Spin pumping by parametrically excited short-wavelength spin waves, *Appl. Phys. Lett.* **99**, 162502 (2011).

239. G. E. Moore, Cramming more components onto integrated circuits, *Electronics* **38**, 114 (1965).

240. K. Bernstein, R. K. Cavin III, W. Porod, A. Seabaugh, and J. Welser, Device and architecture outlook for beyond-CMOS switches, *Proc. IEEE* **98**, 2169 (2010).

241. T. Nozaki, Electric-field-induced ferromagnetic resonance excitation in an ultrathin ferromagnetic metal layer, *Nat. Phys.* **8**, 491 (2012).

242. C.-G. Duan, S. S. Jaswal, and E. Y. Tsymbal, Predicted magnetoelectric effect in Fe/BaTiO$_3$ multilayers: Ferroelectric control of magnetism, *Phys. Rev. Lett.* **97**, 047201 (2006).

243. R. Verba, V. Tiberkevich, I. Krivorotov, and A. Slavin, Parametric excitation of spin waves by voltage-controlled magnetic anisotropy, *Phys. Rev. Appl.* **1**, 044006 (2014).

244. A. Khitun, D. E. Nikonov, M. Bao, K. Galatsis, and K. L. Wang, Feasibility study of logic circuits with a spin wave bus, *Nanotechnology* **18**, 465202 (2007).

245. S. Dutta, D. E. Nikonov, S. Manipatruni, I. A. Young, and A. Naeemi, SPICE circuit modeling of PMA spin wave bus excited using magnetoelectric effect, *IEEE Trans. Magn.* **50**, 1300411 (2014).

246. M. Bao, G. Zhu, K. L. Wong et al., Magneto-electric tuning of the phase of propagating spin waves, *Appl. Phys. Lett.* **101**, 022409 (2012).

247. S. Cherepov, P. Khalili Amiri, J. G. Alzate et al., Electric-field-induced spin wave generation using multiferroic magnetoelectric cells, *Appl. Phys. Lett.* **104**, 082403 (2014).

248. Y. Tabuchi, S. Ishino, A. Noguchi et al., Coherent coupling between a ferromagnetic magnon and a superconducting qubit, *Science* **349**, 405 (2015).

249. D. Gutierrez, H. Chiang, T. Bhowmick et al., Magnonic holographic imaging of magnetic microstructures, *J. Magn. Magn. Mater.* **428**, 348 (2017).

250. Y. Khivintsev, M. Ranjbar, D. Gutierrez et al., Prime factorization using magnonic holographic devices, *J. Appl. Phys.* **120**, 123901 (2016).

251. O. Rousseau, B. Rana, R. Anami et al., Realization of a micrometre-scale spin-wave interferometer, *Sci. Rep.* **5**, 9873 (2015).

7

Spin Torque Effects in Magnetic Systems

Experiment

Maxim Tsoi

7.1 INTRODUCTION

The rapid pace of progress in the computer industry over the past 40 years has been based on the miniaturization of chips and other computer components. Further miniaturization, however, faces serious challenges, e.g., due to increasingly high-power dissipation. To continue at pace, the industry must go beyond incremental improvements and embrace radically new technologies. This book discusses a number of topics from a promising nanoscale technology known as *spintronics* (or magnetoelectronics) in which information is carried not by the electron's charge, as in conventional microchips, but also by the electron's intrinsic *spin*. Changing the spin of an electron is faster and requires less power than moving it. Therefore if a reliable way can be found to control and manipulate spins, spintronic devices could offer higher data processing speeds, lower electric consumption, and many other advantages over conventional chips including, perhaps, the ability to carry out radically new quantum computations.

The present chapter is focused on the spin-transfer-torque (STT) phenomenon which refers to a novel method to actively control and manipulate magnetic moments, or spins, using an electrical current. This method offers unprecedented spatial and temporal control of spin distributions and attracts a great deal of attention because it combines interesting fundamental science with the promise of applications in a broad range of technologies. STT effect is sometimes regarded as inverse to the giant magnetoresistance (GMR) effect, in that the magnetic state of a GMR structure affects its transport properties, while transport currents can alter the magnetic state via STT. This intimate connection can be traced throughout both theoretical and experimental studies of STT and makes GMR, examined in the previous chapters, a most excellent foundation for our discussion. Here we will focus on various aspects of STT experiments in metallic systems including device fabrication and STT detection techniques. In what follows we introduce the basic physics of STT effect in magnetic metallic multilayers. For additional details on STT theory in metallic systems, we refer to the next chapter of this book. STT in magnetic tunnel junctions and semiconductors will be discussed in sections III and IV, respectively.

This chapter is organized as follows: After introducing STT basics, we discuss a variety of STT devices from point contacts to nanopillars and nanowires. Then we focus on experimental methods to detect the effects of STT in such devices. Finally, we conclude by reviewing some of the potential STT applications.

7.2 STT BASIC PRINCIPLES

In the preceding chapters, we have seen that the magnetic state of a ferromagnet can affect its electrical transport properties – the relative orientation of the magnetic moments in magnetic multilayers underlies the phenomenon of GMR [1, 2]. The inverse effect – STT, in which a large electrical current density can perturb the magnetic state of a multilayer has also been predicted [3, 4] and observed [5–8] in systems with magnetic layers. Here the current transfers vector spin between the ferromagnetic layers and induces precession and/or reversal of magnetic moments in the multilayer. Altering a magnetic state with spin currents is based on quantum mechanical exchange interaction and represents a novel method to control and manipulate magnetic moments on nanometer length scales as well as picosecond timescales. In late 1970s Berger [9] was the first to realize that the exchange interaction between conduction electron spins and localized magnetic moments in a ferromagnet should result in a torque on the moments, e.g., in a magnetic domain wall traversed by an electric current. In a series of remarkable but only recently appreciated works [9–15], Berger set the groundwork for one of the most exciting areas in magnetism research today. However, it had to wait until after the discovery of GMR which spawned a major technological change in the information storage industry and, what is of particular interest for STT, provided a simple and efficient means to detect the effects of STT in experiments. A vast majority of such experiments rely on GMR to make STT visible, so it is natural to precede with the discussion of these experiments with GMR basics. However, as both the theory and experiment of GMR were already discussed in the previous chapters, here we only briefly revisit its physical concepts important for the sections to follow.

Both GMR and STT have been observed in magnetic metallic multilayers where several atomic layers of one – ferromagnetic (F) – material alternate with layers of another – nonmagnetic (N) – material (see Figure 7.1). To prepare such F/N multilayers a wide variety of deposition methods have been used, such as electrochemical deposition techniques [16, 17] and various vacuum deposition techniques [18, 19]. The latter shares mainly between two methods using either sputter deposition or molecular beam epitaxy (MBE) systems. Sputter deposition involves the knocking off the atoms of the material of interest from a target by particle bombardment followed by the deposition of highly energetic atoms (~2–30 eV) onto the substrate. A principal advantage of sputter deposition is the ease with which many different materials can be deposited at relatively high deposition rates. In contrast, deposition rates in MBE systems are usually much lower than for sputtering systems, but much lower energies (~0.1 eV) of the evaporated material make this technique favorable for the growth of highly oriented single-crystalline films.

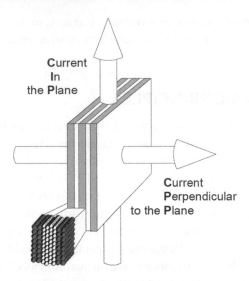

Current
In
the Plane

Current
Perpendicular
to the Plane

FIGURE 7.1 In a magnetic multilayer several atomic layers of magnetic material (shown in gray) alternate with layers of nonmagnetic material (shown in white). GMR occurs in two geometries: (i) when an electric current flows in the plane (CIP geometry) of the layers, or when it flows perpendicular to the layers (CPP geometry). The latter will be of particular interest for STT experiments.

In GMR experiments, a dramatic reduction in resistance of an F/N multilayer film is observed when the film is subject to an external magnetic field. GMR occurs in two different geometries (see Figure 7.1), namely when the current flows in the plane of the layers – CIP geometry, or when current flows perpendicular to the layers – CPP geometry. Most of the experiments on GMR are carried out in the CIP geometry because it is quite easy to measure the fairly large resistance of a thin film (film length is typically in orders of magnitude larger than its thickness). Experiments in the CPP geometry are more difficult [20] and require special techniques for the precision measurements of very small resistances ~10^{-7}–10^{-8} Ω that is the consequence of the "short and wide" geometry of a 1 mm^2 "wide" and 1 μm 'long' sample. In order to increase the resistance to easily observable values, microfabrication techniques can be used to reduce the sample's cross-sectional area [21–23]. Finally, a simple and inexpensive technique as the point-contact technique [24] may be also suitable for the purpose. The samples with the reduced cross-sectional area will be of interest for STT experiments which are mostly carried out in CPP geometry.

GMR originates from the spin-dependent scattering that occurs when electrons move across the multilayer film. The electron scattering and, therefore, the film resistivity depends on the magnetic configuration of the multilayer while the role of the external magnetic field is to change this configuration. The connection between the resistivity and magnetic configuration can be understood most simply by considering electrons traversing an $F_1/N/F_2$ trilayer where two thin-film ferromagnets (F_1 and F_2) are separated by a nonmagnetic spacer layer (N). Figure 7.2 illustrates the case in the simple limit where each F can be considered a perfect spin filter – it transmits

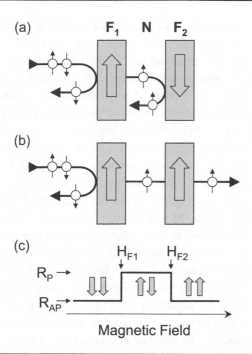

FIGURE 7.2 Qualitative picture of GMR in an $F_1/N/F_2$ spin valve. Right-going electrons are initially polarized by the F_1 layer. If F_2 is antiparallel to F_1 (a), the polarized electrons are filtered by F_2 and no current flows across the spin valve. If F_2 is parallel to F_1 (b), the electrons will be transmitted through both Fs. (c) The resulting spin-valve magnetoresistance (resistance vs magnetic field) for F_1 and F_2 switching at different applied fields (H_{F1} and H_{F2}). The resistance is low (R_P) when F_1 and F_2 are parallel and high (R_{AP}) when they are antiparallel.

only electrons whose spin is aligned with the magnetic moment of F, while electrons with antiparallel spin are completely reflected. When electrons cross such an $F_1/N/F_2$ spin valve from left to right, the electrons transmitted through F_1 will be polarized along F_1. If F_2 is antiparallel to F_1, then all these electrons are reflected off F_2 and no current can flow across the trilayer (Figure 7.2a). The spin-valve resistance R_{AP} in this antiparallel (AP) configuration is thus infinite. In contrast, if F_2 is parallel to F_1 (Figure 7.2b), electrons with like spins will be transmitted through both F layers and result in a finite resistance R_P in this parallel (P) configuration. The size of GMR is defined as $(R_{AP}-R_P)/R_P$ and is infinite for perfect spin-filtering. In real multilayers, the filtering, i.e., transmission and reflection of electrons, is not perfect that gives finite GMR ratios.

The resulting magnetoresistance (MR) curve – resistance R versus external magnetic field – of an $F_1/N/F_2$ trilayer may look like that in Figure 7.2c. Here F_1 and F_2 were made to respond differently to the applied field H and switch at different applied fields H_{F1} and H_{F2}, respectively. Starting up from high negative fields where both Fs are aligned with the field, the resistance is constant at a minimum value R_P below H_{F1} (parallel Fs), rises to a maximum value R_{AP} when F_1 reverses at H_{F1} (antiparallel Fs), and reduces back to R_P as F_2 reverses at H_{F2} (parallel Fs again). The different switching fields of F_1 and F_2 can be achieved either by making them of different materials/thicknesses or by placing an antiferromagnet in contact with one of Fs to effectively "pin" its magnetization through an effect called "exchange bias."

In contrast to this simple picture of GMR where the magnetic configuration of $F_1/N/F_2$ trilayer is assumed to be affected only by external fields, the STT phenomenon is concerned with the effect of transport currents on the magnetic configuration. The basic physical picture of STT relies on conservation of angular momentum as the current propagates through the multilayer. Consider a pedagogically simple case where a conduction electron crosses an interface between a nonmagnetic metal (N) and a ferromagnet (F). We assume that the initial state of the electron's spin **S** in N is not collinear with the F's magnetic moment **M**. Once into F, **S** is subject to an exchange torque caused by **M** which tends to reorient **S**. Therewith, according to Newton's Third Law, there should also exist a reaction torque which acts on **M** – STT torque. Deep into F, **S** is aligned with **M** and the change in angular momentum that occurs from its reorientation has been transferred to **M**. Therefore, the name of the phenomenon – spin-transfer-torque (STT). Of course, the transfer of angular momentum to **M** by a single spin **S** is tiny due to **S** being negligibly small compared to **M**. For a high-density current injected into F, however, the number of such spins can be very large and the effective **S** might become comparable to **M**. This would result in a significant STT torque on **M** and highlights the need for high current densities in experimental demonstrations of STT phenomenon.

The first observation of STT in magnetic multilayers was recorded by Tsoi et al. [5] (see Figure 7.3). In this experiment, the spin-transfer-induced excitations were produced by injecting high-density electric currents into a Co/Cu magnetic multilayer through a mechanical point contact. Point contacts with sizes smaller than 10 nm are formed when a sharpened Cu metal

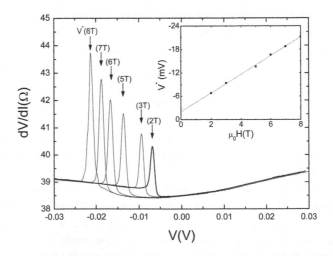

FIGURE 7.3 Differential resistance *dV/dI* of a mechanical point contact to Co/Cu multilayer as a function of bias voltage V (equivalent to current) for a series of magnetic fields. The peak in *dV/dI* indicates the onset of spin-transfer-torque excitations. The inset shows that the threshold current at the peak in *dV/dI* increases linearly with the applied field. (After Tsoi, M. et al., *Phys. Rev. Lett.* 80, 4281, 1998. With permission.)

wire (tip) is carefully brought into contact with the multilayer. The extremely small cross-sectional area of such a contact makes it possible to achieve current densities in excess of 10^{12} A/m^2. Because of its extremely small size (<10 nm), point contact is a very efficient probe of electrical transport properties in extremely small sample volumes yet inaccessible with other techniques (e.g., electron beam lithography patterning). The latter qualifies point contact as the smallest probe of STT today.

A typical experiment on the current-driven excitation of a ferromagnet usually involves two single-domain thin-film magnets separated by a non-magnetic spacer. Here one magnet (F_1) is 'hard' and used to polarize the current while the spacer (N) is thin enough for the polarized current to get through and excite the second 'free' magnet (F_2). This F_1/N/F_2 trilayer structure is similar to the GMR spin valve discussed above. The GMR effect can thus be used to monitor the orientation of F_1 relative to F_2 – GMR varies linearly with cosθ, where θ is the angle between magnetic moments of F_1 and F_2, and a phenomenological description [25] gives the trilayer resistance R(θ) = $R_P + (R_{AP} - R_P)(1 - \cos\theta)/2$. When current flows across F_1/N/F_2, the current-induced torques act on both F_1 and F_2 layers [4], as schematically illustrated in Figure 7.4; here the black arrows indicate the directions of the torques. This qualitative picture of STT assumes that both F_1 and F_2 layers are perfect spin filters [26], so that electron spins aligned with the magnetic moment of, e.g., the F_1 layer are completely transmitted through the layer, while spins aligned antiparallel to the layer moment are completely reflected. When electron current crosses the F_1/N/F_2 trilayer from left to right (Figure 7.4b, and e), electrons transmitted through F_1 will be polarized along F_1. If spin-diffusion length in N is long enough, this spin-polarized current will reach F_2 and exert a torque on F_2 in a direction so as to align F_2 with F_1. Repeating the argument for F_2, we find that electrons reflected from F_2 will be polarized antiparallel to F_2 and, hence, exert a torque on F_1 trying to align F_1 antiparallel with F_2. The net result is a pinwheel-type motion with both F_1 and F_2 rotating in the same direction (anti-clockwise in Figure 7.4b), as described previously by Slonczewski [4]. The higher-order reflections between F_1 and F_2 (Figure 7.4 shows a total of four) can change the magnitudes of torques on F_1 and F_2. Note that all subsequent reflections from F_2 in Figure 7.4b will produce torques opposite to the initial torque acting on F_2 and would reduce it. In contrast, the subsequent reflections from F_1 produce torques in the same direction as the initial torque acting on F_1 and would increase it. However such higher-order contributions from multiple reflections are expected to be small as the electron flux is reduced upon each reflection. When the current crosses the trilayer from right to left, the roles played by F_1 and F_2 are reversed and so are the directions of the initial torques (see Figure 7.4c and f) – the torque on F_1 is trying to align F_1 parallel with F_2, while the torque on F_2 is trying to align F_2 antiparallel with F_1. Figure 7.5e and f illustrate the situation for the antiparallel alignment of F_1 and F_2.

The above discussion implies that there is an asymmetry of STT with respect to the current direction as explained below. Let's fix the orientation of the polarizer F_1; in experiments, this is usually accomplished by making

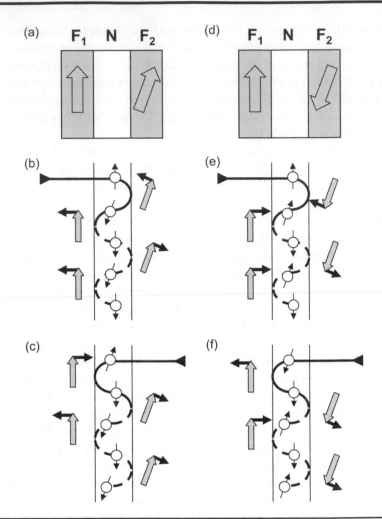

FIGURE 7.4 Qualitative picture of STT in an $F_1/N/F_2$ spin valve. For almost parallel (a) or antiparallel (d) alignment of F_1 and F_2, the right-going electrons (b, e) are first polarized along F_1 and exert a torque on F_2 in a direction so as to align F_2 with F_1. Electrons reflected from F_2 will be polarized antiparallel to F_2 and exert a torque on F_1 trying to align F_1 antiparallel with F_2. Electrons bouncing between F_1 and F_2 (reflections up to 4th are shown) exert higher-order torques on F_1 and F_2. Higher-order contributions from multiple reflections are expected to be small as the electron flux is reduced upon each reflection. Black arrows indicate torque directions. For reverse polarity of the current flow (c, f) the roles played by F_1 and F_2 are reversed and so are the directions of the initial torques.

F_1 very thick (compared to F_2) or by pinning its orientation with an adjacent antiferromagnetic layer via the phenomenon of exchange bias. If initially, F_2 is almost parallel with F_1, the right-going electrons (forward bias) will stabilize this parallel alignment and no STT excitation is present. When current bias is reversed, the torque on F_2 will try to rotate F_2 away from F_1 and will result in STT excitation of the system. If the initial orientation of F_2 is almost antiparallel with F_1, the forward bias will try to switch its orientation while reverse bias will stabilize the antiparallel alignment of F_2 and F_1. This asymmetry with respect to bias (current) polarity is one of the fingerprint-like features of STT

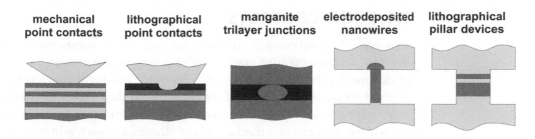

| mechanical point contacts | lithographical point contacts | manganite trilayer junctions | electrodeposited nanowires | lithographical pillar devices |

FIGURE 7.5 Overview of devices for STT observation. High current density needed for STT is achieved by forcing the electric current to flow through a very small constriction: point contact (mechanical or lithographically defined), manganite junction, nanowires, or nanopillar.

in experiments; see, for instance, Figure 7.3 where STT excitations are present only at negative bias.

STT phenomenon currently attracts a great deal of attention because it combines interesting fundamental science with the promise of applications in a broad range of technologies. In high-speed high-density magnetic recording technology, for instance, the spin-transfer-torque could replace the Oersted field currently used for writing magnetic bits in storage media (e.g., in magnetic random access memory or MRAM). This may lead to a smaller and faster magnetic memory. Another possible application is based on the spin-transfer-induced precession of magnetization, which converts a dc current input into an ac voltage output. The frequency of such a precession can be tuned from a few GHz to > 100 GHz by changing the applied magnetic field and/or dc current, effectively resulting in a current controlled oscillator to be used in practical microwave circuits.

7.3 STT DEVICES

Since its prediction by Berger [3] and Slonczewski [4], the STT effect has been observed in a number of experiments, including those with mechanical [5, 27–34] and lithographic point contacts [7, 35–40], lithographically defined nanopillars [41–48], lateral structures [49–52] electrochemically grown nanowires [53–58], thin-film stripes [11, 14, 15, 59–63], manganite junctions [6, 64], tunnel junctions [65–68], and semiconductor structures [69–75]. These different methods all share one characteristic feature – they make it possible to attain extremely high current densities ($\sim10^{11}$-10^{13} A/m^2 for metallic structures) needed to produce sufficiently large spin-transfer torques. This is achieved by forcing the electric current to flow through a very small constriction. The latter can be a mechanical point contact, lithographically defined point contact or nanopillar, or nanowire as illustrated in Figure 7.5. In all cases the maximum current density j_{max}=I/A is defined by the current I flowing through the device and the minimum cross-sectional area A for the current flow. For typical mechanical point contacts I $\sim$1 mA and A $\sim$100 nm^2 gives j_{max} $\sim$10^{13} A/m^2. In lithographically defined structures both I and A are typically larger, I $\sim$10 mA and A $\sim$10000 nm^2, that gives j_{max} $\sim$10^{12} A/m^2. In this section, we review the devices used in STT experiments.

7.4 POINT CONTACTS

Nanoscale electrical contacts are convenient tools for the realization of local injection and detection of conduction electrons in experimental studies of electron kinetics in metals [76–81]. Recently, such point contacts have received an increased amount of attention due to their ability to produce extremely high current densities needed in STT experiments. Two standard types of point contacts are mechanical and lithographic, each with their own advantages. Mechanical contacts, made by gently pressing a metal wire tip onto a sample surface, are relatively simple to make and can be used on samples of arbitrary composition and shape; whereas lithographic techniques offer much greater control of contact geometry and placement, in addition to being more stable and robust, e.g., to temperature variations.

7.4.1 Mechanical Point Contacts

Mechanical point contacts were used for the original observation of STT in metallic multilayers by Tsoi et al. [5]. In the following, we discuss the mechanical setup used in Ref. [5] that provides a means to produce point contacts for a variety of samples of arbitrary composition and shape. Similar systems were used for STT observations in F/Cu multilayers with F = Co [5, 27, 28, 82, 83] and F = $Co_{90}Fe_{10}$, $Ni_{80}Fe_{20}$, $Ni_{40}Fe_{10}Cu_{50}$, or Fe [84], single F = Co layer [29–31], bulk F = Co [30], antiferromagnetic AFM = FeMn, IrMn alloys [32, 85], and Co nanoparticles [33, 34].

Figure 7.6 schematically shows a mechanical point contact between a sharpened metallic wire and a thin-film sample. An electrical contact between the wire tip and the sample is achieved by gently pressing the wire, bent in the usual way to minimize damage to the sample, onto the sample surface. Usually, the wire is moved toward the sample surface by a standard differential screw mechanism, but a more precise piezo-control can also be used. Typical contacts made in this way have an area from ~ 10–10^4 nm^2, estimated from the measured contact resistance [24, 86, 87] assuming a combination of Sharvin (ballistic) [76] and Maxwell (diffusive) [88] scattering. Because the contact is so small, one of the main experimental challenges

FIGURE 7.6   Schematic view of a point contact. A sharpened metal wire is gently pressed onto a sample surface. The resulting electrical contact between the wire tip and sample can be as small as $2a \sim 1$ nm.

is to preserve it throughout an experimental run, as any vibrations may lead to the tip displacements with respect to the sample and, thereby, to contact instability. Such systems have been used to successfully study ballistic transport in many bulk metals and compounds.

Preparation of the contact tip is similar to that used in scanning tunneling microscopy (STM). In Ref. [7] the tip was made of Ag wire (Ø 100–200 μm) electrochemically etched to a sharp point. Other wire materials (Cu, Au, Nb, Fe, Co) have also been used. The tip is moved toward the sample with a standard differential screw mechanism (see Figure 7.7) which consists of a frame (1), a threaded bolt (2), and a piston (3). A room-temperature control knob (not shown) on the rod (4) is used to control the rod rotation and, via fork-blade coupling (5), the threaded bolt (2). The bolt has different pitch threads on opposite bolt ends so that the bolt engages the piston (3) faster than the bolt is disengaged from the frame (1). The resulting vertical displacement of the piston is transferred to the tip (10) via a metallic ball (13). A complete rotation of the room-temperature control knob displaces the tip by 25 μm. The leaf spring (11) minimizes the tip displacements in the horizontal plane and damps unwanted vibrations. A hole in the spacer (9) provides visual control of initial separation ∼ 200 μm between the tip and the sample (8).

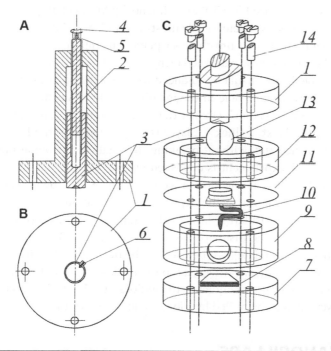

FIGURE 7.7   The mechanical point-contact setup. (a, b) A differential screw mechanism includes frame (1), threaded bolt (2), and piston (3). (c) An exploded view of a block, containing the tip (10) and sample (8). The structural elements of the block – sample table (7), leaf spring (11), and spacers (9, 12) – are tightly fixed to the frame (1) with four screws (14). The vertical displacement of the piston (3), transmitted through a metallic ball (13) to the leaf spring (11), results in tip's motion up/down. A small spring (6) prevents the piston from rotating.

The system just described was used for STT observations both at helium (4.2 K) [5] and room temperature [89] and in magnetic fields up to 20 T [90]. If compared to other types of point contacts, the mechanical contacts have a number of advantages. In addition to being inexpensive and simple to control, the mechanical system provides a unique means to produce a contact of only a few nanometers in size and to change the contact within a single experimental run by simple rotation of the room-temperature control knob.

7.4.2 Lithographic Point Contacts

Although mechanical point contacts have the advantage of being small and simple to control, they also have some disadvantages. First of all, it is a poor control of the contact geometry – the contact size can be defined only indirectly, after establishing the contact, by measuring its resistance. Second, mechanical contacts are too fragile to perform a number of experiments, including temperature dependent measurements, which fail due to different thermal expansions of tip and sample materials.

All of the above limitations can be overcome by microfabricating point contacts with the help of electron beam lithography (EBL). The microfabricated contacts provide excellent stability, e.g., under temperature cycling, and good control over the contact geometry [91]. Such lithographic point contacts were first used in STT experiments by Myers et al. [7]. The device fabrication process is schematically illustrated in Figure 7.8a through c. The process begins with a 50-nm-thick suspended membrane of silicon nitride (Si_3N_4). The membrane is patterned using EBL and reactive ion etching (RIE) to open a hole in the insulator. Reduced RIE rates in constrained geometry result in a bowl-shaped hole in the membrane and an opening 5 to 10 nm in diameter, which is much smaller than a 40–100 nm feature actually patterned by EBL. Next, a thick Cu layer is evaporated onto the bowl-shaped side of the device and, without breaking the vacuum, the metal multilayer needed for STT experiment is deposited on the opposite side of the membrane. An alternative fabrication process (Figure 7.8d through f) starts with the multilayer of interest being deposited on a substrate. The multilayer is covered by a thin insulating layer and a standard EBL patterning is used to create a hole in the insulator. Note that a layer of EBL resist (PMMA= Polymethyl-methacrylate) can be used as the insulator. After in-situ cleaning of the exposed metal surface, a top electrode is deposited to fill the hole and establish an electric contact to the multilayer [35]. The hole in the insulating layer can also be milled by a focused ion beam (FIB) [92].

7.5 NANOPILLARS

Lithographically defined nanopillars of magnetic multilayers are probably the most-studied STT devices to date [93–112]. One of the main reasons for that is the opportunity they provide for realizing the geometry originally suggested by Berger [3] and Slonczewski [4] for STT observation, two ferromagnetic (F) layers separated by a nonmagnetic (N) spacer layer and a

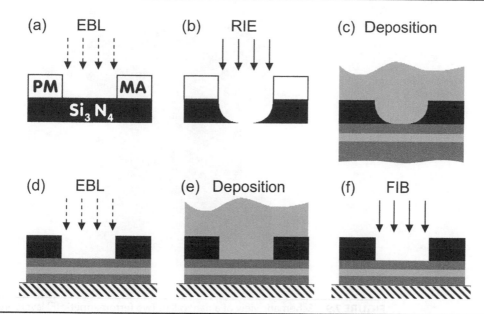

FIGURE 7.8 Illustration of the point-contact fabrication process flow. (a) PMMA resist layer is patterned by EBL. (b) RIE is used to transfer the EBL pattern to the insulating (Si_3N_4) membrane. (c) A thick Cu is evaporated on one side of the membrane, followed by a multilayer deposition on the other side. Alternatively, (d) EBL can be used to create a hole in an insulating layer (black) on top of an existing multilayer, followed by a Cu electrode deposition (e). (f) The hole in the insulator can also be made by focused ion beam (FIB).

uniform current density traversing this F/N/F stuck (see Figure 7.9). The first observation of STT phenomenon with lithographically defined nanopillars was made by Katine et al. [41]. They used electron beam lithography (EBL) to produce pillars consisting of two F = Co layers of different thicknesses separated by a N=Cu layer spacer. The lithographic process allows for greater control of the device geometry while reducing the cross-sectional area of this otherwise standard F/N/F spin valve to ~ 10^4 nm^2. The pillars can be patterned using two different generic approaches: subtractive – the growth of the spin-valve film followed by patterning a hard-mask on top and etching of the film, or additive – metal deposition onto a predefined template. The etching step (typically ion milling) required for subtractive processing can degrade the magnetic behavior of nanoscale patterned elements. On the other hand, the additive process can result in less well-defined edges of the nanopillar. So either approach has to be optimized to avoid potential problems.

7.5.1 SUBTRACTIVE PATTERNING

The fabrication of a nanopillar by a subtractive patterning process as used in [41] is outlined in Figure 7.10. The fabrication begins with buffer layer/multilayer stuck/capping layer deposition onto an oxidized Si substrate. The buffer layer will later serve as the bottom electrode. Next, a Cr dot is patterned on top of the capping layer by EBL: a single dot in resist (PMMA) layer is exposed and developed (Figure 7.10a); after Cr metallization and lift-off, the Cr dot remains on the layer surface (Figure 7.10b). This dot serves as the

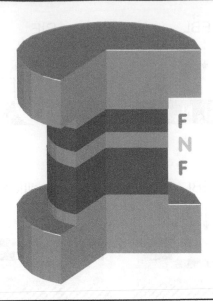

FIGURE 7.9 Schematic view of a nanopillar. Two ferromagnetic (F) layers are separated by a nonmagnetic (N) spacer. One F is typically made much thicker than the other one to minimize current-induced effects on the thick F and focus on STT in the thin F.

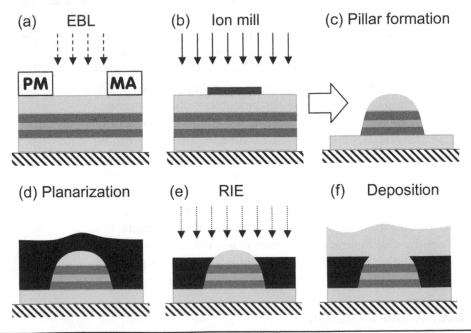

FIGURE 7.10 Illustration of nanopillar fabrication via subtractive patterning. First, a buffer layer (light gray), multilayer stuck, and capping layer (light gray) are deposited on a substrate. (a) A Cr dot is patterned on the surface of multilayer stuck by EBL, Cr deposition (not shown), and lift-off process (not shown). (b–c) Ion milling defines the pillar. (d) Planarization with polyimide seals the pillar with an insulator. (e) RIE opens the pillar cap. (f) The top electrode is formed by final metallization.

mask for incoming ions during the ion-mill step (Figure 7.10b through c) and thus defines the pillar cross-section. The mask and multilayer around it are completely eroded during milling while redeposition of the milled material results in a rounded pillar such as that shown in Figure 7.10c. Following the milling step, planarization with polyimide, reactive ion etching (RIE), and metallization are used to seal the pillar with an insulator (Figure 7.10d), open the pillar cap (Figure 7.10e), and form the top electrode (Figure 7.10f).

7.5.2 ADDITIVE PATTERNING

An alternative way to fabricate nanopillars of magnetic multilayers is the additive patterning method [43] shown schematically in Figure 7.11. Here the fabrication begins with bottom electrode/insulating (SiO_2) spacer/Pt layer deposition onto an oxidized Si substrate. EBL patterning and ion milling are used to create a small hole in the Pt layer which, in the following steps, serves as a mask and defines the pillar size. Next a wet etch is used to open up the insulating spacer, followed by deposition of the multilayer of interest. The nanopillar is formed by the portion of the multilayer which goes through the mask and is laid down directly on the bottom electrode. Finally, without breaking the vacuum, a thick metallic layer is deposited on top to make the top electrode.

To speed up the fabrication process and eliminate the EBL patterning a focused ion beam (FIB) milling can be used to punch a hole in the Pt mask layer [113] similar to the process shown in Figure 7.8f. Another alternative

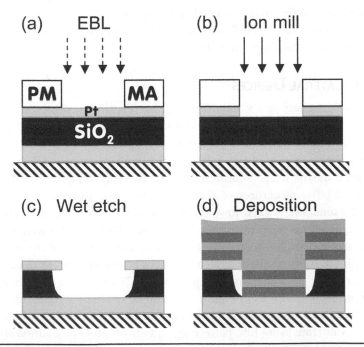

FIGURE 7.11 Additive patterning process for nanopillar fabrication. EBL (a) and ion milling (b) define a small hole in a thin Pt layer. (c) Wet etch opens up the insulating spacer. (d) Deposition of multilayer stack and the top electrode.

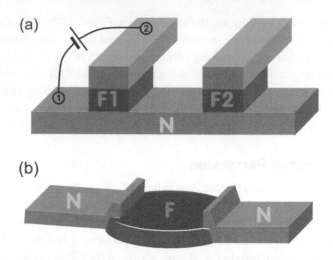

FIGURE 7.12   Lateral devices. (a) Lateral spin valve where two magnets (F1 and F2) are laterally separated by a nonmagnetic (N) spacer. (b) Single F element (thin film disk) contacted by two nonmagnetic electrodes.

may use a layer of PMMA resist as the insulating spacer and an EBL patterned hole in this layer as a template for the nanopillar [114]. This process eliminates the ion mill and wet etch steps – the multilayer is deposited directly into the template hole, similar to Cu electrode deposition in Figure 7.8d and e. A shortcoming of this method lies in the absence of a controlled undercut (see Figure 7.11c) obtained during the wet etch, which means that the nanopillar edges may not be well-defined due to the deposition onto the side-walls of the hole. Yet another method for making a pillar exploits a three-dimensional (3D) FIB milling [115] to sculpt a nanopillar by carving away material from a bulk multilayer film.

7.5.3 LATERAL DEVICES

The two F magnets of a standard nanopillar (Figure 7.9) can be separated laterally [49, 116] in a so-called lateral spin valve (LSV) shown schematically in Figure 7.12a. In addition to providing, e.g., optical access to the N spacer to study the relaxation of spin currents, this multi-terminal LSV geometry provides a means for generating pure spin currents which are impossible in vertical pillar structures of Figure 7.9 where the spin current, needed for STT excitation, always flows together with the charge current. Driving charge current between terminals 1 and 2 (see Figure 7.12a) would result in spin accumulation near the N/F1 interface. The diffusion of the accumulated spins away from the interface generates pure spin current toward the second magnet F2 which, in turn, generates STT on F2 and was shown to switch it [49, 50]. LSV as in Figure 7.12a can be made by a standard multi-level lithography process [117]. Here every level of device elements is laid on a planar substrate using EBL patterning and lift-off process. Caution should be taken to ensure a good electrical contact between elements of different levels as oxidation and contamination during lithographic patterning is known to degrade the contact transparency. A short ion milling is typically used to prepare the element surfaces. To further improve the quality of N/F interfaces, magnetic F elements can be deposited directly onto the N spacer without breaking the vacuum using oblique deposition techniques [50]. Figure 7.12b illustrates a lateral structure with a single F element as used for studying the current-driven dynamics and switching of magnetic vortices in a F=Py (permalloy) disk [51, 52].

7.6 NANOWIRES AND NANOSTRIPES

Currently, it is challenging to make lithographically defined nanopillars with diameters smaller than 50 nm. In this section, we will discuss the template-synthesis method for STT device fabrication which potentially may be used to prepare metallic nanopillars (wires) with diameters down to a few nanometers. Larger wires, however, are still useful for STT experiments with magnetic domain walls. Here we will discuss magnetic nanostripes employed in such experiments prepared by lithographic methods and focused ion beam patterning.

7.6.1 ELECTROCHEMICALLY GROWN NANOWIRES

An alternative method of producing cylindrical pillars of the desired material, known as membrane-based or template-based synthesis, was around for decades and successfully used to prepare nanowires of metals, semiconductors, conducting polymers, and other materials [118]. Here the desired material is deposited within the pores of a template membrane. For cylindrical pores with diameters in the nanometer range, e.g., as in track-etch polymeric membranes or porous aluminas, this results in a nanocylinder of the desired material (nanowire) obtained in each pore.

The template synthesis of metallic nanowires was first demonstrated by Possin [119]. He used electrochemical reduction of metal ions to deposit the appropriate metal within the pores of a template membrane. The process is schematically illustrated in Figure 7.13. Here one (bottom) face of the membrane is covered with an evaporated metallic film. The film serves as a cathode for electroplating when the membrane is immersed in a plating solution containing ions of the desired metal. The filling of membrane pores by electroplating results in wires of a nearly uniform cross-section and reasonable purity. The method can be applied to any metal which can be plated from an acid solution.

The effect of electric current on a magnetic reversal in individual electrodeposited nanowires was evidenced by Wegrowe et al. [8]. They developed a method where contact with a single wire is made in-situ during electrodeposition. The contact is accomplished by continuously monitoring potential between the cathode and a patterned gold electrode on top of the membrane (see Figure 7.13) and by stopping the electrodeposition process as soon as the first nanowire emerges from the membrane pore, attaches to the top electrode, and short-circuits the membrane. Applying this method to track-etch polymeric membranes has led to STT observations in Ni nanowires

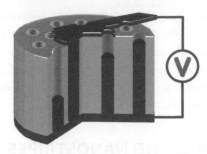

FIGURE 7.13   Electrochemically grown nanowires. A metallic film is evaporated on one (bottom) face of a track-etched membrane to serve as a cathode for electroplating. The filling of membrane pores by electroplating results in wires of the nearly uniform cross-section. Monitoring the potential between the cathode and a patterned top electrode allows for stopping the electrodeposition when a nanowire emerges from membrane pores, attaches to the top electrode, and short-circuits the membrane.

[8, 53, 120, 121], hybrid wires where a Ni half-wire is joined with a Co/Cu multilayer [53–120], nanowires of Co/Cu multilayer [120], and Co/Cu/Co spin valves embedded in Cu nanowires [54, 55]. A relatively low pore density (typically $< 10^9$ cm^{-2}) of polymeric membranes facilitates contacting a single nanowire which can even be made with a sharpened Au tip [56]. However, in templates with higher pore densities, similar to porous aluminas with densities up to $\sim 10^{12}$ cm^{-2}, it is difficult to contact a single nanowire in this way and alternative methods should be used. A nanolithography process based on electrically controlled nanoindentation of alumina templates has been used to study STT properties in nanowires of Py/Cu/Py [57] and Co/Cu/Co [58] spin valves.

7.6.2 LITHOGRAPHICALLY DEFINED NANOSTRIPES

Standard lithographic methods are typically used to prepare nanowires of thin-film materials [121–147]. Figure 7.14a through b illustrates the fabrication process. A nanowire pattern is first defined, e.g., by electron beam lithography (EBL), on MIBK resist layer spanned on an oxidized silicon wafer (Figure 7.14a). A thin-film deposition and lift-off process are then used to transfer the pattern to a thin film of the desired material (Figure 7.14b). The resulting nanowire is rather a nanostripe with a cross-section defined by the film thickness and the width of the lithographic pattern. Scanning electron microscopy (SEM) image in Figure 7.14c shows a 100-nm-wide Py stripe connecting two diamond shaped elements used for nucleation and injection of magnetic domain walls; constrictions in the stripe are typically used to pin the injected domain walls [62].

An alternative method for fabricating similar nanostripes begins with depositing a thin continuous film on a substrate (Figure 7.15a), followed by a selective milling of the film by focused ion beam (FIB) [126, 130, 144]. Removal of selected areas of the film by FIB (Figure 7.15b) results in an arbitrary-shaped thin-film pattern, which can be a nanostripe with desired width and length. The SEM image in Figure 7.15c shows an FIB-prepared

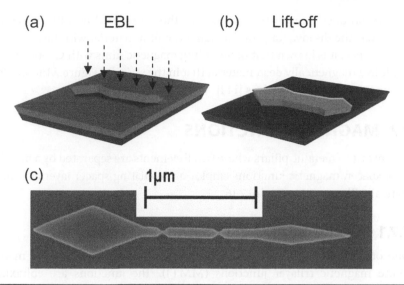

FIGURE 7.14 Lithographically defined nanowires. (a) The desired shape is patterned by EBL in a resist layer on a substrate. (b) The EBL pattern is then transferred to a thin film of the desired material using thin-film deposition and lift-off process. (c) SEM image of a Py thin-film device used in STT experiments where a 100-nm-wide stripe connects two diamond shaped elements and has two narrow constrictions.

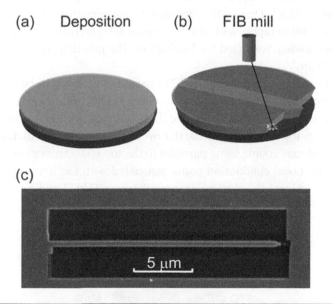

FIGURE 7.15 FIB patterned nanowires. (a) A thin film is evaporated on a substrate. (b) FIB is used to selectively remove the film material. (c) SEM image of a 600-nm-wide Py stripe. Darker color indicates areas were Py film was milled away by FIB.

600-nm-wide 20-μm-long Py nanostripe. The darker color in Figure 7.15c indicates areas were Py film was milled away. One end of the remaining stripe is left connected to the extended unmilled film for domain-wall injection purposes. Although the FIB patterning process is fast and may be used to produce nanostripes as small as those prepared by EBL lithography, it has the disadvantage of damaging the film material with heavy ions. For instance, it is known that bombarding magnetic films with Ga ions may result in a magnetically dead material that in the stripe of Figure 7.15c would be concentrated near its edges [144].

7.7 MAGNETIC JUNCTIONS

In contrast to metallic pillars where two F elements are separated by a metallic N spacer, magnetic junctions employ an insulating spacer layer as schematically illustrated in Figure 7.16.

7.7.1 MANGANITE MAGNETIC TRILAYER JUNCTIONS

One of the first STT experiments was performed by Sun [6] with manganite magnetic trilayer junctions (MMTJ). The junctions are epitaxial LSMO60nm/STO3nm/LSMO40nm thin-film trilayers grown by laser ablation on NGO substrates (LSMO = $La_{0.67}Sr_{0.33}MnO_3$, STO = $SrTiO_3$, NGO = $NdGaO_3$). After deposition, a lithographic process similar to that used for metallic nanopillars was used to form 'pillars' of LSMO/STO/LSMO junctions, where bottom LSMO is only partially patterned and serves as the bottom electrode, and a SiO_2 layer is used to isolate the top (Au) electrode from the bottom LSMO. Due to the significant heat-sensitivity of the junction, only optical lithography, with the maximum temperature reaching ~ 90°C during processing, was used for fabrication. The junction size is, thus, limited to ~10 μm^2.

Electron transport across the STO barrier is believed to be controlled by the morphology of the junction interface. LSMO films grown by laser ablation are known to contain particulates with sizes ranging from 10 nm to more than 100 nm. The STO barrier over such LSMO particulates is nonuniform and may couple some particles to the top LSMO more conductively than others. Local conduction paths associated with the nonuniform coupling can dominate the electron transport across MMTJ and result in high

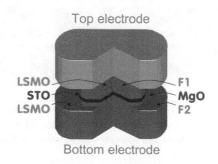

FIGURE 7.16 Magnetic junction. Two thin-film magnets are separated by a nonmagnetic barrier. Top and bottom electrodes are used for electrical access to the junction. The bottom magnet can be a part of the bottom electrode. In MMTJs F1 = F2 = LSMO with STO barrier. In MTJs F1 and F2 are transition-metal magnets separated by AlO_x, MgO, or other tunneling barrier.

densities of local current, thus making STT observable at relatively low current levels [6, 64].

7.7.2 Magnetic Tunneling Junctions

Magnetic tunneling junctions (MTJ) have a similar structure and fabrication process flow to that of the MMTJs discussed in the previous section (see Figure 7.16), except that the F1/barrier/F2 trilayer is typically prepared by sputtering with the F1 and F2 layers made of transition metals or their alloys separated by an oxide barrier. Initial observations of STT in MTJs were done with the barrier = AlO_x [65–67] later replaced by MgO [68] which is more robust to electrical breakdown and produce higher magnetoresistance ratios. More details on STT in MTJs can be found in the following section on spin transport in magnetic tunnel junctions.

Current-induced magnetization reversal has also been reported in magnetic tunnel junctions based on ferromagnetic semiconductors [69–72]. Here the roles of two F electrodes are played by ferromagnetic III-V semiconductors, such as (Ga,Mn)As characterized by carrier-induced ferromagnetism and strong spin-orbit interaction, while GaAs or AlAs is used as a nonmagnetic spacer layer. Such semiconductor-based trilayers are typically grown by molecular beam epitaxy (MBE).

7.8 STT EXPERIMENT

Previous sections introduced the physics of STT and various magnetic nanosystems where STT can be observed. We have seen that a high-density electric current can result in torques on magnetic elements of a system. These torques may be used to control and manipulate the system's magnetic state. However, the resulting behavior of the system can differ significantly from case to case and depends on particular conditions of observation. For instance, in modest external magnetic fields, the magnetization of a small element can be repeatedly reversed between two stable configurations; while at higher fields, where the reversal is energetically unfavorable, the moment can be set into precession at a very high frequency. In order to understand details of what happens with a magnetic moment **S** in a particular situation, one can use Newton's Second Law. For **S** this would be the Landau–Lifshitz-Gilbert equation where the rate of change of S is set equal to the net torque acting on **S**:

$$\dot{\boldsymbol{S}} = \gamma \ \vec{\boldsymbol{S}} \times \vec{\boldsymbol{H}}_{\text{eff}} - \alpha \ \hat{s} \times \dot{\boldsymbol{S}} + \boldsymbol{I} \frac{\eta}{e} \hat{s} \times \left(\hat{s}^* \times \hat{s} \right) \qquad (7.1)$$

Here the first term on the right is the torque on a magnetic moment **S** in an effective magnetic field $\vec{H}_{\text{eff}}$, (including applied, demagnetizing, anisotropy, etc., fields), with γ the gyromagnetic ratio; the second term is a phenomenological damping term introduced by Gilbert, with α the Gilbert damping parameter; the third term is the spin-transfer-torque introduced by Slonczewski [4] where small $\hat{s}$ and $\hat{s}^*$ are unit vectors along $\vec{S}$ and the polarizer $\vec{S}^*$, I is the current, e electron charge and η spin-polarization factor.

The diagram in Figure 7.17 shows the directions of the three torques from Equation 7.1. Note that depending on the polarity of applied current the STT torque can be either in the same direction as the damping torque or opposite to it. In the former case, STT will effectively result in an increased damping for any magnitude of the applied current and suppress any possible excitations of $\vec{S}$ from its equilibrium state along $\vec{H}_{\text{eff}}$. If, however, the STT torque is opposite to the damping we can distinguish two situations. For currents below a critical current, where STT is small compared to damping, $\vec{S}$ spirals toward $\vec{H}_{\text{eff}}$ (gray trajectory in Figure 7.17). For currents larger than the critical current, STT exceeds the damping torque and causes $\vec{S}$ to spiral away from $\vec{H}_{\text{eff}}$, with a steadily increasing precession angle. The ultimate result can be, depending on angular dependence of STT and damping, either stable steady-state precession of $\vec{S}$ around $\vec{H}_{\text{eff}}$ (black trajectory in Figure 7.17) or magnetic reversal of $\vec{S}$ into a state antiparallel to $\vec{H}_{\text{eff}}$

Almost all experimental observations of STT rely on GMR phenomenon to detect the current-induced reorientation of magnetic moments in nanodevices. Typical measurements include (i) measuring static device resistance $R=V/I$ as a function of applied dc bias current I in an applied magnetic field and (ii) measuring R versus applied field at a constant I. Figure 7.3 shows how the differential resistance dV/dI of a Cu point contact to Co/Cu magnetic multilayer varies with the bias voltage V (equivalent to I) applied across the contact. Here the multilayer magnetic moments are saturated out of the plane of the layers by a large external magnetic field (≥ 2 T). The onset of STT-driven magnetic precession is revealed by a peak in the differential dV/dI contact resistance. The peak in dV/dI may indicate that the transition into precession is a reversible process and in a small range of currents one can continuously increase/decrease the angle of precession, although other

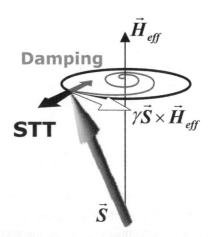

FIGURE 7.17 Torques on a magnetic moment in a magnetic field and subject to an electric current.

scenarios, e.g., fast switching between steady-state precession and static states with current-dependent dwell time, are also possible [112].

If the applied magnetic field is small, the magnetic system can have more than one low-energy state. In the simple case of a magnetic element with uniaxial anisotropy, STT can trigger a transition between two static states which are equally energetically favorable without current. An example of such behavior is shown in Figure 7.18. Here the current is driven across a trilayer Py20nm/Cu12nm/Py4.5 nm spin-valve structure patterned by electron beam lithography (EBL) into a nanopillar with cross-sectional area 40×120 nm^2 [148]. The differential resistance of the nanopillar exhibits a hysteresis as a function of an applied bias current as the magnetization of the thin (free) permalloy (Py) layer is aligned parallel or antiparallel to the thick (hard) Py layer by the current.

The two examples presented above (Figures 7.3 and 7.18) demonstrate how simple dc resistance measurements can be used for STT observation. Here measured resistance of a device provides indirect information about the relative orientation of magnetic elements in the device. However, measured critical currents highlighted by sharp variations in the resistance remain almost the only experimental information which can be used for quantitative comparison with theory. Moreover, the dc measurements of Figure 7.3 provide no information about the fast evolution of magnetization in the device associated with the high-frequency precession of magnetic moments. High-frequency techniques must be employed to provide such capabilities as discussed next.

The first experiment which gave unequivocal evidence that a dc electric current can result in a high frequency (tens of GHz) precession of magnetic

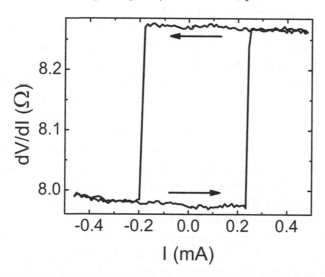

FIGURE 7.18 Spin-torque-driven magnetic switching for a Py20 nm/Cu12 nm/ Py4.5 nm spin valve with cross-sectional area 40×120 nm^2, as the magnetization of the thin (free) magnetic layer is aligned parallel (at high negative currents) and antiparallel (at high positive currents) to the thicker magnetic layer by an applied current. Data are courtesy of Patrick Braganca and Dan Ralph. (After Braganza, P.M. et al., *Appl. Phys. Lett.* 87, 112507, 2005. With permission.)

moments was reported by Tsoi et al. in [27]. Here a STT device – point contact – was placed in a microwave cavity of a high-frequency high-field electron spin resonance (ESR) spectrometer. This arrangement allowed performing dc transport experiments, such as those described above, while the contact was irradiated with high-frequency microwaves. Such irradiation induces microwave currents in the contact and was shown to be equivalent to the case where microwave currents are fed to the contact via electric leads [89]. When the frequency of external microwaves matched the precession frequency excited by dc current, an additional dc voltage was detected across the contact. By detecting this voltage, while varying the external frequency, field, and applied current, the STT-driven frequency could be mapped as a function of these parameters. Rippard et al. [149] fed microwaves to a lithographically defined point contact via electric leads and reported the observation of a similar dc response.

Finally, the high-frequency dynamics of the free-layer magnetization can be measured directly by detecting high-frequency oscillations in voltage across a spin valve under dc bias. Here, the hard magnetic layer is fixed while the free layer is STT-driven into high-frequency precession relative to the hard layer. GMR results in a high-frequency modulation of the spin-valve resistance which, in turn, leads to a high-frequency component of the voltage across the spin valve traversed by a dc current. This voltage can be directly probed with a high-frequency spectrum analyzer as was demonstrated by Kiselev et al. [93] and by Rippard et al. [35]. Moreover, the voltage oscillations due to spin-torque-driven magnetic precession can be directly measured in the time domain using a sampling oscilloscope [94]; an example of the high-frequency oscillations in voltage across an exchange-biased IrMn/NiFe/Cu/NiFe spin valve is shown in Figure 7.19.

The above examples illustrate how broadband instrumentation for measuring voltage in GMR devices may provide important, often unique, information about high-frequency magnetic dynamics driven by spin-transfer torques. However, the detailed understanding of STT is still the subject of debate and requires new experimental techniques that are capable of probing magnetization dynamics on nanometer length scales and sub-nanosecond timescales. In principle, this can be accomplished by the use of synchrotron x-rays that were recently shown [98] to probe interfacial phenomena and directly image the time-resolved response of magnetic nanostructures to sub-nanosecond magnetic field pulses (Oersted switching) and spin-polarized current pulses (STT switching). Figure 7.20 shows scanning transmission x-ray microscopy [150] images of in-plane components of magnetization M – M_x in panel (a) and M_y in panel (b) – in the free CoFe layer (indicated by an ellipse) of a $100 \times 150 \, nm^2$ magnetic nanopillar. The images were obtained by scanning a focused (diameter $\sim 30 \, nm$) circularly polarized x-ray beam across the CoFe layer, with the photon energy tuned to the characteristic Co L3 resonance to provide magnetic contrast through the x-ray magnetic circular dichroism (XMCD) effect [151], and by monitoring transmission of the x-rays as a function of the position x, y with a fast avalanche detector. The M-vector field of the free layer can be reconstructed from the measured

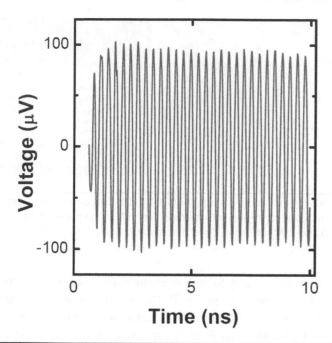

FIGURE 7.19   Oscillatory voltage generated by the precessional motion of the free magnet in IrMn8 nm/NiFe4nm/Cu8nm/NiFe4nm nanopillar, in response to a 335-mV dc voltage step applied to the device at $B = 630$ Oe. Data courtesy of Ilya Krivorotov and Dan Ralph. (After Krivorotov, I.N. et al., *Science* 307, 228, 2005. With permission.)

M_x and M_y components as illustrated in Figure 7.20c and the ultra-fast x-ray microscopy technique provided a means to monitor this field as a function of time with ~100 ps resolution. The spatial resolution of the technique is set by the spot size of the x-ray beam (~30 nm) and currently limits its application to spintronic devices >100 nm in size [98]. Potentially, however, technical development of the ultra-fast x-ray microscopy may lead to an ultimate technique for STT studies which can probe the M-vector field on the nanometer length scale with picosecond time resolution.

7.9 CURRENT-INDUCED DOMAIN-WALL MOTION

Yet another manifestation of STT in metallic ferromagnets is a motion of magnetic domain walls traversed by an electric current. The original prediction of the effect dates back to 1978 when Luc Berger predicted that a spin-polarized current should apply a torque to a magnetic domain wall [9]. In a series of theoretical [9, 10, 12, 13] and experimental [11, 14, 15] works, Berger set the groundwork for current-induced domain-wall motion (CIDWM) which is now documented in materials ranging from superlattices with perpendicular anisotropy [127] to magnetic semiconductors [121]. However, the most widely studied materials by far have been metallic ferromagnets, including Py ($Ni_{81}Fe_{19}$), CoFe, Co, because of their decades-long ubiquity in magnetic storage technology.

The CIDWM effect can be qualitatively understood on the basis of the following arguments. Consider an electric current flowing between two

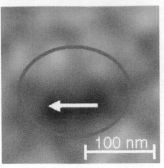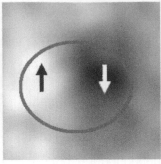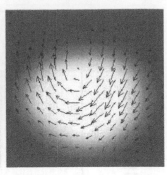

FIGURE 7.20 Scanning transmission x-ray microscopy images of M_x (a) and M_y (b) components of magnetization M combined into the vector field (c), which represents the direction of M in the plane of the CoFe free layer. Data courtesy of Yves Acremann and Joachim Stöhr. (After Acremann, Y. et al., *Phys. Rev. Lett.* 96, 217202, 2006. With permission.)

magnetic domains (A and B) with opposite magnetizations and, thus, traversing a 180° magnetic domain wall. The situation is somewhat similar to the case of a single N/F interface discussed above in conjunction with STT in magnetic multilayers. While in domain A, spins of conduction electrons are preferentially aligned with the magnetic moment of A. But once into B, the spins reverse to align with the magnetic moment of B. In reversing the electron spins, magnetic moments in the domain wall experience a torque associated with the change in angular momentum that occurs from the rotation of spins of transport electrons. This torque can move the domain wall in the direction of the electron flow – CIDWM.

Models which describe the mechanisms responsible for CIDWM fall into two classes, termed "adiabatic" [9, 152, 153] and "non-adiabatic" [10, 152, 154, 155], referring essentially to the way electron spins traverse the magnetization texture of domain walls. The adiabatic torque arises when the spins of transport electrons follow the underlying magnetic landscape adiabatically. On the other hand, when spatial variations in magnetization are too rapid to be followed adiabatically by the transport spins – non-adiabatic contributions appear. The resulting magnetodynamics of domain walls can be resolved by introducing these current-induced torques into the equation of motion for magnetization (see, e.g., Equation 7.1). Most analytical work performs this operation within the framework of one-dimensional (1D) domain walls schematically illustrated in Figure 7.21a. Here we have chosen a Néel type domain wall appropriate for thin films with a large out-of-plane demagnetizing factor that promotes the in-plane orientation of magnetic moments. The in-plane orientation of each spin is defined by angle θ which varies from 0 in one magnetic domain to π in another domain over a characteristic domain-wall width Δ [156]. Experimental realization of such thin-film stripe geometry provides a workbench for testing CIDWM models.

The early experiments on CIDWM by Berger and coworkers [11, 14, 15] used ~3 mm wide and 14–86 nm thick stripes of NiFe and the Faraday effect to observe magnetic domain walls. Relatively large cross-sections of such samples (~10^{-10} m²) required very large currents (up to 45 A) in order to achieve high enough current densities (~10^{11} A/m^2) for CIDWM; and even though the current was pulsed with a pulse duration of ~2 μs most of the samples were destroyed through electrical breakdown or local heating. Those samples which survived the pulses exhibited displacements of domain walls on the scale from 1–10 μm. For films thicker than 40 nm, different walls were observed being moved by the current in different directions as expected for the effect of the Oersted field produced by the current. However, all thinner (14–30 nm) samples exhibited domain-wall displacements in the direction of the electron current flow, which is consistent with the STT mechanism of CIDWM most pronounced in thinner samples. Gan et al. [59] have used 100-nm-thick NiFe stripes with cross-sections almost two orders of magnitude smaller than those used by Berger and coworkers, but still needed currents of the order of 1A to produce μm-scale displacements of domain walls observed with a magnetic force microscope (MFM).

With advances in nanofabrication techniques, magnetic stripes with ~10^{-15} m^2 cross-sections can now be routinely produced where currents of only a few mA and below can trigger CIDWM. Moreover, unlike the initial CIDWM observations of multiple domain walls, which are the consequence of relatively large samples, CIDWM of individual magnetic domain walls is now possible. Most recent experiments [128–147] study domain-wall propagation in thin-film magnetic nanostripes prepared by electron beam lithography (EBL) or focused ion beam (FIB) patterning. Such nanostripes are typically connected to extended magnetic elements as shown in Figures 7.14 and 7.15. These extended areas are used for domain-wall nucleation and subsequent injection into the stripe where it can propagate under the influence of applied current, or magnetic field, or both. The 1D wall geometry shown in Figure 7.21a is an appropriate starting point for analysis of most experiments.

The 1D domain wall is described by two collective coordinates – the wall displacement q and its canting angle ψ (see Figure 7.21b), while (in-plane, out-of-plane) orientation of individual spins is denoted (θ, φ). The motion of the wall requires a torque on θ to align the wall spins along the nanostripe and drive q and ψ motion. For 1D domain wall the equation of motion (see Equation 7.1) reduces to the following equations [152, 155]:

$$\dot{q} = (2\pi M_S)\gamma \ \Delta \sin 2\psi + \alpha \ \Delta \dot{\psi} + \eta u \tag{7.2a}$$

$$\dot{\psi} = \gamma \ \bar{H}_a - (\alpha/\Delta) \ \dot{q} + (\beta u)/\Delta \tag{7.2b}$$

Here current density j is included via $u = -(g\mu_B p/2eM_S)j$, p polarization, and M_S magnetization; Δ is domain-wall width. The last terms of Equations 7.2a and 7.2b describe adiabatic and non-adiabatic interactions, respectively.

Without current (without the last terms in Equations 7.2a and b) the domain wall is driven solely by an applied magnetic field (H_a). If the field is applied along the nanostripe, the field-induced torque $\gamma \ \bar{M} \times \bar{H}_a$ causes the domain-wall spins to cant out of the film plane, as shown in Figure 7.21b, but cannot directly drive the wall motion (cannot change θ). Instead,

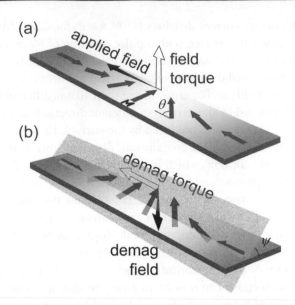

FIGURE 7.21   Schematic of a 1D domain wall in a ferromagnetic nanostripe and the torques involved in riving its motion.

a demagnetizing field H_d, which develops in the opposite direction as the out-of-plane component of M_S, will produce the demagnetizing torque $\gamma \, \vec{M} \times \vec{H}_d$ that drives θ and consequent wall motion $\dot{q}$. For small applied fields, a simple translational motion of the wall occurs with a constant canting ψ. When applied magnetic field increases so does the demagnetizing torque, the canting angle ψ, and the resulting domain-wall velocity $\dot{q}$. But the increase cannot continue indefinitely. At $\psi = \pi/4$, the demagnetizing torque, and thus $\dot{q}$, peaks. If H_a drives ψ past this limit, ψ can no longer remain stationary – a transition known as Walker breakdown [157] occurs. Here ψ advances continually and the demagnetizing torque averages to zero as it changes direction with each quarter period of ψ rotation, i.e., cannot move the wall forward. As ψ rotation becomes more and more rapid with increasing H_a, a small net damping torque $\alpha \, \vec{M} \times \dot{\vec{M}}$ cants the wall spins toward H_a and drives the wall forward in place of the demagnetizing torque. The Walker breakdown of individual domain walls was observed by Beach et al. [126] in NiFe nanostripes using high-bandwidth Kerr polarimetry (see Figure 7.22).

If we now turn the effect of the current on (last terms in Equations 7.2a, and b), one can distinguish two limiting cases – below and far above the Walker breakdown. Below the breakdown ψ is stationary ($\dot{\psi} = 0$) and only the non-adiabatic term can drive wall motion. From Equation 7.2a the wall velocity is

$$\dot{q} = \left(\gamma \Delta / \alpha \right) H_a + \left(\beta u \right) / \alpha \tag{7.3}$$

and the sole effect of adiabatic torque is to shift the steady-state of ψ and vary the Walker breakdown field $H_W = 2\pi \alpha M_S + \left(\alpha \eta - \beta \right) u / \gamma \Delta$.

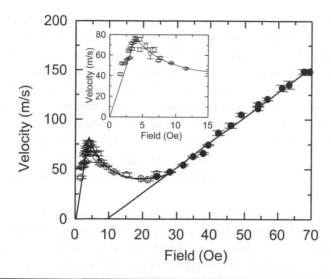

FIGURE 7.22 Average domain-wall velocity in a 600-nm-wide and 20-nm-thick NiFe nanostripe versus the applied magnetic field. The inset shows the detail around the velocity peak. Straight solid lines are linear fits to the data below and far above the Walker breakdown at $H_W \cong 4$ Oe; the curved line is a visual guide. (After Beach, G.S.D. et al., *Nat. Mater.* 47, 741, 2005. With permission.)

Far above the breakdown, it is only the adiabatic torque that can drive the wall motion. For $\psi \gg 0$ and, as a result time-averaged $\langle \sin 2\psi \rangle = 0$, the wall velocity is given by

$$\dot{q} = \gamma \Delta \alpha \, H_a + \eta u \qquad (7.4)$$

In experiments, adiabatic and non-adiabatic interactions may thus be probed independently simply by using an applied magnetic field to select between the regimes of Equations 7.3 or 7.4. Such experiments were performed in permalloy nanowires by Beach et al. [130] using ultra-fast Kerr polarimetry and by Hayashi et al. [158] using real-time resistance measurements. An applied electric current causes the linear parts, at low and high H_a, of the curve in Figure 7.22 to shift vertically as expected from Equations 7.3 and 7.4. This indicates that the equation of motion (Equation 7.2) is capable of describing the effects of field and current at least qualitatively. However, a detailed analysis of the data in Refs. [126] and [130] suggests that this 1D model has difficulties in accounting for quantitative characteristics of both field- and current-driven domain-wall dynamics on an equal footing [159]. Micromagnetic simulations that involve more complex domain-wall textures, e.g., vortex walls, provide a better agreement with experiments and indicate that meaningful comparison between experiment and theory will require models that fully account for realistic, time-variant domain-wall structures.

7.10 STT IN ANTIFERROMAGNETS

The phenomenon of STT is not limited to ferromagnetic materials. In 2006, Núñez et al. [160] predicted that a similar effect could occur in antiferromagnetic (AFM) systems. Such an AFM-STT effect would be indispensable to manipulate antiferromagnetic elements in all-AFM spintronic applications [161]. For example, a simple switching of an antiferromagnetic order parameter could represent the writing operation in a magnetic random access memory (MRAM) based on AFM materials. Such an AFM-MRAM would be insensitive to external magnetic field perturbations and eliminate a cross-talk between its elements thanks to the stray-field-free nature of AFMs. More importantly, it could offer ultra-fast writing schemes thanks to the very high (THz range) natural frequencies of precession in AFMs. The latter may ultimately open the AFM spintronics to a wide range of THz applications.

The first theoretical model of antiferromagnetic spin-transfer torque [160] was soon followed by experiments [162–166]. The experiments were set to probe the effects of spin-polarized currents on the AFM order parameter: a simple ferromagnet/antiferromagnet (F/AFM) bilayer was used as a natural test system where an electrical current can be the first spin-polarized, by being driven across the F layer, and then injected directly into the AFM. It is, however, a challenge to detect the resulting effect of the polarized current on the AFM's magnetic order. Wei et al. [162] used the exchange-bias phenomenon at the F/AFM interface to inquire the information about AFM order near the interface. Indeed, as the exchange-bias phenomenon is known to be associated with the interfacial AFM moments, the observed in [162] current-induced variation of the exchange-bias field can be taken as the first evidence of AFM-STT effect. This experiment exploited the action of a spin-polarized current on AFM in a $F_{free}/N/F_{pinned}/AFM$ spin valve (AFM=FeMn). Here, F_{free}/N served only as a probe of the exchange bias at the F_{pinned}/AFM interface with (F_{free}/N) layer compositions (F_{free} = CoFe/N = Cu) and thicknesses (10 nm/10 nm) designed on purpose to minimize the conventional (ferromagnetic) STT effect on F_{free} and not to perturb the F_{pinned}/AFM interactions. Point contacts were used to achieve high enough densities of the current injected into the spin valve and propagated across the spin-valve stack directly into a back copper electrode to ensure a close to perpendicular-to-plane current flow.

Figure 7.23 shows the key experimental results of Wei et al. [162]. The spin-valve resistance is shown in Figure 7.23a as a function of the applied magnetic field for a fixed value of the bias dc current (10 mA); here black (gray) trace shows a MR sweep from high positive (negative) field to high negative (positive) fields. The shape of MR is typical for spin valves: starting from high positive field, the resistance is constant at a minimum value (magnetizations of F_{free} and F_{pinned} are parallel), rises to a maximum when the magnetization of F_{free} switches near zero field (leading to antiparallel alignment of F_{free} and F_{pinned}), and then decreases to its minimum value beyond the exchange-bias field H_B at which the magnetization of F_{pinned} is finally reversed. The reversed (gray) sweep exhibits a similar behavior. Figure 7.23b shows how the exchange-bias field H_B changes with the applied dc bias; these data were obtained from down MR sweeps (as the one in Figure 7.23a) recorded at different bias currents. The observed

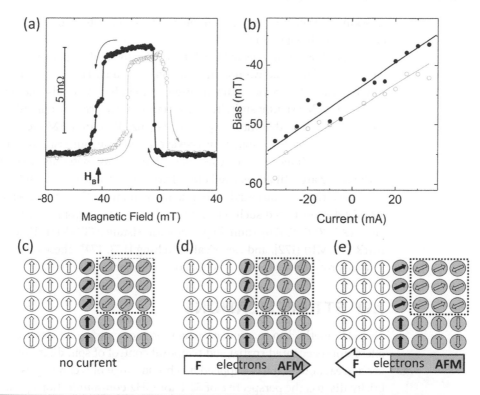

FIGURE 7.23 STT in antiferromagnets. (a) Spin-valve magnetoresistance at a fixed bias current (10 mA). (b) Exchange-bias field H_B vs applied bias current I. Solid (open) symbols show H_B inferred from the magnetoresistance data at 50% (30%) level assuming 0% for minimum resistance and 100% for maximum resistance (see text for details). Linear fits indicate the trend of H_B shift with the current. (c–e) An intuitive picture of AFM-STT at an F/AFM interface. The current gets spin-polarized in F (white/left). The transmitted (reflected) polarized current transfers torque to the uncompensated AFM spins (black) which rotate away from (toward) the F spins (gray/right). It influences the F magnetization reversal and results in a reduction (enhancement) of H_B.

variation of H_B with the applied bias cannot be explained by the standard (ferromagnetic) STT interactions between F_{free} and F_{pinned} – their absence was confirmed by the bias-independent reversal of F_{free} near zero field [162] – but can be associated with the AFM-STT effect. Similar behavior was also found with other AFM materials, e.g. IrMn alloys [167, 168].

The intuitive picture describing the consequences of STT on AFM is shown in Figure 7.23c through e. The cartoon shows an interface between F (white/left) and AFM (gray/right). The surface layer of AFM contains uncompensated magnetic moments (black). A fraction of these uncompensated moments are pinned, thereby inducing an energy barrier for the F-reversal; they do not reverse when F reverses and are responsible for the existence of exchange bias. Electrons flowing from F into AFM (Figure 7.23d) generate spin-transfer torques which alter its magnetic configuration. These torques tend to favor parallel alignment of moments at the F/AFM interface and, thus, increase H_B. Electrons flowing in the opposite direction (Figure 7.23e) will tend to have the opposite effect (decrease H_B). Here the current-induced changes of AFM order are present only as long as the current is present too.

As soon as the current is removed the AFM goes back to its original state just like in an exchange-spring.

It is thus clearly established theoretically and experimentally that AFMs can play an active role in spintronic circuits. However, both theory and experiment were originally focused on metallic materials that limit the choice of AFMs significantly. The materials that theorists had in mind [160] and those used in the initial experiments [162–166] are AFM metals either similar to Cr with its spin-density-wave antiferromagnetism, or the rock salt structure intermetallics used as exchange-bias materials; virtually all of the latter are Mn alloys like FeMn, IrMn, etc. A much larger selection of AFM materials can be found among semiconductors. This type of AFMs has already demonstrated such interesting effects as anisotropic magnetoresistance (AMR) [169, 170], tunneling magnetoresistance (TMR) [171], tunneling AMR (TAMR) [172], and electrical switching [173, 174]. These observations further prove the immense potential of AFMs for spintronics.

7.11 STT APPLICATIONS

The STT method to manipulate magnetic moments by an electric current offers unprecedented spatial and temporal control of spin distributions and has a great potential to be applied in a broad range of technologies. Here we briefly discuss the perspective of STT for GHz communication applications and magnetic recording technology. More details on some of the applications can be found in Section VII of this book.

The STT application in high-frequency technologies is based on the spin-transfer-induced precession of spins. In the previous section, we have seen that the precession of magnetization in GMR devices can convert a dc current input into an ac voltage output. The frequency of this output can be tuned from a few GHz to > 100 GHz by changing the applied magnetic field and/or the dc current, effectively resulting in a current controlled oscillator to be used in practical microwave circuits. Hence, the STT effect in GMR structures provides a means to engineer a nanoscale high-frequency oscillator powered and tuned by dc current. Such an oscillator could potentially have frequency characteristics spanning more than 100 GHz and perhaps into THz range. Linewidths as narrow as 2 MHz were demonstrated [175] leading to quality factors over 18000. Potential applications for such high-frequency sources include integrated transceivers for wireless and wired applications, and chip-to-chip, as well as on-chip communications both wireless and wired. For the latter, logic circuits with a spin-wave bus were proposed [176, 177] as an interface between electronic circuits and integrated spintronics circuits. Here spin waves are used for information transmission and processing and the STT effect can provide a means for efficient spin-wave generation on the nanoscale.

In high-speed high-density magnetic recording technology, STT could replace the Oersted field currently used for writing magnetic bits in a storage media, thus leading to smaller and faster magnetic memory. For instance, the bit cell of a conventional magnetic random access memory (MRAM) is programmed to a "1" or "0" by switching between the parallel and antiparallel states of a GMR-like storage element, e.g., spin valve. The first-generation MRAM utilizes magnetic tunnel junctions (MTJ) as storage elements because of their higher magnetoresistance ratios and impedance-matching constraints. However, in scaling MRAM to small dimensions, the same constraints are expected to drive a transition from MTJs to fully metallic spin-valve storage elements. The switching between "1" and "0" states (writing) relies on the magnetic reversal of the MTJ's free layer achieved by passing electric currents down the 'bit' and 'write' lines that generate a sufficiently strong magnetic field at their intersection, i.e., for a given MTJ. However, as the spatial decay of this Oersted field is rather slow ($\sim 1/r^2$), it may affect neighboring cells. This makes scaling of conventional MRAM to smaller dimensions challenging. So-called STT-MRAM removes the above constraints on scalability. Here the switching of the free-layer magnetic moment is achieved by STT switching when a high-density electric current is driven directly through the storage element (MTJ or spin valve). The writing is thus performed at high current levels, while the reading (measuring the resistance of element) is still done at low currents. Since STT-MRAM eliminates the need for 'write' and 'bypass' lines, a more compact memory can be realized. In 2016, Everspin Technologies announced [178] the production of the first STT-MRAM product (256 Mb) that breaks the record for the highest density commercial MRAM currently available in the market and plans for further density increases with a 1Gb product later the same year.

Finally, CIDWM was proposed as the basis for a new type of magnetic memory called "racetrack" [135]. In contrast to today's hard disc drives (HDD), which rely on the spinning motion of a disk to move its magnetic regions where the data is stored past a read head, the racetrack memory exploits the idea of moving magnetically stored data electronically. The racetrack is a ferromagnetic nanowire, with data encoded as a pattern of magnetic domains along the nanowires. An electric current can drive the entire data-pattern in the racetrack past read and write elements. If a tunnel junction to the racetrack is used as a reading element, the reading operation is achieved simply by measuring the resistance of the junction which depends on the orientation of magnetic domains in the junction area. The same junction can be used to inject a spin-polarized current into the racetrack locally, thus mediating a local reversal of magnetization via STT – the data writing. The reversal (writing) can also be triggered by magnetic fields applied locally. A high-density 3D array of the racetracks would result in a memory chip with storage density higher than that in solid-state memory devices like Flash RAM and similar to conventional HDDs but potentially may have much higher read/write performance than HDD.

What is the future of STT in spintronic applications? We have seen that a number of new STT-based spintronic devices have been proposed. In the foreseeable future, however, a great deal of fundamental work remains to be done before we see the commercial applications of these devices. For the memory industry, this may lead to a universal memory which would combine the cost benefits of DRAM, the speed of SRAM, and the non-volatility of Flash RAM.

Potentially all logic operations on a chip can be carried out by manipulating spins in metallic systems instead of manipulating charges in semiconductor transistors, as in conventional microchips, and, moreover, could be combined on-chip with a universal memory. This would result in a new, scalable, radiation resistant, and more power efficient computers and electronic devices on which we rely in almost every aspect of our everyday life. If STT is the method of choice to control and manipulate spins in such devices, its future will be very bright.

REFERENCES

1. M. N. Baibich, J. M. Broto, A. Fert et al., Giant magnetoresistance of (001)Fe/(001)Cr magnetic superlattices, *Phys. Rev. Lett.* **61**, 2472–2475 (1988).
2. G. Binasch, P. Grünberg, F. Saurenbach, and W. Zinn, Enhanced magneto-resistance in layered magnetic structures with antiferromagnetic interlayer exchange, *Phys. Rev. B* **39**, 4828–4830 (1989).
3. L. Berger, Emission of spin waves by a magnetic multilayer traversed by a current, *Phys. Rev. B* **54**, 9353–9358 (1996).
4. J. C. Slonczewski, Current-driven excitation of magnetic multilayers, *J. Magn. Magn. Mater.* **159**, L1–L7 (1996).
5. M. Tsoi, A. G. M. Jansen, J. Bass et al., Excitation of a magnetic multilayer by an electric current, *Phys. Rev. Lett.* **80**, 4281–4284 (1998).
6. J. Z. Sun, Current-driven magnetic switching in manganite trilayer junctions, *J. Magn. Magn. Mater.* **202**, 157–62 (1999).
7. E. B. Myers, D. C. Ralph, J. A. Katine, R. N. Louie, and R. A. Buhrman, Current-induced switching of domains in magnetic multilayer devices, *Science* **285**, 867–880 (1999).
8. J.-E. Wegrowe, D. Kelly, Y. Jaccard, P. Guittienne, and J.-P. Ansermet, Current-induced magnetization reversal in magnetic nanowires, *Europhys. Lett.* **45**, 626–632 (1999).
9. L. Berger, Low-field magnetoresistance and domain drag in ferromagnets, *J. Appl. Phys.* **49**, 2156–2161 (1978).
10. L. Berger, Exchange interaction between domain wall and current in very thin metallic films, *J. Appl. Phys.* **55**, 1954–1956 (1984).
11. P. P. Freitas and L. Berger, Observation of s-d exchange force between domain-walls and electric-current in very thin permalloy-films, *J. Appl. Phys.* **57**, 1266–1269 (1985).
12. L. Berger, Possible existence of Josephson effect in ferromagnets, *Phys. Rev. B* **33**, 1572–1578 (1986).
13. L. Berger, Exchange interaction between electric-current and magnetic domain-wall containing bloch lines, *J. Appl. Phys.* **63**, 1663–1669 (1988).
14. C.-Y. Hung and L. Berger, Exchange forces between domain-wall and electric-current in permalloy-films of variable thickness, *J. Appl. Phys.* **63**, 4276–4278 (1988).
15. C.-Y. Hung, L. Berger and C. Y. Shih, Observation of a current-induced force on Bloch lines in Ni-Fe thin-films, *J. Appl. Phys.* **67**, 5941–5943 (1990).
16. R. Weil and R. G. Barradas, *Electrocrystallization*, The Electrochemical Society, Pennington, 1981.
17. D. S. Lashmore and M. P. Dariel, Electrodeposited Cu-Ni textured superlattices, *J. Electrochem. Soc.* **135**, 1218–1221 (1988).
18. M. Ohring, *The Materials Science of Thin Films*, Academic, Boston, MA, 1992.
19. J. M. Slaughter, W. P. Pratt, Jr., and P. A. Schroeder, Fabrication of layered metallic systems for perpendicular resistance measurements, *Rev. Sci. Instrum.* **60**, 127–131 (1989).
20. W. P. Pratt, Jr., S. F. Lee, J. M. Slaughter, R. Loloee, P. A. Schroeder, and J. Bass, Perpendicular giant magnetoresistances of Ag/Co multilayers, *Phys. Rev. Lett.* **66**, 3060–3063 (1991).

21. B. Dieny, V. S. Speriosu, S. Metin et al., Magnetotransport properties of magnetically soft spin-valve structures, *J. Appl. Phys.* **69**, 4774–4779 (1991).
22. L. Piraux, J. M. George, J. F. Despres et al., Giant magnetoresistance in magnetic multilayered nanowires, *Appl. Phys. Lett.* **65**, 2484–2486 (1994).
23. A. Blondel, J. P. Meier, B. Doudin, and J.-P. Ansermet, Giant magnetoresistance of nanowires of multilayers, *Appl. Phys. Lett.* **65**, 3019–3021 (1994).
24. M. Tsoi, A. G. M. Jansen, and J. Bass, Search for point-contact giant magnetoresistance in Co/Cu multilayers, *J. Appl. Phys.* **81**, 5530–5532 (1997).
25. B. Dieny, V. S. Speriosu, S. S. P. Parkin, B. A. Gurney, D. R. Wilhoit, and D. Mauri, Giant magnetoresistance in soft ferromagnetic multilayers, *Phys. Rev. B* **43**, 1297–1300 (1991).
26. X. Waintal, E. B. Myers, P. W. Brouwer, and D. C. Ralph, Role of spin-dependent interface scattering in generating current-induced torques in magnetic multilayers, *Phys. Rev. B* **62**, 12317–12327 (2000).
27. M. Tsoi, A. G. M. Jansen, J. Bass, W.-C. Chiang, V. Tsoi, and P. Wyder, Generation and detection of phase-coherent current-driven magnons in magnetic multilayers, *Nature* **406**, 46–48 (2000).
28. W. H. Rippard, M. R. Pufall, T. J. Silva, Quantitative studies of spin-momentum-transfer-induced excitations in Co/Cu multilayer films using point-contact spectroscopy, *Appl. Phys. Lett.* **82**, 1260–1262 (2003).
29. Y. Ji, C. L. Chien, and M. D. Stiles, Current-induced spin-wave excitations in a single ferromagnetic layer, *Phys. Rev. Lett.* **90**, 106601 (2003).
30. I. K. Yanson, Yu. G. Naidyuk, D. L. Bashlakov et al., Spectroscopy of phonons and spin torques in magnetic point contacts, *Phys. Rev. Lett.* **95**, 186602 (2004).
31. O. P. Balkashin, V. V. Fisun, I. K. Yanson, L. Yu. Triputen, A. Konovalenko, V. Korenivski, Spin dynamics in point contacts to single ferromagnetic films, *Phys. Rev. B* **79**, 092419 (2009).
32. Z. Wei, A. Sharma, A. S. Nunez et al., Changing exchange bias in spin valves with an electric current, *Phys. Rev. Lett.* **98**, 116603 (2007).
33. Y. Luo, M. Esseling, M. Münzenberg, and K. Samwer, A novel spin transfer torque effect in Ag_2Co granular films, *New J. Phys.* **9**, 329 (2007).
34. X. J. Wang, H. Zou, and Y. Ji, Spin transfer torque switching of cobalt nanoparticles, *Appl. Phys. Lett.* **93**, 162501 (2008).
35. W. H. Rippard, M. R. Pufall, S. Kaka, S. E. Russek, and T. J. Silva, Direct-current induced dynamics in $Co_{90}Fe_{10}/Ni_{80}Fe_{20}$ point contacts, *Phys. Rev. Lett.* **92**, 027201 (2004).
36. S. Kaka, M. R. Pufall, W. H. Rippard, T. J. Silva, S. E. Russek, and J. A. Katine, Mutual phase-locking of microwave spin torque nano-oscillators, *Nature* **437**, 389–392 (2005).
37. F. B. Mancoff, N. D. Rizzo, B. N. Engel, and S. Tehrani, Phase-locking in double-point-contact spin-transfer devices, *Nature* **437**, 393–395 (2005).
38. O. Ozatay, N. C. Emley, P. M. Braganca et al., Spin transfer by nonuniform current injection into a nanomagnet, *Appl. Phys. Lett.* **88**, 202502 (2006).
39. Q. Mistral, M. van Kampen, G. Hrkac et al., Current-driven vortex oscillations in metallic nanocontacts, *Phys. Rev. Lett.* **100**, 257201 (2008).
40. S. Bonetti, P. Muduli, F. Mancoff, and J. Akerman, Spin torque oscillator frequency versus magnetic field angle: The prospect of operation beyond 65 GHz, *Appl. Phys. Lett.* **94**, 102507 (2009).
41. J. A. Katine, F. J. Albert, R. A. Buhrman, E. B. Myers, and D. C. Ralph, Current-driven magnetization reversal and spin-wave excitations in Co/Cu/Co Pillars, *Phys. Rev. Lett.* **84**, 3149–3152 (2000).
42. J. Grollier, V. Cros, A. Hamzic et al., Spin-polarized current induced switching in Co/Cu/Co pillars, *Appl. Phys. Lett.* **78**, 3663–3665 (2001).
43. J. Z. Sun, D. J. Monsma, D. W. Abraham, M. J. Rooks, and R. H. Koch, Batch-fabricated spin-injection magnetic switches, *Appl. Phys. Lett.* **81**, 2202–2204 (2002).

44. B. Özyilmaz, A. D. Kent, D. Monsma, J. Z. Sun, M. J. Rooks, and R. H. Koch, Current-induced magnetization reversal in high magnetic fields in Co/Cu/Co nanopillars, *Phys. Rev. Lett.* **91**, 067203 (2003).

45. S. Urazhdin, N. O. Birge, W. P. Pratt, Jr., and J. Bass, Current-driven magnetic excitations in permalloy-based multilayer nanopillars, *Phys. Rev. Lett.* **91**, 146803 (2003).

46. M. Tsoi, J. Z. Sun, and S. S. P. Parkin, Current-driven excitations in symmetric magnetic nanopillars, *Phys. Rev. Lett.* **93**, 036602 (2004).

47. M. Covington, A. Rebei, G. J. Parker, and M. A. Seigler, Spin momentum transfer in current perpendicular to the plane spin valves, *Appl. Phys. Lett.* **84**, 3103–3105 (2004).

48. M. R. Pufall, W. H. Rippard, S. Kaka, S. E. Russek, and T. J. Silva, Large-angle, gigahertz-rate random telegraph switching induced by spin-momentum transfer, *Phys. Rev. B* **69**, 214409 (2004).

49. T. Kimura, Y. Otani, and J. Hamrle, Switching magnetization of a nanoscale ferromagnetic particle using nonlocal spin injection, *Phys. Rev. Lett.* **96**, 037201 (2006).

50. T. Yang, T. Kimura, and Y. Otani, Giant spin-accumulation signal and pure spin-current-induced reversible magnetization switching, *Nat. Phys.* **4**, 851–854 (2008).

51. S. Kasai, Y. Nakatani, K. Kobayashi, H. Kohno, and T. Ono, Current-driven resonant excitation of magnetic vortices, *Phys. Rev. Lett.* **97**, 107204 (2006).

52. K. Yamada, S. Kasai, Y. Nakatani et al., Electrical switching of the vortex core in a magnetic disk, *Nat. Mater.* **6**, 269–273 (2007).

53. J.-E. Wegrowe, D. Kelly, T. Truong, P. Guittienne, and J.-P. Ansermet, Magnetization reversal triggered by spin-polarized current in magnetic nanowires, *Europhys. Lett.* **56**, 748–754 (2001).

54. J.-E. Wegrowe, A. Fábián, P. Guittienne, X. Hoffer, D. Kelly, and J.-P. Ansermet, Exchange torque and spin transfer between spin polarized current and ferromagnetic layers, *Appl. Phys. Lett.* **80**, 3775–3777 (2002).

55. A. Fábián, C. Terrier, S. Serrano Guisan et al., Current-induced two-level fluctuations in pseudo-spin-valve (Co/Cu/Co) nanostructures, *Phys. Rev. Lett.* **91**, 257209 (2003).

56. N. Biziere, E. Murè, and J.-P. Ansermet, Microwave spin-torque excitation in a template-synthesized nanomagnet, *Phys. Rev. B* **79**, 012404 (2009).

57. L. Piraux, K. Renard, R. Guillemet et al., Template-grown NiFe/Cu/NiFe nanowires for spin transfer devices, *Nano Lett.* **7**, 2563–2567 (2007).

58. T. Blon, M. Mátéfi-Tempfli, S. Mátéfi-Tempfli et al., Spin momentum transfer effects observed in electrodeposited Co/Cu/Co nanowires, *J. Appl. Phys.* **102**, 103906 (2007).

59. L. Gan, S. H. Chung, K. H. Aschenbach, M. Dreyer, and R. D. Gomez, Pulsed-current-induced domain wall propagation in permalloy patterns observed using magnetic force microscope, *IEEE Trans. Magn.* **36**, 3047–3049 (2000).

60. J. Grollier, D. Lacour, V. Cros et al., Switching the magnetic configuration of a spin valve by current-induced domain wall motion, *J. Appl. Phys.* **92**, 4825–4827 (2002).

61. M. Klaui, C. A. F. Vaz, J. A. C. Blanda et al., Domain wall motion induced by spin polarized currents in ferromagnetic ring structures, *Appl. Phys. Lett.* **83**, 105–107 (2003).

62. M. Tsoi, R. E. Fontana, and S. S. P. Parkin, Magnetic domain wall motion triggered by an electric current, *Appl. Phys. Lett.* **83**, 2617–2619 (2003).

63. T. Kimura, Y. Otani, I. Yagi, K. Tsukagoshi, and Y. Aoyagi, Suppressed pinning field of a trapped domain wall due to direct current injection, *J. Appl. Phys.* **94**, 7226–7229 (2003).

64. J. Z. Sun, Spin-dependent transport in trilayer junctions of doped manginites, *Physica C* **350**, 215–226 (2001).

65. Y. Liu, Z. Zhang, J. Wang, P. P. Freitas, and J. L. Martins, Current-induced switching in low resistance magnetic tunnel junctions, *J. Appl. Phys.* **93**, 8385–8387 (2003).

66. Y. Huai, F. Albert, P. Nguyen, M. Pakala, and T. Valet, Observation of spin-transfer switching in deep submicron-sized and low-resistance magnetic tunnel junctions, *Appl. Phys. Lett.* **84**, 3118–3120 (2004).

67. G. D. Fuchs, N. C. Emley, I. N. Krivorotov et al., Spin-transfer effects in nanoscale magnetic tunnel junctions, *Appl. Phys. Lett.* **85**, 1205–1207 (2004).

68. A. A. Tulapurkar, Y. Suzuki, A. Fukushima et al., Spin-torque diode effect in magnetic tunnel junctions, *Nature* **438**, 339–342 (2005).

69. D. Chiba, Y. Sato, T. Kita, F. Matsukura, and H. Ohno, Current-driven magnetization reversal in a ferromagnetic semiconductor (Ga,Mn)As/GaAs/(Ga,Mn)As tunnel junction, *Phys. Rev. Lett.* **93**, 216602 (2004).

70. R. Moriya, K. Hamaya, A. Oiwa, and H. Munekata, Current-induced magnetization reversal in a (Ga,Mn)As-based magnetic tunnel junction, *Jpn. J. Appl. Phys. Part 2 - Lett. Express Lett.* **43**, L825–L827 (2004).

71. R. Moriya, K. Hamaya, A. Oiwa, H. Munekata, Magnetization reversal by electrical spin injection in ferromagnetic (Ga,Mn)As-based magnetic tunnel junctions, *J. Supercond.* **18**, 3–7 (2005).

72. P. N. Hai, S. Ohya, M. Tanaka, S. E. Barnes, S. Maekawa, Electromotive force and huge magnetoresistance in magnetic tunnel junctions, *Nature* **458**, 489–492 (2009).

73. M. Yamanouchi, D. Chiba, F. Matsukura, and H. Ohno, Current-induced domain-wall switching in a ferromagnetic semiconductor structure, *Nature* **428**, 539–542 (2004).

74. M. Yamanouchi, D. Chiba, F. Matsukura, T. Dietl, and H. Ohno, Velocity of domain-wall motion induced by electrical current in the ferromagnetic semiconductor (Ga,Mn)As, *Phys. Rev. Lett.* **96**, 096601 (2006).

75. M. Yamanouchi, J. Ieda, F. Matsukura, S. E. Barnes, S. Maekawa, H. Ohno, Universality classes for domain wall motion in the ferromagnetic semiconductor (Ga,Mn)As, *Science* **317**, 1726–1729 (2007).

76. Yu. V. Sharvin, A possible method for studying Fermi surfaces, *Sov. Phys. JETP* **21**, 655 (1965).

77. V. S. Tsoi, Focusing of electrons in metal by a transverse magnetic field, *JETP Lett.* **19**, 70–71 (1974).

78. I. K. Yanson, Nonlinear effects in electric-conductivity of point junctions and electron-phonon interaction in normal metals, *Zh. Eksp. Teor. Fiz.* **66**, 1035–1050 (1974).

79. A. G. M. Jansen, A. P. van Gelder, and P. Wyder, Point-contact spectroscopy in metals, *J. Phys. C* **13**, 6073–6118 (1980).

80. V. S. Tsoi, J. Bass, and P. Wyder, Studying conduction-electron/interface interactions using transverse electron focusing, *Rev. Mod. Phys.* **71**, 1641–1693 (1999).

81. Yu. G. Naidyuk and I. K. Yanson, *Point-Contact Spectroscopy*, Springer, New-York, 2004.

82. M. Tsoi, V. Tsoi, J. Bass, A. G. M. Jansen, and P. Wyder, Current-driven resonances in magnetic multilayers, *Phys. Rev. Lett.* **89**, 246803 (2002).

83. M. Tsoi, V. S. Tsoi, and P. Wyder, Generation of current-driven magnons in Co/Cu multilayers with antiferromagnetic alignment of adjacent Co layers, *Phys. Rev. B* **70**, 012405 (2004).

84. M. R. Pufall, W. H. Rippard, and T. J. Silva, Materials dependence of the spin-momentum transfer efficiency and critical current in ferromagnetic metal/Cu multilayers, *Appl. Phys. Lett.* **83**, 323–325 (2003).

85. J. Basset, A. Sharma, Z. Wei, J. Bass, and M. Tsoi, Towards antiferromagnetic metal spintronics, *Proc. SPIE* **7036**, 703605 (2008).

86. G. Wexler, Size effect and non-local Boltzmann transport equation in orifice and disk geometry, *Proc. Phys. Soc. London* **89**, 927 (1966).

87. B. Nikolic and P. B. Allen, Electron transport through a circular constriction, *Phys. Rev. B* **60**, 3963–3969 (1999).

88. J. C. Maxwell, *A Treatise on Electricity and Magnetism*, Clarendon Press, Oxford, 1873.

89. Z. Wei, Spin-transfer-torque effect in ferromagnets and antiferromagnets, Thesis, The University of Texas at Austin, Austin, TX, 2008.

90. M. Tsoi, Electromagnetic wave radiation by current-driven magnons in magnetic multilayers, *J. Appl. Phys.* **91**, 6801–6805 (2002).

91. K. S. Ralls, R. A. Buhrman, and R. C. Tiberio, Fabrication of thin-film metal nanobridges, *Appl. Phys. Lett.* **55**, 2459–2461 (1989).

92. B. O'Gorman and M. Tsoi, Fabrication of point contacts by FIB patterning, *Eur. Phys. J. Appl. Phys.* **49**, 10801 (2010).

93. S. I. Kiselev, J. C. Sankey, I. N. Krivorotov et al., Microwave oscillations of a nanomagnet driven by a spin-polarized current, *Nature* **425**, 380–383 (2003).

94. I. N. Krivorotov, N. C. Emley, J. C. Sankey, S. I. Kiselev, D. C. Ralph, and R. A. Buhrman, Time-domain measurements of nanomagnet dynamics driven by spin-transfer torques, *Science* **307**, 228–231 (2005).

95. T. Devolder, A. Tulapurkar, Y. Suzuki, C. Chappert, P. Crozat, and K. Yagami, Temperature study of the spin-transfer switching speed from dc to 100 ps, *J. Appl. Phys.* **98**, 053904 (2005).

96. N. Smith, J. A. Katine, J. R. Childress, and M. J. Carey, Angular dependence of spin torque critical currents for CPP-GMR read heads, *IEEE Trans. Magn.* **41**, 2935–2940 (2005).

97. S. Mangin, D. Ravelosona, J. A. Katine, M. J. Carey, B. D. Terris, and E. E. Fullerton, Current-induced magnetization reversal in nanopillars with perpendicular anisotropy, *Nat. Mater.* **5**, 210–215 (2006).

98. Y. Acremann, J. P. Strachan, V. Chembrolu et al., Time-resolved imaging of spin transfer switching: Beyond the macrospin concept, *Phys. Rev. Lett.* **96**, 217202 (2006).

99. J. C. Sankey, P. M. Braganca, A. G. F. Garcia, I. N. Krivorotov, R. A. Buhrman, and D. C. Ralph, Spin-transfer-driven ferromagnetic resonance of individual nanomagnets, *Phys. Rev. Lett.* **96**, 227601 (2006).

100. X. Jiang, L. Gao, J. Z. Sun, S. S. P. Parkin, Temperature dependence of current-induced magnetization switching in spin valves with a ferrimagnetic CoGd free layer, *Phys. Rev. Lett.* **97**, 217202 (2006).

101. T. Yang, A. Hirohata, M. Hara, T. Kimura, and Y. Otani, Temperature dependence of intrinsic switching current of a Co nanomagnet, *Appl. Phys. Lett.* **89**, 252505 (2006).

102. O. Boulle, V. Cros, J. Grollier et al., Shaped angular dependence of the spin-transfer torque and microwave generation without magnetic field, *Nat. Phys.* **3**, 492–497 (2007).

103. V. S. Pribiag, I. N. Krivorotov, G. D. Fuchs et al., Magnetic vortex oscillator driven by d.c. spin-polarized current, *Nat. Phys.* **3**, 498–503 (2007).

104. D. Houssameddine, U. Ebels, B. Delaët et al., Spin-torque oscillator using a perpendicular polarizer and a planar free layer, *Nat. Mater.* **6**, 447–453 (2007).

105. O. Ozatay, P. G. Gowtham, K. W. Tan et al., Sidewall oxide effects on spin-torque- and magnetic-field-induced reversal characteristics of thin-film nanomagnets, *Nat. Mater.* **7**, 567–573 (2008).

106. S. Urazhdin and S. Button, Effect of spin diffusion on spin torque in magnetic nanopillars, *Phys. Rev. B* **78**, 172403 (2008).

107. G. Finocchio, O. Ozatay, L. Torres, R. A. Buhrman, D. C. Ralph, and B. Azzerboni, Spin-torque-induced rotational dynamics of a magnetic vortex dipole, *Phys. Rev. B* **78**, 174408 (2008).

108. S. Garzon, L. Ye, R. A. Webb, T. M. Crawford, M. Covington, and S. Kaka, Coherent control of nanomagnet dynamics via ultrafast spin torque pulses, *Phys. Rev. B* **78**, 180401R (2008).

109. S. H. Florez, J. A. Katine, M. Carey, L. Folks, O. Ozatay, and B. D. Terris, Effects of radio-frequency current on spin-transfer-torque-induced dynamics, *Phys. Rev. B* **78**, 184403 (2008).

110. S. Mangin, Y. Henry, D. Ravelosona, J. A. Katine, and E. E. Fullerton, Reducing the critical current for spin-transfer switching of perpendicularly magnetized nanomagnets, *Appl. Phys. Lett.* **94**, 012502 (2009).

111. T. Seki, H. Tomita, A. A. Tulapurkar, M. Shiraishi, T. Shinjo, and Y. Suzuki, Spin-transfer-torque-induced ferromagnetic resonance for Fe/Cr/Fe layers with an antiferromagnetic coupling field, *Appl. Phys. Lett.* **94**, 212505 (2009).

112. B. O'Gorman, S. Dietze, and M. Tsoi, Current-sweep-rate dependence of spin-torque driven dynamics in magnetic nanopillars, *J. Appl. Phys.* **105**, 07D111 (2009).

113. B. Özyilmaz, G. Richter, N. Müsgens et al., Focused-ion-beam milling based nanostencil mask fabrication for spin transfer torque studies, *J. Appl. Phys.* **101**, 063920 (2007).

114. A. Parge, T. Niermann, M. Seibt, and M. Münzenberg, Nanofabrication of spin-transfer torque devices by a polymethylmethacrylate mask one step process: Giant magnetoresistance versus single layer devices, *J. Appl. Phys.* **101**, 104302 (2007).

115. M. C. Wu, A. Aziz, D. Morecroft et al., Spin transfer switching and low-field precession in exchange-biased spin valve nanopillars, *Appl. Phys. Lett.* **92**, 142501 (2008).

116. M. Johnson and R. H. Silsbee, Interfacial charge-spin coupling: injection and detection of spin magnetization in metals, *Phys. Rev. Lett.* **55**, 1790–1793 (1985).

117. F. J. Jedema, A. T. Filip, and B. J. van Wees, Electrical spin injection and accumulation at room temperature in an all-metal mesoscopic spin valve, *Nature* **410**, 345–348 (2001).

118. C. R. Martin, Nanomaterials: A membrane-based synthetic approach, *Science* **266**, 1961–1966 (1994).

119. G. E. Possin, A method for forming very small diameter wires, *Rev. Sci. Instrum.* **41**, 772–774 (1970).

120. J.-E. Wegrowe, D. Kelly, X. Hoffer, P. Guittienne, and J.-P. Ansermet, Tailoring anisotropic magnetoresistance and giant magnetoresistance hysteresis loops with spin-polarized current injection, *J. Appl. Phys.* **89**, 7127–7129 (2001).

121. M. Yamanouchi, D. Chiba, F. Matsukura, and H. Ohno, Current-induced domain-wall switching in a ferromagnetic semiconductor structure, *Nature* **428**, 539–542 (2004).

122. E. Saitoh, H. Miyajima, T. Yamaoka, and G. Tatara, Current-induced resonance and mass determination of a single magnetic domain wall, *Nature* **432**, 203–206 (2004).

123. A. Yamaguchi, T. Ono, S. Nasu, K. Miyake, K. Mibu, and T. Shinjo, Real-space observation of current-driven domain wall motion in submicron magnetic wires, *Phys. Rev. Lett.* **92**, 077205 (2004).

124. N. Vernier, D. A. Allwood, D. Atkinson, M. D. Cooke, and R. P. Cowburn, Domain wall propagation in magnetic nanowires by spin-polarized current injection, *Europhys. Lett.* **65**, 526–532 (2004).

125. C. K. Lim, T. Devolder, C. Chappert et al., Domain wall displacement induced by subnanosecond pulsed current, *Appl. Phys. Lett.* **84**, 2820–2822 (2004).

126. G. S. D. Beach, C. Nistor, C. Knutson, M. Tsoi, J. L. Erskine, Dynamics of field-driven domain-wall propagation in ferromagnetic nanowires, *Nat. Mater.* **4**, 741–744 (2005).

127. D. Ravelosona, D. Lacour, J. A. Katine, B. D. Terris, and C. Chappert, Nanometer scale observation of high efficiency thermally assisted current-driven domain wall depinning, *Phys. Rev. Lett.* **95**, 117203 (2005).

128. L. Thomas, M. Hayashi, X. Jiang, R. Moriya, C. Rettner, and S. S. P. Parkin, Oscillatory dependence of current-driven magnetic domain wall motion on current pulse length, *Nature* **443**, 197–200 (2006).

129. M. Laufenberg, W. Bührer, D. Bedau et al., Temperature dependence of the spin torque effect in current-induced domain wall motion, *Phys. Rev. Lett.* **97**, 046602 (2006).

130. G. S. D. Beach, C. Knutson, C. Nistor, M. Tsoi, and J. L. Erskine, Nonlinear domain-wall velocity enhancement by spin-polarized electric current, *Phys. Rev. Lett.* **97**, 057203 (2006).

131. L. Thomas, M. Hayashi, X. Jiang, R. Moriya, C. Rettner, and S. S. P. Parkin, Resonant amplification of magnetic domain-wall motion by a train of current pulses, *Science* **315**, 1553–1556 (2007).

132. M. Hayashi, L. Thomas, C. Rettner, R. Moriya, and S. S. P. Parkin, Direct observation of the coherent precession of magnetic domain walls propagating along permalloy nanowires, *Nat. Phys.* **3**, 21–25 (2007).

133. G. Meier, M. Bolte, R. Eiselt, B. Krüger, D.-H. Kim, P. Fischer, Direct imaging of stochastic domain-wall motion driven by nanosecond current pulses, *Phys. Rev. Lett.* **98**, 187202 (2007).

134. S. Laribi, V. Cros, M. Muñoz et al., Reversible and irreversible current induced domain wall motion in CoFeB based spin valves stripes, *Appl. Phys. Lett.* **90**, 232505 (2007).

135. S. S. P. Parkin, M. Hayashi, and L. Thomas, Magnetic domain-wall racetrack memory, *Science* **320**, 190–194 (2008).

136. M. Hayashi, L. Thomas, R. Moriya, C. Rettner, and S. S. P. Parkin, Current-controlled magnetic domain-wall nanowire shift register, *Science* **320**, 209–211 (2008).

137. R. Moriya, L. Thomas, M. Hayashi, Y. B. Bazaliy, C. Rettner, and S. S. P. Parkin, Probing vortex-core dynamics using current-induced resonant excitation of a trapped domain wall, *Nat. Phys.* **4**, 368–372 (2008).

138. L. Heyne, M. Klaui, D. Backes et al., Relationship between nonadiabaticity and damping in permalloy studied by current induced spin structure transformations, *Phys. Rev. Lett.* **100**, 066603 (2008).

139. M. Bolte, G. Meier, B. Krüger et al., Time-resolved X-ray microscopy of spin-torque-induced magnetic vortex gyration, *Phys. Rev. Lett.* **100**, 176601 (2008).

140. O. Boulle, J. Kimling, P. Warnicke et al., Nonadiabatic spin transfer torque in high anisotropy magnetic nanowires with narrow domain walls, *Phys. Rev. Lett.* **101**, 216601 (2008).

141. C. Burrowes, D. Ravelosona, C. Chappert et al., Role of pinning in current driven domain wall motion in wires with perpendicular anisotropy, *Appl. Phys. Lett.* **93**, 172513 (2008).

142. T. A. Moore, I. M. Miron, G. Gaudin et al., High domain wall velocities induced by current in ultrathin Pt/Co/AlOx wires with perpendicular magnetic anisotropy, *Appl. Phys. Lett.* **93**, 262504 (2008).

143. L. Bocklage, B. Krüger, R. Eiselt, M. Bolte, P. Fischer, and G. Meier, Time-resolved imaging of current-induced domain-wall oscillations, *Phys. Rev. B* **78**, 180405R (2008).

144. C. Knutson, Magnetic domain wall dynamics in nanoscale thin film structures, Thesis, The University of Texas at Austin, Austin, TX, 2008.

145. S. A. Yang, G. S. D. Beach, C. Knutson et al., Universal electromotive force induced by domain wall motion, *Phys. Rev. Lett.* **102**, 067201 (2009).

146. I. M. Miron, P.-J. Zermatten, G. Gaudin, S. Auffret, B. Rodmacq, and A. Schuhl, DomainWall spin torquemeter, *Phys. Rev. Lett.* **102**, 137202 (2009).

147. A. Fernández-Pacheco, J. M. De Teresa, R. Córdoba et al., Domain wall conduit behavior in cobalt nanowires grown by focused electron beam induced deposition, *Appl. Phys. Lett.* **94**, 192509 (2009).

148. P. M. Braganca, I. N. Krivorotov, O. Ozatay et al., Reducing the critical current for short-pulse spin-transfer switching of nanomagnets, *Appl. Phys. Lett.* **87**, 112507 (2005).

149. W. H. Rippard, M. R. Pufall, S. Kaka, T. J. Silva, S. E. Russek, and J. A. Katine, Injection locking and phase control of spin transfer nano-oscillators, *Phys. Rev. Lett.* **95**, 067203 (2005).

150. A. L. D. Kilcoyne, T. Tyliszczak, W. F. Steele et al., Interferometer-controlled scanning transmission X-ray microscopes at the Advanced Light Source, *J. Synchrotron Radiat.* **10**, 125–136 (2003).

151. J. Stöhr, Y. Wu, B. D. Hermsmeier et al., Element-specific magnetic microscopy with circularly polarized X-rays, *Science* **259**, 658–661 (1993).

152. G. Tatara and H. Kohno, Theory of current-driven domain wall motion: Spin transfer versus momentum transfer, *Phys. Rev. Lett.* **92**, 086601 (2004).

153. Z. Li and S. Zhang, Domain-wall dynamics and spin-wave excitations with spin-transfer torques, *Phys. Rev. Lett.* **92**, 207203 (2004).

154. S. Zhang and Z. Li, Roles of nonequilibrium conduction electrons on the magnetization dynamics of ferromagnets, *Phys. Rev. Lett.* **93**, 127204 (2004).

155. A. Thiaville, Y. Nakatani, J. Miltat, and Y. Suzuki, Micromagnetic understanding of current-driven domain wall motion in patterned nanowires, *Europhys. Lett.* **69**, 990–996 (2005).

156. A. P. Malozemoff and J. C. Slonczewski, *Magnetic Domain Walls in Bubble Materials*, Academic Press, New York, 1979.

157. N. L. Schryer and L. R. Walker, Motion of 180 degrees domain-walls in uniform dc magnetic-fields, *J. Appl. Phys.* **45**, 5406–5421 (1974).

158. M. Hayashi, L. Thomas, Ya. B. Bazaliy et al., Influence of current on field-driven domain wall motion in permalloy nanowires from time resolved measurements of anisotropic magnetoresistance, *Phys. Rev. Lett.* **96**, 197207 (2006).

159. G. S. D. Beach, C. Knutson, M. Tsoi, and J. L. Erskine, Field- and current-driven domain wall dynamics: An experimental picture, *J. Magn. Magn. Mater.* **310**, 2038–2040 (2007).

160. A. S. Núñez, R. A. Duine, P. M. Haney, and A. H. MacDonald, Theory of spin torques and giant magnetoresistance in antiferromagnetic metals, *Phys. Rev. B* **73**, 214426 (2006).

161. A. H. MacDonald and M. Tsoi, Antiferromagnetic metal spintronics, *Philos. Trans. R. Soc.* **369**, 3098 (2011).

162. Z. Wei, A. Sharma, A. Nunez et al., Changing exchange bias in spin valves with an electric current, *Phys. Rev. Lett.* **98**, 116603 (2007).

163. S. Urazhdin and N. Anthony, Effect of polarized current on the magnetic state of an antiferromagnet, *Phys. Rev. Lett.* **99**, 046602 (2007).

164. X.-L. Tang, H.-W. Zhang, H. Su, Z. Y. Zhong, and Y.-L. Jing, Changing and reversing the exchange bias in a current-in-plane spin valve by means of an electric current, *Appl. Phys. Lett.* **91**, 122504 (2007).

165. N. Dai, N. Thuan, L. Hong, N. Phuc, Y. Lee, S. Wolf, and D. Nam, Impact of in-plane currents on magnetoresistance properties of an exchange-biased spin valve with an insulating antiferromagnetic layer, *Phys. Rev. B* **77**, 132406 (2008).

166. J. Bass, A. Sharma, Z. Wei, and M. Tsoi, Studies of effects of current on exchange-bias: A brief review, *J. Magn.* **13**, 1–6 (2008).

167. J. Basset, A. Sharma, Z. Wei, J. Bass, and M. Tsoi, Towards antiferromagnetic metal spintronics, in *Proc. SPIE 7036, Spintronics*, 703605 (4 September 2008) (2008), doi: 10.1117/12.798220; https://doi.org/10.1117/12.798220.

168. Z. Wei, J. Basset, A. Sharma, J. Bass, and M. Tsoi, Spin-transfer interactions in exchange-biased spin valves, *J. Appl. Phys.* **105**, 07D108 (2009).

169. I. Fina, X. Marti, D. Yi et al., Anisotropic magnetoresistance in an antiferromagnetic semiconductor, *Nat. Commun.* **5**, 4671 (2014).

170. C. Wang, H. Seinige, G. Cao, J.-S. Zhou, J. B. Goodenough, and M. Tsoi, Anisotropic magnetoresistance in antiferromagnetic Sr_2IrO_4, *Phys. Rev. X* **4**, 041034 (2014).

171. Y. Y. Wang, C. Song, G. Wang, J. Miao, F. Zeng, and F. Pan, Anti-ferromagnet controlled tunneling magnetoresistance, *Adv. Funct. Mater.* **24**, 6806 (2014).

172. B. G. Park, J. Wunderlich, X. Martí et al., A spin-valve-like magnetoresistance of an antiferromagnet-based tunnel junction, *Nat. Mater.* **10**, 347 (2011).

173. C. Wang, H. Seinige, G. Cao, J. S. Zhou, J. B. Goodenough, and M. Tsoi, Electrically tunable transport in the antiferromagnetic mott insulator Sr_2IrO_4, *Phys. Rev. B* **92**, 115136 (2015).

174. P. Wadley, B. Howells, J. Železný et al., Electrical switching of an antiferromagnet, *Science* **351**, 587–0 (2016).

175. W. H. Rippard, M. R. Pufall, S. Kaka, T. J. Silva, and S. E. Russek, Current-driven microwave dynamics in magnetic point contacts as a function of applied field angle, *Phys. Rev. B* **70**, 100406(R) (2004).

176. A. Khitun, D. E. Nikonov, M. Bao, K. Galatsis, and K. L. Wang, Feasibility study of logic circuits with a spin wave bus, *Nanotechnology* **18**, 465202 (2007).

177. T. Schneider, A. A. Serga, B. Leven, B. Hillebrands, R. L. Stamps, and M. P. Kostylev, Realization of spin-wave logic gates, *Appl. Phys. Lett.* **92**, 022505 (2008).

178. www.everspin.com/pub/press/2016

8
Spin Torque in
Magnetic Systems
Theory

Aurélien Manchon and Shufeng Zhang

M anipulating the magnetization direction of a magnetic layer without the use of an external magnetic field represents an outstanding opportunity and a challenge for spintronic applications [1]. In the past twenty years, vast experimental data and many detailed theoretical results have been produced. In this chapter, we aim to provide a coherent theoretical description of this topic for various phenomena involving the interplay between spin transport and magnetization dynamics. There are already a few excellent review papers [2–6] on this same subject, and we refer to some of the details in these articles. In this chapter, we address the theoretical aspects of current-induced -magnetization control using the so-called the spin torque effect (ST) from both the transport and magnetization dynamics points of view. Although numerous models have been developed, discussing them all is far beyond the scope of this chapter and we rather refer the reader to specific literature. After a brief overview of ST studies in Section 8.1.1, transport characteristics in a number of magnetic systems such as spin valves, magnetic tunnel junctions, and magnetic domain walls are described in Section 8.1.2. At the end of the section, several unconventional spin torques will be briefly discussed. In Section 8.1.3, the ST-driven magnetization dynamics for these magnetic systems is studied. Finally, perspectives and conclusions are given in Section 8.1.4.

8.1 OVERVIEW OF SPIN TRANSFER STUDIES

Despite earlier insights by Berger [7] on magnetic domain walls (DWs) and Slonczewski [8] on magnetic tunnel junctions (MTJs), the concept of *spin transfer torque* (STT) was formalized independently by these authors in 1996 [9, 10] for the case of the metallic spin valve (SV) for quite different reasons. In Berger's picture, the magnetic instability occurs when the current-driven chemical potential splitting for the spin up and down electrons exceeds the spin-wave gap (threshold). Since the chemical potential splitting scales with the spin accumulation, which is proportional to the current, the spin wave is generated at a critical current density. Alternatively, Slonczewski proposed that the absorption of the transverse spin current by a ferromagnetic layer is equivalent to a spin transfer torque exerted on the layer and this torque can be used for rotating the magnetization of the layer. Although Berger addresses spin-wave emission and Slonczewski considers a macrospin representation of the magnetic layers, the two models are largely equivalent. With the above Berger-Slonczewski model, one anticipates that a sufficiently large current applied perpendicular to the spin valve layers can either generate spin waves or rotate the magnetization.

Less than two years after its prediction, evidence of STT was observed in magnetic multilayers [11, 12], metallic spin valves [13–15], magnetic tunnel junctions [16–18], domain walls [19, 20], antiferromagnets [21, 22], semiconductors devices [23, 24], and single ferromagnetic layers in the presence of spin-orbit coupling [25–27]. The details of these experimental achievements are extensively reviewed by M. Tsoi in the previous chapter (Chapter 7) and by A. Manchon and H. Yang in Chapter 9.

These observations raised a number of fundamental questions concerning the microscopic origin of the spin torques in these magnetic structures. Many theoretical approaches have been developed in the past 15 years. For the metallic magnetic structures with non-collinear electrode magnetizations, most of the theories have focused on diffusive scattering in addition to the quantum mechanical interface reflection that induces rotation of spins for reflected and transmitted electrons [2, 28–33]. In MTJs, the spin torques were calculated via quantum mechanical tunneling through using a tight-binding model [34], free electron models [35–37] as well as first principle methods [38]. At high bias voltage for MTJs, inelastic tunneling becomes increasingly important and a generalized form of the transfer Hamiltonian formalism [39–41] for non-collinear structures was developed and its prediction was compared to recent experiments [42–45]. For magnetic domain walls, the spin torques were studied semi-phenomenologically [46–48], and microscopically [49–53]. Several novel concepts, such as spin-motive forces [54–56] and spin-current mediated damping [47, 57] have been developed. In the first part of this chapter, we intend to use a unified approach to formulate the spin torque for all these systems.

The spin torques generated by the current can not be directly measured. Instead, the anticipated consequences of the spin torque, such as the critical current for the magnetization reversal [13, 16], the spin-wave excitations [58–62] or domain walls motion [63–66] were experimentally measured. These measured results not only depend on the nature of the spin torques, but also on other parameters, e.g., the damping coefficient of the materials and detailed magnetization distributions of the film during magnetization dynamics. Furthermore, the influence of the finite temperature is usually quite complex for the current-driven magnetization dynamics, and thus the connection between the theoretically calculated spin torque and experimentally observed dynamic motion becomes a challenging issue [67]. In fact, most of the research efforts were made on the detailed analysis of the magnetization dynamics by assuming a specific form of the spin torques. Usually, either the macrospin model [68, 69] or micromagnetic simulations [70, 71] are used to determine the possible magnetic phases and dynamics driven by the currents; this subject is the focus of the second half of the chapter.

8.2 SPIN TORQUE IN MAGNETIC STRUCTURES

8.2.1 PRINCIPLE OF SPIN TRANSFER TORQUE

The spin transport in heterogeneous magnetic structures with a collinear magnetization configuration gives rise to a longitudinal *spin density* [72] (also called *spin accumulation* in diffusive systems). When the electrodes' magnetizations become non-collinear, as in spin valves or magnetic domain wall structures, a spin accumulation showing a component *transverse* to the background magnetization appears (see Figure 8.1a). This transverse component of the spin accumulation exerts a torque on the background magnetization via the exchange interaction, leading to magnetization reorientation, switching or excitations [9, 10]. Another way to understand the spin transfer torque is through the angular momentum conservation: the net balance of spin current flowing through an enclosed region is the rate of the total angular momentum change, which is defined as a spin transfer torque (see Figure 8.1b). We will show below that these two pictures are equivalent only when one neglects spin-flip scattering and spin-orbit coupling.

Throughout this chapter, we choose an *sd* model to describe the spin transport and magnetization dynamics. The *sd* model artificially separates the itinerant electrons in *sd* bands that are responsible for the spin transport from the localized *d* bands that determine the magnetization [73]. From the microscopic point of view, the *sd* model is excessively simplistic since the Fermi surface for the majority and minority electrons are very different in $3d$ transition metals. This simplified description introduces inaccuracies in the prediction as will be mentioned below. However, the *sd* model remains a very

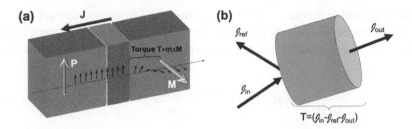

FIGURE 8.1   (a) Schematics of the spin density profile in a metallic spin valve, without spin relaxation: When impinging into the right layer, the spin density m possesses a component perpendicular to the local magnetization **M** and exerts a torque on the form $\mathbf{T} \propto \mathbf{m} \times \mathbf{M}$; (b) Spin transfer picture: the net balance of spin-current density $-\nabla \cdot \mathcal{J}$ is equivalent to a spin transfer to the local magnetization, in the absence of spin relaxation.

useful and pedagogical tool that provides qualitatively valuable results in the context of spin transfer torque. The Hamiltonian of the total spin system is:

$$\hat{H} = \hat{H}^d + \hat{H}^{sp} - J_{sd} \sum_i \hat{\mathbf{S}}_i \cdot \hat{\mathbf{s}}, \tag{8.1}$$

$$\hat{H}^d = -\frac{g\mu_B}{2} \sum_i \hat{\mathbf{S}}_i \cdot \mathbf{H}_{\text{eff}}, \tag{8.2}$$

$$\hat{H}^{sp} = \frac{\hat{\mathbf{p}}^2}{2m} + U(\hat{\mathbf{r}}) + \frac{\hbar}{4m^2c^2}(\nabla V \times \hat{\mathbf{p}}) \cdot \hat{\mathbf{s}}, \tag{8.3}$$

where $\hat{\mathbf{S}}_i$ (s) is the dimensionless spin operator for the i-localized (itinerant) electron, $\mathbf{H}_{\text{eff}}$ is the effective magnetic field (including the anisotropy, dipolar, and exchange fields) that self-consistently depends on the local spin $\mathbf{S}_i$, and J_{sd} is the exchange interaction between localized and itinerant electrons. The itinerant electron Hamiltonian $\hat{H}^{sp}$ includes a kinetic term, the potential profile and the spin-orbit coupling. (In Equation 8.2, the spin-orbit coupling is taken into account within the anisotropy field.) Applying Ehrenfest's theorem, one obtains the local spin density continuity equation for a localized and an itinerant electron, respectively:

$$\frac{\partial \langle \hat{\mathbf{S}}_i \rangle}{\partial t} = -\frac{g\mu_B}{2i\hbar} \left\langle \left[\hat{\mathbf{S}}_i, \hat{\mathbf{S}}_i \cdot \mathbf{H}_{\text{eff}}\right] \right\rangle + \frac{2J_{sd}}{\hbar} \langle \hat{\mathbf{S}}_i \rangle \times \langle \hat{\mathbf{s}} \rangle, \tag{8.4}$$

$$\frac{\partial \langle \hat{\mathbf{s}} \rangle}{\partial t} = \nabla \cdot \mathcal{J}_s + \frac{1}{2m^2c^2} \left\langle (\nabla V \times \hat{\mathbf{p}}) \times \hat{\mathbf{s}} \right\rangle - \frac{2J_{sd}}{\hbar} \langle \hat{\mathbf{S}}_i \rangle \times \langle \hat{\mathbf{s}} \rangle, \tag{8.5}$$

where $\langle \ldots \rangle$ denotes quantum mechanical averaging on either the local or the itinerant electronic states with a non-equilibrium distribution function, $\mathcal{J}_s \equiv -\langle \hat{\mathbf{v}} \otimes \hat{\mathbf{s}} \rangle$ is the spin-current tensor and $\gamma = |g| \mu_B / \hbar$ is the gyromagnetic ratio. In Equation 8.4, the commutator $[\hat{S}_i, \hat{S}_i \cdot \mathbf{H}_{\text{eff}}]$ depends on the detailed dependence of $\mathbf{H}_{\text{eff}}$ on $\hat{\mathbf{S}}_i$. One usually approximates it with the Landau-Lifshitz-Gilbert (LLG) equation, which includes a precessional term (i.e., takes $\mathbf{H}_{\text{eff}}$ as a c-number) and a damping term. There are quite a number of recent studies on the microscopic origins of the Gilbert damping [74–76]. Here we simply assume that the magnetization motion takes the same form except for the additional contribution from the second term of Equation 8.4, i.e.,

$$\frac{\partial \mathbf{M}}{\partial t} = -\gamma \mathbf{M} \times \mathbf{H}_{\text{eff}} + \frac{\alpha}{M_s} \mathbf{M} \times \frac{\partial \mathbf{M}}{\partial t} - \frac{2J_{sd}}{\hbar} \mathbf{M} \times \mathbf{m}, \tag{8.6}$$

where we have defined the magnetization densities for the local and itinerant electron spins $\mathbf{M}(\mathbf{r},t)/M_s = \langle \hat{\mathbf{S}}_i \rangle / S$ and $\mathbf{m}(\mathbf{r},t) = -\langle \hat{\mathbf{s}} \rangle$, M_s being the saturation magnetization, S is the localized spin magnitude and γ is the gyromagnetic ratio*. The first two terms are standard LLG terms and thus the only new term is due to the interaction between the magnetization M and the spin density m, which is usually defined as the spin torque, $\mathbf{T}_{st} = -2SJ_{sd}/\hbar M_s \mathbf{M} \times \mathbf{m}$. To determine the spin density m, a similar equation of motion for the itinerant spins is needed:

$$\frac{\partial \mathbf{m}}{\partial t} = -\nabla \cdot \mathcal{J}_s - \frac{1}{2m^2 c^2} \langle (\nabla V \times \hat{\mathbf{p}}) \times \hat{\mathbf{s}} \rangle + \frac{2SJ_{sd}}{\hbar M_s} \mathbf{M} \times \mathbf{m}$$
$$- \frac{\delta \mathbf{m}}{\tau_{sf}} - \frac{1}{\tau_\varphi M_s^2} \mathbf{M} \times (\delta \mathbf{m} \times \mathbf{M}), \tag{8.7}$$

where we have phenomenologically introduced a spin relaxation time τ_{sf} to model the isotropic spin-flip process by impurities and magnons, and a dephasing time τ_φ to model the relaxation of the spin density component *transverse* to the magnetization due to spin decoherence† [77]. $\delta \mathbf{m} = \mathbf{m} - \mathbf{m}_0$ is the non-equilibrium spin density: $\delta \mathbf{m}$ is actually the quantity responsible for the current-driven torque since the equilibrium spin accumulation $\mathbf{m}_0$ gives rise to a zero-bias interlayer exchange coupling that is usually included in the effective field $\mathbf{H}_{\text{eff}}$ for the sake of simplicity. Therefore, for long spin decoherence time ($\tau_\varphi \gg \hbar/2SJ_{sd}$) the non-equilibrium spin torque has the form: $\mathbf{T}_{st} \equiv -2SJ_{sd}/\hbar M_s \mathbf{M} \times \delta \mathbf{m}$. The coupled Equations 8.6 and 8.7 determine $\mathbf{M}$ and $\delta \mathbf{m}$ as long as one can relate the spin current $\mathcal{J}_s$ to the magnetization and one can properly evaluate the spin-orbit term. One immediately realizes that in a steady state, $\partial \mathbf{m}/\partial t = 0$, and without the spin-orbit coupling and spin relaxation $1/\tau_{sf} = 0$, the above equation reduces to $\mathbf{T}_{st} \equiv -2SJ_{sd}/\hbar M_s \mathbf{M} \times \delta \mathbf{m} = -\nabla \mathcal{J}_s$, i.e., the spin torque can be viewed as the spin current transfer (the processes illustrated by Figure 8.1a and b are equivalent).

Before we proceed in considering the details of the spin torque in several experimental systems, we emphasize that the spin torque introduced here is always transverse to the magnetization $\mathbf{M}$. In principle, there is also a longitudinal component that may change the magnitude of the magnetization. In

---

* The electron spin and magnetic moment are antiparallel. For itinerant holes, we would have $m(r,t) = +\langle s \rangle$.

† This term is hidden in the collision integral of Boltzmann transport equation and emerges upon k-integration in the diffusive limit.

the ferromagnets at low temperature (compared to the Curie temperature), we assume that the magnitude of the magnetization is rigid as it is assumed in the LLG equation. Thus we only consider the two transverse components of the spin torque, which can be formally written as

$$\mathbf{T}_{st} = \frac{a_J}{M_s^2} \mathbf{M} \times (\mathbf{P} \times \mathbf{M}) + \frac{b_J}{M_s} \mathbf{M} \times \mathbf{P}, \tag{8.8}$$

where **P** is any fixed vector that is conveniently chosen as the magnetization direction of the pinned layer in spin valves and magnetic tunnel junctions. The first term is sometimes called the in-plane spin torque or the damping-like term. The second is named the perpendicular spin torque or the field-like term. Depending on the detail of Equation 8.7, a_J and b_J possess different characteristics (angular and bias dependencies) related to the specific system under consideration. The detailed form of the torque $\mathbf{T}_{st}$ will be addressed in the next sections.

8.2.2 METALLIC SPIN VALVES

A metallic spin valve is composed of two ferromagnetic systems (ferromagnetic layers with or without antiferromagnets, laminated layers etc.) separated by a metallic spacer (see Figure 8.2a). Since the voltage drop across the metallic spin valve is rather small, of the order of only 1 mV even for a current density as high as 10^8 A/cm^2, the linear response theory is sufficiently accurate for the calculation of the spin torque. Furthermore, the spin-orbit coupling is usually small compared to the sd interaction and thus (8.7) reduces to

$$\frac{\partial \mathbf{m}}{\partial t} = -\nabla \cdot \mathcal{J}_s - \frac{\delta \mathbf{m}}{\tau_{sf}} - \mathbf{T}_{st}. \tag{8.9}$$

In a steady state ($\partial \mathbf{m} / \partial t = 0$), the spin torque is thus $\mathbf{T}_{st} = -\nabla \cdot \mathcal{J}_s - \delta \mathbf{m} / \tau_{sf}$.

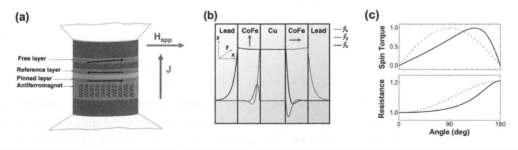

FIGURE 8.2   (a) Schematics of a common metallic spin valve, where the reference layer is antiferromagnetically coupled (usually through Ru) to a ferromagnetic layer pinned by an antiferromagnet (IrMn, PtMn...). The spacer layer that separates the free layer from the reference layer is usually Cu, whereas the ferromagnetic layers can be Co, CoFe, NiFe... (b) Calculated spatial profile of the spin-current densities in a symmetric spin valve: Lead/Co/Cu/Co/Lead. (c) Angular dependence of the normalized spin transfer torque (top) and normalized resistance (bottom) in the same spin valve. The dotted lines represent the purely ballistic case (≡MTJs). (After Manchon, A. et al., *J. Phys. Condens. Matter* 19, 165212, 2007. With Permission.)

In order to explicitly find the spin torque from the above equation, one needs to relate the spin-current tensor $\mathcal{J}_s$ to the non-equilibrium spin accumulation $\delta\mathbf{m}$. In the ballistic transport ($1/\tau_{sf} \equiv 0$), however, the current and the spin accumulation are not directly related because the spin accumulation is not well-defined. In this case, one simply has $\mathbf{T}_{st} = -\nabla \cdot \mathcal{J}_s$. If one integrates the entire magnetic layer, the total spin torque on the layer is $\mathbf{T}_{st} = -\int_{int}^{out} d\Omega \nabla \cdot \mathcal{J}_s = \mathcal{J}_s^{in} - \mathcal{J}_s^{out}$, i.e., the total spin torque is determined by the difference between the incoming and outgoing spin current at two sides of the layer. This simple identification makes first principle calculation possible [78–80].

The assumption of the ballistic transport is, in fact, a very poor approximation in the case of metallic spin valves. It has been well-established that diffusive spin-dependent transport is the proper description for metallic multilayers, as long as the layer thickness is large compared to the mean free path: diffusive scattering in the layers and at interfaces contributes to the spin current and magnetoresistance. For magnetic multilayers with collinear magnetization and current perpendicular to the plane, the most effective model is Valet-Fert's diffusive equation [72]. For the non-collinear magnetization, the extension to the Valet-Fert model is straightforward as long as one replaces the spin up and down channels by the 2 × 2 spinor tensor [31]. Specifically, one defines the spinor form of the current as

$$\hat{j} = \mathbf{j}_e\hat{I} + \mathcal{J}_s \cdot \hat{\sigma}, \tag{8.10}$$

where $\hat{I}$ is the 2 × 2 unity matrix and $\hat{\sigma}$ is the vector of Pauli spin matrices. In the spinor form, the spin-dependent current density is expressed in the basis ($\hat{I}, \hat{\sigma}$); therefore, the charge current vector and spin current tensor are determined consistently. Within this formalism, in the presence of external electric field E and inhomogeneous (spin-dependent) electronic density $\hat{n}$, the spinor form of the current reads

$$\hat{j} = \hat{C}\mathbf{E} - \hat{D}\nabla\hat{n}. \tag{8.11}$$

Here, $\hat{C} = C_0\left(\hat{I} + \beta\hat{\sigma} \cdot \mathbf{M}/M_s\right)$ is the generalized conductivity, $\hat{D} = D_0\left(\hat{I} + \beta'\hat{\sigma} \cdot \mathbf{M}/M_s\right)$ is the generalized diffusion constant and $\hat{n} = n_0\hat{I} + \hat{\sigma} \cdot \mathbf{m}$ is the generalized accumulation accounting for both charge ($2n_0$) and spin (m) accumulations. Both β and β' are the spin asymmetries of the conductivity and diffusion constants, responsible for the spin polarization of the current. All these quantities are 2 × 2 matrices, with a similar form as in Equation 8.10. The diffusion constant and the conductivity are related via the Einstein relation $\hat{C} = e^2\hat{N}\left(E_F\right)\hat{D}$, where $\hat{N}\left(E_F\right)$ is the density of states at the Fermi energy. The spinor form of Equation 8.11 provides one scalar relation for $\mathbf{j}_e$ that must be completed with the steady-state charge density continuity equation $\nabla \cdot \mathbf{j}_e + \partial n / \partial t = \nabla \cdot \mathbf{j}_e = 0$. A formal derivation of the spin-diffusion model can be found, for instance, in Ref. [77].

Equations 8.10 and 8.11 define the spin-dependent Ohm's law. From these relations, the spin current $\mathcal{J}_s = Tr_\sigma\left(\hat{\sigma}\hat{j}\right)$ can be explicitly expressed in terms of the charge current plus a diffusion term, i.e., $\mathcal{J}_s = -\dfrac{\beta}{eM_s}\mathbf{j}_e \otimes \mathbf{M} - 2D_0\nabla\mathbf{m}$, where $e > 0$ is the electric charge. By placing this spin current into Equation 8.9, we have

$$\nabla^2\mathbf{m} = -\frac{1}{\lambda_J^2 M_s}\mathbf{M}\times\mathbf{m} + \frac{\delta\mathbf{m}}{\lambda_{sf}^2} + \frac{1}{\lambda_\varphi^2 M_s^2}\mathbf{M}\times(\mathbf{m}\times\mathbf{M}), \qquad (8.12)$$

where $\lambda_J = \sqrt{\hbar D_0 / 2SJ_{sd}}$ is the transverse precession length, $\lambda_{sf} = \sqrt{2D_0\tau_{sf}}$ is the spin-diffusion length and $\lambda_\varphi = \sqrt{2D_0\tau_\varphi}$ is the spin dephasing length.

The above spin-diffusion equation is the generalization of Valet-Fert's spin-diffusion model to the non-collinear magnetization configurations. While the longitudinal spin accumulation has a length scale determined by the spin-flip length λ_{sf} that is of the order of 10 nm or longer for transition ferromagnets, the transverse spin accumulation has a much shorter length scale λ_J, ranging between a few angstroms to a few tens of angstroms. The solution of Equation 8.12 can be immediately obtained,

$$\delta\mathbf{m}_{||} = \mathbf{A}_{||}\exp\left(x / \lambda_{sf}\right) + \mathbf{B}_{||}\exp\left(-x / \lambda_{sf}\right), \qquad (8.13)$$

$$\delta\mathbf{m}_\perp = \mathbf{A}_\perp\exp\left(\pm x / l_+\right) + \mathbf{B}_\perp\exp\left(\pm x / l_-\right), \qquad (8.14)$$

where $l_\pm^{-1} = \sqrt{\lambda_{sf}^{-2} + \lambda_\varphi^{-2} \pm i\lambda_J^{-2}}$, and the constants of integration are determined by the boundary conditions. In Figure 8.2b, we show the spin-current density in the spin valve when the two ferromagnetic layers are perpendicular to each other. As we have seen, the longitudinal spin current (directly related to the spin accumulation) diffuses in the layers over a substantial distance, whereas the transverse spin-current decays on a much shorter scale, of the order of λ_J. The spin torque $\mathbf{T}_{st} = -2SJ_{sd} / \hbar M_s\mathbf{M}\times\delta\mathbf{m} + 1 / \tau_\varphi M_s^2\mathbf{M}\times(\delta\mathbf{m}\times\mathbf{M})$ occurs at the interface because the transverse spin accumulation $\delta\mathbf{m}_\perp$ is confined near the interface region.

The above treatment leads to two main characteristics of the STT in metallic spin valves. First, it is found that the coefficients a_J and b_J in Equation 8.8 are angular dependent due to the strong influence of the longitudinal spin accumulation on the spin transport, as shown in Figure 8.2c. This angular dependence has been used in the "wavy-structure" proposed by Boulle et al. [81] to generate zero-field magnetization steady precession. Secondly, at metallic interfaces, almost all the Fermi surface contributes to the current so that after averaging over all the electron k-states the effective spin current possesses only a small perpendicular component. As a consequence, the out-of-plane torque is very small and usually neglected in experiments.

A limitation of the diffusive model outlined above is its validity in strong ferromagnets. As a matter of fact, in magnetic transition metals such as Co, Ni, Fe, and their compounds, the rapid alignment of the incoming spins on the magnetization occurs over a few atomic planes away from the interface [29, 78], i.e. over a distance much shorter than the bulk mean free path (~10–20 nm in sputtered films). In such a situation, the drift-diffusion approach is inapplicable and cannot properly account for the spin dynamics close to the interface. To solve this issue, Brataas et al. [2] developed the concept of *spin mixing conductance*. In this approach, the complex interfacial spin dynamics (reflection/transmission, precession, and absorption) is reduced to a set of interfacial parameters that connect the incoming spin current to the local spin density. This concept has been particularly useful to describe interfacial effects such as spin torque and spin pumping [82] and has been recently extended to the case of spin-orbit coupled interfaces [83].

8.2.3 MAGNETIC TUNNEL JUNCTIONS

In magnetic tunnel junctions, the spacer layer that separates the two ferromagnetic layers is a thin insulator (e.g., AlO_x, MgO...) in contrast with the metallic spacer in spin valves. This distinction makes the magnetic tunnel junctions more promising in terms of magnetic memory applications: MTJs have a higher value of magnetoresistance up to several hundred percent and the tunnel resistance is much larger than in metallic spin valves. Unlike metallic spin valves, the resistance and magnetoresistance are solely determined by the electronic states near the barrier. As a consequence, the modification of the electrochemical potential due to the spin accumulation build-up ($\approx$meV) is negligible compared to the barrier height ($\approx$eV). Then, it is convenient to describe the spin transport within the magnetic tunnel junction assuming a ballistic approximation. Indeed, most of the theoretical calculations have been carried out by assuming a perfect tunnel junction within the elastic tunneling regime. When the bias voltage exceeds several hundred milli-volts, however, the inelastic tunneling becomes more important. We will describe below the main characteristics of the elastic and inelastic contributions to the spin torque.

8.2.3.1 Elastic Tunneling

The elastic tunneling process is a textbook example of the elementary application of quantum mechanics. Several approaches, depending on the complexity of the detailed calculation, are free electron [8, 35–37], tight-binding [34], and *ab-initio* methods [38]. These theories all assume the absence of spin relaxation within the barrier and the electrodes. Instead of calculating the spin accumulation, the spin current for a fixed bias voltage is computed using the direct relation between the spin torque and the spin current, $\mathbf{T}_{st} = \mathcal{J}_s^{int}$. In the free electron approach, the Hamiltonian is solved layer-by-layer and the wave functions are connected using the usual continuity principle (the barrier is treated using the Airy functions [35] or WKB approximation [36, 37]).

As mentioned in the previous section, the spin transport in nanodevices is controlled by the spin accumulation within the bulk of the layers and the quantum mechanical reflection/transmission of the spin at the interfaces. Whereas the role of the longitudinal spin accumulation is negligible in MTJs (a_J and b_J in Equation 8.8 are generally independent of the angle, as shown in Figure 8.3b), the spin rotation during the tunneling can be quite significant. Contrary to metallic spin valves where the spin current is averaged over the complete Fermi surface, in MTJs, only electrons close to the perpendicular incidence significantly contribute to the transport yielding a large effective spin rotation [37]. This effective spin rotation is at the origin of a significant out-of-plane torque ($b_J \neq 0$ in Equation 8.8). Furthermore, in the case of a symmetric junction with elastic tunneling, it can be shown that the components of the torque have the following form (see Figure 8.3a)

$$\mathbf{T}_{\parallel} = \left(a_1 V + a_2 V^2\right) \mathbf{M} \times \left(\mathbf{P} \times \mathbf{M}\right), \tag{8.15}$$

$$\mathbf{T}_{\perp} = \left(b_0 + b_2 V^2\right) \mathbf{M} \times \mathbf{P}. \tag{8.16}$$

These bias dependencies have been mentioned by Slonczewski [8, 39] and calculated numerically by others [34–37]. Note that in MTJs the torque is expressed as a function of the bias voltage V, rather than the current density j_e since bias rather than current controls the tunneling transport. One of the important concerns for application is the actual bias dependence of the out-of-plane torque, that significantly modifies the magnetization dynamics [41–43, 83, 84]. In particular, the presence of inelastic scattering, as well as junction structural asymmetries, alters the above bias dependence [45].

8.2.3.2 Inelastic Tunneling

The inelastic scattering of tunneling electrons by phonons or magnons is known to be detrimental to the tunneling magnetoresistance in magnetic tunnel junctions [86]. The interaction Hamiltonian reads

$$\hat{H}_{\text{inel}} = -\frac{J_{sd}}{\sqrt{N}} \sum_{k,q} \left(C^+_{k-q,\uparrow} C_{k,\downarrow} a^+_q + C^+_{k+q,\downarrow} C_{k,\uparrow} a_q\right)$$
$$+ \sum_{k,p,\sigma} M_p \left(C^+_{k-q,\sigma} C_{k,\sigma} b^+_q + C^+_{k+p,\sigma} C_{k,\sigma} b_p\right), \tag{8.17}$$

where $\mathbf{C}_{k,\sigma}$ is the annihilation operator for a $\sigma-$ electron with wave vector k and a_q (b_p) is the magnon (phonon) annihilation operator. In the electron-magnon interaction, an incoming electron relaxes (enhances) its energy by emitting (absorbing) a magnon (see Figure 8.3c). Since the total spin needs to be conserved, the electron flips its spin during the interaction. The electron-phonon interaction is similar except the electron conserves its spin. The influence of electron-magnon scattering on the spin transport has been

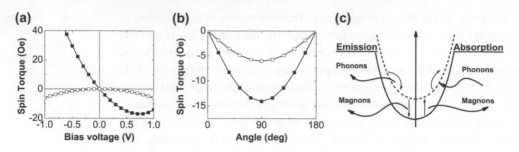

FIGURE 8.3   Free electron calculation of the bias dependence (a) and angular dependence (b) of the in-plane (black) and out-of-plane (red) torques. (c) The principle of the inelastic scattering by phonons and magnons. (After Manchon, A. et al., *J. Phys. Condens. Matter* 20, 145208, 2008. With permission.)

studied within the transfer Hamiltonian formalism for the case of tunneling magnetoresistance [87] and spin torque [40, 41]. The spin-dependent current density is expressed in the spinor form, as a function of the interfacial spin-dependent densities of states matrices $\hat{\rho}$ and of the transfer matrix $\hat{T}_{\mathbf{kp}}$ derived from the tunneling Hamiltonian

$$\hat{J} = \sum_{\mathbf{k},\mathbf{p}} \hat{\rho}_{\mathbf{k}} \hat{T}_{\mathbf{kp}} \hat{\rho}_{\mathbf{p}} (\hat{T}_{\mathbf{kp}})^{+} f_{L} (1 - f_{R}) - \hat{\rho}_{\mathbf{p}} \hat{T}_{\mathbf{pk}} \hat{\rho}_{\mathbf{k}} (\hat{T}_{\mathbf{pk}})^{+} f_{R} (1 - f_{L}). \quad (8.18)$$

The transfer matrix $\hat{T}_{\mathbf{kp}}$ accounts for the elastic tunneling as well as inelastic contributions [40, 41]. One can show that the general form of the in-plane torque is not modified, whereas the out-of-plane torque acquires a symmetric linear component

$$\mathbf{T}_{\parallel} = \left(a_{1}(T)V + a_{2}V^{2} \right) \mathbf{M} \times (\mathbf{P} \times \mathbf{M}), \quad (8.19)$$

$$\mathbf{T}_{\perp} = \left(b_{0} + b_{1}(T)|V| + b_{2}V^{2} \right) \mathbf{M} \times \mathbf{P}. \quad (8.20)$$

Note that the coefficients are now temperature-dependent and that the out-of-plane torque remains an *even* function of the bias voltage since the symmetry of the MTJ is conserved. Interestingly, a bias-dependent field-like effect has been observed by some authors [45, 84, 88]. This field effect is clearly asymmetric against bias and is responsible for *back-hopping* phenomena [45, 88]. Although the interfacial magnon scattering cannot reproduce this bias asymmetry, Li et al. [84] showed that in the case of *bulk* magnons the out-of-plane torque may acquire a component on the form $\propto j_{e}|V|$, that is therefore antisymmetric against the bias. However, other possible mechanisms (temperature effects, non-macrospin behavior, other inelastic interactions) are still under investigation.

8.2.3.3 Structural Asymmetries

The bias dependences given in Equations 8.15–8.16 and 8.19–8.20 assume a symmetric MTJ where the electrodes are equivalent and the barrier has a perfectly symmetric shape at zero bias. However, when structural asymmetries are introduced (the exchange splittings are different in the left and right electrodes or the barrier displays an asymmetric potential profile at zero bias) the transport properties of the junction become asymmetric against bias. In the case of asymmetric barrier or asymmetric exchange splitting in the electrodes, the conductance has the form [35, 89]

$$G(V) = G_0 + (\alpha_\phi \Delta\phi + \alpha_J \Delta J) eV + G_2 V^2, \qquad (8.21)$$

where $\Delta\phi = \phi_R - \phi_L$ and $\Delta J = J_{sd}^R - J_{sd}^L$ are the asymmetries in the barrier height and ferromagnetic exchange splitting. Similarly, the out-of-plane torque is no more an even function of the bias voltage and an antisymmetric linear term appears at the first order [35, 36, 90, 91]. It can be shown that the form of the torque exerted on the *right* electrode is therefore:

$$T_\perp = b_0 + (\beta_\phi \Delta\phi + \beta_J \Delta J) eV + b_2 V^2. \qquad (8.22)$$

The calculation of coefficients $\alpha_{\phi,J}$, $\beta_{\phi,J}$ following Brinkman's method [89] at the lowest order in voltage is a standard derivation that we do not develop here.

Up until now, the free electron and tight-binding models have provided valuable results that have been generally verified experimentally. However, in the case of MgO-based MTJs, the complexity of spin transport in crystalline structures limits the validity of these approaches [92]. The presence of interfacial resonant states [93], as well as the metal-induced gap states [94], are expected to deeply influence the spin transport for very thin insulators. For such systems, an *ab-initio* description of the spin torque is necessary [38].

8.2.3.4 Ferroelectrics and Spin Filters

To conclude this section, we mention the recent exploration of the impact of the tunnel barrier on the spin torque. Since the bias and angular dependences of the spin transfer torque are controlled by the interfacial densities of states on each side of the barrier, it is reasonable to expect dramatic impact when replacing the insulating spacer by a material displaying an additional degree of freedom such as a ferroelectric [95] or even a magnetic insulator [96].

As a matter of fact, in this case, with a ferroelectric spacer (such as $BaTiO_3$ or $BiFeO_3$), the screening of the polarization charges in the metallic electrodes results in a modulation (accumulation or depletion) of the charge density at the interfaces of the barrier [97]. Such a modulation depends on the direction of the ferroelectric polarization. Hence, in the case of (non-magnetic) asymmetric electrodes switching the ferroelectric polarization direction results in a dramatic change of resistance, a phenomenon called *giant electroresistance* [98]. When the electrodes are magnetic, the tunnel junction becomes multiferroic, i.e., the overall resistance can be controlled by tuning both the relative magnetization directions of the electrodes and the ferroelectric polarization of the barrier [99]. Correspondingly, it is also possible to modulate the bias dependence of the spin transfer torque by switching the ferroelectric polarization direction [95].

Replacing the insulating spacer by a magnetic insulator also offers interesting possibilities in terms of system design. In magnetic insulators, also called spin filters, the tunneling transmission differs for majority and minority spins, resulting in very large tunneling polarization [100]. The influence of such spin filters in magnetic tunnel junction has been very recently addressed theoretically [96] and remained to be explored experimentally.

8.2.4 MAGNETIC DOMAIN WALLS

Magnetic domain walls are another example of a magnetic device where the magnetization configuration is not collinear (see Figure 8.4a). At equilibrium, the spin density of itinerant electrons is aligned on the magnetic texture, $\mathbf{m}(\mathbf{r}) = m_0 \mathbf{M}(\mathbf{r}) / M_s$. As an electric current flows in a domain wall, the local spin density aligns away from the texture and adopts the general form

$$\mathbf{m} = \frac{m_0}{M_s} \mathbf{M}(\mathbf{r},t) + \delta\mathbf{m}, \tag{8.23}$$

$$\delta\mathbf{m} \sim \mathbf{M} \times (\mathbf{u} \cdot \nabla)\mathbf{M} + (\mathbf{u} \cdot \nabla)\mathbf{M}. \tag{8.24}$$

Here the current density is $\mathbf{j}_e = j_e \mathbf{u}$, $\mathbf{u}$ being the current direction. The resulting torque reads [47, 48]

$$\mathbf{T}_{st} = b_J (\mathbf{u} \cdot \nabla)\mathbf{M} - \beta \frac{b_J}{M_s} \mathbf{M} \times (\mathbf{u} \cdot \nabla)\mathbf{M}, \tag{8.25}$$

where the first term ($\propto b_J$, odd in magnetization) is called the *adiabatic* torque and the second term ($\propto \beta b_J$, even in magnetization) is called the *non-adiabatic* torque. The dimensionless parameter β is usually called the non-adiabatic coefficient or the non-adiabaticity. Notice that this terminology is

(a)   **(b)**

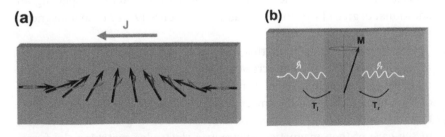

FIGURE 8.4   (a) Schematics of spin transport through a magnetic transverse domain wall. (b) Principle of spin pumping: the temporal oscillation of the magnetization injects a spin current in the neighboring electrodes. Its absorption by the leads gives rise to a torque exerted at the interfaces.

mostly a matter of convention as β is not necessarily related to non-adia-baticity *per se*. In general, both torques are non-local and depend on the magnetic texture.

8.2.4.1 Non-Adiabaticity

As will be discussed in the second part of this chapter, the adiabatic and non-adiabatic torques have a different impact on domain wall dynamics. In a nutshell, the non-adiabaticity β determines the velocity of one-dimensional domain walls. Therefore, major efforts have been made lately to determine the physical origin of β and accurately estimate its magnitude. Theoretical approaches range from the drift-diffusion model [47], the semi-classical Boltzmann equation [50], and quantum kinetics [49, 52, 101, 102] to tight-binding [53, 103] and more recently *ab-initio* calculations [104].

Overall two main mechanisms have been identified. In the presence of spin relaxation (e.g., from spin-orbit coupling or magnetic impurities), the absorption of the non-equilibrium spin current by the local magnetization is not complete, which results in a non-adiabaticity parameter of the form $\beta = \hbar / (2SJ_{sd}\tau_{sf})$ in the limit of large domain walls [47]. For sharp domain walls, the itinerant spins cannot follow the local magnetization and get mis-aligned, a mechanism sometimes called *spin mistracking*. This mistrack-ing enhances the reflection of spin-polarized electrons by the domain wall, resulting in domain wall resistance as well as a momentum transfer leading to a non-adiabatic torque [49, 50, 103]. The latter effect is only significant when the domain wall width is of the order of the spin precession length and decreases exponentially with the wall width. It is therefore vanishingly small in conventional transition metal ferromagnets [50, 103]. However, the presence of spin-independent disorder dramatically increases the non-adi-abaticity by enhancing the spin mixing due to momentum scattering [53]. Moreover, in spin-orbit coupled systems, such as domain walls in (Ga,Mn) As, the additional precession of the itinerant spins about the spin-orbit field enhances the spin mistracking and, thereby, both the domain wall resistance and non-adiabaticity [105, 106].

Spin transport becomes even more intriguing in two- and three-dimen-sional textures such as vortices and skyrmions. In such situations, as dis-cussed further below, the presence of a two- or three-dimensional magnetic texture gives rise to an emergent magnetic field that distorts the trajectory of the flowing electrons. This results in a topological Hall effect as well as a corresponding non-adiabaticity [107].

8.2.4.2 Thermal Activation and Temperature Gradient

The influence of a finite temperature and the interplay with thermal magnons have also been studied theoretically. Duine et al. [108] showed that thermal fluctuations together with the adiabatic spin torque term mimic the effect of the non-adiabatic torque. Due to the thermal noise, the domain is not intrinsically pinned and it is then possible to induce the domain wall motion in the absence of non-adiabatic torque. The influence of electron-magnon scattering on the current-induced domain wall motion has been analyzed and showed to contribute to the non-adiabatic torque [109, 110].

The motion of domain wall induced by temperature gradient or a magnon flow has also attracted some interest lately [111–113]. Different regimes have been identified, depending on the magnons wavelength. When the wavelength is shorter than the domain wall width, a spin transfer occurs and the domain wall flows against the magnon flow (magnons carry a spin angular momentum of $-\hbar$). In contrast, in the long wavelength limit, the domain wall is driven by linear momentum transfer and goes along the magnon flow.

8.2.4.3 Emergent Electromagnetic Field: Hall Effect and Torque

A very instructive way to explore spin transport in magnetic textures is to look at the system in the rotated frame of the local magnetization. Concretely, by rotating the *sd* Hamiltonian given in Equation 8.1 into the moving frame of the magnetization, one can show that each spin projection *s* feels an emergent electromagnetic field ($\mathbf{E}_s, \mathbf{B}_s$) defined [54, 114–117]

$$E_i^s = s\left(\hbar / 2eM_s^3\right)\left(\partial_t\mathbf{M} \times \partial_i\mathbf{M}\right) \cdot \mathbf{M}, \tag{8.26}$$

$$B_i^s = -s\varepsilon^{ijk}\left(\hbar / 2eM_s^3\right)\left(\partial_j\mathbf{M} \times \partial_k\mathbf{M}\right) \cdot \mathbf{M}. \tag{8.27}$$

Here, ε^{ijk} is the Civi-Levita symbol and (i,j,k) are spatial coordinates. In other words, a time- and spatial-dependent magnetic texture induces a spin-dependent electric field on the itinerant electrons, while a two- or three-dimensional texture induces a spin-dependent magnetic field. The former acts as a spin-dependent electromotive force [54, 115–117], while the latter induces so-called topological Hall effect [107]. As a consequence, when spin carriers flow in a moving magnetic texture, the current of spin projection *s* reads

$$\mathbf{j}_e^s = \sigma^s\mathbf{E} + \sigma^s\mathbf{E}^s + \sigma_H^s\mathbf{E}\times\mathbf{B}^s, \tag{8.28}$$

where σ^s (σ_H^s) is the spin-dependent longitudinal (ordinary Hall) conductivity. The implications of such an emergent electromagnetic field is best illustrated through two standard examples.

Let us first consider a one-dimensional moving domain wall in the absence of an external electric field ($\mathbf{E} = 0$). The total spin and charge currents flowing along the direction e_i read

$$j_e^i = \frac{\hbar}{2eM_s^3}\left(\sigma^- - \sigma^\downarrow\right)\left(\partial_t\mathbf{M} \times \partial_i\mathbf{M}\right) \cdot \mathbf{M}, \tag{8.29}$$

$$\mathcal{J}_s^i = -\frac{\hbar}{2e^2M_s^4}\left(\sigma^\uparrow + \sigma^\downarrow\right)\left[\left(\partial_t\mathbf{M} \times \partial_i\mathbf{M}\right) \cdot \mathbf{M}\right]\mathbf{M}. \tag{8.30}$$

The first expression, Equation 8.29, indicates that a moving domain wall pumps a charge current along the domain wall direction [54, 115–117]. This pumped charge current is accompanied by a spin current, Equation 8.30, which in turn exerts a torque on the magnetic texture, $\mathbf{T}_{st} = -\nabla \cdot \mathcal{J}_s$. Since this torque is even in magnetization direction and proportional to the first time-derivative of the magnetization, it enhances the magnetic damping [118].

Alternatively, let us consider a static two-dimensional magnetic texture $(\partial_t \mathbf{M} = 0, \ \nabla = (\partial_x, \partial_y))$, such as a vortex or a skyrmion, under an external electric field ($\mathbf{E} \neq 0$). The total spin and charge currents read

$$j_e^i = \left(\sigma^- + \sigma^\downarrow\right) E_i - \left(\sigma_H^- - \sigma_H^\downarrow\right) B_{em}(\mathbf{E} \times \mathbf{z})_i, \tag{8.31}$$

$$\mathcal{J}_s^{\ i} = -\frac{\hbar}{2e^2 M_s^4} \mathbf{M} \otimes \left[\left(\sigma^\uparrow - \sigma^\downarrow\right) E_i - \frac{\hbar}{e}\left(\sigma_H^\uparrow + \sigma_H^\downarrow\right) B_{em}(\mathbf{E} \times \mathbf{z})_i\right]. \tag{8.32}$$

Here, $B_{em} = \left(\partial_x \mathbf{M} \times \partial_y \mathbf{M}\right) \cdot \mathbf{M}$ is the emergent magnetic field. The charge current, Equation 8.31, experiences the topological Hall effect and is accompanied by a spin current that is even in magnetization direction, and hence also enhances the non-adiabatic torque [107].

8.2.5 Unconventional Spin Torques

In this section, we briefly discuss the nature of three different *unconventional* spin torques that have recently attracted interest from theoretical and experimental perspectives. The first torque is the sibling of the conventional STT discussed in the previous but is induced by a thermal gradient rather than by an electric field. The second is due to the presence of a spin-orbit coupling and does not necessitate a spin polarizer or magnetic texture to occur. The third torque is predicted to occur in antiferromagnets, where there is no overall transfer of spin angular momentum.

8.2.5.1 Thermoelectric and Thermomagnonic Torques

In heterogeneous (non-magnetic) multilayers, the application of a temperature gradient across the device generates a voltage via the Seebeck effect: the spatial dependence of the electronic thermal distribution along the structure is equivalent to the spatial dependence of the electron chemical potential, which in turn gives rise to a bias voltage.

Recently, Hatami et al. [119] proposed a so-called *thermoelectric* spin torque, based on the Seebeck effect in magnetic multilayers. In a heterogeneous magnetic system with non-collinear magnetic electrodes and with a temperature gradient, the thermal distributions of majority and minority electrons are different. The thermalization of the electronic system through inelastic scattering generates a spin current (the total number of electrons is conserved but not the spin polarization) that can be absorbed within the neighboring ferromagnetic layers, inducing a torque. The current spin polarization that drives this torque is no more the conventional polarization of the conductance but is the polarization of the *energy derivative* of the conductance, since the electronic thermal distribution is involved (this comes directly from the Sommerfield expansion of the generalized Ohm's law). This effect has recently been studied by Kovalev and Tserkovnyak [110] for domain walls.

The concept was recently extended to the case of magnetic insulators, where heat and spin transport are mediated by magnons rather than electrons [113, 121, 122]. This *thermomagnonic* torque can be used to move magnetic textures such as domain walls or skyrmions. More recently, it was shown that in the presence of Dzyaloshinskii-Moriya interaction, a magnon flow can exert a torque in magnetic insulators [123], an effect that echoes the spin-orbit torque discussed below.

8.2.5.2 Spin-Orbit Induced Torque

As shown in Equation 8.7, the spin-orbit coupling itself can act as a source for the spin torque (see Figure 8.5a). Assuming a uniformly magnetized single layer ($\nabla \cdot \mathcal{J}_s = 0$) in a steady-state ($\partial \mathbf{m} / \partial t = 0$) condition and neglecting the spin relaxation ($1 / \tau_{sf} = 0$), the out-of-equilibrium torque on the magnetization is directly proportional to the spin density of the conduction electrons,

$$\mathbf{T}_{st} \equiv -\frac{2SJ_{sd}}{\hbar M_s} \mathbf{M} \times \mathbf{m} = -\frac{1}{2m^2c^2} \langle (\nabla V \times \hat{\mathbf{p}}) \times \hat{\mathbf{s}} \rangle. \tag{8.33}$$

Contrary to the spin transfer torque presented in the previous sections, this torque does not come from the spatial variation of the magnetic texture.

While Equation 8.33 involves the general form of the spin-orbit interaction (SOI), the potential V includes all electric potentials in the solids. In the case of a system with a *spatial inversion symmetry* (impurities-induced SOI and Luttinger SOI, for instance) or a *rotational symmetry* (cubic Dresselhaus SOI), the k-integration vanishes and the spin torque is found to be zero at the

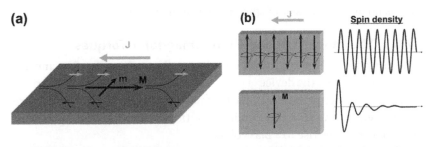

FIGURE 8.5 (a) In a single magnetic layer with spin-orbit interaction, a charge current generates a non-equilibrium spin density **m**, that can be used to reorient the magnetization **M**. (b) Principle of spin transport in an antiferromagnet: in an antiferromagnet (top), the torque exerted by the local magnetization on the itinerant spin is reversed at each layer, which provides a *staggered* non-equilibrium spin density texture (right diagram) in the absence of spin diffusion. This is in sharp contrast with the ferromagnetic case (bottom), where the spin precession remains unchanged along the ferromagnet, allowing for destructive interferences and therefore a damped oscillatory non-equilibrium spin density (right diagram).

first order [123]. However, in the case of *structure inversion asymmetry* and *bulk inversion asymmetry with strain*, the spin-orbit interaction is linear in k and depends on its sign, so that $H_{so}(\mathbf{k}) = -H_{so}(-\mathbf{k})$ and a non-zero torque is found at the first order in SOI [37, 125].

This SOI-induced torque is created by the spin-orbit coupling present in the band structure under a non-equilibrium condition, and thus it does not involve the transfer of the conduction electron spin to the magnetization, in sharp contrast with the spin transfer torque presented in the previous sections. As a result, the SOI-induced torque does not require the presence of a spin polarizer and can excite or switch the magnetization of uniformly magnetized single layers. Such a torque has recently attracted a massive amount of attention and has been studied in details in both model systems [101–126] and from first principles [127, 128]. These investigations have shown that the SOI-induced torque possesses both field-like (odd in magnetization) and damping-like (even in magnetization) components and therefore enables current-driven magnetization switching and excitation of single layers. An overview of this emerging topic is provided in Chapter 9.

8.2.5.3 Spin Torque in Antiferromagnets

In the past ten years, antiferromagnets – up till recently confined to a passive role in conventional magnetic multilayers – have emerged as promising materials for ultrafast spintronics applications [129]. Indeed, the antiferromagnetic order parameter displays ultrafast (THz) inertial dynamics in sharp contrast with the slower (GHz) dynamics of ferromagnets [130]. In these materials, the overall magnetization vanishes due to strong antiferromagnetic interactions within the layer ($J_{ij} < 0$), but the density of states is spin-split on each sublattice due to the strong *sd* exchange coupling. As a consequence, one can expect that an impinging spin current exerts a torque on the antiferromagnetic order parameter, which could be a powerful tool for ultrafast data manipulation.

Careful inspection of the equation of motion of the order parameter suggests that the torque enabling current-driven switching must be *even* upon magnetization reversal [130], i.e. on the form $\sim \mathbf{n} \times (\mathbf{p} \times \mathbf{n})$, where $\mathbf{n}$ is the order parameter direction and $\mathbf{p}$ is an arbitrary vector determined by the source of the spin torque. Several theoretical studies have been conducted to uncover the nature of spin torque in spin valves involving antiferromagnets [131, 132]. An important feature of spin valves is that the spin density acquired in one side of the device needs to be coherently transmitted toward the other part of the device without alteration. This works well in ferromagnetic spin valves but dramatically fails in their antiferromagnetic counterparts: due to the staggered nature of the magnetic configuration in antiferromagnets, the spin torque is very sensitive to disorder [133, 134], which makes its experimental observation a very challenging task [21].

One successful strategy is to consider systems that do not involve the transmission of a spin density, i.e., where the non-equilibrium spin density is generated *locally*. This can be done in two classes of systems: single

antiferromagnets with either bulk or interfacial inversion symmetry breaking where SOI-induced torque is operative [135], or antiferromagnetic textures where spin torque is induced by the magnetization gradient [136, 137].
With the very recent observation of current-driven order parameter switching [138], the field of antiferromagnetic spintronics is expected to expand
dramatically in the coming years.

8.3 CURRENT-INDUCED MAGNETIZATION DYNAMICS

In the previous section, we showed that depending on the structure under
consideration (metallic SVs, MTJs, DWs, single magnetic layers, antiferromagnets), a torque emerges, driven by an external bias (voltage or temperature). This spin torque modifies the dynamics of the local magnetization
and provides a powerful tool for controlling the magnetization direction,
essential for technological applications. In this section, we address the characteristics of the magnetization dynamics in the presence of ST. We do not
discuss the spin torque-driven dynamics of antiferromagnets and invite the
reader to refer to the specific literature [129, 130].

In principle, the magnetization dynamics involves simultaneous time-
dependent solutions for $\mathbf{M}(\mathbf{r},t)$ and $\mathbf{m}(\mathbf{r},t)$, see Equations 8.6 and 8.7.
However, the analysis of the timescale for $\mathbf{M}(\mathbf{r},t)$ and $\mathbf{m}(\mathbf{r},t)$ can reduce
the problem to a steady-state condition for the conduction electrons, i.e.,
$\partial \mathbf{m} / \partial t = 0$. To see this, one notices that it only takes about $1\,fs$ (transport
relaxation time) to establish a steady-state charge current after one applies an
electric field to any heterogeneous conducting system. To establish a steady-
state spin density and spin current in a magnetic system, the time scale
would be $\tau_{sf} = l_{sf} / v_F \approx 10\,fs - 1\,ps$. The time scale of the transverse magnetization described by Equation 8.6 is of the order of nanoseconds, much longer
than τ_{sf} and it is an excellent approximation to assume $\partial \mathbf{m} / \partial t = 0$ if one is
only interested in the nanosecond dynamics of the magnetization $\mathbf{M}(\mathbf{r},t)$.
With this simplification, the spin accumulation $\mathbf{m}(\mathbf{r},t)$ is solely determined
by $\mathbf{M}(\mathbf{r},t)$, i.e., there is no retardation effect. In fact, all the explicit expressions of the spin torques, e.g., Equations 8.14, 8.15, 8.16, 8.25 and 8.33 are
derived under this approximation. Recent self-consistent approaches have
been developed recently [139, 140] and largely confirm this assumption.
Hence, the starting point for the current-driven magnetization dynamics is
the following equation,

$$\frac{\partial \mathbf{M}}{\partial t} = -\gamma \mathbf{M} \times \mathbf{H}_{\text{eff}} + \frac{\alpha}{M_s} \mathbf{M} \times \frac{\partial \mathbf{M}}{\partial t} + \mathbf{T}_{st}(\mathbf{M}), \tag{8.34}$$

where the spin torque $\mathbf{T}_{\text{st}}$ takes a specific form for each magnetic system we
want to discuss below.

8.3.1 Magnetization Dynamics in Spin Valves

As we have shown in the previous sections, the general expressions of the spin torque in spin valves and MTJs take the form of Equation 8.8. Since the perpendicular toque, the second term of Equation 8.8, can be absorbed into the effective field in Equation 8.34, we should neglect this term and solely consider the spin torque

$$\mathbf{T}_{st} = \frac{a_J}{M_s^2} \mathbf{M} \times (\mathbf{P} \times \mathbf{M}).$$
(8.35)

The torque coefficient a_J may depend on the angle between the magnetization directions of the pinned and free layers (e.g., in [81]). In this chapter, we limit our discussions to a constant a_J. While Equation 8.34 can be solved by standard micromagnetic simulations, we shall consider analytical solutions in order to gain physical insight on the role of the spin torque by assuming the magnetization $\mathbf{M}$ is uniform with the layer, i.e., the macrospin model.

8.3.2 Magnetic Phase Diagrams

A useful tool for describing the dynamical properties of a magnetic layer under a spin-polarized current is the magnetic phase diagram that displays the magnetic stability regions of the layer as a function of the external magnetic field and applied bias. The geometry of the system is displayed in Figure 8.2a: the spin valve consists of two ferromagnetic layers separated by a (metallic or insulator) spacer. The system is a pillar along the z direction with an elliptic shape lying in the (x, y) plane (other configurations can be used experimentally as described by Tsoi in the previous chapter). The bottom (reference) layer has a magnetization direction $\mathbf{P} = +\mathbf{z}$ assumed to be fixed, and the top (free) layer has a magnetization direction $\mathbf{M}$, an easy axis $\mathbf{u} = \mathbf{x}$ and a demagnetizing field in the z direction.

8.3.2.1 Switching Threshold

Solving Equation 8.34 requires the precise knowledge of the magnetization distribution as well as the effective field $\mathbf{H}_{\text{eff}}$. In principle, it comprises the exchange field $\mathbf{H}_{ex}$, the anisotropy field $\mathbf{H}_K$, the demagnetization field $\mathbf{H}_d$, and the externally applied field $\mathbf{H}_{\text{ext}}$. The current-induced Oersted field, as well as interlayer exchange coupling (and perpendicular spin torque), can also be included. In the macrospin approach, the spatial variations of $\mathbf{M}$ are neglected so that the exchange field, as well as the Oersted field, are disregarded. Then, the total effective field reduces to $\mathbf{H}_{\text{eff}} = \mathbf{H}_{\text{ext}} + M_x H_K \mathbf{x} - 4\pi M_s M_z \mathbf{z}$. In spherical coordinates, $\mathbf{M} = M_s(\sin\theta\cos\phi, \sin\theta\sin\phi, \cos\theta)$ and (8.34) reduces to [3]:

$$\frac{1+\alpha^2}{\gamma}\dot{\theta} = h_\phi + \alpha h_\theta, \quad \frac{1+\alpha^2}{\gamma}\sin\theta\dot{\phi} = \alpha h_\phi - h_\theta,$$
(8.36)

where h_ϕ and h_θ are functions of (θ,ϕ) that we do not write explicitly (see for example [3]). The conventional method for obtaining stability conditions for the magnetization $\mathbf{M}$ is to study the stability of small deviations from the equilibrium $\delta\mathbf{M}$, under a spin-polarized current j_e. The equilibrium magnetization direction $\mathbf{M}_0$ is defined as $h_{\theta,\phi}(\mathbf{M}_0) = 0$, with $a_J = b_J = 0$ ($j_e = 0$). In the case where the external field is $\mathbf{H}_{ext} = H_x\mathbf{x}$, $\mathbf{M}_0 = M_s(\pm1,0,0)$. Inserting the perturbed magnetization $\mathbf{M} = \mathbf{M}_0 + \delta\mathbf{M}$ in Equation 8.36, and assuming that $|\mathbf{M}|^2 = M_s^2$, one obtains two differential equations for δM_x and δM_y that can be solved analytically. For metallic spin valves ($a_J = aj_e$ and $b_J = 0$), the stability conditions for the parallel (P) and antiparallel (AP) states are [141]:

$$j_e < \alpha\frac{\left(H_x + H_K + \dfrac{H_d}{2}\right)}{a} \qquad j_e > -\alpha\frac{\left(H_x - H_K - \dfrac{H_d}{2}\right)}{a}$$

$$H_x > -H_K \qquad\qquad H_x < H_K$$

(8.37)

As long as the conditions (8.37) are met, either P or AP or even both states are stable, as illustrated in Figure 8.6a. However, when the above conditions are not met, e.g., $j_e > \alpha\left(H_x + H_K + \dfrac{H_d}{2}\right)/a$, the P and AP configurations become unstable: the magnetization jumps to new stable states (denoted as E in Figure 8.6a). These novel states are unique characteristics of the spin torque and are described below.

8.3.2.2 Steady-State Precessions

Investigating the stability diagrams out of the stability regions reveals a rich variety of magnetization regimes, including *in-plane* precessions, *out-of-plane* precessions, and *inhomogeneous* magnetic excitations. The fact that the spin transfer torque term a_J can be either positive or negative, depending on the current direction, enables steady precessional states in which the damping is exactly compensated by the spin torque.

The usual method for determining the current-induced self-sustained precessional motion of the magnetization is based on the work of Bertotti et al. [69]. In the following, we disregard the role of the out-of-plane torque and focus on the influence of the in-plane torque. In the absence of both damping and in-plane torque, the magnetic system is energy conservative and the magnetization trajectories can be found straightforwardly by solving

$$\frac{\partial\mathbf{M}}{\partial t} = -\gamma\mathbf{M}\times\mathbf{H}_{eff}.$$

(8.38)

Equation 8.38 possesses fixed points (e.g., along the anisotropy axis) as well as two types of trajectories: in-plane (IP) and out-of-plane precessions (OP). Details of the derivation of these trajectories are found in [69]. In the following, we assume that we know the detail of such trajectories, their energy and precession period.

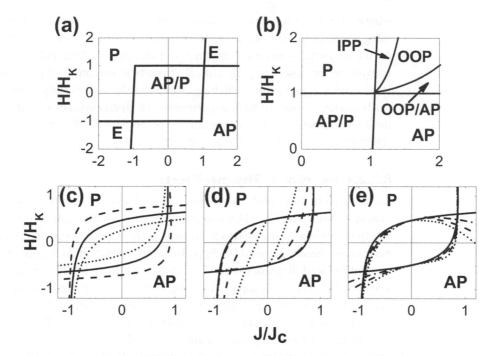

FIGURE 8.6 Stability diagrams for the free layer: (a) at zero temperature, assuming $b_J = 0$. (b) zoom of the zero temperature diagram for positive field and current, including in-plane (IPP) and out-of-plane (OOP) precessions. (c) Finite temperature ($\tau = 1$ s), assuming $b_J = 0$ (black: 300 K, blue: 100 K, red: 600 K). (d) Finite temperature ($\tau = 1$ s), with current heating and $b_J = 0$ ($c_0 = 1 \times 10^4 K^2/V^2$: red, $c_0 = 5 \times 10^5 K^2/V^2$: blue, $c_0 = 5 \times 10^6 K^2/V^2$: green); (e) finite temperature ($\tau = 1$ s), $b_J = b_2 V^2$ (at $V = 1$, $b_J/a_J = 0$: black, $b_J/a_J = 0.6$: red, $b_J/a_J = 1$: blue, $b_J/a_J = 2$: green).

The method introduced by Bertotti et al. consists in determining how the competition between the Gilbert damping and the spin torque stabilizes or destabilizes these unperturbed trajectories and fixed points. This perturbative method is based on Melkinov Integrals [142]

$$\mathcal{M}(E) = \int_0^{T(E)} \frac{\partial \varepsilon}{\partial t} dt, \qquad (8.39)$$

where ε is the total energy of the perturbed magnetic system, E is the conserved energy of the unperturbed trajectory $\mathcal{L}(E)$ and $T(E)$ is the period of the unperturbed cycle. A non-zero integral would indicate that the magnetization $\mathbf{M}$ deviates from its initial trajectory, whereas a zero value means that the competition between the spin torque and the damping produces a self-sustained precessional motion of the magnetization over the unperturbed trajectory. Therefore, a self-sustained precession is described by $\mathcal{M}(E) = 0$ and is stable (unstable) under $\partial \mathcal{M}/\partial E > 0 (< 0)$. Considering the complete LLG equation (Equation 8.34), the Melkinov Integral yields:

$$\mathcal{M}(E) = \int_0^{T(E)} \left[-\alpha |\tilde{\mathbf{M}} \times \mathbf{H}_{eff}|^2 + a_J (\tilde{\mathbf{M}} \times P) \cdot (\tilde{\mathbf{M}} \times \mathbf{H}_{eff}) \right] dt, \qquad (8.40)$$

where M is determined in the *absence* of damping and spin torque. This integral can be understood as the work done by the damping and spin torque on the magnetization on a trajectory over one period. This method is then convenient for describing the critical currents at the onset of precessional states and magnetization switching, revealing the rich variety of dynamical behavior, consistent with experimental observations. For example, this method allows for the determination of the in-plane (IPP) and out-of-plane (OPP) precession states, as illustrated in Figure 8.6b.

8.3.2.3 Inclusion of Thermal Effects

Thermal fluctuations of the magnetization are known to deeply influence the magnetization dynamics [143] and the current densities required to manipulate the magnetization are expected to significantly increase the effective temperature of the sample [67, 144]. Experimental evidence of thermally-assisted current-induced dynamics [145–147] has led to numerous theoretical works and simulations on this topic [148–150]. Note that the thermal fluctuations are unlikely to significantly affect the spin torque amplitude itself since the Fermi level is usually larger than the thermal energy ($E_F \gg k_B T$).

Thermal activation processes are accounted for using the Brown model of thermal magnetic fluctuations at equilibrium [150–153]: thermal fluctuations add a Gaussian stochastic field $\mathbf{h}$, to the effective field $\mathbf{H}_{\text{eff}}$, with

$$\langle h_r^i(t) \rangle_h = 0, \langle h_r^i(t) h_r^j(t') \rangle_h = 2D\delta_{ij}\delta(t - t'), \tag{8.41}$$

where $\langle \ldots \rangle_h$ denotes averaging over all the realizations of the fluctuating field, and $D = \alpha k_B T / \gamma M_s$ is the strength of thermal fluctuations, according to the fluctuation-dissipation theorem.

In the framework of the LLG equation, $|\mathbf{M}|$ is kept constant, so that the random field $\mathbf{h}$ induces a random walk on the $\mathbf{M}$-sphere. The evolution of the probability density for magnetic orientation vectors $\rho(\mathbf{M},t)$ is therefore described by the continuity equation [143]:

$$\frac{\partial \rho(\mathbf{M},t)}{\partial t} + \nabla_{\mathbf{M}} \cdot \mathbf{J}_M(\mathbf{M},t) = 0, \tag{8.42}$$

where $\mathbf{J}_M(\mathbf{M},t)$, the probability current density, contains a convective part (driven by the magnetization dynamics) and a diffusive part (driven by the random diffusion)*

$$J_M(M,t) = \frac{\rho(M,t)}{M_s} \frac{\partial M}{\partial t} - D\nabla \rho(M,t). \tag{8.43}$$

In the case of static states, Equation 8.42 is sufficient for determining the temperature-dependence of the stability diagrams. The probability density has the form [151–153]

---

* Notice the similarity between Equations (8.43) and (8.11).

$$\rho(\mathbf{M},t) \propto \exp\left(-\frac{E_{\text{eff}}}{k_B T}\right), E_{\text{eff}} = E(1 - a_J / a_c), \qquad (8.44)$$

where a_c is the amplitude of the spin torque at the critical switching voltage. For a magnetic layer with uniaxial anisotropy $\mathbf{H}_K = H_K M_x / M_s x$ and in-plane applied field $\mathbf{H}_{\text{ext}} = H_x \mathbf{x}$, the lifetime of the magnetization within a given static state is

$$\tau^{-1} = f_0 \exp\left[-\frac{E_b(1 - a_J / a_c)}{k_B T}\right], \qquad (8.45)$$

Where f_0 is the attempt frequency and E_b is the energy barrier, defined as $E_b = K_u V_0 (1 - H_{\text{ext}} / H_c)^\beta$. Here, $K_u V_0$ is the anisotropy energy and H_c is the switching field defined in Equation 8.37. The exponent β is 2 when the external field is along the easy axis and 1.5 otherwise. The out-of-plane torque can be included as a bias-dependent magnetic field, and one replaces H_{ext} by $H_{\text{ext}} + b_J$. Equation 8.45 relates the lifetime τ of the magnetization in a given stable state. Therefore, assuming a constant τ (e.g., equal to the experimental current-pulse duration), it is possible to analytically determine the boundaries between two stable states, as done in Figure 8.6.

Finally, one has to account for the temperature increase due to Joule heating. From the usual thermal diffusion equations [154], the contribution of the Joule heating to the effective temperature of the free layer is obtained by replacing the temperature of the environment T by $\sqrt{T^2 + c_0 V^2}$ [144], so that the equilibrium thermal factor of the free layer $\Delta = K_u V_0 / k_B T$ is replaced by $\Delta = K_u V_0 / k_B \sqrt{T^2 + c_0 V^2}$ in the presence of an external current.

The approach described above disregards the softening of the magnetic exchange when increasing the temperature (in other words, the saturation magnetization is kept constant). A more comprehensive approach that has recently attracted increasing attention considers such an effect in the frame of the Landau-Lifshitz-Bloch equation [155]. In this context, the dynamic equation reads

$$\partial_t \mathbf{M} = -\gamma \mathbf{M} \times \mathbf{H}_{\text{eff}} - \gamma \frac{\alpha_\perp}{M_s^2} \mathbf{M} \times (\mathbf{M} \times \mathbf{H}_{\text{eff}}) - \gamma \frac{\alpha_\parallel}{M_s^2} (\mathbf{M} \cdot \mathbf{H}_{\text{eff}}) \mathbf{M}. \quad (8.46)$$

Here, $\alpha_\perp = \alpha 2T / 3T_C$ and $\alpha_\parallel = \alpha(1 - T / 3T_C)$ when $T < T_C$ and $\alpha_\perp = \alpha_\parallel = \alpha$ for $T > T_C$. $\alpha_\perp$ and $\alpha_\parallel$ are the transverse and longitudinal damping coefficients, and T_C is the Curie temperature. The effective field $\mathbf{H}_{\text{eff}}$ also depends on temperature. This approach has been particularly successful in modeling (ultrafast) thermally-induced or thermally-assisted magnetization dynamics within the atomistic framework [150, 156].

8.3.2.4 Stability Diagrams

The stability diagrams obtained from the analytical treatment given above are displayed in Figure 8.6. Figure 8.6a shows the stability diagram of a metallic spin valve at 0 K ($a_J = a j_e$ and $b_J = 0$). The diagram shows two stability regions where the perturbation $\delta\mathbf{M}$ is damped and the magnetization relaxes toward either P or AP configuration and two regions where the perturbation $\delta\mathbf{M}$ diverges (regions E). These regions are detailed in Figure 8.6b and show both IPP and OPP. Interestingly, the inclusion of thermal effects modifies the boundaries of the stability diagram. Figure 8.6c shows the same diagram at 100 K, 300 K, and 600 K, assuming a lifetime (of pulse duration) $\tau=1$ s. The critical switching voltage and field both decrease with the temperature as the result of the thermally-assisted magnetization switching. When including the Joule heating of the junction [Figure 8.6d], the effective temperature varies with the bias and the critical switching voltage is then dramatically decreased.

Finally, Figure 8.6e displays the stability diagram in the case of a symmetric magnetic tunnel junction, assuming elastic tunneling. In this case, the out-of-plane torque is no more negligible and reads $b_J = b_2 V^2$, $b_2 < 0$. In this case, the out-of-plane torque favors the P-state, and modifies the stability diagram asymmetrically: for large positive bias and small negative external field, b_J competes with H_{ext} and a_J, then decreasing the stability region AP. The opposite happens for negative bias and positive field.

8.3.2.5 Inhomogeneous Magnetic Excitations

Although the macrospin model provides a simple and powerful tool for the interpretation of a number of experiments [68, 69, 141], some experimental data are not reproduced by this method, indicating the micromagnetic nature of the magnetization [58, 157]. In this case, the Oersted and dipolar fields are taken into account as well as the magnetic exchange arising from the spatially varying magnetization texture. These two terms, Oersted field and magnetic exchange, are responsible for vortex nucleation and collective magnetic excitations (spin waves) that are absent in the macroscopic model and can deeply modify the magnetization dynamics.

As an example, the current-induced magnetization switching shows a much more complex behavior in micromagnetic modeling [139, 158]. Whereas in the macrospin model the magnetization switching occurs after fast coherent magnetization precessions, in micromagnetic modeling, it appears to occur in two steps: First, incoherent spin waves are excited by the current close to the edge of the layer; after a few nanoseconds, these spin waves have extended along the edges and begin to excite the center of the layer, generating inhomogeneous magnetic domains and vortices. Eventually, the magnetization of the free layer switches. The contribution of the Oersted field is essential for distorting the initial magnetic configuration and breaking the symmetry of the system. This type of incoherent magnetization switching has been directly observed by time-resolved x-ray imaging [159].

Most of the magnetic excitation features predicted by the macrospin model are observed within the micromagnetic simulations [160, 161]. In particular, in-plane and out-of-plane precessions (IPP and OPP in Figure 8.6b) are found with both approaches, even if the line width of the magnetic excitations is usually different. However, the presence of spin-wave modes within the ferromagnetic layer can produce magnetic behaviors far from the macrospin predictions [71, 158, 161] up to chaotic dynamics [162, 163]. These modes can be excited either by small perturbations (linear regime) or by a strong spin-polarized current (non-linear regime), resulting in strongly inhomogeneous magnetic excitations [164], as observed both experimentally and theoretically. In the non-linear regime, self-localized modes such as spin-wave bullets [165] or droplets can be observed [62].

Finally, we mention the increasing interest in phase-locked nano-oscillators. Mancoff et al. [59] and Kaka et al. [60] simultaneously demonstrated the mutual phase-locking of the precession frequency of two spin valves in point-contact geometry. A number of theoretical investigations [165–167] addressed this effect within the non-linear regime of spin waves excitations. The current-driven phase-locking of an array of nano-oscillators via an AC current has been recently achieved experimentally [168].

8.3.3 Spin-Wave Dynamics and Domain Wall Motion

Ferromagnetic films display very rich dynamical properties. Most of the theoretical studies are carried out via micromagnetic simulations where standard codes are now available. The dynamic equation in the presence of current and magnetic field is given by

$$
\frac{\partial \mathbf{M}}{\partial t} = -\gamma \mathbf{M} \times \mathbf{H}_{\mathrm{eff}} + \frac{\alpha}{M_s} \mathbf{M} \times \frac{\partial \mathbf{M}}{\partial t} + \frac{b_J}{M_s^2} (\mathbf{u} \cdot \nabla) \mathbf{M}
$$

$$
- \beta \frac{b_J}{M_s} \mathbf{M} \times (\mathbf{u} \cdot \nabla) \mathbf{M}.
$$

(8.47)

As we have discussed earlier, the above model contains two crucial parameters: the damping parameter α and the non-adiabaticity constant β; both parameters are materials dependent, in principle non-local, and have not been well-understood microscopically. The study of the current-driven dynamics given by (8.47) falls into two broadly defined categories: ferro-dynamics and domain wall motion. The former concerns current-induced ferromagnetic instabilities where the initially uniformly magnetized film or nanowire becomes unstable at a large critical value of the current. The latter concerns the current-induced control and transformation of magnetic domain walls when the spin torque exceeds the (extrinsic or intrinsic) pinning potential.

Magnetic domain walls present various structural forms determined by the geometrical and material parameters of the films. In narrow magnetic wires, with a width smaller than the magnetic exchange length, domain walls are well described by one-dimensional models and are referred to as transverse walls (TW – Figure 8.7a). In wider wires, the magnetic texture can adopt a variety of configurations such as vortex walls separating oppositely magnetized domains (VW – Figure 8.7b), or magnetic skyrmions that behave like magnetic topological defects in an otherwise homogeneous magnetic background [107]. When an external force (magnetic field or electric

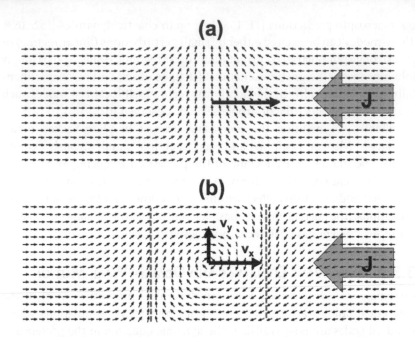

FIGURE 8.7 Magnetization patterns of (a) a transverse wall and (b) a vortex wall. In the presence of an applied current, the transverse wall moves against the current flow. In the case of a vortex or a skyrmion, an additional transverse velocity appears, an effect sometimes referred to as the skyrmion or vortex Hall effect.

current) is applied, TW, VW, and skyrmions are able to move along the wire. The relation between this applied force and the domain wall velocity is determined by the materials parameter, the geometry and the nature of the pinning forces. The dynamics of the walls are generally very complex and micromagnetic simulations are required in order to describe the details of the domain wall motion [107, 169, 170]. Because TW are (mostly) one-dimensional objects, whereas VW and skyrmions are two-dimensional, they display very distinct current-driven response and we choose to treat there separately.

8.3.3.1 Ferrodynamics: Destabilization of Ferromagnetism by Currents

The critical current density for the linear instability of a uniformly magnetized film can be readily obtained by using the conventional linear instability analysis [46]. Consider a small deviation of the uniform magnetization, $\tilde{\mathbf{m}}(x,t) = \mathbf{e}_x + \delta\tilde{\mathbf{m}}e^{i(\omega(k)t + kx)}$ where we have defined $\tilde{\mathbf{m}} = \mathbf{M}/M_s$ as a unit vector, $\delta\tilde{\mathbf{m}}$, k and $\omega(k)$ are the amplitude (a small vector perpendicular to $\mathbf{e}_x$), wave vector and frequency of the spin wave, respectively. Since we assume the wire is infinitely long and the boundary effect is discarded, $\delta\tilde{\mathbf{m}}$ can be treated as a spatially independent constant. By placing $\tilde{\mathbf{m}}(x,t)$ into Equation 8.47 and carrying out the linear instability analysis, we find,

$$\begin{bmatrix} -i(\omega + b_J k) & f_1 \\ f_2 & -i(\omega + b_J k) \end{bmatrix} \begin{bmatrix} \delta\tilde{m}_y \\ \delta\tilde{m}_z \end{bmatrix} = 0, \tag{8.48}$$

where we have defined $f_1 = -\gamma(J_{ex}k^2 + H_1) - i(\alpha\omega + \beta b_J k)$ and $f_2 = \gamma(J_{ex}k^2 + H_2) + i(\alpha\omega - \beta b_J k)$, where $H_1 = H_{ext} + H_K + 4\pi M_s$ and $H_2 = H_{ext} + H_K$, where H_{ext} and H_K are the external field and the anisotropy field along the wire; J_{ex} is the exchange constant. By calculating the determinant of the above secular equation, one could find the critical current $j^c(k)$ at the onset of instabilities ($Im[\omega] = 0$)

$$j^c(k) = \frac{\alpha\gamma}{kb(\alpha - \beta)}\sqrt{(J_{ex}k^2 + H_1)(J_{ex}k^2 + H_2)}, \qquad (8.49)$$

where $b_J = bj_e$. By minimizing $j^c(k)$ with respect to k, i.e., $\partial j^c / \partial k = 0$, we find the minimum critical current density j_{c2} is

$$j_{c2} = \alpha\gamma\sqrt{J_{ex}}\left(\sqrt{H_1} + \sqrt{H_2}\right)/b(\alpha - \beta). \qquad (8.50)$$

Interestingly, only a single spin wave with frequency $\omega_c = j_{c2}(\beta b / \alpha)(H_1 H_2 / J_{ex}^2)^{1/4}$ is excited at j_{c2}. One notes that j_{c2} is infinite when $\beta = \alpha$. The lack of spin-wave excitations, in this case, can be traced back to Equation 8.47 when one makes a Galilean transformation, $x' = x - bjt$ and $t' = t$, Equation 8.47 becomes a usual LLG equation without the spin torque terms. Thus, the spin-wave excitations are prohibited when $\beta = \alpha$ as expected for a Galilean invariant system.

When the current density is larger than j_{c2}, the excited spin-wave frequency expands from a single value to a finite range. Equation 8.48 has two solutions for $\omega(k)$

$$\omega_1 = -bj_e k + \gamma\sqrt{(J_{ex}k^2 + H_1)(J_{ex}k^2 + H_2)}, \qquad (8.51)$$

$$\omega_2 = -\frac{\beta bj_e}{\alpha}k. \qquad (8.52)$$

Equation 8.51 is known as the spin-wave dispersion relation with Doppler shift [171], and Equation 8.52 is an additional solution associated with the non-adiabatic spin torque. To fully describe the ferrodynamic phase for $j > j_{c2}$, it is necessary to quantitatively determine the number of the spin waves excited by the current using micromagnetic simulations. The detailed features, such as fast spin-wave propagation, critical exponents of the order parameters, and generation of localized spin-wave packets have been reported in several references [46, 152, 171–173]. Although a number of experimental results have reported current-driven domain nucleation [20, 174], the relation to ferrodynamics is not clear yet.

Finally, we mention the observation of current-driven magnetic excitations in a single magnetic layer when the current is injected perpendicular to the plane of the layer [175, 176]. These excitations have been associated with the spin transfer torque mediated by the electron diffusion along the interfaces [177, 178] (also called *self-torque*). Another interpretation could be the current-driven instabilities in the framework of the ferrodynamic effect. Up until now, no clear experimental data has been able to determine which phenomenon is more likely responsible for these excitations.

8.3.3.2 Current-Driven Domain Wall Motion: Transverse Walls

The transport regime in which domain walls move under the action of an external force is determined by the relative magnitude of this force with respect to the pinning potential. Two classes of pinning exist: *extrinsic* pinning due to notches or defects, and *intrinsic* pinning, which determines the resistance of the magnetic texture to deformation. In the most common cases, the extrinsic pinning potential is smaller than the intrinsic pinning, which allows to roughly define four main transport regimes [179] (see Figure 8.8).

When the external force (field or current) is smaller than the extrinsic pinning, the domain wall is in the *creep regime* and the velocity is defined as [180]

$$v = v_0 \exp\left[-\frac{T_p}{T}\left(\frac{f_p}{f}\right)^\mu \right],$$

(8.53)

where v_0 is a velocity parameter that depends on the details of the potential profile and energy dissipation, T_p is the depinning temperature and f_p is the depinning force. This law resembles Mott's variable range hopping law for charge transport in strongly disordered systems. The exponent μ depends on the nature of the pinning potential (long range, short range etc.) [181]. Notice that the scaling law (8.53) only applies in the limiting of

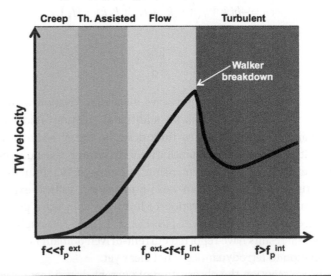

FIGURE 8.8 Schematics of the different transport regime of a transverse domain wall as a function of the applied force f. Here, the ordinate axis represents the domain wall velocity, f_p^{ext} and f_p^{int} are the extrinsic and intrinsic pinning forces, respectively.

a weak driving force ($f \ll f_p$) and small temperature ($T \ll T_p$). Up till now, this regime could not be modeled by micromagnetic simulations due to the excessive computation time demand. This regime is of central importance in the current development of an ultrahigh velocity domain walls and is not fully understood. Upon increasing the external force, the system may transit toward a *thermally-assisted regime* when the velocity is given by Arrhenius law $v \sim v_0 \exp\left[-\dfrac{T_p}{T}\left(1 - \dfrac{f_p}{f}\right)\right]$. When the external force is larger than the extrinsic pinning, but not strong enough to significantly deform the domain wall, the wall is in the *flow regime*. If the external force exceeds the intrinsic pinning (this is the so-called Walker breakdown), the domain wall undergoes periodic transformations and enters the *turbulent regime* [182].

To address the theoretical dynamics in the flow and turbulent regimes, two approaches are commonly used. One is the numerical simulation based on micromagnetic codes [169, 170, 183–185]. This approach can quantitatively address the detailed dynamic features of various domain walls in the presence of defects. Another approach is based on Thiele's method [186] and approximates the domain wall as a rigid object that can be characterized by a few parameters such as the center of the wall, the orientation of the wall plane and the size of the vortex core [183, 185]. For TWs, a simple scenario assumes the domain wall undergoes a uniform motion along the direction of the electron current [184]. This is modeled by $\mathbf{M}(x,t) = \mathbf{M}(x - v_x t)$ and $\partial \mathbf{M}/\partial t = -v_x \partial \mathbf{M}/\partial x$, where v_x is the wall velocity. Placing this into (8.47), we readily find that

$$\gamma \mathbf{M} \times \mathbf{H}_{\text{eff}} = -\frac{v_x \alpha + \beta b_J}{M_s} \mathbf{M} \times \frac{\partial \mathbf{M}}{\partial x} + \left(b_J + v_x\right)\frac{\partial \mathbf{M}}{\partial x}. \tag{8.54}$$

After some algebra (see [184] for details), the TW velocity for a rigid wall is obtained

$$v_x = \frac{\pm \gamma H_{\text{ext}} \Delta - \beta b_J}{\alpha}, \tag{8.55}$$

where the domain wall width Δ is defined as $\Delta \propto \int dV \left|\dfrac{\partial M}{\partial x}\right|^2$. Equation 8.55 demonstrates that the domain wall velocity solely depends on the external magnetic field and on the *non-abiabatic* torque βb_J. Actually, the domain wall width Δ weakly depends on the magnetic field and on the adiabatic torque b_J. Therefore, under an external field, the final domain wall velocity slightly depends on b_J.

In the case where the TW is not rigid ($\Delta = \Delta(t)$), one can parametrize the domain wall by three time-dependent variables: the polar and azimuthal angles $\theta(t)$ and $\phi(t)$, and the domain wall width $\Delta(t)$ [187]. A powerful method is to determine the equation of motion through the Lagrangian formalism [188, 189]. For the sake of concreteness, let us consider a domain wall in a one-dimensional perpendicularly magnetized nanowire. The magnetic energy density reads $W = A|\partial_x \mathbf{M}|^2 - K_\perp M_z^2 - K_D M_y^2$. The first term is the magnetic exchange, the second term is the perpendicular anisotropy and the third term is a (dipolar) anisotropy term that favors Bloch configuration. Assuming the usual ansatz for the polar angle ($\theta(t) = 2\tan^{-1}e^{(x-q)/\Delta}$, $q = q(t)$ being the position of the domain wall center), one obtains the equations of motion

$$\partial_t q - \alpha\Delta\partial_t\phi = -\gamma\Delta H_D\cos\phi\sin\phi - b_J, \tag{8.56}$$

$$\alpha\partial_t q + \Delta\partial_t\phi = -\beta b_J, \tag{8.57}$$

$$\frac{\alpha M_s \pi^2}{12\gamma\Delta}\partial_t\Delta = \frac{A}{\Delta^2} - K_\perp + K_D\sin^2\phi, \tag{8.58}$$

where $H_D = 2K_D / M_s$. Equation 8.56 shows that the adiabatic torque causes a distortion of the wall (via the azimuthal angle ϕ) that absorbs the adiabatic spin angular momentum. In the absence of non-adiabatic torque, once this distortion is complete the net effect of the adiabatic torque vanishes and the wall stops [183, 184]. In contrast, the non-adiabatic torque in (8.57) acts as a magnetic field and induces a steady-state domain wall motion ($\partial_t\phi = 0$): the steady velocity is solely determined by the non-adiabatic torque, as shown in (8.55). This holds true as long as the current does not exceed a critical value above which the azimuthal angle becomes time dependent ($\partial_t\phi \neq 0$) and the wall enters the turbulent regime. In this case, the time-averaged velocity is solely given by the adiabatic torque. Finally, notice that the domain wall width is unlikely to be distorted as long as $K_\perp \gg K_D$.

8.3.3.3 Current-Induced Domain Wall Motion: Vortex Walls and Skyrmions

Two-dimensional magnetic textures such as skyrmions and vortex walls present some striking differences with the one-dimensional transverse walls discussed above [185]. As seen in the previous section, electrons flowing in such textures experience an emergent magnetic field, $\sim [\partial_x \mathbf{M} \times \partial_y \mathbf{M}] \cdot \mathbf{M}$, which induces the topological Hall effect. The transverse electron flow driven by the topological Hall effect, in turn, creates a large non-adiabatic torque. Following Thiele's method [186], one finds that the velocity of such two-dimensional textures reads

$$v_x|_{t=0} \simeq -b_J, \tag{8.59}$$

$$v_y|_{t=0} \simeq -p(\alpha - \beta)\mathcal{F} b_J, \tag{8.60}$$

where $p = \pm 1$ is the vortex polarity and $\mathcal{F}$ is a parameter that depends on the details of the texture.

The VW structure might be schematically understood as the combination of two symmetrical transverse walls diagonally crossing the wire and a central vortex core connecting the two TWs (the vertical red dashed lines in Figure 8.7b). The distorted TWs produce a reacting (or restoring) force to the vortex core during the domain wall motion. Then, two scenarios are possible: the reacting force is strong enough to completely halt the perpendicular wall velocity, or the reacting force is unable to stop the vortex from colliding with the wall edge. In the former case, the perpendicular wall velocity eventually reaches zero and a steady-state wall velocity along the wire is achieved. The final wall velocity is precisely the same as that of the transverse wall i.e., $v_x|_{t=\infty} = -\beta b_J / \alpha$ and $v_y|_{t=\infty} = 0$. In the latter case, the vortex core collides with the wire edge and can either vanish, move out of the wire, or be reflected, i.e., the transformation of the vortex wall to other types of walls occurs. Usually, the vortex core disappears at the edge and the initial VW becomes a TW [63, 64]. However, the above conclusion is not universal and in the presence of the defects or pinning centers, the adiabatic torque plays a critical role in determining the critical depinning current [169, 170, 182–184]. Furthermore, the wall velocity becomes a complicated function of the current and field, depending on the wall structure as well as the strength of currents and fields.

Magnetic skyrmions are quite different from vortices as they are localized in space. In this sense, they behave more like isolated vortex cores. As a result, they react very differently to defects or pinning centers. As a matter of fact, they can easily deform while passing through the defect, thereby avoiding pinning [190, 191]. This robustness is a distinctive characteristic of skyrmions, directly related to their topology. Various situations have been investigated theoretically and numerically, and it is believed that skyrmions offer an interesting platform for ultradense data storage [192].

8.4 CONCLUSION

Since its first theoretical prediction, the spin torque mechanism has attracted a massive amount of attention from the spintronics community. Understanding the interplay between spin currents and magnetization dynamics in spin valves, tunnel junctions, and domain walls still represents one of the most important challenges for applications in magnetic memories, read-heads, and oscillators.

The theory of spin-dependent transport has also made a lot of progress in the understanding of spin currents, and related phenomena in all the structures mentioned above. Even if the main physics is now captured within magnetoelectronics or diffusive theories, quantitative agreement between the theory and the experiment remains to be achieved, and the large interest for numerical studies (first principle calculations and micromagnetic modeling) illustrate the need for parameters-free understanding.

The conventional spin torque is now aiming toward applications, and recent theoretical studies have proposed a different new type of spin torques, using thermokinetics properties, spin-orbit interaction or specific magnetic texture such as antiferromagnets. These last three examples are now fostering massive efforts and excitement as they offer thrilling perspectives for the development of disruptive spin torque-based devices.

REFERENCES

1. A. D. Kent and D. C. Worledge, A new spin on magnetic memories, *Nat. Nanotechnol.* **10**, 187–191 (2015).
2. A. Brataas, G. E. W. Bauer, and P. J. Kelly, Non-collinear magnetoelectronics, *Phys. Rep.* **427**, 157–255 (2006).
3. M. D. Stiles and J. Miltat, Spin transfer torque and dynamics in spin dynamics in confined structures, in *Spin Dynamics in Confined Magnetic Structures III*, Topics in Applied Physics, Vol. 101, B. Hillebrands and A. Thiaville (Eds.), Springer, Berlin, pp. 225–308, 2006.
4. D. C. Ralph and M. D. Stiles, Spin transfer torques, *J. Magn. Magn. Mater.* **320**, 1190–1216 (2008).
5. J. Z. Sun and D. C. Ralph, Magnetoresistance and spin-transfer torque in magnetic tunnel junctions, *J. Magn. Magn. Mater.* **320**, 1227–1237 (2008).
6. A. Brataas, A. D. Kent, and H. Ohno, Current-induced torques in magnetic materials, *Nat. Mater.* **11**, 372–381 (2012).
7. L. Berger, Low-field magnetoresistance and domain drag in ferromagnets, *J. Appl. Phys.* **49**, 2156–2161 (1978).
8. J. C. Slonczewski, Conductance and exchange coupling of two ferromagnets separated by a tunneling barrier, *Phys. Rev. B* **39**, 6995–7002 (1989).
9. L. Berger, Emission of spin waves by a magnetic multilayer traversed by a current, *Phys. Rev. B* **54**, 9353–9358 (1996).
10. J. C. Slonczewski, Current-driven excitation of magnetic multilayers, *J. Magn. Magn. Mater.* **159**, L1–L7 (1996).
11. M. Tsoi, A. G. M. Jansen, J. Bass et al., Excitation of a magnetic multilayer by an electric current, *Phys. Rev. Lett.* **80**, 4281–4284 (1998).
12. E. B. Myers, D. C. Ralph, J. A. Katine, R. N. Louie, and R. A. Buhrman, Current-induced switching of domains in magnetic multilayer devices, *Science* **285**, 867–870 (1999).
13. J. A. Katine, F. J. Albert, R. A. Buhrman, E. B. Myers, and D. C. Ralph, Current-driven magnetization reversal and spin-wave excitations in Co /Cu /Co pillars, *Phys. Rev. Lett.* **84**, 3149–3152 (2000).
14. A. Fert, V. Cros, J.-M. George et al., Magnetization reversal by injection and transfer of spin: experiments and theory, *J. Magn. Magn. Mater.* **272**, 1706–1711 (2004).
15. S. Mangin, D. Ravelosona, J. A. Katine, M. J. Carey, B. D. Terris, and E. E. Fullerton, Current-induced magnetization reversal in nanopillars with perpendicular anisotropy, *Nat. Mater.* **5**, 210–215 (2006).
16. J. Z. Sun, Current-driven magnetic switching in manganite trilayer junctions, *J. Magn. Magn. Mater.* **202**, 157–162 (1999).
17. Y. Huai, F. Albert, P. Nguyen, M. Pakala, and T. Valet, Observation of spin-transfer switching in deep submicron-sized and low-resistance magnetic tunnel junctions, *Appl. Phys. Lett.* **84**, 3118–3120 (2004).
18. G. D. Fuchs, N. C. Emley, I. N. Krivorotov et al., Spin-transfer effects in nanoscale magnetic tunnel junctions, *Appl. Phys. Lett.* **85**, 1205–1207 (2004).
19. A. Yamaguchi, T. Ono, S. Nasu et al., Real-space observation of current-driven domain wall motion in submicron magnetic wires, *Phys. Rev. Lett.* **92**, 077205 (2004).
20. M. Kläui, P.-O. Jubert, R. Allenspach et al., Direct observation of domain-wall configurations transformed by spin currents, *Phys. Rev. Lett.* **95**, 026601 (2005).

21. Z. Wei, A. Sharma, A. S. Nunez et al., Changing exchange bias in spin valves with an electric current, *Phys. Rev. Lett.* **98**, 116603 (2007).

22. S. Urazhdin and N. Anthony, Effect of polarized current on the magnetic state of an antiferromagnet, *Phys. Rev. Lett.* **99**, 046602 (2007).

23. D. Chiba, Y. Sato, T. Kita, F. Matsukura, and H. Ohno, Current-driven magnetization reversal in a ferromagnetic semiconductor (Ga,Mn)As/GaAs/(Ga,Mn)As tunnel junction, *Phys. Rev. Lett.* **93**, 216602 (2004).

24. M. Yamanouchi, D. Chiba, F. Matsukura, and H. Ohno, Current-induced domain-wall switching in a ferromagnetic semiconductor structure, *Nature* **428**, 539–542 (2004).

25. A. Chernyshov, M. Overby, X. Liu, J. K. Furdyna, Y. Lyanda-Geller, and L. P. Rokhinson, Evidence for reversible control of magnetization in a ferromagnetic material by means of spin–orbit magnetic field, *Nat. Phys.* **5**, 656–659 (2009).

26. I. M. Miron, K. Garello, G. Gaudin et al., Perpendicular switching of a single ferromagnetic layer induced by in-plane current injection, *Nature* **476**, 189–194a (2011a).

27. L. Liu, C. F. Pai, Y. Li, H. W. Tseng, D. C. Ralph, and R. A. Buhrman, Spin-torque switching with the giant spin Hall effect of tantalum, *Science* **338**, 555–558 (2012).

28. X. Waintal, E. B. Myers, P. W. Brouwer, and D. C. Ralph, Role of spin-dependent interface scattering in generating current-induced torques in magnetic multilayers, *Phys. Rev. B* **62**, 12317–12327 (2000).

29. M. D. Stiles and A. Zangwill, Anatomy of spin-transfer torque, *Phys. Rev. B* **66**, 014407 (2002).

30. J. C. Slonczewski, Currents and torques in metallic magnetic multilayers, *J. Magn. Magn. Mater.* **247**, 324–338 (2002).

31. S. Zhang, P. M. Levy, and A. Fert, Mechanisms of spin-polarized current-driven magnetization switching, *Phys. Rev. Lett.* **88**, 236601 (2002).

32. J. Barnas, A. Fert, M. Gmitra, I. Weymann, and V. K. Dugaev, From giant magnetoresistance to current-induced switching by spin transfer, *Phys. Rev. B* **72**, 024426 (2005).

33. A. Manchon, N. Ryzhanova, N. Strelkov, A. Vedyayev, and B. Dieny, Modeling spin transfer torque and magnetoresistance in magnetic multilayers, *J. Phys. Condens. Matter* **19**, 165212 (2007).

34. I. Theodonis, N. Kioussis, A. Kalitsov, M. Chshiev, and W. H. Butler. Anomalous bias dependence of spin torque in magnetic tunnel junctions, *Phys. Rev. Lett.* **97**, 237205 (2006).

35. M. Wilczynski, J. Barnas, and R. Swirkowicz, Free-electron model of current-induced spin-transfer torque in magnetic tunnel junctions, *Phys. Rev. B* **77**, 054434 (2008).

36. J. Xiao, G. E. W. Bauer, and A. Brataas, Spin-transfer torque in magnetic tunnel junctions: Scattering theory, *Phys. Rev. B* **77**, 224419 (2008).

37. A. Manchon, N. Ryzhanova, N. Strelkov, A. Vedyayev, M. Chshiev, and B. Dieny, Description of current-driven torques in magnetic tunnel junctions, *J. Phys. Condens. Matter* **20**, 145208 (2008).

38. C. Heiliger and M. D. Stiles, Ab initio studies of the spin-transfer torque in magnetic tunnel junctions, *Phys. Rev. Lett.* **100**, 186805 (2008).

39. J. C. Slonczewski, Currents, torques, and polarization factors in magnetic tunnel junctions, *Phys. Rev. B* **71**, 024411 (2005).

40. P. M. Levy and A. Fert, Spin transfer in magnetic tunnel junctions with hot electrons, *Phys. Rev. Lett.* **97**, 097205 (2006).

41. A. Manchon and S. Zhang, Influence of interfacial magnons on spin transfer torque in magnetic tunnel junctions, *Phys. Rev. B* **79**, 174401 (2009).

42. J. C. Sankey, Y. T. Cui, J. Z. Sun, J. C. Slonczewski, R. A. Buhrman, and D C. Ralph, Measurement of the spin-transfer-torque vector in magnetic tunnel junctions, *Nat. Phys.* **4**, 67 (2008).

43. H. Kubota, A. Fukushima, K. Yakushiji et al., Quantitative measurement of voltage dependence of spin-transfer torque in MgO-based magnetic tunnel junctions, *Nat. Phys.* **4**, 37 (2008).

44. A. M. Deac, A. Fukushima, H. Kubota et al., Bias-driven high-power microwave emission from MgO-based tunnel magnetoresistance devices, *Nat. Phys.* **4**, 803 (2008).

45. S.-C. Oh, S.-Y. Park, A. Manchon et al., Bias-voltage dependence of perpendicular spin-transfer torque in asymmetric MgO-based magnetic tunnel junctions, *Nat. Phys.* **5**, 898–902 (2009).

46. Ya. B. Bazaliy, B. A. Jones, and S.-C. Zhang, Modification of the Landau-Lifshitz equation in the presence of a spin-polarized current in colossal- and giant-magnetoresistive materials, *Phys. Rev. B* **57**, R3213–R3216 (1998).

47. S. Zhang and Z. Li, Roles of nonequilibrium conduction electrons on the magnetization dynamics of ferromagnets, *Phys. Rev. Lett.* **93**, 127204 (2004).

48. A. Thiaville, Y. Nakatani, J. Miltat, and Y. Suzuki, Micromagnetic understanding of current-driven domain wall motion in patterned nanowires, *Europhys. Lett.* **69**, 990 (2005).

49. G. Tatara and H. Kohno, Theory of current-driven domain wall motion: Spin transfer versus momentum transfer, *Phys. Rev. Lett.* **92**, 086601 (2004).

50. J. Xiao, A. Zangwill, and M. D. Stiles, Spin-transfer torque for continuously variable magnetization, *Phys. Rev. B* **73**, 054428 (2006).

51. Y. Tserkovnyak, H. J. Skadsem, A. Brataas, and G. E. W. Bauer, Current-induced magnetization dynamics in disordered itinerant ferromagnets, *Phys. Rev. B* **74**, 144405 (2006).

52. G. Tatara, H. Kohno, J. Shibata, Y. Lemaho, and K.-J. Lee, Spin torque and force due to current for general spin textures, *J. Phys. Soc. Jpn.* **76**, 054707 (2007).

53. C. A. Akosa, W. S. Kim, A. Bisig, M. Kläui, K. J. Lee, and A. Manchon, Role of spin diffusion in current-induced domain wall motion for disordered ferromagnets, *Phys. Rev. B* **91**, 094411 (2015).

54. S. E. Barnes and S. Maekawa, Generalization of Faraday's law to include nonconservative spin forces, *Phys. Rev. Lett.* **98**, 246601 (2007).

55. J.-I. Ohe, A. Takeuchi, and G. Tatara, Charge current driven by spin dynamics in disordered Rashba spin-orbit system, *Phys. Rev. Lett.* **99**, 266603 (2007).

56. R. A. Duine, Effects of nonadiabaticity on the voltage generated by a moving domain wall, *Phys. Rev. B* **79**, 014407 (2009).

57. S. E. Barnes and S. Maekawa, Current-spin coupling for ferromagnetic domain walls in fine wires, *Phys. Rev. Lett.* **95**, 107204 (2005).

58. S. I. Kiselev, J. C. Sankey, I. N. Krivorotov et al., Microwave oscillations of a nanomagnet driven by a spin-polarized current, *Nature* **425**, 380–383 (2003).

59. F. B. Mancoff, N. D. Rizzo, B. N. Engel, and S. Tehrani, Phase-locking in double-point-contact spin-transfer devices, *Nature* **437**, 393–395 (2005).

60. S. Kaka, M. R. Pufall, W. H. Rippard, T. J. Silva, S. E. Russek, and J. A. Katine, Mutual phase-locking of microwave spin torque nano-oscillators, *Nature* **437**, 389–392 (2005).

61. D. Houssameddine, U. Ebels, B. Delaet et al., Spin-torque oscillator using a perpendicular polarizer and a planar free layer, *Nat. Mater.* **6**, 447–454 (2007).

62. S. M. Mohseni, S. R. Sani, J. Persson et al., Spin torque? generated magnetic droplet solitons, *Science* **339**, 1295 (2013).

63. M. Hayashi, L. Thomas, C. Rettner, R. Moriya, Ya. B. Bazaliy, and S. S. P. Parkin, Current driven domain wall velocities exceeding the spin angular momentum transfer rate in permalloy nanowires, *Phys. Rev. Lett.* **98**, 037204 (2007).

64. L. Heyne, M. Kläui, D. Backes et al., Relationship between nonadiabaticity and damping in permalloy studied by current induced spin structure transformations, *Phys. Rev. Lett.* **100**, 066603 (2008).

65. I. M. Miron, T. Moore, H. Szambolics et al., Fast current-induced domain-wall motion controlled by the Rashba effect, *Nat. Mater.* **10**, 419–423 (2011b).

66. S. H. Yang, K. Ryu, and S. S. P. Parkin, Domain-wall velocities of up to 750 m s^{-1} driven by exchange-coupling torque in synthetic antiferromagnets, *Nat. Nanotechnol.* **10**, 221–226 (2015).

67. R. H. Koch, J. A. Katine, and J. Z. Sun, Time-resolved reversal of spin-transfer switching in a nanomagnet, *Phys. Rev. Lett.* **92**, 088302 (2004).

68. J. Z. Sun, Spin-current interaction with a monodomain magnetic body: A model study, *Phys. Rev. B* **62**, 570–578 (2000).

69. G. Bertotti, C. Serpico, I. D. Mayergoyz, A. Magni, M. d'Aquino, and R. Bonin, Magnetization switching and microwave oscillations in nanomagnets driven by spin-polarized currents, *Phys. Rev. Lett.* **94**, 127206 (2005).

70. K.-J. Lee, A. Deac, O. Redon, J.-P. Nozieres, and B. Dieny, Excitations of incoherent spin-waves due to spin-transfer torque, *Nat. Mater.* **3**, 877–881 (2004).

71. D. Berkov and N. Gorn, Transition from the macrospin to chaotic behavior by a spin-torque driven magnetization precession of a square nanoelement, *Phys. Rev. B* **71**, 052403 (2005).

72. T. Valet and A. Fert, Theory of the perpendicular magnetoresistance in magnetic multilayers, *Phys. Rev. B* **48**, 7099 (1993).

73. S. V. Vonsovsky, *Magnetism*, Wiley, New York, 1974.

74. T. L. Gilbert, A phenomenological theory of damping in ferromagnetic materials, *IEEE Trans. Magn.* **40**, 3443–3449 (2004).

75. K. Gilmore, Y. U. Idzerda, and M. D. Stiles, Identification of the dominant precession-damping mechanism in Fe, Co, and Ni by first-principles calculations, *Phys. Rev. Lett.* **99**, 027204 (2007).

76. V. Kambersky, Spin-orbital Gilbert damping in common magnetic metals, *Phys. Rev. B* **76**, 134416 (2007).

77. C. Petitjean, D. Luc, and X. Waintal, Unified drift-diffusion theory for transverse spin currents in spin valves, domain walls, and other textured magnets, *Phys. Rev. Lett.* **109**, 117204 (2012).

78. M. Zwierzycki, Y. Tserkovnyak, P. J. Kelly, A. Brataas, and G. E. W. Bauer, First-principles study of magnetization relaxation enhancement and spin transfer in thin magnetic films, *Phys. Rev. B* **71**, 064420 (2005).

79. P. M. Haney, D. Waldron, R. A. Duine, A. S. Nunez, H. Guo, and A. H. MacDonald, Current-induced order parameter dynamics: Microscopic theory applied to Co/Cu/Co spin valves, *Phys. Rev. B* **76**, 024404 (2007).

80. S. Wang, Y. Xu, and K. Xia, First-principles study of spin-transfer torques in layered systems with noncollinear magnetization, *Phys. Rev. B* **77**, 184430a (2008).

81. O. Boulle, V. Cros, J. Grollier et al., Shaped angular dependence of the spin-transfer torque and microwave generation without magnetic field, *Nat. Phys.* **3**, 492–497 (2007).

82. M. Weiler, M. Althammer, M. Schreier et al., Experimental test of the spin mixing interface conductivity concept, *Phys. Rev. Lett.* **111**, 176601 (2013).

83. K. Chen and S. Zhang, Spin pumping in the presence of spin-orbit coupling, *Phys. Rev. Lett.* **114**, 126602 (2015).

84. Z. Li, S. Zhang, Z. Diao et al., Perpendicular spin torques in magnetic tunnel junctions, *Phys. Rev. Lett.* **100**, 246602 (2008).

85. S. Petit, C. Baraduc, C. Thirion et al., Spin-torque influence on the high-frequency magnetization fluctuations in magnetic tunnel junctions, *Phys. Rev. Lett.* **98**, 077203 (2007).

86. E. Y. Tsymbal, O. N. Mryasov, and P. R. LeClair, Spin-dependent tunnelling in magnetic tunnel junctions, *J. Phys. Condens. Matter* **15**, R109–R142 (2003).

87. S. Zhang, P. M. Levy, A. C. Marley, and S. S. P. Parkin, Quenching of magnetoresistance by hot electrons in magnetic tunnel junctions, *Phys. Rev. Lett.* **79**, 3744–3747 (1997).

88. J. Z. Sun, M. C. Gaidis, G. Hu et al., High-bias backhopping in nanosecond time-domain spin-torque switches of MgO-based magnetic tunnel junctions, *J. Appl. Phys.* **105**, 07D109 (2009).

89. W. F. Brinkman, R. C. Dynes, and J. M. Rowell, Tunneling conductance of asymmetrical barriers, *J. Appl. Phys.* **41**, 1915 (1970).

90. Y.-H. Tang, N. Kioussis, A. Kalitsov, W. H. Butler, and R. Car, Controlling the nonequilibrium interlayer exchange coupling in asymmetric magnetic tunnel junctions, *Phys. Rev. Lett.* **103**, 057206 (2009).

91. A. Manchon, S. Zhang, and K.-J. Lee, Signatures of asymmetric and inelastic tunneling on the spin torque bias dependence, *Phys. Rev. B* **82**, 174420 (2010).

92. W. H. Butler, X.-G. Zhang, T. C. Schulthess, and J. M. MacLaren, Spin-dependent tunneling conductance of Fe|MgO|Fe sandwiches, *Phys. Rev. B* **63**, 054416 (2001).

93. C. Tiusan, J. Faure-Vincent, C. Bellouard, M. Hehn, E. Jouguelet, and A. Schuhl, Interfacial resonance state probed by spin-polarized tunneling in epitaxial Fe/MgO/Fe tunnel junctions, *Phys. Rev. Lett.* **93**, 106602 (2004).

94. P. Mavropoulos, N. Papanikolaou, and P. H. Dederichs, Complex band structure and tunneling through ferromagnet/insulator/ferromagnet junctions, *Phys. Rev. Lett.* **85**, 1088–1091 (2000).

95. A. Useinov, M. Chshiev, and A. Manchon, Controlling the spin-torque efficiency with ferroelectric barriers, *Phys. Rev. B* **91**, 064412 (2015).

96. C. Ortiz Pauyac, A. Kalitsov, A. Manchon, and M. Chshiev, Spin-transfer torque in spin filter tunnel junctions, *Phys. Rev. B* **90**, 235417 (2014).

97. S. Zhang, Spin-dependent surface screening in ferromagnets and magnetic tunnel junctions, *Phys. Rev. Lett.* **83**, 640 (1999).

98. M. Ye Zhuravlev, R. F. Sabirianov, S. S. Jaswal, and E. Y. Tsymbal, Giant electroresistance in ferroelectric tunnel junctions, *Phys. Rev. Lett* **94**, 246802 (2005).

99. E.Y. Tsymbal and H. Kohlstedt, Tunneling across a ferroelectric, *Science* **313**, 181 (2006).

100. J. S. Moodera, T. S. Santos, and T. Nagahama, The phenomena of spin-filter tunneling, *J. Phys. Condens. Matter* **19**, 165202 (2007).

101. I. Garate and A. H. MacDonald, Influence of a transport current on magnetic anisotropy in gyrotropic ferromagnets, *Phys. Rev. B* **80**, 134403 (2009).

102. K. Hosono, J. Shibata, H. Kohno, and Y. Nozaki, Spin torques due to diffusive spin current in magnetic texture, *Phys. Rev. B* **87**, 094404 (2013).

103. O. Wessely, D. M. Edwards, and J. Mathon, Quantum mechanical theory of current-induced domain wall torques in ferromagnetic materials, *Phys. Rev. B* **77**, 174425 (2008).

104. K. Gilmore, I. Garate, A. H. MacDonald, and M. D. Stiles, First-principles calculation of the nonadiabatic spin transfer torque in Ni and Fe, *Phys. Rev. B* **84**, 224412 (2011).

105. A. K. Nguyen, H. J. Skadsem, and A. Brataas, Giant current-driven domain wall mobility in (Ga,Mn)As, *Phys. Rev. Lett.* **98**, 146602 (2007).

106. D. Culcer, M. E. Lucassen, R. A. Duine, and R. Winkler, Current-induced spin torques in III-V ferromagnetic semiconductors, *Phys. Rev. B* **79**, 155208 (2009).

107. N. Nagaosa and Y. Tokura, Topological properties and dynamics of magnetic skyrmions, *Nat. Nanotechnol.* **8**, 900–911 (2013).

108. R. A. Duine, A. S. Nunez, and A. H. MacDonald, Thermally assisted current-driven domain-wall motion, *Phys. Rev. Lett.* **98**, 056605 (2007).

109. J.-I. Ohe and B. Kramer, Dynamics of a domain wall and spin-wave excitations driven by a mesoscopic current, *Phys. Rev. Lett.* **96**, 027204 (2006).

110. Y. Le Maho, J.-V. Kim, and G. Tatara, Spin-wave contributions to current-induced domain wall dynamics, *Phys. Rev. B* **79**, 174404 (2009).

111. D. Hinzke and U. Nowak, Domain wall motion by the magnonic spin Seebeck effect, *Phys. Rev. Lett.* **107**, 027205 (2011).

112. P. Yan, X. S. Wang, and X. R. Wang, All-magnonic spin-transfer torque and domain wall propagation, *Phys. Rev. Lett.* **107**, 177207 (2011).

113. A. A. Kovalev and Y. Tserkovnyak, Thermomagnonic spin transfer and Peltier effects in insulating magnets, *Europhys. Lett.* **97**, 67002 (2012).

114. G. E. Volovik, Linear momentum in ferromagnets, *J. Phys. C* **20**, L83 (1987).

115. W. M. Saslow, Spin pumping of current in non-uniform conducting magnets, *Phys. Rev. B* **76**, 184434 (2007).

116. Y. Tserkovnyak and M. Mecklenburg, Electron transport driven by nonequilibrium magnetic textures, *Phys. Rev. B* **77**, 134407 (2008).

117. R. A. Duine, Spin pumping by a field-driven domain wall, *Phys. Rev. B* **77**, 014409 (2008).

118. S. Zhang and S. S.-L. Zhang, Generalization of the Landau-Lifshitz-Gilbert equation for conducting ferromagnets, *Phys. Rev. Lett.* **102**, 086601 (2009).

119. M. Hatami, G. E. W. Bauer, Q. Zhang, and P. J. Kelly, Thermal spin-transfer torque in magnetoelectronic devices, *Phys. Rev. Lett.* **99**, 066603 (2007).

120. A. A. Kovalev and Y. Tserkovnyak, Thermoelectric spin transfer in textured magnets, *Phys. Rev. B* **80**, 100408 (2009).

121. F. Schlickeiser, U. Ritzmann, D. Hinzke, and U. Nowak, Role of entropy in domain wall motion in thermal gradients, *Phys. Rev. Lett.* **113**, 097201 (2014).

122. X. S. Wang and X. R. Wang, Thermodynamic theory for thermal-gradient-driven domain-wall motion, *Phys. Rev. B* **90**, 014414 (2014).

123. A. Manchon, P. B. Ndiaye, J.-H. Moon, and K.-J. Lee, Magnon-mediated Dzyaloshinskii-Moriya torque in homogeneous ferromagnets, *Phys. Rev. B* **90**, 224403 (2014).

124. A. Manchon and S. Zhang, Theory of spin torque due to spin-orbit coupling, *Phys. Rev. B* **79**, 094422 (2009b).

125. B. A. Bernevig and O. Vafek, Piezo-magnetoelectric effects in p-doped semiconductors, *Phys. Rev. B* **72**, 033203 (2005).

126. H. Li, H. Gao, L. P. Zarbo et al., Intraband and interband spin-orbit torques in noncentrosymmetric ferromagnets, *Phys. Rev. B* **91**, 134402 (2015).

127. P. M. Haney, H. W. Lee, K. J. Lee, A. Manchon, and M. D. Stiles, Current-induced torques and interfacial spin-orbit coupling, *Phys. Rev. B* **88**, 214417 (2013).

128. F. Freimuth, S. Blugel, and Y. Mokrousov, Spin-orbit torques in Co/Pt(111) and Mn/W(001) magnetic bilayers from first principles, *Phys. Rev. B* **90**, 174423 (2014).

129. T. Jungwirth, X. Marti, P. Wadley, and J. Wunderlich, Antiferromagnetic spintronics, *Nat. Nanotechnol.* **11**, 231 (2016).

130. E. V. Gomonay and V. M. Loktev, Spintronics of antiferromagnetic systems, *Low Temp. Phys.* **40**, 17 (2014).

131. A. S. Nunez, R. A. Duine, P. M. Haney, and A. H. MacDonald, Theory of spin torques and giant magnetoresistance in antiferromagnetic metals, *Phys. Rev. B* **73**, 214426 (2006).

132. Y. Xu, S. Wang, and K. Xia, Spin-transfer torques in antiferromagnetic metals from first principles, *Phys. Rev. Lett.* **100**, 226602 (2008).

133. R. A. Duine, P. M. Haney, A. S. Nunez, and A. H. MacDonald, Inelastic scattering in ferromagnetic and antiferromagnetic spin valves, *Phys. Rev. B* **75**, 014433 (2007b).

134. H. B. M. Saidaoui, A. Manchon, and X. Waintal, Spin transfer torque in antiferromagnetic spin valves: From clean to disordered regimes, *Phys. Rev. B* **89**, 174430 (2014).

135. J. Železný, H. Gao, K. Výborný et al., Relativistic néel-order fields induced by electrical current in antiferromagnets, *Phys. Rev. Lett.* **113**, 157201 2014.

136. K. M. D. Hals, Y. Tserkovnyak, and A. Brataas, Phenomenology of current-induced dynamics in antiferromagnets, *Phys. Rev. Lett.* **106**, 107206 (2011).

137. J. Barker and O. A. Tretiakov, Static and dynamical properties of antiferromagnetic skyrmions in the presence of applied current and temperature, *Phys. Rev. Lett.* **116**, 147203 (2016).

138. P. Wadley, B. Howells, J. Železný et al., Electrical switching of an antiferromagnet, *Science* **351**, 587–590 (2016).

139. K.-J. Lee, M. D. Stiles, H.-W. Lee, J.-H. Moon, K.-W. Kim, and S.-W. Lee, Self-consistent calculation of spin transport and magnetization dynamics, *Phys. Rep.* **531**, 89–113 (2013).

140. C. Abert, M. Ruggeri, F. Bruckner et al., A three-dimensional spin-diffusion model for micromagnetics, *Sci. Rep.* **5**, 14855 (2015).

141. J. Grollier, V. Cros, H. Jaffres et al., Field dependence of magnetization reversal by spin transfer, *Phys. Rev. B* **67**, 174402 (2003).

142. V. K. Melkinov, On the stability of the center for time periodic perturbations, *Trans. Moscow Math. Soc.* **12**, 1–57 (1963).

143. W. F. Brown Jr., Thermal fluctuations of a single-domain particle, *Phys. Rev.* **130**, 1677–1686 (1963).

144. G. D. Fuchs, I. N. Krivorotov, P. M. Braganca et al., Adjustable spin torque in magnetic tunnel junctions with two fixed layers, *Appl. Phys. Lett.* **86**, 152509 (2005).

145. A. Fabian, C. Terrier, S. S. Guisan et al., Current-induced two-level fluctuations in pseudo-spin-valve (Co/Cu/Co) nanostructures, *Phys. Rev. Lett.* **91**, 257209 (2003).

146. I. N. Krivorotov, N. C. Emley, A. G. Garcia et al., Temperature dependence of spin-transfer-induced switching of nanomagnets, *Phys. Rev. Lett.* **93**, 166603 (2004).

147. W. L. Lim, N. Anthony, A. Higgins, and S. Urazhdin, Thermal dynamics in symmetric magnetic nanopillars driven by spin transfer, *Appl. Phys. Lett.* **92**, 172501 (2008).

148. S. E. Russek, S. Kaka, W. H. Rippard, M. R. Pufall, and T. J. Silva, Finite-temperature modeling of nanoscale spin-transfer oscillators, *Phys. Rev. B* **71**, 104425 (2005).

149. M. Gmitra and J. Barnas, Thermally assisted current-driven bistable precessional regimes in asymmetric spin valves, *Phys. Rev. Lett.* **99**, 097205 (2007).

150. P. M. Haney and M. D. Stiles, Magnetic dynamics with spin-transfer torques near the Curie temperature, *Phys. Rev. B* **80**, 094418 (2009).

151. Z. Li and S. Zhang, Thermally assisted magnetization reversal in the presence of a spin-transfer torque, *Phys. Rev. B* **69**, 134416a (2004a).

152. Z. Li, J. He, and S. Zhang, Stability of precessional states induced by spin-current, *Phys. Rev. B* **72**, 212411 (2005).

153. D. M. Apalkov and P. B. Visscher, Spin-torque switching: Fokker-Planck rate calculation, *Phys. Rev. B* **72**, 180405(R) (2005).

154. R. Holm, *Electric Contacts: Theory and Applications*, Springer-Verlag, New-York, 1967.

155. D. A. Garanin, Fokker-Planck and Landau-Lifshitz-Bloch equations for classical ferromagnets, *Phys. Rev. B* **55**, 3050 (1997).

156. U. Atxitia, O. Chubykalo-Fesenko, J. Walowski, A. Mann, and M. Münzenberg, Evidence for thermal mechanisms in laser-induced femtosecond spin dynamics, *Phys. Rev. B* **81**, 174401 (2010).

157. I. N. Krivorotov, N. C. Emley, R. A. Buhrman, and D. C. Ralph, Time-domain studies of very-large-angle magnetization dynamics excited by spin transfer torques, *Phys. Rev. B* **77**, 054440 (2008).

158. K.-J. Lee, Micromagnetic investigation of current-driven magnetic excitation in a spin valve structure, *J. Phys. Condens. Matter* **19**, 165211 (2007).

159. J. P. Strachan, V. Chembrolu, Y. Acremann et al., Direct observation of spin-torque driven magnetization reversal through nonuniform modes, *Phys. Rev. Lett.* **100**, 247201 (2008).

160. J. Xiao, A. Zangwill, and M. D. Stiles, Macrospin models of spin transfer dynamics, *Phys. Rev. B* **72**, 014446 (2005).

161. D. V. Berkov and J. Miltat, Spin-torque driven magnetization dynamics: Micromagnetic modeling, *J. Magn. Magn. Mater.* **320**, 1238–1259 (2008).

162. Z. Li, Y. C. Li, and S. Zhang, Dynamic magnetization states of a spin valve in the presence of dc and ac currents: Synchronization, modification, and chaos, *Phys. Rev. B* **74**, 054417 (2006).

163. Z. Yang, S. Zhang, and Y. C. Li, Chaotic dynamics of spin-valve oscillators, *Phys. Rev. Lett.* **99**, 134101 (2007).

164. S. M. Rezende, F. M. de Aguiar, and A. Azevedo, Spin-wave theory for the dynamics induced by direct currents in magnetic multilayers, *Phys. Rev. Lett.* **94**, 037202 (2005).

165. A. N. Slavin and V. S. Tiberkevich, Nonlinear self-phase-locking effect in an array of current-driven magnetic nanocontacts, *Phys. Rev. B* **72**, 092407 (2005).

166. A. N. Slavin and V. S. Tiberkevich, Theory of mutual phase locking of spin-torque nanosized oscillators, *Phys. Rev. B* **74**, 104401 (2006).

167. J.-V. Kim, V. Tiberkevich, and A. N. Slavin, Generation linewidth of an auto-oscillator with a nonlinear frequency shift: Spin-torque nano-oscillator, *Phys. Rev. Lett.* **100**, 017207 (2008).

168. B. Georges, J. Grollier, M. Darques et al., Coupling efficiency for phase locking of a spin transfer nano-oscillator to a microwave current, *Phys. Rev. Lett.* **101**, 017201 (2008).

169. J.-H. Moon, D.-H. Kim, M. H. Jung, and K.-J. Lee, Current-induced resonant motion of a magnetic vortex core: Effect of nonadiabatic spin torque, *Phys. Rev. B* **79**, 134410 (2009).

170. P. Warnicke, Y. Nakatani, S. Kasai, and T. Ono, Long-range vortex domain wall displacement induced by an alternating current: Micromagnetic simulations, *Phys. Rev. B* **78**, 012413 (2008).

171. J. Fernandez-Rossier, M. Braun, A. S. Nunez, and A. H. MacDonald, Influence of a uniform current on collective magnetization dynamics in a ferromagnetic metal, *Phys. Rev. B* **69**, 174412 (2004).

172. J. He and S. Zhang, Magnetic dynamic phase generated by spin currents, *Phys. Rev. B* **78**, 012414 (2008).

173. J. Shibata, G. Tatara, and H. Kohno, Effect of spin current on uniform ferromagnetism: Domain nucleation, *Phys. Rev. Lett.* **94**, 076601 (2005).

174. M. Kläui, M. Laufenberg, L. Heyne et al., Current-induced vortex nucleation and annihilation in vortex domain walls, *Appl. Phys. Lett.* **88**, 232507 (2006).

175. Y. Ji, C. L. Chien, and M. D. Stiles, Current-induced spin-wave excitations in a single ferromagnetic layer, *Phys. Rev. Lett.* **90**, 106601 (2003).

176. B. Ozyilmaz, A. D. Kent, J. Z. Sun, M. J. Rooks, and R. H. Koch, Current-induced excitations in single Cobalt ferromagnetic layer nanopillars, *Phys. Rev. Lett.* **93**, 176604 (2004).

177. M. L. Polianski, and P. W. Brouwer, Current-induced transverse spin-wave instability in a thin nanomagnet, *Phys. Rev. Lett.* **92**, 026602 (2004).

178. M. D. Stiles, J. Xiao, and A. Zangwill, Phenomenological theory of current-induced magnetization precession, *Phys. Rev. B* **69**, 054408 (2004).

179. J. Gorchon, S. Bustingorry, J. Ferré, V. Jeudy, A. B. Kolton, and T. Giamarchi, Pinning-dependent field-driven domain wall dynamics and thermal scaling in an ultrathin Pt|Co|Pt magnetic film, *Phys. Rev. Lett.* **113**, 027205 (2014).

180. P. Chauve, T. Giamarchi, and P. Le Doussal, Creep and depinning in disordered media, *Phys. Rev. B* **62**, 6241 (2000).

181. M. Yamanouchi, J. Ieda, F. Matsukura et al., Universality classes for domain wall motion in the ferromagnetic semiconductor (Ga,Mn)As, *Science* **317**, 1726–1729 (2007).

182. T. Koyama, D. Chiba, K. Ueda et al., Observation of the intrinsic pinning of a magnetic domain wall in a ferromagnetic nanowire, *Nat. Mater.* **10**, 194–197 (2011).

183. Z. Li and S. Zhang, Domain-wall dynamics and spin-wave excitations with spin-transfer torques, *Phys. Rev. Lett.* **92**, 207203 (2004).

184. Z. Li, J. He, and S. Zhang, Effects of spin current on ferromagnets (invited), *J. Appl. Phys.* **99**, 08Q702 (2006).

185. J. He, Z. Li, and S. Zhang, Current-driven vortex domain wall dynamics by micromagnetic simulations, *Phys. Rev. B* **73**, 184408 (2006).

186. A. A. Thiele, Steady-state motion of magnetic domains, *Phys. Rev. Lett.* **30**, 230–233 (1973).

187. N. L. Schryer and L. R. Walker, The motion of 180° domain walls in uniform dc magnetic fields, *J. Appl. Phys.* **45**, 5406 (1974).

188. A. Thiaville, J. M. Garcia, and J. Miltat, Domain wall dynamics in nanowires, *J. Magn. Magn. Mater.* **242–245**, 1061 (2002).

189. O. Boulle, S. Rohart, L. D. Buda-Prejbeanu et al., Domain wall tilting in the presence of the Dzyaloshinskii-Moriya interaction in out-of-plane magnetized magnetic nanotracks, *Phys. Rev. Lett.* **111**, 217203 (2013).

190. J. Sampaio, V. Cros, S. Rohart, A. Thiaville, and A. Fert, Nucleation, stability and current-induced motion of isolated magnetic skyrmions in nanostructures, *Nat. Nanotechnol.* **8**, 839 (2013).

191. J. Iwasaki, M. Mochizuki, and N. Nagaosa, Current-induced skyrmion dynamics in constricted geometries, *Nat. Nanotechnol.* **8**, 742 (2013).

192. A. Fert, V. Cros, and J. Sampaio, Skyrmonics on the track, *J. Nature Nanotech.* **8**, 152 (2013).

9

Spin-Orbit Torques
Experiments and Theory

Aurélien Manchon and Hyunsoo Yang

The electric control of the magnetic order parameter is a powerful means to operate spin devices. While conventional spin transfer torque requires the existence of a spin polarizer and therefore a spin-valve configuration or a magnetic texture, an alternative approach has been intensively investigated in the past ten years. This approach exploits the spin-orbit coupling in inversion asymmetric magnetic structures. In this configuration, a direct transfer of angular momentum between orbital and spin degrees of freedom occurs and a torque is exerted on the magnetization. This torque has been observed experimentally in a wide variety of non-centrosymmetric materials, from magnetic multilayers to the surface of topological insulators and zinc-blende semiconductors (see Figure 9.1). The concept of spin-orbit torque, and its Onsager reciprocal, the charge pumping, are introduced in Section 9.1. The methods used to measure spin-orbit torques are presented in Section 9.2. The nature of the torque observed in metallic multilayers, bulk non-centrosymmetric magnets, and two-dimensional systems are addressed in Sections 9.3 through 9.5. Besides spin-orbit torque, non-centrosymmetric magnets also display chiral magnetic textures with fascinating dynamics. These aspects are addressed in Section 9.6. Finally, Section 9.7 offers conclusion and perspectives.

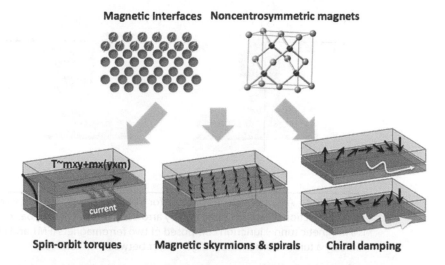

Magnetic Interfaces   Noncentrosymmetric magnets

Spin-orbit torques   Magnetic skyrmions & spirals   Chiral damping

FIGURE 9.1   In magnets lacking inversion symmetry, such as magnetic interfaces or non-centrosymmetric magnetic crystals, the presence of spin-orbit coupling gives rise to a variety of phenomena among which spin-orbit torques, chiral magnetic textures, and chirality-dependent energy dissipation. In this figure, the black arrows refer to the local magnetic moments, the red arrows represent the spin accumulation of the itinerant electrons and the blue arrow denotes the spin-orbit torque. The wavy yellow arrows illustrate the energy dissipation of a moving magnetic texture.

9.1 PRINCIPLES OF SPIN-ORBIT TORQUES

9.1.1 REMINDER ABOUT SPIN TRANSFER TORQUE

Before entering into the physics of spin-orbit torques (SOT), it is useful to go over some basics about spin transfer torque (STT – an in-depth discussion of this phenomenon is provided in Chapters 7 and 8, Volume 1). Originally proposed by Slonczewski [1] and Berger [2], spin transfer can be understood from Newton's angular momentum conservation law. Consider a magnet of volume Ω and (uniform) magnetization direction $\mathbf{m}$. Now imagine that a spin current, i.e., a flow of electrons all possessing the same spin angular momentum direction $\mathbf{p}$, impinges on this magnetic volume. When the spin polarization $\mathbf{p}$ is misaligned with the magnetization $\mathbf{m}$, the spin of the electrons entering the ferromagnet precesses around the magnetization and eventually, due to spin decoherence, gets aligned on the local magnetization $\mathbf{m}$. If the ferromagnet is thicker than the spin decoherence length (i.e. the distance over which an electron spin aligns on the magnetization $\mathbf{m}$), the outgoing spin current has a polarization aligned on $\mathbf{m}$. As a result, a net angular momentum has been transferred to the ferromagnet. This is equivalent to say that a torque $\tau \sim \mathbf{m} \times (\mathbf{p} \times \mathbf{m})$ is exerted by the flowing spins on the ferromagnetic order parameter.

The standard device on which spin transfer torque is investigated is called a spin-valve (see Figure 9.2) and consists of two ferromagnets (the spin polarizer and the free layer) separated by a metallic or tunneling spacer (see Chapters 7, 8, and 15, Volume 1). The existence of the spin polarizer presents some limitations in terms of technological implementation. In fact, the polarizing layer generates stray fields that perturb the free layer and, of course, it also experiences a spin torque from the free layer.

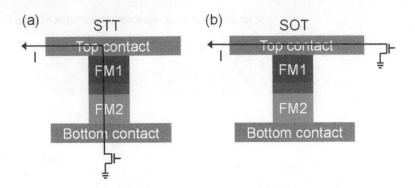

FIGURE 9.2   A typical device structure for STT device (a) and SOT device (b) with current injection directions indicated by arrows. In both cases, the basic structure is a magnetic tunnel junction composed of two ferromagnets (FM1 and FM2) separated by a tunnel barrier (AlOx, MgO – layer between FM1 and FM2).

These considerations led researchers to conceive alternative ways to generate electrically driven spin torque without the need for an external polarizer. To do so, it has been suggested [3–7] to exploit the spin-orbit coupling (SOC), i.e., the intimate connection between spin and orbital angular momenta, to generate spin currents and spin densities and thereby exert an SOT on the ferromagnet. As explained below, to be efficient, the ferromagnet must present some sort of structural symmetry breaking, either at its surface (Sections 9.3 and 9.5) or in its bulk (Section 9.4).

A typical SOT device consists of a bilayer composed of a ferromagnet (FM) and a non-magnetic material (NM) with high spin-orbit coupling [6, 8–10]. An in-plane charge current injected into the bilayer generates a transverse spin density at the interface of the bilayer that exerts a torque on the magnetization of FM. SOT has been proven to switch the magnetization of FM [6–12], move domain walls inside FM [13, 14] and generate oscillations efficiently compared with conventional STT [15, 16]. Figure 9.2 compares the device structure, a magnetic tunnel junction (MTJ), for typical STT and SOT devices. Since the current injection direction in STT is perpendicular to the plane of FMs, a large current density can cause dielectric breakdown of the tunnel barrier. In contrast, in SOT the current direction is in-plane and no current needs to flow throw the tunnel barrier thus improving device stability.

9.1.2 BASICS OF SPIN-ORBIT COUPLING

Spin-orbit coupling is a very powerful interaction that lies at the core of condensed matter physics and magnetism [17]. Its physical origin can be traced back to the Dirac equation for relativistic electrons. Nevertheless, a simple picture can be grasped by considering the motion of spinning objects in a viscous environment. It is well known from classical mechanics that when a spinning ball is thrown in the air, its trajectory depends on the direction of its spin. If the spinning axis is *normal* to the initial direction of the ball, the relative velocity of the air with respect to the ball is different on both sides of the ball (see Figure 9.3a). The difference of pressure produces a force, the Magnus force, that

(a)   **(b)**

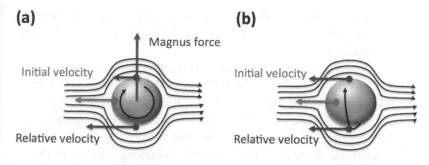

FIGURE 9.3   Spinning ball in airflow, with an angular momentum (a) normal to the initial velocity direction and (b) along the initial velocity direction. The black lines represent the airflow, the central horizontal (red) arrow is the initial velocity while the side horizontal (green) arrows represent the relative velocity of the spinning ball with respect to the air flow. The vertical (blue) arrow is the resulting Magnus force.

distorts the initial trajectory: the spinning ball acquires a transverse velocity. To prevent the spinning ball from diverging from its original trajectory, one can simply make the ball spin about an axis *parallel* to the initial velocity. In this case, the relative velocity of the air with respect to the ball is homogeneous at the surface of the ball and there is no Magnus force (see Figure 9.3b).

This principle, which is nothing other than angular momentum conservation, applies also in a condensed matter when electrons flow in the crystal potential. It produces two important effects: spin Hall and inverse spin-galvanic effects (see Sections 9.1.3 and 9.1.4). More precisely, in the rest frame of the electron, the potential gradient ∇V induces a magnetic field that couples to the spin angular momentum through a Zeeman energy term $\hat{H}_{so} = \xi \hat{\sigma} \cdot (\nabla V \times \hat{p})$.

9.1.3 GENERATING SPIN CURRENT: SPIN HALL EFFECT

Following the process described in Figure 9.3a, when the electron spin direction is normal to the transport plane, the up and down electrons deflect toward opposite directions, creating a spin current transverse to the initial charge current. This effect is called the spin Hall effect [18] and converts an (unpolarized) charge current $\mathbf{J}_c$ into a (chargeless) spin current, $\mathbf{J}_s \sim \sigma \times \mathbf{J}_c$. The rate of conversion between the two currents is usually measured in terms of spin Hall angle, θ_{sh}.

The spin Hall effect, along with its ferromagnetic counterpart, the anomalous Hall effect, has been intensively scrutinized for a number of years. It is not our intention to enter into the details of previous works and we refer the reader to the excellent specialized reviews available [19]. It is, however, instructive to summarize these works from the perspective of experiments: "In order to obtain a large spin Hall conversion efficiency, should the material be more or less resistive?". It turns out that the answer depends on the nature of the spin Hall mechanism. Two main classes of mechanisms have been identified. The first class consists of mechanisms that directly depend on the density of scatterers, such that the Hall conductivity scales with the longitudinal conductivity, $\sigma_H \sim \sigma_{xx}$. In the limit of weak short-range impurities, the main effect is the so-called skew or Mott scattering [19]. It is important in certain compounds such as heavy metals (Bi, Ir) as well as rare-earth-doped light metals [20, 21].

The second class consists of mechanisms that are independent of the density of scatterers, such that the Hall conductivity is independent on the longitudinal conductivity, $\sigma_H \sim O(\sigma_{xx})$. In the limit of weak short-range impurities, two main effects contribute to this second class: the side-jump scattering, which renormalizes the velocity operator in the presence of scatterers, and the intrinsic Berry-curvature induced anomalous velocity. In the latter, the presence of SOC in the band structure bends in the trajectory of the flowing electrons. Distinguishing between side-jump and intrinsic Hall effects remains experimentally difficult. To the best of our knowledge, the spin Hall effect in most of the $4d$ and $5d$ metals (Ta, W, Pt) is attributed to intrinsic spin Hall effect governed by the band structure [22, 23].

9.1.4 GENERATING SPIN DENSITIES: INVERSE SPIN-GALVANIC EFFECT

In Figure 9.3b, we showed that to prevent a spinning ball from deviating from its initial trajectory, the spinning axis needs to point *along* the propagation direction. In materials lacking inversion symmetry, it means that the flowing spins must all align in the same direction. In other words, a current flow is necessarily accompanied by a non-equilibrium spin density. This is the inverse spin-galvanic effect [17], an effect that arises in structures with broken inversion symmetry, such as asymmetric multilayers and quantum wells or non-centrosymmetric bulk materials (strained zinc-blende or wurtzite crystals). In technical terms, in non-centrosymmetric structures, SOC becomes odd in linear momentum [17], i.e. $\hat{H}_{so} \propto \hat{\sigma} \cdot w(\hat{\mathbf{p}})$, where $w(\hat{\mathbf{p}}) = -w(-\hat{\mathbf{p}})$. The simplest version of such SOC is the so-called Rashba SOC, $\hat{H}_R = \alpha_R \hat{\sigma} \cdot (z \times \hat{\mathbf{p}})$, emerging at the interface between two dissimilar materials. [24] This form of SOC enables the direct coupling of the spin density $\mathbf{S} = \langle \hat{\sigma} \rangle$ and charge current $\mathbf{J}_c = (e/m) \langle \hat{\mathbf{p}} \rangle$. Namely, a non-equilibrium spin density generates a charge current, this is the spin-galvanic effect [25]. Reciprocally, a charge current is accompanied by a non-equilibrium magnetization, this is the inverse spin-galvanic effect, also called the Rashba-Edelstein effect [26], recently observed at the surface of transition metals [27]. In magnetic materials lacking inversion symmetry, the spin density induced by inverse spin-galvanic effect can exert a torque on the magnetization [3–5], called the Rashba torque.

9.1.5 SPIN TORQUE AND CHARGE PUMPING

The inverse spin-galvanic effect, present in all non-centrosymmetric materials, generates a non-equilibrium effective magnetic field whose direction is determined by the symmetry of the (bulk or interfacial) structure. For instance, in the case of interfacial Rashba SOC, as depicted in Figure 9.4a, the torque reads $\tau_T \sim \mathbf{m} \times (\mathbf{z} \times \mathbf{J}_c)$, where $\mathbf{z}$ is normal to the plane density. This torque is referred to as the field-like or transverse torque. The spin Hall

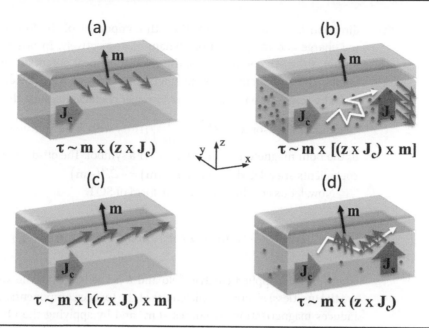

FIGURE 9.4   Illustration of spin-orbit torque mechanisms present in a magnetic bilayer: (a) extrinsic Rashba effect [3–5], (b) spin Hall effect [67], (c) intrinsic Rashba effect [30, 97], and (d) spin swapping effect [31]. Extrinsic Rashba effect and spin swapping both produce field-like torques, while intrinsic Rashba and spin Hall effect both produce damping-like torques.

effect, in turn, is present in the bulk of normal metals. The physics of spin Hall torque is very similar to conventional STT: the spin Hall effect present in the normal metal injects a spin current into the adjacent ferromagnet (see Figure 9.4c), yielding a torque that reads $\tau_{\mathrm{L}} \sim \mathbf{m} \times \left[\left(\mathbf{z} \times \mathbf{J}_c \right) \times \mathbf{m} \right]$. This torque is referred to as a damping-like or longitudinal torque (or also "antidamping torque"). Recent theories have demonstrated that, in fact, the inverse spin-galvanic effect can also give rise to damping-like torque [28–30] (see Figure 9.4b), while spin-orbit scattering mechanisms such as spin swapping can also give rise to a field-like torque [31] (see Figure 9.4d).

It is instructive to relate SOT with its Onsager reciprocal, the spin-orbit-mediated charge pumping [32]. In fact, if a charge current can induce magnetization dynamics, then magnetization dynamics is also expected to induce a charge current. This can be formalized through the Onsager reciprocity principle [33]. The rate of change in a system can be represented in terms of a particle current connected to a force via the so-called Onsager coefficients. In the case of a magnetic system under a charge current, one writes [33].

$$\begin{pmatrix} SJ_{c,i} / e \\ \partial_t m_i \end{pmatrix} = \begin{pmatrix} \mathcal{L}_{ij}^{cc} & \mathcal{L}_{ij}^{cm} \\ \mathcal{L}_{ij}^{mc} & \mathcal{L}_{ij}^{mm} \end{pmatrix} \begin{pmatrix} F_{c,j} \\ F_{m,j} \end{pmatrix}. \tag{9.1}$$

Here, $J_{c,i}$ is the current density of charge e flowing along direction i through a surface S, and $\partial_t m_i$ is the rate of change of the magnetization $\mathbf{m}$ along direction i. $F_{c,j}$ and $F_{m,j}$ are the j-th component of the forces that drive the charge and magnetization dynamics, respectively. In other words, the former is the j-th component of the electrochemical potential and the latter is the j-th component of the magnetic force, defined $F_{m,j} = -\delta_{m_j}W$, where W is the magnetic energy. $\mathcal{L}_{ij}^{\alpha\beta}$ are Onsager coefficients, such that $\mathcal{L}_{ij}^{cc}$ is the conductance tensor and $\mathcal{L}_{ij}^{mm} = \left(\dfrac{\gamma}{\mu_B}\right)\epsilon_{ijk}m_k$, with γ the gyromagnetic ratio, μ_B the Bohr magneton and ϵ_{ijk} Levi-Civita symbol. The off-diagonal Onsager coefficients are related through $\mathcal{L}_{ij}^{mc}(\mathbf{m}) = -\mathcal{L}_{ji}^{cm}(-\mathbf{m})$.

Now, let us consider the generic form of SOT

$$\tau = \left(\frac{\gamma L}{\mu_B}\right)\chi_L \mathbf{m}\times\left[(\mathbf{z}\times e\mathbf{E})\times\mathbf{m}\right] + \left(\frac{\gamma L}{\mu_B}\right)\chi_T \mathbf{m}\times(\mathbf{z}\times e\mathbf{E}), \qquad (9.2)$$

where $\mathbf{E}$ is the applied electric field and L is the length of the system. $\chi_{L,T}$ are the (unitless) electrical efficiencies of the torque components. This torque induces magnetization dynamics $\partial_t\mathbf{m}$ and by applying the Onsager reciprocity principle, one obtains the charge current pumped through magnetization dynamics,

$$J_c^{\text{pump}} = \left(\frac{e}{S}\right)\chi_L \mathbf{z}\times(\mathbf{m}\times\partial_t\mathbf{m}) - \left(\frac{e}{S}\right)\chi_T \mathbf{z}\times\partial_t\mathbf{m}. \qquad (9.3)$$

Notice that the Onsager principle does not provide information about the underlying physics of the spin-charge conversion.

9.2 EXPERIMENTAL METHODOLOGY

9.2.1 MACROSPIN DYNAMIC

In the macrospin model when all the magnetic moments behave coherently, the evolution of the magnetization $\mathbf{m}$ of a FM in presence of an effective magnetic field $\mathbf{H}_{\text{eff}}$ can be modeled using Landau-Lifshitz-Gilbert (LLG) equation,

$$\partial_t\mathbf{m} = -\gamma\mu_0\mathbf{m}\times\mathbf{H}_{\text{eff}} + \alpha\mathbf{m}\times\partial_t\mathbf{m}, \qquad (9.4)$$

where μ_0 is vacuum magnetic permeability and α is the Gilbert damping parameter. On the right-hand side of Equation 9.4, the first term is a precession term that sets $\mathbf{m}$ to rotate around $\mathbf{H}_{\text{eff}}$, while the second term is an energy dissipation term that damps $\mathbf{m}$ to align it along $\mathbf{H}_{\text{eff}}$. In presence of SOT, the modified LLG equation reads

$$\partial_t\mathbf{m} = -\gamma\mu_0\mathbf{m}\times\mathbf{H}_{\text{eff}} + \alpha\mathbf{m}\times\partial_t\mathbf{m} + \tau_L + \tau_T, \qquad (9.5)$$

where $\tau_L = \tau_L\mathbf{m}\times\left[(\mathbf{z}\times\mathbf{J}_c)\times\mathbf{m}\right]$ is the longitudinal (damping-like) torque and $\tau_T = \tau_T\mathbf{m}\times(\mathbf{z}\times\mathbf{J}_c)$ is the transverse (field-like) torque. Conventionally,

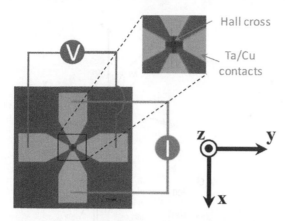

FIGURE 9.5 Optical microscope image of the Hall cross devices structure along with a schematic of measurement configuration for Hall resistance measurements. The center of the image has been zoomed in to show an enlarged view of the Hall cross-device. (After Ramaswamy, R. et al., *Appl. Phys. Lett.* 108, 202406, 2016. With permission.)

these two SOT components are expressed in terms of effective fields: (i) the longitudinal field ($\mathbf{H}_L$) that generates the damping-like torque $\boldsymbol{\tau}_L = -\gamma\mu_0\mathbf{m} \times \mathbf{H}_L$ and (ii) the transverse field ($\mathbf{H}_T$) that generates the field-like torque $\boldsymbol{\tau}_T = -\gamma\mu_0\mathbf{m} \times \mathbf{H}_T$. In presence of $\mathbf{H}_L$ and $\mathbf{H}_T$, $\mathbf{m}$ no longer relaxes towards $\mathbf{H}_{eff}$ in the static regime; instead, it relaxes to the equilibrium position governed by Equation 9.5. Understanding the SOT-induced magnetization dynamics is at the core of the experimental methods used to determine the magnitude of SOT components.

9.2.2 Harmonic Hall Voltage Measurements

To quantify the SOT effective fields, $\mathbf{H}_L$ and $\mathbf{H}_T$, in perpendicularly magnetized systems, harmonic Hall voltage measurements can be utilized in Hall bar device structures [34–37]. Figure 9.5 shows the microscope image of a Hall cross-device structure along with Ta/Cu contacts. The measurement schematic for Hall resistance measurements is also shown in Figure 9.5. To quantify SOT effective fields, a low frequency (e.g., 13.7 Hz) alternating current (I_{ac}) is passed into the NM/FM bilayer, and the first and second harmonic Hall voltages are measured using two lock-in amplifiers [34–37]. This measurement is performed in two configurations, namely longitudinal ($\mathbf{H}\|\mathbf{I}_{ac}$) and transverse configurations ($\mathbf{H}\perp\mathbf{I}_{ac}$) to estimate $\mathbf{H}_L$ and $\mathbf{H}_T$, respectively, as shown in Figure 9.6a, and b. In both configurations, the magnetic field is applied with a small tilting angle (e.g., 4°) to the film plane to ensure the magnetization of the sample is uniformly aligned along $\mathbf{H}$ in order to avoid formation of multi-domain state.

Figures 9.6c, d show the first (V_f) and second harmonic (V_{2f}) Hall voltage signals for longitudinal and transverse configurations, respectively. To determine the magnitude of SOT effective fields, the second harmonic Hall voltage is fitted using a (macrospin) model [37]. From the fitting parameters, one can obtain the values of the anisotropy field ($H_{k,eff}$), H_L and H_T. It must be noted that both anomalous and planar Hall effects should be considered

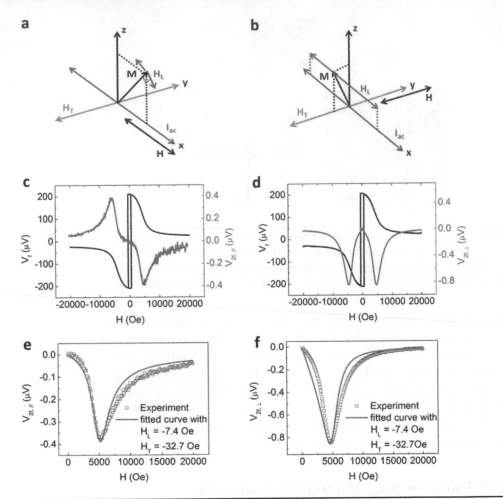

FIGURE 9.6 (a, b) Directions of H_L and H_T for (a) **H** ∥ **I** and (b) **H** ⊥ **I**. (c, d) First and second harmonic voltages of (c) longitudinal and (d) transverse configurations. (e, f) Fitting of the second harmonic voltages. The obtained H_L and H_T values are indicated in the graphs. (After Qiu, X. et al., *Sci. Rep.* 4, 4491, 2014. With permission.)

for fitting as neglecting the latter underestimates the magnitudes of H_L and H_T [36–38]. Figures 9.6e, and f show the fits of second harmonic Hall voltage signals for longitudinal and transverse configurations, respectively. It can be observed that the model (solid line) deviates from the measured data (symbols) for certain magnetic field values. This can be understood by the fact the magnitude of SOT effective fields depends on the angle between the magnetization and the normal to the sample plane [36, 37]. Thus, the magnitude of H_L and H_T changes with increasing magnetic field, whereas the model assumes constant SOT fields. The second harmonic Hall voltages reach their maximum value when the applied field is roughly equal to the anisotropy field.

In the low field regime, the longitudinal and transverse effective fields are calculated by using [35] $\Delta H_{T(L)} = -2 \dfrac{\partial V_{2f}}{\partial H_{T(L)}} / \dfrac{\partial^2 V_f}{\partial H_{T(L)}^2}$. In recent works [39, 40], it was reported that the presence of planar Hall effect (PHE) in the signal can corrupt the 2nd harmonic signal and the effective fields by considering the PHE contribution is given by [40]

$$|H_L| = 2\frac{\left|\dfrac{\partial_X V_{2f}}{\partial_X^2 V_f}\right| + 2\eta\left|\dfrac{\partial_Y V_{2f}}{\partial_Y^2 V_f}\right|}{1 - 4\eta^2} \quad \text{and} \quad |H_T| = 2\frac{\left|\dfrac{\partial_Y V_{2f}}{\partial_Y^2 V_f}\right| + 2\eta\left|\dfrac{\partial_X V_{2f}}{\partial_X^2 V_f}\right|}{1 - 4\eta^2} \quad (9.6)$$

With the Hall bar in the x-y plane and current along the x-direction, $\partial_{X,Y} V_f$ and $\partial_{X,Y} V_{2f}$ are the derivative of the 1st and 2nd harmonic voltage when the field is applied along the x- and y- directions, respectively. $\eta = R_{PHE} / R_{AHE}$ is the ratio between planar and anomalous Hall effects. To measure R_{PHE}, an external field of 2–4 T was applied in the plane of the sample and the sample stage was rotated while measuring the first harmonic Hall voltage. To find R_{PHE}, this curve was fitted with $R_{PHE} \sin 2\theta + R_{AHE} \sin^2 \theta$ where θ is the angle between the current and field direction. By using this low field region fitting method, a large applied magnetic field is not necessary and the measured second harmonic signal is not sensitive to the out-of-plane tilting angle due to the in-plane field. However, the noise level in the low field region has to be well controlled in order to get a clear linear V_{2f} curve.

9.2.3 Spin Torque Ferromagnetic Resonance

In systems where the FM possesses in-plane magnetic anisotropy, the ST-FMR technique can be utilized to determine the SOT magnitude. For ST-FMR measurements, the film stack consisting of FM/NM is patterned into rectangular microstrips and a symmetric coplanar waveguide (CPW) is fabricated to make contact with the device as shown in Figure 9.7a. A ground-signal-ground (GSG) electrical probe is used to pass a microwave electrical current into the device.

Before describing the measurement technique of ST-FMR, the concept of ferromagnetic resonance (FMR) is explained as follows. As described before, $\mathbf{m}$ precesses around $\mathbf{H}_{eff}$ and the amplitude of these precessions continuously decays due to damping. If we supply energy through an alternating magnetic field ($\mathbf{h}_{rf}$), $\mathbf{m}$ undergoes forced precession. When the frequency of $\mathbf{h}_{rf}$, matches the natural precession frequency (ω_{res}) of $\mathbf{m}$ due to $\mathbf{H}_{eff}$, the amplitude of the precessions becomes maximum and forced resonance occurs. This condition is called FMR and the relation between ω_{res} and $\mathbf{H}_{eff}$ is given by Kittel's relation, $\omega_{res} = \gamma\mu_0\sqrt{H_{eff}\left(H_{eff} + 4\pi M_s\right)}$, where M_S is the saturation magnetization of the FM. ST-FMR [41, 42] is an FMR technique where the energy for sustained oscillations for the FM comes from spin torque (STT and SOT), instead of $\mathbf{h}_{rf}$. From Equation 9.5, it is evident that the current generating τ_L and τ_T to compensate damping should be an alternating current. Moreover, ω_{res} is usually in GHz range and hence microwave currents are required to produce ST-FMR in NM/FM.

Figure 9.7b shows a schematic of a ST-FMR setup for an NM/FM bilayer. In the ST-FMR measurements, a microwave current of a fixed frequency (f) is applied to NM/FM bilayer using the signal generator (SG). Simultaneously,

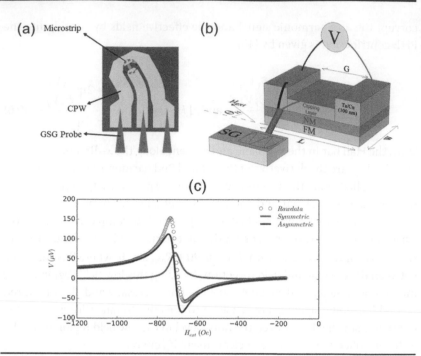

FIGURE 9.7   (a) Microscope image of a typical ST-FMR device with GSG electrical probe. (After Wang, Y. et al., *Appl. Phys. Lett.* 105, 152412, 2014. With permission.) (b) Illustration of a ST-FMR setup. The dimensions of the microstrip devices are *L* (130 μm) × *W* (20 μm), while gap (G) between ground and signal electrodes of the CPW is varied among the devices in order to tune the device impedance close to ~50 Ω. (c) ST-FMR spectrum (blue circles) along with symmetric (red curve) and antisymmetric (green curve) Lorentzian fittings of the spectrum.

an external magnetic field H_{ext} is applied at an angle θ_H with respect to the current channel. The ac charge current generates both an alternating SOT and an alternating Oersted field. These oscillating torques induce magnetization precession in the FM layer and the resistance of the bilayer oscillates at the same frequency as the magnetization precession due to the anisotropic magnetoresistance (AMR) effect. Consequently, a dc mixing voltage V_{mix} is produced by the product of applied alternating charge currents and oscillating resistance [6]. A nanovoltmeter (V) is used to measure the V_{mix} signal across the device. For a given frequency f of the microwave current, H_{ext} is swept to meet the resonance condition given by Kittel's relation. The ST-FMR spectrum is obtained by measuring V_{mix} as a function of H_{ext}.

Figure 9.7c shows a typical ST-FMR spectrum obtained from experiments for $f = 7$ GHz. One can observe that the amplitude of V_{mix} peaks at $H_{ext} \approx 900$ Oe, which is the resonance field for the sample at $f = 7$ GHz. The ST-FMR spectra can be fitted using the equation [6] $V_{mix} = V_S F_S(H_{ext}) + V_A F_A(H_{ext})$ where $F_S(H_{ext})$ is a symmetric Lorentzian function of amplitude V_S and $F_A(H_{ext})$ is an antisymmetric Lorentzian function of amplitude V_A. Figure 9.7c also shows the symmetric and antisymmetric Lorentzian fits of the obtained ST-FMR spectrum.

To understand the origins of the different components of the ST-FMR line shape, the LLG equation with SOT Equation 9.5 can be analyzed to identify the symmetry of the different current-induced torques (τ_L and τ_T). The Oersted field induced torque from the charge current in NM is of the symmetry τ_T and is out-of-phase with the magnetization precession; while the spin Hall torque from the generated spin current is predominantly of the symmetry τ_L and hence is in-phase with the magnetization precession. Therefore, the out-of-phase Oersted field induced torque produces an anti-symmetric Lorentzian spectrum about the resonance field and the in-phase spin Hall torque produces a symmetric Lorentzian spectrum about the resonance field. The (unitless) efficiency of the longitudinal torque, θ_L, can be expressed [6], $\theta_L = \left(V_S / V_A\right)\left(e\mu_0 M_s t d / \hbar\right)\sqrt{1 + 4\pi M_{\text{eff}} / H_{\text{ext}}}$ where M_{eff} is the effective magnetization of the FM layer, and t and d are the thicknesses of the FM and NM layer, respectively. In the case where only spin Hall effect is present in the system, θ_L is the spin Hall angle. However, if the Rashba effect is significant, it can generate an effective transverse field of the same symmetry as the Oersted field and therefore can contribute to V_A. As a result, the value of θ_L may not be accurate from the ST-FMR spectrum. To eliminate this issue, θ_L can be determined by considering only the symmetric component V_S of the ST-FMR spectrum using [43] $\theta_L = \frac{1}{\sigma} \frac{4 V_S M_S t \Delta}{\gamma E I_{rf} \cos\theta_H \left(\partial R / \partial \theta_H\right)}$,

where Δ is the linewidth of the Lorentzian ST-FMR spectrum, E and I_{rf} are the microwave electric field and current through the device, respectively, $\partial R / \partial \theta_H$ is angular dependent magnetoresistance of the device, and σ is the conductivity of the normal metal.

9.3 METALLIC MULTILAYERS

9.3.1 GENERAL OVERVIEW

The first demonstrations of SOT in metallic multilayers have been performed on Pt/Co [8, 34] and Pt/NiFe [6, 44] bilayers using harmonic detection and ST-FMR, respectively. A major breakthrough was the achievement of SOT-induced switching in perpendicularly magnetized Pt/Co/AlOx [9, 45] and in-plane magnetized Ta/CoFeB/MgO [10]. Since then, SOT studies in metallic bilayers have been extended to a large class of $4d$ and $5d$ transition metals (Pt, Ta, W, etc.), but also thick magnetic multilayers such as Co/Pd superlattices.

In general, the observed SOT possesses both damping-like and field-like components (see Table 9.1 for typical values), and displays an angular dependence [36, 37, 46] that can be quite significant (up to ~50% in the case of Pt/Co/AlOx [36] and topological insulators [46]). Several methods have been reported in the literature to manipulate the SOT efficiency such as interface doping [47, 48], change of the NM or FM material [49–53], a variation in NM and FM thickness [35, 54–56] and change of annealing conditions [36, 57]. The search for the physical origin of the damping-like and field-like SOTs (extrinsic and intrinsic Rashba effects, spin Hall and spin swapping effects) is still very active. Angular [36, 37, 46], temperature [37, 58] and materials dependence [49–53] of SOTs might hold the key to this unsettled debate.

TABLE 9.1
Magnitude of SOT Components Measured in Various Classes of Systems

	Pt/Co[36]	Ta/CoFeB[58]	(Co/Pd)n[71]	Bi$_2$Se$_3$/CoFeB[43]	(Ga,Mn)As[98]	NiMnSb[93]	LAOSTO[128]
Field-like θ_T	670 (0.126)	820 (0.3)	5025 (19.1)	110 (0.21)	9.1/1.5 (0.04/0.007)	44.5/10.5 (0.41/0.1)	1.76×10^4 (1.9)
Damping-like θ_T	690 (0.13)	110 (0.04)	1170 (4.5)	330 (0.42)	1.3 (0.006)	~0	–

The torque magnitude is given in Oe/(10^8 A/cm^2) and its efficiency is defined $\theta_{T\perp} = \left(\frac{2e}{\hbar}\right) M_s d H_{T\perp}$. The two values of field-like torque for NiMnSb and (Ga,Mn)As refer to Dresselhaus and Rashba symmetries, respectively. For the Bi$_2$Se$_3$/CoFeB, the torques are reported at 30K.

9.3.2 MAGNETIC BILAYERS

NM/FM bilayers are by far the most studied class of systems. Among the NM underlayers that act as SOT source, Pt is widely used due to its large SOC. Nonetheless, there is currently significant disagreement between the values of SOT in Pt reported by different groups, with a spin Hall angle θ_{sh} ranging from 0.012 to 0.12. These values are collected in Figure 9.8a as a function of the (assumed or measured) spin diffusion length (λ_S) in Pt. A clear inverse correlation between θ_{sh} and λ_S can be observed by the product of $\theta_{sh}\lambda_S \sim 0.13$ nm [56, 59]. The spin diffusion length in a material is generally proportional to its electrical conductivity [60] (σ_{Pt}), as shown in Figure 9.8b. A very wide range of σ_{Pt} and thus λ_S in previous studies might explain the significant disagreement in the reported θ_{sh}. Moreover, it also opens up possibilities to tune θ_{sh} by tuning σ_{Pt} (and thus λ_S) as shown by the inset of Figure 9.8b. Three series of samples with the same structure of Pt(6)/Py(6)/SiO$_2$(3) (thicknesses in nm) were fabricated, where the Pt layer is firstly deposited on the wafers and then is *in-situ* annealed at different temperatures such as 250 and 350°C for 30 minutes. Subsequently, Py and SiO$_2$ capping layers are deposited on top of Pt at room temperature. The measured spin Hall angle in Pt, θ_{sh}, increases as σ_{Pt} decreases.

For commercialization, SOT based spintronic devices should be integrated with CMOS fabrication technologies. However, Pt is not a common metallization element in CMOS processes, in contrast with Cu and Al. Unfortunately, Cu and Al possess very small θ_{sh} due to their smaller atomic number and weaker SOC. One way to enhance the magnitude of θ_{sh} in Cu is by adding non-magnetic impurities with strong SOC that provide extrinsic spin Hall effect, which can be tuned by changing the relative concentrations of host and impurity atoms [20, 61, 62].

Following Pt, the second widely scrutinized underlayer for SOT devices is Ta [10, 35, 37, 57]. Even though Ta is CMOS compatible and a good diffusion barrier, most of the SOT research has been restricted to Ta thicknesses < 5 nm. Moreover, Ta-based structures face difficulty in sustaining the perpendicular magnetic anisotropy (PMA) of the FM for annealing temperatures beyond 300°C. Hence, other elements such as Mo [63] and Hf [52, 64–66] are

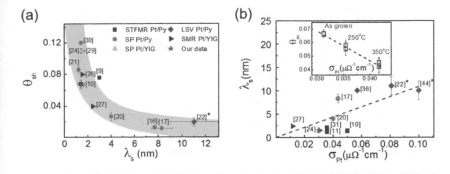

FIGURE 9.8   (a) θ_{sh} as a function of λ_s in Pt from various reports. The ST-FMR, SP, LSV, and SMR represent spin torque ferromagnetic resonance, spin pumping, non-local measurement in lateral spin-valves, and spin Hall magnetoresistance method, respectively. All the data in this figure were measured at room temperature, except for those tagged 22 and 44, which were measured at 10 K, as denoted by small black stars. The blue thick curve shows a correlation of $\theta_{sh}\lambda_s$ ~0.13 nm. (b) λ_s in Pt as a function of its electrical conductivity (σ_{Pt}). The inset shows the measured spin Hall angles in various Pt films with different conductivities. The dashed lines are guides to the eye. (After Wang, Y. et al., *Appl. Phys. Lett.* 105, 152412, 2014. With permission.)

being researched as NM underlayers providing enhanced PMA compared to Ta, W, and Pt.

Hf is particularly interesting, as Hf-based dielectrics are promising gate material for CMOS applications, and its large SOC suggests it could be an efficient source of SOT. An example of a Hf/CoFeB/MgO system is shown in Figure 9.9. Both H_L and H_T are found to increase in magnitude and saturate above a certain Hf thickness, which is a characteristic behavior of spin currents generated by the spin Hall effect [6, 67]. The insets of Figure 9.9 show that there is a significant change of SOT effective fields between 1.75 and 2 nm of the Hf thickness, much larger than in Ta/CoFeB/MgO system where the sign change happens at a Ta thickness of 0.5 nm [35]. This suggests the competition between interfacial (Rashba-like) and bulk (spin Hall-like) SOT mechanims [28–30].

Finally, we mention that the impact of interfacial oxidation has also been highlighted recently. In a Pt/CoFeB/MgO stack, Qiu et al. [48] demonstrated that tuning the amount of oxygen at the CoFeB/MgO interface could lead to a complete change of sign of SOT, associated with a dramatic modification of the interfacial SOT contribution. This recent observation opens appealing avenues to the interfacial engineering of SOT devices.

9.3.3 SUPERLATTICES

In most of the recent reports on SOT, the thickness of the ferromagnetic material was ~1 nm, which is impractical for real applications [9, 34, 45] and renders the theoretical modeling of the experimental conditions rather challenging [31, 67–69]. Furthermore, increasing the thickness of the FM above 1.5 nm results in a decrease in the SOT field by an order of magnitude [35, 70]. To solve these issues, magnetic superlattices such as Co/Pd

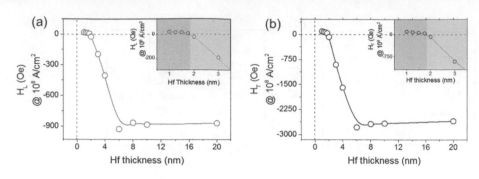

FIGURE 9.9   Extracted values of (a) H_L and (b) H_T measured at a current density of 10^8 A/cm² as a function of Hf thickness (t_{Hf}). The insets show a magnified view of the data for 1 nm $<t_{Hf}<3$ nm. SOT effective fields are positive in the regions shaded by red, while they are negative in the regions shaded by green. (After Ramaswamy, R. et al., *Appl. Phys. Lett.* 108, 202406, 2016. With permission.)

multilayers [71] have been investigated. The torque efficiencies are found to be as large as 1.17 and 5 kOe at 10^8 A/cm² for the in-plane and perpendicular components, respectively [71], which is comparable to the largest value obtained to date in ultrathin bilayers (see Table 9.1). Such large SOT fields, that cannot be attributed to spin Hall effect, can arise if two successive Co/Pd and Pd/Co interfaces are structurally dissimilar; otherwise, their combined effect cancels out. The large lattice mismatch (~9%) between Pd and Co [72] results in strong lattice distortions at their interface. More importantly, this distortion is 30% stronger for Co/Pd than for Pd/Co interfaces [73], which might explain the presence of a Rashba-like SOC in the bulk Co/Pd multilayer. This observation opens promising directions toward the development of SOT in artificial bulk systems.

Recently, SOT-induced domain wall motion in Tb/Co wires with different layered structures has been studied [74]. The magnitude of the damping-like SOT by inner Co/Tb interfaces is tuned by changing the thickness of Co ultrathin sublayers, a number of inner Co/Tb interfaces, and formation of Tb-Co alloy magnetic layer. Co and Tb lattices are antiferromagnetically aligned forming a ferrimagnetic system. The results show that the SOT efficiency is highest for the samples with a larger number of inner Co/Tb interfaces and lowest for the one with Tb-Co alloy, which strongly advocates for a dominant contribution of SOT from the inner Co/Tb interfaces.

9.3.4 SWITCHING

One of the most appealing achievements of SOT is its ability to reverse the magnetization of single magnetic layers, without the need for an external polarizer [9–11, 75, 76]. Since the current is injected in the plane rather than out-of-plane (see Figure 9.2), the critical switching current density does not depend on the lateral extension of the magnet, which enables the reversal of micrometer size samples. In samples larger than the exchange length (~ a few nm in these systems), magnetization reversal is governed by domain walls nucleation and propagation [77–79]. To obtain a hysteretic reversal though, one needs to apply a longitudinal assistive field that ensures the asymmetric

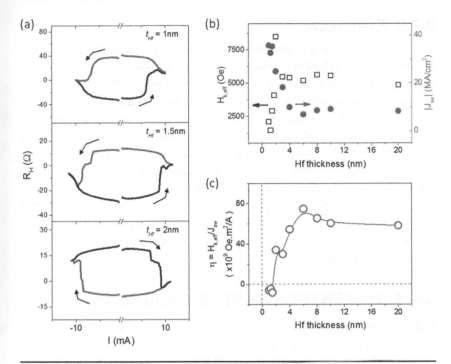

FIGURE 9.10 (a) Anomalous Hall resistance as a function of current for $t_{Hf} = 1$, 1.5, 2 nm. (b) $H_{k,eff}$ and $|J_{sw}|$ plotted as a function of t_{Hf}. J_{sw} is defined as the average value of current densities required to switch from a high to a low state in R_H and vice-versa. J_{sw} is positive for a clockwise switching sequence. (c) SOT switching efficiency plotted as a function of t_{Hf}. (After Ramaswamy, R. et al., *Appl. Phys. Lett.* 108, 202406, 2016. With permission.)

expansion of the reversed magnetic domains driven by the SOT (see Section 9.6 for details). Yet, integrating such an assistive field into nanoscale devices is difficult. Several recent works have endeavored to take advantage of exchange bias or interlayer exchange coupling [80–82] or to break structural symmetry along the current direction [83, 84] in order to overcome this issue. However, the proposed solutions cannot be readily integrated into MTJ structures without structural modifications. Therefore, future works are required to find solutions to integrate the field-free SOT switching scheme in modern MTJs.

Figure 9.10 shows an example of current-driven switching in Hf/CoFeB/MgO [65]. To obtain switching loops, current pulses are injected and the Hall resistance (R_H) of the devices is probed by applying a small assistive field (e.g., 500 Oe) along the current direction. The switching loops and current sweep direction (indicated by black arrows) for different Hf thicknesses are shown in Figure 9.10a, while Figure 9.10b shows the effective anisotropy field ($H_{k,eff}$) and the absolute value of switching current density ($|J_{sw}|$). The SOT switching efficiency $\eta = H_{k,eff}/J_{sw}$, displayed in Figure 9.10c, increases sharply with t_{Hf} and saturates after $t_{Hf} = 6$ nm, which is consistent with SOT arising from the spin Hall effect [56] (see also Figure 9.9). While initial studies on SOT concentrated only on the role of H_L on SOT-induced magnetization switching, recent reports [57, 85–87] emphasize the role of a larger H_T for reducing SOT switching current densities.

9.3.5 EXCITATIONS

SOT can also excite self-sustained GHz oscillations. A limiting factor for achieving such oscillations is the degeneracy of the spin wave modes. If the sample is large, a large number of modes compete with each other to absorb the energy deposited by the SOT, and no self-sustained oscillations can be achieved. To solve this issue, the modes degeneracy must be lifted by reducing the size of the sample and thereby lowering the excitation threshold and bandwidth. SOT-driven GHz excitations were reported in Pt/NiFe bilayers by Liu et al. [11] (nanopillar geometry) and Demidov et al. [15] (point contact configuration). SOT-driven spin wave excitations have also been achieved in the ultralow damping magnetic insulator yttrium iron garnet (YIG) deposited on Pt [88]. Finally, the electrical control of the magnetic damping through SOT has been used to enhance the spin wave propagation length in microwave guides [89, 90].

9.4 NON-CENTROSYMMETRIC MAGNETS

Bulk non-centrosymmetric magnets constitute the first class of materials in which SOT has been predicted [3–5] and observed [7, 91]. They also present a unique platform for the investigation of SOTs: since no spin Hall effect is present, SOT entirely comes from the bulk of the materials, i.e., from inverse spin-galvanic effect [30]. The exact form of the torque can be deduced from the crystal symmetries of the magnet following Neumann's principle [92]. Figure 9.11 presents three examples of non-centrosymmetric magnets: MnSi (Figure 9.11a) is a weak ferromagnet adopting a B20 crystal structure and an appealing candidate for the exploration of chiral magnetism and skyrmions [93]. (Ga,Mn)As (Figure 9.11b) is a low temperature diluted ferromagnetic semiconductor possessing zinc-blende crystal structure [94]. Finally, Mn_2Au (Figure 9.11c) is a newly discovered collinear antiferromagnet [95]. Although its crystal structure is inversion symmetric, the antiferromagnetic sublattices themselves experience a symmetry-broken environment, which enables the emergence of local, sublattice-dependent inverse spin-galvanic effect [96].

(a)   **(b)**   **(c)**

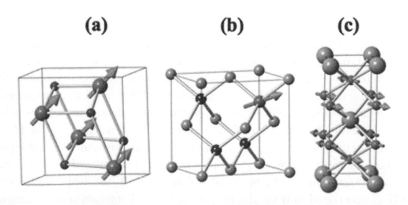

FIGURE 9.11   Various magnetic crystal structures lacking inversion symmetry: (a) MnSi, (b) (Ga,Mn)As, and (c) Mn_2Au.

9.4.1 SOT IN NON-CENTROSYMMETRIC FERROMAGNETS

In their original observation, Chernyshov et al. [7] demonstrated the reversible switching of (Ga,Mn)As thin film under the action of a current. The critical current density required to switch the magnetization was less than 10^6 A/cm^2, much smaller than in magnetic multilayers. This result was confirmed by Endo et al. [91] Further investigations were carried out by Fang et al. [12] who used ST-FMR technique to extract the SOT (see Table 9.1). The results are consistent with an inverse spin-galvanic effect arising from Rashba and Dresselhaus-type SOC [30]. Most interestingly, the authors reported the presence of a damping-like component [97]. This discovery came as a surprise as the spin Hall effect, which is conventionally considered as the sole origin of damping-like torque, is absent in this system. Kurebayashi et al. [97] proposed that this additional damping-like torque could be attributed to electric field-driven interband transitions. In fact, as mentioned in Section 9.1, several theories have shown that in a magnetic material possessing Rashba-like SOC, the SOT possesses two components [28, 30]: one, $\tau_T \sim \mathbf{m} \times (\mathbf{z} \times \mathbf{J}_c)$, odd in magnetization and arising from inverse spin-galvanic effect and the other, $\tau_L \sim \mathbf{m} \times \left[(\mathbf{z} \times \mathbf{J}_c) \times \mathbf{m} \right]$, even in magnetization and arising from the interband transitions (see also Figure 9.4b). This observation suggests that such Rashba-driven damping-like torque should also be present in magnetic multilayers [71]. SOT studies have been recently extended to the room-temperature ferromagnetic Heusler alloy, NiMnSb [92], which also possesses bulk inversion symmetry breaking.

9.4.2 SOT IN NON-CENTROSYMMETRIC ANTIFERROMAGNETS

In a recent breakthrough, Wadley et al. [96] demonstrated SOT-driven switching in the collinear antiferromagnet CuMnAs. This material possesses an intriguing crystal structure, fairly similar to Mn$_2$Au, depicted in Figure 9.11c. Although its (non-magnetic) crystal structure is inversion symmetric, each antiferromagnetic sublattice experiences a non-centrosymmetric environment (they are "inversion partners"). Therefore, upon current injection, each magnetic sublattice experiences a SOT field that changes sign on opposite sublattices [98]. In other words, the current-driven field is spatially *staggered*. As discussed in Chapter 8, Volume 1, such a *staggered* field can exert a torque on the Néel order parameter. In the case of CuMnAs, the material possesses biaxial anisotropy and switching Néel order parameter from one easy axis to the other was demonstrated [96]. Further investigations are expected to better understand the physics at stake and in particular the underlying spin dynamics.

9.5 TWO-DIMENSIONAL MATERIALS

9.5.1 TOPOLOGICAL INSULATORS

The three-dimensional topological insulators (TI) are a new class of materials that have an insulating bulk and spin-momentum-locked metallic surface states [99–101]. They exhibit strong SOC and are expected to show large charge-to-spin current conversion efficiency. The topological surface states (TSS) in TI are immune to the non-magnetic impurities due to the time reversal symmetry protection. While some recent reports have confirmed that the TSS are intact in Bi_2Se_3 covered with Fe [102, 103] or Co [104] adatoms, other investigations point out the extreme sensitity of the Dirac states to transition metal proximate layers [105]. The spin-dependent transport is known to be significant near the Fermi level in the Bi_2Se_3 surface states. However, limited spin-dependent transport experiments have been focused on TI/FM heterostructures. Spin-orbit effects have been recently reported by spin pumping measurements [106–109] and magnetoresistance measurements [110, 111]. Direct charge current-induced SOT on the FM layer has been demonstrated by ST-FMR measurement [43, 112] and magnetization switching at cryogenic temperature [46, 113]. It is known that for Bi_2Se_3 the bulk channel provides an inevitable contribution to transport at room temperature and may diminish the SOT arising from surface states [43]. At low temperatures, however, the surface contribution should become significant [114], and SOT in TI/FM heterostructures should be enhanced [43, 46, 113] (see Table 9.1). The reported amplitude of spin Hall angles (i.e. efficiency of damping-like SOT) is widely distributed from ~0.01 using spin pumping measurements to ~2 using ST-FMR measurements in Bi_2Se_3, and even reaches 140–425 in the $(Bi_{0.5}Sb_{0.5})_2Te_3/(Cr_{0.08}Bi_{0.54}Sb_{0.38})_2Te_3$ bilayer heterostructure [46]. Further research works are required to clarify this discrepancy and understand the interplay between surface states, bulk states, and interfacial Rashba bands.

9.5.2 OXIDE HETEROINTERFACES

Structural inversion asymmetry across the surfaces and interfaces results in a strong Rashba SOC in two-dimensional electron gas (2DEG) systems such as InAs/InGaAs [115], and $LaAlO_3/SrTiO_3$ [116, 117] (LAO/STO). These 2DEG systems are widely explored for high-performance spintronics applications such as spin-transistors, as the gate tunable Rashba field makes such systems an attractive choice [17]. Recently, there has been great interest in LAO/STO systems due to the observation of electric field tunable quasi-2DEG [118]. These oxide systems exhibit unusual properties such as the simultaneous existence of magnetism and superconductivity [119], magnetism and Kondo scattering [120], etc. The existence of these competing phenomena is possible only if SOC is present. Indeed, the electron gas present in LAO/STO heterostructures is confined within a few nm from the interface on the STO side [121] and experiences strong inversion symmetry breaking [122], resulting in large Rashba-like SOC. Due to the dielectric properties of $SrTiO_3$ [116], the magnitude of Rashba SOC can be tuned by a gate voltage.

Probing AMR in a rotating in-plane magnetic field is a powerful tool for detecting possible magnetic ordering in LAO/STO systems [123]. While Rashba SOC in LAO/STO systems has been also probed by AMR [116, 124, 125] and two-dimensional weak (anti)localization experiments [126], these studies have been limited to the electric field effect. Recently, the current-induced Rashba field has been quantified utilizing angular dependent AMR [127], and an induced Rashba field of 1.76×10^4 kOe/(10^8 A/cm^2) has been reported [127]. The observed Rashba field is a few orders of magnitude larger than that in metallic systems (see Table 9.1). The ability to effectively tune this gigantic Rashba field with gate electric fields makes LAO/STO systems very attractive.

9.5.3 Two-Dimensional Hexagonal Lattices

Spin-charge conversion experiments [128] have recently been reported on bilayers consisting in a FM adjacent to a two-dimensional crystal of transition metal dichalcogenides (such as MoS_2, WSe_2 etc.). These crystals possess a honeycomb structure, similar to graphene, with two gapped Dirac cones located at the K and K' valleys of the Brillouin zone [129]. The large band gap together with strong SOC make these materials interesting candidates for optically active spin-valleytronics [129]. Further progress is expected in this direction.

9.6 DOMAIN WALLS

The electrical control of magnetic domain walls (DW) is among the most active topics in spintronics [130]. While earlier experiments mainly focused on NiFe nanowires, it has been recently observed that DWs in asymmetric magnetic multilayers display gigantic velocities [13, 131, 132]. Such large velocities, exceeding the one attainable by conventional STT, arise from the cooperation between two effects unique to non-centrosymmetric magnets: [14, 49, 133, 134] a large damping-like SOT, usually coming from spin Hall effect, and Dzyaloshinskii-Moriya interaction (DMI), an antisymmetric exchange interaction that emerges in magnets lacking inversion symmetry [135, 136]. The latter interaction distorts the DW texture in such a way that the damping-like SOT becomes very efficient, leading to ultrafast domain wall motion [133]. The dynamics of chiral magnetic textures in non-centrosymmetric systems is currently attracting significant attention, with the recent observation of room-temperature skyrmions [137–141] and chiral magnetic damping [142, 143].

9.6.1 Dzyaloshinskii-Moriya and Chiral Magnets

In magnets with inversion symmetry, the collinear alignment of neighboring spins is dictated by Heisenberg exchange coupling, $E_H = -\sum_{ij} J_{ij} \mathbf{S}_i \cdot \mathbf{S}_j$, that promotes ferro- or antiferro-magnetism (see Figure 9.12a). When inversion symmetry is broken, DMI emerges. This interaction, reading $E_{DMI} = -\sum_{ij} \mathbf{D}_{ij} \cdot (\mathbf{S}_i \times \mathbf{S}_j)$, forces the neighboring magnetic moments to align perpendicularly to each other thereby stabilizing spin spiral structures with a unique rotational sense (see Figure 9.12a). The direction and magnitude of

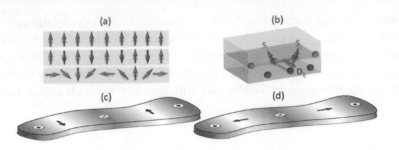

FIGURE 9.12 (a) Ferromagnetic, antiferromagnetic, and spin spiral configurations. (b) Sketch of DMI at the interface between a ferromagnetic metal (red) and a metal with a strong SOC (blue). (c, d) Sketch of Bloch and chiral Néel walls in a perpendicularly magnetized film respectively. Arrows correspond to the magnetization directions.

the DMI vector $\mathbf{D}_{ij}$ is determined by the complex underlying physics and the symmetries of the system [135, 136]. Figure 9.12b illustrates the direction of $\mathbf{D}_{ij}$, $\mathbf{S}_{i,}$ and $\mathbf{S}_j$ in the case of interfacial DMI.

In magnetic nanowires, quasi-one-dimensional DWs can adopt two configurations, Bloch or Néel, depending on the strength of magnetostatic interactions. In most cases, the Bloch wall is favored because of its lower energy (see Figure 9.12c). When DMI is present, however, the wall tends toward an intermediate configuration, between Bloch and Néel. When exceeding the magnetostatic energy, [49, 133] DMI can stabilize Néel DWs of a fixed chirality, [144] either right-handed (for $D > 0$ – ↑⟶↓ or ↓⟵↑) or left-handed (for $D < 0$ – ↑⟵↓ or ↓⟶↑) (see Figure 9.12d). For even stronger DMI, chiral spin textures called skyrmions can form and become stable [93, 145] (see Chapter 10, Volume 1, for details). These "magnetic defects" possess a whirling configuration and a spatial extension of only tens to a few hundred nanometers. They have drawn strong research interest due to their reduced size, stability, and predicted robustness against defects [146, 147]. Most importantly, skyrmions can in principle be moved at a low current density [138, 148] (~10^6 A cm^{-2}) compared to conventional DWs (~10^8 A cm^{-2}), making them a promising choice for high speed, ultrafast data storage device [145].

Néel skyrmions have been experimentally observed at room temperature in a variety of transition metal bilayers [137–140]. In such systems, interfacial DMI creates an effective symmetry breaking in-plane field, H_{DMI}, directed along the magnetization gradient. When it exceeds the magnetic anisotropy, $|H_{DMI}| > (2/\pi)H_K$, this field enables *metastable* skyrmions. These skyrmions have been limited to sizes well over 100 nm, making them ill-suited for practical applications. Furthermore, the total magnetic volume in these systems is below 1.5 nm, posing challenges for thermal stability. Recently, Moreau-Luchaire et al. [141], demonstrated sub-100 nm magnetic skyrmions in dipolar coupled Pt/Co/Ir multilayers, in which the asymmetric Pt and Ir layers result in an enhanced H_{DMI} strong enough to generate skyrmions in films as thick as 6.6 nm. However, to date, no skyrmions have been observed in thick ferromagnetic multilayers in which the disparate magnetic layers are exchange coupled. The discovery of skyrmions in bulk non-centrosymmetric magnets and thin films has led to the proposal of a variety of skyrmionic devices, in which the skyrmion is used as the basic information carrier in next-generation logic and memory devices [145, 146, 149, 150].

9.6.2 Measuring Dzyaloshinskii-Moriya Interaction

There are several methods for engineering DMI, such as changing the bottom NM materials [49, 151, 152], inserting an interfacial dusting layer [134, 153], tuning the underlayer thickness [47], and changing the capping layers. [154] Several measurement methods have been proposed for quantifying the strength of DMI. Ryu et al. proposed a current-induced DW motion method, [134] in which electrical currents are injected into magnetic wires with the presence of a longitudinal magnetic field, and the velocity of the DW is measured as a function of the applied field. The authors found the DW velocity vanishes at some specific in-plane field value equals to H_{DMI}. There are also other methods to quantify the DMI such as DW depinning efficiency measurements [154] and asymmetric magnon dispersion curves measurements by Brillouin light spectroscopy [155, 156] (BLS). In BLS measurements, the in-plane magnetic field H is applied perpendicular to the incident plane of light, corresponding to the Damon-Eschbach (DE) geometry. The counter-propagating spin waves with the same wave vector magnitude, $-k$ and $+k$, are simultaneously recorded in a single BLS spectrum as Stokes and anti-Stokes peaks respectively. The frequency difference between the Stokes and anti-Stokes peaks, Δf, is induced by interfacial DMI and provides a direct measurement of the DMI constant, $D = \pi M_s \Delta f / 2\gamma k$.

Another field-dependent DMI measurement method has been reported by Je et al. [157], in which the DMI effective field, H_{DMI}, can be quantified by studying the DW behavior in the creep regime [153, 157]. The circular DW in the film is driven by the out-of-plane component of the applied field H_z, while the in-plane component H_x breaks the rotational symmetry caused by H_{DMI}, facilitating the anisotropic expansion of the DW. Polar Kerr microscopy is deployed to image the asymmetric DW creep velocity along the direction of an applied in-plane field, illustrated in Figure 9.13a. Blue colored arrows indicate the magnetization directions (either ↑ or ↓) of the film with the presence of a circular domain. The thick yellow arrow indicates the magnetic field (H) applied in-plane with a small out-of-plane tilting angle (~8–10°). Figure 9.13b is obtained by overlapping two DW images recorded by Kerr microscopy at different times from MgO(2)/Pt(4)/CoFeB(1)/MgO(2)/SiO$_2$(3) (thicknesses in nm). The DW distorts along the applied in-plane field, showing that the right edge of the DW moves faster than the left edge. The magnetic moments within the DW are illustrated as green arrows in the DW images in Figure 9.13a. The chirality of DMI stabilizes right-handed Néel walls (↑—→↓ or ↓←—↑) in this case.

In order to quantify the magnitude of DMI, the DW creep velocity under the influence of an external in-plane field H_x is measured as shown in Figure 9.13c. As can be seen, the velocity versus in-plane field curve is asymmetric with respect to the y-axis, which agrees well with the asymmetric domain

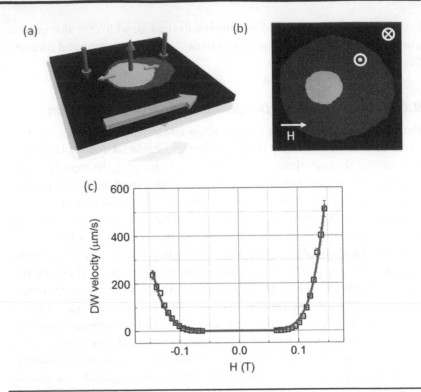

FIGURE 9.13 (a) Illustration of the Kerr microscopy measurement to quantify the DMI effective field (H_{DMI}). The magnetization directions of the domains are shown with the blue arrows, and the H_{DMI} are indicated by green arrows. The thick yellow arrow indicates the applied in-plane field. (After Yu, J. et al., *Sci. Rep.* 6, 32629, 2016. With permission.) (b) The anisotropic domain wall expansion with an in-plane magnetic field. (c) The asymmetric domain wall creep velocity as a function of the applied in-plane magnetic field. The blue symbols are the experimental results and the red solid line shows the fitting curve.

distortion shown in Figure 9.13a and b. The value of H_{DMI} is obtained by fitting the field-driven velocity data with a creep model, [153, 157] $v = v_0 e^{\zeta(H)}$, where the exponent ζ contains information about the disordered energy profile of the system. Phenomenologically, $\zeta = \zeta_0 \left[\sigma(H_x) / \sigma(0) \right]^{1/4}$ where ζ_0 is the scaling fitting parameter, and $\sigma(H_x)$ is the DW energy density as a function of in-plane field H_x [153, 157]. By fitting the curve, the effective DMI constant can be extracted by using $D = \mu_0 H_{DMI} M_s \Delta$. Typical values of DMI range from ~0.3 to 2 mJ m^{-2} in transition metal multilayers [138, 139, 141].

9.6.3 ROLE OF DMI IN SOT-INDUCED MAGNETIZATION SWITCHING

When a perpendicularly magnetized DW is submitted to a damping-like torque, $\tau_L = \tau_L \mathbf{m} \times (\mathbf{y} \times \mathbf{m})$ (e.g., due to spin Hall effect), its velocity depends on the DW azimuthal angle, ϕ. Indeed, the damping-like torque is equivalent to a field $\mathbf{H}_L \sim m_x \mathbf{z}$ that applies either along $+\mathbf{z}$ or $-\mathbf{z}$ depending on $m_x \sim \cos\phi$. Hence, the DW velocity reads $v \propto \tau_L \cos\phi$. In a SOT-driven magnetization reversal experiment, when reversed magnetic domains are nucleated the relative velocity of newly created DWs determines whether reversal occurs or not [133]. When DMI is strong, DWs adopt *homochiral* Néel configuration [$\phi = 0$ – Figure 9.12d] so that under SOT they both propagate along the same direction: the nucleated domain does not expand, and no reversal occurs. However, when applying a large longitudinal field that overcomes DMI, opposite Néel DWs acquire opposite chiralities, and then opposite velocities: the nucleated domain expands, and SOT-driven magnetization reversal occurs [77, 78].

While such a scenario qualitatively explains most switching behaviors [47, 77, 78], it fails to account for some cases. [158] Let us illustrate these aspects by considering two magnetic multilayers MgO(2)/Pt(4)/Co(0.1)/[Ni(0.1)/Co(0.1)]$_4$/[Cu or Ta] (thicknesses in nm) that only differ by their capping layer. In the Cu-capped device, DMI is intermediate and attributed to the bottom Pt layer only, while in the Ta-capped layer, DMI is large as the top Ta interface contributes a large positive DMI [49]. The SOT-driven switching of the Cu-capped device with 400 Oe assist field (see Figure 9.14a and b), as well as the switching of the Ta-capped device with 1000 Oe assistive field (see Figure 9.14e and f) can both be easily explained by the scenario exposed above.

However, at the intermediate external field (400 Oe), the Ta-capped device exhibits an abnormal switching-back behavior that cannot be fully accounted for by the traditional scenario (see Figure 9.14c and d). Indeed, Figure 9.14d clearly shows that the magnetization reversal occurs through the propagation of a *tilted* DW along the electron flow (panels 1 and 2) until it reaches a strong pinning site at the exit of the Hall cross (panel 3). At this stage, the DW tilting almost vanishes and its depinning takes place *against* the electron flow (panel 4). The change of sign of the DW velocity between panel 3 and 4 holds the key to understand the absence of hysteresis for this system and suggests a strong connection between the DW tilting and its direction of motion [158]. The velocity of a tilted magnetic DW can be written as [159] $v \propto \tau_\mathrm{L} \cos\phi / \cos\chi$, where χ is the tilting angle of the wall with respect to the transverse direction. The tilting appears through a competition between DMI and the potential profile of the magnetic nanowire. Hence, the magnetic configuration depends on the position of the DW in the nanowire, as shown in Figure 9.14d and illustrated in Figure 9.14g. Depending on the tilting angle χ and on the azimuthal angle ϕ, the velocity can change sign close to the exit of the nanowire [panel 3 in Figure 9.14d], corresponding to the abnormal switching-back mechanism observed in Figure 9.14c.

9.6.4 Chiral Damping

It has been realized recently that magnetic damping also depends on the chirality of the magnetic texture. Indeed, in magnetic metals, low energy spin waves relax through electron-magnon interaction mediated by SOC [160]. In a system lacking inversion symmetry, it is therefore natural to expect that the energy dissipation depends on the propagation direction of the magnons. In other words, the magnetic damping should fulfill Neumann's principle that

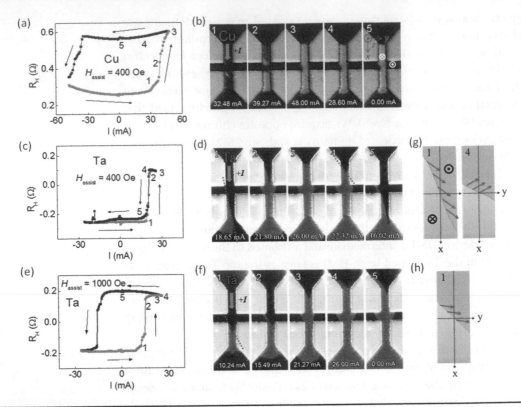

FIGURE 9.14 (a) The SOT current-induced switching curve in the Cu-capped device with a 400 Oe assist field. (b) The DW configurations captured by Kerr microscope simultaneously with the SOT current-induced switching measurement. The magnitude of the applied current is indicated at the bottom. (c, d) The switching curve and Kerr image in the Ta-capped device is displayed in c and d, respectively, with a 400 Oe assist field. (e, f) The switching curve and Kerr image of the Ta-capped device with a 1000 Oe assist field. (g) Domain wall and magnetic moment configurations. (h) Intermediate tilting with a compensated DMI by a strong assist field. (After Yu, J. et al., *Sci. Rep.* 6, 32629, 2016. With permission.)

states, "Any physical properties of a system possesses the symmetry of that system". In a recent experiment, Jué et al. [142] investigated the creep motion of a magnetic bubble under an external field and observed that the magnetic damping depends on the DW chirality, as expected [153, 157]. Measuring the field-driven expansion of the bubble, they analyzed their data following the method proposed by Je et al. [153, 157] and observed that the asymmetry of the DW creeps velocity, $v \approx v_0 e^{\zeta(H)}$, does not lie in the exponent ζ as expected for DMI, but in the prefactor v_0. While the exponent is determined by the energy profile of the magnetic system (and therefore DMI), the prefactor includes a complex (and, to date, undetermined) combination of energy profile and energy dissipation. Since the exponent is not affected by the chirality of the domain wall, the observed asymmetry was attributed to energy dissipation, i.e. chiral damping. The change of magnetic damping can be quite large, of the order of 50% [142].

9.7 CONCLUSION AND PERSPECTIVES

SOTs constitute an outstanding opportunity for the effective electrical control of magnetic domains without the use of an external polarizer. The very fast switching rate, high electrical efficiency, and planar configuration open interesting perspectives from the application standpoint, but several technical challenges such as controllability, reproducibility, and scalability remain to be thoroughly addressed. The fundamental mechanisms taking place in such structures (spin Hall effect, Rashba effect, spin swapping, DMI, and chiral damping) are still under active scrutiny, and systematic theoretical and experimental investigations are required to enable the rational design of robust, technologically relevant SOT-driven devices.

REFERENCES

1. J. C. Slonczewski, Current-driven excitation of magnetic multilayers, *J. Magn. Magn. Mater.* **159**, L1–L7 (1996).
2. L. Berger, Emission of spin waves by a magnetic multilayer traversed by a current, *Phys. Rev. B* **54**, 9353–9358 (1996).
3. B. A. Bernevig and O. Vafek, Piezo-magnetoelectric effects in p-doped semiconductors, *Phys. Rev. B* **72**, 033203 (2005).
4. A. Manchon and S. Zhang, Theory of nonequilibrium intrinsic spin torque in a single nanomagnet, *Phys. Rev. B* **78**, 212405 (2008).
5. I. Garate and A. H. MacDonald, Influence of a transport current on magnetic anisotropy in gyrotropic ferromagnets, *Phys. Rev. B* **80**, 134403 (2009).
6. L. Liu, T. Moriyama, D. C. Ralph, and R. A. Buhrman, Spin-torque ferromagnetic resonance induced by the spin Hall effect, *Phys. Rev. Lett.* **106**, 036601 (2011).
7. A. Chernyshov et al., Evidence for reversible control of magnetization in a ferromagnetic material by means of spin–orbit magnetic field, *Nat. Phys.* **5**, 656–659 (2009).
8. I. M. Miron et al., Current-driven spin torque induced by the Rashba effect in a ferromagnetic metal layer, *Nat. Mater.* **9**, 230–234 (2010).
9. I. M. Miron et al., Perpendicular switching of a single ferromagnetic layer induced by in-plane current injection, *Nature* **476**, 189–193 (2011).
10. L. Liu et al., Spin-torque switching with the giant spin Hall effect of tantalum, *Science* **336**, 555–558 (2012).
11. L. Liu, C.-F. Pai, D. C. Ralph, and R. A. Buhrman, Magnetic oscillations driven by the spin Hall effect in 3-terminal magnetic tunnel junction devices, *Phys. Rev. Lett.* **109**, 186602 (2012).
12. D. Fang et al., Spin-orbit-driven ferromagnetic resonance, *Nat. Nanotechnol.* **6**, 413–417 (2011).
13. I. M. Miron et al., Fast current-induced domain-wall motion controlled by the Rashba effect, *Nat. Mater.* **10**, 419–423 (2011).
14. P. P. J. Haazen et al., Domain wall depinning governed by the spin Hall effect, *Nat. Mater.* **12**, 299–303 (2013).
15. V. E. Demidov et al., Magnetic nano-oscillator driven by pure spin current, *Nat. Mater.* **11**, 1028–1031 (2012).
16. R. Liu, W. Lim, and S. Urazhdin, Spectral characteristics of the microwave emission by the spin hall nano-oscillator, *Phys. Rev. Lett.* **110**, 147601 (2013).
17. A. Manchon, H. C. Koo, J. Nitta, S. M. Frolov, and R. A. Duine, New perspectives for Rashba spin-orbit coupling, *Nat. Mater.* **14**, 871 (2015).
18. M. I. Dyakonov and V. I. Perel, Possibility of orienting electron spins with current, *ZhETF Pis. Red.* **13**, 657 (1971).

19. J. Sinova, S. O. Valenzuela, J. Wunderlich, C. H. Back, and T. Jungwirth, Spin Hall effect, *Rev. Mod. Phys.* **87**, 1213 (2015).
20. Y. Niimi et al., Extrinsic spin Hall effect induced by iridium impurities in copper, *Phys. Rev. Lett.* **106**, 126601 (2011).
21. Y. Niimi et al., Experimental verification of comparability between spin-orbit and spin-diffusion lengths, *Phys. Rev. Lett.* **110**, 016805 (2013).
22. M. Morota et al., Indication of intrinsic spin Hall effect in 4d and 5d transition metals, *Phys. Rev. B* **83**, 174405 (2011).
23. T. Tanaka et al., Intrinsic spin Hall effect and orbital Hall effect in 4d and 5d transition metals, *Phys. Rev. B* **77**, 165117 (2008).
24. Y. A. Bychkov and E. I. Rasbha, Properties of a 2D electron gas with lifted spectral degeneracy, *P. Zh. Eksp. Teor. Fiz.* **39**, 66 (1984).
25. S. D. Ganichev et al., Spin-galvanic effect, *Nature* **417**, 153–156 (2002).
26. V. M. Edelstein, Spin polarization of conduction electrons induced by electric current in two-dimensional asymmetric electron systems, *Solid State Commun.* **73**, 233–235 (1990).
27. H. J. Zhang et al., Current-induced spin polarization on metal surfaces probed by spin-polarized positron beam, *Sci. Rep.* **4**, 4844 (2014).
28. D. A. Pesin and A. H. MacDonald, Quantum kinetic theory of current-induced torques in Rashba ferromagnets, *Phys. Rev. B* **86**, 014416 (2012).
29. X. Wang and A. Manchon, Diffusive spin dynamics in ferromagnetic thin films with a Rashba interaction, *Phys. Rev. Lett.* **108**, 117201 (2012).
30. H. Li et al., Intraband and interband spin-orbit torques in noncentrosymmetric ferromagnets, *Phys. Rev. B* **91**, 134402 (2015).
31. H. B. M. Saidaoui and A. Manchon, Spin-swapping transport and torques in ultrathin magnetic bilayers, *Phys. Rev. Lett.* **117**, 036601 (2016).
32. F. Freimuth, S. Blügel, and Y. Mokrousov, Direct and inverse spin-orbit torques, *Phys. Rev. B* **92**, 064415 (2015).
33. A. Brataas et al., Spin Pumping and Spin Transfer, arXiv 1108.0385v3 (2012).
34. U. H. Pi et al., Tilting of the spin orientation induced by Rashba effect in ferromagnetic metal layer, *Appl. Phys. Lett.* **97**, 162507 (2010).
35. J. Kim et al., Layer thickness dependence of the current-induced effective field vector in Ta|CoFeB|MgO, *Nat. Mater.* **12**, 240–245 (2013).
36. K. Garello et al., Symmetry and magnitude of spin-orbit torques in ferromagnetic heterostructures, *Nat. Nanotechnol.* **8**, 587–593 (2013).
37. X. Qiu et al., Angular and temperature dependence of current induced spin-orbit effective fields in Ta/CoFeB/MgO nanowires, *Sci. Rep.* **4**, 4491 (2014).
38. M. Hayashi, J. Kim, M. Yamanouchi, and H. Ohno, Quantitative characterization of the spin-orbit torque using harmonic Hall voltage measurements, *Phys. Rev. B* **89**, 144425 (2014).
39. H.-R. Lee et al., Spin-orbit torque in a bulk perpendicular magnetic anisotropy Pd/FePd/MgO system, *Sci. Rep.* **4**, 6548 (2014).
40. T. Yang et al., Layer thickness dependence of spin orbit torques and fields in Pt/Co/AlO trilayer structures, *Jpn. J. Appl. Phys.* **54**, 04DM05 (2015).
41. A. A. Tulapurkar et al., Spin-torque diode effect in magnetic tunnel junctions, *Nature* **438**, 339–342 (2005).
42. A. Balocchi et al., Full electrical control of the electron spin relaxation in GaAs quantum wells, *Phys. Rev. Lett.* **107**, 136604 (2011).
43. Y. Wang et al., Topological surface states originated spin-orbit torques in Bi_2Se_3, *Phys. Rev. Lett.* **114**, 257202 (2015).
44. K. Ando et al., Electric manipulation of spin relaxation using the spin Hall effect, *Phys. Rev. Lett.* **101**, 036601 (2008).
45. L. Liu, O. Lee, T. J. Gudmundsen, D. C. Ralph, and R. A. Buhrman, Current-induced switching of perpendicularly magnetized magnetic layers using spin torque from the spin Hall effect, *Phys. Rev. Lett.* **109**, 096602 (2012).

46. Y. Fan et al., Magnetization switching through giant spin-orbit torque in a magnetically doped topological insulator heterostructure, *Nat. Mater.* **13**, 699–704 (2014).

47. J. Torrejon et al., Interface control of the magnetic chirality in CoFeB/MgO heterostructures with heavy-metal underlayers, *Nat. Commun.* **5**, 4655 (2014).

48. X. Qiu et al., Spin–orbit-torque engineering via oxygen manipulation, *Nat. Nanotechnol.* **10**, 333–338 (2015).

49. S. Emori, U. Bauer, S.-M. Ahn, E. Martinez, and G. S. D. Beach, Current-driven dynamics of chiral ferromagnetic domain walls, *Nat. Mater.* **12**, 611–616 (2013).

50. C.-F. Pai et al., Spin transfer torque devices utilizing the giant spin Hall effect of tungsten, *Appl. Phys. Lett.* **101**, 122404 (2012).

51. S. Woo, M. Mann, A. J. Tan, L. Caretta, and G. S. D. Beach, Enhanced spin-orbit torques in Pt/Co/Ta heterostructures, *Appl. Phys. Lett.* **105**, 212404 (2014).

52. C.-F. Pai et al., Enhancement of perpendicular magnetic anisotropy and transmission of spin-Hall-effect-induced spin currents by a Hf spacer layer in W/Hf/CoFeB/MgO layer structures, *Appl. Phys. Lett.* **104**, 082407 (2014).

53. L. M. Loong, P. Deorani, X. Qiu, and H. Yang, Investigating and engineering spin-orbit torques in heavy metal/$Co_2FeAl_{0.5}Si_{0.5}$/MgO thin film structures, *Appl. Phys. Lett.* **107**, 6–11 (2015).

54. X. Fan et al., Observation of the nonlocal spin-orbital effective field, *Nat. Commun.* **4**, 1799 (2013).

55. X. Fan et al., Quantifying interface and bulk contributions to spin-orbit torque in magnetic bilayers, *Nat. Commun.* **5**, 3042 (2014).

56. Y. Wang, P. Deorani, X. Qiu, J. H. Kwon, and H. Yang, Determination of intrinsic spin Hall angle in Pt, *Appl. Phys. Lett.* **105**, 152412 (2014).

57. C. O. Avci et al., Fieldlike and antidamping spin-orbit torques in as-grown and annealed Ta/CoFeB/MgO layers, *Phys. Rev. B* **89**, 214419 (2014).

58. J. Kim et al., Anomalous temperature dependence of current-induced torques in CoFeB/MgO heterostructures with Ta-based underlayers, *Phys. Rev. B* **89**, 174424 (2014).

59. J.-C. Rojas-Sánchez et al., Spin pumping and inverse spin Hall effect in platinum: The essential role of spin-memory loss at metallic interfaces, *Phys. Rev. Lett.* **112**, 106602 (2014).

60. J. Bass and W. P. Pratt, Spin-diffusion lengths in metals and alloys, and spin-flipping at metal/metal interfaces: An experimentalist's critical review, *J. Phys. Condens. Matter* **19**, 183201 (2007).

61. Y. Niimi et al., Giant spin Hall effect induced by skew scattering from bismuth impurities inside thin film CuBi alloys, *Phys. Rev. Lett.* **109**, 156602 (2012).

62. P. Laczkowski et al., Experimental evidences of a large extrinsic spin Hall effect in AuW alloy, *Appl. Phys. Lett.* **104**, 142403 (2014).

63. T. Liu, Y. Zhang, J. W. Cai, and H. Y. Pan, Thermally robust Mo/CoFeB/MgO trilayers with strong perpendicular magnetic anisotropy, *Sci. Rep.* **4**, 5895 (2014).

64. T. Liu, J. W. Cai, and L. Sun, Large enhanced perpendicular magnetic anisotropy in CoFeB/MgO system with the typical Ta buffer replaced by an Hf layer, *AIP Adv.* **2**, 032151 (2012).

65. R. Ramaswamy, X. Qiu, T. Dutta, S. D. Pollard, and H. Yang, Hf thickness dependence of spin-orbit torques in Hf/CoFeB/MgO heterostructures, *Appl. Phys. Lett.* **108**, 202406 (2016).

66. M. Akyol et al., Effect of the oxide layer on current-induced spin-orbit torques in Hf|CoFeB|MgO and Hf|CoFeB|TaOx structures, *Appl. Phys. Lett.* **106**, 032406 (2015).

67. P. M. Haney, H.-W. Lee, K.-J. Lee, A. Manchon, and M. D. Stiles, Current induced torques and interfacial spin-orbit coupling: Semiclassical modeling, *Phys. Rev. B* **87**, 174411 (2013).

68. F. Freimuth, S. Blügel, and Y. Mokrousov, Spin-orbit torques in Co/Pt(111) and Mn/W(001) magnetic bilayers from first principles, *Phys. Rev. B* **90**, 174423 (2014).

69. P. M. Haney, H.-W. Lee, K.-J. Lee, A. Manchon, and M. D. Stiles, Current-induced torques and interfacial spin-orbit coupling, *Phys. Rev. B* **88**, 214417 (2013).

70. T. Suzuki et al., Current-induced effective field in perpendicularly magnetized Ta/CoFeB/MgO wire, *Appl. Phys. Lett.* **98**, 142505 (2011).

71. M. Jamali et al., Spin-orbit torques in Co/Pd multilayer nanowires, *Phys. Rev. Lett.* **111**, 246602 (2013).

72. S. Kim, Y. Koo, V. Chernov, and H. Padmore, Clear evidence for strain changes according to Co layer thickness in metastable Co/Pd(111) multilayers: An extended x-ray absorption fine structure study, *Phys. Rev. B* **53**, 11114 (1996).

73. A. Maesaka and H. Ohmori, Transmission electron microscopy analysis of lattice strain in epitaxial Co − Pd multilayers, *IEEE Trans. Magn.* **38**, 2676–2678 (2002).

74. D. Bang et al., Enhancement of spin Hall effect induced torques for current-driven magnetic domain wall motion: Inner interface effect, *Phys. Rev. B* **93**, 174424 (2016).

75. M. Cubukcu et al., Spin-orbit torque magnetization switching of a three-terminal perpendicular magnetic tunnel junction, *Appl. Phys. Lett.* **104**, 042406 (2014).

76. K. Garello et al., Ultrafast magnetization switching by spin-orbit torques, *Appl. Phys. Lett.* **105**, 212402 (2014).

77. O. J. Lee et al., Central role of domain wall depinning for perpendicular magnetization switching driven by spin torque from the spin Hall effect, *Phys. Rev. B* **89**, 024418 (2014).

78. G. Yu et al., Magnetization switching through spin-Hall-effect-induced chiral domain wall propagation, *Phys. Rev. B* **89**, 104421 (2014).

79. S. Pizzini et al., Chirality-induced asymmetric magnetic nucleation in Pt/Co/AlOx ultrathin microstructures, *Phys. Rev. Lett.* **113**, 047203 (2014).

80. Y.-C. Lau, D. Betto, K. Rode, J. M. D. Coey, and P. Stamenov, Spin−orbit torque switching without an external field using interlayer exchange coupling, *Nat. Nanotechnol.* **11**, 758–762 (2016).

81. A. van den Brink et al., Field-free magnetization reversal by spin-Hall effect and exchange bias, *Nat. Commun.* **7**, 10854 (2016).

82. S. Fukami, C. Zhang, S. Dutta Gupta, and H. Ohno, Magnetization switching by spin-orbit torque in an antiferromagnet/ferromagnet bilayer system, *Nat. Mater.* **15**, 535 (2016).

83. L. You et al., Switching of perpendicularly polarized nanomagnets with spin orbit torque without an external magnetic field by engineering a tilted anisotropy, *Proc. Natl. Acad. Sci.* **112**, 10310–10315 (2015).

84. G. Yu et al., Switching of perpendicular magnetization by spin-orbit torques in the absence of external magnetic fields, *Nat. Nanotechnol.* **9**, 548–554 (2014).

85. W. Legrand, R. Ramaswamy, R. Mishra, and H. Yang, Coherent subnanosecond switching of perpendicular magnetization by the fieldlike spin-orbit torque without an external magnetic field, *Phys. Rev. Appl.* **3**, 1–11 (2015).

86. T. Taniguchi, S. Mitani, and M. Hayashi, Critical current destabilizing perpendicular magnetization by the spin Hall effect, *Phys. Rev. B* **92**, 24428 (2015).

87. J. Park, G. E. Rowlands, O. J. Lee, D. C. Ralph, and R. A. Buhrman, Macrospin modeling of sub-ns pulse switching of perpendicularly magnetized free layer via spin-orbit torques for cryogenic memory applications, *Appl. Phys. Lett.* **105**, 102404 (2014).

88. A. Hamadeh et al., Electronic control of the spin-wave damping in a magnetic insulator, *Phys. Rev. Lett.* **113**, 197203 (2014).

89. V. E. Demidov, S. Urazhdin, A. B. Rinkevich, G. Reiss, and S. O. Demokritov, Spin Hall controlled magnonic microwaveguides, *Appl. Phys. Lett.* **104**, 152402 (2014).

90. K. An et al., Control of propagating spin waves via spin transfer torque in a metallic bilayer waveguide, *Phys. Rev. B* **89**, 140405(R) (2014).

91. M. Endo, F. Matsukura, and H. Ohno, Current induced effective magnetic field and magnetization reversal in uniaxial anisotropy (Ga,Mn)As, *Appl. Phys. Lett.* **97**, 222501 (2010).

92. C. Ciccarelli et al., Room-temperature spin-orbit torque in NiMnSb, *Nat. Phys.* **12**, 855–860 (2016).

93. N. Nagaosa and Y. Tokura, Topological properties and dynamics of magnetic skyrmions, *Nat. Nanotechnol.* **8**, 899–911 (2013).

94. T. Jungwirth et al., Spin-dependent phenomena and device concepts explored in (Ga,Mn)As, *Rev. Mod. Phys.* **86**, 855–896 (2014).

95. V. M. T. S. Barthem, C. V. Colin, H. Mayaffre, M.-H. Julien, and D. Givord, Revealing the properties of Mn_2Au for antiferromagnetic spintronics, *Nat. Commun.* **4**, 2892 (2013).

96. P. Wadley et al., Electrical switching of an antiferromagnet, *Science* **351**, 587–590 (2016).

97. H. Kurebayashi et al., An antidamping spin-orbit torque originating from the Berry curvature, *Nat. Nanotechnol.* **9**, 211–217 (2014).

98. J. Železný, H. Gao, K. Výborný, and J. Zemen, Relativistic Néel-order fields induced by electrical current in antiferromagnets, *Phys. Rev. Lett.* **113**, 157201 (2014).

99. M. Z. Hasan and C. Kane, Colloquium: Topological insulators, *Rev. Mod. Phys.* **82**, 3045–3067 (2010).

100. J. E. Moore, The birth of topological insulators, *Nature* **464**, 194–198 (2010).

101. X.-L. Qi and S.-C. Zhang, Topological insulators and superconductors, *Rev. Mod. Phys.* **83**, 1057–1110 (2011).

102. M. R. Scholz et al., Tolerance of topological surface states towards magnetic moments: Fe on Bi_2Se_3, *Phys. Rev. Lett.* **108**, 256810 (2012).

103. J. Honolka et al., In-plane magnetic anisotropy of Fe atoms on Bi_2Se_3(111), *Phys. Rev. Lett.* **108**, 256811 (2012).

104. M. Ye et al., Quasiparticle interference on the surface of Bi_2Se_3 induced by cobalt adatom in the absence of ferromagnetic ordering, *Phys. Rev. B* **85**, 205317 (2012).

105. J. Zhang, J. P. Velev, X. Dang, and E. Y. Tsymbal, *Phys. Rev. B* **94**, 014435 (2016).

106. P. Deorani et al., Observation of inverse spin Hall effect in bismuth selenide, *Phys. Rev. B* **90**, 094403 (2014).

107. Y. Shiomi et al., Spin-electricity conversion induced by spin injection into topological insulators, *Phys. Rev. Lett.* **113**, 196601 (2014).

108. M. Jamali et al., Giant spin pumping and inverse spin Hall effect in the presence of surface spin-orbit coupling of topological insulator Bi_2Se_3, *Nano Lett.* **15**, 7126 (2015).

109. J. C. Rojas-Sánchez et al., Spin to charge conversion at room temperature by spin pumping into a new type of topological insulator: α -Sn films, *Phys. Rev. Lett.* **116**, 096602 (2016).

110. Y. Ando et al., Electrical detection of the spin polarization due to charge flow in the surface state of the topological insulator $Bi_{1.5}Sb_{0.5}Te_{1.7}Se_{1.3}$, *Nano Lett.* **14**, 6226–6230 (2014).

111. C. H. Li et al., Electrical detection of charge-current-induced spin polarization due to spin-momentum locking in Bi_2Se_3, *Nat. Nanotechnol.* **9**, 218–224 (2014).

112. A. R. Mellnik et al., Spin-transfer torque generated by a topological insulator, *Nature* **511**, 449–451 (2014).

113. Y. Fan et al., Electric-field control of spin-orbit torque in a magnetically doped topological insulator, *Nat. Nanotechnol.* **11**, 352–359 (2016).

114. N. Bansal, Y. S. Kim, M. Brahlek, E. Edrey, and S. Oh, Thickness-independent transport channels in topological insulator Bi_2Se_3 thin films, *Phys. Rev. Lett.* **109**, 116804 (2012).

115. J. Nitta, T. Akazaki, H. Takayanagi, and T. Enoki, Gate control of spin-orbit interaction in an inverted InGaAs/InAlAs hetrostructure, *Phys. Rev. Lett.* **78**, 1335–1338 (1997).

116. A. D. Caviglia et al., Tunable Rashba spin-orbit interaction at oxide interfaces, *Phys. Rev. Lett.* **104**, 126803 (2010).

117. M. Ben Shalom, M. Sachs, D. Rakhmilevitch, A. Palevski, and Y. Dagan, Tuning spin-orbit coupling and superconductivity at the $SrTiO_3$/$LaAlO_3$ interface: A magnetotransport study, *Phys. Rev. Lett.* **104**, 126802 (2010).

118. S. Stemmer and S. James Allen, Two-dimensional electron gases at complex oxide interfaces, *Annu. Rev. Mater. Res.* **44**, 151–171 (2014).

119. L. Li, C. Richter, J. Mannhart, and R. C. Ashoori, Coexistence of magnetic order and two-dimensional superconductivity at $LaAlO_3$/$SrTiO_3$ interfaces, *Nat. Phys.* **7**, 762–766 (2011).

120. Ariando et al., Electronic phase separation at the $LaAlO_3$/$SrTiO_3$ interface, *Nat. Commun.* **2**, 188 (2011).

121. N. Reyren et al., Superconducting interfaces between insulating oxides, *Science* **317**, 1196–1199 (2007).

122. A. Ohtomo and H. Y. Hwang, A high-mobility electron gas at the $LaAlO_3$/$SrTiO_3$ heterointerface, *Nature* **427**, 423–426 (2004).

123. M. Ben Shalom et al., Anisotropic magnetotransport at the $SrTiO_3$/$LaAlO_3$ interface, *Phys. Rev. B* **80**, 140403 (2009).

124. A. Fête, S. Gariglio, A. D. Caviglia, J. M. Triscone, and M. Gabay, Rashba induced magnetoconductance oscillations in the $LaAlO_3$-$SrTiO_3$ heterostructure, *Phys. Rev. B* **86**, 201105 (2012).

125. A. Annadi et al., Fourfold oscillation in anisotropic magnetoresistance and planar Hall effect at the $LaAlO_3$/$SrTiO_3$ heterointerfaces: Effect of carrier confinement and electric field on magnetic interactions, *Phys. Rev. B* **87**, 201102 (2013).

126. H. Nakamura, T. Koga, and T. Kimura, Experimental evidence of cubic Rashba effect in an inversion-symmetric oxide, *Phys. Rev. Lett.* **108**, 206601 (2012).

127. K. Narayanapillai et al., Current-driven spin orbit field in $LaAlO_3$/$SrTiO_3$ heterostructures, *Appl. Phys. Lett.* **105**, 162405 (2014).

128. W. Zhang et al., Research update: Spin transfer torques in permalloy on monolayer MoS_2, *APL Mater.* **4**, 032302 (2016).

129. X. Xu, W. Yao, D. Xiao, and T. F. Heinz, Spin and pseudospins in layered transition metal dichalcogenides, *Nat. Phys.* **10**, 343–350 (2014).

130. O. Boulle, G. Malinowski, and M. Kläui, Current-induced domain wall motion in nanoscale ferromagnetic elements, *Mater. Sci. Eng. R Rep.* **72**, 159–187 (2011).

131. T. A. Moore et al., High domain wall velocities induced by current in ultrathin Pt/Co/AlO$_x$ wires with perpendicular magnetic anisotropy, *Appl. Phys. Lett.* **93**, 262504 (2008).

132. S.-H. Yang, K.-S. Ryu, and S. Parkin, Domain-wall velocities of up to 750 ms^{-1} driven by exchange-coupling torque in synthetic antiferromagnets, *Nat. Nanotechnol.* **10**, 221–226 (2015).

133. A. Thiaville, S. Rohart, É. Jué, V. Cros, and A. Fert, Dynamics of Dzyaloshinskii domain walls in ultrathin magnetic films, *Europhys. Lett.* **100**, 57002 (2012).

134. K.-S. Ryu, L. Thomas, S.-H. Yang, and S. Parkin, Chiral spin torque at magnetic domain walls, *Nat. Nanotechnol.* **8**, 527–533 (2013).

135. I. E. Dzyaloshinskii, Thermodynamic theory of weak ferromagnetism in antiferromagnetic substances, *Sov. Phys. JETP* **5**, 1259 (1957).

136. T. Moriya, Anisotropic superexchange interaction and weak ferromagnetism, *Phys. Rev.* **120**, 91–98 (1960).

137. W. Jiang et al., Blowing magnetic skyrmion bubbles, *Science* **349**, 283 (2015).

138. S. Woo et al., Observation of room-temperature magnetic skyrmions and their current-driven dynamics in ultrathin metallic ferromagnets, *Nat. Mater.* **15**, 501–506 (2016).

139. O. Boulle et al., Room temperature chiral magnetic skyrmion in ultrathin magnetic nanostructures, *Nat. Nanotechnol.* **11**, 449 (2016).

140. G. Chen, A. Mascaraque, A. T. N'Diaye, and A. K. Schmid, Room temperature skyrmion ground state stabilized through interlayer exchange coupling, *Appl. Phys. Lett.* **106**, 242404 (2015).

141. C. Moreau-Luchaire et al., Additive interfacial chiral interaction in multilayers for stabilization of small individual skyrmions at room temperature, *Nat. Nanotechnol.* **11**, 444 (2016).

142. E. Jué, C. K. Safeer, M. Drouard, A. Lopez, and P. Balint, Chiral damping of magnetic domain walls, *Nat. Mater.* **15**, 272 (2015).

143. C. A. Akosa, I. M. Miron, G. Gaudin, and A. Manchon, Phenomenology of chiral damping in noncentrosymmetric magnets, *Phys. Rev. B* **93**, 214429 (2015).

144. M. J. Benitez et al., Magnetic microscopy and topological stability of homochiral Néel domain walls in a Pt/Co/AlOx trilayer, *Nat. Commun.* **6**, 8957 (2015).

145. A. Fert, V. Cros, and J. Sampaio, Skyrmions on the track, *Nat. Nanotechnol.* **8**, 152–156 (2013).

146. J. Sampaio, V. Cros, S. Rohart, A. Thiaville, and A. Fert, Nucleation, stability and current-induced motion of isolated magnetic skyrmions in nanostructures, *Nat. Nanotechnol.* **8**, 839–844 (2013).

147. J. Iwasaki, M. Mochizuki, and N. Nagaosa, Current-induced skyrmion dynamics in constricted geometries, *Nat. Nanotechnol.* **8**, 742–747 (2013).

148. F. Jonietz et al., Spin transfer torques in MnSi at ultralow current densities, *Science* **330**, 1648–1651 (2010).

149. X. Zhang, M. Ezawa, and Y. Zhou, Magnetic skyrmion logic gates: Conversion, duplication and merging of skyrmions, *Sci. Rep.* **5**, 9400 (2015).

150. R. Tomasello et al., A strategy for the design of skyrmion racetrack memories, *Sci. Rep.* **4**, 6784 (2014).

151. G. Chen et al., Tailoring the chirality of magnetic domain walls by interface engineering, *Nat. Commun.* **4** (2013).

152. K.-S. Ryu, S.-H. Yang, L. Thomas, and S. S. P. Parkin, Chiral spin torque arising from proximity-induced magnetization, *Nat. Commun.* **5**, 3910 (2014).

153. A. Hrabec et al., Measuring and tailoring the Dzyaloshinskii-Moriya interaction in perpendicularly magnetized thin films, *Phys. Rev. B* **90**, 020402 (2014).

154. J. H. Franken, M. Herps, H. J. M. Swagten, and B. Koopmans, Tunable chiral spin texture in magnetic domain-walls, *Sci. Rep.* **4**, 5248 (2014).

155. K. Di et al., Direct observation of the Dzyaloshinskii-Moriya interaction in a Pt/Co/Ni film, *Phys. Rev. Lett.* **114**, 047201 (2015).

156. H. T. Nembach, J. M. Shaw, M. Weiler, E. Jué, and T. J. Silva, Linear relation between Heisenberg exchange and interfacial Dzyaloshinskii – Moriya interaction in metal films, *Nat. Phys.* **11**, 825 (2015).

157. S.-G. Je et al., Asymmetric magnetic domain-wall motion by the Dzyaloshinskii-Moriya interaction, *Phys. Rev. B* **88**, 214401 (2013).

158. J. Yu et al., Spin orbit torques and Dzyaloshinskii- Moriya interaction in dual-interfaced Co-Ni multilayers, *Sci. Rep.* **6**, 32629 (2016).

159. O. Boulle et al., Domain wall tilting in the presence of the Dzyaloshinskii-Moriya interaction in out-of-plane magnetized magnetic nanotracks, *Phys. Rev. Lett.* **111**, 217203 (2013).

160. V. Kamberský, Spin-orbital gilbert damping in common magnetic metals, *Phys. Rev. B* **76**, 134416 (2007).

10

All-Optical Switching of Magnetization

From Fundamentals to Nanoscale Recording

Andrei Kirilyuk, Alexey V. Kimel, and Theo Rasing

All-optical switching (AOS) of magnetization has recently emerged as a fascinating discovery with both fundamental and applied aspects. The latter, however, depends on our ability to demonstrate the nanoscale realization of this phenomenon. Apart from the intrinsic material-related behavior of AOS, two more practical problems need to be solved. First of all, this is the focusing of light into a nanometer-sized spot. Second, this is the stability of the written domain. This chapter reviews the various aspects of the nanoscale realization of the AOS, demonstrating reproducible all-optical magnetic recording down to 50 nm.

10.1 INTRODUCTION: OPTICAL MANIPULATION OF MAGNETIC ORDER

The problem with the very fast switching of the magnetization between its metastable states is fundamentally intriguing as well as important for applications [1]. The conventional way to reverse magnetization is based on a precessional motion of the latter in an external magnetic field. A realistic switching time is then determined by the strength and duration of the magnetic field pulse and is of the order of ~100 ps [2, 3, 4]. Related topics to this chapter pertain to discussion of conservation and transfer of angular momentum in Chapters 7–9 and 14, Volume 1, electric-field controlled magnetism in Chapter 13, Volume 2, magnetic skyrmions in Chapter 10, Volume 3, as well as potential applications from magnetic sensors and memories in Chapters 12–14, Volume 3, to spin-lasers and spin-based logic in Chapters 16 and 17, Volume 3 in this book.

While the generation of very short and intense magnetic field pulses remains an experimental challenge, laser technology readily provides light pulses as short as a few tens of femtoseconds. Consequently, after the seminal demonstration of the ultrafast demagnetization of a magnetic thin film by a 60-femtosecond laser pulse [5], laser manipulation and detection of magnetization dynamics has developed into an active research area [6]. At the same time, the studies of the various effects occurring in magnetic media under the action of ultrashort laser pulses rose a number of important fundamental questions. In particular, the laser excitation brings the magnetic medium into a strongly non-equilibrium state [7], where a conventional description of magnetic phenomena in terms of equilibrium thermodynamics and adiabatic approximations is no longer valid. Consequently, ultrafast laser-induced magnetization dynamics is a new and rather unexplored topic at the frontier of modern magnetism [1].

First of all, the laser-induced rapid reduction of the magnetization at the sub-picosecond timescale forced scientists to revise the channels of angular-momentum relaxation in solids. In particular, the role of spin-lattice coupling had to be reconsidered. This is still a subject of hot debates [6]. Whether the angular momentum is dissipated into the lattice via phonons and defects [8, 9], or whether it is carried away by hot electrons [10] or the photons [11] – such questions are still awaiting definitive answers.

Ultrafast laser-induced changes of the magnetization can be exploited to provide an easy experimental method to study magnetic precession, thus studying various magnetic parameters such as the magnetocrystalline or shape anisotropies. As rapid, laser-induced, changes in the temperature and the magnetization value result in a change in the magnetocrystalline and/or shape anisotropies and consequently in a change in the total effective field direction, this change can induce precession if it is faster than the corresponding precessional period. To obtain this, a non-collinear initial configuration of the various components of the net magnetic field is important so that a change in the value of the one component modifies the direction of the sum.

Moreover, a totally new, so-called opto-magnetic mechanism was discovered that depended on the presence of strong spin-orbit interaction. The corresponding magneto-optical susceptibility, responsible for the magneto-optical Faraday effect, i.e., the effect of the magnetization on the polarization of light, was demonstrated to be also responsible for the inverse Faraday effect: the manipulation of the magnetization by circularly polarized light. Via this opto-magnetic effect, an intense femtosecond circularly polarized optical laser pulse was demonstrated to act on spins similarly to an equivalently short effective magnetic field pulse with a strength up to several Teslas [12, 13]. With such an optically induced magnetic field, one may selectively excite different modes of magnetic resonance [14] and realize coherent control of spin oscillations [15].

Ultrafast laser-induced heating, as well as the non-thermal effects of the laser pulses, can also be used to trigger magnetic phase transitions on a sub-picosecond time scale. For example, a spin-reorientation second-order phase transition in weak-ferromagnetic $TmFeO_3$ can be driven at very short time scales, given by the characteristic antiferromagnetic resonance frequencies [16]. In some cases, one can observe the ultrafast emergence of a magnetic moment from an antiferromagnetic phase [17]. Such studies provide the fundamental understanding of the behavior of various types of phase transitions at very short timescales.

Although during the last 15 years the area of ultrafast laser-induced magnetization dynamics has experienced intense research interest, most of these studies have been devoted to single-element metallic ferromagnets or multi-element dielectrics [6]. Moreover, even if the experiments were performed on a multi-sublattice material, this fact has largely been ignored in the interpretation of the results, thus assuming that the laser-induced dynamics of the spins in the different sublattices are similar. However, it appeared that in particular in multi-sublattice ferrimagnets, where an important role is played by the exchange of angular momentum between non-equivalent sublattices, the dynamics of the magnetic moments becomes much more intriguing [16, 18]. In such systems, the time scale of the magnetization dynamics becomes dependent on the exchange interaction and the balance of the angular momentum between the sublattices [19], particularly when these are antiferromagnetically coupled. In pure antiferromagnets, moreover, the dynamic behavior changes its character substantially, demonstrating Newton-like dynamics with acceleration and inertia [20].

In general, many peculiarities of magnetization dynamics are related to the fact that a certain amount of angular momentum is associated with the magnetic moments. In fact, the role of the angular momentum in magnetism has intrigued the magnetic community since 1915, when Einstein and de Haas [21] and Barnett [22] published their experimental observations of the interconnection between the magnetization of a macroscopic body and its rotation. Therefore, as derived by Landau and Lifshitz in 1935 [23], the dynamics of the magnetic moments is described by a simple mechanical relation between the change of angular momentum and the torque applied to the system. The law of the conservation of angular momentum is, therefore, an important issue to be considered when sudden changes of the magnetization are treated.

All-optical switching of magnetization using the shortest possible laser pulses, being the holy grail of these studies for a long time, was finally observed in ferrimagnetic GdFeCo alloys [24], and subsequently also in a larger variety of ferrimagnetic alloys and multilayers [25]. The multi-sublattice character, the angular momentum balance, and the transient non-equilibrium state happen to be the essential ingredients of the switching process [26].

10.2 ALL-OPTICAL SWITCHING IN METALLIC FERRIMAGNETS

10.2.1 MECHANISM: NON-EQUIVALENT FERRIMAGNETIC SUBLATTICES

Note that the basic possibilities for the direct laser manipulation of magnetization have been indicated a long time ago [27, 28, 29], by using inverse opto-magnetic effects and circularly polarized light. The realization of such control in magnetically ordered materials came however only recently [12, 13]. The question was immediately triggered whether it would be possible to use *the same* mechanism for practical switching of the magnetization in for example recording media. A seemingly straightforward answer came very soon afterward, with a direct demonstration of helicity-dependent all-optical magnetic recording in thin films of metallic GdFeCo alloys [24]. In spite of the fact that the switching was clearly reproducible and robust, the understanding of the exact process and mechanism of it remained elusive. The most obvious explanation via the mentioned opto-magnetic effects could only very qualitatively account for the observed features.

Experimentally, single-shot time-resolved magnetic imaging demonstrated that the switching occurs via a strongly non-equilibrium state [7], which could not be described with a macrospin-like approximation. Atomistic-level spin dynamics calculations also failed to explain some essential points quantitatively. In particular, the pulse width required to reverse the magnetization was calculated to be at least 250 fs under the most favorable conditions, while experimentally, 40 fs pulses were used. Optical coherence was invoked, but it seemed unlikely that it would last over 100 fs.

Until this point, the multi-sublattice nature of the samples was not taken into account. New experiments, however, showed that the until now used single-component approach is invalid for these ferrimagnetic alloys. By using element-resolved femtosecond visible–X-ray pump-probe experiments it was shown that the RE and TM sublattices may have very different dynamics, to the point that the FeCo sub-lattice reverses before that of Gd, creating a transient ferromagnetic-like state [18]. This state was directly reproduced with atomistic simulations, even though all exchange constants were assumed to be unchanged by the laser pulse, and only ultrafast heating was taken into account. Most intriguingly, this transient state was shown to be followed by a reversal of the second sub-lattice without any additional stimulus or external magnetic field, creating a fully switched magnetic domain [30]. This very unusual process, that actually happens against the very strong exchange interaction that couples the sublattices, was shown to be polarization-independent (Figure 10.1).

It appeared that the transient state is due to two reasons: (i) considerably different demagnetization times of the RE and TM sublattices, and (ii) angular-momentum conservation and its transfer between the sublattices. First of all, it is the very different magnetic origin of the two sublattices that leads to considerably different demagnetization rates at the time scale of the first few

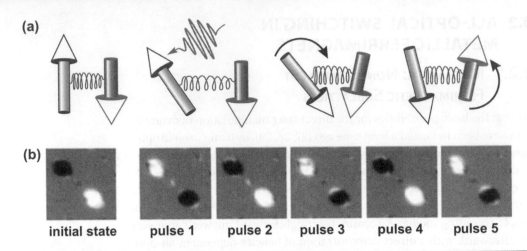

FIGURE 10.1   (a) Upon fs laser heating the ferrimagnetic system reaches a strongly non-equilibrium state within a few hundreds of fs. Successive exchange scattering drives the system into an intermediate ferromagnetic-like state, after which the inter-atomic exchange causes the final switch toward the reversed magnetic state. This scenario leads to all-optical toggle switching of the magnetization from up (white) to down (black) by successive fs laser pulses, as depicted by the experimental data on two ferrimagnetic islands in (b). (After Ostler, T.A. et al., *Nature Comm.* 3, 666, 2012. With permission.)

hundreds of femtoseconds [31]. At this timescale, the sublattices are effectively decoupled because the electronic bath temperature exceeds T_C by far. Later, when the electrons cool down, the angular-momentum transfer to the hot electron bath reduces, and the exchange coupling becomes effective again, preserving the total angular momentum of the two sublattices. The conservation of the angular momentum thus first pushes the TM sublattice through zero, to help further demagnetization of the RE system, as the latter has not yet achieved a thermal equilibrium with the environment because of its slower demagnetization, Then, on the longer timescale, the RE sub-lattice is restored back in the direction against the TM one, to save the exchange energy. This latter process is not angular-momentum conserving. However, it happens much slower, on a time scale of several picoseconds, and the conservation is much less strict then. Thus, each laser pulse with intensity above a certain threshold results in switching the direction of the magnetization [30].

What is the origin of the helicity dependence in the all-optical switching? A careful study [32] showed that it is the helicity-dependent absorption in the RE-TM magnetic layer, caused by the circular magnetic dichroism, that exactly matches the helicity-dependent intensity window as observed in switching experiments [33]. It happened that also multiple-pulse effects, such as observed in the original work [24], could also be explained by the dichroism effects [32].

To summarize, though originally this research was triggered by the observation of the inverse Faraday effect in magnetically ordered media [12, 13], the all-optical switching of the magnetization in ferrimagnetic RE-TM alloys was found to be due to a completely different effect: a combination of ultrafast laser-induced demagnetization with angular-momentum conservation on (sub-)picosecond timescales, assured by the antiferromagnetic exchange coupling between the RE and TM sublattices.

10.2.2 Samples: From Alloys to Multilayers

The typical materials that demonstrate the switching behavior are rare-earth-transition metal alloys (RE-TM), such as GdFe, TbFe, GdCo, or some mixtures. The most usable alloys have amorphous structures, that makes it possible to tune their composition in any desirable range by co-sputter the composition onto a substrate.

RE-TM alloys are ferrimagnets, where the RE and TM sublattices are coupled antiferromagnetically. These alloys were used for the magneto-optical storage technology in the past [34], and are known for their large magneto-optical effects [34,35]. Depending on RE ions concentration, RE-TM alloys can exhibit magnetization (TM) and angular momentum (TA) compensation temperatures, where the magnetizations (angular momenta) of the RE and TM sublattices are equivalent and, consequently, the net magnetization (angular momentum) is zero. For the case of Gd-containing alloys, the magnetization compensation point may be expected below the Curie point T_C if x = 20%,....,30% (see Ref. [36]). The angular momentum compensation point is typically ~50 K above TM. Along with TM, a number of other magnetic properties, such as magnetic anisotropy and coercive field, are defined by the Gd concentration [36].

Replacing Gd with Tb or Dy produces a strong effect on the magnetic anisotropy of the films, because of their large orbital moments. In addition, these orbital moments change the overall ratio between magnetic moments and angular momenta of the RE sub-lattice, thus shifting TA with respect to TM. Otherwise, similar magnetic properties can be expected.

In contrast, replacing heavy RE metals with the light ones produces essentially different magnetic properties, where the total magnetic moment of the RE part is dominated by the orbital moments and is thus aligned in the same direction at the TM magnetization. The compensation points are therefore non-existent. However, a different dynamic peculiarity arises in these samples that is due to their non-collinear fanned out spin structure. In optical pump-probe experiments, there is a certain overshooting effect, where the magnetization temporarily increases above its equilibrium value, for the time period of about 100 ps after the pump pulse [37]. This phenomenon is explained by the laser excitation decrease of the opening angle of the fan.

Apart from various alloys, ferrimagnetically coupled multilayers, or *synthetic ferrimagnets*, are an interesting possibility to expand the area of switchable materials [25, 38]. Atomistic spin dynamics calculations demonstrated the possibility of all-optical magnetic switching in Fe/FePt nanostructures as well as other alloys based on Ni, Fe, and Co ferromagnets such as Fe/CoPt [38]. The interlayer exchange coupling, which drives the switching process, can also be engineered by using different materials such as Ir or Si, though the optimal coupling depends on the thickness of the layers. This allows the tuning of properties such as the Curie temperature, magnetocrystalline anisotropy, and Gilbert damping to achieve the desired dynamic behavior and switching properties and opens the possibility of engineering high-performance magnetic data storage devices.

Experiments basically confirmed these predictions [25], by demonstrating AOS for a range of engineered materials such as RE-TM alloys, multilayers, and heterostructures involving different types of magnetic elements. The common denominator of the diverse structures showing AOS is two ferrimagnetic sublattices showing distinct temperature dependences that can be magnetically compensated at a selected temperature. Included in this study were RE-free heterostructures that mimic the properties of RE-TM alloys, that also showed AOS. These findings open a route for designing complex materials using well-known techniques, of which the magnetization can be controlled by the application of light.

10.2.3 Switching at the Nanoscale

In order to further increase the recording density of magnetic hard drives, alternative solutions to the magnetic recording trilemma arising from the competing requirements between readability, writeability, and stability have to be found [39]. One promising solution is heat-assisted magnetic recording (HAMR), where a laser pulse is used to briefly heat up the magnetic bit to be recorded, to lower its coercivity to a value compatible with the magnetic field produced by the recording head [40, 41, 42]

A possibly simpler solution to the magnetic recording trilemma can be proposed thanks to AOS, involving only a laser heat pulse instead of the currently used magnetic recording head and displaying tremendous recording speed. It is left to be verified, however, how such switching works at the nanoscale, where the initial magnetization states may be strongly modified by magnetostatic fields or by the manufacturing process itself, and where the relaxation from the transient strongly non-equilibrium state could be altered.

10.3 ALL-OPTICAL SWITCHING IN MAGNETIC NANOSTRUCTURES

By carefully tuning the laser fluence and exploiting the threshold barrier that is characteristic for the optically induced reversal [24, 32] it is possible to confine the AOS within a few tens of squared micrometers. Such a dimension is orders of magnitude larger than the current bit size in magnetic data storage. This problem can be circumvented by patterning the magnetic films and realizing micro- and nanosized structures [43, 44]. This approach will allow us to study the feasibility of AOS at the nanoscale in principle, without the need to realize complicated optical schemes.

10.3.1 Demonstration of AOS in Structures of Various Sizes

In this section, we demonstrate the magnetization reversal in GdFeCo nanostructures for dimensions as small as 200 nm after applying a single, linearly polarized femtosecond laser pulse and without applying any magnetic field [44]. The samples considered here were thin films of GdFeCo grown by magnetron sputtering on a silicon substrate, and further structured in squares and disks with sizes ranging from 2×2 μm^2 down to 100×100 nm^2 by electron beam lithography in combination with a lift-off process [43], resulting in isolated magnetic structures. Images of the magnetic domain states in these structures are then obtained using the Elmitec PEEM at the surface/interface microscopy (SIM) beamline [45] at the Swiss Light Source.

The three components of the magnetization vector in the structures were determined with a photoemission electron microscope (PEEM) by performing an azimuthal dependent measurement of the x-ray magnetic circular dichroism (XMCD) at the Fe L3 edge. This azimuthal study showed that the rims of the structures display an in-plane magnetic anisotropy, which seems to be the result of the thinning of the GdFeCo layer at the edge resulting from the fabrication process. Nevertheless, these in-plane rims were found to play no significant role in the reversal of the central out-of-plane domain, as shown below.

A sequence of XMCD images, each recorded after excitation by a single linearly polarized laser pulse, is shown in Figure 10.2a–d and illustrates the efficiency of the laser-induced magnetization switching in these various structures. For the 2 and 1 μm wide structures, the reversal of the magnetization is easily identified as a change from black to white or vice versa with an efficiency approaching 100%. In this specific example, only one case (circled in Figure 10.2c) does not result in a complete switching. The behavior of the smaller structures that are 500, 400, and 300 nm wide is however not as evident. Here, most of them seem to have stable multi-domain states with both black and white contrast in the same structure, in contrast to

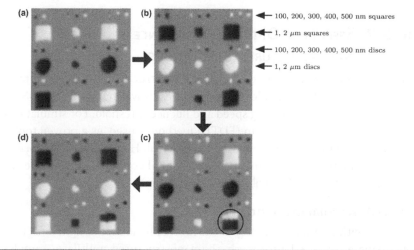

FIGURE 10.2 (a)-(d) Sequence of successive XMCD images from the same region. The arrows indicate the order in which the images were recorded, while a single linearly polarized laser pulse was applied to the sample in-between each of them. The circle indicates the case for which the switching is not complete. (After Le Guyader, L. et al., *Appl. Phys. Lett.* 101, 022410, 2012. With permission.)

the larger structures which look mono-domain. It appeared that this is due to the structures having the aforementioned in-plane magnetization rims, which cannot be distinguished in these images.

It is, however, important to realize that these in-plane rims do not actually play a role in the switching behavior as such. This illustrates experimentally the impetus by which this laser-induced switching occurs. Indeed, simulations have shown that magnetic fields as high as 40 T were not sufficient to prevent the formation of the transient ferromagnetic-like state which leads to the magnetization reversal [30]. This provides an insensitivity of the switched domains to their surroundings, a property which might be exploited to reach a very high recording density. It should be noted, however, that this high magnetic field is defined by the strength of the actual inter-sub-lattice exchange interaction. In realistic alloys and multilayers the fields that trigger spin-flop transitions, cant the sublattices and thus prevent the switching, can be much smaller.

Thus it is clear that the laser-induced magnetization switching is possible in GdFeCo nanostructures down to the sub-micron range. The in-plane edge domains are found to play no particular role in the efficiency of the switching and are likely an artifact of the lift-off structuring method. This specific limitation can be lifted with an improved nanofabrication and optimized material properties, leading to single domain structures with out-of-plane magnetization at sizes much smaller than the ones reported here.

Thus, from the first glance, the nanostructuring does not seem to affect the AOS behavior much. Note, however, that the presence of the boundaries of the nanostructures may also lead to optical interference effects, with a significant redistribution of the light intensity as a consequence. In the following, we shall treat such effects first computationally and then experimentally.

10.3.2 Effects of Optical Interference in Nanostructures: Simulations

This section deals with the question of how structuring down to dimensions comparable to or smaller than the optical wavelength affects AOS key parameters, such as switching speed and fluence threshold. For simulations, a finite difference time domain (FDTD) method is used, as a proven tool to treat the wave interference behavior of light [46]. Here, we have performed FDTD calculations [47] to investigate the lateral confinement and interference of the optical field distribution within the magnetic layer.

10.3.2.1 Normal Incidence

We start with a simple geometry of the normal incidence of light and simulate, using a $5 \times 5 \times 1$ nm^3 mesh, the nanostructured multilayer samples. Optical parameters such as refractive indices were extracted from ellipsometric measurements [32]. The dimension and focusing of the pump beam are chosen to match the values used for experiments on similar structures, pumping at the wavelength of $\lambda = 400$ nm. The simulations were performed

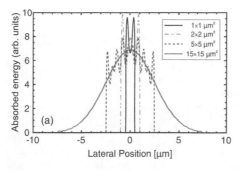
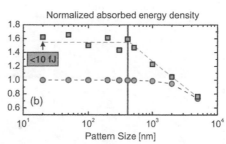

FIGURE 10.3 (a) Energy profiles of the electric-field intensity taken at the center of the structures. The formation of a standing pattern due to the lateral confinement of the incoming radiation, when the dimensions of the structures equal to or smaller than the beam size, is clearly visible. (b) Variation of the simulated energy distribution per unit area as a function of the structure lateral dimension, represented with a squared point, compared with the trend expected in the case of no interference effects present (circles). The expected energy needed to optically switch a $20 \times 20 \, nm^2$ domain is as low as 10 fJ. The vertical line indicates the pump beam wavelength. (After Savoini, M. et al., *Phys. Rev. B* 86, 140404(R), 2012. With permission.)

from a structure size of $20 \times 20 \, nm^2$, basically equivalent to the actual bit size in present-day hard drives, up to $15 \times 15 \, \mu m^2$.

Figure 10.3a shows the resulting intensity profile across a line through the center of the structures. It is apparent that reducing the structure dimension below the beam size (in this case taken as $\approx 5 \, \mu m$ FWHM) leads to the formation of a standing wave pattern due to the scattering of the beam by the structure edges. From these electric-field distributions, we can calculate the integral of the absorbed energy density. Figure 10.3b shows the trend of the simulated absorbed energy density in the structures (squares) compared with the integrated profile of a Gaussian beam within the same area (circles). Regarding the former, we have calculated the integral per unit area of the absorbed energy distribution within the patterned structure, while the latter is simply the integral per unit area of a Gaussian profile matching the absorbed profile of the $15 \times 15 \, \mu m^2$ structure (where no standing pattern is present).

We find two separate behaviors: (i) in structures smaller than $400 \times 400 \, nm^2$, no significant change in the absorption is visible. (ii) Above $400 \times 400 \, nm^2$, a steady increase in the integrated absorbed energy density is visible with decreasing pattern size. The difference between absorption with or without interference effects is remarkable. For structures smaller than $5 \times 5 \, \mu m^2$, the integrated Gaussian distribution (i.e., without interference) quickly reaches a saturation value. In contrast, the simulated energy distribution including interference continues to increase, down to approximately the $400 \times 400 \, nm^2$ structure. Below this point, it remains constant but shows a value which is about 60% higher than the integrated Gaussian distribution. Recall that 400 nm is also the wavelength of the pump beam. This is a clear indication that interference of light in (sub)micrometer structures plays a crucial role; in fact, we have an increase in absorption until the structure size becomes similar to the wavelength. Reducing the pattern size even further does not show any significantly enhanced absorption.

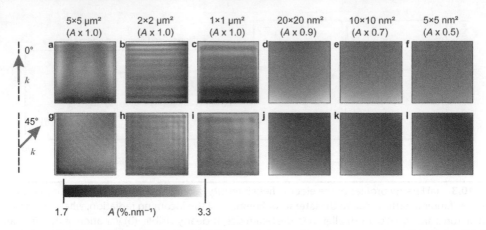

FIGURE 10.4 Simulated light absorption as a function of structure size and incoming laser direction. FDTD-simulated light absorption A inside structures of different sizes ranging from 5×5 μm² (a, g) down to 5×5 nm² (f, l). The simulations are shown for two different incoming laser directions with respect to the structure edge as shown in the inset by the in-plane light wave vector k at 0° for (a–f) and at 45° for (g-l). In both cases, the laser light impinges with a 16° grazing incidence. The absorption profiles A have been scaled by the factor indicated in the parentheses such that they all fall into the same range. (After Le Guyader, L. et al., *Nature Comm.* 6, 5839, 2015. With permission.)

10.3.2.2 Grazing Incidence

Thus even at normal incidence, the interference effects could be very significant, as demonstrated in the previous section. To expand this further, we have simulated the intensity patterns in squares of various sizes and different directions of the laser beam, incident at a nearly grazing angle (about 16° was chosen for practical reasons). For the FDTD modeling, we have chosen a realistic GdFeCo multilayer structure, identical to the one in the previous section. The simulations of the light absorption profiles were performed for different structural sizes ranging from 5×5 μm² down to 5×5 nm². The results are shown in Figure 10.4. At constant 16° grazing incidence geometry, we varied the in-plane incoming laser direction, as indicated in Figure 10.4. The results for 0° and 45° are shown in Figures 10.4a–f and 4g–l, respectively. The first striking feature is that even though the incoming laser pulse is a plane wave with an 800-nm wavelength, that is, orders of magnitude larger than the smallest simulated structure, the light absorption inside the structure is inhomogeneous down to the smallest 5×5 nm² structure. These absorption profiles also depend strongly on the incoming laser direction with respect to the structure edges, as can be seen by a direct comparison between Figure 10.4a–f and 4g–l. In particular, an interesting case occurs at 45° where the absorbed laser energy is mostly confined within a quarter of the structure, as shown in Figure 10.4i in the top right corner, as well as in Figure 10.4j in the opposite bottom left corner. Moreover, these absorption profile inhomogeneities are rather strong, displaying a ratio of ~2 between the highest and lowest absorption regions inside the structures down to the 20×20 nm² structure. This ratio reduces to 1.5 for the 10×10 nm² structure and to 1.1 for the 5×5 nm² structure. Furthermore, the total absorbed energy increases by a factor of 2 from the largest to the smallest structures, making the smaller structures more absorbing and thus more energy efficient as discussed above, in Section 3.2.1.

These focusing and coupling efficiency effects are created by the structure's boundaries and the interference between the waves propagating and absorbing within the sample. Remember that these effects are not only present at 16° grazing incidence as shown here, but also at normal incidence as shown above. As expected, the grazing incidence geometry offers an additional degree of freedom such that, depending on the orientation of the boundary with respect to the propagation wave vector of the light pulse, a variety of patterns can be created. This leads, for example, to the intense side lobes seen in the 5×5 μm^2 structure at 0° incoming azimuthal direction shown in Figure 10.4a. Refraction, reflection, and interference of these waves occurring within the structure, which are best seen in the 45° incoming direction cases shown in Figure 10.4h, create strong intensity variations inside the structure. Strong optical absorption of light leads to the formation of these features on the length scale of a few nanometers.

Thus, using the grazing incidence of the incoming laser pulse, it is possible to focus the laser pulse by the structural geometry into parts of the structure, which are well below the far-field diffraction limit. These results, plus the effective intensity increase demonstrated in the previous section, pose the questions as to whether or not such systems can be realistically employed to all-optically switch a nanoscale region of a magnetic structure. The answer to these questions will now be addressed.

10.3.3 Effects of Optical Interference in Nanostructures: Experiments

10.3.3.1 Normal Incidence

To address the magnetization dynamics in a single structure down to 1×1 μm^2, and later to create and detect sub-micrometer sized domains in a continuous film (see Section 4.1 below) we modified a commercial confocal microscope (α-SNOM, WiTec GmbH, Germany) to obtain femtosecond temporal and sub-micrometer spatial resolution [50]. This setup gives the possibility to work in ambient conditions and with a table-top setup; moreover, it offers several advantages in terms of simplicity and temporal resolution compared with other techniques such as for example Refs. [51, 52, 53].

The samples were a series of patterned square areas with different dimensions fabricated with lift-off techniques starting from a sputtered GdFeCo film, as described above in Section 3.1 [43, 44].

First of all, using the largest 15×15 μm^2 and 50×50 μm^2 structures, we verified that the magnetization dynamics as a function of the fluence is the same as in a continuous layer. While increasing the laser fluence to some extent, only an increasing degree of demagnetization is observed, for fluences higher than about 5.5 mJ/cm^2 we see all-optical switching of magnetization. Below 3 mJ/cm^2, the magnetization relaxes back to the initial state with an exponential behavior with a characteristic time of 100–150 ps. This relaxation becomes slower for demagnetizations higher than 70%. We can explain

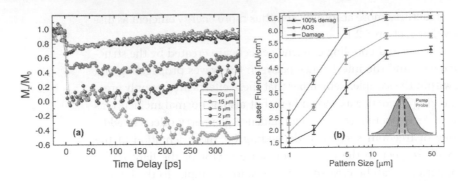

FIGURE 10.5 (a) Magnetization dynamics at a fixed fluence of 1.9 ± 0.2 mJ/cm² for different structure dimensions. A strong dependence is visible as we measure a higher demagnetization upon decreasing the structure dimensions. For the smallest structure, AOS is induced. (b) Energy requirements for 100% demagnetization, all-optical switching, and the damage threshold as a function of the structure size. Inset: Experimental conditions reproducing in scale the dimensions of the pump and probe beams. (After Savoini, M. et al., *Phys. Rev. B* 86, 140404(R), 2012. With permission.)

this trend as a loss of correlation in the local exchange interaction due to the high electronic temperature reached by the system, making the relaxation dynamics slower than exponential. This trend is in qualitative agreement with calculations based on Landau-Lifshitz-Bloch equations [54] and atomistic simulations [55].

Magnetization switching is induced for fluences above 5.5 mJ/cm², and occurs within 50–100 ps, similar to previously reported optical measurements in continuous films [7].

Much more interesting is to study the magnetization dynamics in nanostructures of different dimensions but for the same pump fluence, as presented in Figure 10.5a. For structures with dimensions of 15 × 15 μm² and 50 × 50 μm², both much larger than the pump and probe beam sizes, the effect on the magnetization is practically the same. However, the initial demagnetization considerably increases upon further reducing the structure dimension. Surprisingly, we measured 50% and basically full demagnetization in the 5 × 5 μm² and 2 × 2 μm² structures, respectively. Most intriguing, even a complete all-optical switching is observed in the 1 × 1 μm² structures.

Notably, the different dimensions do not directly affect the initial demagnetization speed which occurs on average within 1 ps. Naively, one would expect that the pump-induced changes should be triggered by a constant energy density, thus we expect no dependence as a function of the dimension. However, we should recall the results of the simulations presented in Section 3.2.1.

The experimental conditions are sketched in the inset of Figure 10.5b, where the pump lateral size is about five times larger than the probe one. Thus the probed area is excited by a homogeneous (practically constant) laser fluence. Figure 10.5b presents, respectively, the trend for the energy densities required to induce 100% demagnetization, AOS, and sample damage, i.e., an irreversible change of the magnetization status of the illuminated area, as a function of the sample size. Indeed, for the larger structures (15 × 15 μm² and 50 × 50 μm²) we do not measure any dependence in the energy requirements as a function of the dimension. Conversely, a clear deviation is present for the smaller ones, and a strong reduction of the required energy densities is apparent. Thus the experiments confirm that optically induced effects, in particular, AOS, become energetically more favorable when reducing the structure size. For example, pattern sizes of $1 \times 1\ \mu m^2$ require three times lower energy compared with larger ones (or continuous films). This finding makes AOS one of the most attractive tools to control the magnetization of a nanoscale sample and may lead to the possibility of its practical applications.

Note that there is one difference between the experimental measurements of Figure 10.5b and the simulated results presented in Figure 10.3b. In the former, a deviation from the expected constant energy density is visible already for the $5 \times 5\ \mu m^2$ structure, while in the latter a clear effect of the interference is present only for patterned areas smaller than $2 \times 2\ \mu m^2$. This difference originates from the different probed area in experiments and simulations. In the experimental results, we are probing the magnetization dynamics locally, with a beam size of approximately 800 nm. In contrast, in the simulations, we compute the absorbed energy integrated over the entire structure.

Note that there is also a certain influence of the thickness of the layers on the interference effects. It happens that the amplification effects are stronger for thinner samples, while they tend to disappear when either the magnetic layer and/or the capping layers become thicker than 100 nm.

10.3.3.2 Grazing Incidence

As simulations indicate, using grazing incidence of light one could create conditions for a passive focusing of light due to its interaction with the structure. It is interesting to verify, whether such predicted partial magnetization switching of the structure is feasible, given the threshold character of the switching. However, another problem is that due to the low coercivity of the GdFeCo alloys, the switched domains are likely to reorganize after switching on the relevant length scale, which here is a few hundred of nanometers. It is, therefore, necessary for the sample investigated to probe the magnetization shortly after the laser pulse. For this, time-resolved XMCD photo-electron emission microscopy (PEEM) imaging was employed, which offers magnetic domain imaging with 70 ps time resolution and ~200 nm spatial resolution in time-resolved mode. By fixing the time delay t between the laser pump and the X-ray probe, the spatially resolved intermediate magnetization state inside a structure at that specific time delay can be recorded.

The time-resolved XMCD images for a $5 \times 5\ \mu m^2$ square microstructure at t = 400 ps after the laser pulse as a function of the incoming laser fluence F, are shown in Figure 10.6a for an incoming laser direction along the structure edge, and in Figure 10.6d for an incoming laser direction of 45° relative to the structure edge. It is important to note that at this relatively long time delay of a few hundred picoseconds, both the Gd and the FeCo sub-lattice magnetization are again in equilibrium with each other, such that measuring only one sub-lattice is enough to characterize the sample magnetization [18]. On the other hand, this time scale is short enough to probe the transient longitudinal magnetization dynamics, in particular, whether

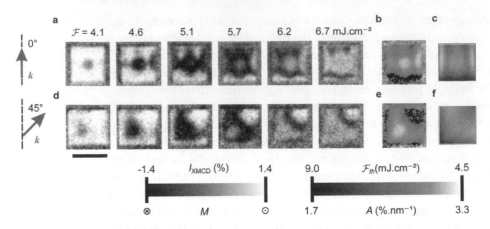

FIGURE 10.6   (a) Time-resolved PEEM images as a function of the incoming laser fluence F and for an incoming laser direction with respect to the structure edge of 0° as indicated by the laser in-plane wave vector k. All images are recorded at a fixed time delay of t = 400 ps after the laser pulse. (d) Same as (a) but for a 45° incoming laser direction. (b and e) The extracted spatially resolved fluence switching threshold F_{th} for the 0° and the 45° incoming laser direction, respectively. (c and f) The FDTD-simulated light absorption A inside the 5×5 μm² structures for the 0° and the 45° incoming laser direction, respectively. The gray scale indicates the relation between the XMCD image contrast IXMCD and the out-of-plane magnetization component M being inward or outward. Scale bar, 5 μm. (After Le Guyader, L. et al., *Nature Comm.* 6, 5839, 2015. With permission.)

partial, total demagnetization or magnetization switching has taken place. The initial state of this microstructure is mono-domain due to a static out-of-plane magnetic field of 50 mT applied along its magnetic easy axis. For the low fluence case of F = 4.1 mJ/cm² shown in Figure 10.6a, one can see that the contrast is no longer uniform and that some partial demagnetization has occurred at the center of the structure. As the fluence is increased, the contrast displayed by each area within the microstructure changes from white to black, and further from black to gray.

From the FDTD simulations shown above in Figure 10.4, we have seen that the structural orientation and size produce a complex combination of refraction and interference, such that the laser energy is focused onto only part of the structure. Comparing the absorption profile A obtained from the FDTD simulations (Figure 10.6c and f for a 0° and 45° incoming laser direction, respectively) with the fluence threshold F_{th} patterns obtained from the experiments (Figure 10.6b and e, respectively), a direct relation between the two is revealed. (Note that the value of F_{th} is determined as the border between switched and non-switched magnetization areas.) Moreover, the fact that the resulting F_{th} pattern completely changes with the incoming laser direction between Figures 10.6b and e rules out an intrinsic inhomogeneous switching threshold due to, for example, chemical inhomogeneities in the structure composition whose effects have been seen in other studies [56]. Therefore, it is not the fluence threshold F_{th} but rather the light absorption A which varies across the structure. A very good qualitative agreement is obtained, as a region with a low F_{th} corresponds to a region with high absorption and vice versa. The agreement is even reasonably quantitative as the ratio between the high and low fluence threshold F_{th} is approximately a factor of 2, in accordance with the ratio between high and low absorption.

Furthermore, these different absorption patterns directly translate to correspondingly different magnetization dynamics [49] and spatially selective ultrafast AOS. Using rare-earth-transition metal alloys with higher magnetic anisotropy, for example, TbFeCo, the reversed patterns induced by the laser pulse inside the microstructure could be stable at smaller sizes [57], see below Section 10.4.1.

Thus, employing the focusing properties of the structure itself, an inhomogeneously absorbed laser energy is obtained at the nanoscale, which in turn triggers inhomogeneous magnetization dynamics. This allows for a selective all-optical magnetization switching inside the microstructure, even though the incoming laser pulse is homogeneously illuminating the structure. These results open novel opportunities for very-high-density data storage media, for example, by either recording several bits of information in a single magnetic structure or by improving the coupling efficiency between the laser pulse and the magnetic structure.

10.4 WRITING MAGNETIC DOMAINS WITH FOCUSED BEAMS

A more universal approach consists of the focusing on the optical beam to smaller sizes onto a continuous magnetic film. While it is impossible to go beyond hundreds of nm using standard far-field optics, much smaller spots sizes can be delivered using near-field optics or plasmonics. There is a question, however, whether the reversed domain will be stable at these small sizes, or what other properties can be essential. This part of our review will deal with these questions.

10.4.1 SKYRMION-LIKE PATTERNS CREATED BY TIGHTLY FOCUSED BEAM

From the first data shown in Ref. [24], and also later in [58], it was clear that AOS may lead to a complicated domain pattern, in particular, if some overheating is involved. Reducing the spot size makes the situation somewhat simpler when the spot becomes comparable with the characteristic domain size in the magnetic film. Indeed, no multi-domain state is then allowed, and the reversed spot would automatically become homogeneous.

From the very first attempts, it has become clear, however, that the widely used GdFeCo film cannot support domains smaller than about 3–5 μm, just because the surface tension of the domain wall overcomes the weak coercivity of these films. Instead, TbFeCo alloys were used to demonstrate the reversal on a sub-micron scale, with their coercive fields of several Teslas. For AOS tightly focused laser pulses were used, with a FWHM spot diameter of not more than 2 μm at the wavelength of $\lambda = 800$ nm. Using the threshold character of AOS [32] and tuning the pulse fluence, it was thus possible to use just the top of pulse profile and create domains much smaller than the spot itself. Varying the fluence thus allowed the adjustment of the size of the switched domains. The magnetization pattern was then read-out with the help of a polarization-conserving hollow pyramid tip of SNOM, as described in Section 3.3.1 above.

The results show that indeed the multi-domain character of the recorded spot vanishes at the smallest sizes, as expected. Instead, a single recorded domain appears, with a diameter around 200–300 nm, that by its topological properties is identical to a so-called *Skyrmion* [59]. In magnetic materials, Skyrmions emerge as soliton-like excitations that cannot be traced back to the ground ferromagnetic state by continuous deformations of the local magnetization field. So far, they have been observed in materials with perpendicular magnetic anisotropy in connection with one of the following stabilization mechanisms: (i) four-spin exchange, leading to the formation of a Skyrmion lattice with each Skyrmion extending over a few lattice sites [60]; (ii) the Dzyaloshinskii-Moriya interaction [61, 62], which is active in non-centrosymmetric helimagnets, where Skyrmions have typical dimensions of few tens of nanometers [63, 64, 65, 66]; (iii) the long-range dipole-dipole interaction (DDI) [67] stabilizing Skyrmions with a typical lateral size of the order of 1 μm [68, 69, 70]. It has been recently found that the DDI can induce a larger and more complex variety of magnetic textures with respect to the ones that are observed in association with four-spin exchange and the Dzyaloshinskii-Moriya interaction [67]. In our case the stabilizing action of the externally applied opposite magnetic field as exploited in Refs. [71, 72] can be effectively replaced by the dipolar field from the magnetic film itself.

At higher laser fluences with respect to those leading to Skyrmion generation (corresponding to an energy density up to 7 mJ/cm^2 for each single laser pulse), we observed the formation of donut-shaped magnetic structures. The patterns became even more complicated above that, up to the damage threshold of 15 mJ/cm^2.

The general evolution of the spin textures that are generated by a single laser pulse illumination with increasing intensity is illustrated in Figure 10.7. For fluences below 4 mJ/cm^2, no magnetization reversal is induced, while single, isolated Skyrmions are generated in the 4–5 mJ/cm^2 fluence range with 100% efficiency. Conversely, fluences between 5 and 7 mJ/cm^2 cause the systematic formation of the donut-shaped structure, that could be represented as a combination of a Skyrmion and anti-Skyrmion, or a "Skyrmionium" atom [57]. The overall diameter of the textures increases with fluence, suggesting that pinning defects might also play a significant role in defining the ultimate size and shape of such structures.

Upon illumination with laser fluences above 7 mJ/cm^2, larger structures are generated, with topological properties that may vary from structure to structure. The texture shown in Figure 10.7e closely resembles those observed in hexaferrite [67]. We interpret it as a combination of two Skyrmions and one anti-Skyrmion, with the topological change Q = +1. It can be viewed as a di-Skyrmionium "cation" molecule, representing the topological analogue of dipositronium molecules. In fact, concentric ring structures, made

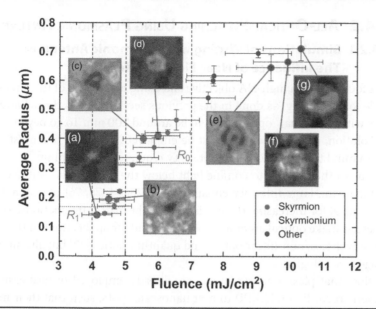

FIGURE 10.7 Evolution of size and topology of magnetic domains generated in TbFeCo after single laser pulse illumination, as a function of laser fluence. The size is estimated by considering the rim of the domains defined by the condition of vanishing Faraday rotation and by averaging the radius of the domain over angle. The error bar on the size also considers the deviations from the axial symmetry of each considered structure. Insets: 2.5×2.5 μm^2 Faraday rotation maps. (After Finazzi, M. et al., *Phys. Rev. Lett.* 110. 177205, 2013. With permission.)

of alternating Skyrmions and anti-Skyrmions nested together, should all be stable [57]. Their Skyrmion number Q is 0 or 1 depending on the number of Skyrmions being larger than or equal to the number of anti-Skyrmions. We find that the values of the zero-order parameter of adjacent (anti-)Skyrmions should always be opposite, in agreement with the alternating helicity reversals inside Skyrmions with multiple-ring structure reported for hexaferrite [67].

Figure 10.7f shows a different di-Skyrmionium texture consisting of one Skyrmion surrounding two anti-Skyrmions, yielding Q=-1. It is still an open question whether this structure is intrinsically stable, or whether a non-uniform distribution of spin-pinning defects in the amorphous film is essential to prevent it from collapsing or expanding. The non-centrosymmetric shape of the structures shown in Figure 10.7 indeed suggests that defects are distorting the spin arrangement. Although we are not able to exactly determine the conditions leading to the creation of these structures, their observation together with the reliable and reproducible production of Skyrmionium represents an important step toward the study of complex topological states in condensed matter.

Thus AOS can be used to locally create stable magnetic configurations characterized by well-defined topological numbers. By tuning the laser fluence one may be able to control the topology of created patterns. Such magnetic excitations are stable without an external magnetic field, which probably is assured in this case by the strong anisotropy and hence the coercive field of the TbFeCo alloys.

10.4.2 ALL-OPTICAL SWITCHING USING PLASMONIC ANTENNAS

10.4.2.1 Simulations of Writing with Plasmonic Antennas: The Influence of Near-Field Interference

Tight focusing with high-NA objective and using just the tip of the intensity profile for recording, as done in the previous section, shows the achievable limits with the "usual" optical tools to be around 200 nm. To go beyond this in resolution, near-field optics must be employed. However, one must also keep in mind the minimum light intensity required for switching.

Due to their ability to confine light below the diffraction limit and the accompanying large intensity enhancement, [73, 74, 75, 76] optical antennas are ideal for increasing the efficiency of light-matter interactions. These properties make optical antennas not only useful for applications in the field of photovoltaics, non-linear optics, and quantum optics [77] but also in data storage technologies.

Note that plasmonic structures are already employed in heat-assisted magnetic recording (HAMR) to heat nanoscale spots, such that their magnetization can then be reversed with a smaller magnetic field, increasing the potential data storage density [42, 78]. With HAMR, an external magnetic field is still necessary. However, exploiting all-optical switching (AOS) magnetic domains can be switched reversibly with femtosecond laser pulses in the absence of any external magnetic field, as discussed above [24, 30] Note, however, that as the actual recording media are usually protected by a dielectric or metallic capping layer of several tens of nanometers, this will severely affect the focusing abilities and subsequent potential data densities for AOS.

To counteract this we have managed to demonstrate that it is possible to deliver energy to the magnetic layer more efficiently by exploiting the near-field interference between the excitation light and the re-emitted light [79]. In particular, we show that due to this interference an off-resonant antenna delivers more energy to a distant plane as compared to a resonant one. Evaluating the spot size and field enhancement that can be obtained inside the magnetic film shows that we have a gain in energy together with a sub-diffraction sized spot, even in the presence of a capping layer. These two quantities are of direct relevance for AOS.

The structure we consider is similar to the samples typically used in the AOS experiments and consists of a glass substrate, a thin film of magnetic material, in our case 20 nm of GdFeCo protected by a capping layer of Si_3N_4. On top of this, a gold cross antenna is positioned.

Here again, we use finite difference time domain (FDTD) simulations to calculate the electromagnetic (EM) fields at different positions in the structure. A circularly polarized plane wave at a wavelength of 800 nm is used to excite the antenna. Note that while the helicity by itself is not crucial for the switching, the ability to manipulate it may be useful for deterministic recording [32]. In the simulations, we assume typical dimensions for real cross antennas: a thickness of 40 nm, an arm width of 50 nm, and a gap size of 35 nm. The total length of two opposite arms including the gap is varied from 110 nm to 300 nm. The antenna is made of gold. In the FDTD simulations, a non-uniform meshing is used with the smallest mesh cells $1 \times 1 \times 1$ nm at the position of the antenna, extending to the GdFeCo film directly underneath the antenna.

In general, when plasmonic antennas are considered, the attention is focused on the intensity enhancement. This intensity enhancement is calculated by normalizing the field intensities obtained in the presence of the antenna structure with the results of a simulation without the antenna. However, we would like to have a measure for the degree of circular polarization as well. For these reasons, we will from here on use the following figure-of-merit (FOM) [80, 81] defined by

$$\text{FOM} = I \cdot C^2$$

where

    I is the intensity and
    C is the polarization degree.

The first finding is that the antenna length at which the maximum energy is delivered to the thin film differs from the resonant antenna length. This difference can be attributed to the interference between the excitation light and the re-emitted light by the antenna.

Next, for the purpose of magnetic switching, both the focusing of the field, as its intensity and polarization are as well important. The patterns presented in Figure 10.8 result from the interference of the field created by the antenna itself with the one reflected from various layers of the structure. These interference patterns give an indication about the size of the magnetic domain that can be switched by using the optical antennas as considered here. The obtained interference patterns for the 60 nm-thick capping

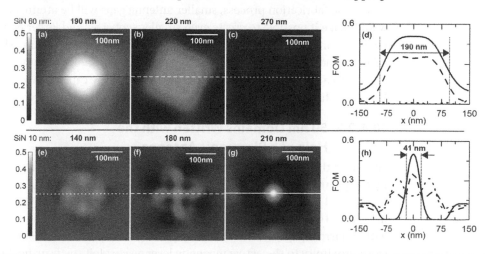

FIGURE 10.8   Figure-of-merit (FOM) distribution in a 2D plane at half height of the GdFeCo layer in case of a 60 nm (a)–(c) and a 10 nm (e)–(g) dielectric layer for different antenna lengths. In (d) and (h), cross-sections through the center of, respectively, (a)–(c) and (e)–(g) are shown. The FWHM of the central spot is indicated for the cross-section of 190 nm (d) and 210 nm (h) antenna length. (After Koene, B. et al., *Appl. Phys. Lett.* 101, 013115, 2012. With permission.)

layer are shown in Figures 10.8a–c for antenna lengths of, respectively, 190 nm (constructive interference), 220 nm (plasmon resonance), and 270 nm (destructive interference). For the antenna with a length of 190 nm, a spot size with a full width at half-maximum (FWHM) of 190 nm is found, as is clearly shown in Figure 10.8d. If we calculate the maximum intensity enhancement in the GdFeCo for this antenna, we find a value of 2.6.

Figures 10.8e–h show the same information as Figures 10.8a–d but now for a 10 nm capping layer. With this thickness, we find an even smaller spot size with a FWHM of 41 nm. This sub-diffraction area is substantially smaller than the spot size of a light beam focused with conventional objectives (typical dimensions with an immersion oil objective are of approximately 350 nm using 800 nm light). We would like to mention that considering only the field intensity would give a spot size FWHM of 52 nm. Hence we have here an advantage due to the dependence of the AOS on the helicity of light. The maximum field enhancement in the 41 nm spot is 3.7. This would mean that the known switching threshold of 2.6 mJ/cm^2 for a GdFeCo sample without any structure on top [32] will reduce to 0.7 mJ/cm^2 in the case that nanoantennas are used.

Although low, the intensity enhancement we find is of the same order of magnitude reported for devices aimed for similar applications [78]. In our case, the low value is caused by the quickly vanishing nature of the near fields together with the presence of a capping layer, while for example in Ref. [78], there is a less efficient coupling due to the high refractive index surrounding the antenna. For the purpose of AOS, this small enhancement factor is not a problem as the main reason to use optical antennas here is to bring down switching to the nanoscale.

Note that here we considered a plasmonic structure that can be made with the current state of the art nanofabrication technologies. With progress in the fabrication process, smaller antenna gaps will be attainable. This would make it possible to create even smaller spot sizes and larger enhancement factors. Moreover, considering data storage technology, in which a writing head (antenna) moves over the medium, we performed some simulations with a 7 nm air slit between the antenna and the capping layer as well. Except in the antenna length, no noticeable changes were observed.

10.4.2.2 50 nm Domains Written with the Help of a Plasmonic Antenna

Following the outcomes of the previous section and abandoning for simplicity the polarization control of the pump beam, we employed two-wire plasmonic gold nanoantennas to obtain localized enhancement of the optical field [82]. The antenna structure consists of two end-to-end aligned wires with a narrow gap between them. By placing such plasmonic resonators in close proximity to the active magnetic layer, we exploit the near-field intensity enhancement in order to confine the area to be magnetically switched to dimensions that are defined by the antenna geometry. The two-wire antennas resonate with linearly polarized light with an electric-field vector parallel to the antenna axis. In order to resolve the all-optically switched magnetic domain structure, we use an X-ray holographic imaging technique, which has a spatial resolution down to 16 nm [83]. The X-ray magnetic contrast is obtained via XMCD [1, 18, 56].

X-ray holography requires the nanoscale integration of the magnetic sample layer and a holography mask. The sample stack, therefore, consisted of a usual 20 nm-thick TbFeCo magnetic layer sandwiched between a 100 nm-thick Si_3N_4 membrane and a 10 nm Si_3N_4 capping layer to prevent oxidation. A gold X-ray holography mask is deposited and fabricated on the other side of the membrane [82]. The optimal antenna geometry is determined by performing finite difference time domain (FDTD) simulations, as discussed in the previous section. Three antenna sets were thus prepared with total lengths of 230, 270, and 310 nm and with gaps between 20 and 30 nm.

The single-crystalline gold antennas were first fabricated by focused ion beam milling on a glass substrate coated with gold before being transferred to the actual samples [84]. This fabrication procedure prevents damage to the magnetic material underneath the antennas by the milling process [85]. Magnetic switching is achieved by directing single femtosecond optical pulses onto the sample surface that contains the antennas. The optical beam focus is 75 μm fullwidth- half-maximum, and we can therefore safely assume that the antennas within the imaging window to be excited by a plane wave.

We will now describe the reversible writing of magnetic information by using plasmonic nanoantennas with AOS. This is done for a laser fluence of 3.7 mJ/cm² that is below the onset of AOS in-between antennas as shown in Figure 10.9f. Figures 10.9a–e shows the actual demonstration of AOS.

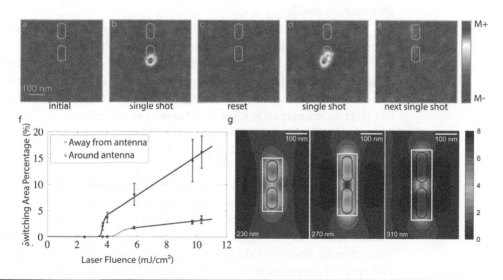

FIGURE 10.9   (a) Initial magnetic contrast around the selected antenna after being magnetically saturated. (b) Magnetic contrast after the first laser pulse. A small domain with a FWHM of 53 nm is switched. (c) The magnetization is reset using an external magnetic field. (d) Magnetic contrast after the first laser pulse on the newly saturated sample. A domain of comparable size as shown in (b) is switched in the same region. (e) Magnetic contrast after a second laser pulse. The magnetization of the region switched in (d) is toggled back to its original state. (f) Switching area as a function of the laser fluence. (g) The calculated near-field enhancement around the three antenna lengths. (After Liu, T.M. et al., *Nano Letters* 15, 6862, 2012. With permission.)

A reversed magnetic domain is written in a uniformly magnetized region below one end of the 230 nm antenna arms with a single optical pulse of 3.7 mJ/cm^2 (Figure 10.9a, b). The diameter of the switched area is 53 nm, almost six times smaller than previous observations [57]. Areas with switched magnetic orientation are observed below the antenna. The domain size compares favorably to the track width of 55 nm used in a recent demonstration of 1+ Tb/in^2 HAMR [86].

The magnetic state can be toggled back-and-forth deterministically, demonstrating that the optically induced switching is controllable on the nanoscale (see Figure 10.9). Following the writing of the initial switched domain in Figure 10.9b, the sample is then magnetically reset ex-situ by applying a magnetic field to the sample (Figure 10.9c), to return the magnetization to the initial state. Subsequently, it receives another laser pulse, resulting in the magnetic reversal in the same region (Figure 10.9d). However, the area and shape of the switched region are not identical. Illuminating the sample again with another laser pulse, the magnetic state is returned back to its original uniformly magnetized state demonstrating the reversibility of the AOS, even on the nanoscale (see Figure 10.9e).

Although the switching is reproducible, we do not yet have full control over the magnetic switching location due to the inhomogeneity in the Tb, Fe, and Co composition in the alloy [56, 82], see also the next section. Such inhomogeneous concentration can lead to strong local variations in the compensation temperature and thus to local variations in the switching fluence threshold. In addition, fabrication imperfections, surface roughness, and a non-perfect positioning of the antennas may also affect the optical illumination profile.

The effect of the nanoantennas on the switching fluence threshold is investigated by examining the fluence dependence of different sample regions as shown in Figure 10.9f. The plot shows the switched area percentage for regions in the near-field of the antennas and regions away from the antennas as a function of fluence. Regions within the near-field region of the antennas start to magnetically reverse at fluences of 3.7 mJ/cm^2. At fluences of 5.8 mJ/cm^2, switching also occurs in areas where no near-field enhancement is expected. These results also confirm the inhomogeneity in the switching threshold across the sample. The switched area for both regions within the near fields and those away from the near fields increase linearly with fluence above a threshold. The slope of this increase is a measure of the switching susceptibility. The slope is 1.98%/mJ/cm^2 for the regions within the near-field enhancement and 0.33%/mJ/cm^2 for regions away from the near field. The ratio of these slopes is 6.0, which is above the modeled average near-field intensity enhancement by the antennas of approximately 4 [82].

Thus by exploiting the field-confining capability of plasmonic nanoantennas, AOS is successfully demonstrated down to a lateral size of about 50 nm, a size that is comparable to what is achieved in the heat-assisted magnetic recording. The switching is shown to be reproducible, and back-and-forth switching is also demonstrated. Further development of the plasmonic antennas is required, including a non-contact geometry to make this a viable alternative technology; it could be achieved using the present two-wire antenna structure atop a flattened atomic force microscopy tip [87].

10.5 IS THERE A LIMIT TO THE MINIATURIZATION?

10.5.1 NON-LOCAL ANGULAR-MOMENTUM TRANSFER BY SPIN CURRENTS

To remind the reader, ultrafast laser techniques have revealed extraordinary spin dynamics in magnetic materials [5, 18, 24] that equilibrium descriptions of magnetism [1] cannot explain. A particularly challenging problem is to access the spatial patterns that occur in these non-equilibrium states. Attempting to reverse a small domain, one notices that the inhomogeneity of the samples plays a role in the reproducibility of the data. From the first experiments on the single-shot imaging of the reversal dynamics [7], it was clear that some pattern smaller than the optical resolution of these experiments must be developing during this non-equilibrium process. In particular, it is critical to understand non-equilibrium (sub-)picosecond spin dynamics on the nanometer length scale, where magnetic order emerges. Given the time and spatial scales of these processes, the resolution of these processes formed a formidable experimental challenge.

To address this challenge, we designed and applied ultrafast diffraction experiments with an X-ray laser that probes the nanoscale spin dynamics following optical laser excitation in the ferrimagnetic alloy GdFeCo, which exhibits macroscopic all-optical switching [56]. Our study revealed that GdFeCo displays nanoscale chemical and magnetic inhomogeneities that affect the spin dynamics. In particular, we observe Gd spin reversal in Gd-rich nanoregions within the first picosecond driven by the non-local transfer of angular momentum from larger adjacent Fe-rich nanoregions.

Although GdFeCo is generally considered to be a homogeneous amorphous alloy, it appeared that at nanometer length scales the material's chemical homogeneity is broken by chemical segregation. The local elemental variations for Gd, Fe, and Co were measured by scanning transmission electron microscopy (STEM) with elemental sensitivity, using energy-dispersive X-ray spectroscopy (EDX). This characterization yielded the surprising result of nanoscale chemical inhomogeneity, and the question remains how this local structure affects the spin subsystems. We address this question using soft X-ray scattering and therefore shift from mapping real-space to probing reciprocal space.

The resulting X-ray diffraction patterns consist of linear combinations of reciprocal space Fourier amplitudes of the sample's chemical and magnetic real-space distributions [88, 89]. These data have shown the correlation length of 10 nm for Gd-rich areas, and about 13 nm for the Fe-rich areas, in good agreement with the real-space data. The sample thus segregates into Gd-rich (~20% of the sample) and Fe-rich (~40% of the sample) areas, which are spatially anti-correlated.

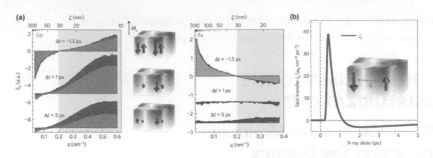

FIGURE 10.10   Magnetic X-ray diffraction, S_q, of Gd 4f and Fe 3d versus wave vector q, for the indicated X-ray delay times, Δt. The darker shaded areas show data with 16 mJ/cm² and lighter shaded areas show data with 24 mJ/cm² optical pump fluence. The thermalized temperature at 1 ps for both fluences is above the critical temperature, T_c. Negative time refers to equilibrium conditions before optical excitation. The data were averaged over 1 ps intervals around the indicated delays and are offset vertically for clarity. (b) Time-resolved angular-momentum flow, J_s, into the Gd-rich regions. The plot shows a time-delayed spike in the angular-momentum transfer to the Gd spins in the Gd-rich nanoregions that lasts 1 ps. Following this spin transfer to the Gd-rich nanoregions, a slower and weaker dissipation of spin from the region occurs. (After Graves, C.E. et al., *Nat. Mater.* 12, 293, 2013. With permission.)

Interestingly, magnetic variations associated with the chemical segregation of the enriched regions dominate at high-q values. Moreover, the local changes in the Gd magnetization were found to be much larger than those in the Fe magnetization. This finding once again underlines the substantial difference in the behavior of the two sublattices.

To probe the dynamic behavior of the magnetization following laser excitation, we measure the S_q distributions, which are proportional to M_z, for different time delays Δt (shown in Figure 10.10a). The femtosecond laser excitation deposits energy directly into the 3d electron system, subsequently generating mobile sp spins [10] without exciting the 4f spins. The low-q and high-$q S_q$ contributions in Figure 10.10a exhibit clearly different time characteristics. Thus, in Fe-rich nanoregions, both sublattices exhibit demagnetization, which is also observed macroscopically [18]. However, the Gd-rich nanoregions respond differently: the Gd part of the magnetization of the Gd-rich regions reverses sign at $\Delta t \sim 1$ ps. As the magnetization of iron maintains its sign, a net ferromagnetic alignment of the two sub-lattice magnetizations occurs within the Gd-rich nanoregions after ~1 ps and persists several picoseconds thereafter.

Analysis of the high-$q S_q$ dynamics quantifies the amount of non-local angular-momentum transfer to the Gd-rich nanoregions. The high-$q \cdot S_q$ measures the local difference of the sub-lattice magnetizations from the average magnetization, and within the enriched regions. Whereas M_{Gd} increases with time, the M_{Fe} becomes smaller, revealing non-local angular-momentum transfer to the Gd-rich regions M_{Gd}, as otherwise, M_{Fe} would necessarily be larger. The non-local transfer is characterized by J_S, the time evolution of the net angular-momentum flow into the Gd-rich nanoregions, or the net spin current, as shown in Figure 10.10b. J_S is the rate of change of M and is found to be dominated by the Gd contribution. Within 1 ps, a net 14 μ_B/nm³ is transferred to the Gd-rich nanoregions. Moreover, the positive sign of J_S during the first picosecond indicates that it is the Fe spins that transport angular momentum into the Gd-rich nanoregions, as only they possess the correct sign. After about 1 ps, J_s becomes negative indicating a slower back-flow of the angular momentum out of the Gd-rich regions. This back-flow continues until between 4 and 6 ps when the Gd magnetization switches back and the ground-state ferrimagnetic order is re-established.

So how does it happen that the non-local transfer of angular momentum leads to Gd spin reversal specifically within the disordered Gd-rich nanoregions? Previous work on GdFeCo focused on the macroscopic angular-momentum transfer observed at relatively low fluences, just sufficient for AOS [18, 31]. In contrast, the measurements of Ref. [56] isolate the nanoscale spatial variations of this transfer. In addition, we observe Gd spin reversal, not Fe, in the Gd-rich areas. Transfer of angular momentum from Fe to Gd occurs through the exchange interaction. This transfer is enhanced within the Gd-rich regions, not only due to local enhancement of the GdFe exchange but also to spin-torque scattering at the interfaces of the Gd-rich regions resulting from non-collinear spins and abrupt chemical variations on the electron scattering length scale [90, 91]. Surprisingly, this spin transfer has a delayed onset, suggesting a threshold-like behavior. Although the observed spin transfer and reversal could be due to spin currents [10], the presence of the delay indicates the need for further theoretical work. The spin transfer to the Gd-rich regions peaks around 1 ps, a timescale characteristic for hot electrons to reach thermal equilibrium with the lattice [92]. The growth of the reversed Gd-rich regions is indicated by the shift of the S_q maximum to lower q values with increasing delay time, characteristic of a size increase of the spin-reversed area. The reversed Gd magnetization relaxes following the spin transfer, proceeding through Fe spin states, as demonstrated by the fact that the increase of Fe and the decrease of Gd magnetizations have the same time constant.

The results discussed above demonstrate that time-resolved X-ray scattering, with its chemical and magnetic sensitivity, opens a new way to study emerging phenomena, such as the order in highly non-equilibrium systems. The obtained results not only link two important present fields in spintronics, namely transport and ultrafast optics, but also point to the general importance of nanoscale inhomogeneities in the study of many fundamental phenomena, such as crystal growth, the role of fluctuations in high-T_c superconductors and the emergence of quantum electronic phases hidden in thermal equilibrium. The observed angular-momentum transfer reverses spins robustly and overcomes the magnetic disorder, suggesting an effective tool for controlling laser-excited spins in microscopically engineered magnetic materials. Furthermore, the observed spin reversal is orders of magnitude faster than reversal by conventional voltage-driven equilibrium spin currents and may offer marked improvements over present spin-torque devices.

10.5.2 THE ROLE OF SAMPLE INHOMOGENEITY IN THE ALL-OPTICAL SWITCHING: MULTISCALE SIMULATIONS

The majorities of the theories that attempt to describe all-optical switching so far have only considered perfectly homogeneous samples, and have not addressed the role of the amorphous structure nor the chemical

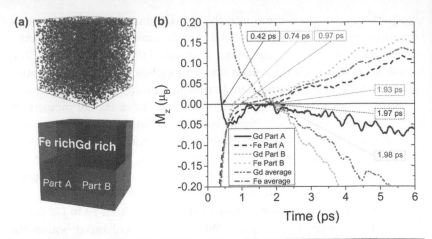

FIGURE 10.11 (a) Schematic figure showing the inhomogeneous samples of $Gd_{0.24}Fe_{0.76}$ plotted in the top part while in the bottom a detailed figure of the amorphous structure is shown with Fe and Gd atoms in red and blue, respectively. Above, Gd-rich regions are called part A and Fe-rich regions are called part B. (b) Magnetization profile for Gd and Fe sublattices for inhomogeneous concentrations. The magnetization of the Gd-rich regions (part A), Gd poor regions (part B), and the sample average is shown in black, green, and red lines, respectively. The strength of the next-nearest-neighbor exchange parameter for the Gd sub-lattice was −0.4 Ry in this simulation. The figure shows clear similarities with experimental data reported in Ref. [18]. (After Chimata, R. et al., *Phys. Rev. B* 92, 094411, 2015. With permission.)

inhomogeneity of the GdFe system and how this may influence the ultrafast switching behavior. Just recently, multiscale simulations addressed this issue, by generating randomly inhomogeneous samples of GdFeCo alloys with various parameters [93]. As before, only temperature effects from the laser pulse were considered, described via a two-temperature model.

A schematic illustration of such a heterogenous sample is shown in Figure 10.11a. First of all, the calculations show that a certain degree of non-collinearity is present in these structures, depending on the strength of the local anisotropy parameter as well as on the next-nearest neighbor exchange, the latter having an antiferromagnetic character, usual for the rare-earth materials [94]. This is supported by the first-principles calculations [93], that show an exchange driven non-collinear configuration as the ground state even at $T = 0$ K.

The results of the simulations of the switching dynamics are briefly summarized in Figure 10.11. The magnetization dynamics of the Gd and Fe sublattices, from the different regions of the sample, i.e., the Gd poor (Fe rich), Gd rich (Fe poor), and regions with an average concentration is shown separately. We can observe that the Gd sub-lattice switches faster (0.42 ps) than the Fe one (1.97 ps) in the Gd-rich regions, while for the Fe-rich regions the process is reversed. This behavior exactly corresponds to what was already observed experimentally for GdFeCo, as discussed in the previous section [56].

Note that the theory put forth in these calculations does not involve any spin current mechanism. Alternatively, this could be the non-collinearity of the magnetic structure, obtained in the calculations, that is responsible for this behavior [93]. It is thus clear that to clarify the exact behavior of this challenging system, further research is imperative. For our purposes of AOS at the nanoscale, such inhomogeneities do put a limit on what minimum domain size could be achieved.

10.6 CONCLUDING REMARKS

The goal of this chapter was to describe the progress in our understanding of ultrafast laser manipulation of magnetic order and its application to the nanoscale switching of isolated magnetic bits. From the first experimental demonstration of AOS by a single 40 fs laser pulse, we have succeeded to realize AOS in various nanostructured samples and to write isolated domains as small as 50 nm. The work also led to a deeper understanding of the exact mechanism, that has fueled a lot of interest in this area. As a result, optical manipulation of magnetic order by femtosecond laser pulses has developed into an exciting and still expanding research field that keeps producing a continuous stream of new and sometimes counterintuitive results.

As an outlook, it is still early to predict whether such all-optical ultra-fast switching of the magnetization could become useful in, for example, magnetic recording devices, and if yes, in which particular realization. The last years have seen a considerable slow-down of the annual growth in the magnetic recording industry, forcing the leading companies to look for novel approaches [95], such as heat-assisted magnetic recording [41] and bit-patterned media [96]. In the former, a laser is introduced in the write-head, to assist the magnetic switching of high-anisotropy media. Our heat-only mechanism would thus be directly compatible with this technology. As for the latter, the continuous media are going to be replaced with arrays of nanoscale magnetic elements, each of them representing a single magnetic bit. It was thus imperative to verify, how the all-optical switching works in nanostructures, where the initial magnetization states may be strongly modified by magnetostatic fields or by the manufacturing process itself, and where the relaxation from the transient strongly non-equilibrium state could be altered. The data have shown that the switching is indeed possible even if the magnetic properties of the elements become inhomogeneous in the manufacturing process [44], and that, due to optical interference effects, a smaller pulse fluence is sufficient for switching [48], decreasing to sub-picoJoule values for the currently used bit sizes. Moreover, the use of plasmonic antenna's can be an essential ingredient in the realization of all-optical recording at the nanoscale [82]. Thus, all the principal ingredients are in place, and we will see in the coming years whether such all-optical recording technology will become the mainstream of the future magnetic recording.

ACKNOWLEDGMENTS

This work was partially supported by The Netherlands Organization for Scientific Research (NWO), the Foundation for Fundamental Research on Matter (FOM), the EU Seventh Framework Program (FP7/2007-2013) Grants

No. NMP3-LA-2010-246102 (IFOX), No. 280555 (Go-Fast), the European Research Council (FP7/2007-2013)/ERC Grant Agreements No. 257280 (Femtomagnetism) and No. 339813 (Exchange).

REFERENCES

1. J. Stöhr and H. C. Siegmann, *Magnetism: From Fundamentals to Nanoscale Dynamics*, Springer Verlag, Berlin, 2006.
2. Th. Gerrits, H. A. M. van den Berg, J. Hohlfeld, L. Bär, and Th. Rasing, Ultrafast precessional magnetization reversal by picosecond magnetic field pulse shaping, *Nature* **418**, 509 (2002).
3. S. Kaka and S. E. Russek, Precessional switching of submicrometer spin valves, *Appl. Phys. Lett.* **80**, 2958 (2002).
4. H. W. Schumacher, C. Chappert, P. Crozat et al., *Phys. Rev. Lett.* **88**, 227201 (2002).
5. E. Beaurepaire, J.-C. Merle, A. Daunois, J.-Y. Bigot, Ultrafast spin dynamics in ferromagnetic nickel, *Phys. Rev. Lett.* **76**, 4250 (1996).
6. A. Kirilyuk, A. V. Kimel, and Th. Rasing, Ultrafast optical manipulation of magnetic order, *Rev. Mod. Phys.* **82**, 2731 (2010).
7. K. Vahaplar, A. M. Kalashnikova, A. V. Kimel et al., Ultrafast path for optical magnetization reversal via a strongly non-equilibrium state, *Phys. Rev. Lett.* **103**, 117201 (2009).
8. B. Koopmans, G. Malinowski, F. Dalla Longa et al., Explaining the paradoxical diversity of ultrafast laser-induced demagnetization, *Nat. Mater.* **9**, 259 (2010).
9. M. Sultan, U. Atxitia, A. Melnikov, O. Chubykalo-Fesenko, and U. Bovensiepen, Electron- and phonon-mediated ultrafast magnetization dynamics of Gd(0001), *Phys. Rev. B* **85**, 184407 (2012).
10. M. Battiato, K. Carva, and P. M. Oppeneer, Superdiffusive spin transport as a mechanism of ultrafast demagnetization, *Phys. Rev. Lett.* **105**, 027203 (2010).
11. G. P. Zhang and T. F. George, Total angular momentum conservation in laser-induced femtosecond magnetism, *Phys. Rev. B.* **78**, 052407 (2008).
12. A. V. Kimel, A. Kirilyuk, P. A. Usachev, R. V. Pisarev, A. M. Balbashov, and Th. Rasing, Ultrafast nonthermal control of magnetization by instantaneous photomagnetic pulses, *Nature* **435**, 655 (2005).
13. F. Hansteen, A. V. Kimel, A. Kirilyuk, and Th. Rasing, Femtosecond photomagnetic switching of spins in ferrimagnetic garnet films, *Phys. Rev. Lett.* **95**, 047402 (2005).
14. A. V. Kimel, C. D. Stanciu, P. A. Usachev et al., Optical excitation of antiferromagnetic resonance in $TmFeO_3$, *Phys. Rev. B.* **74**, 060403R (2006).
15. A. V. Kimel, A. Kirilyuk, F. Hansteen, R. V. Pisarev, and Th. Rasing, Nonthermal optical control of magnetism and ultrafast laser-induced spin dynamics in solids, *J. Phys.: Condens. Matter* **19**, 043201 (2007).
16. A. V. Kimel, A. Kirilyuk, A. Tsvetkov, R. V. Pisarev, and Th. Rasing, Laser-induced ultrafast spin reorientation in the antiferromagnet $TmFeO_3$, *Nature* **429**, 850 (2004).
17. D. Afanasiev, B. A. Ivanov, A. Kirilyuk, Th. Rasing, R. V. Pisarev, and A. V. Kimel, Control of the ultrafast photoinduced magnetization across the morin transition in $DyFeo_3$, *Phys. Rev. Lett.* **116**, 097401 (2016).
18. I. Radu, K. Vahaplar, C. Stamm et al., Transient ferromagnetic-like state mediating ultrafast reversal of antiferromagnetically coupled spins, *Nature* **472**, 205 (2011).
19. C. D. Stanciu, A. V. Kimel, F. Hansteen, A. Tsukamoto, A. Itoh, A. Kirilyuk, and Th. Rasing, Ultrafast spin dynamics across compensation points in ferrimagnetic GdFeCo: The role of angular momentum compensation, *Phys. Rev. B.* **73**, 220402(R) (2006).

20. A. V. Kimel, B. A. Ivanov, R. V. Pisarev, P. A. Usachev, A. Kirilyuk, and Th. Rasing, Inertia-driven spin reorientation in antiferromagnets, *Nat. Phys.* **5**, 727 (2009).

21. A. Einstein W. J. de Haas, Experimenteller nachweis der amperèschen molekülströme, *Verhandl. Deut. Phys. Ges.* **17**, 152 (1915).

22. S. J. Barnett, Magnetization by rotation, *Phys. Rev.* **6**, 239 (1915).

23. L. Landau and E. Lifshitz, On the theory of dispersion of magnetic permeability in ferromagnetic bodies, *Phys. Z. Union* **8**, 153 (1935).

24. C. D. Stanciu, F. Hansteen, A. V. Kimel, A. Tsukamoto, A. Itoh, A. Kirilyuk, and Th. Rasing, All optical magnetic recording with circularly polarized light, *Phys. Rev. Lett.* **99**, 047601 (2007).

25. S. Mangin, M. Gottwald, C. H. Lambert et al., Engineered materials for all-optical helicity-dependent magnetic switching, *Nat. Mater.* **13**, 287 (2014).

26. A. Kirilyuk, A. V. Kimel, and Th. Rasing, Laser-induced magnetization dynamics and reversal in ferrimagnetic alloys, *Rep. Prog. Phys.* **76**, 026501 (2013).

27. L. P. Pitaevskii, Electric forces in a transparent dispersive medium, *J. Exp. Theor. Phys.* **12**, 1008 (1961).

28. J. P. van der Ziel, P. S. Pershan, and L. D. Malmstrom, Optically-induced magnetization resulting from the inverse Faraday effect, *Phys. Rev. Lett.* **15**, 190 (1965).

29. P. S. Pershan, J. P. van der Ziel, and L. D. Malmstrom, Theoretical discussion of the inverse Faraday effect, Raman scattering, and related phenomena, *Phys. Rev.* **143**, 574 (1966).

30. T. A. Ostler, J. Barker, R. F. L. Evans et al., Ultrafast heating as a sufficient stimulus for magnetization reversal in a ferrimagnet, *Nat. Commun.* **3**, 666 (2012).

31. J. H. Mentink, J. Hellsvik, D. V. Afanasiev et al., Ultrafast spin dynamics in multisublattice magnets, *Phys. Rev. Lett.* **108**, 057202 (2012).

32. A. R. Khorsand, M. Savoini, A. Kirilyuk, A. V. Kimel, A. Tsukamoto, A. Itoh, and Th. Rasing, Role of magnetic circular dichroism in all-optical magnetic recording, *Phys. Rev. Lett.* **108**, 127205 (2012).

33. K. Vahaplar, A. M. Kalashnikova, A. V. Kimel et al., All-optical magnetization reversal by circularly polarized laser pulses: Experiment and multiscale modeling, *Phys. Rev. B.* **85**, 104402 (2012).

34. M. Mansuripur, *The Physical Principles of Magneto-Optical Recording*, Cambridge University Press, Cambridge, 1995.

35. X. Jiang, L. Gao, J. Z. Sun, and S. S. P. Parkin, Temperature dependence of current-induced magnetization switching in spin valves with a ferrimagnetic CoGd free layer, *Phys. Rev. Lett.* **97**, 217202 (2006).

36. Y. Mimura, N. Imamura, T. Kobayashi, A. Okada, Y. Kushiro, Magnetic properties of amorphous alloy films of Fe with Gd, Tb, Dy, Ho, or Er, *J. Appl. Phys.* **49**, 1208 (1978).

37. J. Becker, A. Tsukamoto, A. Kirilyuk, J. C. Maan, Th. Rasing, P. C. M. Christianen, and A.V. Kimel, Ultrafast laser-induced dynamics of noncollinear spin structures in amorphous NdFeCo and PrFeCo, *Phys. Rev. B.* **92**, 180407 (2015).

38. R. F. L. Evans, T. A. Ostler, R. W. Chantrell, I. Radu, and Th. Rasing, Ultrafast thermally induced magnetic switching in synthetic ferrimagnets, *Appl. Phys. Lett.* **104**, 082410 (2014).

39. O. Heinonen and K. Z. Gao, Extensions of perpendicular recording, *J. Magn. Magn. Mater.* **320**, 2885 (2008).

40. W. A. Challener, T. W. McDaniel, C. D. Mihalcea, K. R. Mountfield, K. Pelhos, and I. K. Sendur, Light delivery techniques for heat-assisted magnetic recording, *Jpn. J. Appl. Phys.* **42**, 981 (2003).

41. M. H. Kryder, E. C. Gage, T. W. McDaniel et al., Heat assisted magnetic recording, *IEEE Proc.* **96**, 1810 (2008).

42. B. C. Stipe, T. C. Strand, C. C. Poon et al., Magnetic recording at 1.5 Pb/m² using an integrated plasmonic antenna, *Nat. Photonics* **4**, 484 (2010).
43. L. Le Guyader, S. El Moussaoui, E. Mengotti et al., Nanostructuring of GdFeCo thin films for laser induced magnetization switching, *J. Magn. Soc. Jpn.* **36**, 21 (2012).
44. L. Le Guyader, S. El Moussaoui, M. et al., Demonstration of laser induced magnetization reversal in GdFeCo nanostructures, *Appl. Phys. Lett.* **101**, 022410 (2012).
45. U. Flechsig, F. Nolting, A. F. Rodríguez et al., Performance measurements at the SLS SIM beamline, *AIP Conf. Proc.* **1234**, 319 (2010).
46. U. S. Inan and R. A. Marshall, *Numerical Electromagnetics: The FDTD Method*, Cambridge University Press, Cambridge, 2011.
47. *FDTD Solutions v7.0*, Lumerical Solutions Inc., Vancouver, 2010.
48. M. Savoini, R. Medapalli, B. Koene et al., Highly efficient all-optical switching of magnetization in GdFeCo microstructures by interference-enhanced absorption of light, *Phys. Rev. B* **86**, 140404(R) (2012).
49. L. Le Guyader, M. Savoini, S. El Moussaoui et al., Nanoscale sub-100 picosecond all-optical magnetization switching in GdFeCo microstructures, *Nat. Commun* **6**, 5839 (2015).
50. M. Savoini, F. Ciccacci, L. Duo, and M. Finazzi, Apparatus for vectorial Kerr confocal microscopy, *Rev. Sci. Inst.* **82**, 023709 (2011).
51. M. Celebrano, P. Biagioni, M. Zavelani-Rossi et al., Hollow-pyramid based scanning near-field optical microscope coupled to femtosecond pulses: A tool for nonlinear optics at the nanoscale, *Rev. Sci. Inst.* **80**, 033704 (2009).
52. B. A. Nechay, U. Siegner, M. Achermann, H. Bielefeldt, and U. Keller, Femtosecond pump-probe near-field optical microscopy, *Rev. Sci. Inst.* **70**, 2758 (1999).
53. M. R. Freeman and B. C. Choi, Advances in magnetic microscopy, *Science* **294**, 1484 (2001).
54. U. Atxitia, O. Chubykalo-Fesenko, N. Kazantseva, D. Hinzke, U. Nowak, and R. W. Chantrell, Micromagnetic modeling of laser-induced magnetization dynamics using the Landau-Lifshitz-Bloch equation, *Appl. Phys. Lett.* **91**, 232507 (2007).
55. N. Kazantseva, U. Nowak, R. W. Chantrell, J. Hohlfeld, and A. Rebei, Slow recovery of the magnetization after a sub-picosecond heat pulse, *Europhys. Lett.* **81**, 27004 (2008).
56. C. E. Graves, A. H. Reid, T. Wang et al., Nanoscale spin reversal by non-local angular momentum transfer following ultrafast laser excitation in ferrimagnetic gdfeco, *Nat. Mater.* **12**, 293 (2013).
57. M. Finazzi, M. Savoini, A. R. Khorsand et al., Laser-induced magnetic nanostructures with tunable topological properties, *Phys. Rev. Lett.* **110**, 177205 (2013).
58. L. Le Guyader, S. El Moussaoui, M. Buzzi et al., Deterministic character of all-optical magnetization switching in GdFe-based ferrimagnetic alloys, *Phys. Rev. B* **193**, 134402 (2016).
59. T. H. R. Skyrme, A unified field theory of mesons and baryons, *Nucl. Phys.* **31**, 556 (1962).
60. S. Heinze, K. von Bergmann, M. Menze et al., Spontaneous atomic-scale magnetic skyrmion lattice in two dimensions, *Nat. Phys.* **7**, 713 (2011).
61. I. J. Dzyaloshinskii, A thermodynamic theory of "weak" ferromagnetism of antiferromagnetics, *Phys. Chem. Solids.* **4**, 241 (1958).
62. T. Moriya, Anisotropic superexchange interaction and weak ferromagnetism, *Phys. Rev.* **120**, 91 (1960).
63. S. Mühlbauer, B. Binz, F. Jonietz et al., Skyrmion lattice in a chiral magnet, *Science* **323**, 915 (2009).
64. W. Münzer, A. Neubauer, T. Adams et al., Skyrmion lattice in the doped semiconductor $Fe_{1-x}Co_xSi$, *Phys. Rev. B* **81**, 041203 (2010).

65. X. Z. Yu, Y. Onose, N. Kanazawa et al., Real-space observation of a two-dimensional skyrmion crystal, *Nature* **465**, 901 (2010).

66. X. Z. Yu, N. Kanazawa, Y. Onose et al., Near room-temperature formation of a skyrmion crystal in thin-films of the helimagnet fege, *Nat. Mater.* **10**, 106 (2011).

67. X. Yu, M. Mostovoy, Y. Tokunaga et al., Magnetic stripes and skyrmions with helicity reversals, *Proc. Natl. Acad. Sci. U.S.A.* **109**, 8856 (2012).

68. P. J. Grundy and S. R. Herd, Lorentz microscopy of bubble domains and changes in domain wall state in hexaferrites, *Phys. Status Solidi* **20**, 295 (1973).

69. Y. S. Lin, P. J. Grundy, and E. A. Giess, Bubble domains in magnetostatically coupled garnet films, *Appl. Phys. Lett.* **23**, 485 (1973).

70. T. Suzuki, A study of magnetization distribution of submicron bubbles in sputtered Ho-Co thin films, *J. Magn. Magn. Mater.* **31–34**, 1009 (1983).

71. T. Ogasawara, N. Iwata, Y. Murakami, H. Okamoto, and Y. Tokura, Submicron-scale spatial feature of ultrafast photoinduced magnetization reversal in TbFeCo thin film, *Appl. Phys. Lett.* **94**, 162507 (2009).

72. M. Ezawa, Giant skyrmions stabilized by dipole-dipole interactions in thin ferromagnetic films, *Phys. Rev. Lett.* **105**, 197202 (2010).

73. A. A. Mikhailovsky, M. A. Petruska, M. I. Stockman, and V. I. Klimov, Broadband near-field interference spectroscopy of metal nanoparticles using a femtosecond white-light continuum, *Opt. Lett.* **28**, 1686 (2003).

74. M. Celebrano, M. Savoini, P. Biagioni et al., Retrieving the complex polarizability of single plasmonic nanoresonators, *Phys. Rev. B* **80**, 153407 (2009).

75. L. Novotny and N. van Hulst, Antennas for light, *Nat. Photonics* **5**, 83 (2011).

76. P. Biagioni, J.-S. Huang, and B. Hecht, Nanoantennas for visible and infrared radiation, *Rep. Prog. Phys.* **75**, 024402 (2012).

77. P. Bharadwaj, B. Deutsch, and L. Novotny, Optical antennas, *Adv. Opt. Photon.* **1**, 438 (2009).

78. W. A. Challener, C. Peng, A. V. Itagi et al., Heat-assisted magnetic recording by a near-field transducer with efficient optical energy transfer, *Nat. Photonics* **3**, 220 (2009).

79. B. Koene, M. Savoini, A. V. Kimel, A. Kirilyuk, and Th. Rasing, Optical energy optimization at the nanoscale by near-field interference, *Appl. Phys. Lett.* **101**, 013115 (2012).

80. P. Biagioni, J. S. Huang, L. Duo, M. Finazzi, and B. Hecht, Cross resonant optical antenna, *Phys. Rev. Lett.* **102**, 256801 (2009).

81. P. Biagioni, M. Savoini, J.-S. Huang, L. Duo, M. Finazzi, and B. Hecht, Near-field polarization shaping by a near-resonant plasmonic cross antenna, *Phys. Rev. B* **80**, 153409 (2009).

82. T. M. Liu, T. H. Wang, A. H. Reid et al., Nanoscale confinement of all-optical magnetic switching in TbFeCo – competition with nanoscale heterogeneity, *Nano Lett.* **15**, 6862 (2015).

83. D. Zhu, M. Guizar-Sicairos, B. Wu et al., High-resolution X-ray lensless imaging by differential holographic encoding, *Phys. Rev. Lett.* **105**, 043901 (2010).

84. X. Wu, R. Kullock, E. Krauss, and B. Hecht, Single-crystalline gold microplates grown on substrates by solution-phase synthesis, *Cryst. Res. Technol.* **50**, 595, (2015).

85. M. Savoini, A. H. Reid, T. Wang et al., Attempting nanolocalization of all-optical switching through nano-holes in an Al-mask, *Proc. SPIE* **2014**, 9167 (2014).

86. A. Q. Wu, Y. Kubota, T. Klemmer et al., HAMR areal density demonstration of 1+ Tbpsi on spinstand. *IEEE Trans. Magn.* **49**, 779 (2013).

87. J. N. Farahani, H.-J. Eisler, D. W. Pohl et al., Bow-tie optical antenna probes for single-emitter scanning near-field optical microscopy, *Nanotechnology* **18**, 125506 (2007).

88. B. Vodungbo, J. Gautier, G. Lambert et al., Laser-induced ultrafast demagnetization in the presence of a nanoscale magnetic domain network, *Nat. Commun.* **3**, 999 (2012).

89. B. Pfau, S. Schaffert, L. Müller et al., Ultrafast optical demagnetization manipulates nanoscale spin structure in domain walls, *Nat. Commun.* **3**, 1100 (2012).

90. M. D. Stiles and A. Zangwill, Anatomy of spin-transfer torque, *Phys. Rev. B* **66**, 014407 (2002).

91. J. Bass and W. P. Pratt, Spin diffusion lengths in metals and alloys, and spin-flipping at metal/metal interfaces: An experimentalist's critical review, *J. Phys. Condens. Matter.* **19**, 183201 (2007).

92. H. S. Rhie, H. A. Dürr, and W. Eberhardt, Femtosecond electron and spin dynamics in Ni/W(110) films, *Phys. Rev. Lett.* **90**, 247201 (2003).

93. R. Chimata, L. Isaeva, K. Kadas et al., All-thermal switching of amorphous Gd-Fe alloys: Analysis of structural properties and magnetization dynamics, *Phys. Rev. B* **92**, 094411 (2015).

94. J. Jensen and A. R. Mackintosh (Eds.), *Rare Earth Magnetism*, Oxford University Press, New York, 1990.

95. J. van Elk, M. L. Plumer and W. C. Cain, New paradigms in magnetic recording, *Phys. Canada* **67**, 25 (2011).

96. C. Brombacher, M. Grobis, J. Lee et al., $L1_0$FePtCu bit patterned media, *Nanotechnology* **23**, 025301 (2012).

Section III
Magnetic Tunnel
Junctions

11

Tunneling Magnetoresistance
Experiment (Non-MgO)

Patrick R. LeClair and Jagadeesh S. Moodera

Tunneling magnetoresistance (TMR) is a consequence of spin-dependent tunneling, and in a broader sense, another manifestation of spin-dependent transport related to the giant magnetoresistance (GMR) and anisotropic magnetoresistance (AMR) effects. Spin-dependent tunneling between two ferromagnets, an outgrowth of earlier work by Meservey and Tedrow, [1] was first proposed [2] and observed [3] in 1975, but it was reliably demonstrated only in 1995 [4, 5]. In the past two decades, magnetic tunnel junctions (MTJs) have aroused considerable interest due to their suitability for applications in spin-electronic devices and indeed have found applications in hard disk read heads and magnetic random access memories (MRAMs). The diversity of the physical phenomena governing the operation of these magnetoresistive devices also makes MTJs very attractive from the fundamental physics point of view. These facts have recently stimulated tremendous activity in both the experimental and theoretical realms of investigation, with a view to understand and manipulate the electronic, magnetic and transport properties of MTJs.

A magnetic tunnel junction, in essence, consists of two ferromagnetic metal layers separated by a thin insulating barrier layer, as shown in Figure 12.1 of Chapter 12, Volume 1. The insulating layer is sufficiently thin (a few nm or less) that electrons can tunnel through the barrier, provided a bias voltage is applied between the two metal electrodes across the insulator. The most important property of MTJs is that the tunneling current depends on the relative orientation of the magnetizations of the two ferromagnetic layers, which can be changed by an applied magnetic field or more conveniently a spin torque (see Chapter 15, Volume 1). This phenomenon is called tunneling magnetoresistance (sometimes referred to as junction magnetoresistance). Although the tunneling magnetoresistance (TMR) effect has been known from the experiments of Julliere [3] for over 40 years, only a relatively modest number of studies had been performed in this field until the mid-nineties. Partly, the technologically demanding fabrication process was responsible, making it difficult to create robust and reliable tunnel junctions. Also, the fact that the reported values of TMR were small (at most a few percent at low temperatures) did not immediately trigger a considerable interest with regards to sensor/memory applications. In 1995, however Miyazaki et al. [5] and Moodera et al. [4] demonstrated reliable TMR of over 10% which triggered an explosion of research.

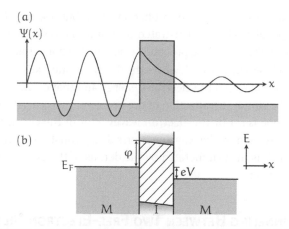

FIGURE 11.1   Tunneling in metal-insulator-metal structures. (a) Electron wave function decays exponentially in the barrier region, and for thin barriers, some intensity remains in the right side. (b) Potential diagram for an M/I/M structure with applied bias *eV*. Shaded areas represent filled states, open areas are empty states, and the hatched area represents the forbidden gap in the insulator.

In this chapter, we will review the history and fundamental physics of spin-dependent tunneling, aiming to provide the reader with a conceptual understanding and an overview of what factors that control the magneto-resistance in magnetic tunnel junctions, and what we view as some of the crucial pedagogical experiments. Starting from early experiments on spin-dependent tunneling which established the basis of the field, we subsequently overview the characteristic features of MTJs and consider experiments and models which highlight the role of the electronic structure of the ferromagnets, the insulating layer, and the ferromagnet/insulator interfaces.

11.1 INTRODUCTION TO ELECTRON TUNNELING

The quantum mechanical tunnel effect is one of the oldest quantum phenomena, dating back to the late 1920s. A general review of electron tunneling and all its nuances is well beyond the scope of this review. We will assume the reader has at least a basic familiarity with the tunneling problem from introductory quantum physics, and for a more complete introduction refer to the excellent overview in Ref. [6] and Chapter 13, Volume 1 (Tunneling Magnetoresistance: Theory) for a discussion of the theory of spin-dependent tunneling. Here we will only briefly introduce the key features of the tunneling problem as applied to metal-insulator-metal structures before embarking on a tour of spin-polarized tunneling.

Electron tunneling is a phenomenon by which an electric current may flow through a thin insulating region. A simple way to understand how tunneling is possible is by considering an electron wave incident on a potential step, as shown in Figure 11.1a. Though most of the wave is reflected, if the potential barrier is sufficiently thin, the evanescent states in the barrier region can emerge on the other side of the barrier, and the electron will have a finite probability of "tunneling" across the potential barrier. Strictly speaking, tunneling is not a purely quantum phenomenon, but a *wave phenomenon*, and has an optical analogy in frustrated total internal reflection. In the step potential model, the tunneling probability depends exponentially on the tunnel barrier thickness d, $P \propto e^{-\kappa d}$, where the decay constant κ, in turn, depends on the difference between the electron's energy and that of the potential barrier. Even in the simplest realization of the problem, two crucial length scales are apparent: one is set by the electron wavevector in the classically allowed region (k), and second by the wavefunction's decay constant in the insulating region (κ), which is in turn dictated by the details of the potential landscape in the barrier region.

11.1.1 TUNNELING BETWEEN TWO FREE-ELECTRON METALS

The most straightforward realization of this problem is in a metal-insulator-metal (M/I/M) trilayer structure (Figure 11.1b), commonly called a tunnel junction, with the insulator typically provided by a metal oxide (e.g., Al_2O_3 or MgO). In this case, a bias voltage (V) is applied between the two metal electrodes. Owing to the enormous difference in resistance between the insulating region and the metal electrodes, this potential difference is developed almost entirely across the tunnel barrier, and this has the effect of raising the Fermi energy of one electrode relative to the other by an amount eV. Thus, in Figure 11.1b, electrons near the Fermi energy in the leftmost electrode can tunnel through the insulating barrier elastically into available states in the rightmost electrode. The two length scales in the tunneling problem now have important implications. First, the electron wavelength at the Fermi energy in a typical metal is extremely short, on the order of a few lattice spacings [7]. This means that tunneling is sensitive to the atomic-scale details of the physical and electronic structure of the metal-insulator interface. Second, the decay constant in the insulating region in a real insulator is dictated by the evanescent states from the metal electrode. The potential barrier height is thus no longer a simple step (except for illustrative purposes), but is in reality governed by the complex electronic structure of the insulator and its coupling to the metal electrodes.

In general, one must consider the electronic structure of the entire trilayer system, which ultimately makes tunneling in realistic systems a subtle problem with a wide variety of rich phenomena exhibited. We should note that our discussion in this chapter focuses almost exclusively on incoherent systems, e.g., those with non-epitaxial electrodes and amorphous barriers. In incoherent systems, k conservation rules are not strict [8], and any analysis of tunneling based on ideal k conservation is generally invalid. For all but a handful of the experiments discussed in this chapter, these considerations play a negligible role, but they will be a primary motivator for the results of Chapters 12 and 13, Volume 1.

In spite of these complications, it is nonetheless illustrative to further consider the idealized M/I/M structure of Figure 11.1b. In the simplest phenomenological picture, [6] the number of electrons tunneling from one electrode to the other is given by the product of the density of states at a given energy in the left electrode, $\rho_l(E)$, and the density of states at the same energy in the right electrode, $\rho_r(E)$, multiplied by the matrix elements $|M|^2$ representing the probability of transmission through the barrier. One must also then multiply by the probabilities that the states in the left electrode are occupied, $f(E)$ and that the states in the right electrode are empty, $1 - f(E - eV)$, where $f(E)$ is the Fermi-Dirac function. This is simply stating the requirement that electrons on one side of the barrier must have empty states to tunnel into on the other side of the barrier. Taking the total tunnel current as the difference between left- and right-going currents, within this picture we have

$$I(V) = I_{l \to r} - I_{r \to l}$$

$$= \int_{-\infty}^{+\infty} \rho_l(E) \cdot \rho_r(E - eV) |M|^2 f(E) \big[f(E - eV) - f(E) \big] dE. \quad (11.1)$$

For a simple square barrier of height φ and thickness d, the matrix elements can be evaluated exactly, and one finds $I \sim e^{-d\sqrt{\varphi}}$. Essentially the same holds for more arbitrarily shaped barriers [9]. The effect of an applied potential difference eV is also already clear in this simple model: the potential difference will develop almost wholly across the insulating barrier, which serves to effectively lower the average barrier height by $eV/2$ for electrons tunneling in one direction, and raise it by the same amount for tunneling in the opposing direction. This leads to a rapidly (and non-linearly) increasing current with increasing potential difference, one hallmark characteristic of tunneling behavior. Extensions to finite temperature further predict an increase in tunnel current a temperature is increased, also commonly used as one criteria for establishing the presence of tunneling [10, 11].

It is often more useful to measure the derivative of the current-voltage characteristic, the *conductance dI/dV*, in order to more readily observe changes in transport. In the limit of low bias voltage, where the tunnel probability matrix elements are essentially independent of energy,

$$G \equiv \frac{dI}{dV} \propto |M|^2 \int \rho_l(E) \rho_r(E - eV) \frac{df(E - eV)}{dV} dE. \quad (11.2)$$

At low temperatures, the derivative of the Fermi-Dirac distribution is similar to a delta function, zero everywhere except where its argument is zero and with a width proportional to $k_B T$. This gives the result that for low temperature and bias voltage, the tunnel conductance should be proportional to the product of the densities of states in the two electrodes: $G \propto \rho_l \rho_r$.

11.2 DENSITY OF STATES IN TUNNELING

In his pioneering experiments, Giaever [12] investigated current-voltage and conductance-voltage characteristics of $Al/Al_2O_3/Pb$ normal-metal – insulator – superconductor (N/I/S) tunnel junctions well below the

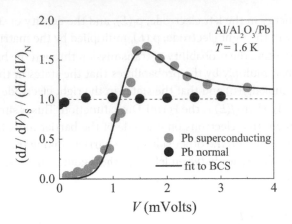

FIGURE 11.2 Conductance vs. voltage curve dI/dV(V) for an Al/Al$_2$O$_3$/Pb junction at $T = 1.6$ K when the Pb is in the normal state (black) and superconducting state (blue). The Al is in the normal state for both curves. The solid line is a fit to a thermally broadened BCS density of states. (After Giaever I., *Phys. Rev. Lett.* 5(4), 147, 1960.)

superconducting transition temperature of the Pb electrode (but above the superconducting transition temperature of the Al electrode). By driving the Pb into its normal state through the application of a magnetic field, Giaever could perform a comparative study of the tunneling characteristics between normal metals, or between a normal metal (Al) and a superconductor (Pb). When the Pb electrode was in the normal state, the tunnel current was linear in voltage and the conductance dI/dV was essentially constant, Figure 11.2, as expected from the previous section. However, when the Pb was in the superconducting state, the tunnel current was dramatically reduced at low voltages, independent of current polarity. Most interestingly, the conductance $dI/dV(V)$, Figure 11.2, very closely resembles the BCS quasiparticle density of states [6, 12]:

$$\frac{\rho_s(E)}{\rho_n(E)} = \text{sgn}(E)\,\text{Re}\left(\frac{E}{\sqrt{E^2 - \Delta^2}}\right). \tag{11.3}$$

Where ρ_n is the density of states in the normal state, and Δ is the energy gap in the quasiparticle excitation spectrum. Giaever interpreted this result [12] as an indication that tunneling conductance is proportional to the density of states in the electrodes, and subsequent measurements confirmed these results shortly thereafter [13]. This result had no adequate theoretical explanation at the time. In fact, several of Giaever's coworkers were initially skeptical of his "naive" experiment to measure directly the superconducting energy gap [14]. This "naive" experiment later led to a shared Nobel prize for Giaever in 1973.

Tunneling into superconductors is now so ubiquitous that it is difficult to appreciate how striking a result it was. A phenomenological explanation, closely following Giaever's original treatment, is now presented in most introductory solid state physics text. However, a proper theoretical justification of Giaever's ideas is subtle, with important consequences for MTJs, and will be treated more rigorously in Chapter 13, Volume 1. Despite the fact that a simple calculation starting with Equation 11.1 seems to include the density of states directly, it has been shown that [15] in a simple independent electron model, the tunneling probability matrix elements are *inversely proportional to the density of states* and thus they exactly cancel the density of states factors in Equation 11.1. Thus, the density of states does *not* directly enter the expression for the tunnel current in a free-electron model [15]. Shortly after Giaever's experiments, Bardeen [16] clarified this apparent contradiction and found that one could indeed recover the result that the tunneling conductance should be proportional to the density of states in the electrodes. The crucial point is that the dependence of the tunnel conductance on the density of states does not come as simply as one might expect from Equation 11.1, as will be pointed out in more detail in Chapter 13, Volume 1. For MTJs, this means that although an explanation of the TMR effect (Julliere's model, Section 11.4.1) seems to come quite easily from (11.1), the proper justification of such an explanation is subtle. In fact, in our view, it is only in recent years that a deeper understanding of the role of electronic structure in tunneling has arisen, and we refer to the reader to Chapter 13, Volume 1 for further discussion.

In the decade following Giaever's groundbreaking results, more sophisticated theoretical treatments by Appelbaum and Brinkman [17] and Zawadowski [18], as well as more recent refinements [19–23], put the role of the electrode density of states on firmer ground. The essential result is that, particularly in incoherent systems, tunneling is specifically sensitive to the *local density of states at the electrode-barrier interface*. Qualitatively, in these more sophisticated treatments, the relevant length scale over which the density of states is "sampled" in the tunneling process is dictated by the Fermi wavelength [17].

In a normal metal, this scale is in the order of a few lattice spacings. However, at distances of only a few lattice spacings from the metal-insulator interface, the wave function in the normal metal is strongly perturbed by the presence of the insulator and the details of the interface bonding. In short, in normal-metal tunneling structures, one can only measure the density of states in a normal metal *in this strongly perturbed region*, rather than measuring bulk-like or even surface-like behavior. Put slightly differently, although one may probe electronic structure effects by normal-metal tunneling, these effects are probed only within a few Fermi wavelengths of the electrode-barrier interface, where the density of states may differ greatly from that in the bulk. One may already imagine that because of this interfacial sensitivity, tunneling characteristics will be a property of both the electrode and barrier *together*. In fact, the junctions with the same electrodes but different barriers *do* behave differently, a phenomenon which will be discussed in Section 11.4.9. Generally speaking, the interfacial sensitivity of tunneling in non-superconducting junctions has extremely important consequences for spin-dependent tunneling [1], particularly in MTJs [24–26] (See Section 11.4.9).

11.3 THE BEGINNINGS OF SPIN-DEPENDENT TUNNELING

Based on Giaever's ideas about tunneling, in the late 1960s and early 1970s Meservey and Tedrow embarked on a series of studies to investigate the influence of spin paramagnetism on high-magnetic-field superconductivity, and in doing so ushered in the era of spin-dependent tunneling. In a simple picture, one can already imagine how spin should play a role in determining the tunneling current: if the tunnel current is proportional to the density of states, and one presumes a two-current model in a ferromagnetic metal, one should have a spin-dependent tunneling current. It was essentially this fact, combined with the unique properties of thin film Al superconductors to serve as spin detectors, which created the field of spin-dependent tunneling. These pioneering experiments are the fundamental basis for the magneto-resistance effect in magnetic tunnel junctions, as well as many other spin-polarized tunneling phenomena discussed in later chapters. Though the full details of spin-polarized tunneling are beyond the scope of this review (and have been excellently reviewed by Meservey and Tedrow [1]), we will briefly describe the concepts behind this powerful technique.

11.3.1 THE SPIN-POLARIZED TUNNELING TECHNIQUE

The spin-polarized tunneling technique (SPT) developed by Meservey and Tedrow [1] utilizes a superconductor as one electrode in a tunneling device to probe the tunneling spin polarization of electrons in the counter-electrode within a few hundreds of μeV of the Fermi level (i.e., on the order of the superconducting energy gap Δ). If the superconductor is sufficiently thin ($\sim$4–5 nm), such that the orbital screening currents in the superconductor are largely suppressed, the effects of the electron spin interaction with the magnetic field may be observed, resulting in Zeeman splitting of the quasiparticle states in the superconductor. However, not only must the critical field of the superconductor be large enough to have observable Zeeman splitting (Zeeman splitting larger than the thermal smearing, i.e., $3.5\,k_BT < 2\mu_BH_{ext}$) at fields below the superconductor's critical temperature, elements with low atomic number must be used to avoid significant spin scattering in the superconductor via spin-orbit scattering. In practice, this usually means temperatures of $\sim$500 mK, magnetic fields of a few teslas, and typically (but not exclusively) the superconductor is pure or doped Al, since ultra-thin Al films can have critical fields > 5 T, and being a light element, spin-orbit scattering is negligible.

As discussed above, when tunneling into a superconductor, the conductance $dI/dV(V)$ is proportional to the BCS density of states (DOS) [6, 12], and the SPT technique achieves spin sensitivity by applying a magnetic field ($\sim$2 – 3 T) parallel to the film plane. If Zeeman splitting is achieved, the total quasiparticle DOS is a superposition of spin-up and spin-down DOSs, each shifted in energy by $\pm\mu_BB$ from the zero-field curve as shown in Figure 11.3a. The total conductance without spin polarization in the normal metal thus

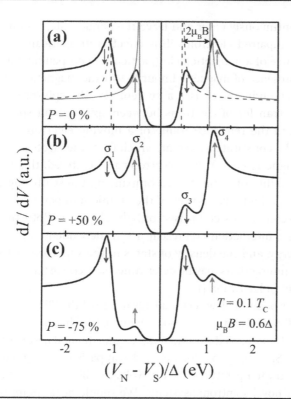

FIGURE 11.3   Conductance (dI/dV) of a S/I/N tunnel junction with Zeeman splitting of the density of states in the superconductor due to a magnetic field B. In (a) the normal metal has zero polarization, $P = +50\%$ in (b), and $P = -75\%$ in (c). The thin solid blue and dashed red lines in (a) show the spin-up and spin-down partial densities of states, respectively. The conductance curves are calculated for $T = 0.1\, T_C$ and $\mu_B B = 0.6\Delta$.

exhibits a four-peaked structure symmetric about $V = 0$. If the tunneling electrons are spin-polarized, however, each of the separate spin channels is weighted by the relative DOS, and $dI/d(V)$ is asymmetric about $V = 0$, Figure 11.3b,c. The heights of the four peaks relate to the spin polarization P. Assuming a proportionality between conductance and the normal metals' densities of states ρ at E_F, the tunneling spin polarization may be *estimated* by the heights of the conductance peaks [27, 28] shown in Figure 11.3, σ_{1-4}:

$$P(E_F) \equiv \frac{\rho_\uparrow - \rho_\downarrow}{\rho_\uparrow + \rho_\downarrow} \approx \frac{(\sigma_4 - \sigma_2) - (\sigma_1 - \sigma_3)}{(\sigma_4 - \sigma_2) + (\sigma_1 - \sigma_3)} . \tag{11.4}$$

The curve in Figure 11.3b has a tunneling spin polarization of $P = +50\%$, while the curve in Figure 11.3c has $P = -75\%$. Thus, not only the magnitude but the sign of the spin polarization may be determined. The difference in conductance peak heights is only an estimate of P, an accurate determination must account for spin-orbit scattering in the superconductor and depairing due to the orbital motion of the carriers [1] (using (0.4) tends to overestimate P). Physically, the depairing corresponds to an interaction between the condensate and unpaired electrons. Due to this interaction, electrons continuously enter and leave the condensate, leading to a finite lifetime for unpaired electrons. This effect is dramatically strengthened by the presence of a magnetic field (particularly perpendicular to the film plane), the presence of magnetic impurities or the magnetic field generated when a superconductor carries a current [1]. The spin-orbit interaction represents a transfer of electrons between spin-up and spin-down subsystems. From the reference frame of a moving electron, the positively charged nuclei constitute a current, which from the electrons' perspective creates a magnetic field. When electrons are scattered, this field leads to a finite probability of a spin-flip accompanying the scattering event. This effect scales as the fourth power of the atomic number, necessitating the use of superconductors comprised of light elements. Both the spin-orbit and depairing interactions represent a spin-dependent correction to the electron energy, and the density of states in the superconductor accounting for both interactions may be determined self-consistently to accurately extract a spin polarization [1].

From our point of view, a crucial advantage of the SPT technique is that a robust, well-tested and microscopic theory of SPT has been developed over the years [1]. This, coupled with the simple, fundamental nature of the experiment makes SPT agreeable for detailed comparison with theory. Though restricted to low temperatures and bias voltages, SPT is a powerful technique for assessing novel spintronic systems. We should again point out that the density of states measured via tunneling (and thus the tunneling spin polarization) is never the "raw" DOS, but always weighted by the barrier transmission probability [29, 30]. For SPT experiments, the transmission factors are essentially evaluated at E_F due to the small bias involved (>1 mVolt), and any energy dependence may be safely ignored. In systems where the spin-dependence of the transmission factors may be ignored, such as disordered Al_2O_3, the tunneling probability is spin-independent and the situation may be somewhat simplified [28]. This is not generally true, however: for ordered MgO, for example (see Chapter 13, Volume 1), the transmission factors are not equal for each spin due to the "symmetry filtering" effect.

For comparison with the theoretical curves, Figure 11.4 shows the first SPT measurement by Meservey and Tedrow [28] using $Al/Al_2O_3/Ni$ junctions. While the zero-field curve is symmetric, the curve for $\mu_o H = 3.37$ T shows an obvious asymmetry, consistent with Figure 11.3c. Though the spin polarization is small in this case, $P_{Ni} = 8.0 \pm 0.2\%$ is obtained from fitting to a model incorporating spin-orbit scattering and orbital depairing in the Al [1], the presence of a finite and *positive* tunneling spin polarization in Ni is clear. Parenthetically, we note that the recent value of tunneling spin polarization for (polycrystalline) Ni (see Table 11.1) is more than a factor of 5 higher, almost certainly due to three decades of improved deposition techniques resulting in cleaner junctions with better interfaces. Table 11.1 lists recently obtained tunneling spin polarization values obtained for several ferromagnetic metals, after correction for spin-orbit scattering. A more extensive list may be found in [31]. Of particular interest is that the sign of the tunneling spin polarization for the $3d$ metals and most alloys is positive. In fact, the

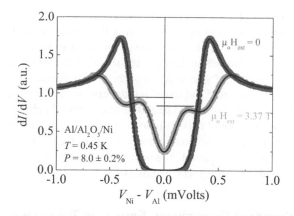

FIGURE 11.4 First observation of spin-polarized tunneling using Ni. Fitting the curves (*lines*) with a model accounting for spin-orbit scattering and orbital depairing [1] gives a spin polarization of $8.0 \pm 0.2\%$. (After Tedrow, P.M. et al., *Phys. Rev. Lett.* 27, 919, 1971.)

TABLE 11.1
Selected tunneling spin polarization values obtained from SPT measurements on FM/I/Al or Al/I/FM junctions, after corrections for spin-orbit scattering and orbital depairing. Crystalline electrodes or barriers are indicated by *

FM	I	P (%)	Ref.
Ni	Al_2O_3	46	[34]
Co	Al_2O_3	42	[35]
$Co_{84}Fe_{16}$	Al_2O_3	55	[35, 36]
$Co_{60}Fe_{40}$	Al_2O_3	50	[36]
$Co_{50}Fe_{50}$	Al_2O_3	55	[35]
$Co_{40}Fe_{60}$	Al_2O_3	51	[36]
Fe	Al_2O_3	44–45	[35]
$Ni_{90}Fe_{10}$	Al_2O_3	36	[36]
$Ni_{80}Fe_{20}$	Al_2O_3	48	[35]
$Ni_{40}Fe_{60}$	Al_2O_3	55	[36]
$SrRuO_3$*	$SrTiO_3$*	−9.5	[32]
$La_{0.67}Sr_{0.33}MnO_3$*	$SrTiO_3$*	78	[37]
CrO_2*	Cr_2O_3	100	[38]
Fe	MgO*	74	[39]
$Co_{70}Fe_{30}$	MgO*	85	[39]

only negative polarizations measured with SPT to date are for $SrRuO_4$ [32] and Co_xGd_{1-x} alloys [33].

Though in retrospect the SPT technique may seem like a natural outgrowth of work on superconducting tunneling in the late 1960s, its development was, in fact, something of a fortuitous accident. Meservey, himself, clarified [40] the original story surrounding the development of SPT:

> What we were doing was actually trying to observe the theoretically predicted paramagnetic critical field of superconducting Al, which seemed to be

attainable because of our ability to make thin enough Al films. It was expected that this result would help in understanding the critical magnetic fields of other superconductors with higher values of spin-orbit scattering.

If you read [41] you will understand the situation. We had spins in mind but had never thought of spin-splitting of the density of states. However, by the time the $x - y$ recorder had traced out the first two peaks, I was calculating what the spin-splitting should be and we knew immediately what we were seeing. I think [41] should make it clear that the paramagnetic critical field was one of the few things not known about superconductors at that time and was the center of much interest. The next and obvious step was to apply the technique to study ferromagnetic metals. [40]

11.3.2 INTERFACE SENSITIVITY, SPIN-FLIP TUNNELING

In the preceding analysis, the determination of P with the SPT technique relied on spin being conserved during the tunneling process. In fact, the Zeeman splitting of the DOS may be also used directly to *prove* that spin flipping does not take place during tunneling. In an elegant experiment [1, 28], Meservey and Tedrow studied spin-polarized tunneling with *two* Zeeman-split superconductors and showed that no spin flipping takes place during tunneling through Al_2O_3, a crucial result for the development of magnetic tunnel junctions. This extension of SPT directly measures the probability of spin-altering events during the tunneling process. Briefly, if no spin flipping or filtering takes place during tunneling, the tunneling $dI/dV(V)$ should display two peaks at $V = (\Delta_1 + \Delta_2)/e$, and will be unaffected by a magnetic field [1, 6, 28]. If the quasiparticle spin changes during the tunneling process, additional conductance peaks are expected at $V = \pm(\Delta_1 + \Delta_2 + 2\mu H)/e$ [1, 28]. The absence of these conductance peaks was used to directly verify that spin is conserved in tunneling through Al_2O_3 barriers.

In another application of this powerful technique, spin polarization measurements on ultra-thin ferromagnetic films [1, 42] beautifully demonstrated the anticipated (in retrospect!) interfacial sensitivity [17, 18, 21] in tunnel junctions with non-superconducting electrodes. In this case, ultra-thin (0–3 nm) films of Fe or Co were deposited on a normal metal (N) layer to form the bottom electrode of the N/F/I/S junctions, on which the Al_2O_3 tunnel barrier and superconducting Al electrodes were grown (Figure 11.5). The spin polarization of electrons tunneling from these ultra-thin layers was then measured as a function of average layer thickness, and compared with the polarization values for thick Fe or Co layers. For Fe and Co backed with (normal-state) Al, show in Figure 11.5 for Co, the polarization increased rapidly with increasing thickness, showing clear spin polarization (and hence, ferromagnetism) for even a single monolayer. The most striking result is that the "bulk" tunneling spin polarization measured on thick layers is nearly reached by only 0.7–1.0 nm, or ~3–5 monolayers. Not only does this experiment clearly indicate that the onset of ferromagnetism is extremely rapid, it definitively demonstrates that tunneling in non-coherent normal-state structures is highly interface sensitive. A more recent experiment using Cu or Au δ-doping layers in Co electrodes corroborates this conclusion [45]: the

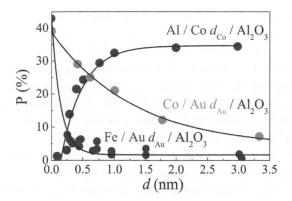

FIGURE 11.5 Blue: spin polarization as a function of thickness for ultra-thin Co inserted at the (normal-state) Al-Al_2O_3 interface. Red: spin polarization of Co as a function of the thickness of an Au layer inserted at the Co-Al_2O_3 interface. Green: spin polarization of Fe as a function of the thickness of an Au layer inserted at the Fe-Al_2O_3 interface. The lines are fits to an exponential decay, which yields length scales ξ of $\xi_{Co-Al} \sim 0.2$ nm, $\xi_{Au-Co} \sim 1.4$ nm, and $\xi_{Au-Fe} \sim 0.2$ nm. (After Tedrow, P.M. et al., *Solid State Commun.* 16, 71, 1975; Moodera, J.S. et al., *Phys. Rev. B* 40(17), 11980, 1989; Gabureac, M.S. et al., *J. Appl. Phys.* 103(7), 07A915, 2008.)

3–5 monolayers adjacent to the insulating barrier appear to dominate the transport properties, in agreement with theoretical expectations [17, 18, 21].

Moodera et al. [43] first probed the opposite side of the question, *viz.*, how far away from the barrier interface may the ferromagnetic layer be to still observe spin polarization? They used SPT to measure the spin polarization in Al/Al_2O_3/Au/Fe junctions as a function of Au interlayer thickness. Consistent with the results of Meservey and Tedrow, they found that the polarization was rapidly quenched for the first two monolayers Au, but a small polarization persisted over even 10 nm Au. A recent follow-up to this experiment by Gabureac et al. [44] probed the spin polarization transmitted across an interfacial Au layer between Co and Al_2O_3, finding a somewhat longer length scale (~1.4 nm). Further studies with both Au and Cu interfacial layers showed that while the decay of spin polarization is rapid, measurable polarization could be observed through even ~6 nm interlayers that were shown by TEM to be continuous [45]. However, as with later experiments involving thin interface layers in MTJs (see Section 11.4.10), we caution that without detailed structural analysis of the interlayers, the absolute numbers should be regarded with caution, since effects such as intermixing and three-dimensional growth may play an important role [24].

11.3.3 WHAT IS TUNNELING SPIN POLARIZATION?

Originally, the positive (i.e., majority) tunneling spin polarization measured for Fe, Ni, and Co was a bit of a puzzle theoretically, since these materials have a density of states which is strongly minority-dominated near the Fermi level. One early explanation was given by Stearns [46]. She recognized that within the independent electron model of tunneling [15], the transmission probability depends on the electron effective mass, which is different for different bands. More generally, the transmission probabilities in Equation 11.1 are different for states of different symmetry and depend on the nature of the evanescent states in the insulator [22, 47–50]. In Fe, Co, and Ni, the relatively localized ("d-like") electrons carry the vast majority of the total magnetic moment, while the itinerant ("s-like") electrons contribute little. However, the localized electrons necessarily have a large effective mass and would be expected to decay extremely rapidly into a vacuum barrier, compared to the mobile hybridized electrons. Since the itinerant bands have a spin polarization opposite in sign, the positive tunneling spin polarization can be rationalized in the case of a vacuum barrier, in spite of a minority-dominated total density of states. Calculations based on a tight binding model through a vacuum barrier [35] essentially reproduce this simple picture (as do different treatments by Mazin [30] and Butler et al. [47]).

However, this picture neglects the electronic structure of the insulating barrier and bonding at the ferromagnet-insulator interface. Based on the discussions above, one expects that the tunnel current is sensitive to the *interfacial* DOS at the ferromagnet-insulator barrier, which is strongly modified by the presence of interface bonding. Tsymbal and Pettifor [20] first addressed the issue, finding that the tunnel conductance strongly depends on the type of bonding present. They showed that different interface bonding between Co or Fe and the insulating barrier could result in either positive or negative tunneling spin polarization. Indeed, experimentally both positive and negative spin polarizations have been measured for Co, depending on the type of insulating barrier used (see Section 11.4.9).

For coherent systems, such as fully epitaxial tunnel junctions, $k_{\parallel}$ conservation must be considered in the tunneling process [8], and the situation is even more complex. As *ab initio* calculations have demonstrated (Chapter 13, Volume 1), the idea of a "tunneling spin polarization" then tends to lose meaning, with the transport characteristics becoming a rather complex function of k, barrier thickness, and the symmetry of the Bloch states involved. In realistic systems, ordered or otherwise, the tunneling spin polarization is not simply a property of the electrode alone or an electrode-barrier combination but is (at least) a property of the electrode-barrier interface, and generally the electronic structure of the entire M-I-M system. We refer to Chapter 13, Volume 1, for more details.

11.4 MAGNETIC TUNNEL JUNCTIONS

The dependence of tunneling current on the density of states, coupled with spin conservation during tunneling, makes it reasonable to anticipate that interesting spin-dependent tunneling effects may be observed without resorting to the use of a superconducting spin detector. In 1975, Julliere [3] and Slonczewski [2] had exactly this idea in mind. Rather than use the Zeeman-split density of states in a superconductor as a spin detector, one should be able to make use of the exchange-split density of states in a second ferromagnetic electrode. In this case, tunneling between two ferromagnets, the tunnel current is expected to depend on the relative magnetization orientation of the two ferromagnetic electrodes, giving rise to a magnetoresistance – the tunneling magnetoresistance (TMR) effect.

In this section, we will introduce the basis for the TMR effect and its characteristic dependencies, for instance, the dependence of the TMR and conductance on applied bias, temperature, and materials choice. Though some of these features may be readily understood from the previous sections (for instance, the magnitude of the TMR and its angular dependence), others are more subtle and their origins have not been definitively determined (for instance, the bias and temperature dependence). Nevertheless, we will outline the main features of MTJs and the fundamental physics behind them as they are now understood.

11.4.1 BASIS FOR THE TMR EFFECT

Following Meservey and Tedrow's model of spin-polarized tunneling, Julliere [3] first proposed a model for tunneling between ferromagnets which we sketch here. First, we consider two identical ferromagnetic electrodes with parallel magnetizations, separated by an insulating barrier. Assuming spin conservation during the tunneling process (Section 11.3.1), we use a two-current model, and tunneling may only occur between bands of the same spin orientation in either electrode. That is, tunneling must proceed from a filled up spin band in one electrode to an empty up spin band in the other, and likewise for the spin-down electrons. In this simplest case, we assume that the tunneling *probability* is itself spin-independent, and we worry only about the spin-dependent density of states in determining the tunneling current (in Chapter 13, Volume 1, we will see how band symmetries may effectively provide a spin-dependent tunneling probability). We also presume a simple proportionality between the tunneling conductance and the product of the spin-dependent densities of states in the two electrodes, with vanishing bias such that the energy dependence of the density of states may be neglected. With these assumptions, one arrives at Julliere's formula for tunneling magnetoresistance, defined as the difference in *conductance* between parallel and antiparallel magnetization states, normalized by the antiparallel conductance:

$$G_\mathrm{p} \propto \rho_{l,\uparrow}\rho_{r,\uparrow} + \rho_{l,\downarrow}\rho_{r,\downarrow}$$

$$G_\mathrm{ap} \propto \rho_{l,\uparrow}\rho_{r,\downarrow} + \rho_{l,\downarrow}\rho_{r,\uparrow}$$

$$\mathrm{TMR}\,|_{V=0} \equiv \frac{G_\mathrm{p} - G_\mathrm{ap}}{G_\mathrm{ap}} = \frac{R_\mathrm{ap} - R_\mathrm{p}}{R_\mathrm{p}} = \frac{2 P_l P_r}{1 - P_l P_r}. \tag{11.5}$$

Where $\rho_{l(r),\uparrow(\downarrow)}$ is the density of states for up ($\uparrow$) or down ($\downarrow$) spins in the left (l) and right (r) electrodes, $P_{l(r)}$ is the tunneling spin polarization in the left (right) ferromagnetic electrode, $G_{p(ap)}$ is the tunneling conductance dI/dV and $R_{p(ap)}$ the tunneling resistance in the parallel (antiparallel) state. The conductances are different for parallel or antiparallel magnetization orientations, and we expect MTJs to display a magnetoresistance, so long as there is a net spin polarization at the Fermi level in both electrodes.

Once again, we must keep in mind that even when the tunneling probability is assumed to be spin- and energy-independent (which ignores any symmetry-based spin filtering, see Chapter 13, Volume 1), P is not simply the difference in total density states at the Fermi level, but is still determined by, e.g., the effective masses of different band electrons, interface bonding, etc. We also note parenthetically that in the tunneling literature "conductance" is defined as $G \equiv dI / dV$, rather than I/V as is common in some fields. The two are not identical except at vanishing bias since tunnel junctions have non-linear current-voltage characteristics.

11.4.2 EARLY MTJ EXPERIMENTS

Before we discuss the characteristics of MTJs in more detail, we will first review some of the early experiments. In 1975, Julliere [3] reported the first observations of tunneling magnetoresistance in Fe/Ge/Co tunnel junctions. By achieving different coercive for the Fe and Co layers, he was able to realize both parallel and antiparallel magnetization orientations. At zero-bias, the maximum observed effects were ~14% but decreased rapidly with increasing bias voltage. Though these results were indeed groundbreaking and stimulated much future research, they were not reproduced by other workers, and their true interpretation is still the subject of debate. Later spin-polarized tunneling experiments showed very little spin conservation when tunneling through amorphous Ge or Si barriers [1]. Whatever the interpretation, Slonczewski [2] quickly realized the potential of the TMR effect, and others soon followed this research.

Maekawa and Gäfvert [51] successfully and unambiguously demonstrated TMR effects in 1982, using Ni/NiO/Ni and Ni/NiO/Co junctions. For Ni/NiO/Co junctions at 4.2 K, ≈2% magnetoresistance was observed, which was, for the first time, clearly correlated with measured hysteresis behavior of the magnetic electrodes. The effects were still small compared to those anticipated, but this work was the first to establish that the magnetoresistance effects were due to the relative magnetization alignment between the two electrodes.

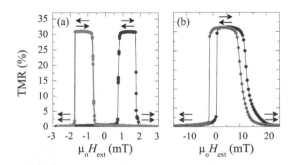

FIGURE 11.6   Magnetoresistance vs. magnetic field for (a) a hard-soft MTJ and (b) an exchange biased MTJ, both at 10 K. The field is swept from negative to positive (blue) and then positive to negative (red). Horizontal arrows represent the relative magnetization orientations of the magnetic layers. Both curves are taken with dc voltage bias well below $k_b T$. (After LeClair, P. et al., in Heinrich, B. et al., Eds., *Ultrathin Magnetic Structures, Vol. III*, Springer-Verlag, Berlin, Germany, 2005.)

In 1995, nearly 20 years after the original "discovery" of the TMR effect, Moodera et al. [4] and Miyazaki and Tezuka [5] solved most of the earlier problems with MTJs and independently demonstrated >10% TMR at room temperature. These demonstrations quickly garnered a great deal of attention and catalyzed many groups to investigate MTJs. The work by Moodera et al. [4] already established many of the basic features associated with MTJs, most notably, the bias and temperature dependence of the TMR, which are discussed in Sections 11.4.7 and 11.4.8.

11.4.3 RESISTANCE VS. FIELD

Julliere's simple model for the TMR effect was discussed previously in this section, with the result that the conductance (resistance) of a MTJ *with spin polarizations of the same sign for both electrodes* is higher (lower) when the magnetizations of the two ferromagnets are parallel. Once said, one must realize experimentally both a parallel and antiparallel magnetization alignment. Perhaps the simplest way to realize this is to use two ferromagnets with different coercive fields (for example by using two different ferromagnets), as shown in Figure 11.6a). Starting from saturation at negative fields, the electrodes have parallel magnetizations, and a low resistance state is realized. As the field is reduced to zero and increased past zero, the electrode with the lower coercive field reverses its magnetization first, yielding an antiparallel alignment of the magnetizations and a high resistance state. As the field is increased further, the electrode with the higher coercive field also reverses its magnetization, and a parallel alignment is again achieved. Were the two electrodes to have opposite spin polarizations (one positive and one negative), the parallel state would be the high resistance state and the antiparallel state the low resistance state.

More relevant for sensor or MRAM technology are exchange biased MTJs. In this case, one of the magnetic electrodes is in direct contact with an antiferromagnetic (AFM) material (e.g., FeMn, IrMn, NiO). The presence of an exchange anisotropy at the FM/AFM interface shifts the entire magnetization-field loop of the ferromagnet away from zero-field. Typical TMR vs. magnetic field behavior for an exchange biased system (at $V = 0$) is shown in Figure 11.6. The exchange biased MTJ is the most commonly studied configuration by far. Technologically, exchange biasing is advantageous because the resistance transition takes place near zero magnetic field, and it generally results in greater magnetic stability [52], while from a fundamental point of view, it allows one to study MTJs with nominally identical electrodes. For more details, we refer to Chapter 2 and Chapter 4 in Volume 1.

11.4.4 FERROMAGNET DEPENDENCE

Using the measured tunneling spin polarization for Co with Al_2O_3 tunnel barriers (see Table 10.1), we may expect a TMR effect of more than 40% for Co/Al_2O_3/Co MTJs, only slightly above the observed (low temperature) value [53]. In general, the most recent spin polarization values with Al_2O_3 barriers [35, 36] obtained via the SPT technique agree well with the maximum

TMR values reported with Al_2O_3 barriers [31, 34, 36]. Clearly, one expects the largest TMR values for materials with the largest tunneling spin polarization. This explains a great deal of the recent interest in so-called "half-metallic" ferromagnets, materials for which only one spin band is occupied at the Fermi level, resulting in perfect 100% spin polarization [55]. Many compounds have been predicted to be half-metallic; however, CrO_2 [38] and $La_{0.67}Sr_{0.33}MnO_3$ [56] are somewhat unique in that they have been shown to be essentially half-metallic by SPT and TMR experiments, respectively. While recently very large TMR effects have been recently observed in MTJs with Heusler alloy electrodes, it is unclear in most cases if the high TMR is due to a high polarization in the Heusler alloy, or a spin filtering effect in the MgO tunnel barriers typically used (see Chapter 12 and Chapter 13 in Volume 1).

Not long after the demonstrations of Moodera and Miyazaki, Lu et al. [57] and Viret et al. [58], observed TMR effects of more than 400% at a low temperature utilizing LSMO with $SrTiO_3$, $PrBaCu_{2.8}Ga_{0.2}O_7$, or CeO_2 barriers. Using Equation 11.5, this implies a spin polarization of more than 80%, in agreement with SPT experiments [37]. More recently, Bowen et al. [56] observed more than 1800% TMR in $LSMO/SrTiO_3/LSMO$ junctions, implying a spin polarization of 95% based on Jullieres model and essentially corroborating photoemission results showing LSMO to be half-metallic. While these TMR results, particularly those of [56], currently exceed anything observed in more traditional MTJs (including MgO-based junctions), the fact that the magnetic ordering temperatures of these oxide ferromagnets are well below room temperature prevents practical applications, and perhaps, as a result, there has been less work in this area compared to Al_2O_3- or MgO-based systems.

In more traditional systems, the search for high spin polarization has often focused on analogies with bulk magnetization (perhaps for lack of a better rationale, owing to the complexity of P). In fact, the relationship between the total moment of the ferromagnetic electrode and the tunneling spin polarization has been a point of controversy since the advent of spin-polarized tunneling. Early experiments by Meservey and Tedrow [1] on various transition-metal alloys apparently showed a simple scaling relationship between the magnetic moment and polarization. More recent studies with improved deposition techniques and cleaner junctions (e.g., [34, 36]) have found essentially no correlation. When the tunnel current is dominated by itinerant electrons, as for Al_2O_3 barriers, while localized electrons provide most of the magnetic moment, relationships between P and, e.g., magnetic moment are not straightforward. Measurements on the NiFe [59] and CoFe systems showed a weak dependence of P on magnetic moment, implying no direct correlation. However, in the CoMn [60] system, polarization and moment were found to be directly related.

Kaiser et al. [61] studied CoV and CoPt alloys and suggested that the weaker composition dependence and higher P in CoPt alloys compared to CoV alloys could be explained by different interface bonding between V and Pt to Al_2O_3, leading to different tunneling rates from different alloy sites. A higher tunneling rate from Pt sites, coupled with the tendency of Pt to polarize in a ferromagnetic matrix, leads to a weaker relative dependence of P with increasing Pt content than the magnetization. On the other hand, CoV alloys showed an even stronger decrease of P with increasing V content than the magnetization, suggesting that the ease of formation of chemical bonds between alloy constituents and the barrier played a primary role in determining P.

In an elegant follow-up study, Kaiser et al. [33] studied ferrimagnetic $Co_{1-x}Gd_x$ alloys, and their work strongly suggested that any relationship between magnetic moment and polarization is purely coincidental. In the CoGd system, one has antiferromagnetically coupled sublattices of Co and Gd, the former giving rise to a much larger tunneling spin polarization, while the later provides a larger magnetic moment. At a critical composition x, the compensation point, the two sublattices give a zero net magnetic moment. However, Kaiser et al. showed that due to the far larger spin polarization of the Co sublattice, *a strongly spin-polarized current could be observed even with zero net magnetization.* Depending on the composition, both positive and negative polarizations could be observed. Using Gd/Co nanolayers, Min et al. [62] came to similar conclusions, finding that the sign of the tunneling spin polarization depended on Gd and Co layer thicknesses, temperature, and applied voltage. After a somewhat colorful history (perhaps highlighted by the recent interest in "zero moment" half metals [63]), it seems that one is forced to retain the original conclusion that magnetic moment and spin polarization have no obvious relationship, and further that the "spin polarization" is not a unique number even for a given metal electrode and tunnel barrier, but a property of the electronic structure of the entire system under study.

11.4.5 CRYSTALLOGRAPHIC ORIENTATION DEPENDENCE

Given the dependence of TMR on the interfacial density of states, one would anticipate a dependence on the crystallographic orientation of the electrodes. However, MTJs with even a single epitaxial layer with different orientations were not realized until 2000. Yuasa et al. [64] prepared Fe(100,110,211)/Al_2O_3/CoFe MTJs with a bottom epitaxial Fe layer to study the effect of the Fermi surface anisotropy on TMR. They observed a strong dependence of the TMR on crystallographic orientation, as shown in Figure 11.7. The largest TMR effect was observed for Fe(211) electrodes, Fe(100) electrodes yielded a TMR effect nearly a factor of 4 lower, while Fe(110) electrodes showed an intermediate TMR value. The fact that the TMR varies so strongly with crystallographic orientation clearly points to the details of the Fe band structure [65] and momentum filtering, which perhaps foreshadows the "Giant" TMR effects [39, 66] first observed in Fe-MgO-Fe MTJs (see Chapter 12 and Chapter 13 in Volume 1).

More recent results by Kim and Moodera [34] compared spin-polarized tunneling into polycrystalline and epitaxial Ni films. For Ni(111), a spin polarization of 25% was measured, despite the fact that bulk band-structure calculations for Ni predict a low or even negative polarization along the (111)

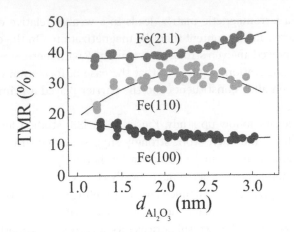

FIGURE 11.7   TMR at 2 K as a function of Al₂O₃ thickness for Fe(211), Fe(110), and Fe(100) epitaxial electrodes in Fe/Al₂O₃/CoFe junctions. Lines are a guide to the eye. (After Yuasa, S. et al., *Europhys. Lett.* 52(3), 344, 2000.)

direction [67]. In addition, a large increase of the polarization compared to previous studies was observed for polycrystalline Ni films (P = +46%) when the Ni surface/interface was contamination free. The dramatically large value of P observed for polycrystalline Ni is in the same range as that for Fe and Co, whereas the magnetic moment is only 0.60 μB for Ni, compared to 2.2 μB for Fe. This again shows the lack of a direct correlation between P or TMR and magnetic moment, and that the factors controlling P (e.g., the *sp*-projected density of states and the nature of the evanescent states in the barrier region [47]) are far subtler than those controlling magnetization.

11.4.6 Conductance vs. Voltage

As we have seen previously in Section 11.1, tunneling between two non-magnetic free-electron metals should result in a conductance (dI/dV) which is quadratic in applied bias (V). Indeed, for non-magnetic junctions such as Al/Al₂O₃/Al, this behavior is often observed [6]. However, in MTJs (or even junctions with one magnetic electrode), the conductance behavior deviates significantly from the expected parabolic dependence, as shown in Figure 11.8a for a Co/Al₂O₃/Co junction (with two polycrystalline electrodes), most noticeably at low voltages. Though the conductance is approximately symmetric with respect to voltage, as expected for (nominally) identical electrodes, there is a pronounced, sharp decrease below ~150 mVolts (the exact value is material-dependent), where the conductance is approximately linear in voltage. This sharp dip in conductance about zero-bias is similar to so-called "zero-bias anomalies" observed by many groups [6], which were attributed to (magnetic) impurities in the tunnel barrier or near the electrode-barrier interfaces. However, in contrast to those zero-bias anomalies, the conductance dip has no strong temperature dependence or magnetic field dependence, persisting even in the cleanest junctions. Further, the energy scale of the anomaly is far greater than what is usually observed in junctions with magnetic impurities. Other mechanisms, such as the presence of metal

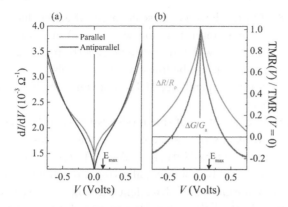

FIGURE 11.8 (a) Conductance vs. applied bias, dI/dV(V), at 5 K for parallel (*red*) and antiparallel (*blue*) magnetizations of a Co/Al$_2$O$_3$/Co MTJ. The vertical arrows indicate the maximum magnon energy (E_{max}) for bulk Co in a mean-field approximation. (b) Bias dependence of the TMR ($\Delta R/R_p$, *orange*) and differential TMR ($\Delta G/G_{ap}$, *purple*) for a Co/Al$_2$O$_3$/Co junction at 5 K. Note that the normal ($\Delta R/R_p$) and "differential" ($\Delta G/G_{ap}$) magnetoresistances are only equivalent at *zero-bias*. (After LeClair, P. et al., *Phys. Rev. Lett.* 88, 107201, 2002 and unpublished data from the authors.)

particles in the barrier (Giaever-Zeller anomalies [6]) are similarly inconsistent. Further, the characteristic (nearly) linear conductance contribution near zero-bias occurs *only* when at least one electrode is magnetic [54, 68, 69]. A linear conductance contribution suggests a continuous spectrum of inelastic excitations, [6] with an upper-energy cutoff of ~150 meV and a lower-energy cutoff below 1 meV, likely in the μeV range.

Moodera et al. [71] initially suggested that magnons excited by tunneling electrons may be at least partly responsible for the conductance anomalies as well as the TMR bias dependence. A theoretical explanation based on this idea was given by Zhang et al. [72] and Bratkovsky [73] shortly thereafter in terms of magnon excitations localized at the FM-Al$_2$O$_3$ interface. Electrons at the Fermi level of one ferromagnet tunnel into the second ferromagnet with an energy eV above the Fermi level (assuming no other inelastic tunneling processes). These "hot" electrons may then lose energy by emitting a magnon of energy $\hbar\omega \leq eV$. A similar process holds for magnon absorption, and as one might expect this leads to a decrease of the TMR with increasing bias. Zhang et al. [72] and Bratkovsky [73] found that this gives an additional inelastic conductance contribution to the conductance $G_{magnon} \propto V$. The slope of the linear contribution is larger in the antiparallel case, with the difference between parallel and antiparallel slopes dictated by the spin polarization. The linearity reflects the fact that a larger bias allows more magnons to be excited, and it is maintained until the eV reaches the maximum magnon energy in the ferromagnetic electrode. Within a mean-field approximation, the maximum magnon energy is given by $E_{max} = 3k_BT_C/(S+1)$, where k_B is Boltzman's constant, and T_C is the Curie temperature of the ferromagnet with spin S. Considering two identical Co electrodes with the bulk T_C, E_{max} ~ 144 meV. Given that the interface T_C is expected to be significantly lower, this is in fairly good agreement with the width of the conductance dip shown in Figure 11.8a. On the low-energy scale, a cutoff arises due to finite size effects (imposing a maximum magnon wavelength) or anisotropy, both in the μeV range and far below the thermal energy at the temperature of typical experiments. For this reason, the low-energy cutoff plays a more important role in the temperature dependence of TMR at vanishing bias as opposed to the bias dependence.

For a detailed application of this model, we refer to Han et al. [54, 69]. The primary conclusions of that study were that by *combining* different MTJ characteristics (such as $dI/dV(V, T)$, $\Delta R/R_p(V, T)$), a consistent set of all model parameters could be obtained, allowing a much more stringent test of the magnon excitation model. Excellent agreement between experimental data and model calculations were demonstrated, suggesting that indeed the additional conductance contribution consistently observed in MTJs (the zero-bias "dip") can be largely, if not completely, explained by interfacial magnons excited by hot tunneling electrons.

One further question with regard to conductance-voltage behavior that frequently arises is the role of the density of states in the tunnel conductance. Given the form of (11.1), which explicitly contains the density of states as a function of energy in each electrode, it is rather surprising that clear observations of the density of states and band-structure effects have *not* been reported until recently [74–76]. One probable reason for this is that in Al_2O_3-based junctions, one is preferentially sensitive to the itinerant density of states, which by its dispersive nature will have little energy dependence [20, 77]. By comparing MTJs with fcc(111)-textured and polycrystalline Co electrodes, LeClair et al. [70] have observed features in the conductance and TMR bias dependence which were attributed to the s-projected density of states of fcc(111)-textured Co electrodes, essentially corroborated by theoretical results.

11.4.7 TMR vs. Voltage

Perhaps the most surprising feature of MTJs initially was the dependence of the TMR effect on applied dc bias, first observed by Julliere [3] and confirmed by Moodera et al. [4]. It was at initially unclear if the bias dependence was an intrinsic effect, or simply due to inelastic tunneling through a non-ideal a barrier and interfaces (e.g., impurity-assisted tunneling). However, subsequent measurements on clean junctions, as well as the observation of this effect by many other groups in the years that followed, led to the conclusion that the bias dependence of the TMR is truly an *intrinsic* effect [68], though its magnitude may vary considerably. A customary figure-of-merit is the voltage at which the TMR ($\Delta R/R_p$) is reduced by a factor of two, usually denoted $V_{1/2}$. Moodera et al. in their initial observation [4] found a "half-voltage" of ~200 mVolts, while now values of >500 mVolts are not uncommon. Moodera et al. [68] also showed early on that the degree of bias dependence is relatively temperature independent.

Figure 11.8b shows the TMR ($\Delta R/R_p$) vs. bias behavior for a $Co/Al_2O_3/Co$ junction at 5 K [29], showing both $\Delta R/R_p$ and $\Delta G/G_{ap}$. Note that these two definitions of the TMR are *not* identical at finite bias since the tunnel current is a non-linear function of voltage. To avoid confusion, we will refer to $\Delta R/R_p$ as simply TMR, following the majority of workers, and to $\Delta G/G_{ap}$ as the *differential* TMR. The differential TMR rapidly decays with bias up to ~0.5 Volts [78], becoming negative at higher biases and then tending to zero. The normal TMR shows a roughly parabolic voltage dependence at intermediate biases (~0.5 Volts) and tends smoothly to zero at higher biases, with a half-voltage of typically ~0.3–0.5 Volts. One insight into its possible origin is the fact that the differential TMR has exactly the same voltage dependence as the magnon-assisted conductance contribution [72, 73] discussed in the previous section.

The aforementioned models of Zhang et al. [72] and Bratkovsky [73] were originally proposed to explain the bias dependence of the TMR. We have seen that the magnon-assisted conductance contribution is linear in voltage, but with a differing slope for parallel and antiparallel magnetization orientations. Thus, the conductance change ΔG from parallel to antiparallel orientations will also be approximately linear in voltage up to $eV > E_{max}$, roughly as observed in $\Delta G/G_{ap}$ (after further analysis to remove other conductance contributions). The fact that both the conductance-voltage and magnetoresistance-voltage agree favorably with model calculations, and further, yield realistic parameters [54, 69], emphasizes that magnon excitations may play a dominant role in the bias dependence. However, several other possible origins have also been proposed and must be considered.

Davis and MacLaren [79] proposed that the bias dependence also has a component resulting from the underlying electronic structure. They considered free-electron tunneling through a square potential barrier, using simple parabolic bands modeled on the itinerant electron bands in Fe determined from *ab initio* calculations. Two primary electronic structure effects were considered, *viz.*, shifting of the Fermi level of one electrode relative to the other, and the altered barrier shape at finite bias. The shifting of the chemical potential allows new states to be accessed for tunneling, which essentially makes the spin polarization a voltage-dependent quantity, whereas the altered barrier shape allows higher-energy states to tunnel more easily. This makes the matrix elements in (11.1) both spin and bias dependent. A strong decrease of the TMR with bias is also found within their model, and also gives good agreement with experimental data with realistic parameter choices. Further, early work by Moodera et al. [4, 68], as well as experiments utilizing composite insulating barriers [74–76] and ultra-thin non-magnetic layers at the FM/Al_2O_3 interface [26], have pointed out the importance of a bias-dependent spin polarization, as will be discussed in Section 11.4.9 and Section 11.4.10. In particular, at large biases ($eV > E_{max}$) one must consider alternative mechanisms to a purely magnon-assisted bias dependence. As one example, the effects of phonon [73] and impurity [80, 81] assisted tunneling on the bias dependence of the TMR have also been considered theoretically. Experimental support exists for both magnon-assisted and intrinsic band-structure mechanisms of bias dependence as well as for an impurity-assisted contribution [82, 83]. The currently accepted view is that all of these mechanisms play a key role to some degree.

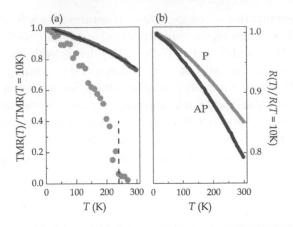

FIGURE 11.9   (a) Temperature dependence of the TMR for a Co/Al$_2$O$_3$/Co MTJ (purple), along with a fit to the model of Shang et al. [84] (*line*) and a Cu$_{38}$Ni$_{62}$/Al$_2$O$_3$/Co junction (orange). Dashed vertical line shows the Curie temperature of Cu$_{38}$Ni$_{62}$ ($T_C \approx 240$ K, note $T_C = 1388$ K for fcc-Co [87]). (b) Temperature dependence of the normalized tunnel resistance for parallel (P, *red*) and antiparallel (AP, *blue*) magnetization orientations, R_p, for a Co/Al$_2$O$_3$/Co MTJ (note the expanded vertical scale). All curves were taken with a dc bias of $V \approx 5$ mV. (After Davis, A.H. et al., *J. Appl. Phys.* 89, 7567, 2001; Hindmarch, A.T. et al., *Phys. Rev. B* 72(10), 100401, 2005.)

11.4.8 TMR vs. Temperature

It was first noticed by Shang et al. [84] that the temperature dependence of the tunnel resistance for magnetic tunnel junctions greatly exceeds that for non-magnetic junctions with nominally identical barriers. Typically, Al/Al$_2$O$_3$/Al junctions showed only a 5–10% [84] change in resistance from 4.2 to 300 K, while magnetic tunnel junctions always exhibited a 15–25% change in resistance, as shown in Figure 11.9 for a Co/Al$_2$O$_3$/Co junction. Further, the TMR ($\Delta R/R_p$) can change as much as 25% or more from 4.2 to 300 K depending on the magnetic electrodes (also shown in Figure 11.9). Shang et al. explained these results within a simple phenomenological model, in which it was assumed that the tunneling spin polarization P decreases with increasing temperature due to spin-wave excitations, as does the surface magnetization. They thus assumed that both the tunneling spin polarization and the interface magnetization followed the same temperature dependence,

the well-known Bloch $T^{\frac{3}{2}}$ law, i.e., $M(T) = M(0)\left(1 - \alpha T^{\frac{3}{2}}\right)$ for low T/T_C.

This temperature dependence holds for surfaces as well as the bulk, though the former has a larger decay constant α [85]. Given the interfacial sensitivity of tunneling, Shang et al. [84] assumed that α was also larger than the bulk value in MTJs, and thus gave a satisfactory explanation for the temperature dependence of the TMR, as demonstrated by the fit in Figure 11.9.

MacDonald et al. [86] provided a more rigorous theoretical justification of these ideas, essentially reproducing the proportionality between $M(T)$ and $P(T)$, though the microscopic origin was slightly different than that considered by Shang et al. The aforementioned model of Zhang et al. [72] also predicts a temperature dependence of TMR, and in contrast to their explanation of the bias dependence of TMR, it is no longer the maximum magnon energy, E_{max}, which sets the relevant energy scale, but rather the lower wavelength cutoff in the spectrum, E_C. This cutoff results physically either from anisotropy, always present for the spins near the FM-I interface, or from a finite coherence length, due to, e.g., a finite grain size. In this model, the zero-bias conductance scales with temperature roughly as $G(T) \approx T \ln(k_B T / E_c)$.

Davis et al. [88] also attempted to put the model of Shang et al. on firmer theoretical ground, somewhat in the spirit of the model of Zhang et al. [72], by assuming that the temperature dependence of the polarization arises from a Stoner-like collapse of the exchange splitting. As with their model of the bias dependence [79], they used free-electron-like bands modeled on the *ab initio* band structure of Co, generating exchange-split parabolic bands with a spin-dependent effective mass. Photoemission data [90] indicate that these itinerant bands are Stoner-like, i.e., the exchange splitting may depend on temperature, and collapses at T_C. Further, citing the fact that the exchange splitting is nearly proportional to $M(T)$, they used the known $M(T)$ behavior of Co to obtain the temperature dependent exchange splitting: $\Delta E_{ex} = \beta M(T)$. Using the bulk $M(T)$ behavior of Co, Davis et al. were able to explain only about a third of the experimental drop in TMR from 0 to 300 K. This is not surprising, given that the interface T_C is expected to be significantly lower [85]. Strikingly, however, they calculated an 18% change in TMR despite the fact that the magnetization changes by only 1.5% over the same temperature range, once again demonstrating that in general P and M are not simply related, though their temperature dependence may share a common origin. The TMR temperature dependence over this range could be well-described by assuming a lower interface ordering temperature for Co, viz., $T_C \approx 982$ K, though this is perhaps unphysically low. As with their work on the TMR bias dependence, the main message is that purely *intrinsic* band effects are able to explain much of the TMR temperature dependence, without resorting to inelastic processes. Of course, in realistic devices, the role of impurity-assisted tunneling must also be considered [82, 83]. The current opinion is generally that both intrinsic (e.g., band effects) and extrinsic (inelastic processes) contributions are of importance.

Several years later, Hindmarch et al. [89, 91] performed a definitive study of the temperature dependence of TMR all the way through the Curie point of a $Cu_{38}Ni_{62}$ alloy, Figure 11.9a. They interpreted their results as being consistent with a Stoner-like collapse of the exchange splitting of the bands responsible for tunneling, lending support to the model of Davis et al. [88]. Further, the normalized TMR versus temperature was found to be independent of applied bias, collapsing onto a "universal" curve. Put another way, the applied bias seems to affect only the *relative magnitude* of the TMR and spin polarization, not the functional form of the temperature dependence. While spin polarization and magnetization versus temperature were clearly *related* in this study, a simple linear relationship was lacking. Their results once again suggest that magnetization and polarization are not simply related. Further, it seems that while the bias and temperature dependence may share a common origin in magnetic excitations, for all intents and purposes they may be treated independently of one another. This does not seem inconsistent with theoretical arguments: the excitation of magnons by finite temperature or hot electrons are independent of one another, and from any bias dependence of the polarization. More bluntly, that these quantities have the same temperature dependence is merely a reflection of the fact that the same physical process causes both M and P to decrease with temperature, viz., magnetic disorder due to thermal excitation of spin waves.

11.4.9 BARRIER DEPENDENCE

The bulk of the early work on MTJs focused almost exclusively on Al_2O_3 tunnel barriers, for a variety of reasons. Perhaps most important were: the ease in fabricating ultra-thin, pinhole-free Al_2O_3 layers; spin conservation has been demonstrated across Al_2O_3 barriers [1]; and the first successful demonstrations of the TMR effect used Al_2O_3 barriers [4, 5]. Significant effort has been made over the last several decades to characterize, understand and improve properties of amorphous alumina barriers. On the other hand, there is little doubt that a myriad of alternative barriers have been explored by many groups (the authors can attest to this) – the lack of evidence for this in literature may simply be due to the fact that successful MTJ tunnel barriers are a rare find, and one tends not to report unsuccessful attempts! Nevertheless, numerous alternative barriers have been successfully employed over the years, some with very distinct behavior compared to Al_2O_3. Of course, the alternative that truly transformed the field of spintronics was epitaxial MgO, discussed in Chapter 12 and Chapter 13 in Volume 1.

As discussed previously, the magnitude and even the sign of the tunneling spin polarization are expected to depend on not only the ferromagnetic electrode but the barrier and nature of the electrode-barrier interface as well. In 1999, two different groups [74–76] independently observed this effect [20]. Later theories confirmed this prediction [48, 77, 92, 93] and demonstrated the possibility of creating novel MTJ properties by not only varying the magnetic electrode but by varying the insulating barrier as well.

In a remarkable series of experiments, de Teresa et al. [74, 75] definitely showed that the tunneling spin polarization depends explicitly on the insulating barrier used. In this case, half-metallic $La_{0.7}Sr_{0.3}MnO_3$ (LSMO) was used as one of the electrodes with barriers of Al_2O_3, $SrTiO_3$ (STO), $Ce_{0.69}La_{0.31}O_{1.845}$ (CLO), or a composite $Al_2O_3/SrTiO_3$ barrier. Since it is known that (ideal) LSMO has only majority states at E_F, the tunneling spin polarization must be positive and close to 100% [37, 94], regardless of the insulating barrier used. Indeed, SPT experiments have measured a tunneling spin polarization of $P = +78\%$, and TMR effects in LSMO-based devices have reached 1800% as noted above. Thus, a LSMO electrode can serve as a spin analyzer for the polarization of a second electrode, in the spirit of the SPT technique of Meservey and Tedrow (see Section 11.3.1). Moreover, a half-metallic analyzer has the added advantage that to some extent the energy dependence of the polarization may be studied.

As expected, de Teresa et al. found that $Co/Al_2O_3/LSMO$ MTJs gave a positive TMR for all biases, not surprising since both LSMO and the Co/Al_2O_3 interface are known to have positive polarizations. Surprisingly, $Co/SrTiO_3/LSMO$ junctions showed *negative* TMR values at zero-bias and

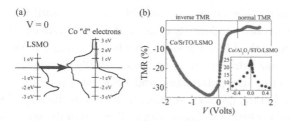

FIGURE 11.10 (a) Schematic of the spin-polarized densities of states of LSMO (as derived from photoemission) and the Co(100) surface (calculated). (b) TMR ratio versus applied bias for a Co/SrTiO$_3$/LSMO junction at T = 5 K. Inverse TMR is observed for V < 0.8 Volts, while normal TMR is observed for V > 0.8 Volts, indicating that the Co/SrTiO$_3$ spin polarization is negative for V < 0.8 Volts. Inset: TMR ratio versus applied bias for a Co/Al$_2$O$_3$/SrTiO$_3$/LSMO junction. In this case, the polarization of Co/Al$_2$O$_3$ and LSMO are both positive, and a normal (positive) TMR is seen. (After De Teresa, J.M. et al., *Phys. Rev. Lett.* 82, 4288, 1999; De Teresa, J.M. et al., *Science* 286, 507, 1999; Tsymbal, E.Y. et al., *J. Phys. Cond. Matt.* 15(4), R109, 2003.)

further displayed a strong bias dependence as shown in Figure 11.10b. In this case, the polarization of the Co/SrTiO$_3$ interface must be *negative*, opposite that of Co/Al$_2$O$_3$ interfaces, in agreement with the d electron polarization of Co at E_F. In order to show this more conclusively, de Teresa et al. investigated Co/Al$_2$O$_3$/SrTiO$_3$/LSMO junctions, with the expectation that since the LSMO and Co/Al$_2$O$_3$ tunneling spin polarizations are both positive, a normal positive TMR would result for all biases. As shown in the inset to Figure 11.10b, a normal positive TMR is observed for all biases, with a bias dependence that is essentially identical to standard Co/Al$_2$O$_3$/Co junctions.

De Teresa et al. argued that the sign change of the Co tunneling spin polarization can be understood in terms of interface bonding, as discussed above (see also Chapter 13, Volume 1). For Co/Al$_2$O$_3$ interfaces, the nature of the bonding at the Co/Al$_2$O$_3$ interface allows the itinerant electrons in Co to tunnel preferentially, and they are *positively* polarized. For Co/SrTiO$_3$ interfaces, however, the interface bonding is sufficiently different that the more localized states tunnel preferentially, and they are *negatively* polarized [20]. Indeed, this conjecture is borne out by first-principles calculations [92, 93, 95] for Co/SrTiO$_3$/Co MTJs. Further experimental proof is evidenced in the bias dependence of Co/SrTiO$_3$/LSMO junctions, as shown in Figure 11.10b. At zero-bias, the Co DOS is dominated by relatively localized minority states, as shown in Figure 11.10a (for the Co(100) surface), and exhibits a large negative tunneling spin polarization. For negative bias (electrons tunneling *from* the LSMO), the LSMO detector predominantly scans the DOS (convoluted with the energy-dependent tunneling probability) above the Fermi level in the Co electrode. As the bias is decreased to −0.4 Volts, in the antiparallel state the Fermi level of LSMO is at the same energy as the peak of the Co minority DOS, giving a larger tunneling spin polarization and, hence, a large (negative) TMR. As the bias becomes more negative, the minority DOS progressively decreases, and the tunneling spin polarization decreases.

For positive bias (electrons tunneling *from* Co), the LSMO detector predominantly scans the DOS *below* the Fermi level in the Co electrode. As the bias is increased, the majority DOS steadily increases, until at approximately 0.8 Volts, the majority and minority densities of states are equal, resulting in a zero crossing of the TMR. For still higher bias, the majority DOS of Co continues to increase (while the minority DOS steadily decreases), resulting in a positive tunneling spin polarization and TMR. At approximately 1.15 Volts, the Fermi level of the LSMO is at the same energy as the *majority* DOS peak in Co, and the positive TMR is maximal. Thus, all of the features in the bias dependence of $Co/SrTiO_3/LSMO$ junctions can be attributed to the structure of the spin-polarized Co DOS, dominated by more localized ("d-like") states. This is in contrast with experiments on Co/Al_2O_3-based tunnel junctions, which demonstrate positive polarization and structure related to the Co itinerant ("s-like") DOS [70].

Subsequent results by Sugiyama et al. [96] and Sun et al. [97] corroborated these results. Experiments by Sharma et al. [76], utilizing Ta_2O_5/Al_2O_3 composite barriers, have indicated that the sign of the spin polarization at Ta_2O_5 interfaces is also negative and have proposed an explanation similar to that of de Teresa et al. However, these results have not yet been independently verified. Overall, these experiments illustrate the rich physics behind spin-polarized tunneling in MTJs, as well as the intriguing possibility of "engineering" MTJs with tailored properties, which leads us naturally to our next example.

11.4.10 INTERFACE DEPENDENCE

As pointed out numerous times, the tunneling current in MTJs is effectively dominated by the electronic structure in the vicinity of the ferromagnet-insulator interface. One powerful way to explore this interface sensitivity is by the insertion of ultra-thin layers (often called "dusting" layers) at the electrode-barrier interfaces. As discussed in Section 11.3.2, Tedrow and Meservey [42] used this technique already in 1975 (Figure 11.5) to probe the onset of spin polarization in ultra-thin ferromagnets. In one view, their experiments established that not only can a few monolayers of Fe, Co, and Ni be ferromagnetic and fully polarized, but that these few monolayers were sufficient to dominate the tunneling current. In retrospect, this was also a direct proof of the interface sensitivity of tunneling.

In 2000, LeClair et al. [24] were among early workers to extend these studies to MTJs, investigating the insertion of ultra-thin Cu layers in $Co/Cu/Al_2O_2/Co$ structures. They were able to demonstrate that much of the early controversy in MTJ studies was morphology-related. In their initial work, they grew Cu interlayers both above and below the Al_2O_3 barrier, which resulted in two different TMR decay lengths, as shown in Figure 11.11a. The longer length scale in the former case simply results from the island-like growth of Cu on Al_2O_3 and hints at the technical challenges that arise in MTJ fabrication. When Cu was grown on Co, atomic-scale smoothness resulted, and a much shorter length scale. In this latter system, LeClair et al. found that the TMR decay was well-described by an exponential decay, with a length scale of $\xi \approx 0.26$ nm, equivalent to just more than one monolayer Cu. Appelbaum and Brinkman [17] first pointed out that tunneling in

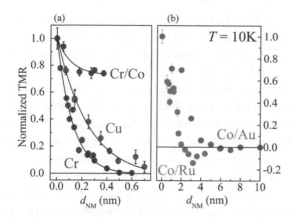

FIGURE 11.11 (a) Normalized TMR at 10 K as a function of Cu thickness for Co/Cu(d_{Cu})/Al$_2$O$_3$/Co, Co/Al$_2$O$_3$/Cu(d_{Cu})/Co, and Co/Cr(d_{Cr})/Co(d_{Co})/Al$_2$O$_3$/Co MTJs. Adding only a few monolayers of Co on Cr almost completely restores the TMR, demonstrating the interfacial sensitivity of MTJs. Lines are fits to exponential decays.
(b) Normalized TMR at 10 K as a function of Cr thickness for Co/Ru(d_{Ru})/Al$_2$O$_3$/Co and Co/Cr(d_{Au})/Al$_2$O$_3$/Ni$_{80}$Fe$_{20}$ tunnel junctions. (After LeClair, P. et al., *Phys. Rev. Lett.* 84(13), 2933, 2000; LeClair, P. et al., *Phys. Rev. Lett.* 86(6), 1066, 2001; LeClair, P. et al., *Phys. Rev. B* 64, 100406, 2001; Moodera, J.S. et al., *Phys. Rev. Lett.* 83(15), 3029, 1999.)

(disordered) normal-metal junctions should be sensitive to the density of states within a few Fermi wavelengths of the electrode-barrier interface, which is consistent with the length scale for TMR decay for smooth Cu in [24] corresponding to about 3.5 Fermi wavelengths. Perhaps more important than the quantitative results of their experiments, however, was the conclusion that detailed characterization of the interlayers must be performed in order to elucidate intrinsic interface effects from morphological effects.

A further demonstration of interface sensitivity was subsequently obtained by LeClair et al. [25] using ultra-thin Cr layers in Co/Cr/Al$_2$O$_3$/Co MTJs. The TMR decay was again approximately exponential, and in this case was even more rapid, (see Figure 11.11b) with $\xi \sim 0.1$ nm, or only about 0.5 monolayers of Cr. In these experiments, they also added an additional Co layer on top of the Cr dusting layer, i.e., Co/Cr(d_{Cr})/Co/Al$_2$O$_3$/Co, shown in Figure 11.11b. Strikingly, the TMR was almost completely restored with only these 3–5 monolayers of Co. This further confirms that only a few monolayers of the electrode adjacent to the ferromagnet-insulator interface dominate MTJ properties, in very good agreement with the earlier SPT work on ultra-thin magnetic layers. Subsequent work by Moodera et al. [99] and Samant and Parkin [100] arrived at essentially the same conclusion, establishing that at least in disordered (i.e., non-epitaxial) systems, the relevant length scale appears to be several monolayers at most. In an elegant experiment discussed briefly above, Gabureac et al. [44] studied the effect of Au interlayers in MTJs and also probed the spin polarization directly by SPT in the spirit of Moodera et al. [43] replacing one of the ferromagnetic layers with an Al spin detector. In this direct comparison between TMR and SPT, they found very good agreement between the decay length of the TMR and spin polarization, though their length scales (~1–1.5 nm) were somewhat longer than in previous work. They further inserted a 0.1 nm Au layer at a variable distance from the ferromagnet-interface, and consistent with LeClair et al. [25] found that the TMR and spin polarization recover their undoped values with length scales of ~0.6–0.7 nm.

The initial failure to observe quantum well states (oscillations of TMR as a function of interface layer thickness), as anticipated by theory and by analogy with GMR, was a source of confusion. Although no quantum well states were observed in the studies above, in *retrospect* one would not expect to observe them except in a nearly perfect epitaxial system, as is the case for quantum well states in metallic multilayers. Although there were observations of a sign inversion of the TMR as a function of Ru thickness in Co/Ru/Al$_2$O$_3$/Co junctions [26, 101] and Au thickness in Co/Au/Al$_2$O$_3$/Ni$_{80}$Fe$_{20}$ junctions [98], the interpretation of these results as quantum well states is unlikely, since none of these studies used single-crystalline electrodes in which quantum well oscillations may reasonably be expected to occur [24, 102]. This is particularly true in the case of Ru interlayers, where one explanation for the observed negative TMR is an interfacial CoRu alloy [26].

However, in 2002, Yuasa et al. [102] reported unambiguous quantum well oscillations of the TMR using single-crystalline Cu interlayers in Co(001)/Cu(001)/Al$_2$O$_3$/Ni$_{80}$Fe$_{20}$ junctions. Figure 11.12 shows the TMR at 2 K as a function of inserted Cu layer thickness. A clear (damped) oscillation of the TMR is evident, and the zero-bias period of 1.14 nm is in good agreement with the Fermi surface of Cu, as is the bias dependence of the oscillation period. Crucially, independent measurements on similarly grown Co/Cu/Co trilayers gave an oscillation of the interlayer exchange coupling with a period of 1.1 nm. This is a key distinction from earlier work with polycrystalline interface layers

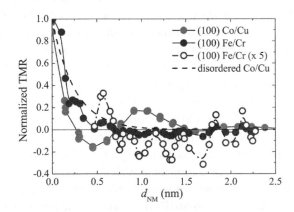

FIGURE 11.12   (red) TMR at $T = 2$ K as a function of Cu interlayer thickness for Co(001)/Cu(001)/Al$_2$O$_3$/Ni$_{80}$Fe$_{20}$ junctions. The period of the oscillation observed, 1.14 nm, is in agreement with the Fermi surface of Cu. For comparison, the dashed line shows the exponential decay of TMR fitted to the data of [44]. (blue) TMR at $T = $ 50 K as a function of Cr interlayer thickness for Fe(100)/Cr(100)/Al$_2$O$_3$/CoFe junctions. A small oscillation of the TMR with 2 ML period is observed due to the layered antiferromagnetic structure of Cr. The open symbols show the same data the vertical axis expanded by a factor of five. (After Yuasa, S. et al., *Science* 297(5579), 234, 2002; Nagahama, T. et al., *Phys. Rev. Lett.* 95(8), 086602, 2005; Matsumoto, R. et al., *Phys. Rev. B* 79(17), 174436, 2009.)

[24], where the polycrystalline nature of the interlayers (or interface mixing) would seem to preclude the observation of quantum well states. The clear correlation between TMR, interlayer coupling, and the Fermi surfaces of Co and Cu suggests that the oscillation indeed arises from spin-dependent reflection at the Co/Cu interface due to the formation of spin-polarized quantum well states within the Cu interlayer. The TMR observed, in this case, is relatively low, even without inserted Cu, but the dependence of spin polarization on crystallographic orientation for fcc-Co is not yet known, and the low TMR value may simply be a result of a low spin polarization for Co(100). Another curiosity is the fact that the overall decay of the TMR, on which the oscillatory behavior is superimposed, occurs at approximately the same rate as observed by LeClair et al. [24], which has still not been adequately explained. Further, the same overall quasi-exponential decay was also observed in Fe/Cr(001)-based junctions (Figure 11.12), where Nagahama et al. [103] also observed an oscillating TMR attributed to the layered antiferromagnetic structure of Cr(001). In Figure 11.13, we illustrate the mechanism for this oscillatory effect in comparison to quantum well states. That Nagahama et al. were able to observe monolayer oscillations of the TMR further confirms the interface sensitivity of tunneling: if it were the *bulk* electronic structure of the electrode that was of importance, no oscillation should be observed. The observation of quantum oscillations by Yuasa et al. [102] were at the time an encouraging first step toward the development of coherent spin-dependent tunneling devices, culminating in the "giant" TMR effects observed in FeMgO-based systems (see Chapters 12 and 13, Volume 1) [39, 66].

11.4.11 THE CURIOUS CASE OF CoFeB

In 2002–2004, very large TMR effects were observed in MTJs based on the amorphous ferromagnet CoFe, [105, 106] as large as 113% (72% at 295 K) at low temperatures, corresponding to $P = 60\%$ in the Julliere model. At

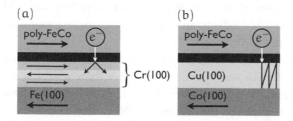

FIGURE 11.13 Dusting layers in epitaxial systems. (a) The magnetic moments in Cr(100) are magnetically aligned within each layer, with each layer aligned opposing adjacent layers. If the TMR ratio is determined by the metal monolayer closest the insulator interface, the TMR should oscillate with a period of 2 ML. If the TMR is determined by bulk electronic structure, there should be no TMR, since there is no net polarization in the Cr(100) layer. (b) If there is suitable matching of the dispersive bands throughout the structure, one spin channel will have relatively easy transmission, while the other will be preferentially reflected at the ferromagnet-dusting layer interface. If multiple scatterings occur within the dusting layer, the down-spin electrons form resonant states (QW states), giving rise to oscillations in the TMR as a function of bias and interlayer thickness. (After Nagahama, T. et al., *Phys. Rev. Lett.* 95(8), 086602, 2005; Yuasa, S. et al., *Science* 297(5579), 234, 2002.)

the time, this was the highest TMR observed at room temperature. More importantly, it was the highest TMR observed in a non-epitaxial junction based on 3d alloys and fabricated by industrially friendly standard sputtering techniques. Further, the bias voltage dependence of the TMR in these MTJs is remarkably weak [106] – even at 0.5 V, the TMR was still >50%, making CoFeB a very interesting system for applications.

However, what is perhaps more interesting in this case is why does the seemingly innocuous addition of B to a standard CoFe alloy increase the TMR (and effectively P) to such a degree? Remarkably, despite the technological importance of these materials, our understanding of their electronic and magnetic behavior is at a surprisingly low level [107]. Further, is the mechanism a general, and exploitable, means for increasing P? Mechanisms for the increased TMR have been proposed. The first assumes that the amorphous CoFeB layer grows more smoothly than standard polycrystalline layers, leading to improved barrier flatness, coverage and quality, thus giving a larger TMR. In this model, the polarization is assumed only weakly affected by the addition of B – it is a structural effect, with the B only serving to make the CoFe amorphous. A second proposition argues that the B hybridizes in such a way with the Co and/or Fe that the itinerant majority density of states is increased near E_F. To an extent, the first argument is borne out by TEM observations [106]. However, one would not expect this to affect the bias dependence so strongly, nor the temperature dependence [106]. While both of these properties are influenced by barrier quality, the amount of TMR(T) and TMR(V) decay usually attributed to barrier imperfections is far too small to explain the differences observed. We believe this lends credence to the hybridization argument, which is borne out to an extent by examining the density of states of the Fe and Co monoborides. On the other hand, the observation of "giant" TMR effects in CoFeB/MgO/CoFeB junctions lends credence to the former hypothesis. In these structures, it is currently believed that the amorphous CoFeB electrodes allow the MgO tunnel barrier to grow in a preferred (100) orientation, promoting "symmetry filtering" [47] which leads to TMR effects an order of magnitude above those observed with amorphous Al_2O_3 barriers [108].

In any case, these results were exciting in retrospect for three reasons: first, the TMR was significantly higher than anything observed previously; second, the TMR observed was nominally higher than what one would expect from the Julliere model [106, 109]; and third, amorphous CoFeB turned out to be crucial to the practical realization of giant TMR effects in MgO-based junctions, which will be explored in the following chapters.

11.5 CONCLUSIONS

In the relatively short time since the first demonstration of large room-temperature magnetoresistance in MTJs in 1995, there has been an enormous interest in the commercial potential of MTJ devices. Some of the main proposed applications are large arrays of magnetic sensors for imaging, ultra-low magnetic field sensors, and most prominently, non-volatile magnetic random access memories and read-head sensors for hard disks. In particular, the latter two have already been commercialized, and TMR-based read sensors have been a mainstay in hard disk drives for several years already. The great demand for improved magnetic devices has certainly played a large role in the "rebirth" of spin-polarized tunneling; a variety of these applications are covered in detail in Volume 3, Section VII, Applications. In particular, Chapter 12 and Chapter 13 in Volume 1 have direct relevance to the present discussion.

More striking from a physics point of view, however, is that despite four decades of research in spin-polarized tunneling, it is only in the last decade or so that the influence of the insulating barrier and the ferromagnet-insulator interface on tunneling spin polarization has been recognized. Some of the experiments mentioned here – e.g., the crystallographic orientation dependence of spin polarization, quantum well oscillations in quasi-epitaxial junctions, and large TMR effects with CoFeB-based junctions – foreshadowed or directly led to the most recent and exciting developments in spin-polarized tunneling, namely the "giant" TMR effects in MgO-based MTJs discussed in Chapter 12 and Chapter 13, Volume 1. Others, such as the spin filtering effects discussed in Chapter 14, Volume 1 are another natural outgrowth of the SPT technique.

REFERENCES

1. R. Meservey and P. M. Tedrow, Spin-polarized electron tunneling, *Phys. Rep.* **238**, 173 (1994).
2. J. C. Slonczewski, Magnetic bubble tunnel detector, *IBM Tech. Discl. Bull.* **19**, 2328–2336 (1976).
3. M. Jullière, Tunneling between ferromagnetic films, *Phys. Lett. A* **54**, 225 (1975).
4. J. S. Moodera, L. R. Kinder, T. M. Wong, and R. Meservey, Large magnetoresistance at room temperature in ferromagnetic thin film tunnel junctions, *Phys. Rev. Lett.* **74**, 3273 (1995).
5. T. Miyazaki and N. Tezuka, Giant magnetic tunneling effect in Fe/Al2O3/Fe junction, *J. Magn. Magn. Mater.* **139**, L231 (1995).
6. E. L. Wolf, *Principles of Electron Tunneling Spectroscopy*, Oxford University Press, London, 1985.
7. C. Kittel, *Introduction to Solid State Physics*, 7th edn., John Wiley and Sons, Inc., New York, 1996.
8. J. E. Dowman, M. L. A. MacVicar, and J. R. Waldram, Selection rule for tunneling from superconductors, *Phys. Rev.* **186**, 452 (1969).
9. J. G. Simmons, Generalized formula for the electric tunnel effect between similar electrodes separated by a thin insulating film, *J. Appl. Phys.* **34**, 1793 (1963).
10. J. M. Rowell, *Tunneling Phenomena in Solids*, Plenum, New York, p. 273, 1969.
11. J. J. Akerman, J. M. Slaughter, and I. K. Schuller, Tunneling criteria for magnetic-insulator-magnetic structures, *Appl. Phys. Lett.* **79**, 3104 (2001).
12. I. Giaever, Energy gap in superconductors measured by electron tunneling, *Phys. Rev. Lett.* **5**, 147–148 (1960).
13. I. Giaever and K. Megerle, Study of superconductors by electron tunneling, *Phys. Rev.* **122**, 1101–1111 (1961).
14. I. Giaever, *Nobel Lectures in Physics, 1971-1980*, World Scientific, Singapore, p. 144, 1992. See also http://nobelprize.org/nobel_prizes/physics/laureates/1973/giaever-lecture.html.

15. W. A. Harrison, Tunneling from an independent-particle point of view, *Phys. Rev.* **123**, 85 (1961).

16. J. Bardeen, Tunneling from a many-particle point of view, *Phys. Rev. Lett.* **6**, 57 (1961).

17. J. A. Appelbaum and W. F. Brinkman, Theory of many-body effects in tunneling, *Phys. Rev.* **186**, 464 (1969).

18. F. Mezei and A. Zawadowski, Kinematic change in conduction-electron density of states due to impurity scattering. II. Problem of an impurity layer and tunneling anomalies, *Phys. Rev. B* **3**, 3127 (1971).

19. S. Zhang and P. M. Levy, Models for magnetoresistance in tunnel junctions, *Eur. Phys. J. B* **10**, 599 (1999).

20. E. Y. Tsymbal and D. G. Pettifor, Modelling of spin-polarized electron tunnelling from 3d ferromagnets, *J. Phys. Condens. Matter* **9**, L411 1997.

21. H. Itoh, J. Inoue, S. Maekawa, and P. Bruno, Tunnel conductance in strong disordered limit, *J. Magn. Soc. Jpn.* **23**, 52 (1999).

22. C. Uiberacker and P. M. Levy, Role of symmetry on interface states in magnetic tunnel junctions, *Phys. Rev. B* **64**, 193404 (2001).

23. H. Itoh and J. Inoue, Interfacial electronic states and magnetoresistance in tunnel junctions, *Surf. Sci.* **493**, 748 (2001).

24. P. LeClair, H. J. M. Swagten, J. T. Kohlhepp, R. J. M. van de Veerdonk, and W. J. M. de Jonge, Apparent spin polarization decay in Cu-dusted Co/Al$_2$O$_3$/Co tunnel junctions, *Phys. Rev. Lett.* **84**, 2933–2936 (2000).

25. P. LeClair, J. T. Kohlhepp, H. J. M. Swagten, and W. J. M. de Jonge, Interfacial density of states in magnetic tunnel junctions, *Phys. Rev. Lett.* **86**, 1066–1069 (2001).

26. P. LeClair, B. Hoex, H. Wieldraaijer, J. T. Kohlhepp, H. J. M. Swagten, and W. J. M. de Jonge, Sign reversal of spin polarization in Co/Ru/Al$_2$O$_3$/Co magnetic tunnel junctions, *Phys. Rev. B* **64**, 100406 (2001).

27. P. M. Tedrow and R. Meservey, Spin-dependent tunneling into ferromagnetic nickel, *Phys. Rev. Lett.* **26**, 192 (1971).

28. P. M. Tedrow and R. Meservey, Direct observation of spin-state mixing in superconductors, *Phys. Rev. Lett.* **27**, 919 (1971).

29. P. LeClair, H. J. M. Swagten, and J. S. Moodera, Spin polarized electron tunneling, in *Ultrathin Magnetic Structures III*, B. Heinrich and J. A. C. Bland (Eds.), Springer-Verlag, Berlin, 2005.

30. I. I. Mazin, How to define and calculate the degree of spin polarization in ferromagnets, *Phys. Rev. Lett.* **83**, 1427 (1999).

31. H. J. M. Swagten, Spin-dependent tunneling in magnetic junctions, in *Volume Seventeen, Handbook of Magnetic Materials*, K. Buschow (Ed.), Elsevier North-Holland, Amsterdam, 2007.

32. D. C. Worledge and T. H. Geballe, Negative spin-polarization of SrRuO$_3$, *Phys. Rev. Lett.* **85**, 5182 (2000).

33. C. Kaiser, A. F. Panchula, and S. S. P. Parkin, Finite tunneling spin polarization at the compensation point of rare-earth-metal–transition-metal alloys, *Phys. Rev. Lett.* **95**, 047202 (2005).

34. T. H. Kim and J. S. Moodera, Large spin polarization in epitaxial and polycrystalline Ni films, *Phys. Rev. B* **69**, 020403 (2004).

35. J. S. Moodera, J. Nassar, and J. Mathon, Spin-dependent tunneling in ferromagnetic junctions, *J. Magn. Magn. Mater.* **200**, 248 (1999).

36. D. J. Monsma and S. S. P. Parkin, Spin polarization of tunneling current from ferromagnet/Al$_2$O$_3$ interfaces using Copper-doped Aluminum superconducting films, *Appl. Phys. Lett.* **77**, 720–722 (2000).

37. D. C. Worledge and T. H. Geballe, Spin-polarized tunneling in La$_{0.67}$Sr$_{0.33}$MnO$_3$, *Appl. Phys. Lett.* **76**, 900 (2000).

38. J. S. Parker, S. M. Watts, P. G. Ivanov, and P. Xiong, Spin polarization of CrO$_2$ at and across an artificial barrier, *Phys. Rev. Lett.* **88**, 196601 (2002).

39. S. S. P. Parkin, C. Kaiser, A. Panchula et al., Giant room-temperature magnetoresistance in single-crystal Fe/MgO/Fe magnetic tunnel junctions, *Nat. Mater.* **3**, 862–867 (2004).

40. R. Meservey, 2009. Private email communications.

41. P. M. Tedrow, R. Meservey, and B. B. Schwartz, Experimental evidence for a first-order magnetic transition in thin superconducting Aluminum films, *Phys. Rev. Lett.* **24**, 1004–1007 (1970).

42. P. M. Tedrow and R. Meservey, Critical thickness for ferromagnetism and the range of spin-polarized electrons tunneling into Co, *Solid State Commun.* **16**, 71 (1975).

43. J. S. Moodera, M. E. Taylor, and R. Meservey, Exchange-induced spin polarization of conduction electrons in paramagnetic metals, *Phys. Rev. B* **40**, 11980–11982 (1989).

44. M. S. Gabureac, K. J. Dempsey, N. A. Porter, C. H. Marrows, S. Rajauria, and H. Courtois, Spin-polarized tunneling with au impurity layers, *J. Appl. Phys.* **103**, 07A915 (2008).

45. M. S. Gabureac, D. A. MacLaren, H. Courtois, and C. H. Marrows, Long-ranged magnetic proximity effects in noble metal-doped cobalt probed with spin-dependent tunneling, *New J. Phys.* **16**, 043008 (2014).

46. M. B. Stearns, Simple explanation of tunneling spin-polarization of Fe, Co, Ni and its alloys, *J. Magn. Magn. Mater.* **8**, 167 (1977).

47. W. H. Butler, X. G. Zhang, T. C. Schulthess, and J. M. MacLaren, Spin dependent tunneling conductance of Fe|MgO|Fe sandwiches, *Phys. Rev. B* **63**, 055416 (2001).

48. P. Mavropoulos, N. Papanikolaou, and P. H. Dederichs, Complex band structure and tunneling through ferromagnet/insulator/ferromagnet junctions, *Phys. Rev. Lett.* **85**, 1088 (2000).

49. J. M. MacLaren, X. G. Zhang, W. H. Butler, and X. Wang, Layer KKR approach to Bloch-wave transmission and reflection: Application to spin-dependent tunneling, *Phys. Rev. B* **59**, 5470 (1999).

50. I. I. Mazin, Tunneling of Bloch electrons through vacuum barrier, *Europhys. Lett.* **55**, 404 (2001).

51. S. Maekawa and U. Gäfvert, Electron tunneling between ferromagnetic films, *IEEE Trans. Magn.* **18**, 707 (1982).

52. S. Gider, B. U. Runge, A. C. Marley, and S. S. P. Parkin, The magnetic stability of spin-dependent tunneling devices, *Science* **281**, 797 (1999).

53. P. LeClair, Fundamental aspects of spin-polarized tunneling, PhD thesis, Technische Universiteit Eindhoven, 2002. See also http://alexandria.tue.nl/extra2/200211695.pdf.

54. X.-F. Han, A. C. C. Yu, M. Oogane, J. Murai, T. Daibou, and T. Miyazaki, Analyses of intrinsic magnetoelectric properties in spin-valve-type tunnel junctions with high magnetoresistance and low resistance, *Phys. Rev. B* **63**, 224404 (2001).

55. W. E. Pickett and J. S. Moodera, Half metallic ferromagnets, *Phys. Today* **5**, 39 (2001).

56. M. Bowen, M. Bibes, A. Barthélémy et al., Nearly total spin polarization in $La_{2/3}Sr_{1/3}MnO_3$ from tunneling experiments, *Appl. Phys. Lett.* **82**, 233–235 (2003).

57. Y. Lu, X. W. Li, G. Q. Gong et al., Large magnetotunneling effect at low magnetic fields in micrometer-scale epitaxial $La_{0.67}Sr_{0.33}MnO_3$ tunnel junctions, *Phys. Rev. B* **54**, R8357 (1996).

58. M. Viret, M. Drouet, J. Nassar, J. P. Contour, C. Fermon, and A. Fert, Low-field colossal magnetoresistance in manganite tunnel spin valves, *Europhys. Lett.* **39**, 545 (1997).

59. B. Nadgorny, R. J. Soulen Jr., M. S. Osofsky et al., Transport spin polarization of Ni_xFe_{1-x}: Electronic kinematics and band structure, *Phys. Rev. B* **61**, R3788, (2000).

60. T. H. Kim and J. S. Moodera, Enhanced ferromagnetism and spin-polarized tunneling studies in Co-Mn alloy films, *Phys. Rev. B* **66**, 104436 (2002).

61. C. Kaiser, S. van Dijken, S.-H. Yang, H. Yang, and S. S. P. Parkin, Role of tunneling matrix elements in determining the magnitude of the tunneling spin polarization of 3d transition metal ferromagnetic alloys, *Phys. Rev. Lett.* **94**, 247203 (2005).

62. B. C. Min, J. C. Lodder, and R. Jansen, Sign of tunnel spin polarization of low-work-function Gd/Co nanolayers in a magnetic tunnel junction, *Phys. Rev. B* **78**, 212403 (2008).

63. N. Thiyagarajah, Y.-C. Lau, D. Betto et al., Giant spontaneous Hall effect in zero-moment Mn_2Ru_xGa, *Appl. Phys. Lett.* **106**, 122402 (2015).

64. S. Yuasa, T. Sato, E. Tamura et al., Magnetic tunnel junctions with single-crystal electrodes: A crystal anisotropy of tunnel magneto-resistance, *Europhys. Lett.* **52**, 344–350 (2000).

65. J. Callaway and C. S. Wang, Energy bands in ferromagnetic iron, *Phys. Rev. B* **16**, 2095 (1977).

66. S. Yuasa, T. Nagahama, A. Fukushima, Y. Suzuki, and K. Ando, Giant room-temperature magnetoresistance in single-crystal Fe/MgO/Fe magnetic tunnel junctions, *Nat. Mater.* **3**, 868–871 (2004).

67. J.-N. Chazalviel and Y. Yafet, Theory of the spin polarization of field-emitted electrons from nickel, *Phys. Rev. B* **15**, 1062–1071 (1977).

68. J. S. Moodera, J. Nowak, and R. J. M. van de Veerdonk, Interface magnetism and spin wave scattering in ferromagnet-insulator-ferromagnet tunnel junctions, *Phys. Rev. Lett.* **80**, 2941 (1998).

69. X.-F. Han, J. Murai, Y. Ando, H. Kubota, and T. Miyazaki, Inelastic magnon and phonon excitations in $Al_{1-x}Co_x/Al_{1-x}Co_x$-oxide/Al tunnel junctions, *Appl. Phys. Lett.* **78**, 2533 (2001).

70. P. LeClair, J. T. Kohlhepp, C. H. van de Vin et al., Band structure and density of states effects in Co-based magnetic tunnel junctions, *Phys. Rev. Lett.* **88**, 107201 (2002).

71. J. S. Moodera and L. R. Kinder, Ferromagnetic - insulator - ferromagnetic tunneling: Spin-dependent tunneling and large magnetoresistance in trilayer junctions (invited), *J. Appl. Phys.* **79**, 4724 (1996).

72. S. Zhang, P. M. Levy, A. C. Marley, and S. S. P. Parkin, Quenching of magnetoresistance by hot electrons in magnetic tunnel junctions, *Phys. Rev. Lett.* **79**, 3744 (1997).

73. A. M. Bratkovsky, Assisted tunneling in ferromagnetic junctions and half-metallic oxides, *Appl. Phys. Lett.* **72**, 2334 (1998).

74. J. M. De Teresa, A. Barthélémy, A. Fert et al., Inverse tunnel magnetoresistance in $Co/SrTiO_3/La_{0.7}Sr_{0.3}MnO_3$: New ideas on spin-polarized tunneling, *Phys. Rev. Lett.* **82**, 4288 (1999).

75. J. M. De Teresa, A. Barthélémy, A. Fert, J. P. Contour, F. Montaigne, and P. Seneor, Role of metal-oxide interface in determining the spin polarization of magnetic tunnel junctions, *Science* **286**, 507 (1999).

76. M. Sharma, S. X. Wang, and J. H. Nickel, Inversion of spin polarization and tunneling magnetoresistance in spin-dependent tunneling junctions, *Phys. Rev. Lett.* **82**, 616 (1999).

77. E. Y. Tsymbal, O. N. Mryasov, and P. R. LeClair, Spin-dependent tunnelling in magnetic tunnel junctions, *J. Phys. Condens. Matter* **15**, R109–R142 (2003).

78. P. LeClair, H. J. M. Swagten, J. T. Kohlhepp, and W. J. M. de Jonge, Tunnel conductance as a probe of spin polarization decay in Cu dusted $Co/Al_2O_3/Co$ tunnel junctions, *Appl. Phys. Lett.* **76**, 3783 (2000).

79. A. H. Davis and J. M. MacLaren, Spin dependent tunneling at finite bias, *J. Appl. Phys.* **87**, 5224 (2000).

80. A. M. Bratkovsky, Tunneling of electrons in conventional and half-metallic systems: Towards very large magnetoresistance, *Phys. Rev. B* **56**, 2344 (1997).

81. A. Vedyayev, D. Bagrets, A. Bagrets, and B. Dieny, Resonant spin-dependent tunneling in spin-valve junctions in the presence of paramagnetic impurities, *Phys. Rev. B* **63**, 064429 (2001).

82. R. Jansen and J. S. Moodera, Influence of barrier impurities on the magnetoresistance in ferromagnetic tunnel junctions, *J. Appl. Phys.* **83**, 6682 (1998).

83. R. Jansen and J. S. Moodera, Magnetoresistance in doped magnetic tunnel junctions: Effect of spin scattering and impurity-assisted transport, *Phys. Rev. B* **61**, 9047 (2000).

84. C. H. Shang, J. Nowak, R. Jansen, and J. S. Moodera, Temperature dependence of magnetoresistance and surface magnetization in ferromagnetic tunnel junctions, *Phys. Rev. B* **58**, R2917 (1998).

85. D. T. Pierce, R. J. Celotta, J. Unguris, and H. C. Siegman, Spin-dependent elastic scattering of electrons from a ferromagnetic glass, $Ni_{40}Fe_{40}B_{20}$, *Phys. Rev. B* **26**, 2566 (1982).

86. A. H. MacDonald, T. Jungwirth, and M. Kasner, Temperature dependence of itinerant electron junction magnetoresistance, *Phys. Rev. Lett.* **81**, 705 (1998).

87. J. Crangle, The magnetic moments of cobalt-copper alloys, *Philos. Mag.* **46**, 499 (1955).

88. A. H. Davis, J. M. MacLaren, and P. LeClair, Inherent temperature effects in magnetic tunnel junctions, *J. Appl. Phys.* **89**, 7567 (2001).

89. A. T. Hindmarch, C. H. Marrows, and B. J. Hickey, Tunneling spin polarization in magnetic tunnel junctions near the Curie temperature, *Phys. Rev. B* **72**, 100401 (2005).

90. T. Kreutz, T. Greber, P. Aebi, and J. Osterwalder, Temperature-dependent electronic structure of nickel metal, *Phys. Rev. B* **58**, 1300 (1998).

91. A. T. Hindmarch, C. H. Marrows, and B. J. Hickey, Tunneling magnetoresistance spectroscopy: Temperature dependent spin-polarized band structure in $Cu_{38}Ni_{62}$, *Phys. Rev. B* **72**, 060406 (2005).

92. J. Velev, K. D. Belashchenko, D. A. Stewart, M. van Schilfgaarde, S. Jaswal, and E. Y. Tsymbal, Negative spin polarization and large tunneling magnetoresistance in epitaxial $Co/SrTiO_3/Co$ magnetic tunnel junctions, *Phys. Rev. Lett.* **95**, 216601 (2005).

93. E. Y. Tsymbal, K. D. Belashchenko, J. P. Velev et al., Interface effects in spin-dependent tunneling, *Prog. Mater. Sci.* **52**, 401−420 (2007).

94. J. H. Park, E. Vescovo, H. J. Kim, C. Kwon, R. Ramesh, and T. Venkatesan, Direct evidence for a half-metallic ferromagnet, *Nature* **392**, 794 (1998).

95. I. I. Oleinik, E. Y. Tsymbal, and D. G. Pettifor, Atomic and electronic structure of $Co/SrTiO_3/Co$ magnetic tunnel junctions, *Phys. Rev. B* **65**, 020401(R) (2001).

96. M. Sugiyama, J. Hayakawa, K. Itou et al., Tunneling magnetoresistance effect and transport properties of tunnel junctions using half-metal ferromagnet, *J. Magn. Soc. Jpn.* **25**, 795−798 (2001).

97. J. Z. Sun, K. P. Roche, and S. S. P. Parkin, Interface stability in hybrid metal-oxide magnetic trilayer junctions, *Phys. Rev. B* **61**, 11244−11247 (2000).

98. J. S. Moodera, J. Nowak, L. R. Kinder et al., Quantum well states in spin-dependent tunnel structures, *Phys. Rev. Lett.* **83**, 3029−3032 (1999).

99. J. S. Moodera, T. H. Kim, C. Tanaka, and C. H. de Groot, Spin-polarized tunnelling, magnetoresistance and interfacial effects in ferromagnetic junctions, *Philos. Mag. B* **80**, 195 (2000).

100. M. G. Samant and S. S. P. Parkin, Magnetic tunnel junctions−principles and applications, *Vacuum* **74**, 705−709 (2004).

101. T. Nozaki, Y. Jiang, Y. Kaneko et al., Spin-dependent quantum oscillations in magnetic tunnel junctions with Ru quantum wells, *Phys. Rev. B* **70**, 172401 (2004).

102. S. Yuasa, T. Nagahama, and Y. Suzuki, Spin-polarized resonant tunneling in magnetic tunnel junctions, *Science* **297**, 234−237 (2002).

103. T. Nagahama, S. Yuasa, E. Tamura, and Y. Suzuki, Spin-dependent tunneling in magnetic tunnel junctions with a layered antiferromagnetic Cr(001) spacer: Role of band structure and interface scattering, *Phys. Rev. Lett.* **95**, 086602 (2005).

104. R. Matsumoto, A. Fukushima, K. Yakushiji et al., Spin-dependent tunneling in epitaxial Fe/Cr/MgO/Fe magnetic tunnel junctions with an ultrathin Cr(001) spacer layer, *Phys. Rev. B* **79**, 174436 (2009).

105. H. Kano, K. Bessho, Y. Higo et al., MRAM with improved magnetic tunnel junction material, in *Magnetics Conference, 2002. INTERMAG Europe 2002. Digest of Technical Papers. 2002 IEEE International*, IEEE, 2002. On pages BB4-.

106. D. Wang, C. Nordman, J. M. Daughton, Z. Qian, and J. Fink, 70% TMR at room temperature for SDT sandwich junctions with CoFeB as free and reference layers, *IEEE Trans. Magn.* **40**, 2269 (2004).

107. J. Hafner, M. Tegze, and Ch. Becker, Amorphous magnetism in Fe-B alloys: First-principles spin-polarized electronic-structure calculations, *Phys. Rev. B* **49**, 285–298 (1994).

108. S. Ikeda, J. Hayakawa, Y. Ashizawa et al., Tunnel magnetoresistance of 604% at 300 k by suppression of Ta diffusion in CoFeB/MgO/CoFeB pseudo-spin-valves annealed at high temperature, *Appl. Phys. Lett.* **93**, 082508 (2008).

109. J. S. Moodera and T. H. Kim, 2004. Unpublished results.

12

Tunnel Magnetoresistance in MgO-Based Magnetic Tunnel Junctions
Experiment

Shinji Yuasa

12.1 INTRODUCTION

12.1.1 TUNNEL MAGNETORESISTANCE EFFECT IN A MAGNETIC TUNNEL JUNCTION

A magnetic tunnel junction (MTJ) consists of an ultra-thin insulating layer (tunnel barrier) sandwiched between two ferromagnetic (FM) metal layers (electrodes), as shown in Figure 12.1a. The resistance of MTJ depends on the relative magnetic alignment (parallel or antiparallel) of the electrodes. The tunnel resistance, R, of MTJ, is lower when the magnetizations are parallel than it is when the magnetizations are antiparallel, as shown in Figure 12.1b. That is, $R_P < R_{AP}$. This change in resistance with the relative orientation of the two magnetic layers, called the tunnel magnetoresistance (TMR) effect, is one of the most important phenomena in spintronics. The size of this effect is measured by the fractional change in resistance $(R_{AP}-R_P)/R_P \times 100(\%)$, which is called the magnetoresistance (MR) ratio. The MR ratio at room temperature (RT) and a low magnetic field (typically < 10 mT) represent a performance index for industrial applications.

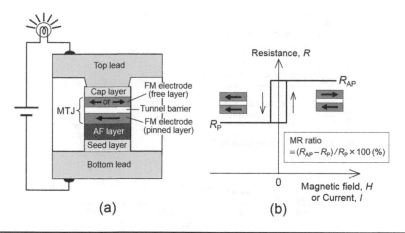

FIGURE 12.1   (a) Typical cross-sectional structure of magnetic tunnel junction (MTJ) with spin-valve structure. (b) Typical magnetoresistance curve of spin-valve-type MTJ and definition of magnetoresistance (MR) ratio.

The TMR effect was first observed in 1975 by Julliere [1], who found that a Fe/Ge-O/Co MTJ exhibited an MR ratio of 14% at 4.2 K.* Although it received little attention for more than a decade because the TMR effect was not observed at RT, it attracted renewed attention after the discovery of giant magnetoresistance (GMR) in metallic magnetic multilayers in the late 1980s [2, 3] (see Chapter 4, Volume 1 for the review on GMR). Because researchers recognized that the GMR effect could be applied to magnetic-sensor devices, such as the read heads of hard disk drives (HDDs), extensive experimental and theoretical efforts were devoted to increasing the MR ratio at RT as well as to understanding the physics of spin-dependent transport. In 1995, Miyazaki et al. [4] and Moodera et al. [5] made MTJs with amorphous aluminum oxide (Al-O) tunnel barriers and 3d ferromagnetic electrodes such as Fe and Co and obtained MR ratios above 10% at RT (Chapter 11, Volume 1). Because these MR ratios were the highest then reported for a practical ferromagnetic/nonmagnetic/ferromagnetic tri-layer structure called a (pseudo-) spin-valve structure, the TMR effect attracted a great deal of attention. Since 1995, extensive experimental efforts have been expended on MTJs with an amorphous Al-O barrier, and the ferromagnetic electrode materials and conditions for fabricating the Al-O barrier have been optimized. As a result, room-temperature MR ratios have been increased to about 70%, as shown in Figure 12.2 (solid circles). These MR ratios of up to 70%, however, are still lower than those needed for many applications of spintronic devices. High-density magnetoresistive random-access-memory (MRAM) cells, for example, will need to have MR ratios that are higher than 150% at RT, and the read head in next-generation ultrahigh-density HDDs will need to have both a high MR ratio and an ultra-low tunneling resistance. The MR ratios of conventional Al-O-based MTJs are simply not high enough for applications to next-generation devices.

---

* In Ref. [1], the tunnel barrier material is described as Ge. However, the Ge layer should have been oxidized because it was exposed to air before the upper electrode layer was deposited.

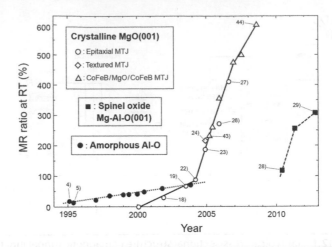

FIGURE 12.2   History of improvement in MR ratio at room temperature (RT) for MTJs with amorphous Al-O barrier (solid circles and dotted line), crystalline MgO(001) barrier (open symbols and solid line), or crystalline spinel Mg-Al-O(001) (solid squares and broken line).

In 2001, theoretical calculations predicted that epitaxial MTJs with a crystalline magnesium oxide (MgO) tunnel barrier would have MR ratios of over 1000% (Chapter 13, Volume 1), and in 2004, huge MR ratios of about 200% were obtained at RT in MTJs with a crystalline MgO(001) barrier. The huge TMR effect in MgO-based MTJs is now called the *giant TMR effect* and is of great importance not only for device applications but also for clarifying the physics of spin-dependent tunneling.

In this chapter, we present an overview of the huge TMR effect in MgO-based MTJs mainly from the experimental point of view. In the following subsection of this introduction, we provide further background information about the TMR effect in MTJs with an amorphous Al-O barrier. In Section 12.2, we briefly explain the theoretical basis for the TMR effect in MgO-based MTJs, and in Sections 12.3 and 12.4, we summarize experimental studies done on fully epitaxial MgO-based MTJs. In Section 12.5, we explain the structure and fabrication of CoFeB/MgO/CoFeB MTJs that are important for applications, and in Section 12.7, we report recent progress in applications of the giant TMR effect and those in MgO-based MTJs.

In other chapters, other topics related to MgO-based MTJ are explained in more detail. For example, the TMR effect in Al-O-based MTJs and measurement of tunneling spin polarization (TSP) are explained in Chapter 11, Volume 1. The theory of the TMR effect in MgO-based MTJs is explained in Chapter 13, Volume 1. Spin-transfer torque in MTJs is explained in Chapter 15, Volume 1. The application of MTJs to magnetic sensors is explained in Chapter 12, Volume 3 and that to MRAMs is explained in Chapter 13, Volume 3.

12.1.2 Incoherent Tunneling Through Amorphous Tunnel Barrier

Before going into the details of coherent tunneling through a crystalline MgO(001) barrier, we will explain the tunneling process through an amorphous tunnel barrier and the phenomenological model for the TMR effect. Because this subject is explained in detail in Sections 10.3, 10.4, and 2.2, we discuss it only briefly in this chapter. Julliere proposed a simple phenomenological model in which the TMR effect is due to spin-dependent electron tunneling [1]. According to this model, the MR ratio of an MTJ could be expressed in terms of the tunneling spin polarizations (TSP), P, of the ferromagnetic electrodes

$$MR = 2P_1P_2 / (1 - P_1P_2), \qquad (12.1)$$

where

$$P_\alpha \equiv [D_{\alpha\uparrow}(E_F) - D_{\alpha\downarrow}(E_F)]/[D_{\alpha\uparrow}(E_F) + D_{\alpha\downarrow}(E_F)]; \quad \alpha = 1, 2. \qquad (12.2)$$

Here P_α is the spin polarization of a ferromagnetic electrode, and $D_{\alpha\uparrow}(E_F)$ and $D_{\alpha\downarrow}(E_F)$ correspond to the densities of states (DOSs) of the electrode at the Fermi energy (E_F) for the majority-spin and minority-spin bands. In Julliere's model, spin polarization is an intrinsic property of an electrode material. When an electrode material is nonmagnetic, $P = 0$. When the DOS of the electrode material is fully spin-polarized at E_F, $|P| = 1$.

The spin polarization of a ferromagnet at low temperature can be directly measured using ferromagnet/Al-O/superconductor tunnel junctions (the so-called Tedrow-Meservey method: see Chapter 11, Volume 1) [6]. Measured this way, the spin polarization of 3d ferromagnetic metals and alloys based on iron (Fe), nickel (Ni), and cobalt (Co) are always positive and are usually between 0 and 0.6 at low temperatures below 4.2 K [6, 7]. The MR ratios estimated from Julliere's model (Equation 12.1) using these measured P values agree relatively well with the MR ratios observed experimentally in MTJs, but the theoretical values of P (Equation 12.2) obtained from band calculations do not agree with the measured spin polarizations and the MR ratios observed experimentally. Even the signs of P often differ between theoretical predictions and experimental results. This paradox in spin polarization is considered to be due to the incorrect definition of spin polarization in Equation 12.2, as explained below.

Tunneling in an MTJ with an amorphous Al-O barrier is schematically illustrated in Figure 12.3a, where the top electrode layer is Fe(001), is an example of a 3d ferromagnet. Various Bloch states with different symmetries of wave functions exist in the electrode. Because the tunnel barrier is amorphous, there is no crystallographic symmetry in the tunnel barrier. Because of this non-symmetric structure, Bloch states with various symmetries can couple with evanescent states in Al-O and therefore have finite tunneling probabilities. This tunneling process can be regarded as 'incoherent' tunneling. In 3d ferromagnetic metals and alloys, Bloch states with Δ_1 symmetry (*spd* hybridized states) usually have a large positive spin polarization at E_F, whereas Bloch states with Δ_5 and Δ_2 symmetry (*d* states) usually have a much smaller or even negative spin polarization at E_F. Julliere's model assumes that tunneling probabilities are equal for all the Bloch states in the

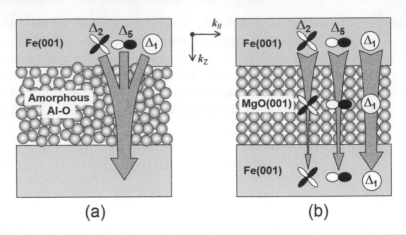

FIGURE 12.3   Schematic of electron tunneling through (a) amorphous Al-O barrier and (b) crystalline MgO(001) barrier.

electrodes. This assumption corresponds to completely incoherent tunneling, in which none of the momentum or coherency of Bloch states is conserved. This assumption, however, is not valid even for an amorphous Al-O barrier. Although the spin polarization, P, defined by Equation 12.2 is negative for Co and Ni, the P experimentally observed for these materials when they are combined with an Al-O barrier is positive [6, 7]. This discrepancy indicates that the tunneling probability in actual MTJs depends on the symmetry of each Bloch state. Even for amorphous barriers, the Δ_1 Bloch states that usually have a large positive P are considered to have higher tunneling probabilities than the other Bloch states [8, 9]. This results in a positive tunneling spin polarization for the ferromagnetic electrode. Because the other Bloch states that have smaller spin polarizations also contribute to the tunneling current, the tunneling spin polarization of the electrode is reduced below 0.6 for standard 3d ferromagnetic metals and alloys. If only highly spin-polarized Δ_1 states coherently tunnel through a barrier (Figure 12.3b), a very high spin polarization of tunneling current and thus a very high MR ratio are expected. Such ideal coherent tunneling is theoretically expected in an epitaxial MTJ with a crystalline MgO(001) tunnel barrier (see the next section). It should be noted here that the actual tunneling process through the amorphous Al-O barrier is considered to be an intermediate process between the completely incoherent tunneling represented by Julliere's model and the coherent tunneling illustrated in Figure 12.3b.

Although Julliere's model is simply a phenomenological model, we risked the use of Equation 12.1 with experimentally observed P here to roughly estimate the maximum possible MR ratio for Al-O-based MTJs. Equation 12.1 with the spin polarizations measured experimentally ($0 < P < 0.6$) yields a maximum MR ratio of about 100% at low temperatures. An MR ratio is reduced to about 70% at RT likely due to thermal spin fluctuations. One way to obtain an MR ratio significantly higher than 70% at RT is to use special kinds of ferromagnetic materials called *half metals* as electrodes, which have full tunneling spin polarization ($|P| = 1$) and are therefore theoretically expected to yield MTJs with huge MR ratios (up to ∞, according to Julliere's model). Some candidate half metals are CrO_2, Heusler alloys such as Co_2MnSi and Fe_3O_4, and manganese perovskite oxides such as $La_{1-x}Sr_xMnO_3$. Very high MR ratios, above several hundred percent, have been obtained at low temperatures in $La_{1-x}Sr_xMnO_3/SrTiO_3/La_{1-x}Sr_xMnO_3$ MTJs [10] and $Co_2MnSi/Al-O/Co_2MnSi$ MTJs [11]. However, such high MR ratios have never been observed at RT for half-metal electrodes (for details, see Section 12.6). Another way to obtain a very high MR ratio is to use coherent spin-dependent tunneling in an epitaxial MTJ with a crystalline tunnel barrier such as MgO(001). That is explained in the following sections.

12.2 THEORY OF COHERENT TUNNELING THROUGH CRYSTALLINE MgO(001) BARRIER

A crystalline MgO(001) barrier layer can be epitaxially grown on a bcc Fe(001) layer with a relatively small lattice mismatch of about 3% because this amount of lattice mismatch can be absorbed by lattice distortions in the Fe and MgO layers and/or by dislocations formed at the interface. Coherent tunneling transport in an epitaxial Fe(001)/MgO(001)/Fe(001) MTJ is schematically illustrated in Figure 12.3b. In ideal coherent tunneling, Fe Δ_1 states are theoretically expected to dominantly tunnel through the MgO(001) barrier using the following mechanism [12, 13]. For the $k_\parallel = 0$ at which the tunneling probability is the highest, there are three kinds of evanescent states (tunneling states) in the band gap of MgO(001): Δ_1, Δ_5, and $\Delta_{2'}$. It should be noted that although free-electron theories often assume tunneling states to be plane waves, they actually have specific orbital symmetries. When the symmetries of tunneling wave functions are conserved, Fe Δ_1 Bloch states couple with MgO Δ_1 evanescent states, as schematically illustrated in Figure 12.4b. Of the three kinds of evanescent states in MgO, the Δ_1 evanescent state has the slowest decay (i.e., the longest decay length) as shown in Figure 12.3 of Chapter 13, Volume 1. The dominant tunneling channel for a parallel magnetic state is Fe $\Delta_1 \longleftrightarrow$ MgO $\Delta_1 \longleftrightarrow$ Fe Δ_1. The band dispersion of bcc Fe for the [001] ($k_\parallel = 0$) direction is shown in Figure 12.4a. It is seen that the Fe Δ_1 band is fully spin-polarized at E_F ($P = 1$). A very large TMR effect in the epitaxial Fe(001)/MgO(001)/Fe(001) MTJ is therefore expected when Δ_1 electrons dominantly tunnel through the MgO(001) barrier. It should also be noted that a finite tunneling current flows even for antiparallel magnetic states. However, the tunneling conductance in the P state is much larger than that in the AP state, making the MR ratio very high [12, 13].

It should be noted that the Δ_1 Bloch states are highly spin-polarized not only in bcc Fe(001) but also in many other bcc ferromagnetic metals and alloys based on Fe and Co (e.g., bcc FeCo, bcc CoFeB, and some of the Heusler alloys). The band dispersion of bcc Co(001) (a metastable structure) has been shown in Figure 12.4b as an example. The Δ_1 states in bcc Co, like those in bcc Fe, are fully spin-polarized at E_F. According to first-principles calculations, the TMR of a Co(001)/MgO(001)/Co(001) MTJ is even larger than that of an Fe(001)/MgO(001)/Fe(001) MTJ [14]. A very large TMR should

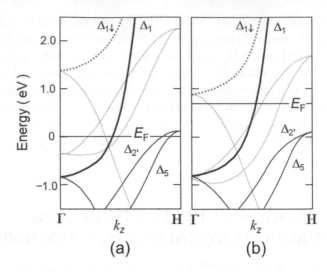

FIGURE 12.4 Band dispersions of (a) bcc Fe in [001] (Γ-H) direction and (b) bcc Co in [001] (Γ-H) direction. (Redrawn from Yuasa, S. et al., *Appl. Phys. Lett.* 89, 042505, 2006.) Black solid and gray dotted lines correspond to majority-spin and minority-spin bands. Thick black solid and gray dotted lines correspond to majority-spin and minority-spin Δ_1 bands. E_F denotes Fermi energy.

be characteristic of MTJs with 3d-ferromagnetic-alloy electrodes with a bcc(001) structure based on Fe and Co. Note also that very large TMR is theoretically expected not only for the MgO(001) barrier but also for other crystalline tunnel barriers such as ZnSe(001) [15] and SrTiO$_3$(001) [16]. Large TMR has, however, never been observed in MTJs with crystalline ZnSe(001) or SrTiO$_3$(001) barriers because of experimental difficulties in fabricating high-quality MTJs without pinholes and interdiffusion at the interfaces.

12.3 GIANT TMR EFFECT IN MAGNETIC TUNNEL JUNCTIONS WITH CRYSTALLINE MgO(001) BARRIER

12.3.1 DEVELOPMENT OF MgO-BASED MAGNETIC TUNNEL JUNCTIONS

Since the theoretical predictions of very large TMR effects in Fe/MgO/Fe MTJs [12, 13], there have been several experimental attempts to fabricate fully epitaxial Fe(001)/MgO(001)/Fe(001) MTJs [17–19]. Bowen et al. were the first to obtain a relatively high MR ratio in Fe(001)/MgO(001)/Fe(001) MTJs (30% at RT and 60% at 30 K, Figure 12.5a) [18], but the RT MR ratios obtained in MgO-based MTJs did not exceed the highest of 70% obtained in Al-O-based MTJs. The main difficulty in the early stages of these experimental attempts was the fabrication of an ideal interface structure. It was experimentally observed that Fe atoms at the Fe(001)/MgO(001) interface were easily oxidized [20]. According to first-principles calculations, tunneling processes through the ideal interface and the oxidized interface are significantly different [21]. At the ideal interface, where there are no O atoms

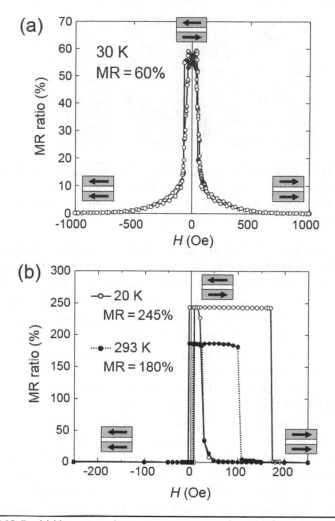

FIGURE 12.5   (a) Magnetoresistance curve measured at 30 K for epitaxial Fe(001)/ MgO(001)/Fe(001) MTJ. (Redrawn from Bowen, M. et al., *Appl. Phys. Lett.* 79, 1655, 2001.) (b) Magnetoresistance curves measured at RT and 20 K for epitaxial Fe(001)/ MgO(001)/Fe(001) MTJ. (a) (Redrawn from Yuasa, S. et al., *Nat. Mater.* 3, 868, 2004.)

in the first Fe monolayer at the interface, the Fe Δ_1 Bloch states effectively couple with the MgO Δ_1 evanescent states in the $k_{//} = 0$ direction, which results in the coherent tunneling of Δ_1 states and thus a very large TMR effect. At the oxidized interface, where there are excess oxygen atoms in the interfacial Fe monolayer, the Fe Δ_1 states do not effectively couple with the MgO Δ_1 states. This prevents coherent tunneling of Δ_1 states and significantly reduces the MR ratio. Coherent tunneling is thus very sensitive to the structure of barrier/electrode interfaces. Oxidation of even a monolayer at the interface significantly suppresses the TMR effect.

In 2004, Yuasa et al. fabricated high-quality fully epitaxial Fe(001)/ MgO(001)/Fe(001) MTJs with a single-crystal MgO(001) tunnel barrier by using MBE growth under an ultrahigh vacuum and attained RT MR ratios that were significantly higher than those of conventional Al-O-based MTJs [22, 23]. There is a cross-sectional transmission electron microscope (TEM)

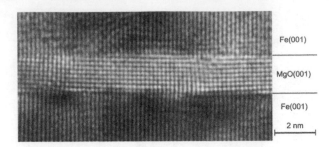

FIGURE 12.6   Cross-sectional transmission electron microscope (TEM) image of epitaxial Fe(001)/MgO(001)(1.8 nm)/Fe(001) MTJ. (Redrawn from Yuasa, S. et al., *Nat. Mater.* 3, 868, 2004.) Vertical and horizontal directions correspond to MgO[001] (Fe[001]) axis and that of MgO[100] (Fe[110]).

image of an epitaxial MTJ in Figure 12.6, where single-crystal lattices of MgO(001) and Fe(001) are clearly evident. The magnetoresistance curves of the epitaxial Fe(001)/MgO(001)/Fe(001) MTJ are shown in Figure 12.5b. They obtained giant MR ratios of up to 188% at RT (see also Figure 12.2) [22, 23]. During the same period, Parkin et al. fabricated highly oriented poly-crystal (or textured) FeCo(001)/MgO(001)/FeCo(001) MTJs with a textured MgO(001) tunnel barrier by using sputtering deposition on an SiO$_2$ substrate with a tantalum nitrate seed layer that was used to orient the entire MTJ stack in the (001) plane [24]. They obtained a giant MR ratio up to 220% at RT (Figure 12.2) [24]. These were the first RT MR ratios that were higher than the previous highest MR ratio obtained with an Al-O-based MTJ. It should be noted that fully epitaxial MTJs and textured MTJs are basically the same from a microscopic point of view if structural defects such as grain boundaries do not strongly influence the tunneling properties. It is therefore considered that the giant MR ratios observed in the epitaxial and textured MTJs originate from the same mechanism. This very large TMR effect in MgO-based MTJs is called the *giant TMR effect*.

The key to obtaining such high MR ratios is thought to be the production of clean Fe/MgO interfaces without oxidized Fe atoms. X-ray absorption spectroscopy (XAS) and x-ray magnetic circular dichroism (XMCD) studies have revealed that none of the Fe atoms adjacent to the MgO(001) layer are oxidized, indicating that there are no oxygen atoms in the interfacial Fe monolayer [25]. Since the experimental achievements of a giant TMR effect in 2004, even higher MR ratios of up to 410% at RT have been demonstrated in epitaxial Co/MgO/Fe and Co/MgO/Co MTJs with metastable bcc Co(001) electrodes (Figure 12.2) [26, 27]. The fact that the MR ratio of Co/MgO/Co MTJ is even larger than that of Fe/MgO/Fe MTJ qualitatively agrees with the theoretical prediction made by Zhang et al. [14].

After the achievement of giant MR ratios with MgO(001) barrier in 2004, extensive experimental efforts have been made for synthesizing crystalline tunnel barriers with other materials. Recently, Sukegawa et al. successfully fabricated fully epitaxial Fe(-Co)(001)/Mg-Al-O(001)/Fe(-Co)(001) MTJs with spinel-type crystalline Mg-Al-O(001) tunnel barrier and attained MR ratios of up to 300% at room temperature (Figure 12.2), which is due to the coherent tunneling of Δ_1 electrons [28, 29]. With advanced epitaxial growth techniques, it will be possible to develop crystalline tunnel barrier with various kinds of new materials, some of which may outperform the MgO(001) tunnel barrier in the future.

12.4 OTHER PHENOMENA OBSERVED IN TUNNEL JUNCTIONS WITH CRYSTALLINE MgO(001) BARRIER

It is difficult to investigate the detailed mechanisms of the TMR effect by using an amorphous Al-O barrier because of its structural uncertainty, and no significant progress has been made toward the further understanding of the physics of the TMR effect since Julliere proposed his phenomenological model in 1975. Unlike Al-O-based MTJs, epitaxial MTJs with a single-crystal MgO(001) barrier is a model system for studying the physics of spin-dependent tunneling because of their well-defined crystalline structure with atomically flat interfaces. Other than the giant TMR effect, several interesting phenomena have been observed in epitaxial MgO-based MTJs. For example, epitaxial Fe(001)/MgO(001)/Fe(001) MTJs were observed to exhibit an oscillatory TMR effect with respect to MgO barrier thickness [23, 30], and complex spin-dependent tunneling spectra [31]. It should be noted that these phenomena have never been observed in conventional MTJs with an amorphous Al-O barrier. Clarifying their underlying mechanisms should lead to a deeper understanding of the physics of spin-dependent tunneling. Of all the novel phenomena that have been observed in epitaxial MgO-based MTJs, we will discuss the interlayer exchange coupling mediated by tunneling electrons and injection of spin-polarized electrons through the MgO(001) barrier in this section.

12.4.1 INTERLAYER EXCHANGE COUPLING MEDIATED BY TUNNELING ELECTRONS

The epitaxial Fe(001)/MgO(001)/Fe(001) MTJ is also a model system used in studying interlayer exchange coupling (IEC) between two ferromagnetic (FM) layers separated by an insulating nonmagnetic (NM) spacer. IEC in an FM/NM/FM structure with a metallic NM spacer has been extensively studied and is well known to exhibit oscillations as a function of the spacer thickness [30, 32]. IEC for a metallic spacer is induced by conduction electrons at E_F. Although similar IEC is also expected in an FM/NM/FM structure with an insulating NM spacer (i.e., an MTJ) [33, 34], such intrinsic IEC has not been observed in MTJs with an amorphous Al-O barrier. Faure-Vincent et al. were the first to observe intrinsic antiferromagnetic IEC on which extrinsic ferromagnetic magnetostatic coupling was superposed in an epitaxial Fe(001)/MgO(001)/Fe(001) MTJ structure, as shown in Figure 12.7a [35]. Katayama et al. obtained a more refined experimental result that exhibited little extrinsic magnetostatic coupling (Figure 12.7b) [36]. Antiferromagnetic coupling is seen at $t_{MgO} < 0.8$ nm. With increasing t_{MgO} the IEC changes sign at $t_{MgO} = 0.8$ nm and then gradually approaches zero. Because of the atomically flat

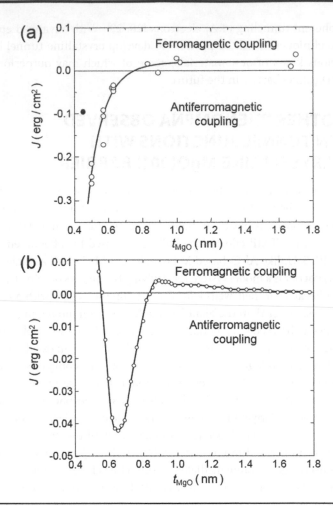

FIGURE 12.7 Interlayer exchange coupling (IEC) in epitaxial Fe(001)/MgO(001)/Fe(001) at RT. (a) Redrawn from Faure-Vincent et al. [35] and (b) redrawn from Katayama et al. [36].

barrier/electrode interfaces (see Figure 12.6), no extrinsic magnetostatic coupling was observed. The IEC for a MgO(001) spacer is apparently mediated by spin-polarized tunneling electrons.

12.4.2 SPIN INJECTION THROUGH CRYSTAL MgO(001) BARRIER

Spin injection into semiconductors is the key to developing semiconductor spintronic devices such as Datta-Das-type spin-dependent field-effect transistors [37] and spin-dependent metal-oxide-semiconductor field-effect transistors (spin MOSFETs) [38]. However, the direct contact of ferromagnetic metal on semiconductors results in very small spin polarization because of the so-called *conductivity mismatch* problem [39]. It was theoretically proposed that this problem can be solved by inserting a tunnel barrier between the ferromagnetic metal and semiconductor [40, 41] (for details, see Chapter 2, Volume 2). A bcc FeCo(001) electrode combined with a crystalline MgO(001) tunnel barrier is especially useful for injecting

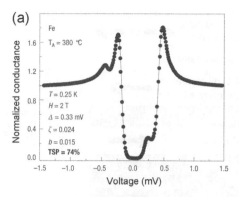 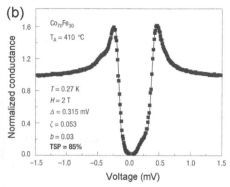

FIGURE 12.8   Measurement of tunneling spin polarization (TSP) by using Tedrow-Meservey method. Symbols denote experimental data and curves denote fitting. The superconducting counter electrode is $Al_{96}Si_4$. Tunnel barrier is crystalline MgO(001). Ferromagnetic electrode is (a) Fe and (b) $Co_{70}Fe_{30}$ ferromagnetic. (After Parkin, S. S. P. et al., *Nat. Mater.* 3, 862, 2004. With permission.)

spin-polarized electrons into a semiconductor because the tunneling current is highly spin-polarized.

Parkin et al. [24] first injected tunneling electrons from a bcc FeCo(001) electrode into a superconducting counter electrode of $Al_{96}Si_4$ through the MgO(001) barrier at low temperature to evaluate the magnitude of tunneling spin polarization P using the Tedrow-Meservey method (see Section 9.3.1) [6]. Typical results are plotted in Figure 12.8. They observed $P = 74\%$ for a Fe electrode and 85% at 0.3 K for a $Co_{70}Fe_{30}$ electrode. Such high P values have never been observed in Al-O-based tunnel junctions. Although these values do not agree precisely with the MR ratio of a CoFe/MgO/CoFe MTJ when Julliere's model is applied, the fact that a $Co_{70}Fe_{30}$ electrode has a higher P than a Fe electrode agrees qualitatively with the theory on MgO-based MTJs [14]. The high spin polarization is closely related to the high MR ratio of an MgO-based MTJ.

Jiang and Parkin et al. used the CoFe/MgO(001) structure to inject spins into a III-V semiconductor based on GaAs [42]. They fabricated the CoFe/MgO(001) spin injector on a GaAs/AlGaAs quantum well structure that acted as a light emitting diode (LED). When a spin-polarized current with the quantization axis perpendicular to the plane was injected into the quantum well structure, circularly polarized light was emitted. By measuring the circular polarization of the light, the spin polarization of the electrons injected into the semiconductor could be estimated. They observed a circular polarization of 57% at 100 K and 47% at 290 K. It should be noted that such a high degree of circular polarization by spin injection through an amorphous Al-O tunnel barrier has never been observed. It was thus proved that the MgO(001) tunnel barrier is very useful for obtaining spin injection into semiconductors.

12.5 CoFEB/MgO/CoFeB MAGNETIC TUNNEL JUNCTIONS FOR DEVICE APPLICATIONS

12.5.1 MTJ STRUCTURE FOR PRACTICAL APPLICATIONS

As explained in Section 12.3, epitaxial MTJs with a single-crystal MgO(001) barrier or textured MTJs with a poly-crystal MgO(001) barrier exhibit the

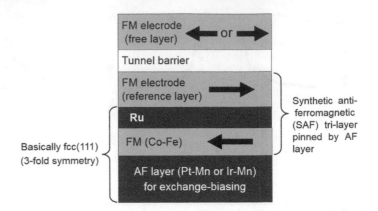

FIGURE 12.9 Cross-sectional structure of MTJ for practical applications.

giant TMR effect at RT. This is an attractive property for device applications such as MRAMs and the read heads of HDDs. These MTJ structures, however, cannot be applied to practical devices because of the following problem. MTJs for practical applications need to have the stacking structure shown in Figure 12.9. That is, they need to have an antiferromagnetic (AF) layer for exchange biasing, a synthetic antiferromagnetic (SAF) tri-layer structure that acts as a pinned layer (reference layer), a tunnel barrier, and a ferromagnetic layer that acts as a free layer. Ir-Mn or Pt-Mn is used as the antiferromagnetic layer. The SAF structure, which consists of an antiferromagnetically coupled FM/NM/FM tri-layer such as CoFe/Ru/CoFe, is exchange biased by the AF layer and acts as the pinned layer of a spin valve. This type of pinned-layer structure is indispensable for device applications because of its robust exchange bias and small stray magnetic field acting on the top free layer. A reliable AF/SAF pinned layer is based on an fcc structure with (111)-orientation. A fundamental problem with this structure is that a NaCl-type MgO(001) tunnel barrier and a bcc(001) ferromagnetic electrode, both of which have fourfold in-plane crystallographic symmetry, cannot be grown on the fcc(111)-oriented AF/SAF structure that has threefold in-plane symmetry because of the mismatch in structural symmetry. We could theoretically solve this problem by developing a new pinned-layer structure having fourfold in-plane symmetry. This solution, however, is not acceptable to the electronics industry because they have spent more than a decade in developing a reliable pinned-layer structure. It should be noted that the reliability of practical devices is determined by the reliability of the pinned layer.

12.5.2 Giant TMR Effect in CoFeB/MgO/CoFeB MTJs

To solve this problem concerning crystal growth, Djayaprawira et al. developed a new MTJ structure, i.e., CoFeB/MgO/CoFeB, by using a sputtering-deposition technique [43]. There is a cross-sectional TEM image of the MTJ in an *as-grown* state in Figure 12.10a. As can be seen in the figure, the bottom and top CoFeB electrode layers have an amorphous structure in the as-grown state. Surprisingly, the MgO barrier layer grown on the amorphous

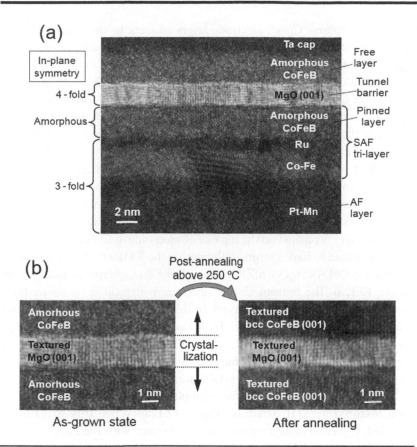

FIGURE 12.10   (a) Cross-sectional TEM image in an as-grown state of CoFeB/MgO/CoFeB MTJ with practical AF/SAF structure underneath MTJ part. (b) Cross-sectional TEM image of MTJ part in as-grown state and after annealing (courtesy of the Canon-Anelva Corp.).

CoFeB is (001)-oriented poly-crystalline. Because the bottom CoFeB electrode is amorphous, this CoFeB/MgO/CoFeB MTJ can theoretically be grown on all kinds of underlayers. For practical applications, the CoFeB/MgO/CoFeB MTJ can be grown on the standard AF/SAF pinned-layer structure shown in Figure 12.10a. It is possible to grow the MgO(001) barrier with fourfold symmetry on the practical pinned layer with threefold symmetry by inserting the amorphous CoFeB bottom electrode layer between them. After annealing at 360°C, the CoFeB/MgO/CoFeB MTJ exhibited an MR ratio of 230% at RT (Figures 12.2 and 12.11) [43]. Up to now, MR ratios above 600% have been obtained at RT in CoFeB/MgO/CoFeB MTJs [44]. The giant TMR effect in CoFeB/MgO/CoFeB MTJs is closely related to the crystallization of the CoFeB electrode layers during the post-annealing process, which is explained below.

As described in Section 12.2, crystallographic symmetry in both the MgO barrier and the electrodes is essential for the coherent tunneling of Δ_1 states. Because the fully spin-polarized Δ_1 band in bcc(001) ferromagnetic electrodes is essential for the giant TMR effect, amorphous CoFeB electrodes are not expected to result in a giant MR ratio. It has, however, been observed that the amorphous CoFeB electrode layers adjacent to the MgO(001) barrier layer crystallized in the bcc(001) structure during annealing above 250°C (Figure 12.10b) [45–47]. It should be noted that a $Co_{60}Fe_{20}B_{20}$ layer adjacent to an MgO(001) layer crystallizes in a bcc(001) structure although the stable structure of $Co_{60}Fe_{20}B_{20}$ is not bcc but fcc [45]. This clearly indicates that the MgO(001) layer acts as a template for crystallizing the amorphous CoFeB layers because of the good lattice matching between MgO(001) and bcc CoFeB(001). This type of crystallization process is known as *solid phase epitaxy*. The giant TMR effect observed in CoFeB/MgO/CoFeB MTJs can therefore be interpreted within the framework of the theory on epitaxial MTJs because the microscopic structure of bcc CoFeB(001)/MgO(001)/bcc CoFeB(001) MTJs is basically the same as that of epitaxial Fe(001)/MgO(001)/Fe(001) MTJs.

A cap layer deposited on the top CoFeB electrode in CoFeB/MgO/CoFeB MTJs was found to have a strong influence on the TMR effect in some cases. In standard CoFeB/MgO/CoFeB MTJs, a Ta or Ru cap layer is usually used (Figure 12.12a). The bottom CoFeB electrode is deposited on an Ru layer (the spacer layer of the SAF tri-layer). In this standard structure, MR ratios of over 200% are obtained at RT. Tsunekawa et al. deposited various cap-layer materials on CoFeB/MgO/CoFeB MTJs and found that the MR ratios were significantly reduced for some cap-layer materials [48]. This is because the cap layer can influence the crystallization of the top CoFeB electrode. A $Ni_{0.8}Fe_{0.2}$ (permalloy) cap layer, for example, grown on amorphous CoFeB has a textured fcc(111) structure (Figure 12.12a). When this structure is annealed, CoFeB crystallization from the interface with the $Ni_{0.8}Fe_{0.2}$ layer takes place at about 200°C before crystallization from the MgO interface takes place at about 250°C. Consequently, the top CoFeB layer crystallizes in a textured fcc(111) structure (Figure 12.12a) [45]. Because the top CoFeB electrode does not have a bcc(001) structure in this case, coherent tunneling of Δ_1 electrons does not take place. This results in a significant decrease in

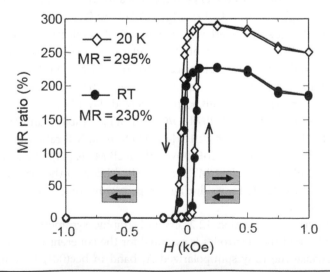

FIGURE 12.11 Magnetoresistance curves at RT and 20 K for CoFeB/MgO/CoFeB MTJ. (Redrawn from Djayaprawira, D. D. et al., *Appl. Phys. Lett.* 86, 092502, 2005.)

the MR ratio. Thus, careful optimizations of the free-layer stacking is necessary to obtain giant MR ratios in CoFeB/MgO/CoFeB MTJs. As shown in Figure 12.12c, for example, a Ta layer inserted between amorphous CoFeB and $Ni_{0.8}Fe_{0.2}$ effectively prevents the crystallization of CoFeB from $Ni_{0.8}Fe_{0.2}$ side. Reducing the B concentration in CoFeB near MgO(001) layer is effective for promoting the crystallization of CoFeB from MgO side. When the Ta insertion layer is ultra-thin, the CoFeB and $Ni_{0.8}Fe_{0.2}$ layers are ferromagnetically coupled, making the free-layer magnetization soft enough for magnetic-sensor applications.

As mentioned above, the CoFeB/MgO/CoFeB MTJs can be fabricated by sputtering deposition at RT followed by *ex situ* annealing. This makes it possible to fabricate the MTJ films on large substrates with high throughput. Djayaprawira et al. actually deposited CoFeB/MgO/CoFeB MTJ films on 8-inch-diameter Si substrates by using a standard sputtering-deposition chamber to mass-produce HDD read heads and MRAMs [43]. Because this fabrication process is highly compatible with high-throughput mass-production processes, current research and development efforts devoted to producing spintronic devices such as MRAMs and HDD read heads are based on this MTJ structure. The CoFeB/MgO/CoFeB MTJs are also commonly used in basic researches on spin-torque-induced phenomena such as magnetization switching and microwave oscillation. These applications and researches are explained more fully in Section 12.7.

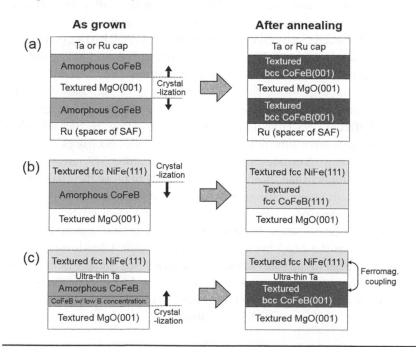

FIGURE 12.12 Schematic of the structure of CoFeB/MgO/CoFeB MTJ in as-grown state and after annealing above about 250°C. (a) Cap layer is Ta or Ru. (b) Cap layer is $Ni_{0.8}Fe_{0.2}$ (permalloy). (c) Cap layer is $Ni_{0.8}Fe_{0.2}$, and ultra-thin Ta layer is inserted between amorphous CoFeB and $Ni_{0.8}Fe_{0.2}$ to prevent crystallization of CoFeB from $Ni_{0.8}Fe_{0.2}$ side. B concentration is reduced near MgO(001) layer to promote crystallization of CoFeB from MgO side.

12.6 MTJS WITH HEUSLER-ALLOY ELECTRODES

Heusler alloys have an ordered bcc structure with $L2_1$-type chemical ordering. According to band calculations [49], some Heusler alloys are *half metals*, which have full tunneling spin polarization ($|P| = 1$) and are therefore theoretically expected to yield giant MR ratios even in Al-O-based MTJs and current-in-plane (CPP) GMR junctions. Experimentally, a very high MR ratio of 570% at low temperatures has been obtained in $Co_2MnSi/Al-O/Co_2MnSi$ MTJs with an amorphous Al-O barrier [11]. However, the giant MR ratio was observed to disappear at RT as a result of a rapid decrease in the MR ratio with temperature. This giant TMR effect at low temperature should originate from the half-metallic nature of Heusler-alloy electrodes because the amorphous Al-O barrier does not have a function for orbital-symmetry filtering, unlike the crystalline MgO(001) barrier.

A MTJ with an MgO(001) barrier and (001)-oriented Heusler-alloy electrodes can be fabricated experimentally because of the relatively small lattice mismatch of a few percent. High-quality epitaxial MTJs with an MgO(001) barrier and Heusler-alloy electrodes have been recently developed and have been observed to demonstrate giant MR ratios of up to about 380% at RT (about 830% at low temperatures) [50–53]. This giant TMR effect, however, is not proof of the half-metallic nature of Heusler-alloy electrodes because the MgO(001) barrier can yield a giant TMR effect if the Δ_1 Bloch states of the electrodes are fully spin-polarized at E_F. It should be noted that the Δ_1 band of half-metallic Heusler alloys is also fully spin-polarized at E_F. Note also that, as described in Sections 12.3 and 12.5, when combined with a crystalline MgO(001) barrier, even simple ferromagnetic electrodes such as bcc Fe, Co, and CoFeB yield MTJs with MR ratios of up to 600% at RT (1000% at low temperatures) [23, 27, 43, 44]. In both Al-O-based and MgO-based MTJs, RT MR ratios of Heusler-alloy electrodes have never exceeded those of CoFe or CoFeB electrodes since the discovery of RT TMR effect in 1995. From the engineering point of view, Heusler-alloy electrodes are much more difficult to synthesize than CoFeB electrodes.

Heusler-alloy electrodes, therefore, need to exhibit MR ratios that are substantially higher than those of CoFeB not only to prove their half-metallic nature but also to establish their technological importance. Theoretically, Heusler-alloy electrodes have an advantage over CoFeB especially in the ultra-low-resistance range because the MgO's orbital-symmetry filtering weakens as the MgO barrier becomes ultra-thin. Ultra-low-resistance MTJs are crucial in magnetic-sensor applications (see Section 12.7.1). Thus, Heusler alloys still have a great deal of potential for device applications especially in the ultra-low-resistance range either as the electrodes of MTJs or as the ferromagnetic layers of CPP-GMR junctions.

12.7 DEVICE APPLICATIONS OF MgO-BASED MTJS

From the viewpoint of device applications, the history of spintronics is a history of magnetoresistance at RT. The history of magnetoresistance and

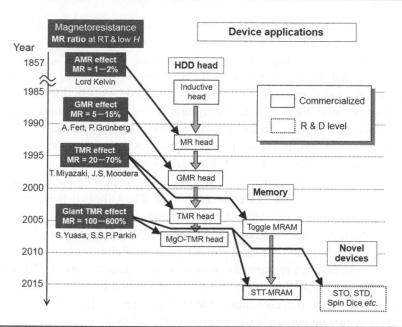

FIGURE 12.13 History and perspectives of magnetoresistive effects, MR ratios at RT and applications of MR effects in spintronic devices.

its applications is summarized in Figure 12.13. GMR spin-valve devices have MR ratios of 5–15% at RT and have been used in the read heads of HDDs. Al-O-based MTJs have MR ratios of 20%–70% at RT and have been used not only in HDD read heads but also in MRAM cells. MgO-based MTJs have MR ratios of 100%–1000% at RT and are expected to be used in various spintronic devices such as HDD read heads, spin-transfer MRAM cells, and novel microwave devices. In this section, we briefly describe recent developments in these applications. Further details on the HDD read heads and MRAM are explained in Chapters 12 and 13, Volume 3.

12.7.1 READ HEADS OF HDDS

As explained in Section 12.5, CoFeB/MgO/CoFeB MTJs with the giant TMR effect are compatible with the mass-manufacturing processes for spintronics devices because they can be fabricated on a practical pinned-layer structure by sputtering deposition at RT followed by post-annealing. Apart from requiring giant MR ratios and compatibility with manufacturing processes, industrial applications require MTJs that have MR ratios that have little dependence on bias voltage, that are robust and have high breakdown voltages, that can be manufactured uniformly and reproducibly, that have a free layer with small magnetostriction, and that have appropriate resistance-area (*RA*) products. It is now possible to satisfy all these requirements by using CoFeB/MgO/CoFeB MTJs. At the early stage of research and development, however, an ultra-low *RA* product that was required for read head applications was not easy to accomplish. In this subsection, we explain how the ultra-low *RA* product was attained and how MgO-based MTJs have been applied to the read heads of ultra-high-density HDDs.

Because impedance matching in electronic circuits is crucial for the high-speed operation of electronic devices, the RA product of MTJs should be adjusted to satisfy impedance-matching conditions. MRAM applications require an RA in a range from several $\Omega \cdot \mu m^2$ to several $k\Omega \cdot \mu m^2$, depending on the lateral MTJ size (i.e., the areal density of an MRAM). In this RA range, MR ratios of over 200% at RT can easily be obtained using MgO-based MTJs. A very low RA product, on the other hand, is required for the read heads of high-density HDDs. MTJs with an amorphous Al-O or Ti-O barrier have been used in the TMR read heads for HDDs with areal recording densities of 100–150 Gbit/inch^2 since 2004 [54]. These MTJs have low RA products (about 3 $\Omega \cdot \mu m^2$) and MR ratios of 20–30% at RT. Although these properties are sufficient for recording densities of 100–150 Gbit/inch^2, even lower RA products and much higher MR ratios are needed for recording densities above 200 Gbit/inch^2. For example, RA products below 2 $\Omega \cdot \mu m^2$ and MR ratios above 50% are required for areal recording densities above 250 Gbit/inch^2 (Figure 12.14). Such ultra-low RA products and high MR ratios have never been obtained in conventional MTJs with an amorphous Al-O or Ti-O barrier. A current perpendicular to plane (CPP) GMR device, which is one of the candidates for the next-generation of HDD read heads, has an ultra-low RA product (below 1 $\Omega \cdot \mu m^2$), but the MR ratio of a CPP-GMR device is too low (<10% for a practical spin-valve structure) for a device that is used as an HDD read head (Figure 12.14).

To reduce the RA product, Tsunekawa et al. made CoFeB/MgO/CoFeB MTJs with an ultra-thin textured MgO(001) barrier (t_{MgO} = 1.0 nm, which corresponds to only 4–5 ML) [55]. Because the textured growth of MgO(001) on amorphous CoFeB deteriorates below an MgO thickness of 1.0 nm [45], it is necessary to carefully optimize the conditions for growing a 1.0-nm-thick textured MgO(001) barrier. For example, the textured growth of MgO layers

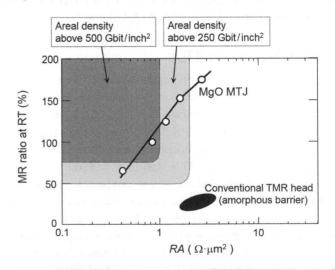

FIGURE 12.14   MR ratio at RT *vs.* resistance-area (*RA*) product. Open circles are values for CoFeB/MgO/CoFeB MTJs. (Redrawn from Nagamine, Y. et al., *Appl. Phys. Lett.* 89, 162507, 2006.) Light gray and dark gray areas are zones required for HDDs with recording densities above 250 Gbit/inch^2 and 500 Gbit/inch^2.

was found to be sensitive to the base pressure of the sputtering chamber. Residual H_2O molecules in the chamber were found to degrade the crystalline orientation of MgO(001), reducing the MR ratio. By carefully removing residual H_2O molecules by using a Ta getter technique, Nagamine et al. were able to grow a highly textured 1.0-nm-thick MgO(001) barrier [56]. The removal of residual H_2O molecules was also found to effectively prevent the CoFeB/MgO interface from becoming oxidized [57]. The MR ratios for these CoFeB/MgO/CoFeB MTJs are indicated by the open circles in Figure 12.14. Nagamine et al. obtained both ultra-low RA products (<1 $\Omega \cdot \mu m^2$) and high MR ratios (>100% at RT). More recently, Maehara et al. obtained MR ratios above 170% for RA products below 1 $\Omega \cdot \mu m^2$ [58]. These properties satisfy the requirements for areal recording densities well above 500 Gbit/inch^2.

By using ultra-low-resistance MgO-based MTJs, HDD manufacturers were able to commercialize TMR read heads for ultrahigh-density HDDs in 2007. For example, Fujitsu Corp. commercialized TMR read heads for ultra-high-density HDDs in 2007. There is a cross-sectional TEM image of a TMR read head in Figure 12.15. MgO-TMR heads have already been used for HDDs with recording densities of 250–800 Gbit/inch^2. This means that the recording density of HDDs has tripled thanks to MgO-TMR heads combined with perpendicular magnetic recording media. MgO-TMR heads currently represent the mainstream technology for HDD read heads and are also expected to be applied to recording densities of up to about 1–2 Tbit/inch^2.

12.7.2 Spin-Transfer-Torque MRAMs

The giant TMR effect in MgO-based MTJs is also useful in developing MRAMs. In conventional MRAMs, the writing process (i.e., magnetization reversal of a free layer) uses a magnetic field generated by pulse currents, and the readout process uses a resistance change between parallel and antiparallel magnetic states (i.e., the TMR effect). The giant TMR effect enables high-speed readout because a giant MR ratio yields a very high output signal [59].

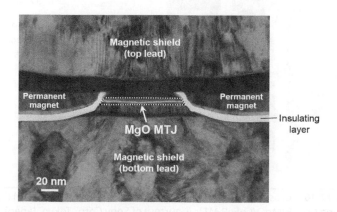

FIGURE 12.15 Cross-sectional TEM image of MgO-TMR read head for HDD with a recording density of 250 Gbit/inch^2 (courtesy of the Fujitsu Corp.).

In conventional MRAMs, however, the writing pulse currents increase when the lateral size of MTJs is reduced, which makes it difficult to develop Gbit-scale high-density MRAMs. In the new types of MRAMs called spin-transfer-torque MRAMs (STT-MRAMs), on the other hand, the writing process uses the magnetization switching induced by spin-transfer torque (for details, see Chapter 15, Volume 1). This phenomenon, called spin-transfer-torque switching (STT switching) [60], is especially important in developing high-density MRAMs because the writing pulse current flowing through the MTJ can be reduced by reducing the MTJ's lateral size. STT switching was experimentally demonstrated first in CPP-GMR devices [61] and later in Al-O-based MTJs [62]. STT switching in MgO-based MTJs, which is especially important for MRAMs, has been demonstrated using CoFeB/MgO/CoFeB MTJs [63–66].

STT-MRAM cells based on the giant TMR effect and STT switching in MgO-based MTJs have been developed. The 4-kbit STT-MRAM (or Spin-RAM) developed by Sony Corp. (shown in Figure 12.16), for example, provides reliable readout and write operations [67]. In CoFeB/MgO/CoFeB MTJs, the intrinsic critical current density, J_{c0}, or switching pulse-current density when the pulse duration is 1 ns, is about 2×10^6 A/cm^2 [66, 67]. This J_{c0} is, however, not small enough for high-density STT-MRAMs. The J_{c0} value needs to be smaller than about 1×10^6 A/cm^2 and high thermal stability for free-layer magnetization is necessary to retain data for more than 10 years to develop Gbit-scale STT-MRAMs. High thermal stability is especially difficult when the lateral size of MTJ cells is smaller than about 50 nm, which is a typical cell size in Gbit-scale STT-MRAMs.

A solution to achieving Gbit-scale STT-MRAMs is to use MTJs with perpendicularly magnetized electrodes. It is theoretically possible to simultaneously have high thermal stability in 50 nm-sized MTJ cells and a J_{c0} smaller than 1×10^6 A/cm^2 by using perpendicularly magnetized electrodes with a

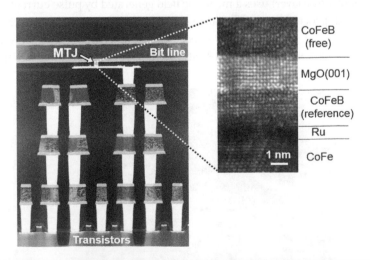

FIGURE 12.16   Cross-sectional TEM images of 4-kbit spin-torque MRAM (Spin-RAM) using CoFeB/MgO/CoFeB MTJs (Courtesy of Sony Corp., Tokyo, Japan) (Redrawn from Hosomi, M. et al., A novel nonvolatile memory with spin torque transfer magnetization switching: Spin-RAM, *Technical Digest of IEEE International Electron Devices Meeting (IEDM)*: 19.1, Washington, DC, 2005.)

high perpendicular magnetic anisotropy, K_u. Toshiba Corp. recently developed a proto-type STT-MRAM using 50 nm-sized MgO-based MTJs with perpendicularly magnetized electrodes (Figure 12.17) [68]. They successfully demonstrated a small J_{c0}, thermal stability that was high enough to retain data for more than a decade, and high-speed read/write operations. More recently, Ikeda et al. developed perpendicularly magnetized CoFeB/MgO/CoFeB MTJs with very thin CoFeB electrode layers [69]. Here, the perpendicular magnetization originates from interfacial perpendicular magnetic anisotropy (PMA) energy at the MgO/CoFeB interfaces. When the CoFeB layer is thicker than about 2 nm, the CoFeB layer has in-plane magnetization because the demagnetization energy, which prefers the in-plane magnetization, is larger than the interfacial PMA energy (Figure 12.18a). When the CoFeB layer is thinner than about 1–1.5 nm, on the other hand, the interfacial PMA energy can be larger than the demagnetization energy, resulting in the perpendicularly magnetized CoFeB. The PMA energy of CoFeB layer can be reinforced by laminating it with a bulk PMA material layer (Figure 12.18b) or by covering it with a MgO cap layer and doubling the interfacial PMA energy. The perpendicularly magnetized MgO-based MTJs are a promising solution to accomplishing Gbit-scale STT-MRAMs.

12.7.3 MICROWAVE APPLICATIONS

MgO-based MTJs are also potentially useful for microwave device applications. Tulapurlar et al. demonstrated that a DC voltage is generated between the two electrodes when an AC current with a microwave frequency flows through a CoFeB/MgO/CoFeB MTJ with a lateral size of about 100 nm [70]. This phenomenon called the *spin-torque diode (STD) effect* or *spin-torque ferromagnetic resonance (ST-FMR)*, results from a combination of the giant TMR effect and resonant precession of the free-layer magnetic moment induced by spin-transfer torque. Because of this effect, MgO-based MTJs can thus act as microwave detectors. A spin-torque diode effect is also a powerful tool for measuring the magnitude of spin-transfer torque. Using

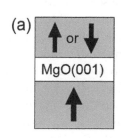
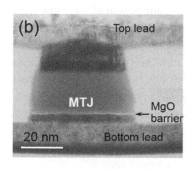
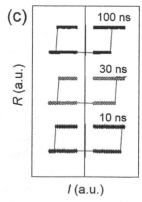

FIGURE 12.17   MgO-based MTJ with perpendicularly magnetized electrodes (a) Schematic and (b) cross-sectional TEM image of MTJ. (c) Spin-torque switching at pulse durations of 10 - 100 ns. (Redrawn from Kishi, T. *et al.*, Lower-current and fast switching of a perpendicular TMR for high speed and high density spin-transfer-torque MRAM, *Technical Digest of IEEE International Electron Devices Meeting (IEDM)*: 12.6, San Francisco, CA, 2008.)

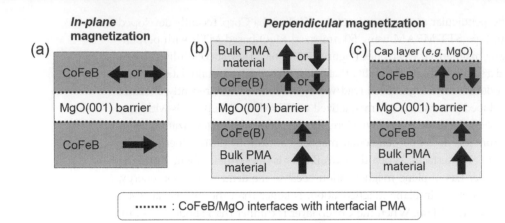

FIGURE 12.18 Basic structure of practical CoFeB/MgO/CoFeB MTJs. (a) MTJ with in-plane magnetization. (b)–(c) MTJs with perpendicular magnetization. Horizontal dotted lines represent CoFeB/MgO interfaces with interfacial perpendicular magnetic anisotropy (PMA) energy.

the spin-torque diode effect, Sankey et al. [71] and Kubota et al. [72] carried out quantitative measurements of the magnitude of spin-transfer torque as a function of the applied bias voltage. More recently, Miwa et al. demonstrated that spin-torque diodes under optimal conditions can outperform p-n junction diodes [73].

An inverse of the spin-torque diode effect is the microwave emissions that occur when a DC current flows through a 100-nm-scale magnetoresistive device. Spin-transfer torque acting on the free-layer magnetic moment can induce steady precession of the free-layer moment at an FMR frequency. The steady precession of the free-layer moment in magnetoresistive device induces an AC current or AC voltage at an FMR frequency (i.e., a microwave frequency). This microwave emission was first demonstrated using a CPP-GMR device with an MR ratio of about 1% at RT [74]. Such a microwave oscillator is called a spin-torque oscillator (STO). The microwave power from a CPP-GMR-based STO, however, is only of the order of nanowatts and is too small for practical use. However, because the microwave power is theoretically proportional to the square of the MR ratio, MgO-based MTJs with giant MR ratios are potentially able to emit high-power microwaves. We recently obtained microwave emissions of microwatt order from a 100-nm CoFeB/MgO/CoFeB MTJ with an MR ratio of about 100% [75]. More recently, emission power higher than a few μW, as well as very narrow linewidth, has been demonstrated in MgO-MTJ-based STOs [76–78]. Moreover, a practical STO stabilized by a phase-locked loop (PLL) was also developed [79]. MgO-MTJ-based STOs are expected to be the third major application of spintronics technology.

As previously described, MgO-based MTJs can be used in various spintronic devices. In these applications, output performance is roughly proportional to the MR ratio at RT. The giant TMR effect in MTJs is therefore expected not only to extend the applications of existing devices but also to help achieve novel spintronic applications (Figure 12.13).

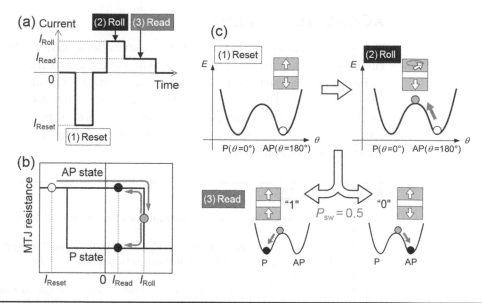

FIGURE 12.19 Principles underlying physical random number generator, Spin Dice. (a) Operation sequence and (b) corresponding R–I curve. (c) Schematics for energy diagram.

12.7.4 Physical Random Number Generator: "Spin Dice"

Another novel application of MgO-based MTJs is in nanosized physical random number generators (RNGs) called Spin Dice [80]. The working principles of Spin Dice are outlined in Figure 12.19. (1) First, reset current (I_{reset}) is passed through the MTJ to reset the initial state to an antiparallel (AP) state. (2) Then, roll current (I_{roll}) is applied to the MTJ to induce STT switching with a probability (P_{SW}) of approximately 0.5. This process is analogous to rolling classical dice. (3) Last, the final state is detected by passing sense current (I_{read}) through the MTJ. By adjusting I_{roll}, we can easily achieve $P_{SW} \approx 0.5$ and as a result have a sequence of random bits. By applying the "exclusive or" (XOR) operation to two or more independent sequences of random bits generated from two or more MTJs, we can have high-quality random bits, which satisfy the ideal equiprobability and unpredictability.

Among all types of physical RNGs, Spin Dice is the only RNG, which is scalable with silicon LSI technology. Note that Spin Dice is a kind of STT-MRAM with a switching probability of 0.5. Note also that the fabrication processes and digital circuits for Spin Dice are basically the same as those of STT-MRAM. Therefore, we can easily embed a Spin Dice circuit in an STT-MRAM chip at little additional cost. Spin Dice can theoretically generate enormous amounts of random numbers using large parallel data processing. Random numbers are essential building blocks in modern cryptographic systems to keep information communication technologies highly secure and confidential. Physical random numbers are superior to pseudo-random numbers from the viewpoint of secure data encryption. Spin Dice can, therefore, provide a novel security functionality to Si LSI technology, especially to STT-MRAM.

ACKNOWLEDGMENTS

The author would like to acknowledge the contributions made by Messrs. A. Fukushima, T. Nagahama, H. Kubota, K. Ando (AIST), D. D. Djayaprawira, K. Tsunekawa, H. Maehara, Y. Nagamine, S. Yamagata, N. Watanabe (Canon-Anelva Corp.), R. Matsumoto, A. M. Deac, and Y. Suzuki (Osaka University) for their valuable discussions, and the financial support given by the New Energy and Industrial Technology Development Organization (NEDO) and the Japan Science and Technology Corporation (JST).

REFERENCES

1. M. Julliere, Tunneling between ferromagnetic films, *Phys. Lett.* **54A**, 225–226 (1975).
2. M. N. Baibich, J. M. Broto, A. Fert, F. Nguyen Van Dau, and F. Petroff, Giant magnetoresistance of (001)Fe/(001)Cr magnetic superlattices, *Phys. Rev. Lett.* **61**, 2472–2475 (1988).
3. G. Binasch, P. Grünberg, F. Saurenbach, and W. Zinn, Enhanced magneto-resistance in layered magnetic structures with antiferromagnetic interlayer exchange, *Phys. Rev. B* **39**, 4828 (1989).
4. T. Miyazaki and N. Tezuka, Giant magnetic tunneling effect in Fe/Al$_2$O$_3$/Fe junction, *J. Magn. Magn. Mater.* **139**, L231–L234 (1995).
5. J. S. Moodera, L. R. Kinder, T. M. Wong, and R. Meservey, Large magnetoresistance at room temperature in ferromagnetic thin film tunnel junctions, *Phys. Rev. Lett.* **74**, 3273–3327 (1995).
6. R. Meservey and P. M. Tedrow, Spin-polarized electron-tunneling, *Phys. Rep.* **238**, 173–243 (1994).
7. S. S. P. Parkin et al., Magnetically engineered spintronic sensors and memory, *Proc. IEEE* **91**, 661–680 (2003).
8. S. Yuasa, T. Nagahama, and Y. Suzuki, Spin-polarized resonant tunneling in magnetic tunnel junctions, *Science* **297**, 234–237 (2002).
9. T. Nagahama, S. Yuasa, E. Tamura, and Y. Suzuki, Spin-dependent tunneling in magnetic tunnel junctions with a layered antiferromagnetic Cr(001) spacer: Role of band structure and interface scattering, *Phys. Rev. Lett.* **95**, 086602 (2005).
10. M. Bowen et al., Nearly total spin polarization in La$_{2/3}$Sr$_{1/3}$MnO$_3$ from tunneling experiments, *Appl. Phys. Lett.* **82**, 233–235 (2003).
11. Y. Sakuraba et al., Giant tunneling magnetoresistance in Co$_2$MnSi/Al-O/Co$_2$MnSi magnetic tunnel junctions, *Appl. Phys. Lett.* **88**, 192508 (2006).
12. W. H. Butler, X.-G. Zhang, T. C. Schulthess, and J. M. MacLaren, Spin-dependent tunneling conductance of Fe/MgO/Fe sandwiches, *Phys. Rev. B* **63**, 054416 (2001).
13. J. Mathon and A. Umersky, Theory of tunneling magnetoresistance of an epitaxial Fe/MgO/Fe(001) junction, *Phys. Rev. B* **63**, 220403R (2001).
14. X.-G. Zhang and W. H. Butler, Large magnetoresistance in bcc Co/MgO/Co and FeCo/MgO/FeCo tunnel junctions, *Phys. Rev. B* **70**, 172407 (2004).
15. P. Mavropoulos, N. Papanikolaou, and P. H. Dederichs, Complex band structure and tunneling through ferromagnet/insulator/ferromagnet junctions, *Phys. Rev. Lett.* **85**, 10881091 (2000).
16. J. P. Velev et al., Negative spin polarization and large tunneling magnetoresistance in epitaxial Co vertical bar SrTiO$_3$ vertical bar Co magnetic tunnel junctions, *Phys. Rev. Lett.* **95**, 216601 (2005).
17. W. Wulfhekel et al., Single-crystal magnetotunnel junctions, *Appl. Phys. Lett.* **78**, 509–511 (2001).

18. M. Bowen et al., Large magnetoresistance in Fe/MgO/FeCo(001) epitaxial tunnel junctions on GaAs(001), *Appl. Phys. Lett.* **79**, 1655–1657 (2001).

19. J. Faure-Vincent et al., High tunnel magnetoresistance in epitaxial Fe/MgO/Fe tunnel junctions, *Appl. Phys. Lett.* **82**, 4507–4509 (2003).

20. H. L. Meyerheim et al., Geometrical and compositional structure at metal-oxide interfaces: MgO on Fe(001), *Phys. Rev. Lett.* **87**, 07102 (2001).

21. X.-G. Zhang, W. H. Butler, and A. Bandyopadhyay, Effects of the iron-oxide layer in Fe-FeO-MgO-Fe tunneling junctions, *Phys. Rev. B* **68**, 092402 (2003).

22. S. Yuasa, A. Fukushima, T. Nagahama, K. Ando, and Y. Suzuki, High tunnel magnetoresistance at room temperature in fully epitaxial Fe/MgO/Fe tunnel junctions due to coherent Spin-Polarized tunneling, *Jpn. J. Appl. Phys.* **43**, L588–L590 (2004).

23. S. Yuasa, T. Nagahama, A. Fukushima, Y. Suzuki, and K. Ando, Giant room-temperature magnetoresistance in single-crystal Fe/MgO/Fe magnetic tunnel junctions, *Nat. Mater.* **3**, 868–871 (2004).

24. S. S. P. Parkin et al., Giant tunnelling magnetoresistance at room temperature with MgO (100) tunnel barriers, *Nat. Mater.* **3**, 862–867 (2004).

25. K. Miyokawa et al., X-ray absorption and X-ray magnetic circular dichroism studies of a monatomic Fe(001) layer facing a single-crystalline MgO(001) tunnel barrier, *Jpn. J. Appl. Phys.* **44**, L9–L11 (2005).

26. S. Yuasa et al., Giant tunneling magnetoresistance in fully epitaxial body-centered-cubic Co/MgO/Fe magnetic tunnel junctions, *Appl. Phys. Lett.* **87**, 222508 (2005).

27. S. Yuasa, A. Fukushima, H. Kubota, Y. Suzuki, and K. Ando, Giant tunneling magnetoresistance up to 410% at room temperature in fully epitaxial Co/MgO/Co magnetic tunnel junctions with bcc Co(001) electrodes, *Appl. Phys. Lett.* **89**, 042505 (2006).

28. H. Sukegawa et al., Tunnel magnetoresistance with improved bias voltage dependence in lattice-matched Fe/spinel $MgAl_2O_4$/Fe(001) junctions, *Appl. Phys. Lett.* **96**, 212505 (2010).

29. H. Sukegawa et al., Enhanced tunnel magnetoresistance in a spinel oxide barrier with cation-site disorder, *Phys. Rev. B* **86**, 184401 (2012).

30. R. Matsumoto, A. Fukushima, T. Nagahama, Y. Suzuki, K. Ando, and S. Juasa, Oscillation of giant tunneling magnetoresistance with respect to tunneling barrier thickness in fully epitaxial Fe/MgO/Fe magnetic tunnel junctions, *Appl. Phys. Lett.* **90**, 252506 (2007).

31. Y. Ando et al., Spin-dependent tunneling spectroscopy in single-crystal Fe/MgO/Fe tunnel junctions, *Appl. Phys. Lett.* **87**, 142502 (2005).

32. S. S. P. Parkin, N. More, and K. Roche, Oscillations in exchange coupling and magnetoresistance in metallic superlattice structures - Co/Ru, Co/Cr, and Fe/Cr, *Phys. Rev. Lett.* **64**, 2304–2307 (1990).

33. P. Bruno, Theory of interlayer magnetic coupling, *Phys. Rev. B* **52**, 411–439 (1995).

34. J. C. Slonczewski, Currents, torques, and polarization factors in magnetic tunnel junctions, *Phys. Rev. B* **71**, 024411 (2005).

35. J. Faure-Vincent et al., Interlayer magnetic coupling interactions of two ferromagnetic layers by spin polarized tunneling, *Phys. Rev. Lett.* **89**, 107206 (2002).

36. T. Katayama et al., Interlayer exchange coupling in Fe/MgO/Fe magnetic tunnel junctions, *Appl. Phys. Lett.* **89**, 112503 (2006).

37. S. Datta and B. Das, Electronic analog of the electro-optic modulator, *Appl. Phys. Lett.* **56**, 665 (1990).

38. S. Sugahara, Spin metal-oxide-semiconductor field-effect transistors (spin MOSFETs) for integrated spin electronics, *IEE Proc.-Circuits Devices Syst.* **152**, 355–365 (2005).

39. G. Schmidt, D. Ferrand, L. W. Molenkamp, A. T. Filip, and B. J. van Wees, Fundamental obstacle for electrical spin injection from a ferromagnetic metal into a diffusive semiconductor, *Phys. Rev. B* **62**, R4790 (2000).

40. E. I. Rashba, Theory of electrical spin injection: Tunnel contacts as a solution of the conductivity mismatch problem, *Phys. Rev. B* **62**, R16267 (2000).

41. A. Fert and H. Jaffres, Conditions for efficient spin injection from a ferromagnetic metal into a semiconductor, *Phys. Rev. B* **64**, 184420 (2001).

42. X. Jiang et al., Highly spin-polarized room-temperature tunnel injector for semiconductor spintronics using MgO(100), *Phys. Rev. Lett.* **94**, 056601 (2005).

43. D. D. Djayaprawira et al., 230% room-temperature magnetoresistance in CoFeB/MgO/CoFeB magnetic tunnel junctions, *Appl. Phys. Lett.* **86**, 092502 (2005).

44. S. Ikeda et al., Tunnel magnetoresistance of 604% at 300 K by suppression of Ta diffusion in CoFeB/MgO/CoFeB pseudo-spin-valves annealed at high temperature, *Appl. Phys. Lett.* **93**, 082508 (2008).

45. S. Yuasa, Y. Suzuki, T. Katayama, and K. Ando, Characterization of growth and crystallization processes in CoFeB/MgO/CoFeB magnetic tunnel junction structure by reflective high-energy electron diffraction, *Appl. Phys. Lett.* **87**, 242503 (2005).

46. S. Yuasa and D. D. Djayaprawira, Giant tunnel magnetoresistance in magnetic tunnel junctions with a crystalline MgO(001) barrier, *J. Phys. D Appl. Phys.* **40**, R337–R354 (2007).

47. Y. S. Choi et al., Transmission electron microscopy study on the polycrystalline CoFeB/MgO/CoFeB based magnetic tunnel junction showing a high tunneling magnetoresistance, predicted in single crystal magnetic tunnel junction, *J. Appl. Phys.* **101**, 013907 (2007).

48. K. Tsunekawa et al., *Digests of the IEEE International Magnetics Conference (Intermag)*: HP-08, Nagoya, Japan, 2005.

49. I. Galanakis, P. H. Dederichs, and N. Papanikolaou, Slater-Pauling behavior and origin of the half-metallicity of the full-Heusler alloys, *Phys. Rev. B* **66**, 174429 (2002).

50. N. Tezuka et al., Giant tunnel magnetoresistance at room temperature for junctions using full-heusler $Co_2FeAl_{0.5}Si_{0.5}$ electrodes, *Jpn. J. Appl. Phys.* **46**, L454–L456 (2007).

51. S. Tsunegi et al., Large tunnel magnetoresistance in magnetic tunnel junctions using a Co_2MnSi Heusler alloy electrode and a MgO barrier, *Appl. Phys. Lett.* **93**, 112506 (2008).

52. N. Tezuka et al., Improved tunnel magnetoresistance of magnetic tunnel junctions with Heusler $Co_2FeAl_{0.5}Si_{0.5}$ electrodes fabricated by molecular beam epitaxy, *Appl. Phys. Lett.* **94**, 162504 (2009).

53. W. H. Wang et al., Giant tunneling magnetoresistance up to 330% at room temperature in sputter deposited Co_2FeAl/MgO/CoFe magnetic tunnel junctions, *Appl. Phys. Lett.* **95**, 182502 (2009).

54. J.-G. Zhu and C.-D. Park, Magnetic tunnel junctions, *Mater. Today* **9–11**, 36–45 (2006).

55. T. Tsunekawa et al., Giant tunneling magnetoresistance effect in low-resistance CoFeB/MgO(001)/CoFeB magnetic tunnel junctions for read-head applications, *Appl. Phys. Lett.* **87**, 072503 (2005).

56. Y. Nagamine et al., Ultralow resistance-area product of 0.4 Ω (µm)2 and high magnetoresistance above 50% in CoFeB/MgO/CoFeB magnetic tunnel junctions, *Appl. Phys. Lett.* **89**, 162507 (2006).

57. Y. S. Choi et al., Effect of Ta getter on the quality of MgO tunnel barrier in the polycrystalline CoFeB/MgO/CoFeB magnetic tunnel junction, *Appl. Phys. Lett.* **90**, 012505 (2007).

58. H. Maehara et al., Tunnel magnetoresistance above 170% and resistance–area product of 1 Ω (µm)^2 attained by in situ annealing of ultra-thin MgO tunnel barrier, *Appl. Phys. Express* **4**, 033002 (2011).

59. J. M. Slaughter et al., High speed toggle MRAM with MgO-based tunnel junctions, *Technical Digest of IEEE International Electron Devices Meeting (IEDM)*: 35.7, 2005.

60. J. C. Slonczewski, Current-driven excitation of magnetic multilayers, *J. Magn. Magn. Mater.* **159**, L1–L7 (1996).
61. J. A. Katine et al., Current-driven magnetization reversal and spin-wave excitations in Co/Cu/Co pillars, *Phys. Rev. Lett.* **84**, 3149–3152 (2000).
62. Y.-M. Huai et al., Observation of spin-transfer switching in deep submicron-sized and low-resistance magnetic tunnel junctions, *Appl. Phys. Lett.* **84**, 3118–3120 (2004).
63. H. Kubota et al., Evaluation of spin-transfer switching in CoFeB/MgO/CoFeB magnetic tunnel junctions, *Jpn. J. Appl. Phys.* **44**, L1237–L1240 (2005).
64. H. Kubota et al., Dependence of spin-transfer switching current on free layer thickness in Co-Fe-B/MgO/Co-Fe-B magnetic tunnel junctions, *Appl. Rev. Lett.* **89**, 032505 (2006).
65. J. Hayakawa et al., Current-driven magnetization switching in CoFeB/MgO/CoFeB magnetic tunnel junctions, *Jpn. J. Appl. Phys.* **44**, L1267–L1270 (2005).
66. Z. Diao et al., Spin transfer switching and spin polarization in magnetic tunnel junctions with MgO and AlOx barriers, *Appl. Phys. Lett.* **87**, 232502 (2005).
67. M. Hosomi et al., A novel nonvolatile memory with spin torque transfer magnetization switching: Spin-RAM, *Technical Digest of IEEE International Electron Devices Meeting (IEDM)*: 19.1, 2005.
68. T. Kishi et al., Lower-current and fast switching of a perpendicular TMR for high speed and high density spin-transfer-torque MRAM, *Technical Digest of IEEE International Electron Devices Meeting (IEDM)*: 12.6, 2008.
69. S. Ikeda et al., A perpendicular-anisotropy CoFeB–MgO magnetic tunnel junction, *Nat. Mater.* **9**, 721 (2010).
70. A. A. Tulapurkar et al., Spin-torque diode effect in magnetic tunnel junctions, *Nature* **438**, 339–342 (2005).
71. J. C. Sankey et al., Measurement of the spin-transfer-torque vector in magnetic tunnel junctions, *Nat. Phys.* **4**, 67–71 (2008).
72. H. Kutoba et al., Quantitative measurement of voltage dependence of spin-transfer torque in MgO-based magnetic tunnel junctions, *Nat. Phys.* **4**, 37–41 (2008).
73. S. Miwa et al., Highly sensitive nanoscale spin-torque diode, *Nat. Mater.* **13**, 50 (2014).
74. S. I. Kiselev et al., Microwave oscillations of a nanomagnet driven by a spin-polarized current, *Nature* **425**, 380–383 (2003).
75. A. M. Deac et al., Bias-driven high-power microwave emission from MgO-based tunnel magnetoresistance devices, *Nat. Phys.* **4**, 803–809 (2008).
76. H. Kubota et al., Spin-torque oscillator based on magnetic tunnel junction with a perpendicularly magnetized free layer and in-plane magnetized polarizer, *Appl. Phys. Express* **6**, 103003 (2013).
77. H. Maehara et al., Large emission power over 2 mW with high Q factor obtained from nanocontact magnetic-tunnel-junction-based spin torque oscillator, *Appl. Phys. Express* **6**, 113005 (2013).
78. H. Maehara et al., High Q factor over 3000 due to out-of-plane precession in nano-contact spin-torque oscillator based on magnetic tunnel junctions, *Appl. Phys. Express* **7**, 023003 (2014).
79. S. Tamaru et al., Extremely coherent microwave emission from spin torque oscillator stabilized by phase locked loop, *Sci. Rep.* **5**, 18134 (2015).
80. A. Fukushima et al., Spin dice: A scalable truly random number generator based on spintronics, *Appl. Phys. Express* **7**, 083001 (2014).

Tunneling Magnetoresistance
Theory

Kirill D. Belashchenko and Evgeny Y. Tsymbal

13.1 INTRODUCTION

Tunneling magnetoresistance (TMR) is a change of the electrical resistance of a magnetic tunnel junction (MTJ) under the influence of an external magnetic field. An MTJ is a tunnel junction with two magnetic metals serving as electrodes; the resistance change occurs when the orientation of the electrode magnetizations is switched between parallel and antiparallel configurations (see Chapters 11 and 12, Volume 1). In this respect, TMR is similar to the giant magnetoresistance (GMR) effect, where the magnetic electrodes are separated by a nonmagnetic metal, rather than an insulating barrier (see Chapters 4 and 5 for the reviews on GMR).

As reviewed in Chapters 11 and 12, TMR strongly depends on the choice of the electrode and barrier materials, as well as on the fabrication technique and quality of the interfaces. The discoveries of high and reproducible TMR values at room temperature in Al_2O_3-based MTJs (Chapter 11) and later in high-quality epitaxial Fe/MgO/Fe MTJs (Chapter 12) have stimulated a surge of interest to this phenomenon due to its many potential applications, and a variety of electrode/barrier combinations have been explored. These discoveries have also stimulated significant progress in the theory of spin-dependent tunneling.

The simplest models of tunneling based on the semiclassical approximation [1–3] are often used for the analysis of the tunneling I-V curves. However, these models are not able to describe TMR because they completely ignore the nature of the electrodes and consequently cannot take into account the changes associated with the magnetization configuration. Over the years, the TMR has usually been interpreted in terms of models based either on Bardeen's view of tunneling as being driven by overlap of wavefunctions penetrating into the barrier from the electrodes [4] or on the one-dimensional picture of tunneling across a potential barrier with exchange-split free-electron bands of the ferromagnetic electrodes [5]. These models are physically transparent and, though simple, capture some important physics of TMR. However, they have no predictive power for the quantitative description of TMR because they do not incorporate an accurate band structure of MTJs. Significant progress has been achieved based on the band theory, which takes into account the particular material-specific form of the crystal potential and the multi-orbital electronic structure.

In this chapter, we describe the current theoretical understanding of spin-dependent tunneling. Starting from the general background and simple formulations of electron tunneling, we then consider more sophisticated approaches capturing the important physics related to a realistic spin-dependent electronic band structure of MTJs. We also discuss the influence of imperfections on TMR and the phenomenon of tunneling anisotropic magnetoresistance (TAMR), which is induced by spin–orbit coupling (SOC) and is closely related to TMR. This chapter complements previous review articles that include the theory of TMR [6–9].

13.2 EXPLORING TUNNELING MAGNETORESISTANCE USING SIMPLE MODELS

13.2.1 BARDEEN'S DESCRIPTION OF TUNNELING

In Bardeen's approach [4], the transmission probability of a tunnel barrier is assumed to be small, so that the coupling between the electrodes can be regarded as a small perturbation. At $t = 0$, one considers the many-electron eigenfunctions of the unperturbed potential corresponding to the left or right electrode with the barrier extended to infinity. Since these states are coupled by tunneling, they are not eigenstates of the full Hamiltonian. The perturbation that couples the electrodes is called the transfer Hamiltonian. Using the standard time-dependent perturbation theory, one can then find the rates of tunneling transitions, that is, those involving an electron transfer between the electrodes [4], which are given by Fermi's golden rule. Multiplying by the electron charge e, one obtains for the tunneling current:

$$J = \frac{2\pi e}{\hbar} \sum_n P_n |M_{0n}|^2 \, \delta(E_n - E_0) \tag{13.1}$$

where:

- 0   labels the many-body ground state that has N_L electrons in the left and N_R in the right electrode
- n enumerates the excited states with $N_L \mp 1$ electrons in the left and $N_R \pm 1$ in the right electrode
- M_{0n} is the tunneling matrix element
- $P_n = \pm 1$ counts left-to-right and right-to-left tunneling rates with appropriate signs

Since the tunneling Hamiltonian is a one-particle operator, its matrix elements M_{0n} are nonzero only for states n in which one single quasiparticle is removed from the ground state of the left electrode and one single quasiparticle is added to the ground state of the right electrode (or vice versa). Bardeen has shown [4] that this matrix element is equal (up to the factor $-i\hbar$) to the matrix element of the current density operator between the states 0 and n taken at a plane inside the barrier.

Therefore, if we represent the many-body wavefunctions as Slater determinants of quasiparticle states occupied according to the Fermi distribution $f(\varepsilon-\mu)$ for the two electrodes (with different chemical potentials μ_L and μ_R), we find

$$J = \frac{2\pi e}{\hbar}\sum_{ij}|b_{ij}|^2\left[f(\varepsilon_i-\mu_L)-f(\varepsilon_j-\mu_R)\right]\delta(\varepsilon_i-\varepsilon_j) \qquad (13.2)$$

where:

i and j label the single-quasiparticle states of the left and right electrodes, respectively

b_{ij} is the tunneling matrix element between these states (which can be calculated using Bardeen's prescription)

The result (Equation 13.2) for the tunneling current is not exact, because it only contains the first order of the perturbation theory, neglecting the interference of multiple scatterings from the interfaces.

13.2.2 JULLIÈRE'S MODEL OF TUNNELING MAGNETORESISTANCE

Jullière's model for TMR [10] may be viewed as an application of Bardeen's approach with two assumptions. First, the electron spin is assumed to be conserved in the tunneling process. It follows then, that tunneling of up- and down-spin electrons are two independent processes. Such a two-current model has also been used to interpret the closely related phenomenon of GMR (see Chapter 4, Volume 1), but spin-flip corrections are sometimes needed to include the effects of SOC or thermal spin disorder. The electrons from the filled quasiparticle states of the left ferromagnetic electrode are thus accepted by unfilled states of the same spin of the right electrode. If the two ferromagnetic electrodes are magnetized parallel, the minority spins tunnel to the minority states and the majority spins tunnel to the majority states. If, however, the magnetizations are antiparallel, the identity of the majority- and minority-spin electrons is reversed in one of the electrodes. Second, Jullière's model assumes that the matrix elements b_{ij} are the same for all single-particle states. The corresponding factor in Equation 13.2 can then be moved outside the sum. Since the difference in the chemical potentials is equal to electron charge times the voltage drop across the barrier, at $T = 0$ one then immediately obtains

$$G_P = \frac{2\pi e^2}{\hbar}|b|^2\left[\rho_L^\uparrow(\varepsilon_F)\rho_R^\uparrow(\varepsilon_F)+\rho_L^\downarrow(\varepsilon_F)\rho_R^\downarrow(\varepsilon_F)\right]$$

$$G_{AP} = \frac{2\pi e^2}{\hbar}|b|^2\left[\rho_L^\uparrow(\varepsilon_F)\rho_R^\downarrow(\varepsilon_F)+\rho_L^\downarrow(\varepsilon_F)\rho_R^\uparrow(\varepsilon_F)\right]$$

$$(13.3)$$

where:

G_P and G_{AP} are the conductances for parallel and antiparallel electrode magnetizations

ρ_n^σ is the density of states (DOS) of the electrode n ($n = L$ or $n = R$) for spin channel σ ($\sigma = \uparrow$ for majority spin, and $\sigma = \downarrow$ for minority spin)

If we define a TMR ratio as the conductance difference between parallel and antiparallel magnetizations normalized by the antiparallel conductance, we arrive at Jullière's formula:

$$\text{TMR} \equiv \frac{G_P - G_{AP}}{G_{AP}} = \frac{2P_L P_R}{1 - P_L P_R} \tag{13.4}$$

which expresses the TMR via the spin polarization of the two ferromagnetic electrodes

$$P_n = \frac{\rho_n^\uparrow - \rho_n^\downarrow}{\rho_n^\uparrow + \rho_n^\downarrow} \tag{13.5}$$

Jullière's formula (13.4) allows one to link the values of spin polarization obtained from the Tedrow–Meservey experiments [11] to TMR (see also Chapter 11, Volume 1). As long as both measurements are made for the same kinds of the ferromagnet/insulator interfaces, this relation usually holds reasonably well, at least qualitatively. However, any attempts to interpret Equation 13.5 literally by using bulk DOS in Equation 13.5 almost invariably fail. For example, positive spin polarizations have been measured for tunneling from Fe, Co, and Ni into Al_2O_3 [11], while the DOS at the Fermi level is much larger for minority-spin electrons in Co and Ni. This failure led to the notion of the "tunneling DOS," which is supposed to appear in Equation 13.5 instead of the bulk DOS. The idea is that there is supposedly a subset of states that tunnel easily and have similar tunneling matrix elements, while other states barely tunnel at all. In this case, it is the density of the first group of states that matters in Equation 13.4. It has been common in the literature to explain the measured TMR and positive Tedrow–Meservey spin polarization based on the notion that "s-electrons" dominate in the tunneling. However, one should consider this separation with caution because s and d-electrons are strongly hybridized. In particular, band structure calculations show that transport in transition metals is dominated by 3d electrons. It is often true that most of the current in the diffusive regime is carried by electrons on just one of the Fermi surface sheets that is characterized by large Fermi velocities, but meaningful predictions require detailed calculations, and also the situation may change at the interface compared to the bulk. Therefore, a quantitative understanding of TMR requires calculations based on realistic electronic band structure. It is, however, helpful to consider a free-electron model as a starting point for a qualitative description of spin-dependent tunneling.

13.2.3 Free-Electron Model in One Dimension

The transmission can be easily calculated for a one-dimensional asymmetric rectangular barrier by matching the wavefunction and its derivative:

$$T = \frac{(2k_L\kappa)(2k_R\kappa)}{(k_L + k_R)^2\kappa^2 + (k_L^2 + \kappa^2)(k_R^2 + \kappa^2)\sinh^2\kappa d} \tag{13.6}$$

where:

d is the barrier thickness

$\kappa = \sqrt{2m(U - E)}/\hbar$ is the decay parameter in the barrier

U is the barrier height

k_L, k_R are the wavevectors in the left and right electrode

This free-electron model can also be used for a qualitative description of a spin-polarized metal by adding an exchange splitting to the potential profile. In this case, all wavevectors in Equation 13.6 (but not κ, unless the barrier is also ferromagnetic) carry a spin index, which we omit to simplify the notation.

The conductance can be calculated exactly using the Landauer–Büttiker (LB) formula [12], which says that the conductance of a two-terminal device can be written as

$$G_\sigma = \frac{e^2}{h}\sum_{\mu\nu} T_{\mu\nu}^\sigma \tag{13.7}$$

where:

σ is the spin index, the sum is taken over the conducting channels (μ, ν) in the two leads

$T_{\mu\nu}$ is the transmission probability from channel μ in lead L to channel ν in lead R (see Ref. [12] for details)

In our one-dimensional free-electron problem there is only one conduction channel, and the sum in Equation 13.7 contains only one term e^2T/h with T given by Equation 13.6. Let us make some general observations, which will be very useful later when we consider more realistic models.

Observation 1: The transmission probability is not equal to a product of two factors characterizing the two sides of the tunnel junction. This is because the coefficients of the propagating and evanescent waves in the wavefunction are determined from a coupled set of matching conditions; a tunneling electron can reflect off of the interfaces multiple times and interfere with itself.

Observation 2: If the barrier is sufficiently strong and tunneling weak ($\kappa d \gg 1$), then with exponential accuracy the transmission probability reduces to a product of two factors, T_L and T_R, characterizing the two interfaces:

$$T = \frac{4k_L\kappa e^{-\kappa d}}{k_L^2 + \kappa^2} \cdot \frac{4k_R\kappa e^{-\kappa d}}{k_R^2 + \kappa^2} = T_L T_R \qquad (\text{at } e^{-2\kappa d} \ll 1) \tag{13.8}$$

This approximation can be understood as follows: In the wavefunction matching at the left interface, we neglect the barrier eigenstate that is exponentially growing to the right, whose amplitude is exponentially small at this interface. Then, T_L describes the "transmission probability" of the electron with k_L into the evanescent barrier eigenstate $e^{-\kappa x}$ (with the barrier extended to infinity). T_R is, likewise, the transmission probability of the decaying barrier eigenstate into the right electrode's outgoing eigenstate k_R. The word "transmission" may be a little confusing here because evanescent waves do not carry current, but T_L and T_R are simply analytic continuations of the transmission probability from the range of positive kinetic energies in the barrier into the range of negative kinetic energies. The reciprocity relation persists into this region (as seen from the symmetry of Equation 13.8) due to the analytic properties of the scattering amplitude. Two exponential factors in Equation 13.8 show the decay of the evanescent barrier state between the two interfaces, which may be extracted in a separate spin-independent factor. We will refer to T_L and T_R as "interface transmission probabilities."

Observation 3: Equation 13.8, which can be directly generalized to the spin-split case, resembles the formulas (13.3) used in the derivation of Jullière's formula, but T_L and T_R are by no means proportional to the electrode DOS, which in our one-dimensional case is proportional to the inverse wavevector. The relation with DOS holds only in the limit of a very low barrier height $\kappa \ll k_L$. We will refer to Equations 13.4 and 13.5 in which the DOS is replaced by some more appropriate quantity as the "generalized Jullière's formula" (which is also, as we will see below, not universally valid).

Observation 4: Instead of a single free electron, consider a free-electron metal in which the levels are filled according to Fermi statistics (still in one dimension). The expectation value of the density operator for one of the electrodes (with the barrier extended to infinity) over the ground state (filled Fermi sphere) is $\rho(x) = \langle 0 | \Sigma_i \delta(x - x_i) | 0 \rangle = \Sigma_k | \psi_k(x) |^2$, where $\psi_k(x)$ are the normalized eigenfunctions, and $k \le k_F$ with k_F the Fermi wavevector. If the electrode has a large but finite length L, the values of k are quantized with an interval of π/L. Converting the sum over k to an integral and changing the integration variable to energy ε, we find

$$\rho(x) = (L/\pi) \int | \psi_k(x) |^2 \, dk = (L/\pi) \int v_\varepsilon^{-1} | \psi_\varepsilon(x) |^2 \, d\varepsilon,$$ where $v_\varepsilon = d\varepsilon/dk = \hbar k/m$ is the electron velocity. On one hand, the integrand $\rho(x,\varepsilon)$ represents the energy-resolved charge density at point x, or, in other words, the DOS projected on the eigenstate of the coordinate operator at x. On the other hand, if x is placed inside the barrier, one can easily verify that the integrand is proportional to T_L from Equation 13.8. Thus, the interface transmission probability can be interpreted as the electrode-induced DOS inside the barrier. This conclusion is analogous to the result of the Tersoff–Hamann theory of scanning tunneling microscopy [13]. It is the spin polarization of this quantity that determines TMR through Equations 13.4 and 13.5 in the weak tunneling limit.

Observation 5: It is obvious from wavefunction continuity that, as long as the factorization (Equation 13.8) holds, the amplitude of the evanescent wave in the barrier is proportional to the amplitude of the wavefunction at the interface. Therefore, the quantities T_L or, equivalently, $\rho(x,\varepsilon)$ calculated for x in the middle of the barrier and right at the interface differ only by the spin-independent exponential decay factor. Therefore, instead of T_L and T_R in Equation 13.8, we can just as well calculate the spin polarization of the *interface DOS* for use in Equation 13.4.

Observation 6: In the limit of weak tunneling, the Bardeen and LB approaches produce identical results. In Bardeen's approach, we need to calculate the matrix element b_{LR} of the current density operator

$$\hat{j} = \frac{i\hbar}{2m}\left[\overline{\nabla}\delta(x - x_b) - \delta(x - x_b)\overrightarrow{\nabla}\right] \tag{13.9}$$

between the states $\psi_L(x) = C_L \exp(-\kappa x)$ and $\psi_R(x) = C_R \exp[\kappa(x-d)]$, where x_b is a point inside the barrier, and the gradient operator $\nabla = d/dx$ acts in the direction shown by the arrow. The matrix element then is $b_{LR} = \langle \psi_L | \hat{j} | \psi_R \rangle = -(i\hbar/m)C_L^*C_R\kappa\exp(-\kappa d)$. The amplitudes are found from interface matching: $C_L/A_L = 2k_L/(k_L + i\kappa)$, where A_L is the amplitude of the incoming wave whose magnitude is found from the normalization of the scattering state, $|A_L|^2 = 1/(2L_L)$, and L_L is the length of the left electrode. Substituting this matrix element in Equation 13.2 and introducing the DOS of the left and right electrodes, after some cancellations we obtain the conductance identical to the LB Equations 13.7 and 13.8.

13.2.4 New Features in Three Dimensions

Now consider an electron moving in three-dimensional space in an external potential $V(z)$, which depends only on one coordinate z and again corresponds to a rectangular tunnel junction. The Schrödinger equation for this potential is separable in Cartesian components, and the wavefunction can be represented in the following form: $\psi(\mathbf{r}) = \exp(i\mathbf{k}_{\|}\mathbf{r}_{\|})\psi(z)$, where $\mathbf{k}_{\|}$ and $\mathbf{r}_{\|}$ are the projections of $\mathbf{k}$ and $\mathbf{r}$ on the xy plane, and $\Psi(z)$ is the solution of the one-dimensional Schrödinger equation with potential $V(z)$ and energy $E_z = E - \hbar^2 k_{\|}^2/(2m)$. At $T = 0$, the electrons in each electrode fill a Fermi sphere. In the magnetic case, the Fermi wavevector depends on spin. Since the barrier decay parameter κ increases if the electron energy is decreased, it is clear that the electrons with $\mathbf{k}_{\|} = 0$ have the highest tunneling probability. In the limit of a very thick barrier, the tunneling current is, therefore, confined to the narrow vicinity of the point $\mathbf{k}_{\|} = 0$.

Let us estimate the size of this region for large but finite κd. Introducing κ_0 according to $\hbar^2\kappa_0^2 = 2m(U - E)$, assuming that $k_{\|}^2 \ll \kappa_{\|}^2$ in the important range of $k_{\|}$, and expanding $\kappa(k_{\|}^2)$ to first order in $k_{\|}^2$, we approximate the barrier decay factor by $\exp(-2\kappa d) \approx \exp(-2\kappa_0 d)\exp(-k_{\|}^2 d/\kappa_{\|})$. The prefactors in Equation 13.8 depend weakly on $k_{\|}$, so the tunneling is suppressed at $k_{\|}^2 d \sim \kappa_0$, which can be rewritten as $k_{\|}a \sim \sqrt{\kappa_0 d}(a/d)$, where a will have the meaning of the lattice parameter when we consider crystalline electrodes. Typical tunnel junctions have barriers that are a few monolayers thick, and values of $\kappa_0 d$ usually do not exceed 10 (otherwise the tunneling current is no longer measurable). We, therefore, see that even for relatively thick barriers $k_{\parallel} a$ is never very small, and the current is carried by electrons in a significant portion of the Brillouin zone rather than just those with $k_{\parallel} \approx 0$.

Each $\mathbf{k}_{\parallel}$ allowed by the boundary conditions in the xy plane serves as an independent channel in the LB formula (Equation 13.7), and the transmission probability is diagonal in the $\mathbf{k}_{\parallel}$ representation because $\mathbf{k}_{\parallel}$ is conserved for the entire system. Thus, we find that the conductance per spin is given by

$$G = \frac{e^2}{h} \sum_{k_{\parallel}} T(\mathbf{k}_{\parallel}) \tag{13.10}$$

where $T(\mathbf{k}_{\parallel})$ is the transmission probability obtained from the solution of the one-dimensional scattering problem given by Equation 13.6 at energy E_z. All observations about the features of the transmission probability made in the previous section are carried over *verbatim* for tunneling at the given $\mathbf{k}_{\parallel}$ point. However, the total current in each spin channel is now given by a sum over $\mathbf{k}_{\parallel}$ in Equation 13.10, so that, even if the transmission probability is factorized as in Equation 13.8 in the weak tunneling limit at each $\mathbf{k}_{\parallel}$, the entire integral can no longer be represented as a product of two factors. Therefore, the generalized Jullière's formula, in general, does not hold in three dimensions even in the weak tunneling limit.

As an illustration, consider the limit of a very thick barrier when the important region of integration in Equation 13.10 is confined to the vicinity of the $\mathbf{k}_{\parallel} = 0$ point. In this case, one can neglect the variation of the prefactors in Equation 13.8 and expand the decay parameter in the exponential as we just did at the beginning of this section. Then the only $k_{\parallel}$-dependent factor under the integral is $\exp(-k_{\parallel}^2 d / \kappa_0)$, which is spin-independent. The generalized Jullière's formula is then restored, and the relevant quantity in Equation 13.5 for the given electrode (say, L) is the value $T_L(\mathbf{0})$. Taking these values from Equation 13.8, we obtain after a little algebra:

$$P_L = \frac{k_L^{\uparrow} - k_L^{\downarrow}}{k_L^{\uparrow} + k_L^{\downarrow}} \cdot \frac{\kappa_0^2 - k_L^{\uparrow} k_L^{\downarrow}}{\kappa_0^2 + k_L^{\uparrow} k_L^{\downarrow}} \tag{13.11}$$

and a similar expression for the right electrode. Here $k_L^{\uparrow}$ and $k_L^{\downarrow}$ are the wavevectors at $\mathbf{k}_{\parallel}=0$ for spin-up and spin-down electrons in the left electrode, respectively. This result was derived by Slonczewski [5].

Equation 13.11 has a very peculiar and suggestive feature. At κ_0 (very low barrier), P_L is equal to the spin polarization of the *inverse* wavevector, which is the same as the spin polarization of DOS at $\mathbf{k}_{\parallel} = 0$. We have mentioned this in *Observation 3* of the previous section. On the other hand, in the opposite high-barrier limit ($\kappa^2 \gg k^{\uparrow} k^{\downarrow}$) the spin polarization has the same magnitude but an *opposite sign*. This is because in the $\kappa \to \infty$ limit the scattering wavefunction develops a node at the interface, while in the $\kappa \to 0$ limit it has an antinode there. In the first case, the amplitude of the wavefunction at the interface is suppressed by a factor k/κ, so that the transmission probability is proportional to k. In the second case, this amplitude is just twice the incoming amplitude, and the transmission probability is proportional to k^{-1}. The spin polarization (Equation 13.5) changes sign if the underlying quantity is replaced by its inverse. This simple example [5] shows the futility of "plausible" arguments based on Jullière's formula in predicting even the sign of the spin polarization and TMR without taking into account the details of the electronic structure and interfacial scattering.

13.3 BAND THEORY OF TUNNELING

The current in a tunnel junction is directly related to transmission probabilities, which, as we have seen from previous sections, depend in a delicate way on the matching of wavefunctions at the interfaces. Solids, and especially transition metals that are usually used as electrodes in MTJs, have complicated electronic band structures; moreover, metal-oxide interfaces almost never look as a simple concatenation of bulk materials. Surfaces and interfaces develop qualitatively new features due to the reduction of the coordination number, severing of bonds, charge transfer, and metal-oxide hybridization; in reality, there are also defects, intermixing, interfacial strain, and other imperfections. The TMR is sensitive to all these details.

Until 2001, the only barrier material that produced a sizable TMR at room temperature was amorphous Al_2O_3. Over the years, the fabrication techniques have been perfected, so that junctions with high TMR of about 70%–80% can now be routinely produced. Tedrow–Meservey experiments for the ferromagnet/Al_2O_3 interfaces show positive spin polarization of the tunneling current (see Chapter 11, Volume 1), which, however, is still not completely understood from the theoretical point of view due to the extreme complexity of the atomic structure. The situation has changed dramatically with the prediction [14, 15] and successful fabrication [16, 17] of epitaxial Fe(001)/MgO/Fe tunnel junctions exhibiting very high TMR values. In spite of the fact that the details of the interfacial structure are almost never perfectly controlled or even known, it has become clear that certain simple selection rules are instrumental in analyzing the TMR in epitaxial tunnel junctions. In this section, we discuss the tunneling of Bloch electrons across epitaxial interfaces.

13.3.1 COMPLEX BAND STRUCTURE OF THE TUNNELING BARRIER

Since the potential of a crystal is periodic, the Hamiltonian commutes with any lattice translation, which also commute between themselves. The Bloch theorem immediately follows, stating that the eigenstates of the crystal Hamiltonian can be taken to be the eigenstates of the lattice translation operators, namely, $\psi_i(\mathbf{r}) = \exp(i\mathbf{k}\mathbf{r})u_i(\mathbf{r})$, where $\mathbf{k}$ is the quasi-wavevector, and $u(\mathbf{r})$ has the periodicity of the lattice. $\mathbf{k}$ is defined up to a reciprocal lattice

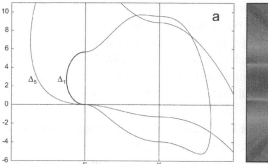
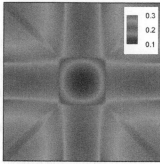

FIGURE 13.1   Illustration of the complex band structure of MgO. (a) Complex band structure at $\mathbf{k}_{||} = 0$ (energy in eV). The Δ_1 complex band is shown by a thick line straddling the insulating gap. For clarity, complex bands with variable real and imaginary parts, which attach to the band extrema at −5 and at 9.5 eV, are not shown. (b) Right panel: Brillouin zone plot of the lowest value of Im k_z taken at mid-gap.

vector and can, therefore, be taken inside the first Brillouin zone; the index i labels the eigenstates with the same $\mathbf{k}$, of which there is an infinite number.

In a bulk crystal, only the wavefunctions with real quasi-wavevectors $\mathbf{k}$ are allowed in the Hilbert space; those with complex $\mathbf{k}$ blow up at infinity and must be discarded. However, in the tunneling problem, the wavefunction matching across the interfaces involves the general solutions of the Schrödinger equation including all evanescent eigenstates in the barrier and the decaying evanescent states in the electrodes. These evanescent eigenstates obey the Bloch theorem, but their quasi-wavevectors are complex. The analytic properties of these states as a function of complex $\mathbf{k}$ were studied by Kohn [18] and Heine [19]. Only the states with real energies have physical significance. The set of all these states is called "complex band structure," which extends the standard band structure which includes only real-$\mathbf{k}$ solutions.

An example below illustrates this concept. For an epitaxial junction, the projection of the quasi-wavevector on the plane parallel to the interfaces ($\mathbf{k}_{||}$) must be real. For each $\mathbf{k}_{||}$ there is an infinite number of eigenstates with complex k_z, whose imaginary part determines the decay parameter. Figure 13.1b shows the smallest decay parameter as a function of $\mathbf{k}_{||}$ for MgO with the high-symmetry (001) interface and the energy taken in the middle of the insulating gap. Note that $\mathbf{k}_{||}$ lies in the *surface Brillouin zone* defined with respect to the two-dimensional lattice translations parallel to the interface, while Re k_z lies in the one-dimensional "Brillouin zone" defined by the translational period in the z direction. The notion of conserved $\mathbf{k}_{||}$ is meaningful only if the interface lies parallel to one of the low-index planes of the crystal. Points in the surface Brillouin zone are usually denoted with a bar on top of the letter; for example, the $\mathbf{k}_{||} = 0$ points is denoted as $\overline{\Gamma}$.

As it was pointed out by Mavropoulos [20], tunneling should be more efficient through barrier eigenstates with the smallest decay rates. In the limit of large barrier thickness, the state with the smallest decay rate controls the tunneling current, similar to Slonczewski's result (Equation 13.11) for the free-electron case. As seen in Figure 13.1b, for MgO(001) the minimum decay rate is reached at the $\bar{\Gamma}$ point. This is similar to the case of free-electron tunneling through a rectangular barrier, where the smallest decay parameter κ is reached at $\mathbf{k}_\parallel = 0$. The main difference is that the crystalline barrier has an infinite number of decaying eigenstates at each $\mathbf{k}_\parallel$ within the Brillouin zone, while in the free-electron case there is only one eigenstate per $\mathbf{k}_\parallel$, which is unrestricted.

Let us now set $\mathbf{k}_\parallel = 0$ and look at the complex bands of MgO shown in Figure 13.1a. These bands are plotted using three panels: the middle panel shows the real part of k_z, and the two side panels show its imaginary part. It is useful to use two side panels for the following reason. The complex bands can be classified into four categories [21]: (1) ordinary real bands with Im k_z = 0, (2) imaginary bands with Re k_z = 0, (3) imaginary bands with the maximum allowed Re $k_z = k_{max}$, and (4) complex bands for which both Im k_z and Re k_z depend on energy. Complex bands are continuous; therefore, bands of type 2 either go to infinity or connect to real bands at k_z = 0. Likewise, bands of type 3 can connect to real bands at $k_z = k_{max}$. Bands of type 2 and 3 are therefore plotted in the left and right panels, respectively; in this way the connections to the real bands are obvious. Type 4 bands may be shown in two panels at the same time [21]; they attach to real bands at their extremal points. Two such bands, which are far from the Fermi level and irrelevant for tunneling, were omitted from Figure 13.1b to avoid clutter.

The point k_z = 0 in Figure 13.1a corresponds to the bulk Γ point. The degenerate triplet at E = 0 is the valence band maximum (VBM), and the singlet at $E \approx 5.5$ eV is the conduction band minimum (CBM). (The band gap is underestimated due to the use of the local density approximation [LDA] in the calculation.) We see that VBM and CBM are connected by a Δ_1 band with imaginary k_z, and the decay constant is lower for the state at this loop than anywhere else in the surface Brillouin zone. This decay constant appears in the center of Figure 13.1b. Note that in the free-electron case the "band gap" corresponds to negative kinetic energies, and the only "complex band" $\kappa(E)$ is just an inverted parabola.

The crucial point is that all bands, be they real or complex, can be classified by the irreducible representations of the group of the quasi-wavevector $\mathbf{k}$. The $\bar{\Gamma}$ point, which is the most important for tunneling, has a high degree of symmetry. Since the full quasi-wavevector of all states with $\mathbf{k}_\parallel = 0$ has only the z component, the symmetry group is that of the [001] direction in the bulk Brillouin zone, that is C_{4v} (or $4mm$ in the international notation). The [001] direction in a cubic crystal is customarily denoted as Δ.

The C_{4v} group has five irreducible representations: four one-dimensional and one two-dimensional [22]. For the electronic states on the Δ line, they are usually denoted as Δ_1, Δ_1', Δ_2, Δ_2', and Δ_5. This notation was introduced by Bouckaert et al. [23] and differs a little from the standard notation for the C_{4v} group [22]. Index 1 or 2 distinguishes the one-dimensional representations with even or odd symmetry with respect to the C_4 rotation; unprimed representations are even with respect to reflections in the planes parallel to the given Δ direction and normal to one of the other two cubic axes; primed representations are odd with respect to these reflections. Δ_1 is the identity -representation; Δ_5 is the two-dimensional representation. If the character of the wavefunctions is described by the s, p, d basis, the Δ_1 bands can only contain s, p_z, and d_{z^2} character; Δ_5 can only have p_x, p_y, d_{xy}, and d_{yz} character; Δ_2 is $d_{x^2-y^2}$ only, and Δ_2' is d_{xy} only. These rules can be used for the assignment of symmetry labels to the electronic states. The Δ_1' wavefunction must have four node lines in the xy plane and is not representable by s, p, d functions.

In MgO, the triplet at VBM has Γ_{15} symmetry, and the singlet at CBM has Γ_1 symmetry (notation of Ref. [23]). As **k** is moved away from Γ along the Δ line, the degeneracy is partially lifted, and the Γ_{15} triplet splits in one Δ_1 band and a doubly degenerate Δ_5 band. These bands are labeled in Figure 13.1a; we see that the complex band connecting VBM and CBM has Δ_1 symmetry. (By continuity, only bands of compatible symmetry can connect to each other.)

If we move away from the $\bar{\Gamma}$ point in a general $\mathbf{k}_\parallel$ direction, all symmetry operations (except the identity) are lost, and all eigenstates belong to the trivial identity representation. However, if $\mathbf{k}_\parallel$ is small, the properties of the eigenstates will still strongly resemble those at the $\bar{\Gamma}$ point. For example, the matrices of the C_{4v} group operators in the basis of the new eigenstates will be almost diagonal at small $\mathbf{k}_\parallel$ (except for the matrix elements in the subspace "lifted" from Δ_5). Therefore, at small $\mathbf{k}_\parallel$ one can talk about, say, "Δ_1-like" or "Δ_5-like" states, according to the representation to which they are adiabatically connected by sending $\mathbf{k}_\parallel$ to zero. However, this terminology loses its meaning at larger $\mathbf{k}_\parallel$, especially when the different sheets of the $\kappa(\mathbf{k}_\parallel)$ function intersect with each other at finite $\mathbf{k}_\parallel$.

13.3.2 TUNNELING SPIN FILTERING DUE TO SYMMETRY SELECTION RULES

If the junction is epitaxial, there is a common translational periodicity for both the electrodes and the barrier in the directions parallel to the interfaces, and $\mathbf{k}_\parallel$ is a conserved quantum number for the eigenstates of the whole system. At $\mathbf{k}_\parallel = 0$ the scattering eigenstates may be classified by the irreducible representation of the symmetry group characterizing the junction. If we now view the whole problem in terms of wavefunction matching, it is clear that a Δ_p scattering eigenstate is obtained by matching the Δ_p states in the electrodes with the Δ_p states in the barrier. In addition to the conventional real bands, the metallic electrodes also have an infinite number of complex bands, which have a given symmetry, and all of them will be represented in all the scattering states of this symmetry (of course, only those that decay *into* the metal are allowed) [24]. However, the current is only carried by real bands, of which there are usually at most one or two.

As an example, consider an epitaxial Fe/MgO/Fe tunnel junction with both MgO and Fe grown in the [001] direction. The relevant real bands in Fe for $\mathbf{k}_\parallel = 0$ cross the Fermi level along the Δ line connecting the Γ and H points in the bulk Brillouin zone. The band structure of Fe along this line

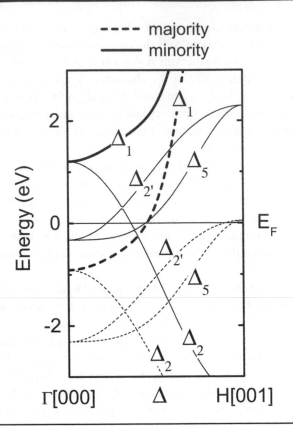

FIGURE 13.2 Spin-resolved band structure of Fe along the [001] direction. Bands are labeled by their symmetry group representation. Thick lines show the Δ_1 bands, which decay the slowest in MgO.

is shown in Figure 13.2. We see that there is one Δ_1 band for majority-spin electrons and *none* for minority-spin electrons.

For a certain state to carry tunneling current across the barrier, it should have a finite amplitude of at least one real band on *each* side of the barrier. If the magnetizations of the two Fe electrodes are parallel, only the majority-spin Δ_1 eigenstate of the junction can carry the tunneling current across the barrier. However, when the electrodes are magnetized antiparallel, *none* of the Δ_1 states of the junction can carry the tunneling current. Therefore, one can expect that the tunneling conductance is larger in the parallel configuration, and hence the TMR defined by Equation 13.4 is positive. This feature of epitaxial MTJs, which was elucidated in Refs. [20, 25], can be called symmetry-enforced spin filtering.

These qualitative considerations are confirmed by explicit calculations [14, 15]. Figure 13.3 illustrates the spin filtering effect. The four panels show the tunneling DOS of the scattering states of the given symmetry incoming from the left electrode plotted as a function of the position in the system. Different decay rates of the states of different symmetry are apparent in the barrier region. We also see that the rightward propagating Δ_1 tunneling state exists only for a majority-spin electron in the parallel configuration (upper left panel). In all other cases, such state either does not exist (upper and lower

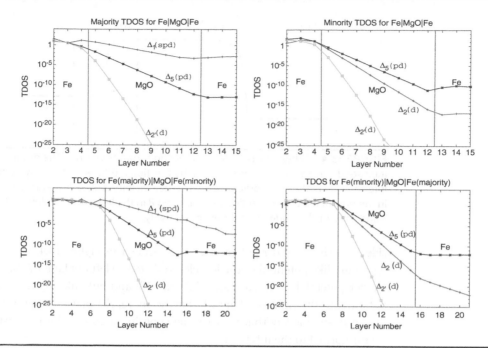

FIGURE 13.3   Tunneling DOS (TDOS) for $k_\parallel = 0$ for Fe(100)|MgO|Fe(100) tunnel junction for majority (a), minority (b), and antiparallel alignment of the magnetizations in the two electrodes (c,d). Each curve is labeled by the symmetry of the incident Bloch state in the left Fe electrode. (After Butler, W.H. et al., *Phys. Rev. B* 63(5), 054416, 2001. With permission.)

right panels) or decays exponentially into the right electrode and carries no current (lower left panel).

Symmetry-enforced spin filtering qualitatively explains large values of TMR observed experimentally in crystalline Fe/MgO/Fe tunnel junctions [16, 17, 26]. In fact, symmetry-enforced spin filtering should be quite common as long as the given junction can be grown epitaxially with low-index interfaces. In addition to bcc Fe, bcc Co and Fe-Co alloys have a Δ_1 symmetry band crossing the Fermi energy only in the majority-spin channel. Therefore, large values of TMR are expected for junctions with these electrodes and MgO barriers stacked along the [001] direction. This was confirmed by first-principles calculations [27], and high TMR was observed experimentally [28, 29].

13.3.3 Factorization of Transmission

In Section 13.2.3 (see *Observation 2*), we have explained that in the one-dimensional free-electron case, when the barrier is sufficiently thick ($e^{-2\kappa d} \ll 1$), the transmission probability is factorized in a product of two interface transmission probabilities. This factorization holds at each $k_\parallel$ for free electrons in three dimensions. We now show that the situation for Bloch electrons is, in fact, quite similar [30]. Let us consider an epitaxial tunnel junction and treat the scattering problem for the given $k_\parallel$.

First, we assume that the barrier is thick enough to have a region where the potential is bulk-like. We can then draw an imaginary plane on each

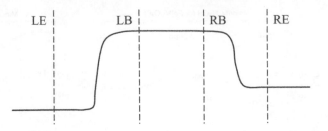

FIGURE 13.4   Schematic picture of tunneling through a barrier. The dashed lines indicate the imaginary planes showing the positions sufficiently far away from the interfaces so that the periodic potential in their vicinity is already identical to that in the bulk of the corresponding region. L and R in the labels refer to left and right; E and B mean electrode and barrier, respectively.

side of both left and right interfaces, as shown in Figure 13.4, so that we have bulk-like potential of the left electrode to the left of plane LE, bulk-like barrier potential between planes LB and RB, and bulk-like right electrode potential to the right of plane RE. Furthermore, the planes LB and RB are drawn in such a way that they can be superimposed by a bulk translation vector normal to the interface.

The general solution of the Schrödinger equation to the left of plane LE is given by a linear combination of the Bloch eigenstates of the left electrode for the given $\mathbf{k}_\parallel$, and similarly for the right electrode. Although there are an infinite number of evanescent states, we are only interested in real bands. There is an equal number of incoming and reflected waves. The general solution in the barrier (between LB and RB) is given by a linear combination of all evanescent barrier eigenstates. Similar to regular Bloch waves, the evanescent waves can be represented as $\psi(\mathbf{r}) = \exp[i\mathbf{k}_\parallel \mathbf{r}_\parallel + ik_z z]u(\mathbf{r})$, where $u(\mathbf{r})$ is periodic, and $k_z = k' + i\kappa$, where k' is the real part of k_z. We define the amplitudes of the evanescent waves in such a way that $z = 0$ at plane LB (or RB for the right interface).

We now treat the problem by the standard method of combining S-matrices [12]. We will denote the barrier eigenstates decaying to the right as "outgoing," and those decaying to the left as "incoming" with respect to the left interface, and vice versa with respect to the right interface. (Outgoing eigenstates remain finite at infinity when a small positive imaginary part is added to the energy; these are the boundary conditions for the retarded Green's function.) In order to limit the number of subscripts, we restrict ourselves to the case when the barrier has a symmetry plane normal to the axis of the junction so that the incoming and outgoing barrier eigenstates come in degenerate pairs with complex conjugate k_z. The eigenstate of the junction must satisfy the Schrödinger equation everywhere. This condition introduces a linear relation between the amplitudes A_i^L of incoming and B_i^L of outgoing waves (referenced from planes LE and LB, respectively), which is described by an S-matrix: $B_i^L = \sum_j S_{ij}^L A_j^L$ $\bar{B}_L = \hat{S}_L \bar{A}_L$ in matrix notation). There is a similar relation at the right interface: $\bar{B}_R = \hat{S}_R \bar{A}_R$, where the incoming and outgoing waves are referenced from planes RE and RB.

Due to our assumption about the existence of a translation vector connecting planes LB and RB, the amplitudes of the barrier eigenstate b at these planes are related by the corresponding Bloch translation factor: $A_b^R = \exp(ik_b'd - \kappa_b d)B_b^L, A_b^L = \exp(ik_b'd - \kappa_b d)B_b^R$, where d is the distance between the planes LB and RB (which is a multiple of the translation vector along the z direction). We can now eliminate the amplitudes of the barrier waves from the linear relationships given by the left and right S-matrices, and find the transmission matrix connecting the amplitudes of the outgoing waves of the right electrode with those of the incoming waves of the left electrode. Following the standard convention [12], we denote the subblock of $\hat{S}_L$ connecting the outgoing LB states with incoming LE states as $\hat{t}_L$, and that connecting the outgoing LB states with incoming LB states as $\hat{r}_L'$. In a similar way, for the right interface, the outgoing RE states are connected to the incoming BR states by the matrix $\hat{t}_R'$, and the outgoing RB states to incoming RB states by the matrix $\hat{r}_R'$. Elimination of the barrier amplitudes then results in the following matrix relation:

$$\hat{t} = \hat{t}_R' e^{i\hat{k}_z d/2}\left[1 - e^{i\hat{k}_z d/2}\hat{r}_L' e^{i\hat{k}_z d}\hat{r}_R' e^{i\hat{k}_z d/2}\right]^{-1} e^{i\hat{k}_z d/2}\hat{t}_L \tag{13.12}$$

where the dependence on $\mathbf{k}_{||}$ is omitted for brevity. Here $\hat{k}_z$ is a diagonal matrix whose eigenvalues are the complex wavevectors of the barrier eigenstates with negative imaginary parts. Equation 13.12 has an obvious interpretation in terms of the Feynman paths going back and forth between the two interfaces [12]. It is exact under our assumption that the region between LB and RB is bulk-like. In fact, it is correct even in the case of a metallic spacer layer that has real k_z eigenvalues.

For a sufficiently thick barrier, the factors $\exp(-\kappa d)$ for all evanescent states become small. In this limit, the matrix in square brackets in Equation 13.12 is close to a unit matrix, and we obtain (with exponential accuracy):

$$\hat{t} = \hat{t}_R' e^{i\hat{k}_z d}\hat{t}_L \quad \text{(thick barrier, Im } k_z < 0) \tag{13.13}$$

The generalization of the Landauer–Büttiker formula (Equation 13.10) for an epitaxial crystalline junction is

$$G = \frac{e^2}{h}\sum_{\mathbf{k}_{||}}\sum_{ij}\left|t_{ij}(\mathbf{k}_{||})\right|^2 \tag{13.14}$$

Using (13.13), for a thick barrier we obtain $G = (e^2/h)\sum_{\mathbf{k}_{||}} T(\mathbf{k}_{||})$, where

$$T(\mathbf{k}_{||}) = \mathrm{Tr}\left[\hat{T}_L e^{-i\hat{k}_z^\dagger d}\hat{T}_R e^{i\hat{k}_z d}\right] \tag{13.15}$$

where the Hermitian matrices $T_{bc}^L = \sum_i t_{bi}t_{ci}^* = \left(\hat{t}_L \hat{t}_L^\dagger\right)_{bc}$ and $T_{bc}^R = \sum_j t_{jc}' t_{jb}'^* = \left(\hat{t}_R'^\dagger \hat{t}_R'\right)_{bc}$ describe the matching between the propagating electrode states and the barrier eigenstates, the trace is taken over the barrier eigenstates with positive $\kappa = \text{Im } k_z$, and we omitted the $\mathbf{k}_{||}$ arguments. In contrast to the free-electron case where the transmission amplitude factorizes for each given $\mathbf{k}_{||}$, we now have a double sum over the barrier eigenstates. Of course, only those barrier eigenstates that have a sufficiently small decay parameter are important. Note that $T(\mathbf{k}_{||})$ in Equation 13.15 is automatically real, as it should be.

Using the condition of current continuity, one can derive a reciprocity relation between $\hat{t}$ and $\hat{t}'$ matrices for the same interface:

$$\hat{t}\,\hat{t}^+ = \hat{t}'^+\hat{t}' \tag{13.16}$$

which holds as long as the propagating states in the electrodes are normalized to unit flux, and the evanescent barrier eigenstates are normalized in such a way that $\sqrt{2}$ times a sum of the right and left-going evanescent states differing in the sign of κ carries unit flux. This implies that the matrix $\hat{T}$ for the given interface in Equation 13.15 does not depend on whether this interface forms the left or the right side of the system. In particular, if the junction is symmetric, we have $\hat{T}_L = \hat{T}_R$.

Equation 13.15 has an interesting consequence. If there are at least two barrier eigenstates for which $\text{Re } k_z^b \neq \text{Re } k_z^c$ (while one of them may be zero), then $T(\mathbf{k}_{||})$ contains a part that oscillates as a function of d (and $\mathbf{k}_{||}$) due to the interference of these eigenstates. This feature was pointed out by Butler et al. [14], who found oscillations of the transmission probability as a function of $\mathbf{k}_{||}$ and exponentially damped oscillations as a function of d in Fe/MgO/Fe junctions. This feature is often invoked to explain the experimentally observed oscillations of the tunneling conductance (in both parallel and anti-parallel configurations) and TMR as a function of barrier thickness [17, 31]. However, the evanescent states in MgO have purely imaginary wavevectors in a large region of $\mathbf{k}_{||}$ around the $\bar{\Gamma}$ point [32, 33]. This implies that any conductance oscillations should decay much faster than the conductance itself, and therefore the TMR oscillations should be strongly damped. This behavior is different from the experimentally observed almost [17] or completely [31] undamped oscillations. It was argued in Ref. [34] that disorder in the barrier, counterintuitively, can restore the conductance oscillations through a mechanism involving scattering of the states at $\mathbf{k}_{||} = 0$ into the pair of interfering states at a large-$\mathbf{k}_{||}$ "hot spot." However, this argument implicitly assumes that the phase relation between the interfering states near the hot spot depends only on the distance from the interface and not on the location of the scattering impurity, which is unphysical.

Based on the above arguments, we believe that the undamped oscillations of the tunneling conductance and TMR observed in Fe/MgO/Fe MTJs are likely unrelated to the interference of evanescent states. An alternative explanation is suggested by the measurements of the anisotropy constant for MgO-based MTJs with Co/Pt multilayer electrodes, which show undamped oscillations as a function of MgO thickness with a period identical to that for

TMR oscillations [35]. These oscillations are, furthermore, reduced under annealing, pointing toward an extrinsic morphology-related effect.

For a $\mathbf{k}_{||}$ point of a general position, the evanescent barrier eigenstates are nondegenerate. Moreover, it often happens that the tunneling conductance is dominated by a region of the interface Brillouin zone where the smallest-κ evanescent state is nondegenerate and widely separated from the next-smallest κ state. This region is usually located around the $\bar{\Gamma}$ point, as occurs, for example, in Fe/MgO/Fe. In this case, one can neglect all terms in Equation 13.15 except one in which k_z in both exponential factors has the smallest imaginary part κ (to be denoted as κ_0) for the given $\mathbf{k}_{||}$. Then, the sums are eliminated and we find

$$T(\mathbf{k}_{||}) = T_L(\mathbf{k}_{||})e^{-2\kappa_0(\mathbf{k}_{||})d}T_R(\mathbf{k}_{||}), \tag{13.17}$$

where $T_L(\mathbf{k}_{||}) = \Sigma_i |t_{i0}|^2$, and likewise for T_R. Note that T_L and T_R are always real; these quantities can be called *interface transmission functions* [30].

Since we are working at fixed $\mathbf{k}_{||}$, the problem is essentially one-dimensional, and the DOS for each propagating electrode band is inversely proportional to the projection of its group velocity on the z direction, or to the flux carried by a Bloch wave in that direction. We can, therefore, use arguments invoked in *Observation 4* of Section 13.3 to associate $T_L(\mathbf{k}_{||})$ with the metal-induced $\mathbf{k}_{||}$-resolved DOS at the plane LB in the barrier; this relation is meaningful as long as the conditions of validity of Equation 13.17 are satisfied.

13.3.4 ASYMPTOTIC DEPENDENCE ON BARRIER THICKNESS

Let us now consider the asymptotic behavior of TMR when the barrier thickness is increased. Although this limit is almost never reached in practice, it is important from the methodological point of view. We assume for definiteness that the Δ_1 state at $\mathbf{k}_{||} = 0$ has the lowest decay rate in the barrier, as it happens in MgO and in many other crystalline barriers.

If the electrodes have Δ_1 bands crossing the Fermi level in both spin channels (or if one electrode has Δ_1 bands in both spin channels, and the other electrode has them in only one spin channel), the situation is qualitatively similar to the free-electron case considered in Section 13.2.4, which led to Equations 13.11 and 13.4. At $d \to \infty$, the TMR tends to a finite constant, which is determined by Jullière's formula with the spin polarizations calculated for the interface transmission functions taken at $\mathbf{k}_{||} = 0$.

However, if both electrodes have Fermi-crossing Δ_1 bands in only one spin channel, the TMR tends to infinity at $d \to \infty$, and it is interesting to consider its asymptotic behavior at large d. We assume that the Δ_1 bands are in the "up" spin channel for both electrodes so that the conductance in the parallel configuration is dominated by the Δ_1 band near $\mathbf{k}_{||} = 0$. The barrier decay factor can be expanded near $\mathbf{k}_{||} = 0$ as explained in Section 13.4, and we have

$$G_P = \frac{e^2}{h}e^{-2\kappa_0 d}\int T_L^\uparrow e^{-\alpha k_{||}^2 d}T_R^\uparrow d\mathbf{k}_{||} \sim T_L^\uparrow T_R^\uparrow \frac{e^{-2\kappa_0 d}}{\alpha d}, \tag{13.18}$$

where $T_L^\uparrow$ and $T_R^\uparrow$ are taken at $\mathbf{k}_{||} = 0$, and $\alpha = d^2\kappa / d\mathbf{k}_{||}^2$.

In the antiparallel configuration, the Δ_1 channel does not carry any current, but the Δ_1-like states with nonzero $\mathbf{k}_{\parallel}$ do, and it is these states that dominate in G_{AP} at large d [36]. For a general $\mathbf{k}_{\parallel}$, the coefficient of the Δ_1-like barrier eigenstate in the scattering state is proportional to $k_{\parallel}$ (which can be seen, for example, using the $\mathbf{k} \cdot \mathbf{p}$ method [37]), and the interface transmission function is proportional to $k_{\parallel}^2$. It is possible that tunneling through the Δ_1-like state is forbidden along some high-symmetry directions in the $\mathbf{k}_{\parallel}$ plane. For example, if there is a symmetry plane normal to the interface and containing $\mathbf{k}_{\parallel}$, all states will be classified as even or odd with respect to reflection in this plane. If the electrode Δ_1-like eigenstate is, say, odd with respect to this plane and the barrier eigenstate is even, then these states do not mix. If this is the case, then the interface transmission function will have an additional positive-definite factor depending on the direction of $\mathbf{k}_{\parallel}$, which we denote $f(\phi)$, that is $T(\mathbf{k}_{\parallel}) = Q k_{\parallel}^2 f(\phi)$. Thus, we have for G_{AP}:

$$G_{AP} = \frac{e^2}{h} e^{-2\kappa_0 d} \int \left[T_L^{\uparrow} Q_R^{\downarrow} f_R(\phi) + T_R^{\uparrow} Q_L^{\downarrow} f_L(\phi) \right]$$

$$\times k_{\parallel}^2 e^{-\alpha k_{\parallel}^2 d} \, d\mathbf{k}_{\parallel} \sim \left[T_L^{\uparrow} Q_R^{\downarrow} \overline{f_R} + T_R^{\uparrow} Q_L^{\downarrow} \overline{f_L} \right] \frac{e^{-2\kappa_0 d}}{(\alpha d)^2}. \tag{13.19}$$

Comparing Equations 13.18 and 13.19, we see that the asymptotic thickness dependence of G_P and G_{AP} is dominated by the same exponential factor, but G_{AP} has an additional factor ($\propto d$)$^{-1}$. Thus, the ratio G_P/G_{AP} increases linearly in d in this asymptotic regime, and the coefficient depends on the curvature of the dominant complex band at the $\mathbf{k}_{\parallel}$ point where k_0 reaches its minimum.

Thus, we see that the TMR increases rather slowly in the asymptotic regime, which is consistent with the fact that the important range of $\mathbf{k}_{\parallel}$ also shrinks slowly, as explained in Section 13.2.4. (The important range is $k_{\parallel} \sim (\alpha d)^{-1/2}$.) This behavior was illustrated in Ref. [32] for Fe/MgO/Fe junctions.

13.3.5 TUNNELING THROUGH INTERFACE STATES

Surfaces and interfaces between different materials often have spectral features that are absent in the bulk. Such features may include *surface states* (or interface states), which are localized at the surface (or interface) and are orthogonal to all bulk states [19, 38, 39]. Surface states appear as bands that have energy dispersion in $\mathbf{k}_{\parallel}$; if they cross the Fermi level, they produce Fermi contours in the $\mathbf{k}_{\parallel}$ plane. *Surface resonances* are similar, but they are not strictly orthogonal to bulk states. Due to coupling to bulk states, surface resonances have a finite lifetime; the corresponding bands and the Fermi contours are, accordingly, broadened. Of course, it only makes sense to talk about resonant states if their lifetime is larger than the relevant inverse bandwidth. Surface resonances often form on the (001) surfaces of transition metals; in particular, the presence of minority-spin resonant states on the Fe(001) surface is well established experimentally [40].

The formation of interface resonant states is largely controlled by the atomic structure and bonding at the interface. This can already be seen within a simple tight-binding model [41], which shows that the interplay between the atomic level shifts and changes in bonding strength at the interface may result in the appearance of an interface resonance. The same parameters determine the coupling of the resonance to bulk states, and hence the width of this resonance.

Interface resonances may appear in MTJs and influence their transport properties [7, 14, 42]. The minority-spin interface resonances are present at the Fe(001)/MgO interface, as was found from first-principle calculations [14] and indirectly observed in tunneling experiments [43, 44]. Within a two-dimensional Brillouin zone, these resonant states represent a surface band sharply peaked near the Fermi energy. Calculations show that the interface Fe(001)/MgO band is extremely sensitive to electron energy [45]. This is seen from Figure 13.5, which shows the $\mathbf{k}_{\parallel}$-resolved DOS for different energies. Interface (resonant) bands show up as (broadened) contours in the surface Brillouin zone. A very small change in energy of 0.02 eV leads to a significant change in the location of these bands.

In the presence of interface resonant states, the tunneling conductance is strongly enhanced for the values of $\mathbf{k}_{\parallel}$ lying within a linewidth of their Fermi contour. This happens simply because for these $\mathbf{k}_{\parallel}$ the scattering states have a large amplitude at the interface; the electron approaching the interface gets "trapped" in the resonant state for a time of the order of its inverse linewidth and therefore has a larger probability to tunnel across the barrier.

While there is no doubt that interface (resonant) states can significantly affect the tunneling conductance and TMR in MTJs, it is extremely difficult to describe their contribution quantitatively for particular junctions. First, as it is clear from the symmetry considerations of Section 13.3.2, this contribution strongly depends on the locus of the Fermi contours of the interface states, which, in turn, is usually very sensitive to the atomic structure of the interface. In most cases, this structure is complicated and not known in sufficient detail. Second, even if the surface structure is known, metal/insulator interfaces are difficult to treat using density functional theory due to the well-known problems with the description of band gaps, band offsets, and correlation effects.

Apart from these serious problems, two issues related to tunneling through interface states are worth mentioning. The first one appears when the conductance of a *symmetric* MTJ is calculated theoretically. If interface resonant states exist in one of the spin channels, their dispersions at the two interfaces are identical. For the parallel magnetization of the electrodes, the interface states then appear at exactly the same contour in the $\mathbf{k}_{\parallel}$ plane and "match" with each other. This matching greatly enhances the conductance in the corresponding spin channel. For example, if the width of the surface resonance is γ, and the coupling between the interface states localized on the two electrodes is much smaller than γ, the $\mathbf{k}_{\parallel}$-integrated tunneling rate is proportional to γ^{-1} [46]. On the other hand, for *mismatched* surface resonances on the two electrodes, the integrated conductance does not depend on γ. For sharp interface resonances (i.e., those with small γ compared to bandwidth), any structural disorder, the difference in the surface structure due to the different growth kinetics for the bottom and top electrodes, as well as the bias voltage will tend to destroy their matching. On the other hand, the conductance in the antiparallel configuration should not be as strongly affected by these factors. In practice, this means that the conductance calculated for a symmetric MTJ with parallel magnetization may contain a large fictitious (unobservable) contribution due to surface resonant states, which usually needs to be properly subtracted [45, 46]. The suppression of this contribution was found using first-principles calculations including interface roughness or disorder [47, 48].

A particularly peculiar feature occurs when the coupling between the "matched" resonant states on the two interfaces exceeds their intrinsic width due to coupling to bulk states [42]. In this case a "resonance of resonances" occurs, when in the first approximation one can neglect the coupling to the bulk and find the exact surface eigenstates to be the symmetric and antisymmetric linear combinations of the degenerate surface states on the two sides of the barrier. When the energy splitting of these states is large compared to the coupling to the bulk, the electrons transmit across the barrier with unit probability near the corresponding energies. This feature is generally regarded as an artifact of the perfectly symmetric junction, which cannot realistically be observed in practice.

Although the situation with perfectly matched interface states on the two sides of the barrier is likely unphysical, interface resonant states often play an important role in the transport properties of MTJs. Examples of first-principles calculations for specific electrode/barrier combinations can be found in the literature (see, for example, Refs. [30, 45]).

The second issue is related to the neglect of all inelastic processes within the simple one-electron picture. Consider a region of $\mathbf{k}_\parallel$ in the interface Brillouin zone where there are no bulk states in one of the electrodes (like the dark blue regions in Figure 13.5). According to the single-electron Landauer–Büttiker formalism, this region does not contribute at all to the tunneling current. However, surface states may form in this region (see the pairs of arcs extending into the dark blue regions in Figure 13.5; they are revealed by adding a small imaginary part to energy). If the problem is treated using Bardeen's approach, which considers the current at $t = 0$ when all states below the Fermi level are filled (including the surface states, if present), the surface states do contribute to the tunneling current. The Landauer–Büttiker approach correctly gives zero for the steady-state current, unless there is a mechanism allowing the surface states to be dynamically refilled to compensate for the tunneling outflux across the barrier. Since the rate of this refilling process should only keep up with the (slow) tunneling, it is reasonable to expect that in any practical situation it will be made possible through inelastic processes. Thus, it may be argued that Bardeen's approach correctly includes the contribution of pure surface states, while the Landauer–Büttiker approach does not. This argument was advanced by Wortmann et al. [49], who have also developed an elegant computational

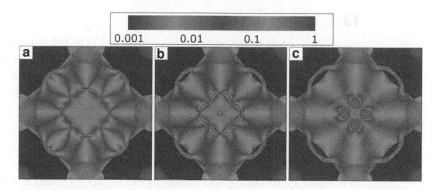

FIGURE 13.5 $\mathbf{k}_{\|}$-Resolved minority-spin density of states (in arbitrary units) for the Fe(001)/MgO interface calculated for three different energies near the Fermi energy (E_F): (a) $E = E_F - 0.02$ eV, (b) $E = E_F$, (c) $E = E_F + 0.02$ eV. (After Belashchenko, K.D. et al., *Phys. Rev. B* 72(14), 140404, 2005. With permission.)

technique designed to implement Bardeen's formula for the tunneling current within a full-potential linear augmented plane wave (FLAPW) method. This technique is based on the embedded Green function formulation [50] for the electronic structure of an active layer sandwiched between semi-infinite electrodes, which was previously used to implement the Landauer–Büttiker calculation of the conductance in FLAPW [51]. Unfortunately, Bardeen's approach is not universal, because it is limited to the case when the tunneling coupling is weak, which allows one to treat tunneling as a first-order process in this coupling.

As it was mentioned above, bonding at the interface may be very different from the bulk, which often leads to the formation of interface resonant states. Additional complications occur due to interface reconstruction, interdiffusion, disorder, etc. Apart from the possible formation of surface states, there are, obviously, other ways in which interface bonding and atomic rearrangements can strongly influence the spin polarization and TMR. For example, excess oxidation at the interface can tend to quench or reduce the magnetic moments near the surface [30, 52] and remove some of the surface spectral weight from the Fermi level. Combined with the symmetry selection rules for the surface states, this may effectively lead to the formation of additional spin-dependent tunneling barriers inside the surface layer. These changes may sometimes even lead to a reversal of the spin polarization of the tunneling current.

13.4 INFLUENCE OF INTERFACE ROUGHNESS AND DEFECTS IN THE BARRIER

Interface imperfections and defects in the barrier may have a strong effect on spin-dependent tunneling due to the mixing of the tunneling channels, as well as resonant transmission through the localized electronic states created by these defects. In this section, we will briefly discuss the influence of point defects on the TMR within the single-electron picture.

13.4.1 INTERFACE ROUGHNESS

Since high TMR is achieved in high-quality MTJs exhibiting symmetry-enforced spin filtering (see Section 13.3.2), defects usually should reduce the TMR. It appears that atomic-scale interface roughness or disorder is the most detrimental because it can introduce scattering between otherwise independent tunneling channels and thereby reduce the efficiency of the symmetry-enforced spin filtering [33, 48]. On the other hand, sparsely spaced steps between atomically flat terraces can be expected to be of lesser importance. The characteristic scale here is the inverse of the size of the area in $\mathbf{k}_\parallel$ space contributing appreciably to the tunneling current. According to Sections 13.2.4 and 13.3.4, this size is of order $\Delta r \sim \sqrt{\alpha d}$, where α is typically a few lattice parameters; Δr can be viewed as the transverse size of the tunneling wave packet. Thus, if the typical distance L between the atomically flat steps is much larger than Δr, they will make a contribution proportional to the areal density of the steps (i.e., this contribution contains a factor $\Delta r/L$). However, this contribution activates the tunneling channel with the lowest decay parameter and therefore takes over at a large barrier thickness. In any case, surface roughness limits the efficiency of spin filtering and imposes the upper bound for TMR, which saturates as a function of barrier thickness.

13.4.2 NONRESONANT DEFECTS IN THE BARRIER

The behavior of the tunneling transparency is very different in the cases when the point defects do or do not introduce localized levels close to the Fermi level [53, 54]. In this context, "close to the Fermi level" means that the distance from the Fermi level is not large compared the level's linewidth due to coupling with the electrodes. For example, a continuum of localized levels should always be present in an amorphous barrier (like Al_2O_3), while isolated point defects, such as oxygen vacancies, would be the most common in high-quality crystalline barriers.

The effect of oxygen vacancies in Fe/MgO/Fe MTJs was considered by Velev et al. [55]. Figure 13.6 shows the calculated majority-spin transmission as a function of energy for an ideal Fe/MgO/Fe MTJ (solid line) and for an MTJ with O vacancies (dashed line). For the ideal MTJ, the majority-spin transmission has a characteristic energy point (denoted by the arrow in Figure 13.6) above which the transmission increases significantly due to the occurrence of the Δ_1 symmetry band in the electronic structure of Fe(001). O vacancies produce a pronounced peak in the transmission due to resonant tunneling of electrons via the O vacancy s state. The width of the peak depends strongly on the vacancy density and on the coupling to the electrodes determined by MgO thickness.

The transmission at resonance is not of practical interest in this case, however, because the resonant level lies at about 1 eV below the Fermi energy. With correction for the self-interaction, the resonant level moves even deeper down in energy as corroborated by the experimental data [56]. Therefore, for a moderate bias, the transport mechanism is controlled by nonresonant scattering at O vacancies. It is evident from Figure 13.6 that at and above the Fermi energy the transmission of an MTJ with vacancies is reduced by

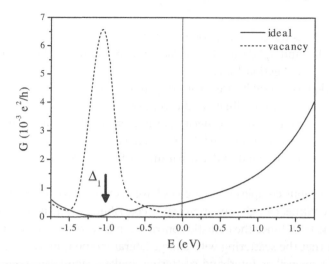

FIGURE 13.6   Majority-spin conductance per unit cell area in Fe/MgO/Fe(001) tunnel junction with five monolayers of MgO for ideal MgO and MgO with O vacancies (1/8 cell). The arrow indicates the bottom of the majority-spin Δ_1 band. The Fermi energy lies at zero. (After Velev, J.P. et al., *Appl. Phys. Lett.* 90, 072502, 2007. With permission.)

a factor of 5–7 depending on energy, as compared to the transmission of a perfect MTJ. This detrimental effect of O vacancies on the majority-spin transmission is due to scattering of tunneling electrons between states with different transverse wave vectors $\mathbf{k}_{\parallel}$. In a perfect Fe/MgO/Fe junction, the tunneling probability in the majority-spin channel is dominated by the Δ_1 evanescent state at the $\bar{\Gamma}$ point ($\mathbf{k}_{\parallel} = 0$), because it has the longest attenuation length in MgO. Scattering to states with $\mathbf{k}_{\parallel} \neq 0$ reduces the transmission coefficient due to a shorter decay length of these states.

For the nonresonant case, Lifshitz and Kirpichenkov [54] developed a rigorous technique allowing one to calculate the tunneling transparency by performing the configurational averaging in a "subbarrier kinetic equation" that describes the decay of the electronic density of a particular incident wave inside the barrier. The "collision integral" in this equation appears as a power series in the defect concentration. In the first order, the kinetic equation is linear; a similar equation was used by Zhang et al. [34] to discuss the effect of nonresonant point defects on TMR in Fe/MgO/Fe tunnel junctions. The effect of point defects is to scatter electrons from one evanescent state $\mathbf{k}_{\parallel}$ to another $\mathbf{k}'_{\parallel}$. (It may be more physically appealing to imagine an outgoing spherical evanescent wave emitted by the defect, on which a plane evanescent wave is an incident. On each side of the defect, the evanescent spherical wave can be expanded in $\mathbf{k}_{\parallel}$ states, which decay in the corresponding direction.)

Generalizing the approach of Ref. [54], single defect scattering leads to the appearance of new $\mathbf{k}_{\parallel}$-nonconserving transmission channels, each contributing to the tunneling current as

$$T(\mathbf{k}_{\parallel}, \mathbf{k}'_{\parallel}, x) = c e^{-2\kappa_0 d} T_L(\mathbf{k}_{\parallel}) e^{-2(\kappa-\kappa_0)x}$$
$$\times |t(\mathbf{k}_{\parallel}, \mathbf{k}'_{\parallel})|^2 \, e^{-2(\kappa'-\kappa_0)(d-x)} T_R(\mathbf{k}'_{\parallel}),$$

(13.20)

where:

c	is the concentration of defects
$T_L(\mathbf{k}_{\parallel})$	and $T_R(\mathbf{k}_{\parallel}')$ are the surface transmission functions introduced in Section 13.3.3
$t(\mathbf{k}_{\parallel}, \mathbf{k}_{\parallel}')$	is the subbarrier scattering amplitude
x	is the coordinate of the impurity
κ	and κ' are the decay parameters of the evanescent barrier states with $\mathbf{k}_{\parallel}$ and $\mathbf{k}_{\parallel}'$, respectively
κ_0	is the smallest decay parameter for all evanescent bands

The obvious complex band indices have been omitted. It is seen that the $\mathbf{k}_{\parallel} \rightarrow \mathbf{k}_{\parallel}'$ channel can give an appreciable contribution only if $x(\kappa - \kappa_0) + (d - x)(\kappa' - \kappa_0) \lesssim 1$. Since the ballistic contribution is restricted to $(\kappa - \kappa_0)d \sim 1$, it is seen that the scattering with a large lateral momentum transfer (either $\mathbf{k}_{\parallel}$ or $\mathbf{k}_{\parallel}'$), as well as interband scattering, make a significant contribution only if the scattering impurity is located close to one of the interfaces, that is, within a distance $(\kappa - \kappa_0)^{-1}$ from the left interface, or within $(\kappa' - \kappa_0)^{-1}$ from the right interface. Averaging over the position of the impurity in (13.20) yields, therefore, a factor $(\kappa - \kappa_0)^{-1}$, that is,

$$T(\mathbf{k}_{\parallel}, \mathbf{k}_{\parallel}') = \frac{c}{2} e^{-2\kappa_0 d} T_L(\mathbf{k}_{\parallel}) \, | \, t(\mathbf{k}_{\parallel}, \mathbf{k}_{\parallel}') \, |^2 \left[\frac{1}{\kappa - \kappa_0} + \frac{1}{\kappa' - \kappa_0} \right] T_R(\mathbf{k}_{\parallel}'). \quad (13.21)$$

The first (second) term in square brackets comes from impurity scattering near the left (right) interface. The behavior of $T(\mathbf{k}_{\parallel}, \mathbf{k}_{\parallel}')$ in the region of $\mathbf{k}_{\parallel}$ where $(\kappa - \kappa_0)d \lesssim 1$ (and similar for $\mathbf{k}_{\parallel}'$) needs a more detailed consideration. As shown in Ref. [54] within their simplified model, the unphysical divergence at $\mathbf{k}_{\parallel}, \mathbf{k}_{\parallel}' \rightarrow 0$ is an artifact of the perturbative treatment.

Since the $\mathbf{k}_{\parallel}$-nonconserving channels produce a contribution decaying with thickness at a similar rate to the symmetry-filtered ballistic contribution, nonresonant defect scattering in the barrier leads to the saturation of TMR at a value that depends on impurity concentration (similar to the case of interface roughness discussed above), and fitting to the experimental thickness dependence of the parallel and antiparallel conductances is possible [34]. The case of very thick barriers, for which the ballistic tunneling is fully replaced by diffusive tunneling, requires a separate treatment along the lines of Ref. [54].

13.4.3 RESONANT DEFECTS IN THE BARRIER

If localized levels in the barrier are found close to the Fermi level, and as long as the temperature is sufficiently low to preserve the wavefunction coherence across the barrier, the tunneling current is dominated by resonant channels through which an electron can tunnel with no attenuation [53, 54]. If the defect concentration is sufficiently low and the barrier is thin enough, the resonant channels are formed by defects located close to the middle of the barrier, so that their coupling to both electrodes is similar. For thicker barriers or larger impurity concentrations, the resonant channels are formed either by percolating resonant paths connecting the defects lying close to a straight line perpendicular to the interfaces or by even more complicated arrangements of defects. It is typical in this situation that the channels carrying most of the current are statistically rare, and the conductance does not behave as a self-averaging quantity. In particular, large fluctuations of the conductance, as well as irregularities in the current–voltage curves, are expected in junctions of small transverse sizes depending on the specific arrangement of defects [53, 57].

Let us consider an isolated defect in the barrier, and assume that in an infinite barrier it would have one nondegenerate eigenstate ψ_0 at energy ε_0. Neglecting direct tunneling, and applying time-dependent perturbation theory in the spirit of Bardeen's approach (see Section 13.2.1), where the role of perturbation is played by the hopping matrix elements TL and TR between the defect level and the Bloch states in the left and right electrodes (respectively), one obtains the rate of tunneling between the states ψ_l and ψ_r in the electrodes [57]:

$$w_{lr}(\varepsilon) = \frac{2\pi}{\hbar} \frac{|t_l|^2 |t_r|^2}{(\varepsilon - \varepsilon_0)^2 + (\Gamma_L + \Gamma_R)^2} \delta(\varepsilon_r - \varepsilon_l) \tag{13.22}$$

where the quantities

$$\Gamma_L = \pi \sum_l |t_l|^2 \delta(\varepsilon - \varepsilon_l); \quad \Gamma_R = \pi \sum_r |t_r|^2 \delta(\varepsilon - \varepsilon_r) \tag{13.23}$$

play the role of the self-energy of the impurity level due to its coupling to the electrodes. The zero-temperature conductance for the given spin channel is obtained by taking the trace:

$$G(\varepsilon) = \frac{4e^2}{h} \frac{\Gamma_L \Gamma_R}{(\varepsilon - \varepsilon_0)^2 + (\Gamma_L + \Gamma_R)^2} \tag{13.24}$$

It is seen that the conductance has a sharp maximum when the impurity level ε_0 lies close to the Fermi energy. This behavior is an example of resonant conductance.

For a planar tunnel junction, the indices l and r in the hopping matrix elements t_l and t_r denote $\mathbf{k}_\parallel$ and the band index. If for each $\mathbf{k}_\parallel$, only one evanescent state is important, these matrix elements are related to the interface transmission functions T_L and T_R, that is,

$$|t_l|^2 \rightarrow |t_L(\mathbf{k}_\parallel)|^2 = T_L(\mathbf{k}_\parallel) e^{-2\kappa(\mathbf{k}_\parallel)x} |V(\mathbf{k}_\parallel)|^2 \tag{13.25}$$

where L stands for the left electrode, the electrode band index is absorbed in $\mathbf{k}_\parallel$, and $V(\mathbf{k}_\parallel) = \langle \psi_0 | V | u(\mathbf{k}_\parallel) \rangle$, where V is the perturbing defect potential, and $u(\mathbf{k}_\parallel)$ the periodic part of the evanescent wave (from which the decaying Bloch exponent has been excluded). The coordinate of the impurity is denoted as x; a similar expression for t_R contains $d - x$ instead.

The conductance (Equation 13.24) at the given energy depends only on Γ_L and Γ_R, which are equal to the surface Brillouin zone integrals of Equation 13.25.

The situation is very different from nonresonant tunneling. In particular, in the resonant case $\varepsilon = \varepsilon_0$, the conductance is dominated by impurities with $\Gamma_L \approx \Gamma_R$, which set up transparent channels with $G \approx e^2/h$. These impurities are located close to the *middle* of the barrier, contrary to the dominant nonresonant impurities, which lie close to the interfaces, as discussed above. Let us consider particular cases.

In the nonresonant regime, that is, at $\varepsilon_0 - \varepsilon_F \gg (\Gamma_L + \Gamma_R)$, the situation is rather similar to the direct nonresonant tunneling considered in the previous section. Whether direct or impurity-assisted nonresonant tunneling dominates depends on the relation between the matrix elements and on the distance of the impurity level from the Fermi level.

For a narrow (point) junction, the conductance of a particular sample may be dominated by a single resonant impurity. In this case, the sign of the TMR may be reversed compared to the nonresonant one. Indeed, consider the case $\varepsilon_0 - \varepsilon_F = 0$ and an asymmetric impurity position. If the impurity is closer to the left electrode, then $\Gamma_L \gg \Gamma_R$, and $G = (4e^2/h)(\Gamma_L/\Gamma_R)$. In the opposite case $\Gamma_L \ll \Gamma_R$; Γ_R we have $G = (4e^2/h)(\Gamma_L/\Gamma_R)$. In both cases, the conductance is inversely proportional to the self-energy due to coupling with one of the electrodes, which results in the sign reversal of TMR [58]. This behavior was observed experimentally in narrow nanojunctions [58] and in gate-field-controlled carbon nanotubes connecting two ferromagnetic electrodes through tunnel barriers [59]. This latter effect occurs due to resonant tunneling through quantum well states in the nanotube, whose energy is changed by the gate voltage.

For a wide tunnel junction, the averaging over the random impurity positions should be performed. For simplicity, let us again take $\varepsilon_F = \varepsilon_0$. If the system consisted of independent one-dimensional threads, Equation 13.23 would contain only one term with the x dependence from Equations 13.25, and 13.24 can then be easily integrated over x. Neglecting for simplicity the discreteness of x due to the lattice periodicity, and assuming $\kappa d \gg 1$, we find with these approximations that the conductance $G = ce^2/(h\kappa)$ does not depend on Γ_L, Γ_R at all. This means that in this case, the current does not depend on the magnitude of the coupling to the bulk states, and therefore the TMR vanishes. Note that the -dominant impurities are always those that maximize G in Equation 13.24; for -different configurations of the electrode magnetizations and different spin channels, this maximum is achieved at different x.

The case when the impurity states are not all pinned to ε_F but are rather distributed with a constant density may be appropriate for an amorphous barrier. In the quasi-one-dimensional case, one then has to perform an additional integration over $\varepsilon_0 > \varepsilon_F$, which results in $\langle G \rangle \propto \sqrt{T_L T_R}$ [8].

The situation for a three-dimensional system is more complicated. The conductance looks similar to Equation 13.24, but it now contains a double sum over $\mathbf{k}_\parallel$, $\mathbf{k}'_\parallel$, while Γ_L, Γ_R are replaced by quantities given in Equation 13.25. If x is integrated over, the resonance for different $\mathbf{k}_\parallel \to \mathbf{k}'_\parallel$ channels is achieved at different values of x (i.e., different impurities are at resonance).

Therefore, the total conductance cannot, in general, be represented as a product of two ("left" and "right") factors in the three-dimensional case.

13.5 SPIN–ORBIT COUPLING AND TUNNELING ANISOTROPIC MAGNETORESISTANCE

Until now, we have treated the transport of electrons with two different spin projections as being completely independent. This is a good approximation as long as the magnetizations of the two electrodes are collinear, the temperature is low enough so that scattering on spin fluctuations is not important, and SOC is negligible. Spin transport in noncollinear configurations is of main interest for problems involving spin torque effects, which are considered in Chapters 7–9, and 15, Volume 1.

SOC may naturally be expected to affect the TMR appreciably if it can generate a significant amount of spin-flip scattering at the interfaces. In addition, in the presence of SOC, the evanescent barrier eigenstates do not have a definite spin quantum number, which weakens the selection rules leading to tunneling spin filtering. In this case, SOC should normally lead to the reduction of TMR due to the reduced spin selectivity of the tunneling process. The details of this problem were studied for a Fe/GaAs/Fe MTJ using a fully relativistic Green's function formalism [60]. In this method, the current operator is approximated to be spin diagonal in the leads, while the spins are mixed in the "active region"; thus the Landauer–Büttiker transmission function has both spin-conserving and spin-flip components, the latter representing the effect of SOC. For Fe/GaAs/Fe, the main effect was found to be on the conductance for the antiparallel configurations, where SOC is able to open efficient spin-flip conducting channels that were otherwise forbidden; the TMR is thereby reduced.

However, under favorable conditions, SOC can also lead to the appearance of an entirely new effect called tunneling anisotropic magnetoresistance in which the resistance of a tunnel junction with only *one* ferromagnetic electrode depends on the magnetization direction of that electrode. First experimental demonstrations of TAMR utilized a diluted magnetic semiconductor (Mn-doped GaAs) as the magnetic electrode [61]. In general, the TAMR effect occurs because the electronic structure of the magnetic side of the junction depends on the orientation of the magnetization with respect to the crystallographic axes, and this dependence is reflected in the tunneling conductance. A similar effect exists even in bulk resistivity (it is called anisotropic magnetoresistance), but the tunneling setup can greatly increase its magnitude because SOC is often strongly enhanced at interfaces.

Consider the effect of SOC on the interface states, whose role in tunneling was discussed in Section 13.3.5. Suppose that the -tunneling current has a large contribution from such states. Having the largest amplitude at the surface, these states are the most sensitive to enhanced interfacial SOC. (In other words, the expectation value of the SOC operator is usually larger for the interface resonant states.) The effect of SOC is to shift these states up or down in energy depending on their in-plane wavevector $\mathbf{k}_\parallel$ and on the magnetization direction (Rashba effect) [62]. (Note that the spin degeneracy is already lifted by the exchange splitting; the SOC does not lead to any additional symmetry breaking.) In turn, these shifts deform the Fermi contour of the interface states. Since the tunneling probabilities strongly depend on $\mathbf{k}_\parallel$ (see Section 13.3.2), the tunneling current, in turn, depends on the direction of magnetization, leading to the TAMR effect [63]. A similar effect occurs in an MTJ with two magnetic electrodes, that is, the tunneling current depends on the orientation of the magnetization of both electrodes, which remain parallel to each other [64, 65]. In this latter case, however, the situation is complicated by the presence of two interfaces, which may both support interface states.

One can distinguish out-of-plane and in-plane TAMR effects, with respect to the mode in which the magnetization is rotated. The out-of-plane (in-plane) TAMR refers to the resistance variation for the rotation of the magnetization in a plane normal -(parallel) to the interfaces. The angular dependence of the resistance reflects the underlying (nonmagnetic) symmetry of the junction. The resistance is symmetric with respect to the change of sign of the magnetization. (This follows from the time-reversal symmetry of the Hamiltonian; it is assumed that the external field is transformed along with the magnetization.) Obviously, if the junction has a rotation axis (normal to the interfaces), the TAMR automatically has the same axis of symmetry. Further, if the junction has a reflection plane normal to the interfaces, the TAMR also has the same reflection plane. (This property follows from the consideration of the response of the axial $\mathbf{M}$ vector to the mirror reflection combined with time reversal.)

A particular case of interest is the Fe/GaAs/Au tunnel junction, whose in-plane TAMR has a C_{2v} angular symmetry [66]. The junction itself cannot have a fourfold axis, because zinc blende GaAs only has screw fourfold axes, which are destroyed in a junction setup. For example, for an As-terminated interface, the normal planes containing the GaAs bonds next to the interface are oriented in a certain unique way. For this reason, even without SOC, the electronic structure of the Fe/GaAs interface at most has a C_{2v} symmetry (see, e.g., Figure 2 of Ref. [67]). Therefore, in the presence of SOC the in-plane TAMR cannot have a fourfold axis.

A further discussion of TAMR is given in Chapter 4, Volume 2.

13.6 CONCLUDING REMARKS

In this chapter, we have discussed the main properties of TMR, which follow from basic quantum-mechanical considerations. In many cases, these considerations (along with model or first-principles calculations) are sufficient to understand the qualitative features of measured TMR and its dependence on variable parameters. However, it is worth reemphasizing that the tunneling currents are very sensitive to the electronic structure of the electrodes, barrier, and interfaces. Consequently, quantitative comparison with experiment requires that this electronic structure is known in sufficient detail. Although the epitaxial relations are known for some high-quality epitaxial interfaces (like Fe/MgO), there always remains some uncontrolled variation in such details as the degree of disorder, interdiffusion, impurity segregation, and other imperfections dependent on film growth conditions. For example, the presence of excess oxygen at the Fe/MgO interface, which likely incorporates itself in the surface Fe layer, was found to have a large effect on spin-dependent tunneling across this interface [68]. Likewise, it was found that the presence of excess oxygen at the Co/Al_2O_3 interface can lead to a reversal of the sign of TMR in $Co/Al_2O_3/Co$ MTJs [52]. The influence of surface structure on TMR is especially strong if interface resonant states are involved. Tunneling across a -vacuum gap from a high-quality surface of a sufficiently simple metal, for which the spin polarization can be measured using spin-polarized STM, is an important example where the surface structure may be known in full detail.

Even if the atomic structure of the interfaces is known (or hypothesized), the standard methods used to describe the electronic structure based on density functional theory often lead to large errors in insulating band gaps, band offsets at the interfaces, as well as in the dispersion relations of the interface states. It is, therefore, necessary to establish whether the behavior of the calculated TMR is robust against such errors; this is, unfortunately, not always possible. The development of new methods to redress deficiencies of the local density approximation (such as the GW method) is desirable. An application of the GW method to a Fe/MgO/Fe tunnel junction was recently reported [69].

ACKNOWLEDGMENTS

This work was supported by the National Science Foundation through the Nebraska Materials Research Science and Engineering Center (Grant No. DMR-1420645) and the Nebraska Research Initiative.

REFERENCES

1. W. F. Brinkman, R. C. Dynes, and J. M. Rowell, Tunneling conductance of asymmetrical barriers, *J. Appl. Phys.* **41**, 1915–1921 (1970).
2. W. A. Harrison, Tunneling from an independent-particle point of view, *Phys. Rev.* **123**, 85–89 (1961).
3. J. G. Simmons, Generalized formula for the electric tunnel effect between similar electrodes separated by a thin insulating film, *J. Appl. Phys.* **34**, 1793–1803 (1963).
4. J. Bardeen, Tunnelling from a many-particle point of view, *Phys. Rev. Lett.* **6**, 57–59 (1961).
5. J. C. Slonczewski, Conductance and exchange coupling of two ferromagnets separated by a tunneling barrier, *Phys. Rev. B* **39**, 6995–7002 (1989).
6. J. S. Moodera, J. Nassar, and G. Mathon, Spin-tunneling in ferromagnetic junctions, *Ann. Rev. Mater. Sci.* **29**, 381–432 (1999).
7. E. Y. Tsymbal, K. D. Belashchenko, J. P. Velev et al., Interface effects in spin-dependent tunneling, *Prog. Mater. Sci.* **52**, 401–420 (2007).
8. E. Y. Tsymbal, O. N. Mryasov, and P. R. LeClair, Spin-dependent tunnelling in magnetic tunnel junctions, *J. Phys.: Condens. Matter* **15**, R109–R142 (2003).

9. X.-G. Zhang and W. H. Butler, Band structure, evanescent states, and transport in spin tunnel junctions, *J. Phys.: Condens. Matter* **15**, R1603–R1639 (2003).

10. M. Julliére, Tunneling between ferromagnetic films, *Phys. Lett. A* **54**, 225–226 (1975).

11. P. M. Tedrow and R. Meservey, Spin polarization of electrons tunneling from films of Fe, Co, Ni, and Gd, *Phys. Rev. B* **7**, 318–326 (1973).

12. S. Datta, *Electronic Transport in Mesoscopic Systems*, Cambridge University Press, Cambridge, 1997.

13. J. Tersoff and D. R. Hamann, Theory of the scanning tunneling microscope, *Phys. Rev. B* **31**, 805–813 (1985).

14. W. H. Butler, X.-G. Zhang, T. C. Schulthess, and J. M. MacLaren, Spin-dependent tunneling conductance of Fe/MgO/Fe sandwiches, *Phys. Rev. B* **63**, 054416 (2001).

15. J. Mathon and A. Umerski, Theory of tunneling magnetoresistance of an epitaxial Fe/MgO/Fe(001) junction, *Phys. Rev. B* **63**, 220403 (2001).

16. S. S. P. Parkin, C. Kaiser, A. Panchula et al., Giant tunnelling magnetoresistance at room temperature with MgO (100) tunnel barriers, *Nat. Mater.* **3**, 862–867 (2004).

17. S. Yuasa, T. Nagahama, A. Fukushima, Y. Suzuki, and K. Ando, Giant room-temperature magnetoresistance in single-crystal Fe/MgO/Fe magnetic tunnel junctions, *Nat. Mater.* **3**, 868–871 (2004).

18. W. Kohn, Analytic properties of Bloch waves and Wannier functions, *Phys. Rev.* **115**, 809–821 (1959).

19. V. Heine, On the general theory of surface states and scattering of electrons in solids, *Proc. Phys. Soc.* **81**, 300–310 (1963).

20. Ph. Mavropoulos, N. Papanikolaou, and P. H. Dederichs, Complex band structure and tunneling through ferromagnet/insulator/ferromagnet junctions, *Phys. Rev. Lett.* **85**, 1088–1091 (2000).

21. Y.-C. Chang, Complex band structures of zinc-blende materials, *Phys. Rev. B* **25**, 605–619 (1982).

22. M. Tinkham, *Group Theory and Quantum Mechanics*, Dover, New York, 2003.

23. L. P. Bouckaert, R. Smoluchowski, and E. Wigner, Theory of Brillouin zones and symmetry properties of wave functions in crystals, *Phys. Rev.* **50**, 58–67 (1936).

24. G. C. Osbourn and D. L. Smith, Transmission and reflection coefficients of carriers at an abrupt GaAs-GaAlAs (100) interface, *Phys. Rev. B* **19**, 2124–2133 (1979).

25. J. M. MacLaren, X.-G. Zhang, W. H. Butler, and X. Wang, Layer KKR approach to Bloch-wave transmission and reflection: Application to spin-dependent tunneling, *Phys. Rev. B* **59**, 5470–5478 (1999).

26. J. Faure-Vincent, C. Tiusan, E. Jouguelet et al., High tunnel magnetoresistance in epitaxial Fe/MgO/Fe -tunnel junctions, *Appl. Phys. Lett.* **82**, 4507–4509 (2003).

27. X.-G. Zhang and W. H. Butler, Large magnetoresistance in bcc Co/MgO/Co and FeCo/MgO/FeCo tunnel junctions, *Phys. Rev. B* **70**, 172407 (2004).

28. T. Moriyama, C. Ni, W. G. Wang, X. Zhang, and J. Q. Xiao, Tunneling magnetoresistance in (001)-oriented FeCo/MgO/FeCo magnetic tunneling junctions grown by -sputtering deposition, *Appl. Phys. Lett.* **88**, 222503 (2006).

29. S. Yuasa, A. Fukushima, H. Kubota, Y. Suzuki, and K. Ando, Giant tunneling magnetoresistance up to 410% at room temperature in fully epitaxial Co/MgO/Co magnetic tunnel junctions with bcc Co(001) electrodes, *Appl. Phys. Lett.* **89**, 042505 (2006).

30. K. D. Belashchenko, E. Y. Tsymbal, M. van Schilfgaarde, D. A. Stewart, I. I. Oleynik, and S. S. Jaswal, Effect of interface bonding on spin-dependent tunneling from the oxidized Co surface, *Phys. Rev. B* **69**, 174408 (2004).

31. R. Matsumoto, A. Fukushima, T. Nagahama, Y. Suzuki, K. Ando, and S. Yuasa, Oscillation of giant tunneling magnetoresistance with respect to tunneling barrier thickness in fully epitaxial Fe/MgO/Fe magnetic tunnel junctions, *Appl. Phys. Lett.* **90**, 252506 (2007).

32. C. Heiliger, P. Zahn, B. Yu. Yavorsky, and I. Mertig, Thickness dependence of the tunneling current in the coherent limit of transport, *Phys. Rev. B* **77**, 224407 (2008).

33. J. Mathon and A. Umerski, Theory of tunneling magnetoresistance in a disordered Fe/MgO/Fe(001) junction, *Phys. Rev. B* **74**, 140404 (2006).

34. X.-G. Zhang, Y. Wang, and X. F. Han, Theory of nonspecular tunneling through magnetic tunnel junctions, *Phys. Rev. B* **77**, 144431 (2008).

35. C.-M. Lee, J.-M. Lee, L.-X. Ye, S.-Y. Wang, Y.-R. Wang, and T.-H. Wu, Effects of MgO barrier thickness on magnetic anisotropy energy constant in perpendicular magnetic tunnel junctions of (Co/Pd)n/MgO/(Co/Pt)n, *IEEE Trans. Magn.* **44**, 2558–2561 (2008).

36. I. I. Mazin, Tunneling of Bloch electrons through vacuum barrier, *Europhys. Lett.* **55**, 404–410 (2001).

37. W. A. Harrison, *Solid State Theory*, Dover, New York, 1980.

38. J. Bardeen, Surface states and rectification at a metal semi-conductor contact, *Phys. Rev.* **71**, 717–727 (1947).

39. W. Shockley, On the surface states associated with a periodic potential, *Phys. Rev.* **56**, 317–323 (1939).

40. J. A. Stroscio, D. T. Pierce, A. Davies, R. J. Celotta, and M. Weinert, Tunneling spectroscopy of bcc (001) surface states, *Phys. Rev. Lett.* **75**, 2960–2963 (1995).

41. E. Y. Tsymbal and K. D. Belashchenko, Role of interface bonding in spin-dependent tunneling, *J. Appl. Phys.* **97**, 10C910 (2005).

42. O. Wunnicke, N. Papanikolaou, R. Zeller, P. H. Dederichs, V. Drchal, and J. Kudrnovský, Effects of resonant interface states on tunneling magnetoresistance, *Phys. Rev. B* **65**, 064425 (2002).

43. C. Tiusan, J. Faure-Vincent, C. Bellouard, M. Hehn, E. Jouguelet, and A. Schuhl, Interfacial resonance state probed by spin-polarized tunneling in epitaxial Fe/MgO/Fe tunnel junctions, *Phys. Rev. Lett.* **93**, 106602 (2004).

44. C. Tiusan, M. Sicot, J. Faure-Vincent et al., Static and dynamic aspects of spin tunnelling in crystalline magnetic tunnel junctions, *J. Phys.: Condens. Matter* **18**, 941–956 (2006).

45. K. D. Belashchenko, J. Velev, and E. Y. Tsymbal, Effect of interface states on spin-dependent tunneling in Fe/MgO/Fe tunnel junctions, *Phys. Rev. B* **72**, 140404 (2005).

46. J. P. Velev, K. D. Belashchenko, and E. Y. Tsymbal, Comment on Destructive effect of disorder and bias voltage on interface resonance transmission in symmetric tunnel junctions, *Phys. Rev. Lett.* **96**, 119601 (2006).

47. Y. Ke, K. Xia, and H. Guo, Disorder scattering in magnetic tunnel junctions: Theory of nonequilibrium vertex correction, *Phys. Rev. Lett.* **100**, 166805 (2008).

48. P. X. Xu, V. M. Karpan, K. Xia, M. Zwierzycki, I. Marushchenko, and P. J. Kelly, Influence of roughness and disorder on tunneling magnetoresistance, *Phys. Rev. B* **73**, 180402 (2006).

49. D. Wortmann, H. Ishida, and S. Blügel, Embedded Green-function formulation of tunneling conductance: Bardeen versus Landauer approaches, *Phys. Rev. B* **72**, 235113 (2005).

50. J. E. Inglesfield, A method of embedding, *J. Phys. C: Solid State Phys.* **14**, 3795–3806 (1981).

51. D. Wortmann, H. Ishida, and S. Blügel, Embedded Green-function approach to the ballistic electron transport through an interface, *Phys. Rev. B* **66**, 075113 (2002).

52. K. D. Belashchenko, E. Y. Tsymbal, I. I. Oleynik, and M. van Schilfgaarde, Positive spin polarization in Co/Al$_2$O$_3$/Co tunnel junctions driven by oxygen adsorption, *Phys. Rev. B* **71**, 224422 (2005).

53. I. M. Lifshitz, S. A. Gredeskul, and L. A. Pastur, *Introduction to the Theory of Disordered Systems*, Wiley, New York, 1988.

54. I. M. Lifshitz and V. Y. Kirpichenkov, On tunneling transparency of disordered systems, *Sov. Phys. JETP* **50**, 499–511 (1979).

55. J. P. Velev, K. D. Belashchenko, S. S. Jaswal, and E. Y. Tsymbal, Effect of oxygen vacancies on spin-dependent tunneling in Fe/MgO/Fe magnetic tunnel junctions, *Appl. Phys. Lett.* **90**, 072502 (2007).

56. P. G. Mather, J. C. Read, and R. A. Buhrman, Disorder, defects, and band gaps in ultrathin (001) MgO tunnel barrier layers, *Phys. Rev. B* **73**, 205412 (2006).

57. A. I. Larkin and K. A. Matveev, Current-voltage characteristics of mesoscopic semiconductor contacts, *Sov. Phys. JETP* **66**, 580–584 (1987).

58. E. Y. Tsymbal, A. Sokolov, I. F. Sabirianov, and B. Doudin, Resonant inversion of tunneling magnetoresistance, *Phys. Rev. Lett.* **90**, 186602 (2003).

59. S. Sahoo, T. Kontos, J. Furer et al., Electric field control of spin transport, *Nat. Phys.* **1**, 99–102 (2005).

60. V. Popescu, H. Ebert, N. Papanikolaou, R. Zeller, and P. H. Dederichs, Influence of spin-orbit coupling on the transport properties of magnetic tunnel junctions, *Phys. Rev. B* **72**, 184427 (2005).

61. C. Gould, C. Rüster, T. Jungwirth et al., Tunneling anisotropic magnetoresistance: A spin-valve-like tunnel magnetoresistance using a single magnetic layer, *Phys. Rev. Lett.* **93**, 117203 (2004).

62. O. Krupin, G. Bihlmayer, K. Starke et al., Rashba effect at magnetic metal surfaces, *Phys. Rev. B* **71**, 201403 (2005).

63. A. N. Chantis, K. D. Belashchenko, E. Y. Tsymbal, and M. van Schilfgaarde, Tunneling anisotropic magnetoresistance driven by resonant surface states: First-principles calculations on an Fe(001) surface, *Phys. Rev. Lett.* **98**, 046601 (2007).

64. L. Gao, X. Jiang, S.-H. Yang, J. D. Burton, E. Y. Tsymbal, and S. S. P. Parkin, Bias voltage dependence of tunneling anisotropic magnetoresistance in magnetic tunnel junctions with MgO and Al_2O_3 tunnel barriers, *Phys. Rev. Lett.* **99**, 226602 (2007).

65. M. N. Khan, J. Henk, and P. Bruno, Anisotropic magnetoresistance in Fe/MgO/Fe tunnel junctions, *J. Phys.: Condens. Matter* **20**, 155208 (2008).

66. J. Moser, A. Matos-Abiague, D. Schuh, W. Wegscheider, J. Fabian, and D. Weiss, Tunneling anisotropic magnetoresistance and spin-orbit coupling in Fe/GaAs/Au tunnel junctions, *Phys. Rev. Lett.* **99**, 056601 (2007).

67. A. N. Chantis, K. D. Belashchenko, D. L. Smith, E. Y. Tsymbal, M. van Schilfgaarde, and R. C. Albers, Reversal of spin polarization in Fe/GaAs (001) driven by resonant surface states: First-principles calculations, *Phys. Rev. Lett.* **99**, 196603 (2007).

68. X.-G. Zhang, W. H. Butler, and A. Bandyopadhyay, Effects of the iron-oxide layer in Fe-FeO-MgO-Fe tunneling junctions, *Phys. Rev. B* **68**, 092402 (2003).

69. S. V. Faleev, O. N. Mryasov, and M. van Schilfgaarde, Effect of electron correlations on spin polarized transport in Fe/MgO tunnel junctions, arXiv: 1010.4086.

14

Spin Filter Tunneling

Tiffany S. Santos and Jagadeesh S. Moodera

14.1   INTRODUCTION

Over the last two decades or more, condensed matter physics, and in particular the field of magnetism, is overwhelmingly active in the area of spintronics: the effects of spin-polarized current, its generation, injection, transport, and dynamics. The amount of information generated and the discoveries that have resulted are unprecedented, reshaping not only the direction of magnetism research but also driving technological advances to dizzying heights. Thousands of articles, countless reviews, and several books have been written that cover the field as it is evolving. The field is developing so rapidly and becoming so diverse that it is hard to keep up with. The driving force is the spin-polarized current generated from ferromagnetic elements, compounds and alloys (metallic or semiconducting) due to their intrinsic exchange energy split conduction band. Generally, the degree of spin polarization (P) is limited to about 50% in conventional ferromagnetic elements and alloys materials. It can reach higher values in a special set of materials called half-metallic ferromagnets [1], which by design are quite difficult to achieve. In this chapter we deal with another phenomenon called spin filter tunneling, currently hotly pursued for its capability to provide tunable P all the way to near 100% and can be interfaced with metals, semiconductors or superconductors, giving the versatility for exploring new effects.

14.2   THE SPIN FILTER EFFECT

Spin polarization of the tunnel current in conventional ferromagnet/insulator/ferromagnet magnetic tunnel junctions, or in the Meservey-Tedrow experiments on Al/Al_2O_3/ferromagnet tunnel junctions, arises from the spin imbalance of the spin-up and spin-down density of states (DOS) of the conduction electrons at the Fermi energy $[N_\uparrow(E_F) > N_\downarrow(E_F)]$ in the ferromagnet (F). It is also dependent on the F/barrier interface. An entirely different way to create a spin-polarized tunnel current is to have spin-selective tunneling through barriers that are sensitive to spin orientation. For example, EuS is a barrier material that has a different tunneling probability $|T|^2$ for the two spin directions: $|T\uparrow|^2 \gg |T\downarrow|^2$. EuS is a magnetic semiconductor having a band gap of 1.65 eV and a ferromagnetic Curie temperature, $T_C = 16.6$ K. The exchange splitting of the conduction band ($2\Delta E_{ex}$) gives rise to two barrier heights for spin-down and spin-up electrons: $\Phi_{\downarrow,\uparrow} = \Phi_0 \pm \Delta E_{ex}$ where Φ_0 is the average barrier height at a temperature above T_C. In the case of EuS, the

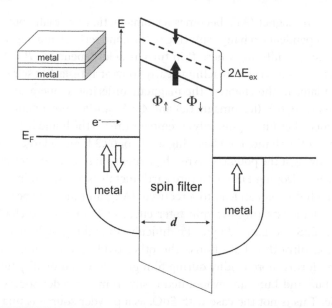

FIGURE 14.1 Schematic of the spin filter effect. Unpolarized electrons from the non-magnetic metal electrode tunnel through the ferromagnetic spin filter. Exchange splitting creates a lower barrier height for spin-up electrons and higher barrier height for spin-down electrons, generating a highly spin-polarized current.

barrier for spin-up electrons ($\Phi_\uparrow$) is lower than the one for spin-down electrons ($\Phi_\downarrow$). Since the tunneling probability depends exponentially on barrier height, the spin-up current can greatly exceed the spin-down current, yielding a highly spin-polarized current in a metal/EuS/metal tunnel junction where the metal electrodes are non-magnetic. This phenomenon is called the spin filter effect, shown schematically in Figure 14.1. For a given barrier thickness d, the tunnel current density J varies exponentially with the corresponding barrier height [2, 3]:

$$J_{\uparrow(\downarrow)} \propto \exp(-\Phi_{\uparrow(\downarrow)}^{1/2}d). \tag{14.1}$$

Therefore, even with a modest separation of spin-up and spin-down barrier heights, the tunneling probability for spin-up electrons is much higher than for spin-down electrons, resulting in polarization (P) of the tunnel current:

$$P = \frac{J_\uparrow - J_\downarrow}{J_\uparrow + J_\downarrow} \tag{14.2}$$

The exchange splitting of the conduction band in europium chalcogenides (EuX, where X is O or S) is substantial, the largest being 0.54 eV for EuO, making them ideal candidates as spin filter materials. Thus they could completely filter out spin-down electrons, leading to total spin polarization, $P = 100\%$.

The europium chalcogenide tunnel barriers EuS [4] and EuSe [5] have demonstrated the spin filter effect in spectacular fashion. EuS barriers have shown P as high as 85%, even with no magnetic field applied. EuSe, which is an antiferromagnet (AF), becomes ferromagnetic in a small applied field, and field-dependent exchange splitting of the conduction band appears. This gives rise to spin filtering where P, in turn, is field-dependent: $P = 0$ in zero field and increases with H, reaching nearly 100% at 1 Tesla. These results are described later in the chapter. The magnetic ordering temperatures of EuS and EuSe are 16.6 K (ferromagnetic) and 4.6 K (antiferromagnetic), respectively. Hence, they filter spins only at temperatures in the liquid helium range.

On the other hand, EuO has a higher T_C of 69.3 K and a larger exchange splitting, giving it the potential to reach greater spin filter efficiency at higher temperatures. Doping EuO with rare earth metals is known to increase the T_C, although doping can lead to a reduced $2\Delta E_{ex}$ and metallic behavior [6]. However, demonstrating the spin filter effect in EuO is more challenging than with EuS and EuSe, due to the difficulty in making high-quality, stoichiometric, ultra-thin EuO films. The other oxide phase, Eu_2O_3, is more stable and forms more readily during film growth. Good quality ultra-thin films of EuS and EuSe are evaporated easily from a powder source of EuS and EuSe. This is not the case with EuO, as a powder source is unavailable due to its metastable form. Despite this challenge, EuO tunnel barriers have displayed large exchange splitting and near 30% polarization by direct measurement [7].

Other promising spin filter candidates have arrived on the scene in more recent times, namely ferrites and perovskites, mainly due to advances in the synthesis of complex oxide thin film heterostructures. The Curie temperature of the ferrites is well above room temperature. Thus, they could potentially filter spins at a technologically useful temperature range. However, with their complex structure, the materials aspects are complicated. As will be described in this chapter, some degree of spin filtering has been achieved in ferrimagnetic, insulating $NiFe_2O_4$ and $CoFe_2O_4$ [8], even at room temperature in $CoFe_2O_4$, though there is much room for improvement. Among the perovskites, some interesting tunneling results have been observed with insulating, ferromagnetic $BiMnO_3$, especially since this material is multiferroic, such that the barrier profile is modified due to both magnetic and ferroelectric order (see Chapter 17, Volume 1).

Spin filtering in GdN barriers has been shown more recently in Josephson junctions with NbN electrodes, unlocking some new phenomena in these structures. Table 14.1 lists the properties of the known spin filter materials.

14.3 ELECTRONIC STRUCTURE OF EU-CHALCOGENIDES

In the 1960s and 1970s, the magnetic, optical, and electronic properties of the EuX compounds in the form of bulk crystals and thick films were extensively studied. The earlier work has been well covered in several past reviews; see for example Refs. [13, 14, 15, 16]. We briefly mention here some properties that are relevant for the discussion of these materials as spin filters. The europium chalcogenides are Heisenberg ferromagnets. The magnetic moment originates from the half-filled $4f$ states, giving rise to a large saturation moment of

TABLE 14.1
Spin Filter Materials

Material	Magnetic Behavior	T_C (K)	Moment (μ_B)	Structure, a(nm)	Eg (eV)	$2\Delta E_{ex}$ (eV)	P (%)	Spin Filter Reference
EuO	F	69.3	7	Fcc, 0.514	1.12	0.54	29	Santos et al. [7]
EuS	F	16.6	7	Fcc, 0.596	1.65	0.36	86	Moodera et al. [4]
EuSe	AF	4.6	7	Fcc, 0.619	1.80		100	Moodera et al. [5]
BiMnO$_3$	F	105	3.6	perovskite			22	Gajek et al. [9]
NiFe$_2$O$_4$	ferri	850	2	spinel	1.2		22	Lüders et al. [10]
CoFe$_2$O$_4$	ferri	796	3	spinel	0.80		−25	Ramos et al. [11]
GdN	F	60		Fcc, 0.497				Pal et al. [12]

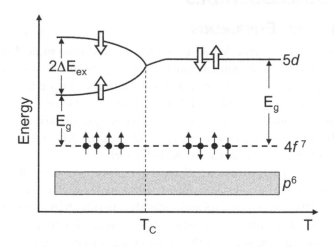

FIGURE 14.2 Simple schematic of the energy level diagram as a function of temperature for ferromagnetic, semiconducting EuS and EuO. (Modified from Wachter, P., in *Handbook on the Physics and Chemistry of Rare Earths*, vol. 2, Eds. K. A. Schneider and L. Eyring, North-Holland Publishing Co., New York, 1979, pp. 507–574.)

$g\mu_B J = 7 \mu_B$ per Eu^{2+} ion. This has been experimentally observed in bulk single crystals and thin films and shown in band structure calculations [17, 18, 19]. A simple schematic of the energy gap is shown in Figure 14.2. The $4f^7$ states are localized within the energy gap between the valence band and conduction band. The energy gap between the $4f^7$ states and the $5dt_{2g}$ states at the bottom of the conduction band defines the optical band gap (E_g).

When cooled below the T_C, the ferromagnetic order of the $4f^7$ spins causes Zeeman splitting of the conduction band, splitting it into two levels: the energy level for spin-up electrons is lowered by an amount ΔE_{ex}, whereas the spin-down level is raised by the same amount, thus lifting the spin degeneracy. The exchange interaction H_{exch} is given by the following Heisenberg exchange relation: $H_{exch} = 2\Delta E_{ex} = -2J_c \sum \vec{s} \cdot \vec{S}_n$, where $\vec{s}$ is the spin of a conduction electron, $\vec{S}_n$ is the spin of neighboring Eu^{2+} ions and J_c is the space-dependent exchange constant between the electron spin and the ion spin [6]. This exchange splitting of the conduction band causes a reduction in

E_g, and Busch et al. [20] measured this shift in the optical band gap to lower energy in a series of absorption measurements using single crystal samples as they were cooled below T_C. In this way, they determined the magnitude of the exchange splitting: 0.36 eV for EuS and 0.54 eV for EuO. Exchange splitting of the conduction band in EuO was reported again more recently [21] in spin-resolved x-ray absorption spectroscopy measurements on a thin film of Eu rich-EuO. They concluded that the doped electrons (due to oxygen vacancies) in the exchange-split conduction band were 100% spin-polarized, forming a half-metal.

14.4 SPIN FILTERING IN THE EUROPIUM CHALCOGENIDES

14.4.1 EARLY EXPERIMENTS

The spin filter effect was demonstrated for the first time in field-emission experiments with EuS by Müller et al. [22] P as high as 89 ± 7% was measured for field-emitted electrons from an EuS-coated tungsten tip at low temperature. Similar work later confirmed that electrons tunnel from the Fermi level of the tungsten into the exchange-split conduction band of the EuS [23, 24, 25]. The high P of the emitted electrons resulted from the different spin-up and spin-down barrier potentials at the W-EuS interface.

Transport experiments of EuS and EuSe films by Esaki et al. [26], performed previous to the field-emission experiments, showed indirect evidence of the spin filter effect. The EuS and EuSe films in this study were 20 and 60 nm thick, between either Al or Au electrodes. With this trilayer structure, they observed tunneling across the potential barrier between the metal and the EuX layer, as shown in Figure 14.3a. This potential barrier was formed by the energy difference between the Fermi level of the metal electrode and the bottom of the conduction band in the EuX. When measuring the temperature dependence of the I–V behavior of these structures, a significant drop in voltage (measured with constant current) occurred below the T_C for both EuS and EuSe junctions, shown in Figure 14.3b. This drop resulted from the exchange splitting of the conduction band, thereby lowering the barrier height. Applying a magnetic field shifted the drop to a temperature slightly higher than the T_C. However, the 20% decrease in voltage below ~10 K in zero field for EuSe is surprising given that it is an antiferromagnet at zero field and thus no exchange splitting was expected (see below the discussion for EuSe). Spin polarization of the current was not quantified in this early experiment, whereas it showed indirect evidence of the spin filter effect. Similar behavior of tunnel junction resistance can be seen for all Eu chalcogenide junctions described later. A few years after Esaki's work, Thompson et al. [27] also measured an increase in conductance at low temperature due to conduction band splitting in a Schottky barrier contact between a EuS single crystal and indium metal.

Subsequently, the spin polarization resulting from spin filtering in Eu chalcogenide tunnel barriers was directly measured via the Meservey-Tedrow

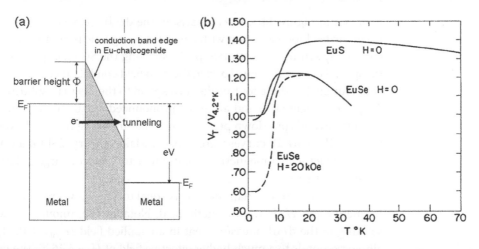

FIGURE 14.3   (a) Energy diagram for metal/EuS or EuSe/metal junction of Esaki *et al.* Φ is the barrier height for tunneling from the metal into the bottom of the conduction band of EuS or EuSe. (b) Normalized voltage across the junction (measured with constant current) versus temperature, with and without applying a magnetic field. (After Esaki, L., *Phys. Rev. Lett.* 19, 852, 1967. With permission.)

technique, whereby a superconducting Al electrode was used to detect the spin of the tunneling electrons, and we describe this elegant set of experiments in this section. Refer to Chapter 11, Volume 1 for a detailed description of this measurement technique. Tunnel junctions for these studies were fabricated *in situ* in a high vacuum chamber on glass substrates, with the general structure Al/EuX/M (where M = Al, Au, Ag, or Y). EuS and EuSe tunnel barriers (about 1 to 4 nm thick) were deposited by thermal evaporation from EuS and EuSe powder sources. EuO was grown by reactive evaporation of Eu metal in the presence of a small oxygen flow. The metal electrodes were deposited by thermal evaporation and patterned by shadow masks into a cross configuration. The Al electrode, typically 4.2 nm thick, was deposited onto a liquid-nitrogen-cooled substrate. Conductance was measured at 0.4 K, well below the critical temperature of the Al superconducting electrode (~2.7 K). The degree of spin polarization of the tunnel current was calculated from the dynamic tunnel conductance (dI/dV) measured within a few millivolts of the Fermi level.

The EuS, EuSe, and EuO thin films were polycrystalline, having saturation magnetization close to bulk values even for thickness down to 2 nm. For EuS bulk T_C for was maintained in this thickness range, whereas by ~1 nm thickness the T_C came down to ~10 K. Reduction of T_C with thickness for ultra-thin EuO is discussed in more detail later in this section. Because the magnetic moment of Eu^{2+} is rather high (~7 μ_B), the magnetic properties of ultra-thin EuX films can be easily measured using a SQUID magnetometer.

14.4.2 EuS

The initial demonstration of spin filter tunneling using the Meservey-Tedrow technique was performed with a ferromagnetic EuS barrier [4]. Dynamic tunnel conductance versus voltage at 0.4 K for an Au/EuS/Al tunnel junction is shown in Figure 14.4. In comparison, the dI/dV curve at a zero applied magnetic field for a junction with a non-ferromagnetic barrier (i.e., Al_2O_3) is completely symmetric with two peaks showing the SC energy gap in Al. In the special case of this EuS barrier, the zero field conductance is asymmetric with four peaks. These four peaks correspond to the Zeeman splitting of the Al quasi-particle DOS and are more pronounced at 0.07 T. Asymmetry of the spin-up and spin-down peaks is the signature of a spin-polarized tunnel current. These dI/dV curves were fit using Maki's theory [28] for a thin film superconductor in a magnetic field, in order to extract a large polarization value P = 80 ± 5%.

The amount of Zeeman splitting is equal to 2 $\mu_B H_0$, where μ_B is the Bohr magneton and H_0 is the magnetic field. Notably, the amount of Zeeman splitting in the dI/dV measurement in an applied field H_{appl} = 0.07 T actually corresponds to a much higher effective field of H_0 = 3.46 T. The internal exchange field of the ferromagnetically ordered Eu^{2+} ions in the EuS barrier, acting on the quasi-particles of the superconducting Al (discussed in detail later) causes this phenomenally enhanced Zeeman splitting. This internal field was even large enough to drive the Al to the normal state at H_{appl} = 0.15 T, as seen in Figure 14.4, which otherwise occurs at about 5 T (without EuS; with Al_2O_3 barriers, for example). Zeeman splitting and polarization was observed even at zero applied field, resulting from significant remanent magnetization in the EuS ultra-thin layer. Because the electrodes are not ferromagnetic, and thus cannot be a source of polarized spins, this large P is generated clearly by spin filtering in the EuS barrier. Polarization as high as 85% was found for Al/3.3 nm EuS/Al superconductor-superconductor tunnel junctions [4].

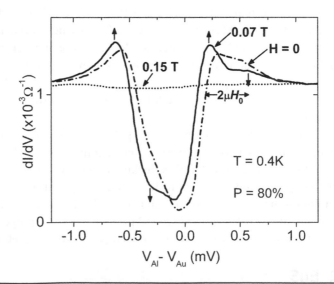

FIGURE 14.4 Dynamic conductance of an Au/EuS/Al junction at 0.4 K in the indicated applied magnetic field. The arrows indicate the spin of the Zeeman split DOS for the 0.07 T curve, showing a polarization of P = 80±5%. The amount of Zeeman splitting (2 $\mu_B H_0$) is indicated for the 0.07 T curve. (Modified from Moodera, J.S. et al., *Phys. Rev. Lett.* 61, 637, 1988; Hao, X. et al., *Phys. Rev.* B 42, 8235, 1990.)

The temperature dependence of the junction resistance (R_J) is another indicator of the spin filter effect in the EuS. When the temperature decreases below the T_C of EuS, the lowering of the tunnel barrier height for spin-up electrons results in a significant decrease of R_J. This is contrary to the typical behavior for a non-magnetic barrier, for which R_J increases continuously when cooled (by a small amount for high quality, insulating barriers, and by a larger amount for semiconducting barriers.) The amount of the drop in R_J below T_C can be used to determine the amount of exchange splitting, which is discussed later for EuO barriers.

14.4.3 EuSe

Because EuSe is an antiferromagnet at zero field, the spin filter behavior of EuSe is quite different from EuS. EuSe has an interesting magnetic phase diagram [29] such that it transitions from the antiferromagnetic state into a ferrimagnetic state at a low field or a ferromagnetic state at high field. Thus in an applied field such that EuSe is ferromagnetic, exchange splitting develops in the EuSe conduction band and the splitting increases as the field increases. This results in spin polarization that increases as the field increases, such that P becomes a function of H_{appl}. For the Ag/EuSe/Al tunnel junction shown in Figure 14.5 at zero field, the EuSe was in the antiferromagnetic state. Thus, no Zeeman splitting was observed and conductance was symmetric about $V = 0$. As the field increased and EuSe entered the ferromagnetic state, polarization also increased. This is evident in the increasing Zeeman splitting and asymmetry as the field increased. The Zeeman splitting is much higher than the applied field, as with the EuS barriers. The increasing asymmetry means higher value of P. In fact with an optimum interface, P even reached as high as

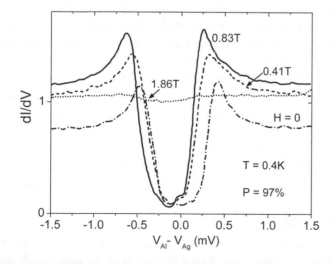

FIGURE 14.5 Conductance versus voltage for an Ag/EuSe/Al junction at 0.4 K taken at the applied fields indicated. Note the highly symmetric curve at $H = 0$, whereas the peaks shift to lower voltage as H increases, corresponding to the Zeeman energy of the spin-up DOS of the Al. This indicates that the polarization is nearly 100%. (Modified from Moodera, J.S. et al., *Phys. Rev. Lett.* 70, 853, 1993.)

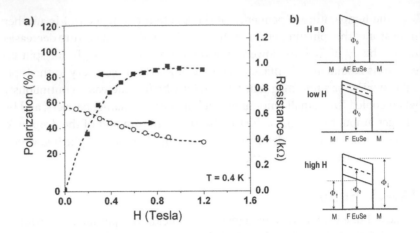

FIGURE 14.6   (a) Polarization and R_J as a function of applied field for a EuSe junction. (b) Schematic of how the exchange splitting of the EuSe conduction band develops as a field is applied. (Modified from Moodera, J.S. et al., *Phys. Rev. Lett.* 70, 853, 1993.)

$97 \pm 3\%$. Because polarization is near 100%, the dI/dV shows only the spin-up peaks: two peaks on either side of $V = 0$, shifted left of center. There is nearly zero conductance in the spin-down DOS and thus they are not seen in the dI/dV curve. This was the first demonstration of a fully polarized tunnel current in a tunnel junction. The $H_{appl} = 1.86$ T curve shows that this applied field plus the large internal field in the EuSe barrier drove the Al to a normal state.

Clearly displayed in Figure 14.5 is an increase in conductance as the applied field increases. In fact, this magnetoresistance is quite high, as much as 400%. This is the first observation of a resistance change of such magnitude for a tunnel junction in an applied field. This increase in conductance corresponds to a significant drop in R_J with increasing magnetic field, shown in Figure 14.6a. This decrease in R_J resulted from the lowering of the spin-up barrier height as the field increased, as shown schematically in Figure 14.6b. At zero field EuSe is in the antiferromagnetic state. When a field is applied, EuSe transitions to the ferromagnetic state and the conduction band undergoes exchange splitting, reaching a constant value at ~1 T in this junction. Accompanying the increase of exchange splitting with the field is an improvement in spin filter efficiency, resulting in increasing P until a constant value is reached. This dependence of polarization on the applied field is unique to the EuSe spin filter. This study also confirmed that at zero field, when the EuSe is antiferromagnetic, the polarization is indeed zero.

14.4.4 EuO

Stoichiometric EuO is quite difficult to grow at the few monolayers scale necessary for tunneling, due to the formation of the more stable and non-magnetic Eu_2O_3 which is not a spin filter. Therefore, compared to EuS and EuSe, demonstrating spin filter tunneling with EuO barriers is much more challenging. With careful control of the Eu metal evaporation and oxygen flow, good quality EuO was grown in an Al/EuO/Y/Al tunnel junction, showing 29% polarization [7] using superconducting Al as the spin detector. Similar to EuS and EuSe barriers, EuO showed a large enhancement of Zeeman splitting, amounting to an effective magnetic field of 3.9 T in a small applied field of just 0.1 T. Thus, the exchange interaction in just 1.4 nm thick EuO was strong enough to create a large internal exchange field of 3.8 T. The exchange interaction of the EuO/Al bilayer had been observed in an earlier tunneling study as well [30]. Even though EuO has a larger $2\Delta E_{ex}$ than EuS, which should result in a higher P, spin filtering efficiency of EuO is not as high. This was explained as due to the non-stoichiometry at the EuO/electrode interfaces, which is quite difficult to control during sample growth. The best polarization values were obtained using yttrium as the top electrode. Yttrium prevented over-oxidation of EuO into the undesirable Eu_2O_3 phase. Confirmation of this came from an x-ray absorption spectroscopy study of the chemical nature of the EuO interface with Y, Al and Ag capping layers [31]. The result from this study was found to agree with the tunneling result.

One must be aware that when a magnetic insulator is made thin enough to act as a tunnel barrier, it may have a lower T_C and a lower exchange splitting than the bulk. Consequently, this could limit efficient spin filtering to temperatures much less than T_C and reduce the spin filter efficiency. A reduction in T_C with thickness was found to be the case for EuO. As shown in Figure 14.7a, T_C for thickness <3 nm is significantly reduced from bulk T_C = 69 K. This thickness dependent study of the T_C was the first experimental confirmation of a theoretical calculation by Schiller *et al.* [32] showing reduced T_C for a few monolayers of EuO.

Measuring the exchange splitting in ultra-thin films is more difficult. The optical absorption measurements used to measure the exchange splitting of the conduction band for crystals of EuO cannot be as easily applied to these ultra-thin films. As an alternative, spin filter tunneling was used to quantitatively determine the exchange splitting in ultra-thin EuO [33]. With a tunnel junction comprised of a EuO barrier sandwiched between

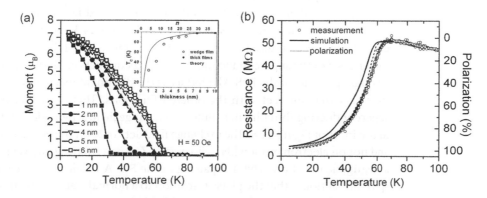

FIGURE 14.7 (a) $M(T)$ for different thicknesses of EuO, 1 nm–6 nm, showing a lower T_C for films <4 nm. This agrees with a calculation by Schiller et al. [32] shown in the inset. (b) Resistance vs temperature for a Y/EuO/Al junction, plotted along with a simulation of the measurement based on the $M(T)$ curve for the 3 nm film in (a) and calculated polarization resulting from the large exchange splitting that causes the decrease in junction resistance. (Modified from Santos, T.S. et al., *Phys. Rev. Lett.* 101, 147201, 2008.)

non-ferromagnetic electrodes, the junction resistance as a function of temperature was measured when varying the temperature across the T_C of EuO. For $T > T_C$, the junction resistance increased as T decreased, as expected for a non-magnetic semiconducting barrier. When the Eu^{2+} ions ferromagnetically ordered at T_C, exchange splitting lowered the spin-up tunnel barrier height, and the junction resistance decreased dramatically as the temperature was lowered further below T_C, as shown in Figure 14.7b. In accordance with an exponential dependence on barrier height, the resistance dropped by several orders of magnitude, especially for larger applied bias at which Fowler-Nordheim tunneling occurs. The barrier height Φ_0 (before splitting) at low bias was found by applying the Brinkman-Dynes-Rowell [3] relation to the current-voltage characteristics of the junction for $T > T_C$, then $R_J(T)$ was used to quantify the amount of exchange splitting for $T < T_C$. Equation 14.2 was used to calculate P, assuming no spin scattering. The amount of resistance drop corresponded to an exchange splitting of 0.31 eV for the 3 nm thick EuO. Even though this is less than the bulk value, this splitting is still large enough to produce a nearly fully spin-polarized current, though polarization could not be measured directly without a spin detector. A complete study of exchange splitting as a function of thickness was not performed. However, it was demonstrated that even though thinner barriers may have smaller exchange splitting, the high spin filter efficiency can still be maintained by the lower average barrier height of the thinner barrier.

14.4.5 EXCHANGE INTERACTION AT FI/SC INTERFACES

A novel feature of ferromagnetic insulator/superconductor (FI/SC) interface is the large internal exchange field of the ferromagnet acting on the superconductor. Sarma [34] and de Gennes [35] had predicted this phenomenon over 40 years ago as an exchange interaction between the ferromagnetically ordered ions in the FI and the conduction electrons of the SC. This is dramatically displayed in the enhanced Zeeman splitting of the Al superconducting DOS when in contact with the EuS, EuSe and EuO barriers in the spin-polarized tunneling measurements (see Figures 14.4 and 14.5). As discussed above, this internal exchange field is as much as ~4 Tesla. For small thickness d of the superconducting film during conduction, the scattering of Al quasi-particles at the EuS/Al interface is the dominant mechanism, where they exchange interact with the ordered Eu^{2+} ions. This is analogous to a superconducting thin film in a uniform exchange field, schematically shown in the inset of Figure 14.8. The exchange field acts only on the electron spins, and not on electron motion [34, 35], causing Zeeman splitting and yet having a negligible effect on orbital depairing in the SC. A result of a purely field-spin interaction is that the phase transition to a normal state occurs through a first-order transition when the critical field is reached.

de Gennes predicted that for d less than the superconducting coherence length ξ, the magnitude of the field is inversely proportional to d. This was experimentally verified using an $EuS/Al/Al_2O_3/Ag$ junction structure [36], with the EuS layer under only half of the patterned Al electrode strip. The

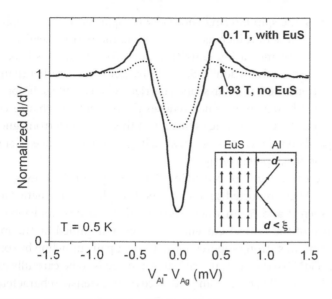

FIGURE 14.8 Comparison of the tunnel conductance for an $Al/Al_2O_3/Ag$ junction with and without an EuS layer beneath the Al superconducting electrode. The EuS layer creates an internal field in the Al, resulting in enhanced Zeeman splitting with minimal orbital depairing in a small applied field. Inset: Schematic of the interaction between the Al quasi-particles with the ferromagnetically ordered ions in the EuS, as they scatter at the FI/SC interface. (Modified from Hao, X. et al., *Phys. Rev. Lett.* 67, 1342, 1991.)

conductance of the junctions for the two configurations was compared as shown in Figure 14.8. For the $EuS/Al/Al_2O_3/Ag$ junction, enhanced Zeeman splitting, corresponding to a field of 1.9 T, was observed in an applied field of 0.1 T. For the $Al/Al_2O_3/Ag$ trilayer without the EuS underlayer, a much higher conductance at zero bias and broadening of the peaks is observed with $H_{appl} = 1.93$ T, showing significant orbital depairing and no Zeeman splitting in the Al quasi-particle DOS. It was also shown in this study that the internal field drove the Al to the normal state via a first-order transition, and the internal saturation field was inversely proportional to Al film thickness. The exchange interaction only occurs for an FI and SC in immediate proximity to each other. The effect was lost when a thin layer of Al_2O_3, even two monolayers, formed at the FI/SC interface.

14.5 GdN SPIN FILTER

The spin filter effect in the ferromagnetic semiconductor GdN shows a different behavior to that observed for EuS and EuO. GdN has a Curie temperature of around 60 K for the bulk and a lower Curie temperature ~35 K for thin film barrier samples. Pal et al. [37] have reported a spin filter efficiency >90% for GdN barriers in between NbN electrodes that have a superconducting critical temperature of 14 K. The spin filter efficiency was determined from the temperature dependence of the junction resistance, using a similar method as described in Section 14.4.4. There was no spin detector for directly probing the polarization in this experiment. In contrast to EuS and

EuO barriers, the spin filtering with GdN is suppressed as the applied voltage across the junction increases. In addition, the barrier height decreases significantly with increasing GdN thickness. To explain this behavior the junction is described as a double Schottky barrier system, rather than having the conventional, rectangular-shaped barrier. When a voltage is applied, the depletion width increases, which reduces the magnetic exchange coupling and consequently the exchange splitting. In this way, the ferromagnetism in GdN depends on the carrier density, leading to a dependence of spin filter efficiency on the applied voltage.

In a separate study of NbN/GdN/NbN Josephson junctions, Senapati et al. [12, 38] have reported supercurrents through ferromagnetic GdN barriers between 2 and 5 nm thick, showing that the field and temperature dependence of the critical current was strongly modified by the magnetic barrier. There was strong suppression of Cooper pair tunneling by spin filtering in the GdN barrier. This is a rich area and needs to be carefully explored with a well-controlled barrier and interfaces with extensive characterization.

14.6 QUASI-MAGNETIC TUNNEL JUNCTIONS

The early experiments with spin filter tunnel barriers showed the spin filter effect by using the Meservey-Tedrow technique. In more recent studies, a quasi-magnetic tunnel junction (QMTJ) has been used to observe spin filtering, by measuring tunnel magnetoresistance (TMR) in junctions where the spin filter barrier is sandwiched between a non-magnetic electrode (NM) and a F electrode. If the polarization of the ferromagnet (P_F) is known, the spin filter efficiency (P_{SF}) can be calculated from the TMR measurement by using Jullière's formula [39]:

$$\text{TMR} = \frac{\Delta R}{R} = \frac{2P_F P_{SF}}{(1 - P_F P_{SF})} \tag{14.3}$$

Resistance (R_J) of the junction depends on the relative alignment of the magnetization of the spin filter and the ferromagnet, as shown in Figure 14.9. For a tunnel barrier that preferentially allows spin-up electrons to tunnel, R_J is low (high) for parallel (antiparallel) alignment. Just as in conventional MTJs, independent switching with well-separated coercivities for the two ferromagnetic layers is necessary for observing TMR and quantifying P_{SF}. In order to have a clear effect, magnetic coupling between the spin filter barrier and the F electrode must be prevented. For this reason, a non-magnetic spacer layer, such as Al_2O_3 or $SrTiO_3$ is sometimes placed between the spin filter and F electrode.

The high spin polarization (~90%) found for EuS spin filter barriers in the Meservey-Tedrow measurements described earlier motivated the incorporation of EuS in QMTJs. LeClair et al. [40] showed the results for a QMTJ. They measured MR ~ 100% at 2 K for a 5 nm EuS barrier between Al and Gd electrodes. There was considerable noise in the MR signal, attributed to instabilities in the EuS magnetization. The TMR decreased as the temperature was raised closer to the T_C of EuS (16 K) and was absent at above T_C.

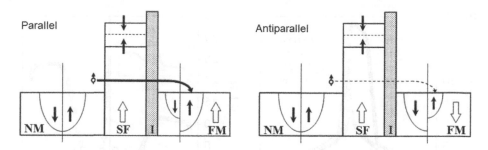

FIGURE 14.9 (Left) Low resistance state (high tunnel current) for a QMTJ, with parallel alignment of magnetization of the spin filter barrier (SF) and F electrode. (Right) High resistance state (low tunnel current) for antiparallel alignment. In some junction structures, there is a very thin, non-magnetic, insulating layer between SF and F, shown here, in order to prevent magnetic coupling between them.

As discussed earlier, a clear decrease of junction resistance was observed for these QMTJs when cooled below T_C, due to the lowering of the barrier height.

Much more robust and reproducible measurements of QMTJs with EuS barriers were made by inserting an AlO_x spacer layer between the EuS and a Co electrode, by Nagahama et al. [41], shown in Figure 14.10. The AlO_x effectively decoupled with EuS and Co layers, and the antiferromagnetic CoO acted as an exchange biasing layer to pin the Co magnetization. Figure 14.10a shows the typical shape of the MR for these QMTJs, having an MR value of 33% at 4.2 K with a bias of 800 mV. The magnetic configuration is well-defined and stable, with no instability in junction resistance.

In contrast to the behavior of conventional F/insulator/F MTJs, these QMTJs display a novel bias dependence of MR, in which the MR actually *increases with bias* (Figure 14.10b). This unique bias dependence for a spin filter MTJ was predicted theoretically by Saffarzadeh [42]. The increase in MR at higher bias was attributed to Fowler-Nordheim (FN) tunneling into the fully spin-polarized conduction band of the EuS barrier. Direct tunneling via the spin filter effect occurred in the low bias regime, and the MR decreased monotonically with bias (see Figure 14.10b inset), similar to conventional MTJs. As the applied bias was increased further, a rise in MR marked the transition from direct tunneling to FN tunneling via the spin-up conduction band for the spin-up electrons. Upon even further increase of bias, a subsequent decrease in TMR occurred as the spin-down electrons also underwent FN tunneling via the higher spin-down conduction band level, resulting in a lower polarization of the tunnel current. Thus, in contrast to conventional MTJs, a sizable TMR effect in this spin filter MTJ is not limited to low bias. As a consequence of having two tunnel barriers (EuS + AlO_x), the junction resistance is very high, especially in the low bias regime (few $G\Omega$). Noise in the measurements at low bias (below ±400 mV in Figure 14.10b) made the measurement in this regime uncertain. Measurement of a EuS QMTJ in a low noise setup [43] confirmed high MR, up to 70% (see inset), and the familiar bias dependence behavior (decreasing MR with bias) before the onset of FN tunneling.

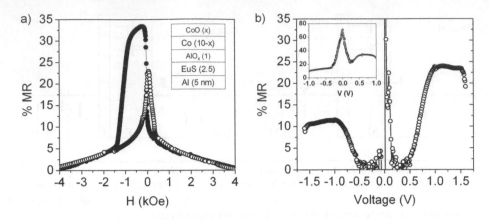

FIGURE 14.10 (a) MR of an Al/EuS/AlO$_x$/Co/CoO junction at 4.2 K with 800 mV bias. Filled (open) circles correspond to decreasing (increasing) field during measurement. (b) Bias dependence for a similar junction, showing an abrupt *increase* in MR at finite bias. Noise in the measurement for V less than ±400 mV prevented reliable data in this range. Inset: Full bias dependence measurement at 2 K in a low noise setup [43] of a similar junction with 3 nm EuS.

14.7 OXIDE SPIN FILTERS

The phenomenal spin filtering effects described up to this point were achieved using EuX compounds that were polycrystalline, although in some cases near epitaxy was seen depending on the substrate. Correspondingly, with the latter films, better properties were observed. More effort is needed toward complete epitaxial EuX layer systems. Fully epitaxial structures can be prepared with a radically different material selection—perovskites or ferrites. While good structural quality has been shown for these complex oxide spin filters, on either elemental metal or oxide electrodes, the spin filtering is not yet optimized, as to show similar spectacular effects that are seen with EuX compounds. However, the performance of these oxide spin filters continues to improve, and the role of oxygen vacancies and defects comes to light. This section focuses on epitaxial spin filter materials.

The spin filter effect has been successfully demonstrated by Gajek et al. in the perovskite BiMnO$_3$ [9] and La-doped BiMnO$_3$ [44], which is ferromagnetic with a bulk T_C = 105 K. In addition to being ferromagnetic, this material is also ferroelectric, whereby its electric dipole moment can be switched by applying an electric field. The electric polarization of the barrier modifies the barrier profile, causing a different average barrier height for the two polarization states. Thus, four stable resistance states can be obtained with this barrier having two ferroic parameters at play [45]. Multiferroic tunnel junctions are the subject of Chapter 17, Volume 1.

14.7.1 FERRITES: LEADING THE WAY TO ROOM-TEMPERATURE SPIN FILTERING

All of the spin filter materials described thus far have beautifully demonstrated the spin filter phenomenon, but all at low temperature due to the low Curie temperature of the spin filter. Having a Curie temperature far above room temperature, the ferromagnetic, spinel ferrites are the best candidates for achieving spin filtering at room temperature. We present in this section the progress toward high-temperature spin filtering.

The spinel ferrites have the general formula XFe_2O_4 where X is a divalent transition metal cation such as Fe^{2+}, Mn^{2+}, Ni^{2+} or Co^{2+}. With the exception of Fe_3O_4, all are insulating and ferrimagnetic with a Curie temperature above 700 K. While the conductive Fe_3O_4 has been studied as a half-metallic electrode in MTJ structures [46], its insulating relatives $NiFe_2O_4$ and $CoFe_2O_4$ have the potential to function as spin filter tunnel barriers.

The ferrites ideally have an inverse spinel structure consisting of a face-centered cubic oxygen sublattice and a distribution of cations in both the tetrahedral (A) and octahedral (B) interstitial sites, resulting in a net moment of 2 μ_B per formula unit in $NiFe_2O_4$ and 3 μ_B in $CoFe_2O_4$. Theoretical calculations on the inverse spinel structure yield band gaps of 0.98 eV for $NiFe_2O_4$ and 0.80 eV for $CoFe_2O_4$, while the exchange splitting of the conduction band in both materials is quite large, around 1.2 eV [47]. The exchange splitting is such that the lowest energy level of the conduction band corresponds to *spin-down* states. The result should, therefore, be a lower tunnel barrier height for spin-down electrons, thus leading to preferential tunneling of the spin-down electrons and a *negative* spin polarization. One important factor to consider for these spinel ferrites is that they rarely exhibit a perfect inverse spinel structure. Migration of X^{2+} cations from B to A sites readily occurs [48], leading to "mixed" spinel having a moment and electronic band structure that can vary significantly from the inverse scenario. Theoretical calculations have demonstrated the effect of cation-site inversion on the electronic band structure and magnetic moment [48]. A site inversion of the Ni^{2+} and Co^{2+} cations results in a large increase in moment to 8 μ_B for $NiFe_2O_4$ and 7 μ_B for $CoFe_2O_4$. As a consequence, a largely amplified exchange splitting in the conduction band results, and therefore the possibility of even higher spin filtering efficiencies. While theoretical predictions confirmed the high potential of $NiFe_2O_4$ and $CoFe_2O_4$ to filter spins, only a few groups have successfully fabricated tunnel junctions containing spinel ferrite barriers and measured the spin filter effect.

14.7.1.1 NiFe$_2$O$_4$

Ultra-thin $NiFe_2O_4$ films indeed show a dramatic enhancement of the saturation moment compared to the bulk value, indicating a large degree of cation-site inversion [49, 50]. In addition, under some deposition conditions, the films are metallic, making them unsuitable as a tunnel barrier. As is the case for $CoFe_2O_4$ discussed later, the oxygen content plays a critical role in the spin-polarized transport properties of $NiFe_2O_4$, including simply the difference between insulating and metallic behavior. Using insulating $NiFe_2O_4$ barriers, the spin filter effect was demonstrated in TMR measurements with Au and $La_{2/3}Sr_{1/3}MnO_3$ electrodes at low temperature [10], shown in Figure 14.11. With one ferromagnetic electrode and one normal metal electrode, the observed TMR signal indicates that $NiFe_2O_4$ acts as a spin filter. One may apply the Jullière relation in Equation 14.3 to this TMR measurement,

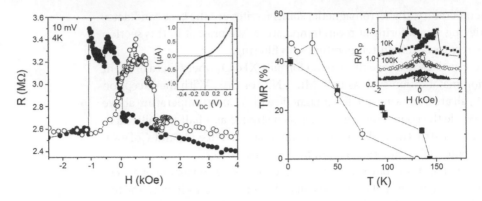

FIGURE 14.11   (a) TMR for a LSMO/NiFe$_2$O$_4$/Au tunnel junction at 4 K. The current-voltage curve is shown in the inset. (b) Temperature dependence of the TMR with (circles) and without (squares) SrTiO$_3$ at the LSMO/NiFe$_2$O$_4$ interface. (After Lüders, U. et al., *Appl. Phys. Lett.* 88, 082505, 2006. With permission.)

using $P_F = 90\%$ for La$_{2/3}$Sr$_{1/3}$MnO$_3$ (LSMO). In this way, a spin filter efficiency $P_{SF} = 19\%$ was found for the NiFe$_2$O$_4$ spin filter. Further studies showed that this value was improved slightly to 22% by inserting a thin SrTiO$_3$ spacer layer in between the LSMO and NiFe$_2$O$_4$.

The positive sign of P_{SF} is contrary to the negative spin polarization expected from the band structure calculations for NiFe$_2$O$_4$ mentioned earlier. One proposed explanation attributes the positive polarization to the different effective masses for spin-up and spin-down electrons ($m^*_\uparrow < m^*_\downarrow$) [47]. Another possibility could be defects in the NiFe$_2$O$_4$ barrier that provide alternative tunneling pathways other than direct tunneling across the electronic band gap.

Despite the prediction of TMR at room temperature, the TMR rapidly vanishes by 150 K, well below the T_C of 850 K for NiFe$_2$O$_4$, as shown in Figure 14.11b. In addition, there was a systematic decrease in TMR with increasing bias for the NiFe$_2$O$_4$-based tunnel junctions, which is not the theoretically expected behavior for a spin filter. The loss of TMR could be due to a decrease in coercivity for the NiFe$_2$O$_4$ barrier as temperature increases, or to a drop in the T_C of LSMO caused by non-stoichiometry at the LSMO/NiFe$_2$O$_4$ interface. Furthermore, in the case of even a small non-stoichiometry in either of the oxide films, defect states in the barrier and/or at the interfaces can certainly lead to spin scattering events or spin-independent tunneling. These effects become more prominent at higher temperatures and bias.

14.7.1.2 CoFe$_2$O$_4$

Ramos et al. [51] performed a detailed study on the film growth and magnetic properties of epitaxial CoFe$_2$O$_4$, including CoFe$_2$O$_4$/Fe$_3$O$_4$ and CoFe$_2$O$_4$/Co heterostructures. A fully epitaxial, quasi-magnetic tunnel junction structure was adopted for their TMR experiments: Pt(111)/CoFe$_2$O$_4$(111)/γ-Al$_2$O$_3$(111)/Co(0001) deposited by oxygen plasma-assisted molecular beam epitaxy onto a sapphire substrate. Note that the Al$_2$O$_3$ spacer layer used to decouple the CoFe$_2$O$_4$ barrier and Co electrode was crystalline in this case, which is rare for this type of tunnel barrier. The high quality of this fully epitaxial system,

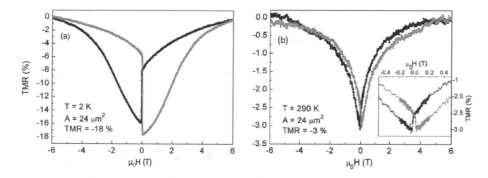

FIGURE 14.12 TMR in a 20 nm Pt/3 nm $CoFe_2O_4$/1.5 nm Al_2O_3/10 nm Co tunnel junction at 2 K (a) and at room temperature (b) with V = 200 mV. (After Ramos, A.V. et al., *Appl. Phys. Lett.* 91, 122107, 2007. With permission.)

especially the interfaces, largely contributed to the successful spin transport measurements, even at room temperature. Ramos et al. [11] observed a small TMR signal of –3% at room temperature, which grew to –18% at 2 K, as shown in Figure 14.12. Using Jullière's formula with P_{Co} = 40%, this yields polarization values for $CoFe_2O_4$ of –4% and –25%, respectively. This is the first demonstration of spin filtering at room temperature. Furthermore, unlike the $NiFe_2O_4$ junction, this $CoFe_2O_4$ junction showed the expected bias dependence behavior of a spin filter, with an increase of TMR at high bias.

Though the TMR signal at room temperature for this $CoFe_2O_4$ junction was small, it showed the importance of having a barrier of high structural quality for these spinel ferrites in order to see room-temperature TMR and robust junctions with the expected bias dependence. It is known that antiphase boundaries in thinner films and oxygen vacancies can affect the magnetic properties and hence the spin filtering performance. Ramos, in collaboration with the authors [52], performed tunneling experiments using the Meservey-Tedrow technique that demonstrated the influence of oxidation during film growth on the spin polarization produced by the $CoFe_2O_4$. Barriers grown in lower oxygen pressure had a stronger dependence of junction resistance on temperature compared to barriers grown in higher oxygen pressure, as shown in Figure 14.13. This indicated the presence of defect states in the band gap and a lowering of the effective barrier height as a result of oxygen vacancies. Direct measurement of spin polarization in the Meservey-Tedrow experiments (with no ferromagnetic electrodes) revealed a higher spin polarization for the barrier grown in higher oxygen pressure, ranging from P = 6% for oxygen pressure of 0.20 Torr to P = 26% for 0.26 Torr. The sign of polarization measured in these experiments is not in agreement with the negative sign of polarization in the TMR measurements previously described. The origin of this discrepancy is unclear, but may be related to the different spin detectors or different bias regimes used in the two cases: superconducting Al spin detector and <2 mV bias regime for the Meservey-Tedrow experiments, and Co spin detector in the higher, FN tunneling regime for the TMR experiments.

It is interesting to note that the oxygen vacancies detected in these transport experiments were undetectable by an extensive investigation of

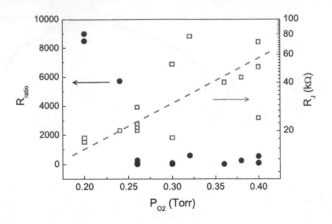

FIGURE 14.13 Junction resistance at 4 K (right axis) for junctions with $CoFe_2O_4$/Al_2O_3 barriers deposited in different oxygen pressures. R_{ratio} on the left axis is the temperature resistance ratio R_J (4 K)/R_J (300 K). The dashed line is a guide for the eyes. (Modified from Ramos, A.V. et al., *Phys. Rev. B* 78, 180402(R), 2008.)

this material system by standard structural, magnetic and chemical characterization techniques. This proves the sensitivity of spin-polarized tunneling to such minute defects. Even though oxygen plasma was used during film growth, enough oxygen vacancies were incorporated to degrade the spin filter efficiency. This result emphasizes the importance and challenge of obtaining high-quality films for achieving high spin filter efficiency with complex oxide barriers.

14.8 ORGANIC SPIN FILTERS

Altogether another exotic type of spin filtering has been seen more recently in molecular devices: organic molecules sitting over a ferromagnetic surface develop this property. For example, non-magnetic Zinc methyl phenalenyl (ZMP) molecules placed on, for example, a Co surface, serving as a tunnel barrier or spacer layer (the counter electrode being Cu), showed not only ferromagnetic coupling but also the spin filtering property [53]. When these molecules are grown on a ferromagnetic surface, interface spin transfer causes a hybridized organometallic, supramolecular magnetic layer to develop and show spin filter properties. As a result, this interface layer creates a spin-dependent resistance, displaying a large interface magnetoresistance effect. In the schematic shown in Figure 14.14, the first coupled molecule over the Co surface has magnetic information while the second dimer molecule on top of the first one develops exchange-split energy levels and spin filtering features (shown on the right). This ferromagnet/molecule interface effect, discussed further in Refs 54 and 55, opens up a vast area for exploration.

14.9 DOUBLE SPIN FILTER JUNCTIONS

Worledge and Geballe [56] proposed another way to reach high TMR in a junction having two spin filter barriers. In a double spin filter junction,

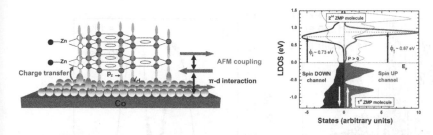

FIGURE 14.14   Left: Schematic of the coupling between the ZMP molecules and the Co surface. Right: The spin-resolved density of states for the two ZMP molecules, showing the spin-splitting in the second molecule.

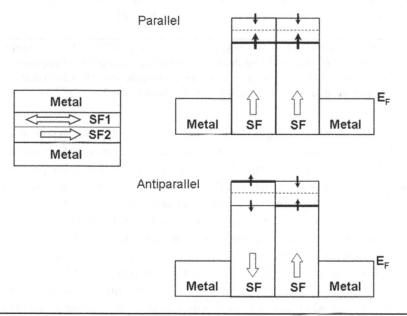

FIGURE 14.15   Schematic of the double spin filter junction proposed by Worledge and Geballe [56]. The magnetization of one spin filter (SF1) is free to switch in a small applied field while the other (SF2) is effectively pinned. The spin-up and spin-down barrier heights for the two spin filters are shown for the parallel and antiparallel alignment. In this schematic, the two SF barriers are assumed to have identical band property.

shown schematically in Figure 14.15, one spin filter acts as the polarizer and the other as the analyzer, and the tunnel current depends on the relative alignment of magnetization between them. When they are magnetized parallel to each other, the spin-up electrons have a lower barrier height for both spin filters, in contrast to antiparallel alignment for which the spin-up electrons preferentially tunnel through the first spin filter (the polarizer) but are blocked by the higher barrier height in the analyzer. The TMR is generated by the two magnetic barriers, without the need for a ferromagnetic electrode. In order to reach both parallel and antiparallel alignment, the two spin filter barriers must be magnetically decoupled and have separate coercive fields.

Worledge and Geballe estimated an MR value as high as 10^5. However, one can easily anticipate the difficulties in experimentally realizing a double

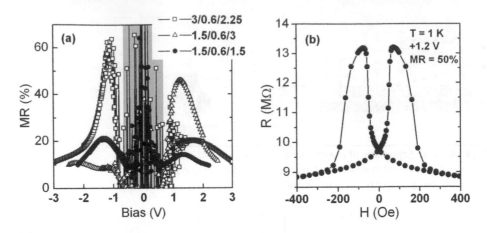

FIGURE 14.16 (a) The bias dependence measured for 10 nm Al/EuS/0.6 nm AlO$_x$/EuS/10 nm Al double spin filter junction, for the various EuS thicknesses indicated. The shaded region at low bias contains noisy data due to the high impedance of the junction. (b) Magnetoresistance curve for the junction with 1.5 nm and 3 nm thick EuS barriers. (Modified from Miao, G.-X. et al., *Phys. Rev. Lett.* 102, 076601, 2009.)

spin filter junction. Because the junction resistance depends exponentially on barrier thickness (Equation 14.1), a junction with two spin filter barriers would have very high resistance. In addition, it is likely that the two spin filters would be magnetically coupled so that the antiparallel alignment state could not be reached. Any coupling would significantly reduce the MR ratio. One could insert a non-magnetic, insulating spacer layer between the two spin filters to decouple them. A spin-independent spacer layer would not decrease the MR ratio, but it would increase the junction resistance even further.

Miao et al. gave the first experimental demonstration of high TMR in a double spin filter junction, using two EuS barriers and *non-magnetic* electrodes [57]. An Al$_2$O$_3$ spacer layer was used to prevent coupling between the two EuS barriers. These junctions produced high TMR and an unconventional bias dependence, in which the TMR *increased* at high bias, reaching some peak value, then decreased as the bias was increased further, as shown in Figure 14.16. Similar to the behavior of quasi-magnetic tunnel junctions described earlier, this bias dependence is a result of a progression from direct tunneling for both spin orientations, to FN tunneling for spin-up electrons only, to FN tunneling for both spin orientations, as the bias is increased. The asymmetry of the peaks and the peak positions for positive and negative applied bias is especially pronounced and is caused by the different thickness of the two EuS layers. With three barrier layers, the impedance was very high (several GΩ) so as to complicate measurement at low bias, below the onset of FN tunneling (see Figure 14.16).

For Miao et al. the trick to fabricating a double spin junction was to deposit the two EuS layers using different growth conditions. Deposition of the first EuS layer onto a liquid-nitrogen-cooled substrate yielded a lower coercive field and sharper switching compared to the second EuS layer subsequently deposited at room temperature. Another important ingredient was the Al$_2$O$_3$ spacer of optimal thickness, thick enough to sufficiently decouple

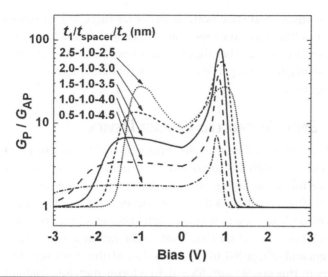

FIGURE 14.17   Calculated bias dependence for a double spin filter junction in which the relative thicknesses of the two spin filter layers is varied while the total thickness remains constant. (Modified from Miao, G.-X. et al., *Phys. Rev. Lett.* 106, 023911, 2009.)

the two layers but not too thick so as to prevent single-step tunneling and/or make the junction impedance too high. Only then could the EuS layers have a large enough separation of coercive field so that a well-defined, antiparallel resistance state was reached [58].

For a double spin filter junction, there are many parameters with which to tune the magnitude of the MR and bias dependence. Relative thickness, exchange splitting, and average barrier height all play a role in the dramatic bias dependence behavior of these junctions, as shown by the numerical calculations of Miao et al. based on WKB approximation [59]. As can be expected, the TMR is enhanced by increasing the thickness and exchange splitting, or by reducing the average barrier height. In qualitative agreement with the experiment in Ref. [57], the maximum TMR value is obtained when electrons first tunnel through the more efficient spin filter, meaning the one that is thicker or has the larger exchange splitting, whereas a lower TMR is obtained for the opposite current flow direction. This is shown in Figure 14.17, in which the relative thickness of the two layers is varied while keeping the total thickness constant. Barrier thickness is the parameter most easily varied experimentally. For optimal spin filter efficiency, a thicker spin filter layer is preferred [56]. Layers that are too thin are likely to have a reduced Tc and consequently a lower exchange splitting. This was shown to be the case for EuO [33], and the opposite is expected to occur in the ferrites. However, on the other hand, inherent problems associated with having a thick barrier is the increased probability of multi-step tunneling via defect states, which are sites for spin-flip scattering, causing a lower MR ratio than expected. Exchange splitting and average barrier heights are parameters that are determined by the material, though these parameters can be influenced by layer thickness as well [33]. Exchange splitting is more effective for a lower average barrier height.

At this time, we are not aware of any other successful demonstration of a double spin filter junction. One potential candidate is a $CoFe_2O_4/NiFe_2O_4$ double spin filter. Given that these ferrites have the same crystal structure and similar magnetic properties, it is likely that an epitaxial, non-magnetic spacer layer is needed.

14.9.1 SPIN-CONTROLLED VOLTAGE DEVICE

As we observed in earlier sections, magnetic insulators can provide large effective Zeeman fields that are confined at an interface, making them especially powerful in modifying adjacent one- or two-dimensional electronic structures. When combining this phenomenon with the spin filtering property, it is possible to see a new effect—spin-assisted charge transfer across a EuS/Al/EuS heterostructure that leads to the generation of a spontaneous spin current and voltage [60, 61]. The very thin Al film in between the two EuS spin filters in this double spin filter device forms metallic Coulomb islands within the magnetic barrier. The large interfacial exchange field acting on these islands, as large as tens of Tesla in this device, causes spin-splitting of the energy levels in the islands. The internal exchange field and the spin filter effect work together in the EuS/Al/EuS device to produce this unique energy profile in which there is an asymmetry of chemical potentials for the two spin channels, leading to a method of harvesting spin information.

The EuS/Al_2O_3/Al/EuS device structure utilized for this observation is shown in Figure 14.18a, having two 10 nm thick Al electrodes on either side of the heterostructure and a thin Al layer only 0.3 nm thick in the middle, which forms isolated, metallic islands within the barrier. The two EuS layers serve as a double barrier with the Al islands in between. A natural oxide (about 0.3 nm thick) is formed on one side of the Al layer to block the exchange field at that interface. The electrical conduction of the device was analyzed based on the energy diagram in Figure 14.18b, using the two-spin current model arising from the spin filter barriers acting as a set of spin-dependent resistors for the spin channels facing the higher/lower energy barrier. For no net charge current flows across the structure in the equilibrium state, the measured voltage shift between the parallel and antiparallel

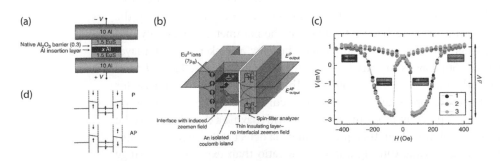

FIGURE 14.18 (a) Schematic of the EuS/AlOx/Al/EuS spin-dependent voltage device. (b) Electron energy diagram across the junction. (c) Spin-dependent voltage measurement. (d) Energy diagram for the ideal condition with perfect spin filtering. (After Miao, G.-X. et al., *Nat. Commun.* 5, 3682, 2013. With permission.)

states of the spin filter layers was expressed in terms of the spin filtering efficiency α and exchange splitting Δ as $V = \alpha \cdot 2\Delta$.

The spontaneous, spin-dependent voltage driven by this energy scheme is shown in Figure 14.18c. The occurrence of this voltage has been interpreted as follows: the spin filter barrier preferentially allows electrons of one spin-type to communicate across it, and the detector electrode is therefore only probing the energy level of this spin-type. Reversing the magnetization orientation of the spin filter layer allows for the detection of the opposite spin-type. The spin-controlled spontaneous voltage measured while tuning the system with a small magnetic field between parallel and antiparallel alignment of the two spin filters is shown in Figure 14.18c, having well-defined voltage states corresponding to the up and down spin channels, respectively. This voltage was largest for the thinnest Al insertion layer, about 3.8 mV corresponding to an effective Zeeman field of >60 T operating only on the electrons' spin degrees of freedom, not on their orbitals. Thus we see that a spin-dependent voltage, albeit quite small, is spontaneously generated across a multilayer device, making it desirable to utilize the spin current in such a system. However, the available power is extremely small in the fW range.

14.10 SPIN INJECTION INTO SEMICONDUCTORS

Sugahara and Tanaka proposed a spin filter transistor (SFT) device, based on the successful, experimental demonstration of spin filter tunneling [62]. Their model device, shown in Figure 14.19, consisted of a semiconductor base separated from the non-magnetic, semiconducting emitter and collector by two spin filter tunnel barriers. When an emitter-base bias is applied, the emitter barrier filters the hot electron spins as they tunnel into the base, allowing only spin-up electrons into the base. The collector barrier acts as an analyzer for the spin-polarized transport through the base. The width of the base must be less than the spin-flip scattering length. The relative magnetic alignment of the emitter and collector tunnel barriers determines the output of the SFT. The collector current for parallel configuration can be significantly higher than that for antiparallel configuration. Depending on the spin filter barrier parameters, the mean free path of the hot electrons in the base layer and width of the base, Sugahara and Tanaka showed that the SFT could achieve large magneto-current ratio, current gain, and power gain.

Filip et al. [63] proposed a similar device, consisting of two EuS spin filter tunnel barriers separated by a non-magnetic, semiconducting quantum well. The spin injection device was modeled using the EuS/PbS system because it could be grown epitaxially as well as the high spin filter efficiency of EuS. According to their model, when the EuS layers are aligned antiparallel, a non-equilibrium spin accumulation takes place in the PbS quantum well. No spin accumulation takes place when they are aligned parallel. The calculation showed that this could result in high magnetoresistance for a range of EuS and PbS layer thicknesses.

The above theoretical models are yet to be explored experimentally due to the degree of difficulty in integrating spin filters with semiconductors.

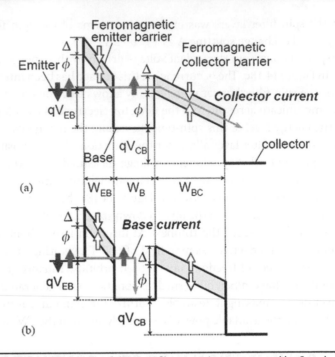

FIGURE 14.19 Schematic of the spin filter transistor proposed by Sugahara and Tanaka. (a) The parallel magnetic orientation of the emitter barrier and collector barrier, resulting in a high collector current. (b) Antiparallel magnetic orientation, resulting in no collector current. (After Sugahara, S. et al., *Physica E* 21, 996, 2004. With permission.)

There have been some studies in this direction: the metal/EuS Schottky contact [64] and EuS/GaAs semiconductor heterojunction [65]. For the metal/EuS Schottky contact (100 nm thick EuS layer), exchange splitting of the EuS conduction band was determined by measuring the I-V characteristics at temperatures above and below the T_C of EuS, and then extracting (from the forward bias I-V) the Schottky barrier height change as it lowers for $T < T_C$. The exchange splitting lowers the Schottky barrier height. Similarly, using forward bias I-V (electron injection from EuS into GaAs), the EuS/GaAs heterojunction showed an exchange splitting of 0.48 eV (much higher than bulk value). In addition, under reverse bias (injection from GaAs into EuS) a shift of the I-V with temperature was observed, suggesting filtering of the unpolarized electrons from GaAs. In this manner, the performance of the EuS/GaAs heterojunction as an injector and detector was probed individually. The observed shifts in the I-V characteristics with temperature, caused by exchange splitting of the EuS conduction band, implied spin-polarized injection and detection, although no spin detector was explicitly used.

There has also been a successful integration of EuS with carbon nanotubes, which shows a mechanism for optical injection of spin-polarized electron-hole pairs in carbon nanotubes [66]. In an exotic experiment, Mohite et al. demonstrated the optical excitation and electrical detection of the triplet exciton in samples of single-walled carbon nanotubes coated with EuS thin film. Triplet excitons are optically inactive in carbon nanotubes. However, in

vertical   or   in-plane

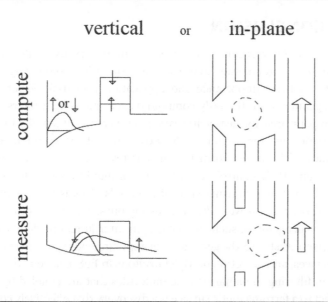

compute

measure

FIGURE 14.20 Proposed scheme of a spin filter quantum measurement. Possible vertical and in-plane structures are shown. In the computing phase, the electron wave function in the quantum dot is held far from the spin filter. In the measurement phase, the gate potentials are changed so that the electron wave function is pressed up against the spin filter. (After DiVincenzo, D.P., *J. Appl. Phys.* 85, 4785, 1999. With permission.)

this case, an exchange interaction between the nanotube and the magnetic Eu^{2+} ions causes the mixing of singlet and triplet states, thus enabling photon absorption into the triplet exciton state and measurement of the triplet contribution to the photocurrent for the first time in carbon nanotubes. This method of using a spin filter increases the efficiency of photocurrent generation, as it is no longer limited to excitation of just the singlet exciton.

Spin filter tunneling is even being considered as an approach to achieving quantum computing, where the electron spin is used as a qubit [67, 68] as shown schematically in Figure 14.20. The spin filter property of the barrier provides the method for the quantum measurement. The correlation of the electron wave function from one quantum dot to the adjacent one is by tunneling through the spin-selective barrier. The energy level of the trapped electron is varied by an applied gate voltage, allowing it to tunnel or not through the barrier placed beside a quantum dot. The tunneling probability depends on the magnetic orientation of the spin filter. In this way, the spin state of the trapped electron can be read, allowing measurement of the state of each qubit individually and reliably by electrical means, rather than the extremely difficult task of magnetically measuring a single spin. Thus spin information is converted into charge information. This is a very simple scheme, and of course, there are many materials and other parameters that must be investigated to realize it. Another group has proposed a spin filter and a spin memory using quantum dots. It is realized by the quantum dots in the Coulomb blockade regime in a magnetic field [69].

14.11 CONCLUSION

Although spin filter studies mostly began in the 1970s while exploring spin-polarized photoemission and field emission, it has come of age in later years both in basic importance and application potential. Ferromagnetic Eu-chalcogenides were the only compounds serving as spin filters, limited to low temperatures, for nearly 40 years. However, recent activity in the ferrites shows promise for developing this novel property at room temperature. Understanding the defect formation in ferrites, and other complex oxide candidates and their control needs to be thoroughly investigated in order to benefit from this promising class of materials. This is true to an extent for Eu-chalcogenides as well. Very interesting physics continues to come out of the Eu-chalcogenides, such as double spin filter junctions demonstrating large magnetoresistance, the generation of a spontaneous spin current in EuS/Al/EuS devices, and detection of triplet excitons in EuS-coated carbon nanotubes. Spin filtering in planar organic molecules that are coupled by a proximity effect to a ferromagnetic surface is a discovery that effectively combines multidisciplinary areas and is very rich for future exploration. With the availability of large interfacial exchange fields, exchange splitting, and quantum well states, unraveling the physics is both challenging and exciting. This can eventually lead to spin filter transistors, opening up this field to various application possibilities. Other engrossing topics that can benefit from spin filter tunneling would be the electrical information readout scheme for quantum computing and spin injection into semiconductors or carbon nanotubes, not to mention more new phenomena waiting to be discovered.

14.12 ACKNOWLEDGMENTS

It is a pleasure to acknowledge the past extensive collaboration and fruitful discussions with Drs. Robert Meservey and Paul Tedrow. In recent times, we have largely benefited from the input from Drs. Guoxing Miao, Taro Nagahama, Karthik V. Raman, Ana Ramos, Martina Mueller, Jean-Baptiste Moussy, Aditya Mohite, Raghava Punguluri, and Boris Nadgorny. The collaborations over the years with Prof. Yves Idzerda and his team as well as with Drs. Julie Borchers and Shannon Watson have greatly contributed to our understanding of the spin filter materials and interfaces. Research funding during many years of research from National Science Foundation, Office of Naval Research and the KIST-MIT joint program has made this all possible.

REFERENCES

1. W. E. Pickett and J. S. Moodera, Half metallic magnets, *Phys. Today* **54**, 39–44 (2001).
2. J. G. Simmons, Generalized formula for the electric tunnel effect between similar electrodes separated by a thin insulating film, *J. Appl. Phys.* **34**, 1793–1803 (1963).
3. W. F. Brinkman, R. C. Dynes, and J. M. Rowell, Tunneling conductance of asymmetrical barriers, *J. Appl. Phys.* **41**, 1915–1921 (1970).

4. J. S. Moodera, X. Hao, G. A. Gibson, and R. Meservey, Electron-spin polarization in tunnel junctions in zero applied field with ferromagnetic EuS barriers, *Phys. Rev. Lett.* **61**, 637–640 (1988); X. Hao, J. S. Moodera, and R. Meservey, Spin-filter effect of ferromagnetic europium sulfide tunnel barriers, *Phys. Rev. B* **42**, 8235–8243 (1990).

5. J. S. Moodera, R. Meservey, and X. Hao, Variation of the electron-spin polarization in EuSe tunnel junctions from zero to near 100% in a magnetic field, *Phys. Rev. Lett.* **70**, 853–856 (1993).

6. J. Schoenes and P. Wachter, Exchange optics in Gd-doped EuO, *Phys. Rev. B* **9**, 3097–3105 (1974).

7. T. S. Santos and J. S. Moodera, Observation of spin filtering with a ferromagnetic EuO tunnel barrier, *Phys. Rev. B* **69**, 241203(R) (2004).

8. M. Pénicaud, B. Siberchicot, C. B. Sommers, and J. Kübler, Calculated electronic band structure and magnetic moments of ferrites, *J. Magn. Magn. Mater.* **103**, 212–220 (1992).

9. M. Gajek, M. Bibes, S. Barthélémy et al., Spin filtering through ferromagnetic BiMnO$_3$ tunnel barriers, *Phys. Rev. B* **72**, 020406(R) (2005).

10. U. Lüders, M. Bibes, K. Bouzehouane et al., Spin filtering through ferrimagnetic NiFe$_2$O$_4$ tunnel barriers, *Appl. Phys. Lett.* **88**, 082505 (2006).

11. A. V. Ramos, M. J. Guittet, J. B. Moussy et al., Room temperature spin filtering in epitaxial cobalt-ferrite tunnel barriers, *Appl. Phys. Lett.* **91**, 122107 (2007).

12. K. Senapati, M. G. Blamire, and Z. H. Barber, Spin-filter Josephson junctions, *Nat. Mater.* **10**, 849–852 (2011).

13. C. Hass, Magnetic Semiconductors, *CRC Crit. Rev. Solid State Sci.* **1**, 47–98 (1970).

14. P. Wachter, Europium chalcogenides: EuO, EuS, EuSe, and EuTe, in *Handbook on the Physics and Chemistry of Rare Earths*, Vol. 2, K. A. Schneider and L. Eyring (Eds.), North-Holland Publishing Co, New York, pp. 507–574, 1979.

15. F. Holtzberg, S. von Molnar, and J. M. D. Coey, Rare earth magnetic semiconductors, in *Handbook on Semiconductors*, Vol. 3, S. P. Keller (Ed.), North-Holland Publishing Co, New York, pp. 803–856, 1980.

16. A. Mauger and C. Godart, The magnetic, optical, and transport properties of representatives of a class of magnetic semiconductors: The europium chalcogenides, *Phys. Rep.* **141**, 151–176 (1986).

17. É. T. Kulatov, Y. A. Uspenskiĭ, and S. V. Khalilov, Electronic structure and magnetooptical properties of europium monochalcogenides EuO and EuS, *Phys. Solid State* **38**, 1677–1678 (1996).

18. M. Horne, P. Strange, W. M. Temmerman, Z. Szotek, A. Svane, and H. Winter, The electronic structure of europium chalcogenides and pnictides, *J. Phys. Condens. Matter* **16**, 5061–5070 (2004).

19. D. B. Ghosh, M. De, and S. K. De, Electronic structure and magneto-optical properties of magnetic semiconductors: Europium monochalcogenides, *Phys. Rev. B* **70**, 115211 (2004).

20. G. Busch, P. Junod, and P. Wachter, Optical Absorption of ferro- and anitferromagnetic europium chalcogenides, *Phys. Lett.* **12**, 11–12 (1964).

21. P. G. Steeneken, L. H. Tjeng, I. Elfimov et al., Exchange splitting and charge carrier spin polarization in EuO, *Phys. Rev. Lett.* **88**, 047201 (2002).

22. N. Müller, W. Eckstein, W. Heiland, and W. Zinn, Electron spin polarization in field emission from EuS-coated tungsten tips, *Phys. Rev. Lett.* **29**, 1651–1654 (1972).

23. E. Kisker, G. Baum, A. H. Mahan, W. Raith, and K. Schröder, Conduction-band tunneling and electron-spin polarization in field emission from magnetically ordered europium sulfide on tungsten, *Phys. Rev. Lett.* **36**, 982–985 (1976).

24. G. Baum, E. Kisker, A. H. Mahan, W. Raith, and B. Reihl, Field emission of monoenergetic spin-polarized electrons, *Appl. Phys.* **14**, 149–153 (1977).

25. E. Kisker, G. Baum, A. H. Mahan, W. Raith, and B. Reihl, Electron field emission from ferromagnetic europium sulfide on tungsten, *Phys. Rev. B* **18**, 2256–2275 (1978).

26. L. Esaki, P. J. Stiles, and S. von Molnar, Magnetointernal field emission in junctions of magnetic insulators, *Phys. Rev. Lett.* **19**, 852–854 (1967).

27. W. A. Thompson, W. A. Holtzberg, T. R. McGuire, and G. Petrich, Tunneling study of EuS magnetization *Magnetism and Magnetic Materials*, in *Proc. of the 17th* Annual Conf. on Magnetism and Magnetic Materials (Chicago, 1971) (*AIP Conf. Proc. No. 5*). C. D. Graham Jr. and J. J. Rhyne (Eds.), AIP, New York, pp. 827–836, 1972.

28. K. Maki, *Superconductivity*, Vol. 2, R. D. Parks (Ed.), Marcel Dekker, New York, p. 1035, 1969; P. Fulde and K. Maki, Theory of superconductor combining magnetic impurities, *Phys. Rev.* **141**, 275 (1966).

29. R. Griessen, M. Landolt, and H. R. Ott, A new ferromagnetic phase in EuSe below 1.8K, *Solid State Commun.* **9**, 2219–2223 (1971).

30. P. M. Tedrow, J. E. Tkaczyk, and A. Kumar, Spin-polarized electron tunneling study of an artificially layered superconductor with internal magnetic field: EuO-Al, *Phys. Rev. Lett.* **56**, 1746 (1986).

31. E. Negusse, J. Holroyd, M. Liberati et al., Effect of electrode and EuO thickness on EuO-electrode interface in tunneling spin filter, *J. Appl. Phys.* **99**, 08E507 (2006).

32. R. Schiller and W. Nolting, Prediction of a surface state and a related surface insulator-metal transition for the (100) surface of stoichiometric EuO, *Phys. Rev. Lett.* **86**, 3847 (2001).

33. T. S. Santos, J. S. Moodera, K. V. Raman et al., Determining exchange splitting in a magnetic semiconductor by spin-filter tunneling, *Phys. Rev. Lett.* **101**, 147201 (2008).

34. G. Sarma, On the influence of a uniform exchange field acting on the spins of the conduction electrons in a superconductor, *J. Phys. Chem. Solids* **24**, 1029–1032 (1963).

35. P. G. de Gennes, Coupling between ferromagnets through a superconducting layer, *Phys. Lett.* **23**, 10–11 (1966).

36. X. Hao, J. S. Moodera, and R. Meservey, Thin-film superconductor in an exchange field, *Phys. Rev. Lett.* **67**, 1342–1345 (1991).

37. A. Pal, K. Senapati, Z. H. Barber, and M. G. Blamire, Electric-field-dependent spin polarization in GdN spin filter tunnel junctions, *Adv. Mater.* **25**, 5581–5585 (2013).

38. A. Pal, Z. H. Barber, J. W. A. Robinson, and M. G. Blamire, Pure second harmonic current-phase relation in spin-filter Josephson junctions, *Nat. Commun.* **5**, 3340 (2014).

39. M. Julliere, Tunneling between ferromagnetic films, *Phys. Lett.* **54A**, 225–226 (1975).

40. P. LeClair, J. K. Ha, H. J. M. Swagten, J. T. Kohlhepp, C. H. van de Vin, and W. J. M. de Jonge, Large magnetoresistance using hybrid spin filter devices, *Appl. Phys. Lett.* **80**, 625–627 (2002).

41. T. Nagahama, T. S. Santos, and J. S. Moodera, Enhanced magnetotransport at high bias in quasimagnetic tunnel junctions with EuS spin-filter barriers, *Phys. Rev. Lett.* **99**, 016602 (2007).

42. A. Saffarzadeh, Spin-filter magnetoresistance in magnetic barrier junctions, *J. Mag. Mag. Mater.* **269**, 327–332 (2004).

43. Measurement by T. S. Santos, T. Nagahama and J. S. Moodera, in collaboration with P. Señeor, M. Bibes, and A. Barthélémy.

44. M. Gajek, M. Bibes, M. Varela et al., $La_{2/3}Sr_{1/3}MnO_3$-$La_{0.1}Bi_{0.9}MnO_3$ heterostructures for spin filtering, *J. Appl. Phys.* **99**, 08E504 (2006).

45. M. Gajek, M. Bibes, S. Fusil et al., Tunnel junctions with multiferroic barriers, *Nat. Mater.* **6**, 296 (2007).

46. G. Hu and Y. Suzuki, Negative spin polarization of Fe_3O_4 in magnetite/magnanite-based junctions, *Phys. Rev. Lett.* **89**, 276601 (2002).

47. Z. Szotek, W. M. Temmerman, D. Ködderitzsch, A. Svane, L. Petit, and H. Winter, Electronic structures of normal and inverse spinel ferrites from first principles, *Phys. Rev. B* **74**, 174431 (2006).

48. G. Hu, J. H. Choi, C. B. Eom, V. G. Harris, and Y. Suzuki, Structural tuning of magnetic behavior in spinel-structure ferrite thin films, *Phys. Rev. B* **62**, 779 (2000).

49. U. Lüders, M. Bibes, J.-F. Bobo, M. Cantoni, R. Bertacco, and J. Fontcuberta, Enhanced magnetic moment and conductive behavior in $NiFe_2O_4$ spinel ultrathin films, *Phys. Rev. B* **71**, 134419 (2005).

50. F. Rigato, S. Estradé, J. Arbiol et al., Strain-induced stabilization of new magnetic spinel structures in epitaxial oxide heterostructures, *Mater. Sci. Eng. B* **144**, 43 (2007).

51. A. V. Ramos, J.-B. Moussy, M.-J. Guittet et al., Influence of a metallic or oxide top layer in epitaxial magnetic bilayers containing $CoFe_2O_4$(111) tunnel barriers, *Phys. Rev. B* **75**, 224421 (2007).

52. A. V. Ramos, T. S. Santos, G. X. Miao, M.-J. Guittet, J.-B. Moussy, and J. S. Moodera, Influence of oxidation on the spin-filtering properties of $CoFe_2O_4$ and the resultant spin polarization, *Phys. Rev. B* **78**, 180402(R) (2008).

53. K. V. Raman, A. M. Kamerbeek, A. Mukherjee et al., Interface-engineered templates for molecular spin memory devices, *Nature* **493**, 509 (2013).

54. K. V. Raman, N. Atodiresei, and J. S. Moodera, Tailoring ferromagnetic-molecule interfaces: Towards molecular spintronics, *SPIN* **4**, 1440014 (2014).

55. J. S. Moodera, B. Koopmans, and P. M. Oppeneer, On the path toward organic spintronics, *MRS Bull.* **39**, 578–581 (2014).

56. D. C. Worledge and T. H. Geballe, Magnetoresistive double spin filter tunnel junction, *J. Appl. Phys.* **88**, 5277–5279 (2000).

57. G.-X. Miao, M. Müller, and J. S. Moodera, Magnetoresistance in double spin filter tunnel junctions with nonmagnetic electrodes and its unconventional bias dependence, *Phys. Rev. Lett.* **102**, 076601 (2009).

58. G.-X. Miao and J. S. Moodera, Controlling magnetic switching properties of EuS for constructing double spin filter magnetic tunnel junctions, *Appl. Phys. Lett.* **94**, 182504 (2009).

59. G.-X. Miao and J. S. Moodera, Numerical evaluations on the asymmetric bias dependence of magnetoresistance in double spin filter tunnel junctions, *J. Appl. Phys.* **106**, 023911 (2009).

60. G.-X. Miao, J. Chang, B. A. Assaf, D. Heiman, and J. S. Moodera, Spin regulation in composite spin-filter barrier devices, *Nat. Commun.* **5**, 3682 (2013).

61. G. X. Miao and J. S. Moodera, Spin manipulation with magnetic semiconductor barriers, *Phys. Chem. Chem. Phys.* **17**, 751 (2015).

62. S. Sugahara and M. Tanaka, A novel spin transistor based on spin-filtering in ferromagnetic barriers: A spin-filter transistor, *Physica E* **21**, 996–1001 (2004).

63. A. T. Filip, P. LeClair, C. J. P. Smits et al., Spin-injection device based on EuS magnetic tunnel barriers, *Appl. Phys. Lett.* **81**, 1815–1817 (2002).

64. C. Ren, J. Trbovic, P. Xiong, and S. von Molnár, Zeeman splitting in ferromagnetic Schottky barrier contacts based on doped EuS, *Appl. Phys. Lett.* **86**, 012501 (2005).

65. J. Trbovic, C. Ren, P. Xiong, and S. von Molnár, Spontaneous spin-filter effect across EuS/GaAs heterojunction, *Appl. Phys. Lett.* **87**, 082101 (2005).

66. D. Mohite, T. S. Santos, J. S. Moodera, and B. W. Alphenaar, Observation of the triplet exciton in EuS-coated single-walled nanotubes, *Nat. Nanotechnol.* **4**, 425 (2009).

67. D. P. DiVincenzo, Quantum computing and single-qubit measurements using the spin-filter effect, *J. Appl. Phys.* **85**, 4785 (1999).

68. C. H. Bennett and D. P. DiVincenzo, Quantum information and computation, *Nature* **404**, 247 (2000).

69. P. Recher, E. V. Sukhorukov, and D. Loss, Quantum dot as spin filter and spin memory, *Phys. Rev. Lett.* **85**, 1962–1965 (2000).

15

Spin Torques in Magnetic Tunnel Junctions

Yoshishige Suzuki and Hitoshi Kubota

15.1 SPIN TORQUES IN MAGNETIC TUNNEL JUNCTIONS

15.1.1 MAGNETIC TUNNEL JUNCTION FOR SPIN-INJECTION

As discussed in Chapters 7 and 8, Volume 1, an electric current in a magnetic multilayer is spin-polarized and yields a torque exerted on a local spin momentum. Such an electric current-induced spin-torque in magnetic multilayers was first predicted theoretically [1, 2] and subsequently observed experimentally in metallic nanojunctions by excitation of spin waves [3] and spin-injection magnetization switching (SIMS) [4, 5]. The effect of the spin-torque was also claimed to be observed in a perovskite system [6]. Further, SIMS was also observed in magnetic tunnel junctions (MTJs) with an Al-O barrier [7] and an MgO barrier [8, 9] and in magnetic semiconductor systems [10].

Figure 15.1 shows a schematic of an MTJ with in-plane magnetization for spin-injection switching. The MTJ consists of two ferromagnetic layers (e.g., Co and Fe) separated by an insulating barrier layer (e.g., AlO and MgO). Lateral shape of an MTJ pillar with in-plane magnetization is an ellipse or a rectangle with dimensions of about 200 nm × 100 nm or less. The angular momentum in the fixed layer, $\vec{S}_1$, which is opposite to the magnetization, is fixed along the long axis of the ellipse through an exchange interaction with an antiferromagnetic layer (e.g., PtMn). Without current injection, the angular momentum in the free layer, $\vec{S}_2$, also lies along the long axis of the ellipse because of magnetostatic shape anisotropy and is either parallel (P) or antiparallel (AP) with respect to $\vec{S}_1$. To induce asymmetry between the two magnetic layers, the thickness of the free layer is often less than that of the fixed layer.

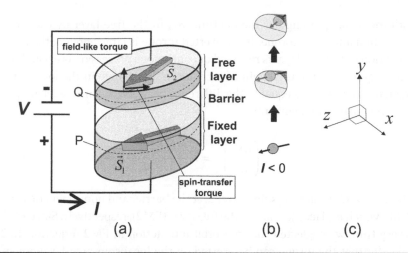

FIGURE 15.1   (a) Typical structures of magnetic nanopillars designed for spin-injection magnetization switching (SIMS) experiments. The upper magnetic layer in the bird's-eye-view images acts as a magnetically free layer, whereas the lower magnetic layer is thicker than the free layer and act as a spin polarizer. The spins in the lower layers are usually pinned to an antiferromagnetic material, which is placed below the pinned layer (not shown), as a result of exchange interaction at the bottom interface. Therefore, the spin polarizer layer is often called the "pinned layer," "fixed layer," or "reference layer." The layer between the two ferromagnetic layers is made of insulators such as MgO. Here, the diameter of the pillar is around 100 nm. The free layer is typically a few nanometers thick. The large arrows indicate the direction of the total spin moment in each layer, and the two small arrows indicate two different spin-torques. (b) The flow and precession of conduction electrons. The injected electrons from the bottom are spin-polarized by the thick ferromagnetic layer (FM1: spin polarizer) and then injected into the thin ferromagnetic layer (FM2: free layer) through the non-magnetic layer (NM1). The injected spins undergo precession in FM2 and lose their transverse components. The lost spin components are transferred to the local spin moment in FM2, $\vec{S}_2$ (details are provided in the text). (c) Coordinate system.

The electric conductance, G, and resistance, R, of the junction vary with the relative angle, θ_{12}, between the two magnetization directions, as already discussed in previous chapters [11],

$$G(\theta) = R^{-1}(\theta) = \frac{G_P + G_{AP}}{2} + \frac{G_P - G_{AP}}{2} \cos\theta_{12}. \tag{15.1}$$

Here, G_P ($=R_P^{-1}$) and G_{AP} ($=R_{AP}^{-1}$) are the tunnel conductances (resistances) for the parallel ($\theta_{12} = 0$) and antiparallel ($\theta_{12} = \pi$) configurations of magnetization, respectively. To inject enough current to switch the magnetization, the resistance-area product ($(RA)_P = A/G_P$ where A is the junction area) is set to below 10 Ω μm^2.

15.1.2 Spin-Transfer Torque and Torquance

When current is passed through a magnetic nanopillar, the electrons are first polarized by the thick fixed layer (spin polarizer) and then injected into the free layer through the barrier. The spins of the injected electrons interact with the local spin-angular momentum, $\vec{S}_2$, in the free layer by exchange interaction and exert torque. If the exerted torque is large enough, the magnetization in the free layer is reversed or continuous precession is excited. As a consequence of the total spin conservation, the change in the local spin-angular momentum is equal to the difference between the injected spin current and gushed spin current:

$$\left(\frac{d\vec{S}_2}{dt}\right)_{ST} = \vec{I}_1^S - \vec{I}_2^S, \tag{15.2}$$

where $\vec{I}_1^S$ and $\vec{I}_2^S$ are the spin currents at the barrier and non-magnetic layer (NM), which is placed just above the free layer (FM2), respectively. Since FM2 is very thin, we neglected the spin-orbit interaction in FM2. Equation 15.2 indicates that the torque can be exerted on the local spin-angular momentum as a result of spin-transfer from the conduction electrons.

Slonczewski showed an intuitive way to evaluate the spin-transfer torque in MTJs [11, 12] by evaluating the spin currents inside ferromagnetic layers. We assume that FM1 is sufficiently thick; therefore, at cross-section P in FM1 (see Figure 15.1a and Figure 15.1b), the conducting spins are relaxed and aligned parallel to $\vec{S}_1$. Those spin-polarized electrons are injected into FM2. The injected spins are subjected to an exchange field made by the local magnetization and show precession motion. Here, we also assume that at cross-section Q inside FM2, the spins of the conducting electrons have already lost their transverse spin component on average because of the decoherence of the precessions and the spins have aligned parallel to $\vec{S}_2$ on average. Therefore, the spin currents at P and Q, $\vec{I}_1'^S$ and $\vec{I}_2'^S$, are parallel to $\vec{S}_1$ and $\vec{S}_2$, respectively. Since the spins of the conduction electrons at P and Q are either the majority or minority spins of the host material, the total charge current in the MTJ can be expressed as a sum of the following four components as shown in Figure 15.2:

$$I^C = I_{++}^C + I_{+-}^C + I_{-+}^C + I_{--}^C. \tag{15.3}$$

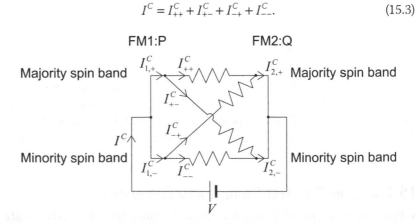

FIGURE 15.2 Circuit model of a magnetic tunnel junction (After Slonzewski, J.C., *Phys. Rev. B* 71, 024411, 2005. With permission.)

Here, the suffixes + and − indicate the majority and minority spin channels, respectively. For example, I_{+-}^C represents a charge current flow from the FM1 minority spin band into the FM2 majority spin band. These charge currents are expressed using the conductance for each spin subchannel, $G_{\pm\pm}$,

$$
\begin{cases}
I_{\pm\pm}^C = V\, G_{\pm\pm} \cos^2 \dfrac{\theta_{12}}{2} \\[2mm]
I_{\mp\pm}^C = V\, G_{\mp\pm} \sin^2 \dfrac{\theta_{12}}{2}
\end{cases}.
\tag{15.4}
$$

Here, V is the applied voltage. The angle dependence of the conductions can be derived from the fact that the spin functions in the FM1 are $|\text{maj}\rangle = \cos(\theta_{12}/2)|\uparrow\rangle + \sin(\theta_{12}/2)|\downarrow\rangle$ for the majority spins and $|\text{min}\rangle = \sin(\theta_{12}/2)|\uparrow\rangle - \cos(\theta_{12}/2)|\downarrow\rangle$ for the minority spins, and those in the FM2 are $|\uparrow\rangle$ and $|\downarrow\rangle$, respectively. Since the spin quantization axes at P and Q are parallel to $\vec{S}_1$ and $\vec{S}_2$, respectively, the spin currents at P and Q are obtained easily as follows:

$$
\begin{cases}
\vec{I}_1^{\prime S} = \dfrac{\hbar}{2}\dfrac{1}{(-e)}\left(I_{++}^C + I_{+-}^C - I_{-+}^C - I_{--}^C\right)\vec{e}_1 \\[3mm]
\vec{I}_2^{\prime S} = \dfrac{\hbar}{2}\dfrac{1}{(-e)}\left(I_{++}^C - I_{+-}^C + I_{-+}^C - I_{--}^C\right)\vec{e}_2
\end{cases},
\tag{15.5}
$$

where unit vectors $\vec{e}_1$ and $\vec{e}_2$ are parallel to the majority spins in the FM1 and FM2, respectively. Now, we apply the total angular momentum conservation between the P and Q planes, i.e.,

$$
\left(\frac{d}{dt}\left(\vec{S}_1 + \vec{S}_2\right)\right)_{ST} = \vec{I}_1^{\prime S} - \vec{I}_2^{\prime S}.
\tag{15.2'}
$$

Then, after a straightforward calculation, the total current and spin-transfer torque are obtained as follows:

$$
\begin{cases}
I^C = \left(\dfrac{G_{++} + G_{--} + G_{+-} + G_{-+}}{2} + \dfrac{(G_{++} + G_{--}) - (G_{+-} + G_{-+})}{2}\,\vec{e}_2 \cdot \vec{e}_1\right)V \\[3mm]
\quad = \left(\dfrac{G_P + G_{AP}}{2} + \dfrac{G_P - G_{AP}}{2}\cos\theta_{12}\right)V \\[3mm]
\left(\dfrac{d\vec{S}_2}{dt}\right)_{ST} = \dfrac{\hbar}{2}\dfrac{1}{(-e)}\left(\dfrac{G_{++} - G_{--}}{2} + \dfrac{G_{+-} - G_{-+}}{2}\right)\left(\vec{e}_2 \times \left(\vec{e}_1 \times \vec{e}_2\right)\right)V \\[3mm]
\quad = T_{ST}\left(\vec{e}_2 \times \left(\vec{e}_1 \times \vec{e}_2\right)\right)V
\end{cases},
\tag{15.6}
$$

The first equation in Equation 15.6 shows the $\cos\theta_{12}$ dependence of the tunnel conductance. The second equation shows the $\sin\theta_{12}$ dependence of the spin-torque (note that $|\vec{e}_2 \times (\vec{e}_1 \times \vec{e}_2)| = \sin\theta_{12}$). Slonczewski called T_{ST}/V the "torquance", which is an analogue of "conductance." In particular, in MTJs, the spin-torque should be bias voltage dependent because $G_{\pm\pm}$ is bias voltage dependent. The direction of the spin-transfer torque is shown in Figure 15.1a for $T_{ST} < 0$. This direction is the same as that in junctions utilizing current perpendicular to the plane (CPP) giant magneto-resistance (GMR) junctions, as was explained in Chapter 4, Volume 1.

15.1.3 FIELD-LIKE TORQUE IN MTJS

The direction of $\vec{S}_2$ may change in two different ways. One is along the direction parallel to the spin-transfer torque, $(\vec{e}_2 \times (\vec{e}_1 \times \vec{e}_2))$. The other is the direction parallel to $(\vec{e}_1 \times \vec{e}_2)$ (see Figure 15.1a). If the torque is parallel to $(\vec{e}_1 \times \vec{e}_2)$, it has the same symmetry as a torque exerted by an external field. Therefore, the latter torque is called a field-like torque. It can also be called a "perpendicular torque" based on its direction in the plane expanded by $\vec{e}_1$ and $\vec{e}_2$.

It has been pointed out that one of the important origins of the field-like torque in MTJs is the change in the interlayer exchange coupling through the barrier layer at a finite biasing voltage [13, 14].

The field-like torque could originate from other mechanisms that are similar to those responsible for the "β-term" in magnetic nanowires. Several mechanisms, such as spin relaxation [15, 16], Gilbert damping itself [17], momentum transfer [18], or a current-induced ampere field have been proposed for the origin of the β-term.

15.2 SPIN-TORQUE DIODE EFFECT AND VOLTAGE DEPENDENCE OF THE TORQUE

15.2.1 LINEARIZED LLG EQUATION AND SPIN-TORQUE DIODE EFFECT

Spin-torques have been quantitatively measured from the ferromagnetic resonance (FMR) excited by them [19–23]. To observe the spin-torque FMR, Tulapurkar et al. [19] developed a homodyne detection technique and named it as "spin-torque diode effect measurement" wherein they used the rectification function of the MTJ. To observe the spin-torque diode effect, we may apply an external field to set a specific relative angle between the free layer and fixed layer magnetizations. In Figure 15.3b, we show a case in which the free layer and fixed layer magnetizations are in-plane but perpendicular to each other. We then apply an alternating current to the junction. A negative current induces a preferential parallel configuration of the spins. Thus, the resistance of the junction becomes smaller and we observe only a small negative voltage across the junction for a given current (Figure 15.3a). A positive current of the same amplitude induces a preferential antiparallel configuration and the resistance becomes higher. We observe a larger positive voltage appearing across the junction (Figure 15.3c). As a result, we observe a positive voltage on average. This is the spin-torque diode effect. This effect can be

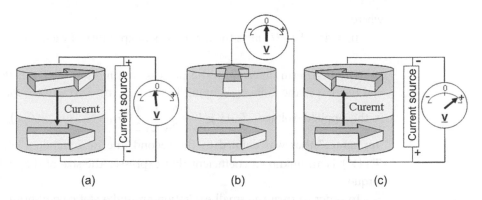

(a)   (b)   (c)

FIGURE 15.3   Schematic explanation of the spin-torque diode effect: (a) negative current; (b) null current; (c) positive current. (After Suzuki, Y. et al., *J. Phys. Soc. Jpn.* 77, 031002, 2008. With permission.)

large if the frequency of the applied current matches the FMR frequency of the free layer. In other words, this effect provides a sensitive FMR measurement technique for the nanopillar moment excited by the spin-torque and provides a quantitative measure of the spin-torques.

To treat the dynamic property of the free-layer spin-angular momentum, here, we introduce the Landau-Lifshitz-Gilbert (LLG) equation including the spin-transfer torque and field-like torque as follows [1, 12, 24]:

$$\frac{d\vec{S}_2}{dt} = \gamma\vec{S}_2 \times \vec{H}_{\text{eff}} - \alpha\vec{e}_2 \times \frac{d\vec{S}_2}{dt} + T_{ST}(V)\vec{e}_2 \times (\vec{e}_1 \times \vec{e}_2)$$
$$+ T_{FT}(V)(\vec{e}_2 \times \vec{e}_1), \tag{15.7}$$

The first term is the effective field torque; second, Gilbert damping; third, spin-transfer torque; and fourth, field-like spin-torque. $\vec{S}_2 = S_2\vec{e}_2$ is the total spin-angular momentum of the free layer and is opposite to its magnetic moment, $\vec{M}_2$. $\vec{e}_2\,(\vec{e}_1)$ is a unit vector that expresses the direction of the spin-angular momentum of the free layer (fixed layer). For simplicity, we neglect the distribution of the local spin-angular momentum inside the free layer and assume that the local spins within each magnetic cell are aligned in parallel and form a coherent "macrospin" [25, 26]. This assumption is not strictly valid since the demagnetization field and current-induced Oersted field inside the cells are not uniform. Such non-uniformities introduce incoherent precessions of the local spins and cause domain and/or vortex formation in the cell [26–28]. Despite the predicted limitations, the macrospin model is still useful, because of both its transparency and its validity for small excitations. γ is the gyromagnetic ratio, where $\gamma<0$ for electrons ($\gamma = -2.21 \times 10^5 \ m/A \cdot \sec$ for free electrons). The effective field, $\vec{H}_{\text{eff}}$, is the sum of the external field, demagnetization field, and anisotropy field. It should be noted that the demagnetization field and the anisotropy field depend on $\vec{e}_2$. $\vec{H}_{\text{eff}}$ is derived from the magnetic energy, E_{mag}, and the total magnetic moment, M_2, of the free layer:

$$\vec{H}_{\text{eff}} = \frac{1}{\mu_0 M_2} \frac{\partial E_{\text{mag}}}{\partial \vec{e}_2}, \tag{15.8}$$

where

$\mu_0 = 4\pi \times 10^7 \ H/m$ is the magnetic susceptibility of vacuum.

The first term determines the precessional motion of $\vec{S}_2$. In the second term, α is the Gilbert damping factor (e.g., $\alpha > 0$, $\alpha \approx 0.007$ for Fe). V is the applied voltage, $T_{ST}(V)/V = \dfrac{\hbar}{2} \dfrac{1}{(-e)} \dfrac{1}{2} (G_{++} - G_{--} + G_{+-} - G_{-+})$ is the "torquance" that was defined in the second line in Equation 15.6, while $T_{FT}(V)$ is an unknown coefficient that expresses the size of the "field-like torque."

In order to treat the small excitation around a static equilibrium point of $\vec{S}_2$, we first introduce the following Hermitiane conjugate variables for a macrospin system;

$$\begin{cases} x^1 \equiv \phi, \\ x^2 \equiv S_2(\cos\theta - 1). \end{cases} \tag{15.9}$$

Here, the coordinate (θ, ϕ) comprises the polar and azimuthal angle of a spherical coordinate system. Using this curvilinear coordinate system, Equation 15.7, can be rewritten as,

$$\frac{dx^i}{dt} = F^i \quad (i = 1, 2), \tag{15.10}$$

where

$$\begin{cases} F^i(t) \cong \displaystyle\sum_{j=1}^{2} \varepsilon^{ij} \partial_j E_{\text{mag}+FT} - \alpha S_2^{-1} \partial^i E_{\text{mag}+ST} \\[2mm] E_{\text{mag}+FT} \equiv E_{\text{mag}} - T_{FT}(\vec{e}_2 \cdot \vec{e}_1) \\[2mm] E_{\text{mag}+ST} \equiv E_{\text{mag}} - \alpha^{-1} T_{ST}(\vec{e}_2 \cdot \vec{e}_1) \\[2mm] \partial_i \equiv \dfrac{\partial}{\partial x^i}, \partial^i \equiv \displaystyle\sum_{j=1}^{2} g^{ij} \partial_j \\[2mm] (\varepsilon^{ij}) = (\varepsilon_{ij}) \equiv \begin{pmatrix} 0 & 1 \\ -1 & 0 \end{pmatrix} \\[2mm] (g_{ij}) = \begin{pmatrix} \sin^2\theta & 0 \\ 0 & \dfrac{1}{S_2^2 \sin^2\theta} \end{pmatrix} = (g^{ij})^{-1} \end{cases} \tag{15.11}$$

Here, ε is the antisymmetric Levi–Civita symbol, and g is the metric tensor. We note that in Equation 15.10, terms proportional to α^2, αT_{ST} and αT_{FT}, are neglected and that we have also assumed that $\partial_j T_{FT} = \partial_j T_{ST} = 0$.

A further point to note is that in principle, the torquance does not have an angular dependence [12].

In Equation 15.10, we clearly see that the spin-transfer torque term, which is a consequence of the spin current, takes the role of a damping term. This Hamilton-type equation of motion using a curvilinear coordinate is a useful tool for obtaining an analytic understanding of the spin-transfer torque dynamics.

Equation 15.10 describes the non-linear dynamics of a macrospin in the MTJ under an applied magnetic field and a given voltage. Before we discuss non-linear behaviors of the system, such as switching, we first derive a linearized equation of motion and discuss the stability of the system with respect to infinitesimal perturbations.

The equilibrium point of the macrospin, (x_0^1, x_0^2), under a static external field and dc bias voltage, (H_0^{ext}, V_0), can be obtained by solving,

$$F^i\left(x_0^1, x_0^2; H_{\text{ext},0}, V_0\right) = 0. \tag{15.12}$$

The linearized equation of motion is obtained by considering the deviation from the equilibrium point as new coordinates, i.e., $(\delta x^1(t), \delta x^2(t)) = (x^1(t) - x_0^1, x^2(t) - x_0^2)$. Then, the linear expansion of Equation 15.12′ yields,

$$\frac{d\delta x^i(t)}{dt} = \sum_{j=1}^{2}\left(\partial_j F^i\right)\delta x^j(t) + \left(\frac{\partial F^i}{\partial H_{ext}}\right)\delta H_{\text{ext}}(t) + \left(\frac{\partial F^i}{\partial V}\right)\delta V(t), \tag{15.13}$$

where $\delta H_{\text{ext}}(t)$ and $\delta V(t)$ are the time-dependent parts of the external field and bias voltage, respectively. All derivatives appearing in Equation 15.13 are evaluated at the equilibrium point (x_0^1, x_0^2) and (H_0^{ext}, V_0).

The solution of the linearized LLG (i.e., Equation 15.13) is a forced oscillatory motion around the equilibrium point, driven by a small rf voltage and/or rf field. Since Equation 15.13 is a linear equation, the solution can be easily obtained using Fourier transformation, i.e. $f(t) = \int_{-\infty}^{+\infty} f(\omega)e^{-i\omega t}d\omega$, as follows:

$$\delta x^i(\omega) \cong \sum_{j=1}^{2} \frac{i\omega g_j^i - \sum_{k=1}^{2}\varepsilon^{ik}\Omega_{kj}}{\omega^2 - \omega_0^2 + i\omega\Delta\omega}$$

$$\left(\sum_{m=1}^{2}\varepsilon^{jm}\partial_m\left(\frac{\partial E_{\text{mag}}}{\partial H_{ext}}\delta H_{\text{ext}}(\omega) + \frac{\partial E_{\text{mag}}}{\partial V}\delta V(\omega) - (\vec{e}_2 \cdot \vec{e}_1)T_{FT}'\delta V(\omega)\right)\right.$$

$$\left. + S_2^{-1}\partial^j(\vec{e}_2 \cdot \vec{e}_1)T_{ST}'\delta V(\omega) \right),$$

$$\tag{15.14}$$

$$\begin{cases} \hat{\Omega} = \left(\Omega_{ij}\right) \equiv \left(\partial_i \partial_j E_{\mathrm{mag}}\right) \\[4pt] \omega_0^2 \equiv \det\left[\hat{\Omega}\right] \\[4pt] \Delta\omega \equiv \alpha S_2^{-1}\left(\partial_1 \partial^1 + \partial_2 \partial^2\right)\left(E_{\mathrm{mag}} - \alpha^{-1} T_{ST}\left(\vec{e}_2 \cdot \vec{e}_1\right)\right). \qquad (15.15) \\[4pt] \quad = S_2^{-1}\left(\alpha\left(\partial_1 \partial^1 + \partial_2 \partial^2\right) E_{mag} + 2 T_{ST}\left(\vec{e}_2 \cdot \vec{e}_1\right)\right) \\[4pt] T_{ST}' \equiv \dfrac{\partial T_{ST}}{\partial V}, T_{FT}' \equiv \dfrac{\partial T_{FT}}{\partial V} \end{cases}$$

Here, ω_0 is the ferromagnetic resonance (FMR) frequency of the system. A positive $\Delta\omega$ provides the line width of the magnetic resonance, while a negative $\Delta\omega$ implies that the system is unstable. The equation above shows that the spin-transfer term may linearly increase or decrease the magnitude of $\Delta\omega$, depending on the sign of the applied voltage.

There are four terms appearing in the parentheses on the left-hand-side of Equation 15.14. The first term describes a magnetic field driven FMR; the second term corresponds to an FMR excitation due to the voltage control of magnetic anisotropy (VCMA); while third and fourth terms show that both the spin-transfer torque and the field-like torque can excite a uniform mode (FMR mode) in the free layer. However, the phases of the FMR excitations differ by 90°. This difference in the precessional phase is a result of the different directions of the respective torques (Figure 15.1). In addition, the width of the resonance, $\Delta\omega$, is affected only by the spin-transfer torque exerted by the direct voltage, V_0. This is the (anti)damping effect of the spin-transfer torque, which was already discussed in the previous section. The field-like torque exerted by the direct voltage, V_0, changes the resonance frequency, ω_0, in a manner similar to an external field.

When the rf voltage across the MTJ is $\delta V \cos\omega t$, Equations 15.1 and 15.14 can be used to determine the precessional motion and oscillating part of the junction resistance, $\delta R\cos(\omega t + \phi)$, which is linear in δV. The rf current through the junction can be approximated as $\delta V G(\theta_0)\cos\omega t$. Thus from Ohm's law, the following additional voltages appear across the junction,

$$\delta R\cos(\omega t + \phi)\times \delta V G(\theta_0)\cos\omega t = \delta R G(\theta_0)\frac{\delta V}{2}\left(\cos\phi + \cos(2\omega t + \phi)\right).$$

Here, the frequencies of the additional voltages are zero (dc) and 2ω. This implies that, under spin-torque FMR excitation, the MTJs may possess a rectification function and a mixing function. Because of these new functions, Tulapulkar et al. referred to these MTJs as spin-torque diodes and to these effects as spin-torque diode effects [19]. These are non-linear effects that result from two linear responses, that is, the spin-torque FMR and Ohm's law.

When the MTJ is placed at the end of a waveguide, the explicit expression of the rectified dc voltage under a small bias voltage is given as follows:

$$V_{d.c.} \cong \frac{\eta}{4}\sin^2\theta_0 \frac{(G_P - G_{AP})}{S_2 G(\theta_0)}$$

$$\mathrm{Re}\left[\frac{(\omega_{\theta\varphi} - \omega_{ST} - i\omega)T'_{ST} + (\omega_{\varphi\varphi} + \omega_{FT})T'_{FT}}{\omega^2 - \omega_0^2 + i\omega\Delta\omega}\right]V_\omega^2, \tag{15.16}$$

where V_ω is the rf voltage amplitude applied to the emission line and η is the coefficient used to correct the impedance matching between the MTJ and the waveguide with a characteristic impedance of Z_0, where

$$\eta = \left(\frac{2R(\theta_0)}{R(\theta_0) + Z_0}\right)^2. \tag{15.17}$$

If the emission line and the MTJ include some parasitic impedances (capacitance in most cases), we should employ an appropriate value of η to correct this effect [22, 24].

This is a type of homodyne detection and is, thus, phase-sensitive. The motion of the spin, illustrated in Figure 15.3, corresponds to that excited by the spin-transfer torque at the resonance frequency. However, the motion of the spin excited by the field-like torque shows a 90° difference in phase. As a consequence, only the resonance excited by the spin-transfer torque can rectify the rf current at the resonance frequency. In Figure 15.4, the dc voltage spectra predicted for the spin-transfer torque excitation and for the field-like torque are both shown. The spectrum excited by the spin-transfer torque exhibits a single bell-shaped peak (dashed line), whereas that excited by the field-like torque is of a dispersion type (dotted line). This very clear difference provides us with an elegant method to distinguish a spin-transfer torque from a field-like torque [19–24].

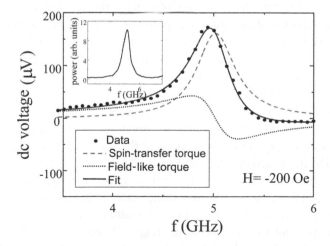

FIGURE 15.4 Spin-torque diode spectra for a CoFeB/MgO/CoFeB MTJ. Data (closed dots) are well-fitted by a theoretical curve that includes contributions from both the spin-transfer torque and the field-like torque. The inset shows an rf noise spectrum obtained for the same MTJ. (After Tulapurkar, A.A. et al., *Nature* 438, 339, 2005. With permission.)

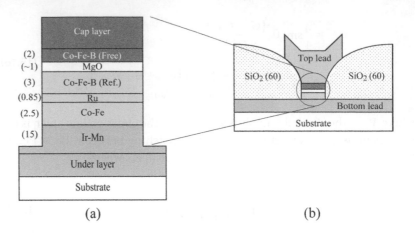

FIGURE 15.5 The multilayer stack (a) and cross-sectional structure (b) of the magnetic tunnel junction.

15.2.2 EXPERIMENTS AND BIAS VOLTAGE DEPENDENCE OF THE TORQUE

The multilayer stack is shown in Figure 15.5a. The numbers in parentheses indicate layer thicknesses in nm units. CoFe/Ru/CoFeB is a synthetic ferrimagnet structure, in which the magnetizations of CoFe and CoFeB align in an antiparallel configuration. The magnetization of CoFe is pinned unidirectionally by an exchange-biasing field from the antiferromagnetic Ir-Mn layer. The thin MgO layer is a crystalline tunnel barrier with a (001) plane [29, 30].

The thin CoFeB on the MgO is a free layer whose magnetization can reverse freely by applying a magnetic field or current. The underlayer consists of a Ta/Cu/Ta trilayer, which has a low resistance as a bottom lead. The multilayer was microstructured into submicron-sized tunnel junctions. The cross-sectional structure of the prepared magnetic tunnel junction is shown in Figure 15.5b. The junction with dimensions of the order of 100 nm is sandwiched between a top lead and a bottom lead. The top and bottom leads are isolated by a thick SiO_2 layer so that a current flows through only the magnetic multilayer.

The sequence of the microfabrication process is shown in Figure 15.6. The top and side views of one tunnel junction are illustrated in the process steps. The side views represent cross-sections of the center part of the sample as indicated by the dotted line A–B in (Figure 15.6a). The first step was the fabrication of a bottom structure. On the surface of the multilayer, the resist pattern for the bottom lead was formed using optical lithography (Figure 15.6a). The resist pattern was transferred to the multilayer using an Ar ion etching technique and a lift-off technique (Figure 15.6b). The minimum width of the bottom lead was about 4 μm. The next step was the fabrication of a submicron-sized junction and electrical contacts. On the bottom electrode, resist patterns for a small junction and large contact pads were prepared using electron beam lithography. The sizes of the junction and contact pads were 50 nm × 250 nm and 50 μm × 50 μm, respectively (Figure 15.6c).

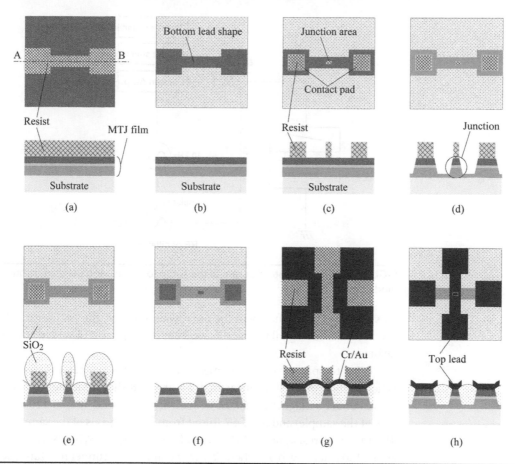

FIGURE 15.6   The sequence of steps in the microfabrication process of the magnetic tunnel junctions.

Ar ion etching was terminated at a predetermined time to stop in the Ir-Mn layer (Figure 15.6d). A thick SiO_2 layer, which passivated the sample surface (Figure 15.6e), was partially removed using the lift-off technique (Figure 15.6f). The final step was the fabrication of a top lead. A Cr(5 nm)/Au(20 nm) double layer was sputtered. Then, the resist patterns for a top lead and two contact pads for the bottom lead were prepared using photolithography (Figure 15.6g). The shapes of the top lead and contact pads were transferred to the Cr/Au double layer using an Ar ion etching technique. Then, the remaining resist patterns were removed (Figure 15.6h). Finally, we obtained about 200 MTJs on a 20 mm × 20 mm substrate chip. The substrate chip was annealed in vacuum at 330°C under a magnetic field of 1T. The magnetic field direction was parallel to the long axis of the junction.

The measurement setup for the spin-torque diode effect is shown in Figure 15.7. A microwave current of −15 dBm was applied to an MTJ sample through a bias-T. The microwave currents (50 MHz~15 GHz) were modulated at 10 kHz. The diode signal, that is, the AC voltage across the sample that was synchronized with the modulation, was measured using a lock-in amplifier at a dc terminal of the bias-T. A dc bias current between −1.5 mA and 1.5 mA was applied to the sample and the dc bias voltage was measured.

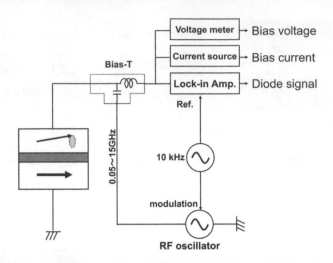

FIGURE 15.7 Schematic diagram of the setup for measuring the spin-torque diode effect.

By sweeping the frequency of the microwave current, the spin-torque diode spectrum (diode signal vs. microwave frequency (f)) was observed at various dc biases.

Prior to spin-torque diode measurements, the magnetoresistive properties of the sample were measured. Figure 15.8 shows magneto-resistance curves under a magnetic field, H, with an angle, θ_H, of 0° (parallel to the long axis of the sample) and θ_H of 45° (tilted from the long axis of the sample) [22]. The magneto-resistance (MR) ratio defined by $(R_{AP} - R_P)/R_P \times 100$ (%) was about 150% at a low bias, where R_{AP} (R_P) denotes the tunnel resistance when the magnetizations of the CoFeB layers are antiparallel (parallel) to each other, corresponding to the resistance value of point A (C) in the figure. The resistance-area product of the sample was about 2 $\Omega\mu m^2$ at room temperature (RT). At θ_H of 0°, the resistance shows abrupt jumps at low fields because of the magnetization switching of the top CoFeB free layer. At θ_H of 45°, in addition to the resistance jumps at low fields, a resistance decrease was observed at negative magnetic fields because of the rotation of the bottom CoFeB magnetization. To measure the spin-torque diode effect, a finite θ_{12} is required: when θ_{12} is equal to 0 or π, the diode signal cancels completely. Therefore, we applied −400 Oe, in which θ_{12} is equal to 137° (point B), for the spectrum measurements. The value of θ_{12} was evaluated using Equation 15.1. To check the deviation of θ_{12} under finite biases, the bias dependence of the tunnel resistance was measured at several points. It was confirmed that the change in θ_{12} with bias was small in the range of ±400 mV.

The observed spectra are shown in Figure 15.9 [22]. At zero bias, the diode signal shows a peak at around 6.7 GHz. The peak shape is symmetric with respect to frequency. This suggests that the spin-transfer torque oscillating in phase with the microwave current was dominant. When negative (positive) biases were applied to the sample, the diode signals became larger (smaller) with the bias. The change in the peak height corresponded well with our expectation: spin precession was enhanced with a negative bias and

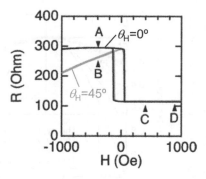

FIGURE 15.8 Magneto-resistance curves for the Ir-Mn/Co-Fe/Ru/CoFeB/MgO/ CoFeB magnetic tunnel junction. θ_H is the angle between the magnetic fields and the long axis of the sample.

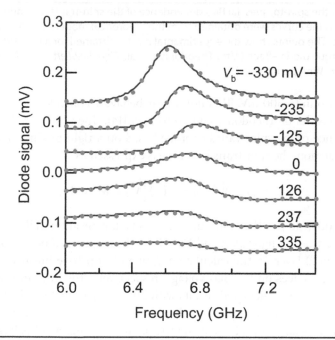

FIGURE 15.9 Spin-torque diode spectra obtained for various dc biases.

dampened with a positive bias because of the voltage-dependent spin-transfer torque. The peak shape also displayed remarkable changes: it became asymmetric with a frequency under large positive and negative biases. This asymmetry was caused by the field-like torque that oscillated out-of-phase with the microwave current. The lines in Figure 15.9 represent Equation 15.16 with appropriate parameters, which fit the experimental results very well. From the parameters, we evaluated the magnitudes of both the spin-transfer and field-like torques. Figure 15.10a shows the spin-transfer torque as a function of the bias voltage [22]. The solid circles and lines represent the experimental results and calculations, respectively. The experimental result for the spin-transfer torque shows a linear dependence around a zero bias. However, at high biases, it shows a minimum around +300 mV and a slope

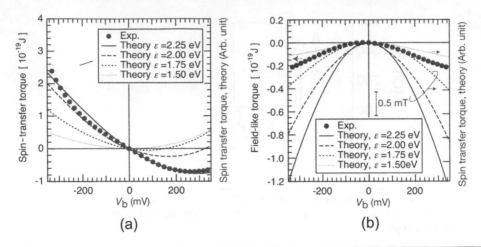

FIGURE 15.10   Bias dependence of the spin-torques. (a) Bias dependence of the spin-transfer torque. (b) Bias dependence of field-like torque. Fine curves were obtained from the theoretical model calculations with different spin splitting parameters, ε. The points show the experimental results obtained for a CoFeB/MgO/CoFeB MTJ by exploiting the spin-torque diode effect. (After Theodonis, I. et al., *Phys. Rev. Lett.* 97, 237205, 2006; Kubota, H. et al., *Nat. Phys.* 4, 37, 2008. With permission.)

change around −300 mV. This non-linear behavior coincides with the calculated results of Theodonis and coworkers [14]. They calculated the bias dependence of the spin-transfer torque based on a tight binding model. In the calculation, the spin-transfer torque is expressed as

$$T_{||} = (1/2)\big[J_s(\pi) - J_s(0)\big]\mathbf{M}_2 \times (\mathbf{M}_1 \times \mathbf{M}_2), \tag{15.18}$$

where $J_s(\pi)$ ($J_s(0)$) is the spin-current density for an antiparallel (parallel) magnetization configuration and $\mathbf{M}_1$ ($\mathbf{M}_2$) is the magnetization vector of the free (pinned) layer. In the calculation, $J_s(0)$ and $J_s(\pi)$ have linear and parabolic dependences on the bias voltage, respectively. The lines in the figure correspond to the calculated results with different parameter values for the minority spin onsite energy, ε, in the tight binding calculation. The linear component is large for a large ε and the parabolic component is large for a small ε. One of the calculations with ε = 2.25 eV is very close to the experimental result. Although the calculation is based on a simple model disregarding detailed electronic structures of the materials, the calculated results can explain the observed dependence well.

Similar experimental results were reported by a Cornell University group [21, 23]. They reported a basically similar bias dependence for the spin-transfer torque in a wider bias range up to about ±500 mV with a θ_{12} of close to 90 degrees. However, the non-linear voltage dependence term was smaller in their result. In our analysis, since the contribution of the non-linear magnetic response [23, 31] was ignored, the non-linear voltage dependence could be overestimated.

Figure 15.10b shows the field-like torque as a function of the bias voltage [22]. The observed field-like torque shows a quadratic bias dependence, which agrees with the calculations [14]. The calculated results again can reproduce the qualitative nature and order of the intensity of the field-like torque. In this case, smaller ε values show a better fit, but it is due to the oversimplified model. Such a quadratic dependence of the field-like torque was also observed by other groups [21, 32]. In contrast, some groups have reported that the field-like torque changes its sign when the bias current direction is reversed. Devolder et al. analyzed an asteroid loop displacement with bias and separated the contributions of the field-like torque, heating and spin-transfer torque. They concluded that the field-like torque has a linear dependence on the bias [33]. Li et al. also reported that the sign of the field-like torque depends on the bias direction [34]. They measured spin-transfer switching under various magnetic fields and current pulse widths. The magnitude of the observed field-like torque was proportional to the product of the current density and absolute value of the voltage. They explained the dependence qualitatively by considering the energy-dependent inelastic tunneling. It should be noted here that because these results are affected by a charge current-induced Oersted field and Joule heating, the results may vary depending on the sample geometry. By contrast, our experiment uses a phase-sensitive detection technique; therefore, such artifacts do not affect the observed results. It should be noted that a recent finding concerning the voltage-induced magnetic anisotropy change in the metallic system (see Section 15.5) [35, 36], which was not included in the previous analysis, may provide a coherent explanation to those contradictions. Further investigations should be performed carefully to conclude the bias dependence of the field-like torque.

15.3 SPIN-INJECTION MAGNETIZATION SWITCHING IN MTJs

15.3.1 CRITICAL CURRENT AND SWITCHING TIME

In MTJs, SIMS may occur that are similar to that in CPP-GMR nanopillars (see Chapter 7, Volume 1). Minor but important differences are in the angular dependences of the torque intensity and in the junction resistances. The much larger resistance in the MTJ results in a much larger temperature rise during switching because of Joule heating. The critical current (instability current) at zero temperature is obtained easily by setting $\Delta \omega$ in Equation 15.15 to zero.

$$I_{c,0} = \alpha S_2 \left(\frac{\omega_{\theta\theta} + \omega_{\varphi\varphi}}{2} \right) \frac{G(\theta)}{T_{ST}}. \tag{15.19}$$

Many efforts have been paid to reduce $I_{c,0}$. Since above expression is essentially the same as in the CPP-GMR nanopillars, the same techniques that were effective are also useful in MTJs. The first attempt is to reduce total spin-angular momentum, S_2, in the free layer. SIMS requires effective injection of spin-angular momentum that is equal to that in the free layer. Therefore, reduction in S_2 results in a reduction in $I_{c,0}$. Reduction in the free layer thickness reduces S_2 and $I_{c,0}$. S_2 can be also reduced by reducing the magnetization of the ferromagnetic material. Especially in the nanopillar with in-plane magnetization, since magnetization also affects the size of the anisotropy field ($-\omega_{\phi\phi} / \gamma$), $I_{c,0}$ is a quadratic function of the magnetization. Therefore, reduction in the magnetization is very effective [37]. The second attempt is to use double spin-filter structure. This method was originally proposed by L. Berger [38]. By using this structure, Huai et al. observed a substantial reduction of the threshold current to 2.2×10^6 A/cm² [39]. The third attempt is to use perpendicular magnetic anisotropy, which can reduce the size of the anisotropy field [40]. Yakata et al. pointed out that a free layer with a Fe rich composition in a FeCoB/MgO/FeCoB stacking possesses a perpendicular crystalline anisotropy and results in a reduction of switching current [41]. The perpendicular crystalline anisotropy partially cancels the demagnetization field and reduces the size of the anisotropy field. Nagase et al. obtained a significant reduction of the threshold current under the required thermal stability factor for the MTJ nanopillars with the CoFeB/[Pd/Co]x2/Pd free layer and FePt/CoFeB pinned layer [42]. In this structure, the films have perpendicular magnetization [43]. Relatively small anisotropy field, small magnetization can be the origin of the reduction.

The switching time can be obtained by integrating the LLG equation with the spin-transfer torque (Equation 15.7). For an MTJ with orthogonal symmetry, the equation of motion for the opening angle of precession, θ, is derived from Equation 15.7 as follows:

$$\frac{d\theta}{dt} \cong -\alpha(-\gamma)\left(\langle H_{\text{ani}}\rangle\cos\theta - H_{\text{ext}}\right)\sin\theta$$

$$-\frac{T_{ST}V}{S_2}\sin\theta + (-\gamma)H_{\text{stochastic}}(t), \tag{15.20}$$

where $H_{\text{stochastic}}$ is a stochastic field that expresses thermal agitation. In the present treatment, we neglect $H_{\text{stochastic}}$. H_{ext} is a z-direction external field. $\langle H_{\text{ani}}\rangle = \dfrac{H_{\text{ani},x} + H_{\text{ani},y}}{2}$ is an averaged anisotropy field. For an in-plane magnetized film, as shown in Figure 15.1, those fields are the in-plane coercive force ($= H_{\text{ani},x}$) and the out-of-plane demagnetization field ($= H_{\text{ani},y}$). To obtain the above equation, so-called "one-cycle average" of the LLG equation was taken. The approximation is exact if $H_{\text{ani},x} = H_{\text{ani},y}$, that is, the uniaxial symmetry case. The switching time is obtained by integrating the equation [25]:

$$\tau_{sw} \cong \frac{I_{c,0}}{I - I_{c,0}}\frac{1}{\alpha(-\gamma)\langle H_{\text{ani}}\rangle}\log\left(\frac{2}{\delta\theta_0}\right), \tag{15.21}$$

where $\delta\theta_0$ is the angle between the direction of magnetization and the easy axis at the beginning of the current application. For the thermal equilibrium, the average value of $\delta\theta_0$ is estimated as,

$$\delta\theta_0 = \sqrt{\frac{k_BT}{\mu_0M_2H_{\text{ani},x}}} = \sqrt{\frac{1}{2\Delta}}, \tag{15.22}$$

where k_B is the Boltzmann constant and T is the absolute temperature. Δ is the thermal stabilization factor.

Koch et al. [44] and Li et al. [45] explained the reduction in the switching current for long pulse durations by considering the thermal excitation of the macrospin fluctuation [44] induced by the stochastic field, $H_{\text{stochastic}}(t)$. Because of the stochastic field, the direction of the free-layer spin-angular momentum, $\vec{e}_2$, becomes also a stochastic variable. Therefore, we can only discuss its probability density, $p(\vec{e}_2,t)$, by considering the statistical nature of $H_{\text{stochastic}}(t)$. From the fluctuation-dissipation theorem [46], $H_{\text{stochastic}}(t)$ in Equation 15.20 is related to the Gilbert damping constant.

$$\alpha \cong \frac{(-\gamma)\mu_0 M_2}{k_B T} \frac{1}{2} \int_{-\infty}^{\infty} d\tau \left\langle H_{\text{stochastic}}(t) H_{\text{stochastic}}(t+\tau) \right\rangle, \quad (15.23)$$

where $\langle\ \rangle$ expresses a thermodynamical ensemble average. Using this, the time evolution of the probability density is expressed as follows,

$$\begin{cases} \dfrac{\partial}{\partial t} p(z,t) + \dfrac{\partial}{\partial z} \left\{ -\dfrac{\alpha}{S_2}(1-z^2) \left[\dfrac{dE_{\text{eff}}(z)}{dz} + k_B T \dfrac{\partial}{\partial z} \right] p(z,t) \right\} = 0 \\[2mm] E_{\text{eff}}(z) = \mu_0 M_2 H_{\text{ext}} z - \dfrac{1}{2}\mu_0 M_2 \left(\dfrac{H_{\text{ani},x} + H_{\text{ani},y}}{2} \right) z^2 - \dfrac{z}{\alpha} T_{ST} V \end{cases}, \quad (15.24)$$

where $E_{\text{eff}}(z)$ is the effective magnetic energy, and $z = \cos\theta$ is the z-component of $\vec{e}_2$. The term that includes $dE_{\text{eff}}(z)/dz$ is the contribution from the drift driven probability current and is proportional to the sum of the three torques, namely the precession, damping, and spin-transfer torques. The term with $k_B T (\partial / \partial z)$ expresses the contribution from the thermal diffusion current of the probability density and is proportional to the gradient of the probability density. This is the Fokker-Planck equation [46, 47] adapted for the LLG equation including the spin-transfer torque [48, 49].

Without a charge current, Equation 15.24 has the Boltzmann distribution, $p(\vec{e}_2) \propto e^{\frac{E_{\text{mag}}(\vec{e}_2)}{k_B T}}$, as a solution for the thermal equilibrium. If the system is out of equilibrium because of a charge current application, the probability density currents start to flow to approach a steady state.

The probabilistic distribution of the switching time, $p_{sw}(t)$, and its integration, $P_{sw}(t) = \int_0^t p_{sw}(t_1) dt_1$, in the thermal activation regime can be obtained by applying Kramers's method [47] to Equation 15.24, for a case with a high thermal barrier, which corresponds to a large thermal stability factor, small current and small external field, as follows:

$$\begin{cases} p_{sw}(t) = \tau_{sw}^{-1} e^{-\frac{t}{\tau_{sw}}} \\[2mm] P_{sw}(t) = 1 - e^{-\frac{t}{\tau_{sw}}} \\[2mm] \tau_{sw}^{-1} \cong \tau_0^{-1} e^{-\Delta\left(1 - \frac{I}{I_{c,0}} - \frac{H_{\text{ext}}}{H_c}\right)^2} \end{cases}, \quad (15.25)$$

where t is the elapsed time, τ_{sw} is the average switching time, and $\tau_0^{-1} = \alpha(-\gamma)\langle H_{ani} \rangle \sqrt{\dfrac{\Delta}{\pi}}$ is the attempt frequency. Δ is a thermal stability factor calculated for zero current and zero external field. The reader can find that Néel-Brown's exponential law is valid under a current injection. Li et al. solved the Fokker-Plank equation for in-plane magnetization with a small current and proposed a slightly different formula to also include parameter β [45]. In their paper, the term $\left(1 - \dfrac{I}{I_{c0}} - \dfrac{H_{ext}}{H_c}\right)^2$ in Equation 15.25 is replaced with $\left(1 - \dfrac{H_{ext}}{H_c}\right)^\beta \left(1 - \dfrac{I}{I_{c,0}}\right)$. Taniguchi et al. proposed a theory which includes another exponent parameter on the current term, $\left(1 - \dfrac{I}{I_{c,0}}\right)^{\beta'}$ [50]. The theoretical curve, derived on the basis of the above-mentioned theories, fits the experimental data well when reasonable parameter values are used.

In the thermal activation regime, SIMS can occur at currents smaller than $I_{c,0}$ because of thermal activations. Moreover, the switching time is distributed widely following an exponential law for a given applied current. In another way, if we apply current pulses with constant widths but different heights (current), we will see the following switching current distribution:

$$p_{sw}\left(I\right) = \frac{2\Delta}{I_{c,0}} \frac{\tau_{pulse}}{\tau_{sw}} e^{-\frac{\tau_{pulse}}{\tau_{sw}}}, \tag{15.26}$$

where τ_{pulse} is the width of the applied pulse current, and τ_{sw} is a function of the applied current Equation 15.25. Experimentally, $I_{c,0}$ and Δ can be determined either from the pulse width dependence of the average switching current or from the switching current distribution measured using pulses with constant width.

15.3.2 PROBABILISTIC SWITCHING UNDER LONG DURATION PULSE CURRENT

In a magnetic recording system such as a hard disk drive (HDD), data retention is guaranteed based on the probabilistic magnetization reversal model with a thermal activation process [51]. Spin-transfer switching is also probabilistic and governed by the thermal activation, which was discussed experimentally by Koch et al. [44]. and theoretically by Li and Zhang [45]. When the current pulse duration is longer than the inverse of the attempt frequency, thermal activation plays an important role in the switching. The observed switching current I_c at a finite temperature is always reduced from the intrinsic switching current I_{c0} that corresponds to a switching current without thermal activation. In the spin dynamics, including the spin-transfer torque, I_{c0} is the current at which the effective damping constant becomes zero. Since I_c depends on the measurement condition, I_{c0} is a very important quantity for characterizing samples. The thermal stability index, Δ, is another important parameter of the switching, which is defined as $\Delta = K_u V/(k_B T)$. Here, K_u, V, k_B, and T correspond to the uniaxial magnetic anisotropy, free layer cell volume, Boltzmann constant, and temperature, respectively. A higher Δ corresponds to a lower switching probability, resulting in a longer data retention time in memory applications. In the thermally activated spin-transfer switching model proposed by Li and Zhang [45], the switching probability is expressed as

$$P_{sw}\left(I\right)=1-\mathrm{Exp}\left[-\frac{t}{\tau_0}\mathrm{Exp}\left[-\Delta\left(1-\frac{I}{I_{c0}}\right)\left(1-\frac{H}{H_{c0}}\right)^2\right]\right], \quad (15.25')$$

where t, τ_0, I, H, and H_{c0} are the switching time, the inverse of the attempt frequency, current, external magnetic field, and intrinsic coercivity, respectively. With this model, two methods are used to evaluate I_{c0} and Δ. The first method is the observation of the pulse duration dependence of the switching current [52]. The switching current is plotted as a function of $\ln(t_p/\tau_0)$, where t_p usually ranges from 10^{-1} s to 10^{-5} s. By extrapolating to τ_0 (=1 ns), we can obtain I_{c0}. From the slope of the extrapolating line, we can obtain Δ. Another method is the observation of the switching current distribution at a constant pulse duration [53, 54]. By measuring the switching current repeatedly many times, we can obtain the switching current distribution. Theoretically, this distribution is expressed as dP_{sw}/dI, which has a peak at around a mean value. By fitting the function to experimental values, we can evaluate I_{c0} and Δ.

Figure 15.11a shows the measurement setup of the switching current [55]. A lock-in amplifier is used to measure the junction resistance at a zero dc bias. The pulse generator produces sequential pulse currents as shown in Figure 15.11b, which were applied to a sample. The sample was a CoFeB/MgO/CoFeB MTJ, which had a structure similar to the one used in the previous section (Figure 15.5). Figure 15.12a shows the magneto-resistance loop of the tunnel junction. The resistance jumps at low fields correspond to the magnetization reversal of the top 2-nm-thick CoFeB free layer. The bottom 3-nm-thick CoFeB layer is a reference layer, whose magnetization does not reverse due to strong exchange biasing. When the free-layer magnetization is parallel (P) to that of the reference layer, the resistance is as low as 320 Ω. When the free-layer magnetization is antiparallel (AP) to that of the reference layer, the resistance is as high as 750 Ω. Figure 15.12b shows a

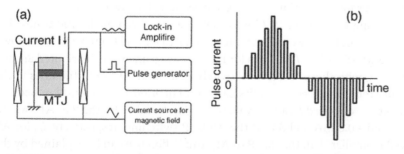

FIGURE 15.11   Schematic diagram of the setup for observing spin-transfer switching (a) and the pulse current sequence. (After Kubota, H. et al., *Oyo Buturi* 78, 232, 2009. With permission.)

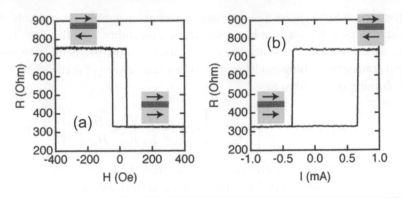

FIGURE 15.12 A *R-H* loop (a) and an *R-I* loop (b) of the CoFeB/MgO/CoFeB magnetic tunnel junction. (After Kubota, H. et al., *Oyo Buturi* 78, 232, 2009. With permission).

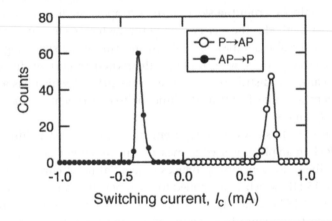

FIGURE 15.13 Distribution of switching currents for the spin-transfer switching in CoFeB/MgO/CoFeB magnetic tunnel junction. (After Kubota, H. et al., *Oyo Buturi* 78, 232, 2009. With permission.)

resistance-current (*R-I*) loop of the sample. In this plot, the resistance was measured after each pulse current. The resistance jumps abruptly at 0.6 mA and −0.4 mA. The resistance change by current agrees well with that in the magneto-resistance loop as shown in Figure 15.12a. This proves that the free-layer magnetization is completely switched by the current.

Using this sample, *R-I* loops were measured repeatedly 100 times. The switching currents were distributed from P to AP and from AP to P, as shown in Figure 15.13. The I_c's values for P to AP switching were distributed around 0.7 mA, with those for AP to P switching around −0.4 mA. The lines in the figure represent theoretical fits based on Equation 15.19. From the theoretical fit, we obtained I_{c0} and Δ values of 1.3 mA and 41 for the P to AP switching and −0.8 mA and 33 for the AP to P switching, respectively. I_{c0} for AP to P is smaller than that for P to AP. This difference can be explained by the difference of $G(\theta)$ between the AP and P state as shown in Equation 15.19. The difference of Δ between P and AP states probably originates in the different effective fields in the free layer cell. During *R-I* switching experiment,

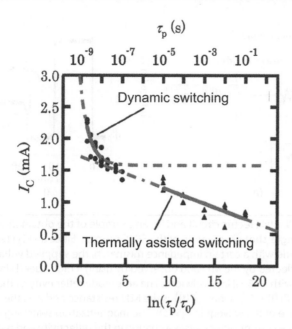

FIGURE 15.14 Pulse width (τ_p) dependence of the average critical current for the spin-injection magnetization switching (SIMS) in a CoFeB/MgO/CoFeB MTJ. The junction area, free layer thickness, resistance-area product, and MR ratio are 110 nm × 330 nm, 2 nm, 17 $\Omega\mu m^2$, and about 115%, respectively. Measurements were performed at room temperature using electric current pulses with durations ranging from 2 nsec to 100 msec. Dotted curved line and a straight line are fittings using Equation 15.21 and 15.25, respectively. (After Aoki, T. et al., *J. Appl. Phys.* 103, 103911, 2008. With permission.)

we adjusted the external field to the center of hysteresis in Figure 15.12a. However, due to magnetostatic coupling at cell edges or through rough interfaces, local fields would be non-uniform and the effective field can be different depending on the magnetization arrangement.

In our experience, the above two methods give almost the same value for I_{c0} and Δ. The advantage of the latter method, measuring the switching current distribution, is the simplicity of the measurement setup. Since a rather long pulse can be used, impedance matching in a circuit is not very important. On the other hand, when we use the former method, the pulse duration dependence of the switching current, the short pulse measurement requires impedance matching in the circuit. Otherwise, the pulse current waveform is degraded, resulting in indistinct current values.

15.3.3 High-Speed Measurements

To analyze high-speed dynamics of SIMS in MTJs, time-domain high-speed electrical observations were performed [56–58]. An example of the pulse width dependence of the switching current for a CoFeB/MgO/CoFeB magnetic nanopillar is shown in Figure 15.14 [56]. A clear transition from the dynamic switching regime to the thermally assisted switching regime can be seen.

Figure 15.15 shows the circuit used to observe the real-time switching signal (Figure 15.15a) and an example of a switching signal obtained by

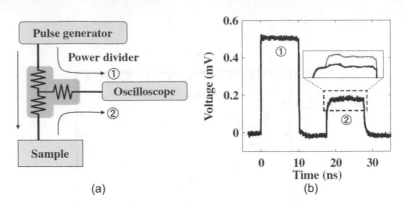

(a)   (b)

FIGURE 15.15 (a) Electric circuit and (b) an example of the obtained signal for a real-time single-shot observation of the SIMS process. Since an MTJ terminates an rf waveguide with a certain impedance mismatch, the supplied voltage pulse is partially reflected by the MTJ and observed by using a high-speed single-shot oscilloscope with a 16-GHz bandwidth. The amplitude reflectivity of the MTJ is $\left(R(\theta)-Z_0\right)/\left(R(\theta)+Z_0\right)$, where $R(\theta)$ is the MTJ resistance and Z_0 is the characteristic impedance of the waveguide (50 Ω). The magnetization switching during an application of a current pulse causes a change in the reflectivity and is observed as a jump in the reflected voltage wave, as shown in the inset in (b). In the inset gray (black) signal shows reflected pulse signal with(without) jump because of magnetization switch (P to AP) (After Tomita, H. et al., *Appl. Phys. Express* 1, 061303, 2008. With permission.)

Tomita et al. [58]. For this measurement, a CoFeB/MgO/CoFeB nanopillar was placed at the end of an rf waveguide. A 10 ns pulse from a pulse generator was supplied to the MTJ through a power divider. The pulse was partially reflected by the MTJ depending on its resistance states and was provided to a high-speed storage oscilloscope with a 16-GHz bandwidth after being made to pass through the power divider again. Figure 15.15b shows both the direct pulse signal and the reflected pulse signal. Since the pulse reflectivity is dependent on the sample resistance, a magnetization switching event can be observed as a step in the reflected signal, as shown in the inset of Figure 15.15b. To cancel the parasitic signals generated by pulse reflections from the pulse generator and the cable connections, the obtained signal was numerically divided by the signal obtained in the absence of switching; the resulting signals are shown in Figure 15.16 for a current slightly below $I_{c,0}$. The signal shows a long waiting time followed by a short transition time (about 500 psec) as expected for a thermally assisted SIMS. The waiting time seems to vary considerably. Tomita et al. [58] also analyzed 1000 single-shot data and obtained the distribution of the non-switched probability, $1-P_{sw}(t)$, which should be a simple exponential function of the elapsed time according to Néel-Brown's law. The obtained distribution was not explained by simple Néel-Brown's law but was well-fitted by considering a certain initial period in which the switching probability was negligibly small, as shown in Figure 15.17. Tomita et al. [58] called this initial period a "non-reactive time (t_{NR})" and explained that this was the time required to complete a transition from

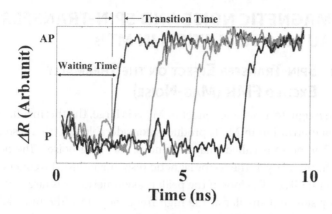

FIGURE 15.16 Real-time, single-shot observation of the spin-injection magnetization switching (SIMS) process in a CoFeB/MgO/CoFeB nanopillar. All the traces are obtained at the same current but show differences because of the probabilistic nature of switching. The current magnitude is 2.7 mA (2.4×10^7 A/cm²), which is slightly smaller than $I_{c,0}$. The signals are normalized by the signal obtained without switching (After Tomita, H. et al., *Appl. Phys. Express* 1, 061303, 2008. With permission.)

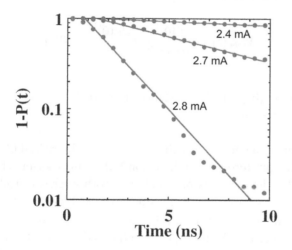

FIGURE 15.17 Non-switched probability, $1 - P_{sw}(t)$, as a function of elapsed time. The data deviate from Néel-Brown's law at the initial stages and are well-fitted by the function $1 - P_{sw}(t) = e^{-(t-t_{NR})}$, where t_{NR} is the non-reactive time. (After Tomita, H. et al., *Appl. Phys. Express* 1, 061303, 2008. With permission.)

the initial thermal equilibrium without current to a quasi-thermal equilibrium state for a finite current. Considering this point Equation 15.25 is rewritten as follows:

$$P_{sw}(t) = 1 - e^{-\frac{t-t_{NR}}{\tau_{sw}}}. \tag{15.25''}$$

15.4 MAGNETIC NOISE AND SPIN-TRANSFER AUTO-OSCILLATION IN MTJs

15.4.1 SPIN-TRANSFER EFFECT ON THE THERMALLY EXCITED FMR (MAG-NOISE)

Since the magneto-resistance effect in MTJs is large, the thermal fluctuation in the magnetization in MTJs produces considerable electric noise under a dc bias. The noise is called "magnetic noise" or "mag-noise." The noise has its peak in intensity at the ferromagnetic resonance (FMR) frequency (typically several GHz). The foot of the peak is asymmetric, it is larger at the low frequency side and smaller at the high frequency side. The most advanced magnetic read heads in HDDs require a wide signal band in order to assure a high data transfer rate, and they are subject to mag-noise. When the system noise is governed by the mag-noise, an enhancement in the MR ratio does not enhance the signal to noise ratio (S/N) because both the signal and the noise are proportional to the MR ratio. Therefore, understanding the mechanism of generation and control of mag-noise in MTJs is important [59–69].

For a given fluctuation of the relative angle between two spin moments, $\delta\theta(t)$, around its average angle θ_0, the output noise voltage is

$$V_{\text{noise}}(t) = \frac{R(\theta_0)}{R(\theta_0) + Z_0} I \left.\frac{dR(\theta)}{d\theta}\right|_{\theta=\theta_0} \delta\theta(t), \tag{15.27}$$

where

$R(\theta_0) = G^{-1}(\theta_0)$ is the resistance of the MTJ
I is the bias current

For a small excitation of the thermal noise, the linearized LLG equation, including the spin-torque term (Equation 15.14) can be adapted by replacing the external excitation fields δH_θ and δH_φ with stochastic field $H_{\text{stochastic},\theta}$ and $H_{\text{stochastic},\varphi}$, respectively. If the stochastic field has a very short correlation time and a white spectrum in the GHz region, $\delta\theta(t)$ and $V_{\text{noise}}(t)$ show a resonance type Fourier spectrum that is determined by the response function. From Equation 15.14, the power spectrum density (PSD) of $\delta\theta(t)$ is expressed as

$$PSD_{\delta\theta}(\omega) = \frac{\left(\omega_{\varphi\varphi} + \omega_{FT}\right)^2 + \left(\omega_{\theta\varphi} - \omega_{ST}\right)^2 + \omega^2}{\left(\omega^2 - \omega_0^2\right)^2 + \left(\omega\Delta\omega\right)^2} \tag{15.28}$$

$$\int d\tau \left\langle (-\gamma) H_{\text{Stochastic}}(0)(-\gamma) H_{\text{Stochastic}}(\tau) \right\rangle.$$

The PSD of the noise voltage, $V_{\text{noise}}(t)$, is obtained by multiplying

$$\frac{1}{Z_0}\left(\frac{R(\theta_0)}{R(\theta_0) + Z_0} I \left.\frac{dR(\theta)}{d\theta}\right|_{\theta=\theta_0}\right)^2 \text{ with the above equation. In the inset of}$$

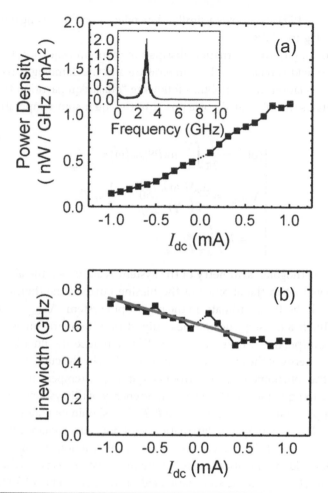

FIGURE 15.18 Rf noise properties observed in Fe/MgO/Fe single crystalline MTJs in the parallel configuration. The junction size is 220 × 420 nm². (a) Peak power density as a function of the biasing current. The power is normalized by I^2. (b) FWHM as a function of the biasing current. Inset: A typical noise power spectrum ($I = 0.8$ mA). (After Matsumoto, R. et al., *Phys. Rev. B* 80, 174405, 2009. With permission.)

Figure 15.18a, a typical "mag-noise" spectrum in Fe/MgO/Fe epitaxial MTJ is shown [62]. The peak frequency, $\omega_0/(2\pi)$, and the FWHM of the peak, $\Delta\omega/(2\pi)$, are expressed by Equation 15.15. In the linear regime, the width of the peak is a linear function of the bias voltage:

$$\Delta\omega = \alpha Tr\left[\hat{\Omega}\right] + 2\frac{T_{ST}V}{S_2}\cos\theta_0. \tag{15.29}$$

The peak width at a zero bias voltage is determined by the Gilbert damping factor, α. Since the injected spin-polarized current under a finite bias voltage causes a forced damping or anti-damping depending on the polarity of the current, the bias voltage contributes to an increase or decrease in the peak width depending on its sign. In Figure 15.18b, the linear dependence of $\Delta\omega/(2\pi)$ on the biasing current is shown for a parallel state. The peak height in the PSD also shows a similar dependence on the biasing current, as shown in Figure 15.18a.

According to the fluctuation-dissipation theorem (Equation 15.23), the stochastic field is related to the temperature and the damping constant. By using this relation, we can obtain following "quasi-equipartition theorem," which relates the integrated noise power to the "quai-spin-temperature":

$$
\left\{
\begin{aligned}
&\left\langle \delta\theta^2 \right\rangle = \frac{1}{2\pi} \int_{-\infty}^{+\infty} d\omega \, PSD_{\delta\theta}(\omega) = \frac{k_B T^*}{S_2 \omega_{\theta\theta}} \\
&\Delta E_{mag,\theta} = \frac{\mu_0 M_2 \omega_{\theta\theta}}{2(-\gamma)} \left\langle \delta\theta^2 \right\rangle = \frac{1}{2} k_B T^* \\
&T^* = T \frac{I_{c,0}}{I_{c,0} - I}
\end{aligned}
\right.
\tag{15.30}
$$

Here, the orthogonal symmetry of the system is assumed. The above equation shows that in the absence of the biasing current, the thermal energy shared with the θ–coordinate in the macrospin system, $\Delta E_{mag,\theta}$, is equal to $k_B T/2$. This is a consequence of the equipartition theorem. Since the injection of spin-polarized current can amplify or reduce the fluctuations, the magnetic energy of the free layer can be controlled by the current. In order to express this phenomenon, we introduce quasi-spin-temperature, T^*, which is defined in Equation 15.30. The magnetic energy of the system is dependent on the applied current and is equal to $k_B T^*/2$. It should be noted that in this case, the change in magnetic fluctuation does not cause a change in the S/N. Since the effects of the current on the response functions to the stochastic and signal fields are identical the S/N remains constant, as far as the current has no direct effect on the stochastic field. If there is an effect like the spin-Peltier effect, we can control the S/N by current injection.

15.4.2 SPIN-TRANSFER AUTO-OSCILLATION IN MTJs

A large current injection may result not only in magnetization switching but also in a continuous precession of magnetization. This spin-transfer oscillation was first observed in CPP-GMR nanopillars and nanocontacts (see Chapter 7 and Chapter 1, Volume 1). The maximum output power was limited to about 1 nW because of its small MR ratio. In order to obtain higher output power, it is important to use pillars with a large MR ratio and with a resistance that matches the impedance of the waveguide (usually 50 Ω). By replacing a CPP-GMR pillar with an MTJ, we can expect a significant increase in the output power [60, 61]. Deac et al. reported a large output power of 0.14 μW from a single CoFeB/MgO/CoFeB nanopillar of cross-section 70×160 nm^2 [60]. The junction was specially designed to have a high MR ratio of about 100% and a very low resistance-area product ($RA = 4$ $\Omega\mu$m^2) [70]. Figure 15.19 shows the output power spectra obtained from a CoFeB/MgO/CoFeB nanopillar. With an increase in current, a steep increase in the power of the first harmonic peak at around 6 GHz can be seen. Such MTJs showing auto-oscillation are named as "spin-torque oscillator (STO)."

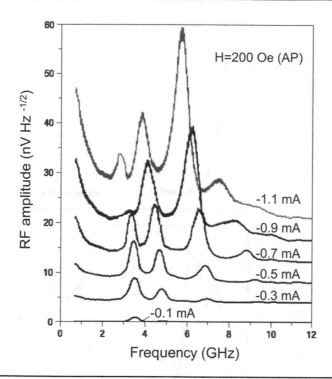

FIGURE 15.19 Rf power spectrum observed for a nanopillar prepared from a low resistance CoFeB/MgO/CoFeB MTJ with in-plane magnetization (AP state). Large rf power emission and a steep increase in the first harmonic (around 6 GHz) intensity with an increase in the injected current were observed. (After Deac, A.M. et al., *Nat. Phys.* 4, 803, 2008. With permission.)

Clearer evidence of the on-set of auto-oscillation is shown in Figure 15.20 [62]. When the injection current is less than $I_{c,0}$, an increase in the injection current results in a linear reduction in the peak width, as in the previous case (Figure 15.18b and Figure 15.20b). The threshold current ($I_{c,0}$), which is indicated by an arrow in Figure 15.20b, corresponds to the current at which the peak width reduces to zero if a leaner reduction holds until $I_{c,0}$. In practice, when the injection current is around the threshold current, there is a sudden increase in the peak width. The peak width has its maximum value slightly below $I_{c,0}$. Further increase in the injection current reduces the peak width and results in a sudden increase in the output power. These observations provide clear evidence of the threshold properties, which are in good agreement with the theory developed by Kim et al. [64]. The width of the spectral lines are, however, very wide when compared to the width of the spectral lines for the CPP-GMR nanopillars and nanocontacts. This is explained by the findings of Houssameddine et al.: they observed the time-domain oscillation signals and found spontaneous mode jumps that may significantly increase the averaged line width [69]. The mechanism of the mode jump was discussed by considering the current distribution, instability mechanism, and inhomogeneities in MTJs.

Another method for obtaining a higher output power is to synchronize the number of oscillators. Kaka et al. [71] and Mancoff et al. [72] demonstrated

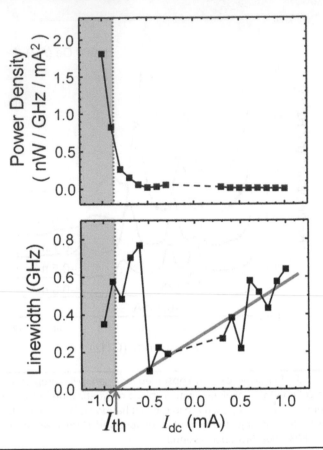

FIGURE 15.20   Rf auto-oscillation properties observed in Fe/MgO/Fe single crystalline MTJs (AP state). The junction size is 220×420 nm^2. (a) Peak power density as a function of the biasing current. The power is normalized by I^2. (b) FWHM as a function of the biasing current. (After Matsumoto, R. et al., *Phys. Rev. B* 80, 174405, 2009. With permission.)

the synchronization of two metal-based oscillators by exchange coupling. For this purpose, they developed two magnetic oscillators that were very close to each other (separation was about 100 nm). On the other hand, George et al. proposed an electric coupling among the oscillators [73]. Such a coupling would be feasible if each oscillator has sufficient power output. Here, Tamaru et al. have demonstrated excellent progress. By introducing an MTJ into a phase-locked-loop circuit, they succeeded in locking the oscillation phase using a feedback on the biasing current [74]. As a result, the STO showed an extremely narrow linewidth that was comparable to the reference oscillator.

15.5 SPIN-ORBIT TORQUES

15.5.1 Spin-Orbit Torque

The structure of the 3-terminal device in which a magnetic tunnel junction has been installed on a non-magnetic metal wire is shown in Figure 15.21 [75]. The wire is made of a heavy metal such as Ta or Pt. Owing to the

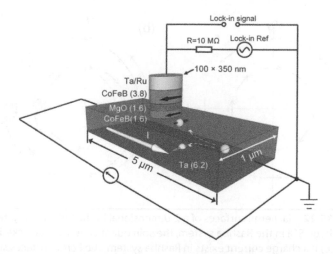

FIGURE 15.21 Structure of a 3-terminal device comprising a magnetic tunnel junction and a non-magnetic metal wire below. The electron flow inside the wire (charge current: white arrow) is converted to a vertical spin current in the wire and injected into the lower magnetic layer of the MTJ. (After Liu, L. et al., *Science* 336, 555, 2012. With permission.)

spin-asymmetry of electron scattering in heavy metals with large spin-orbit interaction (SOI), the electrons which are upward are spin-polarized as shown in Figure 15.21a. As a result, by passing an electric current through the wire, one can inject pure spin current from the wire to the lower magnetic electrode in the magnetic tunnel junction. The generation of the spin current due to such spin-dependent scattering is called the "spin-Hall effect" [76–79]. The orientation of the injected spins is in-plane and normal to the direction of the charge current. A spin-transfer torque is exerted to the lower ferromagnetic electrode in MTJ when it absorbs the spin current. In 2012, Liu et al. found a surprisingly large charge-to-spin conversion efficiency, i.e., a spin-Hall angle of 0.3 [75]. One may use such large spin current in order to switch the magnetization [75] or to conduct it toward an auto-oscillation [80, 81]. Since the spin current is induced by the SOI in this device, the generated torque is regarded as spin-orbit torque (SOT). For 3-terminal devices which utilize the SOT, one may separate the read circuit from the write circuit and thus avoid the read disturb problem.

Part of the charge current penetrates into the interfaces between the ferromagnetic and the adjacent non-magnetic layers where the inversion symmetry is broken. The electric current flowing through such interface states produces a torque with a different symmetry [82]. This torque is also called an SOT, or Rashba-torque since the torque was predicted from the Rashba Hamiltonian [83, 84]. Eigenstates of the Rashba Hamiltonian demonstrate spin-locking to the k-vectors as shown in Figure 15.22a because of a considerable SOI in the system. Here, for simplicity, we show an example of a non-magnetic system. In this system, if a charge current exists, the Fermi surfaces shift to opposite the current direction and spin accumulation takes

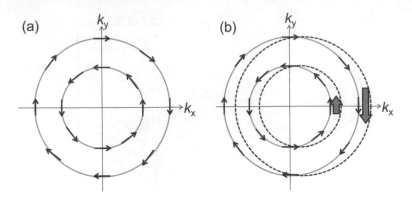

FIGURE 15.22 (a) Fermi surfaces of a 2-dimensional Rashba electron system. Due to a large SOI in the Rashba system, the spin quantization axis is locked in k-space. (b) If a charge current exists in Rashba system, the Fermi surfaces shift to oppose the current flow and an imbalance between the numbers of electrons with momenta $+k$ and $-k$ appears. The imbalance results in a spin accumulation due to spin-locking.

place (the Edelstein effect), as shown in Figure 15.22b. This accumulated non-equilibrium spin may exert a torque on the ferromagnetic spins. In order to distinguish between the two types of torques mentioned above, the former can be regarded as a damping-like torque, while the latter can be considered as a field-like torque.

15.5.2 VOLTAGE CONTROL OF MAGNETIC ANISOTROPY (VCMA)

In this section, a method of controlling magnetization without the use of an electric current, or spin current, will be shown. Instead, one may use an applied voltage to control the magnetization. Such voltage-based devices are expected to exhibit ultralow power consumption since such ultralow energy consumption has already been proven in voltage-driven CMOS circuits, in which electric field transistors (FETs) are utilized.

The spin-orbit interaction (SOI) is, in principle, an interaction between the electron's spin and the electric field. Therefore, it is expected that we may control the SOI by the application of a voltage. Indeed, the magnetization in materials with an inverse magnetostriction effect can be controlled with a voltage, by making multilayered stacks with piezoelectric materials [85]. Already, the Curie temperature and magnetic anisotropy of ferromagnetic semiconductors have been deliberately altered or controlled by an electric field effect, through the mechanism of carrier control [86, 87]. The magneto-electric effect has been observed in multiferroic and composite materials[88]. Moreover, voltage control of magnetic anisotropy (VCMA) was also observed in metallic ferromagnetic films steeped in a liquid electrolyte [89] and in a solid-state junction [90]. Further results for ferromagnetic metals include the voltage control of the Curie temperature [91]; asymmetric exchange interaction (i.e., the Dzyaloshinskii-Moriya Interaction) [92]; and magnetic domain size [93, 94]; while the large electric field, which exists under an STM tip, has made it possible to control magnetic phase transition [95] and skyrmionic state [96].

Among many other kinds of magneto-electric effects, VCMA in simple 3d transition metal ferromagnets attracts much attention because of its potential for room temperature operation and its expected unlimited endurance. In Figure 15.22a and b, the effect of voltage on the magnetic hysteresis loop of a FeCo ultrathin film is shown. The hysteresis clearly shows a transition from an in-plane to out-of-plane magnetization on the application of ±200 V at room temperature [97]. The magnitude of this effect was measured quantitatively by considering the VCMA in typical MTJs from the voltage dependence of the MR-curve [98] and from the bias voltage dependence of the voltage-driven FMR frequency [99]. The size of the VCMA was approximately 20 to 30 fJ/(Vm) and was determined by considering the voltage change in the surface magnetic anisotropy energy, K_s. This value roughly agrees with that expected from first-principle calculations [100–103]. The magnetic energy is expressed in terms of the surface anisotropy energy,

$E_{\text{mag}} = K_s(\mathcal{E})\dfrac{v}{d}\sin^2\theta$, where v is the volume of the magnetic cell, d is the ferromagnetic film thickness. θ is the angle between the film's normal and the magnetization and $\mathcal{E}$ is the electric field. From Equation 15.8, the induced effective magnetic field is,

$$\vec{H}_{\text{eff}} = \frac{1}{\mu_0 M_2} K_s \frac{v}{d}\sin 2\theta\vec{e}_0. \tag{15.8'}$$

Assuming an example case with the values, $\dfrac{dK_s}{d\mathcal{E}} = 30\text{fJ/Vm}$, $\mathcal{E} = 1$ V/nm, $\mu_0 M_2 / v = 1$ T, $d = 0.5$ nm, $\theta = \pi/4$, H_{eff} is estimated to be 60kA/m $\simeq$ 750Oe, which though not very large, has considerable size.

One of the main obstacles for VCMA as a means of magnetization switching is the fact that though the electric field may change perpendicular magnetizations to in-plane magnetizations, it does not change the orientation from up to down. To overcome this problem, short pulse voltages can be employed. Shiota et al. have reported on the observation of coherent magnetization switching induced by the application of pulse voltages to an MTJ comprising an Au (50 nm)/$Fe_{80}Co_{20}$ (0.70 nm) free layer/MgO (1.5 nm)/Fe (10 nm) epitaxial layer stack [104]. The size of their junction was 800 × 200 nm². Without a bias magnetic field and at zero bias voltage the free layer magnetization is aligned in-plane and parallel to the long axis of the junction due to the shape anisotropy of the junction.

Figure 15.24 shows LLG simulation results for the above-mentioned MTJ under a perpendicular bias field of 700 Oe. Contours of the magnetic energy are shown in Figures 15.24a, b, c. Under zero bias voltage, the magnetization stays at the energy minima, which are indicated by the red and blue circles in a and c, respectively. These points are the initial state (I. S.) and the final state (F. S.) of the switching process (see Figure 15.24d and f).

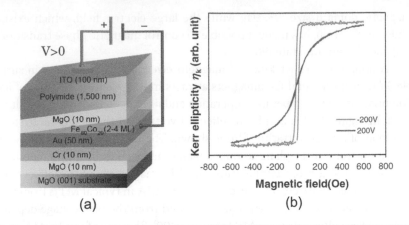

(a)   (b)

FIGURE 15.23   (a) Structure of the multilayer used to observe VCMA. (b) Under a bias voltage of +200 V (−200 V), the film shows in-plane (out-of-plane) magnetic anisotropy. (After Shiota, Y. et al., *Appl. Phys. Express* 2, 063001, 2009. With permission.)

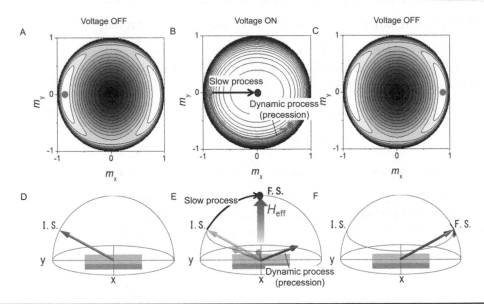

FIGURE 15.24   Simulation of dynamic magnetization switching using a short voltage pulse. A and C show stereographic-projections of magnetic energy contours under zero bias voltage. Applying a small perpendicular bias field stabilizes a tilted state of magnetization, as shown in D and F (state I. S.). The application of a voltage induces a perpendicular magnetic anisotropy, as shown in B, and generates a new energy minimum at the north pole. A slow process brings the magnetization to the lowest energy state (north pole) in a dissipative process. However, immediate voltage application causes the precession of the magnetization, as shown in B and E. A sudden cut in the voltage at F. S. causes the magnetization to stop at F. S., as shown in C and F. (After Shiota, Y. et al., *Nat. Mater.* 11, 39, 2012. With permission.)

Application of a voltage creates a new polar energy minimum position (see Figure 15.24b). If the voltage was increased slowly, the magnetization would follow the trajectory indicated as "slow process" to reach the polar position. This process naturally includes some energy dissipation because of magnetic damping. However, if the voltage were applied instantaneously,

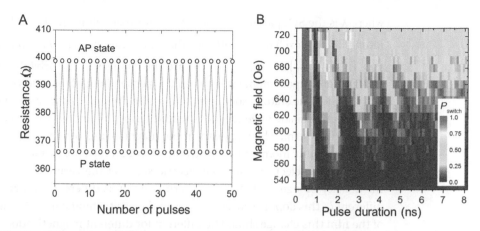

FIGURE 15.25 Experimental dynamic magnetization switching using a short voltage pulse. (a) Repeated application of pulses causes toggle-type switching of magnetization. (b) Switching probabilities as a function of bias field and pulse duration. Oscillation of the switching probability as a function of the pulse duration is a consequence of precessional switching. (After Shiota, Y. et al., *Nat. Mater.* 11, 39, 2012. With permission.)

the magnetization cannot lose its magnetic energy immediately. Therefore, it would follow the iso-energy contour marked "dynamic process." This is a precession of the magnetization. After a half period of the precession, the magnetization reaches the F. S. At this same moment, if we turn off the applied voltage, this F. S. point becomes an energy minimum again and the magnetization precession stops at this point. If we then apply the next pulse, the magnetization may be switched back from the F. S. to the I. S. This process demonstrates the dynamic switching of the magnetization by the application of a voltage pulse.

Experimental results of such a process are shown in Figure 15.25. The dynamic voltage pulse driven switching simulated above was experimentally demonstrated by applying short pulses of 0.55 ns duration and –1.5 V height. The applied voltage is much smaller than that shown in the previous experiment of Figure 15.23 since the voltage was applied through the ultrathin MgO barrier layer (1.5 nm). Figure 15.25b shows the switching probability as a function of the pulse duration and bias field strength. In the figure, a clear oscillation indicative of precession-driven switching can be seen. Such voltage-driven dynamic switching was also observed in an MTJ with perpendicular magnetization [105]. Shiota et al. demonstrated that by employing a perpendicularly magnetized MTJ and an even larger VCMA, a very small writing error of the order of 10^{-5} could be achieved [106].

Laan has discussed the second order perturbation energy of the SOI in a ferromagnetic system with C4v symmetry and has derived following anisotropy energy [107]:

$$\Delta E \cong \frac{\lambda}{4\mu_B}\left(\Delta m_{L,\downarrow} - \Delta m_{L,\uparrow}\right) - \frac{21}{2\mu_B}\frac{\lambda^2}{E_{ex}}\Delta m_T, \qquad (15.31)$$

where λ is the SOI parameter, μ_B is the Bohr magneton, and E_{ex} is the magnitude of the exchange splitting. The quantities, $\Delta m_{L,s} \left(= m_{L,s}^{\perp} - m_{L,s}^{\parallel} \right)$ and $\Delta m_T \left(= m_T^{\perp} - m_T^{\parallel} \right) \cong -3\mu_B Q_{zz} \langle S \rangle / 7\hbar$, are the changes in the orbital moment and the spin-dipole moment between the perpendicularly magnetized state and the in-plane state, respectively. Q_{zz} is the dimensionless zz-component of an electric quadrupole tensor. The first term in Equation 15.31 is the so-called Bruno term, which appears in the magnetic anisotropy appears if the orbital magnetic moment is anisotropic. Through the application of voltage, one may dope electrons into the electric states at the Fermi surface. Since the density of states (DOS) at the Fermi energy is different for states with different l_z, this doping causes changes in the orbital moment. At the surface of the film this change should be different for different magnetization directions, and therefore, one may obtain the VCMA.

The second term in Equation 15.31 contributes to the magnetic anisotropy if the magnetic dipole moment is anisotropic. Miwa et al. reported that the term is large for atoms with a large SOI, and for certain induced moments such as those found in Pt [108]. The spin dipole can be approximated as a product between the electric quadrupole and spin moment. Since the electric field at the surface is extremely inhomogeneous due to the strong shielding effect present in the metals, such an inhomogeneous electric field may couple with the electric quadrupole, and one may thus control the spin dipole at the surface by the application of a voltage. This is the second mechanism used in order to achieve the VCMA.

Much effort has been made in order to increase the effect of VCMA. It has shown that the inclusion of heavy metals like Pt [108], Pd [109], and Ir [110] results in an increased voltage effect of up to 300 fJ/(Vm) at room temperature. Such contributions can be considered as an increase in the orbital moment of the $3d$ transition metal due to adjacent heavy metal atoms or as a contribution from the second mechanism mentioned here. Duan et al. have also predicted a large VCMA effect at the ferromagnetic metal/ferroelectric insulator interface[111]. Similar kinds of attempts were performed by Bi et al. [112] and Bauer et al. [113]. Their results showed an extremely large VCMA that corresponds to more than 5000 fJ/(Vm) in a GdO/Co/Pt system. Here, the application of a voltage controls the oxygen migration in the multilayer system, resulting in such a huge VCMA effect. However, the response speed is as slow as 25 ms at room temperature, since it requires thermal activation.

APPENDIX 15A

15A.I   Elliptical Nanopillar with an In-Plane Magnetization

For the in-plane magnetization case, we assume that $\vec{e}_1$ is fixed parallel to the easy axis of the free layer ($\vec{e}_1 = (0, 0, 1)$) and an external field is applied in-plane ($\vec{H}_{ext} = -H_{ext}(\sin\theta_{ext}, 0, \cos\theta_{ext})$). Here, the coordinate system in Figure 15.1 is used. Then, $\vec{e}_2^{(0)}$ is aligned almost opposite to the external-field direction, $\vec{e}_2^{(0)} = (\sin\theta_0,\, 0,\, \cos\theta_0)$. The magnetic energy and the effective field are

$$E_{mag} = \mu_0 M\left(\vec{H}_{ext} \cdot \vec{e}_2^{(0)} + \frac{1}{2}H_{//,H}\left(\hat{e}_x \cdot \vec{e}_2^{(0)} \right)^2 \right.$$

$$\left. + \frac{1}{2}H_{\perp}'\left(\hat{e}_y \cdot \vec{e}_2^{(0)} \right)^2 + \frac{1}{2}H_{//,E}\left(\hat{e}_z \cdot \vec{e}_2^{(0)} \right)^2 \right)$$

$$\vec{H}_{eff}^{(0)} = \frac{1}{\mu_0 M}\frac{\partial E_{dipole\_and\_anisotropy}}{\partial \hat{e}_2^{(0)}} \qquad (A1)$$

$$= \vec{H}_{ext} + \begin{pmatrix} H_{//,H} & 0 & 0 \\ 0 & H_{\perp}' & 0 \\ 0 & 0 & H_{//,E} \end{pmatrix} \vec{e}_2^{(0)},$$

where $H_{//,H}$, $H_{\perp}'$, and $H_{//,E}$ are the demagnetization field for the x-, y-, and z-directions, respectively.

From Equation 15.13, the $\hat{\Omega}$ matrix elements in the $\left(x^1, x^2\right)$ coordinate system are calculated as follows:

$$\hat{\Omega} = \left(\partial_i\partial_j E_{mag} \right) \cong -\,\gamma H_{ext} \sin\theta_{ext} \begin{pmatrix} S_2 \sin\theta_0 & 0 \\ 0 & \dfrac{1}{S_2 \sin^3\theta_0} \end{pmatrix}$$

$$-\gamma\begin{pmatrix} S_2\left(H_{\perp} - H_{\parallel} \right)\sin^2\theta_0 & 0 \\ 0 & -\dfrac{1}{S_2}H_{\parallel} \end{pmatrix}, \qquad (A2)$$

where $H_{\parallel} \equiv H_{//,H} - H_{//,E}$ and $H_{\perp} = H_{\perp}' - H_{//,E}$ are in-plane and out-of-plane effective anisotropy fields, respectively. Be noticed that the expression is not applicable for $\theta_0 = 0, \pi$ where (x^1, x^2) coordinate has singular behavior. For an application of large in-plane field toward $-x$-direction, the resonance frequency is estimated to be $\omega_0^2 = \det\left[\hat{\Omega}\right] \cong \gamma^2\left(H_{ext} + H_{\perp} - H_{\parallel} \right)\left(H_{ext} + H_{\parallel} \right)$. This is a Kittel's formula.

15A.II   OUT-OF-PLANE MAGNETIZATION CASE

For a cylindrical pillar with the out-of-plane magnetization case, we assume that $\vec{e}_1$ is fixed perpendicular to the film plane ($\vec{e}_1 = (0,\, 0,\, 1)$) and that an external field is applied with an angle ($\vec{H}_{ext} = -H_{ext}(\sin\theta_{ext},\, 0,\, \cos\theta_{ext})$). Here, the x-axis and y-axis are taken in-plane and z-axis perpendicular to the film plane. The free-layer magnetization, $\vec{e}_2^{(0)} = (\sin\theta_0,\, 0,\, \cos\theta_0)$, is assumed to have a perpendicular easy axis and to be tilted because of the in-plane component in the external field. The magnetic energy and the effective field are,

$$E_{\text{mag}} = \mu_0 M \left(\vec{H}_{\text{ext}} \cdot \vec{e}_2^{(0)} + \frac{1}{2} H_{\parallel}'' \left(\left(\hat{e}_x \cdot \vec{e}_2^{(0)} \right)^2 + \left(\hat{e}_y \cdot \vec{e}_2^{(0)} \right)^2 \right) \right.$$

$$\left. + \frac{1}{2} \left(H_{\perp}'' - H_u'' \right) \left(\hat{e}_z \cdot \vec{e}_2^{(0)} \right)^2 \right)$$

$$\vec{H}_{\text{eff}}^{(0)} = \frac{1}{\mu_0 M} \frac{\partial E_{\text{dipole\_and\_anisotropy}}}{\partial \hat{s}_2^{(0)}} \tag{A3}$$

$$= \vec{H}_{\text{ext}} + \begin{pmatrix} H_{\parallel}'' & 0 & 0 \\ 0 & H_{\parallel}'' & 0 \\ 0 & 0 & H_{\perp}'' - H_u'' \end{pmatrix} \vec{e}_2^{(0)},$$

where $H_{\parallel}''$, $H_{\perp}''$ and H_u'' are in-plane demagnetization field, out-plane demagnetization field, and uniaxial anisotropy fields, respectively.

From Equation 15.13, the $\hat{\Omega}$ matrix elements in (x^1, x^2) the coordinate system are calculated as follows:

$$\hat{\Omega} = -\gamma H_{\text{ext}} \sin \theta_{\text{ext}} \begin{pmatrix} S \sin \theta_0 & 0 \\ 0 & \dfrac{1}{S \sin^3 \theta_0} \end{pmatrix} - \gamma \begin{pmatrix} 0 & 0 \\ 0 & -\dfrac{1}{S} H_u \end{pmatrix}, \tag{A4}$$

where $H_u = H_u'' + H_{\parallel}'' - H_{\perp}''$ is the effective uniaxial anisotropy field. It should be noticed that the expression is not applicable for $\theta_0 = 0, \pi$ where (x^1, x^2) coordinate has singular behavior. For an application of large in-plane field toward a $-x$ direction, the resonance frequency is estimated to be $\omega_0^2 = \det\left[\hat{\Omega}\right] \cong \gamma^2 H_{\text{ext}} \left(H_{\text{ext}} - H_u \right)$. This is a Kittel's formula.

REFERENCES

1. J. C. Slonczewski, Current-driven excitation of magnetic multilayers, *J. Magn. Magn. Mater.* **159**, L1–L7 (1996).
2. L. Berger, Emission of spin waves by a magnetic multilayer traversed by a current, *Phys. Rev. B* **54**, 9353–9358 (1996).
3. M. Tsoi, A. G. M. Jansen, J. Bass et al., Excitation of a magnetic multilayer by an electric current, *Phys. Rev. Lett.* **80**, 4281 (1998); ibid. **81**, 492 (1998).
4. E. B. Myers, D. C. Ralph, J. A. Katine, R. N. Louie, and R. A. Buhrman, Current-induced switching of domains in magnetic multilayer devices, *Science* **285**, 867–870 (1999).
5. J. A. Katine, F. J. Albert, R. A. Buhrman, E. B. Myers, and D. C. Ralph, Current-driven magnetization reversal and spin-wave excitations in Co/Cu/Co pillars, *Phys. Rev. Lett.* **84**, 3149–3152 (2000).
6. J. Z. Sun, Current-driven magnetic switching in manganite trilayer junctions, *J. Magn. Magn. Mat.* **202**, 157–162 (1999).

7. Y. Huai, F. Albert, P. Nguyen, M. Pakala, and T. Valet, Observation of spin-transfer switching in deep submicronsized and low-resistance magnetic tunnel junctions, *Appl. Phys. Lett.* **84**, 3118–3120 (2004).

8. H. Kubota, A. Fukushima, Y. Ootani et al., Evaluation of spin-transfer switching in CoFeB/MgO/CoFeB magnetic tunnel junctions, *Jpn. J. Appl. Phys.* **44**, L1237–L1240 (2005).

9. Z. Diao, D. Apalkov, M. Pakala, Y. Ding, A. Panchula, and Y. Huai, Dependence of spin-transfer switching current on free layer thickness in Co–Fe–B/MgO/Co–Fe–B magnetic tunnel junctions, *Appl. Phys. Lett.* **87**, 232502 (2005).

10. D. Chiba, Y. Sato, T. Kita, F. Matsukura, and H. Ohno, Current-driven magnetization reversal in a ferromagnetic semiconductor (Ga,Mn)As/GaAs/(Ga,Mn)As tunnel junction, *Phys. Rev. Lett.* **93**, 216602 (2004).

11. J. C. Slonczewski, Conductance and exchange coupling of two ferromagnets separated by a tunneling barrier, *Phys. Rev. B* **39**, 6995–7002 (1989).

12. J.C. Slonczewski, Currents, torques, and polarization factors in magnetic tunnel junctions, *Phys. Rev. B* **71**, 024411 (2005).

13. D. M. Edwards, F. Federici, J. Mathon, and A. Umerski, Self-consistent theory of current-induced switching of magnetization, *Phys. Rev. B* **71**, 054407 (2005).

14. I. Theodonis, N. Kioussis, A. Kalitsov, M. Chshiev, and W. H. Butler, Anomalous bias dependence of spin torque in magnetic tunnel junctions, *Phys. Rev. Lett.* **97**, 237205 (2006).

15. S. Zhang and Z. Li, Roles of nonequilibrium conduction electrons on the magnetization dynamics of ferromagnets, *Phys. Rev. Lett.* **93**, 127204 (2004).

16. A. Thiaville, Y. Nakatani, J. Miltat and Y. Suzuki, Micromagnetic understanding of current-driven domain wall motion in patterned nanowires, *Europhys. Lett.* **69**, 990 (2005).

17. S.E. Barnes and S. Maekawa, Current-spin coupling for ferromagnetic domain walls in fine wires, *Phys. Rev. Lett.* **95**, 107204 (2005).

18. G. Tatara and H. Kohno, Theory of current-driven domain wall motion: Spin transfer versus momentum transfer, *Phys. Rev. Lett.* **92**, 086601 (2004).

19. A. A. Tulapurkar, Y. Suzuki, A. Fukushima, H. Kubota, H. Maehara, K. Tsunekawa, D. D. Djayaprawira, N. Watanabe, and S. Yuasa, Spin-torque diode effect in magnetic tunnel junctions, *Nature* **438**, 339 (2005).

20. J. C. Sankey, P. M. Braganca, A. G. F. Garcia, I. N. Krivorotov, R. A. Buhrman, and D. C. Ralph, Spin-transfer-driven ferromagnetic resonance of individual nanomagnets, *Phys. Rev. Lett.* **96**, 227601 (2006).

21. J. C. Sankey, Y.-T. Cui, J. Z. Sun, J. C. Slonczewski, R. A. Buhrman, and D. C. Ralph, Measurement of the spin-transfer-torque vector in magnetic tunnel junctions, *Nat. Phys.* **4**, 67–71 (2008).

22. H. Kubota, A. Fukushima, K. Yakushiji, et al., Quantitative measurement of voltage dependence of spin-transfer torque in MgO-based magnetic tunnel junctions, *Nat. Phys.* **4**, 37–41 (2008).

23. C. Wang, Y. T. Cui, J. Z. Sun, J. A. Katine, R. A. Buhrman, and D. C. Ralph, Bias and angular dependence of spin-transfer torque in magnetic tunnel junctions, *Phys. Rev. B* **79**, 224416 (2009).

24. Y. Suzuki and H. Kubota, Spin-torque diode effect and its Application, *J. Phys. Soc. Jpn.* **77**, 031002 (2008).

25. J. Z. Sun, Spin-current interaction with a monodomain magnetic body: A model study, *Phys. Rev. B* **62**, 570 (2000).

26. M. D. Stiles and J. Miltat, Spin dynamics in confined magnetic structures III, in *Topics in Applied Physics*, Vol. 101, B. Hillebrands and A. Thiaville (Eds.), Springer, Heidelberg, pp. 225–308, 2006.

27. J. Miltat, G. Albuquerque, A. Thiaville, C. Vouille, Spin transfer into an inhomogeneous magnetization distribution, *J. Appl. Phys.* **89**, 6982–6984 (2001).

28. K.-J. Lee, A. Deac, O. Redon, J.-P. Nozières, and B. Dieny, Excitations of incoherent spin-waves due to spin-transfer torque, *Nat. Mater.* **3**, 877 (2004).

29. S. Yuasa, T. Nagahama, A. Fukushima, Y. Suzuki, and K. Ando, Giant room-temperature magnetoresistance in single-crystal Fe/MgO/Fe magnetic tunnel junctions, *Nat. Mater.* **3**, 868–871 (2004).

30. S. S. P. Parkin, C. Kaiser, A. Panchula, P. M. Rice, B. Hughes, M. Samant, and S. H. Yang, Giant tunneling magnetoresistance at room temperature with MgO (100) tunnel barriers, *Nat. Mater.* **3**, 862–867 (2004).

31. S. Miwa, S. Ishibashi, H. Tomita, et al., Highly sensitive nanoscale spin-torque diode, *Nat. Mater.* **13**, 50–56 (2014).

32. C. Heiliger and M. D. Stiles, Ab initio studies of the spintransfer torque in magnetic tunnel junctions, *Phys. Rev. Lett.* **100**, 186805 (2008).

33. T. Devolder, J. V. Kim, C. Chappert, et al., Direct measurement of current-induced field like torque in magnetic tunnel junctions, *J. Appl. Phys.* **105**, 113924 (2009).

34. Z. Li, S. Zhang, Z. Diao et al., Perpendicular spin torques in magnetic tunnel junctions, *Phys. Rev. Lett.* **100**, 246602 (2008).

35. T. Maruyama, Y. Shiota, T. Nozaki, et al., Large voltage-induced magnetic anisotropy change in a few atomic layers of iron, *Nat. Nanotechnol.* **4**, 158 – 161 (2009).

36. T. Nozaki, Y. Shiota, S. Miwa, et al., Electric-field-induced ferromagnetic resonance excitation in an ultrathin ferromagnetic metal layer, *Nat. Phys.* **8**, 491–496 (2012).

37. K. Yagami, A. A. Tulapurkar, A. Fukushima, Y. Suzuki, Low-current spin-transfer switching and its thermal durability in a low-saturation-magnetization nanomagnet, *Appl. Phys. Lett.* **85**, 5634–5636 (2004).

38. L. Berger, Multilayer configuration for experiments of spin precession induced by a dc current, *J. Appl. Phys.* **93**, 7693 (2003).

39. Y. Huai, M. Pakala, Z. Diao, and Y. Ding, Spin transfer switching current reduction in magnetic tunnel junction based dual spin filter structures, *Appl. Phys. Lett.* **87**, 222510 (2005).

40. S. Mangin, D. Ravelosona, J. A. Katine, M. J. Carey, B. D. Terris, and E. E. Fullerton, Current-induced magnetization reversal in nanopillars with perpendicular anisotropy, *Nat. Mater.* **5**, 210–215 (2006).

41. S. Yakata, H. Kubota, Y. Suzuki et al., Influence of perpendicular magnetic anisotropy on spin-transfer switching current in CoFeB/MgO/CoFeB magnetic tunnel junctions, *J. Appl. Phys.* **105**, 07D131 (2009).

42. T. Nagase, K. Nishiyama, M. Nakayama et al., Spin transfer torque switching in perpendicular magnetic tunnel junctions with Co based multilayer, C1.00331, *American Physics Society March Meeting*, New Orleans, LA. 2008.

43. S. Ikeda, K. Miura, H. Yamamoto, et al., A perpendicular-anisotropy CoFeB–MgO magnetic tunnel junction, *Nat. Mater.* **9**, 721–724 (2010).

44. R. H. Koch, J. A. Katine, and J. Z. Sun, Time-resolved reversal of spin-transfer switching in a nanomagnet, *Phys. Rev. Lett.* **92**, 088302 (2004).

45. Z. Li and S. Zhang, Thermally assisted magnetization reversal in the presence of a spin-transfer torque, *Phys. Rev. B* **69**, 134416 (2004).

46. W. Coffey, Yu. P. Kalmykov and J. T. Waldron, *The Langevin Equation: With Applications to Stochastic Problems in Physics, Chemistry and Electrical Engineering*, 2nd edn, World Scientific Series in Contemporary Chemical Physics, Vol. 14, World Scientific, Singapore, 2006.

47. W. F. Brown, Jr., Thermal fluctuations of a single-domain particle, *Phys. Rev.*, **130**, 1677 (1963).

48. D. M. Apalkov, P. B. Visscher, Spin-torque switching: Fokker–Planck rate calculation, *Phys. Rev. B* **72**, 180405(R), 274–275 (2005).

49. D. M. Apalkov, P. B. Visscher, Slonczewski, Spin-torque as negative damping: Fokker–Planck computation of energy distribution, *J. Magn. Magn. Mater.* **286**, 370 (2005).

50. T. Taniguchi, Y. Utsumi, M. Marthaler, D. S. Golubev, and H. Imamura, Spin torque switching of an in-plane magnetized system in a thermally activated region, *Phys. Rev. B* **87**, 054406 (2013).

51. M. P. Sharrock, Time dependence of switching fields in magnetic recording media, *J. Appl. Phys.* **76**, 6413–6418 (1994).
52. K. Yagami, A. A. Tulapurkar, A. Fukushima, and Y. Suzuki, Low-current spin-transfer switching and its thermal durability in a low-saturation-magnetization nanomagnet, *Appl. Phys. Lett.* **85**, 5634–5636 (2004).
53. M. Pakala, Y. M. Huai, T. Valet, Y. F. Ding, and Z. T. Diao, Critical current distribution in spin-transfer-switched magnetic tunnel junctions, *J. Appl. Phys.* **98**, 056107 (2005).
54. M. Morota, A. Fukushima, H. Kubota, K. Yakushiji, S. Yuasa, and K. Ando, Dependence of switching current distribution on current pulse width of current-induced magnetization switching in MgO-based magnetic tunnel junction, *J. Appl. Phys.* **103**, 3 (2008).
55. H. Kubota, Y. Suzuki, and S. Yuasa, Development of high density MRAM with spin-torque writing, *Oyo Buturi* **78**, 232 (2009).
56. T. Aoki, Y. Ando, D. Watanabe, M. Oogane, and T. Miyazaki, Spin transfer switching in the nanosecond regime for CoFeB/MgO/CoFeB ferromagnetic tunnel junctions, *J. Appl. Phys.* **103**, 103911 (2008).
57. T. Devolder, J. Hayakawa, K. Ito, et al., Single-shot timeresolved measurements of nanosecond-scale spin-transfer induced switching: Stochastic versus deterministic aspects, *Phys. Rev. Lett.* **100**, 057206 (2008).
58. H. Tomita, K. Konishi, T. Nozaki, et al., Single-shot measurements of spin-transfer switching in CoFeB/MgO/CoFeB magnetic tunnel junctions, *Appl. Phys. Express* **1**, 061303 (2008).
59. S. Petit, C. Baraduc, C. Thirion, et al., Spin-torque influence on the high-frequency magnetization fluctuations in magnetic tunnel junctions, *Phys. Rev. Lett.* **98**, 077203 (2007).
60. A. M. Deac, A. Fukushima, H. Kubota, et al., Bias-driven high-power microwave emission from MgO-based tunnel magnetoresistance devices, *Nat. Phys.* **4**, 803 (2008).
61. D. Houssameddine, S. H. Florez, J. A. Katine et al., Spin transfer induced coherent microwave emission with large power from nanoscale MgO tunnel junctions, *Appl. Phys. Lett.* **93**, 022505 (2008).
62. R. Matsumoto, A. Fukushima, K. Yakushiji, et al., Spintorque-induced switching and precession in fully epitaxial Fe/MgO/Fe magnetic tunnel junctions, *Phys. Rev. B* **80**, 174405 (2009).
63. J.-V. Kim, Stochastic theory of spin-transfer oscillator linewidths, *Phys. Rev. B* **73**, 174412 (2006).
64. J.-V. Kim, Q. Mistral, C. Chappert, V. S. Tiberkevich, and A. N. Slavin, Line shape distortion in a nonlinear auto-oscillator near generation threshold: Application to spintorque nano-oscillators, *Phys. Rev. Lett.* **100**, 167201 (2008).
65. J.-V. Kim, V. Tiberkevich, and A. N. Slavin, Generation linewidth of an auto-oscillator with a nonlinear frequency shift: Spin-torque nano-oscillator, *Phys. Rev. Lett.* **100**, 017207 (2008).
66. V. S. Tiberkevich, A. N. Slavin, J.-V. Kim, Temperature dependence of nonlinear auto-oscillator linewidths: Application to spin-torque nano-oscillators, *Phys. Rev. B* **78**, 092401 (2008).
67. A. N. Slavin and P. Kabos, Approximate theory of microwave generation in a current-driven magnetic nanocontact magnetized in an arbitrary direction, *IEEE Trans. Magn.* **41**, 1264 (2005).
68. V. Tiberkevich, A. Slavin, and J.-V. Kim, Approximate theory of microwave generation in a current-driven magnetic nanocontact magnetized in an arbitrary direction, microwave power generated by a spin-torque oscillator in the presence of noise, *Appl. Phys. Lett.* **91**, 192506 (2007).
69. D. Houssameddine, U. Ebels, B. Dieny et al., Temporal coherence of MgO based magnetic tunnel junction spin torque oscillators, *Phys. Rev. Lett.* **102**, 257202 (2009).

70. Y. Nagamine, H. Maehara, K. Tsunekawa et al., Ultralow resistance-area product of 0.4 Ω (μm)2 and high magnetoresistance above 50% in CoFeB/MgO/CoFeB magnetic tunnel junctions, *Appl. Phys. Lett.* **89**, 162507 (2006).

71. S. Kaka, M. R. Pufall, W. H. Rippard, T. J. Silva, S. E. Russek, and J. A. Katine, Mutual phase-locking of microwave spin torque nano-oscillators, *Nature* **437**, 389 (2005).

72. F. B. Mancoff, N. D. Rizzo, B. N. Engel, and S. Tehrani, Phase-locking in double-point-contact spin-transfer devices, *Nature* **437**, 393 (2005).

73. B. Georges, J. Grollier, V. Cros, and A. Fert, Impact of the electrical connection of spin transfer nano-oscillators on their synchronization: An analytical study, *Appl. Phys. Lett.* **92**, 232504 (2008).

74. S. Tamaru, H. Kubota, K. Yakushiji, S. Yuasa, and A. Fukushima, Extremely coherent microwave emission from spin torque oscillator stabilized by phase locked loop, *Sci. R.* **5**, 18134 (2015).

75. L. Liu, C.-F. Pai, Y. Li, H. W. Tseng, D. C. Ralph, and R. A. Buhrman, Spin-torque switching with the giant spin Hall effect of tantalum, *Science* **336**, 555 (2012).

76. M. I. Dyakonov and V. I. Perel, Current-induced spin orientation of electrons in semiconductors, *Phys. Lett. A.* **35**, 459 (1971).

77. Y. Kato, R. C. Myers, A. C. Gossard, and D. D. Awschalom, Observation of the spin Hall effect in semiconductors, *Science* **306**, 1910 (2004).

78. S. O. Valenzuela and M. Tinkham, Direct Electronic measurement of the spin Hall effect, *Nature* **442**, 176 (2006).

79. T. Kimura, Y. Otani, T. Sato, S. Takahashi, and S. Maekawa, Room-temperature reversible spin Hall effect, *Phys. Rev. Lett.* **98**, 156601(2007).

80. V. E. Demidov, S. Urazhdin, H. Ulrichs et al., Magnetic nano-oscillator driven by pure spin current, *Nat. Mater.* **11**, 1028 (2012).

81. L. Liu, C.-F. Pai, D. C. Ralph, and R. A. Buhrman, Magnetic oscillations driven by the spin Hall effect in 3-terminal magnetic tunnel junction devices, *Phys. Rev. Lett.* **109**, 186602 (2012).

82. I. M. Miron, K. Garello, G. Gaudin et al., Perpendicular switching of a single ferromagnetic layer induced by in-plane current injection, *Nature* **476**, 189 (2011).

83. V. M. Edelstein, Spin polarization of conduction electrons induced by electric current in two-dimensional asymmetric electron system, *Solid State Communications*, **73**, 233 (1990).

84. A. Manchon S. Zhang, Theory of spin torque due to spin-orbit coupling, *Phys. Rev. B* **79**, 094422 (2009).

85. V. Novosad, Y. Otani, A. Ohsawa et al., Novel magnetostrictive memory device, *J. Appl. Phys.* **87**, 6400 (2000).

86. H. Ohno, D. Chiba, F. Matsukura et al., Electric-field control of ferromagnetism, *Nature* **408**, 944 (2000).

87. D. Chiba, M. Yamanouchi, F. Matsukura, and H. Ohno, Electrical manipulation of magnetization reversal in a ferromagnetic semiconductor, *Science* **301**, 943 (2003).

88. For example, W. Eerenstein, N. D. Mathur, and J. F. Scott, Multiferroic and magnetoelectric materials, *Nature* **442**, 759 (2006) and references therein.

89. M. Weisheit, S. Fahler, A. Marty, Y. Souche, C. Poinsignon, and D. Givord, Electric field-induced modification of magnetism in thin-film ferromagnets, *Science* **315**, 349 (2007).

90. T. Maruyama, Y. Shiota, T. Nozaki et al., Large voltage-induced magnetic anisotropy change in a few atomic layers of iron, *Nat. Nanotechnol.* **4**, 158 (2009).

91. D. Chiba, S. Fukami, K. Shimamura, N. Ishiwata, K. Kobayashi, and T. Ono, Electrical control of the ferromagnetic phase transition in cobalt at room temperature, *Nat. Mater.* **10**, 853 (2011).

92. K. Nawaoka, S. Miwa, Y. Shiota, N. Mizuochi and Y. Suzuki, Voltage induction of interfacial Dzyaloshinskii–Moriya interaction in Au/Fe/MgO artificial multilayer, *Appl. Phys. Exp.* **8**, 063004 (2015).

93. F. Ando, H. Kakizakai, T. Koyama et al., Modulation of the magnetic domain size induced by an electric field, *Appl. Phys. Lett.* **109**, 022401 (2016).

94. T. Dohi, S. Kanai, A. Okada, F. Matsukura, and H. Ohno, Effect of electric-field modulation of magnetic parameters on domain structure in MgO/CoFeB, *AIP Adv.* **6**, 075017 (2016).

95. L. Gerhard, T. K. Yamada, T. Balashov et al., Magnetoelectric coupling at metal surfaces, *Nat. Nanotechnol.* **5**, 792 (2010).

96. N. Romming, C. Hanneken, M. Menzel et al., Writing and deleting single magnetic skyrmions, *Science* **341**, 636–639 (2016).

97. Y. Shiota, T. Maruyama, T. Nozaki, T. Shinjo, M. Shiraishi, and Y. Suzuki, Voltage-assisted magnetization switching in ultrathin Fe80Co20 alloy layers, *Appl. Phys. Express* **2**, 063001 (2009).

98. T. Nozaki, Y. Shiota, M. Shiraishi, T. Shinjo, and Y. Suzuki, Voltage-induced perpendicular magnetic anisotropy change in magnetic tunnel junctions, *Appl. Phys. Lett.* **96**, 022506 (2010).

99. T. Nozaki, Y. Shiota, S. Miwa et al., Electric-field-induced ferromagnetic resonance excitation in an ultrathin ferromagnetic metal layer, *Nat. Phys.* **8**, 491 (2012), and its supplementary information.

100. X. Nie and S. Bluegel, Electric field for the change of the magnetization in thin films, patented at the patent office Munich Amtl. Az. 19841034.4, 2000. European Patent Nr. 1099217.

101. C.-G. Duan, J. P. Velev, R. F. Sabirianov et al., Surface Magnetoelectric effect in ferromagnetic metal films, *Phys. Rev. Lett.* **101**, 137201 (2008).

102. K. Nakamura, R. Shimabukuro, Y. Fujiwara, T. Akiyama, T. Ito, and A. J. Freeman, Giant Modification of the magnetocrystalline anisotropy in transition-metal monolayers by an external electric field, *Phys. Rev. Lett.* **102**, 187201 (2009).

103. M. Tsujikawa and T. Oda, Finite electric field effects in the large perpendicular magnetic anisotropy surface Pt/Fe/Pt(001): A first-principles study, *Phys. Rev. Lett.* **102**, 247203 (2009).

104. Y. Shiota, T. Nozaki, F. Bonell, S. Murakami, T. Shinjo, and Y. Suzuki, Induction of coherent magnetization switching in a few atomic layers of FeCo using voltage pulses, *Nat. Mater.* **11**, 39 (2012).

105. S. Kanai, M. Yamanouchi, S. Ikeda, Y. Nakatani, F. Matsukura, and H. Ohno, Electric field-induced magnetization reversal in a perpendicular-anisotropy CoFeB-MgO magnetic tunnel junction, *Appl. Phys. Lett.* **101**, 122403 (2012).

106. Y. Shiota, T. Nozaki et al., Evaluation of write error rate for voltage-driven dynamic magnetization switching in magnetic tunnel junctions with perpendicular magnetization, *Appl. Phy. Exp.* **9**, 013001 (2016), and private communication.

107. G. van der Laan, Microscopic origin of magnetocrystalline anisotropy in transition metal thin films, *J. Phys.: Condens. Matter* **10**, 3239 (1998).

108. S. Miwa, M. Suzuki, M. Tsujikawa et al., Voltage controlled interfacial magnetism through platinum orbits, *Nat. Commun.* **8**, 15848 (2017).

109. A. Obinata, Y. Hibino, D. Hayakawa et al., Electric-field control of magnetic moment in Pd, *Sci. Rep.* **5**, 14303 (2015).

110. T. Nozaki, private communication.

111. C.-G. Duan, J. P. Velev, R. F. Sabirianov, W. N. Mei, S. S. Jaswal, and E. Y. Tsymbal, Tailoring magnetic anisotropy at the ferromagnetic/ferroelectric interface, *Appl. Phys. Lett.* **92**, 122905 (2008).

112. C. Bi, Y. Liu, T. Newhouse-Illige et al., Reversible Control of Co magnetism by voltage-induced oxidation, *Phys. Rev. Lett.* **113**, 267202 (2014).

113. U. Bauer, L. Yao, A. J. Tan et al., Magneto-ionic control of interfacial magnetism, *Nat. Mater.* **14**, 174 (2015).

16

Phase-Sensitive Interface and Proximity Effects in Superconducting Spintronics

Matthias Eschrig

16.1 INTRODUCTION

Spintronics with superconductors has been developing speedily since a number of pivotal experiments established its feasibility at the end of the 1990s and beginning of the 2000s. It is based on a number of theoretical predictions and has proved of interest for both the fundamental point of view in realizing new states of matter as well as for the practical point of view via the current urge to develop an energy-saving technology with large-scale computing and data storage centers.

The main driving forces are along three directions. The first is control of the phase of the order parameter via frustration between two types of Cooper pairs, spin-singlet and spin-triplet Cooper pairs. The second consists of utilizing equal-spin Cooper pairs in ferromagnetic devices, which allows for long-ranged proximity effects with penetration depths comparable to those that would occur in normal metals. The third is the production and control of pure spin currents utilizing the spin-polarized single-particle excitation spectrum in superconductors.

A number of spin-offs are of strong interest for fundamental research. Among those is the possibility of studying the so-called odd-frequency pairing, an elusive type of pairing that has never been observed in bulk superconductors, and, indeed, arguments have been put forward that it is not thermodynamically stable in a bulk superconductor.

The emerging field of superconducting spintronics has the goal of developing memory and logic devices based on the unique properties resulting from the combination of superconductivity with spin-ordered states. In

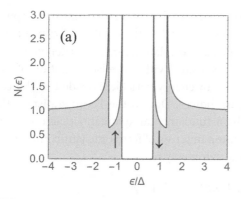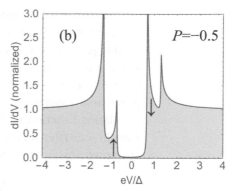

FIGURE 16.1 (a) Density-of-states $N(\varepsilon)$ in a superconductor with a Zeeman splitting of $|\mu B| = 0.3\Delta$. The quasiparticle states near the gap edges are nearly 100% spin-polarized. (b) Tunneling conductance vs bias voltage for an S-F tunnel junction, where the ferromagnet has a spin-polarization of $P = (N_\uparrow - N_\downarrow)/(N_\uparrow + N_\downarrow) = -0.5$.

complementing this material, several chapters on the normal state properties are directly relevant, for example, on spin injection and its nonlocal detection (Chapter 5, Volume 1; Chapters 7 and 9, Volume 3), spin relaxation (Chapter 1, Volume 2), proximity effects (Chapters 5, Volume 3), and spin torque (Chapters 7–9, Volume 3). In the superconducting state, Chapter 2, Volume 3, addresses Andreev reflection, the microsocopic mechanisms of supercoducting proximity effect, while Chapter 13, Volume 2, shows how elusive spin-triplet pairing could enable topologically protected states.

16.2 SINGLET-TRIPLET MIXING

16.2.1 SPIN-POLARIZED PAIR CORRELATIONS: BULK

When a magnetic field is applied parallel to a thin superconducting film, superconductivity may survive to field strengths that give rise to an appreciable Zeeman splitting of the electronic bands in the superconductor. This Zeeman splitting of the spin bands by $2\mu|B|$, where μ is the effective magnetic moment of the electron,† leads to a shift of the excitation spectrum of quasiparticles in the superconductor, an effect that was studied in great detail and exploited by Tedrow and Meservey [1, 2]. In particular, the Zeeman splitting leads to a spin-split quasiparticle density-of-states in the superconductor, which can be used as a spin detector. Tunneling from a ferromagnetic material into a spin-split superconductor allows to determine the spin-polarization of the tunneling electrons emerging from the ferromagnet.

As shown in Figure 16.1, tunneling from a ferromagnet results in an asymmetric tunneling conductance as a function of the bias voltage, which is due to the fact that the tunnel current due to the spin-up and due to the spin-down electron states in the ferromagnet have different magnitude, and the spin-up electronic quasiparticle states in the superconductor are shifted with respect to the spin-down quasiparticle states. Using such devices, it is possible to produce almost fully spin-polarized tunnel currents.

---

† For free electrons the magnetic moment is $\mu = \mu_e < 0$.

Apart from this, another aspect of the Zeeman splitting of spin bands in a superconductor is of interest, which is the creation of triplet superconducting pair correlations. The pair correlation function is denoted $F_{\alpha\beta}(\boldsymbol{p},z)$, where $\boldsymbol{p}$ is the momentum associated with the relative motion of the electrons in a pair, $z = \varepsilon_1 + i\varepsilon_2$ is an energy variable extended to the complex plane, and α and β are spin indices. Due to the Fermi statistics, the pair correlation function fulfills a fundamental symmetry relation, which follows directly from its definition in terms of fermionic (anticommuting) field operators. This fundamental symmetry reads

$$F_{\alpha\beta}(\boldsymbol{p},z) = -\, F_{\beta\alpha}(-\boldsymbol{p},-\,z). \tag{16.1}$$

We use the spin-antisymmetric and spin-symmetric combinations

$$F_s = \frac{1}{2}(F_{\uparrow,\downarrow} - F_{\downarrow,\uparrow}), \quad F_t = \frac{1}{2}(F_{\uparrow,\downarrow} + F_{\downarrow,\uparrow}), \tag{16.2}$$

where:

- F_s is the spin-singlet component, and
- F_t is one of the three spin-triplet components.

Furthermore, in the bulk homogeneous superconducting state with an isotropic gap Δ, the pair correlation function is independent of $\boldsymbol{p}$. In this case, the fundamental symmetry relation translates into the following equations,

$$F_s(z) = F_s(-z), \quad F_t(z) = -F_t(-z). \tag{16.3}$$

This means, the singlet correlations are even in z and the triplet correlations are odd in z. The latter are therefore called *odd-frequency* spin-triplet pair correlations [3]. They exist because the momentum symmetry is s-wave, or even, in relative momentum. Explicitly, they are given by the expressions

$$F_{s,t}(z) = \frac{\pi\Delta}{\sqrt{|\Delta|^2 - (z - |\mu\boldsymbol{B}|)^2}} \pm \frac{\pi\Delta}{\sqrt{|\Delta|^2 - (z + |\mu\boldsymbol{B}|)^2}}, \tag{16.4}$$

where the upper sign is for singlet and the lower sign for triplet.

Figure 16.2 illustrates the pair correlation functions $F_s(z)$ and $F_t(z)$ in the complex z-plane for $|\mu\boldsymbol{B}| = 0.5\Delta$.[†] For $\boldsymbol{B} = 0$ the triplet correlations vanish, for $\boldsymbol{B} \neq 0$ they are nonzero, and their magnitude increases linearly in $|\mu\boldsymbol{B}|$ for small fields. The plot in Figure 16.2 shows $F_{s,t}(z)$ in the complex z-plane. The correlation functions coincide in the upper open half plane, where $\mathrm{Im}(z) > 0$, with the retarded correlation function, and in the lower open half plane, where $\mathrm{Im}(z) < 0$, with the advanced correlation function.[‡] Note that changing z to $-z$ changes retarded into advanced functions and vice versa, see Figure 16.3a. Of particular interest is the imaginary z-axis, as equilibrium properties can be expressed via the values of the correlation

---

[†] This value of is $|\mu B|$ below the Chandrasekhar-Clogston limit [4, 5] of $\Delta_0/\sqrt{2}$.

[‡] They can be analytically continued to a Riemann surface with branch cuts on the real axis. In this case, all singularities are situated on the real axis or in lower Riemann sheets.

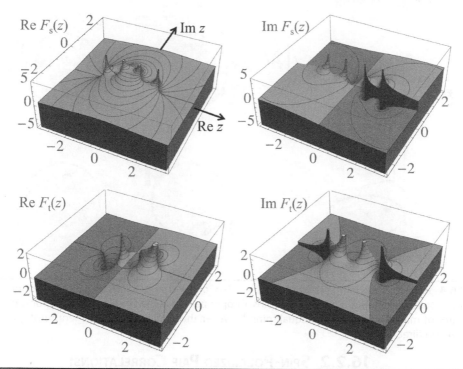

FIGURE 16.2 Pair correlation functions $F_s(z)$ and $F_t(z)$ for $|\mu B| = 0.5\Delta$, plotted in the complex z plane. Energies are in units of Δ. Positive and negative function values are indicated by different brightness. The symmetries $F_s(z) = F_s(-z)$ and $F_t(z) = -F_t(-z)$ hold.

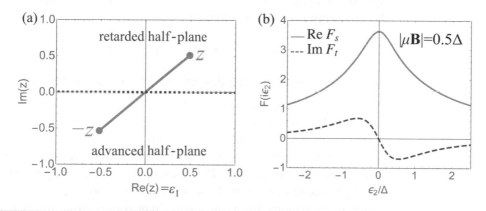

FIGURE 16.3 (a) Retarded and advanced half-planes in the complex z-plane. Pair correlation functions have a symmetry between points at z and at $-z$, resulting from Fermi statistics. (b) Pair correlation functions $F_s(i\varepsilon_2)$ and $F_t(i\varepsilon_2)$ for $|\mu B| = 0.5\Delta$. The symmetries $F_s(i\varepsilon_2) = F_s(-i\varepsilon_2)$ and $F_t(i\varepsilon_2) = -F_t(-i\varepsilon_2)$ hold. F_t is a so-called odd-frequency component.

functions on this axis alone. As can be seen from the figure, only $\mathrm{Re}(F_s)$ and $\mathrm{Im}(F_t)$ are nonzero for $z = i\varepsilon_2$ (for the real gauge of the gap Δ we are using here). The former is an even function of ε_2, and the latter an odd function of ε_2, as shown in Figure 16.3b.

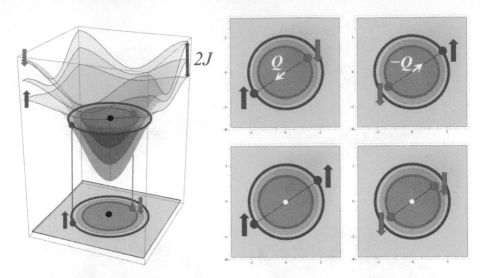

FIGURE 16.4   Due to an exchange splitting $H_{exch} = -J \cdot \sigma$ of the energy bands by $\pm J$, a corresponding Fermi-surface splitting appears. For opposite spins, Cooper pairing at the Fermi surfaces can only occur with a finite center-of-mass momentum. For equal spins, however, the center-of-mass momentum stays zero. Spins are quantized in direction of J.

16.2.2 Spin-Polarized Pair Correlations: Proximity Effect

A spin splitting of the quasiparticle bands in the normal state, $\varepsilon(p) \rightarrow \varepsilon(p) - J(p) \cdot \sigma$, leads to a splitting between the Fermi surface for spin-up and spin-down quasiparticles. For $J \equiv |J|$ much smaller than the Fermi energy, the latter is defined by

$$\hbar v_F \cdot (p_F) \equiv v_F \cdot (p_{F\uparrow} - p_{F\downarrow}) = 2|J(p_F)|, \qquad (16.5)$$

obtained by linearizing the dispersion relation around the Fermi energy.[†]

This, in turn, can cause Cooper pairs $\{p_{F\uparrow}, -p_{F\downarrow}\}$ and $\{p_{F\downarrow}, -p_{F\uparrow}\}$ built from Landau quasiparticles close to the Fermi energy to develop a finite center-of-mass momentum $\pm \hbar Q / 2$. Such Cooper pairs can give rise to a spatial variation of the order parameter, which is the so-called Fulde-Ferrell-Larkin-Ovchinnikov effect [6, 7]. However, even with a homogeneous singlet order parameter present in the superconductor, Cooper pairs with a finite center-of-mass momentum can develop in proximity structures that involve a superconductor and a ferromagnet (Figure 16.4). From Equation 16.5, one can construct a length scale $\xi_{F0} = \hbar v_F / 2J$, the magnetic coherence length, which plays an important role in proximity-induced pair correlations.

In a spin-polarized system, the pair correlations amplitudes are 2×2 matrices in spin space. In a superconductor in proximity to a spin-polarized

---

[†] The Fermi velocity is assumed to be approximately equal for the two spin bands due to the small splitting, and $p_F = (p_{F\uparrow} + p_{F\downarrow})/2$.

system, it is useful to decompose the pair amplitude into a spin-singlet and a spin-triplet contribution,

$$f_{\alpha\beta} = f_s (i\sigma_2)_{\alpha\beta} + \mathbf{f}_t \cdot (\sigma i\sigma_2)_{\alpha\beta}. \tag{16.6}$$

where $\sigma = (\sigma_1, \sigma_2, \sigma_3)$ refers to the vector of Pauli spin matrices. The component f_s describes spin-singlet pair correlations, whereas the component $\mathbf{f}_t$ refers to spin-triplet pair correlations. They both depend on mixed Wigner coordinates energy z, momentum $\mathbf{p}_F$, spatial center-of-mass coordinate $\mathbf{R}$, and (only in nonequilibrium) time t. The Fermi statistics leads to the fundamental symmetry relations

$$f_s(z, \mathbf{p}_F, \mathbf{R}, t) = f_s(-z, -\mathbf{p}_F, \mathbf{R}, t), \quad \mathbf{f}_t(z, \mathbf{p}_F, \mathbf{R}, t) = -\mathbf{f}_t(-z, -\mathbf{p}_F, \mathbf{R}, t). \tag{16.7}$$

The transport equation of the pair amplitudes f_s and $\mathbf{f}_t$ depend on the energy variable z. It is useful to introduce a variable s_z equal to the sign of the imaginary part of z, $s_z = \mathrm{sign}\left[\mathrm{Im}(z)\right]$. For retarded correlation functions $s_z = 1$, for advanced correlation functions $s_z = -1$. Equilibrium properties can be expressed via Matsubara pair amplitudes, which are obtained for $z = i\varepsilon_n$, where $\varepsilon_n \equiv (2n+1)\pi k_B T$ is called Matsubara energy.

In ferromagnetic materials, a hierarchy of length scales exists, which determine the physical properties of superconducting proximity structures involving ferromagnets. For ballistic systems they are

$$\xi_{F0} = \frac{\hbar v_F}{2J}, \quad \xi_{T0} = \frac{\hbar v_F}{2\pi k_B T}, \quad \xi_{n0} = \frac{\hbar v_F}{2|\varepsilon_n|} = \frac{\xi_{T0}}{|2n+1|}, \quad \ell_F = v_F \tau_F, \tag{16.8}$$

where:

τ_F is the electronic scattering time in the ferromagnet

ℓ_F the corresponding mean free path

Typically, $\xi_{F0} < \xi_{T0}$, whereas both $\ell_F < \xi_{F0}$ and $\ell_F > \xi_{F0}$ are possible, depending on the magnitude of the exchange splitting and the amount of impurity scattering. For diffusive systems, when $\ell_F \ll \xi_{F0}$, one has a corresponding hierarchy

$$\xi_{F1} = \sqrt{\frac{\hbar D}{2J}}, \quad \xi_{T1} = \sqrt{\frac{\hbar D}{2\pi k_B T}}, \quad \xi_{n1} = \sqrt{\frac{\hbar D}{2|\varepsilon_n|}} = \frac{\xi_{T1}}{\sqrt{|2n+1|}}. \tag{16.9}$$

Here, D is the diffusion constant $D = v_F^2 \tau / 3 = v_F \ell_F / 3$.

For a clean ballistic ferromagnet, in the case of a weak proximity effect, when all pair amplitudes are small, the pair of equations

$$\left(\frac{1}{2} i\hbar \mathbf{v}_F \cdot \nabla_R + z\right) f_s + \mathbf{J} \cdot \mathbf{f}_t = i\pi s_z \Delta \tag{16.10}$$

$$\left(\frac{1}{2} i\hbar \mathbf{v}_F \cdot \nabla_R + z\right) \mathbf{f}_t + \mathbf{J} f_s = 0, \tag{16.11}$$

together with the appropriate boundary conditions determine the pair amplitudes. Homogeneous solutions of these equations for a given direction of the Fermi velocity $\mathbf{v}_F$ are proportional to $e^{\pm ix'/\xi_1}e^{-x'/\xi_2}$, where x' is the coordinate in direction of $\mathbf{v}_F$. There are two types of solutions. The first ones have eigenvectors with $\mathbf{f}_t = \pm f_s \mathbf{J}/J$, i.e., $\mathbf{f}_t$ is collinear with $\mathbf{J}$. For $z = i\varepsilon_n$ with $\varepsilon_n > 0$, one obtains for these solutions $\xi_1 = \xi_{F0}$ and $\xi_2 = \xi_{n0}$. They correspond to an equal-weight superposition of the singlet amplitude f_s and the triplet amplitude $f_{t0} \equiv \mathbf{f}_t \cdot \mathbf{J}/J$ with zero spin-projection to $\mathbf{J}$, oscillating spatially out of phase with respect to each other. Thus, at an interface with a singlet superconductor, the induced pair amplitudes in the ferromagnet (where $\Delta = 0$) have in the limit of small transmission amplitude t the form

$$f_s(x') = |t|^2 f_0 \cos(x'/\xi_{F0})e^{-x'/\xi_{n0}}, \quad f_{t0}(x') = i|t|^2 f_0 \sin(x'/\xi_{F0})e^{-x'/\xi_{n0}}$$

$$(16.12)$$

(at the interface $x' = 0$), or when choosing the quantization axis along $\mathbf{J}$,

$$f_{\uparrow\downarrow}(x') = |t|^2 f_0 e^{iQx'}e^{-x'/\xi_{n0}}, \quad f_{\downarrow\uparrow}(x') = |t|^2 f_0 e^{-iQx'}e^{-x'/\xi_{n0}}, \quad (16.13)$$

with $Q = \xi_{F0}^{-1}$ according to Equation 16.5.

The strong oscillations of these solutions on the length scale $\xi_{F0} = \hbar v_F/2J$ can lead to a strong reduction when averaging over all directions, with algebraic spatial decay, before exponential decay sets in on the scale ξ_{n0}. In general, a Fermi-surface mismatch between the superconductor and the ferromagnet at an interface leads to a contact resistance. When the Fermi wave vector in the ferromagnet is much larger than in the superconductor, then momentum conservation parallel to the interface ensures that pair amplitudes propagating in the ferromagnet are induced only for Fermi velocity vectors almost normal to the interface. In this case, no algebraic decay due to averaging over directions takes place. In the opposite limit, however, when the Fermi wave vector in the ferromagnet is smaller than in the superconductor, then for three-dimensional isotropic materials the Fermi surface averaged quantity

$$f_s(x)/f_0 = \int_0^1 d\mu \left[\mu \cos\left(\frac{x}{\mu\xi_{F0}}\right)e^{-\frac{x}{\mu\xi_{n0}}} \right] \tag{16.14}$$

determines the pair amplitude in the ferromagnet, where μ is the cosine of the angle between the Fermi velocity and the surface normal (such that $x' = x/\mu$). In Figures 16.5a and 16.6a, this case is illustrated. In addition, for large exchange splitting, the magnitude of these pair amplitudes is of the order of Δ/J, and thus vanishes in the limit of large J.

The second type of solution for Equations 16.10–16.11 have no oscillations, $\xi_1^{-1} = 0$, and decays on the length scale $\xi_2 = \xi_{n0}$. These solutions correspond to eigenvectors with $f_s = 0$ and $\mathbf{f}_t \cdot \mathbf{J} = 0$ and represent two equal-spin ($\uparrow\uparrow$ and $\downarrow\downarrow$) pair amplitudes $f_{t,\pm 1}$ with respect to the quantization axis $\mathbf{J}$.

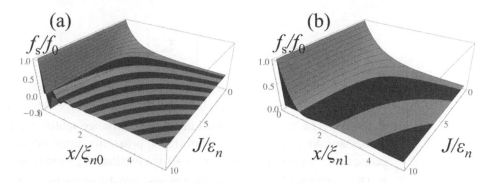

FIGURE 16.5 Oscillating (short-ranged) solutions for pair amplitudes in a ferromagnet in contact with a super-conductor, as a function of spatial coordinate x and the exchange energy J normalized to $\varepsilon_n = \pi k_B T(2n+1)$. (a) For a ballistic system, with $\xi_{n0} = \hbar v_F / 2\varepsilon_n$, calculated from Equation 16.14; (b) For a diffusive system, with $\xi_{n1} = \sqrt{\hbar D / 2\varepsilon_n}$, calculated from Equation 16.17. Note the strongly damped oscillations as a function of x for the diffusive case.

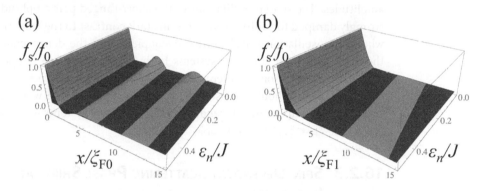

FIGURE 16.6 Oscillating (short-ranged) solutions for pair amplitudes in a ferromagnet in contact with a superconductor, as a function of spatial coordinate x and temperature ($\varepsilon_n / J = \pi k_B T(2n+1) / J$). (a) For a ballistic system, with $\xi_{F0} = \hbar v_F / 2J$, calculated from Equation 16.14; (b) For a diffusive system, with $\xi_{F1} = \sqrt{\hbar D / 2J}$, calculated from Equation 16.17. Note the temperature dependence of the spatial oscillation period in the diffusive case.

For diffusive systems, in the case of a weak proximity effect, the corresponding equations are

$$\left(\frac{\hbar s_z}{2i} \nabla_j D_{jk} \nabla_k + z \right) f_s + \mathbf{J} \cdot \mathbf{f}_t = i\pi s_z \Delta \tag{16.15}$$

$$\left(\frac{\hbar s_z}{2i} \nabla_j D_{jk} \nabla_k + z \right) \mathbf{f}_t + \mathbf{J} f_s = 0. \tag{16.16}$$

For isotropic systems ($D_{ij} \to D\delta_{ij}$) the fundamental solutions again come in two short-ranged and two long-ranged varieties. The short-range solutions are proportional to $e^{ix/\xi_1} e^{-x/\xi_2}$, and for $z = i\varepsilon_n$ one finds [8]

$$\xi_1 = \left(\frac{\hbar D}{\sqrt{\varepsilon_n^2 + J^2} - |\varepsilon_n|}\right)^{\frac{1}{2}}, \quad \xi_2 = \left(\frac{\hbar D}{\sqrt{\varepsilon_n^2 + J^2} + |\varepsilon_n|}\right)^{\frac{1}{2}}. \tag{16.17}$$

showing strongly damped spatial oscillations. This is illustrated in Figures 16.5b and 16.6b. For $J \gg |\varepsilon_n|$ both length scales merge and approach $\xi_{1,2} = \sqrt{2}\xi_{F1}\left(1 \pm |\varepsilon_n| / 2J\right)$ [9]. With increasing exchange splitting, the decay length shrinks to zero and the short-range proximity amplitudes are suppressed to unmeasurable size. For $J \to 0$ the oscillation period goes to infinity ($\xi_1^{-1} \to 0$), and $\xi_2 \to \xi_{n1}$. The long-ranged equal-spin solutions $f_{t,\pm 1}$ have no oscillations, $\xi_1^{-1} = 0$, and decay on the length scale $\xi_2 = \xi_{n1}$. They are thus insensitive to J.

In hybrid structures with collinear spin alignment throughout the structure, the long-range equal-spin solutions $f_{t,\pm 1}$ are not induced. A noncollinear spin arrangement is crucial for the creation of equal-spin pair amplitudes. The spatial oscillations of the short-ranged pair amplitudes are strongly damped for diffusive systems, in stark contrast to the ballistic case, where the oscillation wavelength is decoupled from the decay length. On the other hand, for diffusive systems both length scales are temperature dependent. Thus, if one is interested in observing many oscillation periods, ballistic systems should be used. If, on the other hand, one would like to study highly temperature-dependent effects, diffusive systems with small J (such that $\ell_F \ll \xi_{F0}$) are preferable.

16.2.3 SPIN-DEPENDENT SCATTERING PHASE SHIFTS AT INTERFACES

Scattering of quasiparticles at interfaces is described by an interface scattering matrix, which connects incoming and outgoing Bloch waves. In the presence of spin-polarized materials, this scattering matrix will be spin-dependent as well. For a superconductor in contact with a ferromagnetic insulator, the interface scattering matrix has only a reflection amplitude, which is a 2×2 spin matrix. For collinear spin arrangements, this matrix can be chosen diagonal (by choosing the spin quantization axis appropriately). The diagonal elements have in this case unit modulus (as the scattering matrix must be unitary), and the scattering is entirely described by the (spin-dependent) scattering phases. It can be parameterized by a scalar phase ψ and a "spin-mixing angle" ϑ,

$$R = e^{i\psi}\begin{pmatrix} e^{\frac{i}{2}\vartheta} & 0 \\ 0 & e^{-\frac{i}{2}\vartheta} \end{pmatrix}. \tag{16.18}$$

For small pair amplitudes, the reflected amplitudes at such an interface have the form $f_{out} = f_0 R(i\sigma_2)R^*$. Thus, the $f_{\uparrow\downarrow}$ amplitude acquires a scattering phase factor $e^{i\vartheta}$, whereas the $f_{\downarrow\uparrow}$ component acquires a scattering phase

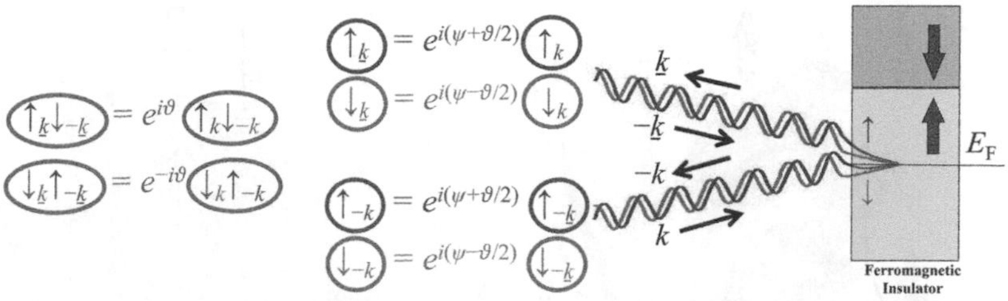

FIGURE 16.7   Singlet-triplet pair mixing mechanism at an interface between a superconductor and a ferromagnetic insulator. Spin mixing leads to spin-dependent scattering phases $\psi \pm \vartheta/2$, which are indicated for spin-up and spin-down quasiparticles. This leads to a dephasing of $f_{\uparrow\downarrow}$ pair amplitudes with respect to $f_{\downarrow\uparrow}$ pair amplitudes.

factor $e^{-i\vartheta}$, see Figure 16.7. As a result, the scattered pair amplitude takes the form

$$\hat{f}_{out} = f_0\left[\cos(\vartheta) + i\sin(\vartheta)\sigma_z\right]i\sigma_2. \tag{16.19}$$

Thus, a spin-triplet component $f_{t0} = i f_0 \sin(\vartheta)$ is induced by the spin-dependent scattering phase shifts. This effect was first described by Tokuyasu, Sauls, and Rainer [10].

Spin-dependent scattering phases have been employed extensively in various models of interfaces between a superconductor and a strongly spin-polarized ferromagnet [11, 12, 13]. The spin-mixing parameter $R_{\uparrow\uparrow}R_{\downarrow\downarrow}^* = |R_{\uparrow\uparrow}R_{\downarrow\downarrow}|e^{i\vartheta}$ plays also a prominent role in spintronics devices where it governs the spin-mixing conductance [14].

16.2.4 SPIN-POLARIZED ANDREEV BOUND STATES AT INTERFACES

One consequence of spin-dependent scattering at interfaces between a superconductor and a ferromagnet is the appearance of spin-polarized Andreev bound states [15]. If a superconductor is in contact with a ferromagnetic insulator, Andreev bound states appear below the gap and show complete spin-polarization. Accompanied with this is a suppression of the singlet superconducting order parameter near the interface due to the appearance of the Andreev bound states. This suppression is complete for $\vartheta = \pi$, when the bound states for spin-up and spin-down are degenerate and appear at zero-energy. Neglecting this order parameter suppression (as one can for example if one considers a point contact between a superconductor and a ferromagnetic insulator), the density-of-states can be obtained analytically and is [15]

$$N_\uparrow(\varepsilon,\vartheta) = \frac{\varepsilon\cos(\vartheta/2) + \Omega\sin(\vartheta/2)}{\Omega\cos(\vartheta/2) - \varepsilon\sin(\vartheta/2)}, \quad N_\downarrow(\varepsilon,\vartheta) = \frac{\varepsilon\cos(\vartheta/2) - \Omega\sin(\vartheta/2)}{\Omega\cos(\vartheta/2) + \varepsilon\sin(\vartheta/2)},$$

$$\tag{16.20}$$

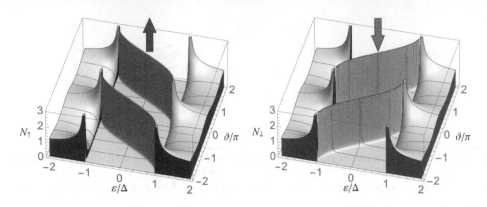

FIGURE 16.8 Spin-polarized Andreev bound states at an interface between a superconductor and a ferromagnetic insulator as a function of energy ε and spin-mixing angle ϑ. On the left the density-of-states at the interface for spin-up, on the right for spin-down.

where $\Omega = \sqrt{|\Delta|^2 - \varepsilon^2}$. It is plotted in Figure 16.8 for both spin projections, showing the complete spin-polarization of the bound states for $|\varepsilon| < |\Delta|$ (the states above the gap are unpolarized).

The Andreev bound states are obtained as

$$\varepsilon_\uparrow = |\Delta|\cos(\vartheta/2)\cdot\text{sign}\left[\sin(\vartheta/2)\right], \quad \varepsilon_\downarrow = -\varepsilon_\uparrow. \tag{16.21}$$

The Fermi-surface averaged pair amplitudes are obtained for $z = i\varepsilon_n$ within the same approximation as

$$f_s(\varepsilon_n,\vartheta) = \frac{\pi\Omega_n\Delta\cos(\vartheta/2)^2}{\Omega_n^2 - |\Delta|^2\sin(\vartheta/2)^2}, \quad f_{t0}(\varepsilon_n,\vartheta) = \frac{1}{2}\frac{i\pi\varepsilon_n\Delta\sin(\vartheta/2)}{\Omega_n^2 - |\Delta|^2\sin(\vartheta/2)^2},$$

$$(16.22)$$

where $\Omega_n = \sqrt{|\Delta|^2 + \varepsilon_n^2}$. The triplet component is an odd-frequency amplitude [16]. There are also odd-parity amplitudes, e.g. an even-frequency odd-parity spin-triplet proportional to $\sin(\vartheta)$ and an odd-frequency odd-parity spin-singlet proportional to $\sin(\vartheta/2)^2$ [17, 18].

Andreev bound states at spin-active interfaces have been verified experimentally [19, 20]. They also play an important role in the interpretation of point contact spectra of spin-polarized materials [21, 22, 23, 24, 25], a technique often used to determine spin-polarization [26, 27, 28], further discussed in Chapter 2, Volume 3.

An interesting interaction between spin-polarized Andreev states arises when one considers a superconductor contacted with ferromagnets at two points separated by a distance of the order of a coherence length. Such a geometry gives rise to so-called crossed Andreev reflection and elastic co-tunneling processes [29, 30], observed first experimentally by Beckmann et al. [31]. In this case, at each point contact spin-polarized Andreev bound states appear, and a nonlocal interaction between these bound states takes place [32]. As can be seen in Figure 16.9, avoided

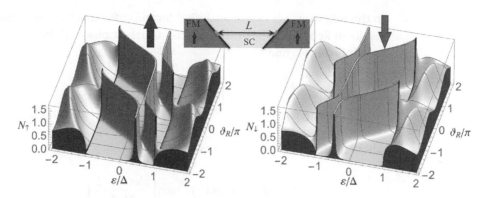

FIGURE 16.9 Spin-polarized Andreev bound states in a superconductor contacted with two ferromagnetic point contacts. The reflection amplitudes, $(r_\uparrow)_{L,R} = (\rho_\uparrow e^{i(\psi+\vartheta/2)})_{L,R}$ and $(r_\downarrow)_{L,R} = (\rho_\downarrow e^{i(\psi-\vartheta/2)})_{L,R}$, in both contacts have equal magnitude, with $(\rho_\uparrow\rho_\downarrow)_{L,R} = 0.95$. The spin-mixing angles are $\vartheta_{L,R}$. Shown is the spin-resolved density-of-states for quasiparticles traveling between the contacts, as a function of ε and ϑ_R in the superconductor, in the midpoint between the two contacts, for fixed $\vartheta_L = 0.7\pi$. The distance between the contacts is $L = 2\xi_{\Delta 0}$ with $\xi_{\Delta 0} = \hbar v_{F,SC} / |\Delta|$ the coherence length in the superconductor. Left for spin-up, right for spin-down.

crossings between Andreev bound states with an equal spin-polarization play an important role in such structures when the two contacts are characterized by comparable spin-mixing angles. At the avoided crossing the bound states have equal weight at both contacts, and a strong hybridization of the two Andreev states originating from the two contacts takes place. The spin-polarization and weight of the bound states determine the magnitude of the nonlocal currents due to crossed Andreev reflection and elastic co-tunneling in such structures [32].

Spin-polarized Andreev bound states arising from spin-dependent scattering phases at the interface are responsible for the transport of spin-polarized supercurrents across magnetically active point contacts [15] and across strongly spin-polarized ferromagnets and half-metals [11] in superconductor-ferromagnet Josephson devices. The Josephson current across a magnetically active point contact shows complex current-phase relations. Current is carried in such junctions by spin-polarized Andreev bound states at energies

$$\varepsilon_\pm = |\Delta|\cos\left(\frac{1}{2}[\chi_t \pm \vartheta]\right), \quad \text{with} \quad \sin\left(\frac{1}{2}\chi_t\right) = |t|\sin\left(\frac{1}{2}\chi\right), \quad (16.23)$$

for spin-up quasiparticles, and $-\varepsilon_\pm$ for spin-down quasiparticles. For $|t| = 1$ one has $\chi_t = \chi$, and for $|t| \to 0$ it follows $\chi_t = 2|t|\sin\left(\frac{1}{2}\chi\right)$. Here, χ is the Josephson phase difference across the junction and t the transmission amplitude (the transmission probability is $|t|^2$). When the spin-mixing angle ϑ exceeds a critical value $\vartheta_{crit} = 2\arccos(|t|)$, then there exists a Josephson phase χ_{jump} at which for zero temperature a jump occurs in the current-phase relation [33, 34, 35]. This happens when one of the bound states crosses zero-energy [15], i.e., for $\chi_t = \pi \pm \vartheta$, corresponding to

$$\chi_{jump} = 2\arcsin\left(|t|^{-1}\cos\left(\frac{1}{2}\vartheta\right)\right). \quad (16.24)$$

In the following, we derive analytical formulas for much of the numerical results in Refs. [15, 33]. For the Josephson current-phase relation for general transmission, we find

$$I(\chi,\vartheta) = \frac{\pi|\Delta|}{2|e|R_N} \frac{F\left(\frac{1}{2}[\chi_t+\vartheta]\right) + F\left(\frac{1}{2}[\chi_t-\vartheta]\right)}{\sin(\chi_t)} \sin(\chi), \quad (16.25)$$

where the function F is defined as

$$F(x) = \sin(x) \ \tanh\left(\frac{|\Delta|\cos(x)}{2k_BT}\right), \quad (16.26)$$

with R_N the normal state resistance of the junction. For $\vartheta = 0$ it reduces in the limit $|t| \to 0$ to the Ambegaokar-Baratoff formula [36], and for $|t| = 1$ the Kulik-Omel'yanchuk formula [37] is recovered, see also Ref. [38]. At zero temperature, $F(x) = \sin(x) \cdot \text{sign}[\cos(x)]$. For $0 \le \chi \le \pi$ and $0 \le \vartheta \le \pi$, the positive and negative zero-temperature branches of the current-phase relation are at

$$I_0^{(+)}(\chi,\vartheta) = I_0 \frac{\cos\left(\frac{1}{2}\vartheta\right)\sin(\chi)}{\sqrt{1 - |t|^2 \sin\left(\frac{1}{2}\chi\right)^2}}, \quad I_0^{(-)}(\chi,\vartheta) = -I_0 \frac{2\sin\left(\frac{1}{2}\vartheta\right)\cos\left(\frac{1}{2}\chi\right)}{|t|},$$

$$(16.27)$$

with $I_0 = \pi|\Delta| / 2|e|R_N$. An example of the current-phase relation is shown in Figure 16.10a. At the position of the jump,

$$|I_0^{(-)} / I_0^{(+)}|_{\chi=\chi_{\text{jump}}} = \tan\left(\frac{1}{2}\vartheta\right)^2, \quad |I_0^{(+)} - I_0^{(-)}|_{\chi=\chi_{\text{jump}}} = 2\frac{\sqrt{|t|^2 - \cos\left(\frac{1}{2}\vartheta\right)^2}}{|t|^2\left|\sin\left(\frac{1}{2}\vartheta\right)\right|}.$$

$$(16.28)$$

The maximal possible value on the positive branch $I_0^{(+)}$ occurs at χ_{max} with

$$\chi_{\text{max}} = 2\arcsin\left(\left[1+\sqrt{1-|t|^2}\right]^{-\frac{1}{2}}\right), \quad I_0^{(+)}(\chi_{\text{max}},\vartheta) = I_0 \frac{2\cos\left(\frac{1}{2}\vartheta\right)}{1+\sqrt{1-|t|^2}},$$

$$(16.29)$$

whereas the negative branch has its minimum at $\chi = 0$ with value $-2|t|^{-1}\sin\left(\frac{1}{2}\vartheta\right)$. The critical Josephson current is given by

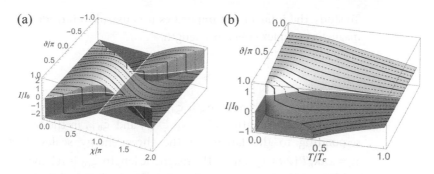

FIGURE 16.10 (a) Current-phase relation according to Equation 16.25 for $|t|^2 = 0.7$ at zero temperature. (b) Critical Josephson current $I_c(\vartheta, T)$ according to Equation 16.30 for $|t|^2 = 0.2$. T is normalized to the critical temperature T_c, and for $\Delta(T)$ the BCS relation was used.

$$I_c(\vartheta) = \max\left\{ I_0^{(+)}\left(\min\left[\chi_{\text{jump}}, \chi_{\text{max}}\right], \vartheta\right), \left|I_0^{(-)}\left(\chi_{\text{jump}}, \vartheta\right)\right| \right\}. \qquad (16.30)$$

The equality $\chi_{\text{jump}} = \chi_{\text{max}}$ holds for $\cos\left(\dfrac{1}{2}\vartheta\right) = |t|\left[1 + \sqrt{1 - |t|^2}\right]^{\frac{1}{2}}$. An example for $I_c(\vartheta, T)$ is shown in Figure 16.10b. The jump in the current-phase relation at χ_{jump} caused by the zero-energy crossing of spin-polarized Andreev bound states leads to a complex behavior of the critical Josephson current, I_c, as a function of temperature, with possible first-order transitions at a finite temperature between two different global maxima of the current-phase relation [15, 33, 34].

16.3 PAIR-AMPLITUDE OSCILLATIONS

16.3.1 0–π OSCILLATIONS IN JOSEPHSON JUNCTIONS

The idea of Josephson junctions with a π phase difference in its ground state when a magnetic impurity is inserted in the junction was introduced by Bulaevskiĭ, Kuziĭ, and Sobyanin [38], based on the previous work of Kulik [39]. A π-Josephson junction is a Josephson junction with a current-phase relation (CPR) taking the form $I(\chi) = |I_c|\sin(\chi + \pi) = -|I_c|\sin(\chi)$, where χ is the superconducting phase difference across the junction. Thus, for $0 \leq \chi \leq \pi$, the critical Josephson current in a π-junction is opposite to that of a zero-junction. Buzdin, Bulaevskiĭ, and Panyukov showed that the pair amplitude oscillates in ferromagnets in contact with a superconductor [40]. The study of the π-Josephson junction attracted increasing attention after this experiment [41, 42, 43]. For a review see Ref. [44]. Starting with the pioneering work of Ryazanov and co-workers [45, 8], in the 2000s all theoretical predictions were impressively confirmed in a series of experiments [46, 47, 48, 9].

Consider a Josephson junction made from two superconductors coupled via a ferromagnetic layer of thickness d_F. We assume a symmetric junction.

To study the influence of impurities it is useful to introduce for the ferromagnet a complex coherence length,

$$\xi_n^{-1} = \ell_F^{-1} + \xi_{0,\varepsilon_n}^{-1} + i\xi_{0,J}^{-1}. \tag{16.31}$$

This complex coherence length ξ_n for given ε_n is dominated by the smallest of the three length scales, $\ell_F, \xi_{0,\varepsilon_n}$, and $\xi_{0,J}$, in the ferromagnet, corresponding to the largest of the three energy scales $\hbar/\tau_F = \hbar v_F/\ell_F$, $\varepsilon_n = 2\pi k_B T (2n+1)$, and J. The magnetic length ξ_{F0} is relatively short in typical ferromagnets, such that often $\ell_F > \xi_{F0}$. In this case, the impurities in the ferromagnet cannot be treated in the diffusive limit, and one must resort to a more general approach. The formalism of Eilenberger [49], and of Larkin, and Ovchinnikov [50] is appropriate for this case.

The equations linearized in the pair amplitudes in the presence of normal isotropic impurity scattering with electronic scattering time τ reads

$$\left(\frac{1}{2}i\hbar v_F \cdot \nabla_{\mathbf{R}} + z\right)f_s + \mathbf{J}\cdot\mathbf{f}_t + \frac{i\hbar s_z}{2\tau}\left(f_s - \langle f_s\rangle_{p_F}\right) = i\pi s_z \Delta \tag{16.32}$$

$$\left(\frac{1}{2}i\hbar v_F \cdot \nabla_{\mathbf{R}} + z\right)\mathbf{f}_t + \mathbf{J} f_s + \frac{i\hbar s_z}{2\tau}\left(\mathbf{f}_t - \langle \mathbf{f}_t\rangle_{p_F}\right) = 0, \tag{16.33}$$

where $\langle\cdots\rangle_{p_F}$ denotes a (density-of-states weighted) Fermi-surface average. This differs from Equations 16.10 and 16.11 by the impurity term proportional to $\hbar/\tau$.

The linearized problem for the case of moderately strong spin-polarized ferromagnets has been solved theoretically in Ref. [51]. The expression for the critical Josephson current for general impurity scattering is [51, 52]

$$J_c\left(d_F, T\right) = \frac{\pi T_{SF}}{4|e|R_N} k_B T \sum_{\varepsilon_n>0} \frac{\Delta^2}{\Delta^2 + \varepsilon_n^2} \text{Re}\left[\frac{\xi_n}{d_F} \sum_{m=-\infty}^{\infty} (-1)^m H_{\ell_F/\xi_n}\left(m\pi\frac{\xi_n}{d_F}\right)\right], \tag{16.34}$$

where a spherical Fermi surface was assumed (the Fermi wavevector in the ferromagnet is assumed to be smaller than that in the superconductor), T_{SF} is a tunneling probability, and R_N is the normal state resistance of the junction (dominated by the boundary resistances). The function entering the expression (16.34) is given by $H_\kappa(z) = \left[1 - a(z)\right]z^{-2} + \kappa^{-1}a(z)^2 / \left[z^2 a(z) + 3(\kappa - 1)\right]$, where $a(z) = 3[z - \arctan(z)]/z^3$. A tunneling characteristics appropriate for a potential barrier due to Fermi velocity mismatch is assumed. Results for J_c as a function of temperature and thickness of the ferromagnetic layer for various degrees of disorder are shown in Figure 16.11.

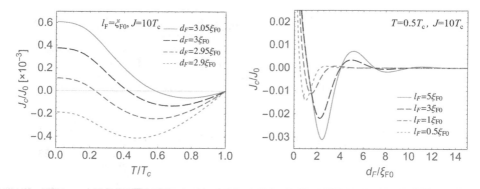

FIGURE 16.11 (a) Temperature dependence of the critical Josephson current across a ferromagnet of thickness $\ell_F = \xi_{F0}$, for various ferromagnet thicknesses d_F. A transition from a $0-$ to a π–junction takes place with increasing temperature. For the temperature dependence of the gap, the approximation $\Delta(T) \approx \Delta(0)\tanh\left(1.74\sqrt{(T_c/T)-1}\right)$ was used. (b) Dependence of the critical Josephson current across a ferromagnet for fixed temperature and various degrees of disorder. All curves in (a) and (b) were calculated with Equation 16.34.

The temperature dependence in Figure 16.11a shows a π–junction behavior near T_c, and a transition to a zero-junction at lower temperatures. For a larger thickness of the ferromagnet, one can also achieve a situation where a π–junction is stabilized at low temperature and a 0–junction near T_c.

Such a temperature-dependent switching between zero and π states can only be achieved for sufficiently small J in the range where the mean free path is not much larger than ξ_{F0}. This can already be inferred qualitatively from Figure 16.6, which shows that in clean systems the oscillation period of the pair amplitudes in the ferromagnet do not depend on temperature, whereas in the diffusive limit they do so. Thus, the best-suited ferromagnets to observe this effect are ferromagnetic alloys, which show a rather small spin splitting of the electronic conduction bands and are intrinsically diffusive. Typical ferromagnetic alloys that have been used are $Cu_{1-x}Ni_x$(CuNi) and $Pd_{1-x}Ni_x$ (PdNi).

The dependence on the thickness of the ferromagnet is shown in Figure 16.11b. For diffusive systems ($\ell_F = 0.5\xi_{F0}$ in the figure), the effect becomes very small already after the first half oscillation period. In contrast, several oscillations can be observed in clean materials, where the mean free path exceeds ξ_{F0} considerably. This is consistent with the behavior seen in Figures 16.5 and 16.6. Due to the strong damping, it is common to plot the critical current in a logarithmic plot. In Figure 16.12, a logarithmic plot of these same calculations, as in Figure 16.11b, is shown on the left. The $0-\pi$ -transitions are here visible in a large range of thicknesses. Note that the solutions of Equation 16.34 show that there is an offset at small thicknesses. In the experimentally accessible range, the asymptotic behavior of the oscillations is usually not reached yet. In Figure 16.12, on the right, the experimental results of Ref. [9] for $0-\pi$ -oscillations as a function of the ferromagnet thickness are shown.

Studies of strongly spin-polarized interlayers were performed, using various materials, among those Ni [53, 54, 55], Co [56], Py [57], and all three

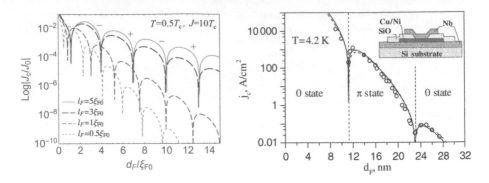

FIGURE 16.12   Left: Theoretical predictions for critical Josephson current as a function of the thickness of the ferromagnetic layer for various degrees of disorder in a logarithmic plot. Right: The ferromagnet-layer thickness dependence of the critical current density for Nb/Cu$_{0.47}$Ni$_{0.53}$/Nb junctions at temperature 4.2 K. Open circles represent experimental results; solid and dashed lines show model calculations. The inset shows a schematic cross-section of the SFS junctions. (After Oboznov, V.A. et al., *Phys. Rev. Lett.* 96, 197003, 2006. With permission.)

in Ref. [58]. In these cases, the oscillation period of the pair amplitudes in the ferromagnet is very short, such that it has been an experimental challenge to resolve the separate oscillations.

The π-Josephson junctions based on superconductor/ferromagnet proximity structures are being implemented in superconducting digital and quantum circuits as π-phase shifters in a d.c. SQUID geometry [59].

In so-called double-proximity structures, two superconductors are connected by a weak link of a normal-metal/ferromagnet bilayer or a ferromagnet/normal-metal/ferromagnet trilayer forming a bridge between the superconducting banks. The superconductors provide pair correlations to the bridge, which then at the interface between the normal metal and the ferromagnet develop spin-triplet components [60, 61, 62, 63].

16.3.2 T_c OSCILLATIONS

Proximity-induced pairs in a superconductor-ferromagnet bilayer lead to spatially oscillating pair amplitudes in the ferromagnetic regions. These act back on the superconducting singlet pair potential in the superconducting region. For a sufficiently thin superconducting layer, the oscillating nature of the pair amplitudes in the ferromagnet ultimately leads to oscillations in the modulus of the pair potential of the superconductor (and thus in the transition temperature) as a function of the geometric dimensions of the hybrid structure.

Motivated by experimental results on V/Fe superlattices [64], an oscillatory behavior in the dependence of the critical temperature $T_c(d_F)$ on the ferromagnetic layer thickness d_F in an S/F bilayer or multilayer was discovered [65, 66]. Further theoretical work showed that the oscillatory behavior could even give rise to a re-entrant superconductivity with

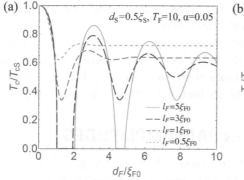
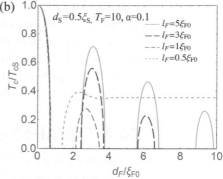

FIGURE 16.13 Oscillations of the superconducting critical temperature T_c of a superconductor/ferromagnet bilayer, for a superconductor of thickness $d_S = 0.5\xi_S$, as a function of the thickness of the ferromagnet d_F. The various curves are for a different mean free path ℓ_F in the ferromagnet. In (a) $\alpha = 0.05$, and in (b) $\alpha = 0.1$, where the parameter is defined in the text.

increasing d_F after T_c is suppressed to zero ("extinction") in a certain range of d_F [67, 68, 69].

Tagirov developed an approximate theory [67, 68] that also covers the range when the mean free path ℓ_F in the ferromagnet becomes comparable to or longer than the (usually short) coherence length $\xi_{F0} = \hbar v_F / 2J$. In contrast, the mean free path ℓ_S in the superconductor is usually much shorter then $\hbar v_S / 2\pi k_B T_{cS}$, such that the diffusive coherence length $\xi_S = \sqrt{\hbar D_S / 2\pi k_B T_{cS}}$ is appropriate, where $D_S = v_S \ell_S / 3$ is the diffusion constant in the superconductor, and T_{cS} is the bulk transition temperature of the superconductor.

Using the dimensionless parameters $\delta_F = d_F / \xi_{F0}$, $\delta_S = d_S / \xi_S$, $\lambda_F = \xi_{F0} / \ell_F$, $\alpha = (N_F v_F)/(N_S v_S) \cdot (3\xi_S / \ell_S)$, and a transmission parameter T_F (ranging between 0 and ∞), one obtains $t_c = T_c / T_{cS}$ from

$$\ln t_c = \mathrm{Re}\left[\Psi\left(\frac{1}{2}\right) - \Psi\left(\frac{1}{2} + \frac{\varphi^2}{2t_c \delta_S^2}\right) \right], \qquad (16.35)$$

with φ resulting from the solution of

$$\varphi \tan(\varphi) = \alpha \cdot \frac{2\delta_S \tanh(i\Lambda_F \delta_F)}{\Lambda_F + 2T_F^{-1} \tanh(i\Lambda_F \delta_F)}, \qquad (16.36)$$

where $\Lambda_F = \sqrt{1 - i\lambda_F}$ for $\xi_{F0}\ell_F$. In the extreme diffusive limit, for $\xi_{F0} \gg \ell_F$, the replacement $\Lambda_F \to \sqrt{3}\Lambda_F$ should be made. Solutions of these equations are shown in Figure 16.13.

In Figure 16.13a, the transition temperature for large thicknesses d_F is still finite for all shown curves, and oscillatory behavior occurs for smaller thicknesses. This oscillatory behavior is more pronounced for ferromagnets where the mean free path is comparable to or larger than ξ_{F0}. In the limit of $\ell_F \to \infty$, the function $t_c(\delta_F)$ becomes periodic with period π, such that T_c oscillates as a function of d_F with period $\pi\xi_{F0}$. In Figure 16.13b, superconductivity is entirely suppressed at a sufficiently large d_F for all curves except the one with the smallest ℓ_F.

These theoretical predictions have been confirmed in a number of experimental studies [70, 71, 72, 73, 74]. In particular, the phenomenon of re-entrant superconductivity with increasing d_F has attracted attention [75, 76], where even double extinction has been demonstrated [77].

16.4 EQUAL-SPIN PAIR AMPLITUDES

In Equations 16.10, 16.11, 16.15 and 16.16 two of the spin-triplet solutions are equal-spin pair amplitudes with spins aligned parallel or antiparallel with **J**. These solutions have a large penetration length scale into the ferromagnet, governed only by temperature, mean free path, and Fermi velocity, however, not by the exchange splitting of the electronic bands *J*. In order to induce these components in a ferromagnet coupled by proximity effect to a superconductor, the spin rotational symmetry around **J** must be broken. This means one needs a noncollinear spin arrangement.

For this it is useful to consider the spin coordinate transformation from one quantization axis (the *z*-axis) to another one, the direction of which is given by polar and azimuthal angles, α and ϕ, respectively, measured from the *z*-axis in spin space. Let us define the unit vector $\mathbf{n} = (n_x, n_y, n_z) = (\sin\alpha\cos\phi, \sin\alpha\sin\phi, \cos\alpha)$. Then the eigenvectors of $\mathbf{n} \cdot \mathbf{S}$ for eigenvalues $\pm\dfrac{\hbar}{2}$, where $\mathbf{S} = \dfrac{\hbar}{2}\boldsymbol{\sigma}$, define the spin-up and spin-down states with respect to the new quantization axis **n**. The well-known transformation formulas for basis vectors quantized along the direction $\mathbf{n}(\alpha,\phi)$ in terms of basis vectors quantized along the *z*-axis read

$$\uparrow_n = \cos\frac{\alpha}{2}\uparrow_z + \sin\frac{\alpha}{2}e^{i\phi}\downarrow_z, \quad \downarrow_n = -\sin\frac{\alpha}{2}e^{-i\phi}\uparrow_z + \cos\frac{\alpha}{2}\downarrow_z. \quad (16.37)$$

We note here that if R, defined in Equation 16.18, is applied to a spin state $\uparrow_n$ or $\downarrow_n$, it leads to a spin state that differs from the original one (apart from a phase) by the replacement $\mathbf{n}(\alpha,\phi) \to \mathbf{n}(\alpha,\phi-\vartheta)$. Thus, the spin will, after reflection, be rotated by an angle ϑ around the *z*-axis. The spin-mixing angle ϑ has consequently also the interpretation of a spin precession angle around the spin quantization axis of the interface.

Using Equation 16.37, one can express a Cooper-pair amplitude with respect to the new quantization axis, which leads to

$$(\uparrow\downarrow - \downarrow\uparrow)_\mathbf{n} = (\uparrow\downarrow - \downarrow\uparrow)_z \qquad (16.38)$$

$$(\uparrow\downarrow + \downarrow\uparrow)_\mathbf{n} = -\sin\alpha\left[e^{-i\phi}(\uparrow\uparrow)_z - e^{i\phi}(\downarrow\downarrow)_z\right] + \cos\alpha\,(\uparrow\downarrow + \downarrow\uparrow)_z. \quad (16.39)$$

Two conclusions arise from this formula. First, once triplet pair amplitudes of the form $(\uparrow\downarrow + \downarrow\uparrow)_\mathbf{n}$ with respect to some axis **n** are created in some

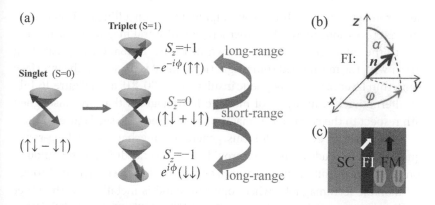

FIGURE 16.14 (a) Mechanism for the creation of a long-range pair amplitudes in a proximity structure. The spin-mixing effect provides mixing between spin-singlet and spin-triplet pair amplitudes, creating a short-range triplet amplitude with $S_z = 0$. If a noncollinear spin arrangement is present in the system, e.g., by inserting a ferromagnetic insulating (FI) thin layer between the superconductor (SC) and the metallic ferromagnet (FM), then long-range amplitudes with $S_z = \pm 1$ are generated in FM. The structure is shown in (c), with the spheric angles defined in (b).

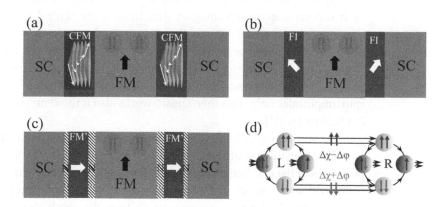

FIGURE 16.15 (a)–(c) Various geometries appropriate for the generation of long-range spin-triplet pair amplitudes in a ferromagnet (FM, e.g., Co) when used in a Josephson junction with a conventional singlet superconductor (SC, e.g., Nb). In (a) two layers of a conical ferromagnet (CFM, e.g., Ho) are used as singlet-triplet converters. In (b) insulating spin-polarized barriers (or thin layers of a ferromagnetic insulator FI) function as a converter, with misaligned moments e.g. due to spin-orbit coupling. In (c) two misaligned ferromagnets (FM', e.g., Cu-Ni or Pd-Ni alloys) are used instead. (d) shows the indirect nature of the Josephson effect, involving a conversion from opposite-spin pairs to equal-spin pairs on either side of the junction [93].

spatial region, they can give rise to long-range equal-spin pair amplitudes $(\uparrow\uparrow)_z$ and $(\downarrow\downarrow)_z$ that can penetrate into a ferromagnet with J aligned with the z-axis. Second, these equal-spin pair amplitudes carry a relative phase of $\pm(\pi + 2\phi)$ with respect to each other.

In Figure 16.14, the mechanism for generating long-range equal-spin components in a superconductor/ferromagnet heterostructure is shown [78]. The crucial step is to insert a region between the superconductor and the ferromagnet in which the spin alignment is noncollinear. This can be achieved by various methods. Either a region of spiral magnetism is present next to the interface [16, 79], as shown in Figure 16.15a. Or an insulating barrier with the misaligned spin magnetic moment is present at the interface (e.g., a thin layer of a ferromagnetic insulator) [11, 80, 81], as in Figure 16.15b. Or, alternatively, thin layers of metallic ferromagnets that are misaligned with respect to the central ferromagnet can be used [82], see Figure 16.15c.

Experimental evidence for this generation mechanism of long-range equal-spin amplitudes has been provided by several different techniques. Long-range proximity coupling as well as spin-triplet supercurrents through half-metallic ferromagnets, where one spin band is metallic and the other spin band not, were observed various times [83, 84, 85, 86, 87, 88, 89]. Other experiments used layers of spiral magnets (e.g., holmium) on either side of a bulk ferromagnet (e.g., cobalt) in order to generate a noncollinear spin-order, and found long-range proximity effects [90] and spin-triplet supercurrents across a cobalt layer [91]. Yet other experimental designs are based on fabricating multilayer structures, with control over the magnetization profile across the layers, demonstrating long-range spin-triplet supercurrents as well [92].

Figure 16.15d shows the indirect nature of the Josephson effect in such superconductor-ferromagnet structures if the central ferromagnet is strongly spin-polarized [93]. In this case, a singlet-triplet conversion takes place at either interface between the superconductor and the ferromagnet (with spin-noncollinearity either artificially engineered or intrinsic), and the coupling between the superconducting electrodes is provided by the equal-spin amplitudes only. This mechanism works also if the central ferromagnet is a half-metal [11].

The slow-decay of long-range equal-spin pair amplitudes in a ferromagnet allows for the use of rather thick ferromagnetic layers. In this case, the properties of the long-range pair amplitudes are typically governed by diffusive transport, as the mean free path, although it may exceed the penetration depth of the short-range components, is typically much shorter than the penetration depth of the long-range components. As diffusive pair amplitudes are Fermi-surface averaged quantities, they are even in parity. Thus, long-range triplet pairs are odd-frequency [16]. The possibility of spatially separating such odd-frequency components from the generating superconductor due to the superconducting proximity effect allows them to be studied and utilized. Odd-frequency pair amplitudes have a paramagnetic Meissner response [94], and their experimental observation has important consequences for the hypothetical existence of bulk odd-frequency superconductors [95]. Bulk odd-frequency pairing has not been observed in nature to date.

16.5 SPIN-VALVE JUNCTIONS

The spin-valve effect consists of a superconductor-ferromagnet device in which a change in the magnetization profile across the device leads to

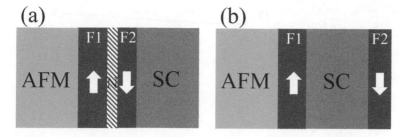

FIGURE 16.16 Spin-valve geometries: (a) F1/N/F2/S spin-valve; (b) F1/S/F2 spin-valve. F1 is pinned by an antiferromagnetic layer (AFM), whereas F2 can be rotated by an externally applied magnetic field. For the thin normal-conducting layer N between F1 and F2 in (a) often the same material is used as for the superconductor S, in which case the thickness of N is chosen such that it is below the critical thickness for superconductivity.

a change of its superconducting properties. In the ideal case, it allows to switch on or off superconductivity in the device by controlling its magnetic state. The physical origin of the spin-valve effect is the pair-breaking and pair-amplitude oscillation effect ferromagnetism has on conventional singlet superconductivity. Thus, at an interface between a superconductor and a ferromagnet, singlet superconductivity is strongly affected. This leads to a modification of the transition temperature T_c.

The device was first suggested by Oh, Youm, and Beasley [96] in an F1/N/F2/S geometry, with two ferromagnets separated by a thin normal-conducting layer that can be brought into a parallel or antiparallel magnetized state, and a superconductor deposited on the second ferromagnet, as illustrated in Figure 16.16a. The thickness of F2 is such that singlet superconducting pair correlations can penetrate into the N layer, and be modified by the F1 layer. Typically, the F1 layer is pinned by an insulating antiferromagnetic layer, and the F2 layer is able to rotate within an externally applied magnetic field.

An alternative device is to place a superconductor between two ferromagnets, to obtain an F1/S/F2 structure, illustrated in Figure 16.16b. This was proposed first by Tagirov [68], and by Buzdin, Vedyaev, and Ryzhanova [97].

Two questions are typically addressed for spin-valves. The first is about the difference of T_c between the parallel (P) and the antiparallel (AP) configuration of the magnetizations in F1 and F2. One defines $\Delta T_c = T_c^{AP} - T_c^{P}$. The case when $\Delta T_c > 0$ is called the *standard spin-valve effect*, and the case where $\Delta T_c < 0$ is called the *inverse spin-valve effect*. The second question is about the so-called *triplet spin-valve effect*, in which case the angle between the magnetizations in F1 and F2 takes a value α different from zero or π. In this case, $T_c(\alpha)$ is the quantity of interest, and in particular the value of T_c at $\alpha = \pi/2$. The new physics arising in the case when the magnetization is noncollinear ($\alpha \neq 0, \pi$) is the appearance of long-range equal-spin pairs, in addition to the short-range pairs in the ferromagnet. Thus, an additional channel for leakage of Cooper pairs into the ferromagnets, leading to a local minimum in $T_c(\alpha)$ for $\alpha \approx \pi/2$. An example of an F1/S/F2 structure is shown in Figure 16.17.

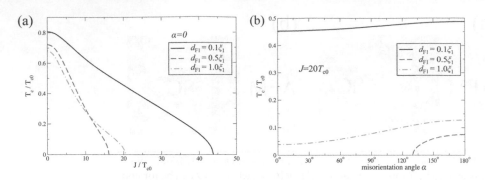

FIGURE 16.17 F1/S/F2 spin-valve effect for a trilayer with dimensions $d_S = 2\xi_1$, $d_{F2} = 0.5\xi_1$, and d_{F1} as indicated, with $\xi_1 = \sqrt{\hbar D / 2\pi k_B T_{c0}}$ (D is assumed as equal in S, F1, and F2), and T_{c0} is the critical temperature of the bulk superconducting material. (a) $T_c(J)$ for of the trilayer for parallel magnetization in F1 and F2, as a function of exchange splitting J (assumed to be equal in F1 and F2). (b) $T_c(\alpha)$ for $J = 20T_{c0}$ as a function of magnetization misorientation angle α between F1 and F2. (After Eschrig, M. et al., *Phys. Rev. B* 75, 014512, 2007. With permission.)

The two geometries shown in Figure 16.16 have been theoretically studied in detail. Near T_c the pair amplitudes are small, and the gap function is given by the self-consistency equation

$$\Delta(x)\ln\frac{T}{T_{c0}} = T\sum_n \left(\left\langle f_s(\varepsilon_n, \boldsymbol{p}_F, x) \right\rangle_{\boldsymbol{p}_F} - \frac{\pi\Delta(x)}{|\varepsilon_n|} \right), \qquad (16.40)$$

where the angle brackets denote a Fermi-surface average, x is the coordinate perpendicular to the layers, T_{c0} is the bulk transition temperature of the superconducting material, and f_s is itself a functional of the function $\Delta(x)$. If one writes

$$\left\langle f_s(\varepsilon_n, \boldsymbol{p}_F, x) \right\rangle_{\boldsymbol{p}_F} = \int d x' K_n(x, x')\Delta(x'), \qquad (16.41)$$

then the determination of T_c reduces to an eigenvalue problem, with the eigenvectors yielding the profile of the function $\Delta(x)$ near T_c across the structure.[†] In order to obtain the relation (16.41), it is necessary to find the solutions for the amplitudes f_s and $\mathbf{f}_t$ in the entire structure, by solving Equations 16.32 and 16.33, or in the diffusive limit Equations 16.15 and 16.16, together with appropriate boundary conditions.

It was found that for the F1/N/F2/S spin-valve, T_c can depend nonmonotonically on the angle between the magnetizations in F1 and F2, and T_c can be higher either in the parallel or in the antiparallel configuration [99, 100], as shown in Figure 16.18.

For the F1/S/F2 spin-valve, where all layers are in the diffusive limit [102], T_c is monotonically increasing from the parallel to the antiparallel configuration. In the clean limit, the same holds in various limiting cases [103, 104]. The general case for intermediate impurity scattering requires more theoretical work.

---

† The magnitude of Δ cannot be obtained from the linearized equation.

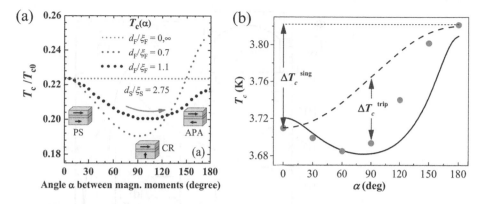

FIGURE 16.18 (a) Triplet spin-valve effect for a F1/N/F2/S spin-valve according to the model in Ref. [99]. The various magnetization configurations of the F1/N/F2 part are shown in the pictograms. (After Zdravkov, V. et al., *Phys. Rev. B 87*, 144507, 2013. With permission.) (b) Triplet spin-valve effect for a CoO$_x$/Py/Cu/Py/Cu/Pb structure. Shown is the angular dependence of T_c measured at a field of $H_0 = 150$ Oe (circles) and the theoretical prediction according to Ref. [99] (solid line). The dashed line illustrates the spin-valve effect that would appear in the absence of long-range amplitudes. ΔT_c^{sing} and ΔT_c^{trip} quantify the short-range and the long-range contributions to the spin-valve effect. (After Leksin, P.V. et al., *Phys. Rev. B 93*, 100502, 2016. With permission.)

The standard spin-valve effect was observed in both F1/N/F2/S structures [105] and F1/S/F2 structures [106, 107, 108, 109], including full angular dependence [110]. The inverse spin-valve effect arises from the pair-amplitude oscillations in the ferromagnets, which leads to an interference effect that can be constructive or destructive. Thus, in F1/N/F2/S spin-valves, depending on the layer thickness of F2, a positive or a negative value of $\Delta T_c'$ can be obtained [99]. These interference effects are sensitive to disorder in F1/S/ F2 spin-valves and allow for an inverse spin-valve effect only in sufficiently clean systems. Inverse spin-valve effects were experimentally observed in Fe/Cu/Fe/In and Fe/Cu/Fe/Pb structures [111, 112], as well as in a number of F1/S/F2 structures [113, 114, 115, 116, 117]. The triplet spin-valve effect was observed in both F1/N/F2/S structures [112, 118, 119, 120] and F1/S/F2 structures [120]. The effect can be strongly enhanced when half-metallic ferromagnets, such as CrO$_2$, are used [121].

In Figure 16.19, the leakage mechanism for the F1/S/F2 spin-valve is illustrated. As a measure for the singlet amplitudes is taken $\Phi_s(x) = T\sum_n \left(f_s(\varepsilon_n, x) - \pi\Delta/\varepsilon_n\right)$, for the triplet amplitudes $\Phi_t(x) = T\sum_{n>0} f_t(\varepsilon_n, x)$. The exchange fields are aligned perpendicularly with respect to each other, $J_1 = Je_z$ and $J_2 = Je_y$. The singlet amplitude $\Phi_s(x)$ (dashed-dotted line) leaks into the F regions and oscillates, decaying on a short length scale. The triplet amplitudes $\Phi_{ty}(x)$ (full line) and $\Phi_{tz}(x)$ (dashed line) are short-range or long-range, depending on their respective projections on J_1 or J_2. Note that the long-range component in F1 is due to the triplet component created at the S/F2 interface, and the long-range component at F2 due to the triplet component created at the F1/S interface. The long-range pair amplitudes are absent for parallel or antiparallel alignment of J_1 and J_2.

Similarly to a spin-valve, a spin-valve Josephson junction is based on two ferromagnetic layers F1 and F2 misaligned with respect to each other,

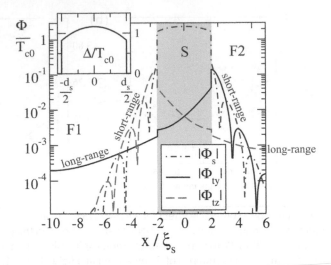

FIGURE 16.19 Leakage of short-range and long-range pair amplitudes into the ferromagnetic regions for an F1/S/F2 triplet spin-valve with perpendicular relative alignment of the magnetizations in F1 and F2, $\alpha = \pi/2$, $J_1 = Je_z$, $J_2 = Je_y$. The profile of $\Delta(x)$ is shown in the inset. Parameters are $T = 0.1T_{c0}$, $J = 20T_{c0}$. (After Eschrig, M. et al., *Phys. Rev. Lett.* 95, 187003, 2005. With permission.)

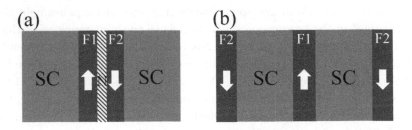

FIGURE 16.20 Spin-valve Josephson junction geometries: (a) S/F1/N/F2/S; (b) F2/S/F1/S/F2. Here, F1 and F2 should be dissimilar ferromagnets, such that they have different magnetization curves, allowing for control by an external magnetic field. For the thin normal-conducting layer N between F1 and F2 in (a) often the same material is used as for the superconductor S, in which case the thickness of N is chosen such that it is below the critical thickness for superconductivity.

as shown in Figure 16.20. The structure is of the form S/F1/N/F2/S or F2/S/F1/S/F2, with contacts at the superconductors to allow for a Josephson current to flow. Various of the ferromagnets can also be replaced by ferromagnetic insulating layers [123].

It is important to control the influence of stray fields, Josephson vortices, and domain switching in such devices. Such fully controllable spin-valve junctions have been produced and characterized [124].

Another type of spin-valve can be produced by utilizing spiral ferromagnets in which the magnetization vector lies parallel to the interface and rotates along a direction parallel to the interface. If the spiral pitch of such a cycloidal spiral is comparable to the superconducting coherence length then 0-π transitions can be induced depending on the spiral wavelength

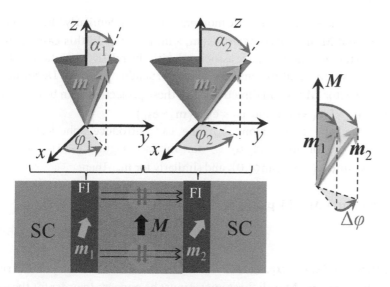

FIGURE 16.21 S/FI1/F/FI2/S Josephson junction that can act as a phase battery. Equal-spin Cooper pairs are transmitted through the ferromagnet F, which has magnetization **M**. The triple of vectors ($\mathbf{m}_1$, $\mathbf{m}_2$, **M**) is noncoplanar, as seen in the picture to the right. The directions of $\mathbf{m}_1$ and $\mathbf{m}_2$ are described by spherical angles (α_1, φ_1) and (α_2, φ_2), respectively, where the z-axis is in the direction of **M**. The equilibrium phase difference between the superconductors can take on a nonzero value depending on the angle $\Delta\varphi = \varphi_2 - \varphi_1$.

[125]. In such a structure only short-range triplet components are present. In a more general case, with magnetic domains separated by Néel walls, long-range triplet components are present and arise at the domain walls, decaying inside the domains [126, 127].

A new type of spin-valve constitutes the *re-orientation spin-valve*, in which a spiral magnetic wave vector can switch between two orientations, using, e.g., magnets from the B20-family of itinerant cubic helimagnets, with potentially giant spin-valve behavior [128].

16.6 PHASE BATTERIES

An interesting aspect of combining superconducting phase coherence with ferromagnetic spin-selectivity is the possibility to modify and control the current-phase relation in a ferromagnetic Josephson junction via the spin-magnetization. Consider an S/FI1/F/FI2/S junction, with two ferromagnetic insulating barriers FI1 and FI2, and an itinerant ferromagnet F. The barriers are characterized by magnetic moments $\mathbf{m}_1$ and $\mathbf{m}_2$, whereas the ferromagnet has magnetization **M**.

Geometric properties of the magnetization profile across a device are expressed in terms of collinearity and coplanarity. The role of noncollinearity has already been discussed in the previous sections in detail, and it is necessary in order to induce long-range superconducting components in a ferromagnet. Two vectors $\mathbf{m}_1$ and $\mathbf{m}_2$ are called *noncollinear* if $\mathbf{m}_1 \times \mathbf{m}_2 \neq 0$. The magnetization profile of a device is called noncollinear if it contains at least two noncollinear magnetic moment directions. Similarly, three vectors $\mathbf{m}_1$, $\mathbf{m}_2$, and $\mathbf{M}$ are *noncoplanar* if $(\mathbf{m}_1 \times \mathbf{m}_2) \cdot \mathbf{M} \neq 0$. In this case, there is an important geometric phase associated with the relative orientation of the projections of $\mathbf{m}_1$ and $\mathbf{m}_2$ onto the plane perpendicular to $\mathbf{M}$. Denoting $\mathbf{e}_\mathbf{M}$ the unit vector $\mathbf{M}/|\mathbf{M}|$ in direction of $\mathbf{M}$, these projections can be expressed as $\boldsymbol{\mu}_1 \equiv \mathbf{e}_\mathbf{M} \times (\mathbf{m}_1 \times \mathbf{e}_\mathbf{M})$ and $\boldsymbol{\mu}_2 \equiv \mathbf{e}_\mathbf{M} \times (\mathbf{m}_2 \times \mathbf{e}_\mathbf{M})$.[†]

Taking the z-axis along $\mathbf{e}_\mathbf{M}$, one can introduce spherical coordinates α_1, φ_1 for $\mathbf{m}_1$, and α_2, φ_2 for $\mathbf{m}_2$ (see Figure 16.21), such that $\boldsymbol{\mu}_1 = |\mathbf{m}_1|\sin\alpha_1 \cdot (\cos\phi_1, \sin\phi_1, 0)$, and similarly for $\mathbf{m}_2$. Then,

$$(\mathbf{m}_1 \times \mathbf{m}_2) \cdot \mathbf{M} = |\mathbf{M}| \, \boldsymbol{\mu}_1 \times \boldsymbol{\mu}_2 = |\mathbf{M}| |\mathbf{m}_1| |\mathbf{m}_2| \sin(\alpha_1)\sin(\alpha_2)\sin(\varphi_2 - \varphi_1).$$

(16.42)

The three angles α_1, α_2, and $\varphi_2 - \varphi_1$ fix the relative orientation of the triple of vectors $(\mathbf{m}_1, \mathbf{m}_2, \mathbf{M})$; all three angles must be nonzero in order for the spin arrangement to be noncoplanar. In particular, these angles do not depend on the coordinate system (i.e., they are independent of the choice of the global spin quantization axis), as they are expressed solely in terms of scalar and vector products.

The difference $\Delta\varphi = \varphi_2 - \varphi_1$ is of similar fundamental importance as the phase difference $\Delta\chi = \chi_2 - \chi_1$ of the superconducting pair potentials $\Delta_1 = |\Delta_1|e^{i\chi_1}$ and $\Delta_2 = |\Delta_2|e^{i\chi_2}$ on the two sides of a Josephson junction. In fact, the geometric angle $\Delta\varphi$ enters the current-phase relation of an S/FI1/F/FI2/S Josephson junction for each spin band, and precisely with opposite sign for the two spin projections: the transmission of an $\uparrow\uparrow$ -pair involves a Josephson phase of $\Delta\chi - \Delta\varphi + \pi$, whereas the transmission of a $\downarrow\downarrow$ -pair involves a Josephson phase of $\Delta\chi + \Delta\varphi + \pi$. The Cooper-pair current for a phase-coherent process that involves the transmission of m $\uparrow\uparrow$ -pairs and n $\downarrow\downarrow$ -pairs involves a phase $m(\Delta\chi - \Delta\varphi + \pi) + n(\Delta\chi + \Delta\varphi + \pi)$. The corresponding Josephson current for each spin-projection can be written as a sum over all transmission processes as [81]

$$I_{\uparrow\uparrow} = \frac{1}{2}\sum_{mn} m(-1)^{m+n} I_{mn} \sin\left[(m+n)\Delta\chi - (m-n)\Delta\varphi\right] \quad (16.43)$$

$$I_{\downarrow\downarrow} = \frac{1}{2}\sum_{mn} n(-1)^{m+n} I_{mn} \sin\left[(m+n)\Delta\chi - (m-n)\Delta\varphi\right]. \quad (16.44)$$

The coefficients fulfill the relation $I_{-m,-n} = I_{mn}$. The charge current I_c and the spin current I_s result from

$$I_c = 2e(I_{\uparrow\uparrow} + I_{\downarrow\downarrow}) \quad (16.45)$$

$$I_s = \hbar(I_{\uparrow\uparrow} - I_{\downarrow\downarrow}). \quad (16.46)$$

[†] In the coplanar but noncollinear case there still can be an important geometric phase of introduced by the triple of magnetic vectors if $(\mathbf{m}_1 \times \mathbf{M}) \cdot (\mathbf{m}_2 \times \mathbf{M}) < 0$.

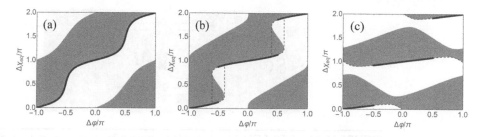

FIGURE 16.22 Examples for ϕ-Josephson-junctions resulting from Equation 16.47 and 16.48. For these plots, $I_{1,0} = 0.7$, $I_{0,1} = 0.3$, and (a) $I_{1,1} = 0.05$, (b) $I_{1,1} = 0.25$, (c) $I_{1,1} = 0.4$. The full lines correspond to the equilibrium solution $\Delta\chi_{eq}$ with lowest free energy for given $\Delta\varphi$. Dashed lines indicate metastable solutions (local, but not global minima in the free energy). The shaded regions correspond to a positive, the white regions to a negative Josephson current.

The current-phase relation fulfills the symmetry $I_{c/s}(-\Delta\chi, \Delta\phi) = -I_{c/s}(\Delta\chi, -\Delta\phi)$ following from the behavior of the current density under time reversal.

A symmetric high-transmissive junction involves typically only processes with $m - n = 0, \pm 1$. For this case, one obtains up to $|m + n| \leq 2$ the terms

$$I_{\uparrow\uparrow} = I_{1,1}\sin(2\Delta\chi) - I_{1,0}\sin(\Delta\chi - \Delta\phi) \qquad (16.47)$$

$$I_{\downarrow\downarrow} = I_{1,1}\sin(2\Delta\chi) - I_{0,1}\sin(\Delta\chi + \Delta\phi). \qquad (16.48)$$

As shown in Figure 16.22, the equilibrium superconducting phase $\Delta\chi_{eq}$ is neither zero nor π, however, takes a value in between, depending on the value of $\Delta\varphi$. For sufficiently small $I_{1,1}$ the variation of $\Delta\chi_{eq}$ with $\Delta\varphi$ is continuous, allowing to tune $\Delta\chi_{eq}$ in the full range between zero to π. Thus, the system can act as a phase battery. The first terms involving $I_{1,1}$ are *crossed pair transmission* processes [93], which involve the transmission of two Cooper pairs with opposite Cooper-pair spin through the two spin bands of the ferromagnet. The equilibrium configuration carries a nonzero spin current $I_s = 2\hbar I_{\uparrow\uparrow}(\Delta\chi_{eq}, \Delta\phi) \neq 0$. In a loop geometry, i.e., for zero phase difference, a spontaneous current

$$I_c(\Delta\chi = 0) = (I_{1,0} - I_{0,1})\sin(\Delta\phi) \qquad (16.49)$$

appears.

In the case of a strongly asymmetric Josephson junction, where the charge current is much smaller than the spin current, the leading terms are for $m + n = 0, \pm 1$, and in this case, one obtains up to $|m - n| \leq 2$ the terms

$$I_{\uparrow\uparrow} = I_{1,-1}\sin(2\Delta\varphi) - I_{1,0}\sin(\Delta\chi - \Delta\phi) \qquad (16.50)$$

$$I_{\downarrow\downarrow} = -I_{1,-1}\sin(2\Delta\varphi) - I_{0,1}\sin(\Delta\chi + \Delta\phi). \qquad (16.51)$$

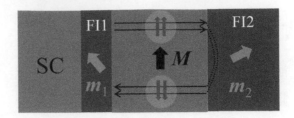

FIGURE 16.23 S/FI1/F/FI2 device carrying a pure spin-supercurrent I_s. Under the condition that **M**, $\mathbf{m}_1$, and $\mathbf{m}_2$ are noncoplanar (linear independent), equal-spin Cooper pairs, generated at the S/FI1/F interface, are transmitted through the ferromagnet F and transformed into each other at the ferromagnetic insulator FI2. Manipulating the spin-magnetization in FI2 (e.g., by ferromagnetic resonance) allows for the manipulation of I_s.

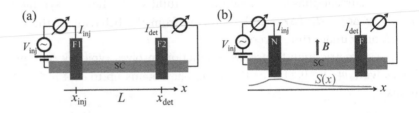

FIGURE 16.24 Spin injection and nonlocal detection. In (a) a spin accumulation is generated at the contact when a spin-polarized current flows into the superconductor from F1. In (b) a current is injected from a normal metal N into a superconductor with spin-polarized density-of-states due to an externally applied field B. In both cases, the spin accumulation S generated at the injector contact is transferred to a ferromagnetic detector contact, F2 in (a) and F in (b), where it leads to a charge current that can be measured.

In this case, the first term involving $I_{1,-1}$ contributes only to the spin current, however, not to the charge current.

For a half-metallic ferromagnet, only terms with $n = 0$, i.e., $I_{m0} \equiv I_m$ are nonzero, and the relation

$$I_{\uparrow\uparrow} = \sum_{m>0} m(-1)^m I_m \sin\left[m(\Delta\chi - \Delta\varphi)\right] \qquad (16.52)$$

holds. In this case, the effective Josephson phase is $\Delta\chi - \Delta\varphi$ [17, 80, 13].

Finally, one can also consider a S/FI1/F/FI2 junction, which contains only one superconductor, as illustrated in Figure 16.23. In such a junction $I_c = 0$, however, the spin current is given by

$$I_s = -\hbar \sum_{m>0} 2m\, I_{m,-m} \sin(2m\Delta\varphi). \qquad (16.53)$$

Such a junction thus carries a pure spin-supercurrent [93].

16.7 SPIN INJECTION, SPIN ACCUMULATION, AND NONLOCAL DETECTION

Superconducting spintronics devices exploit various concepts of the superconductors, such as nonlocal transport, charge- and spin-imbalance, spin-split density-of-states, as well as the coupling between the Cooper-pair condensate and Bogoliubov quasiparticles. In a typical spin-injection experiment [129, 130, 131, 132, 133], a spin-polarized current passes from a ferromagnet into a superconductor, see Figure 16.24a, or from a normal metal into a spin-polarized superconductor with Zeeman-split density-of-states, see Figure 16.24b. This leads to the nonequilibrium distribution of quasiparticle spin in the injection region, which propagates on a spin-relaxation length scale to a detector a certain distance away from the injection contact, outside of the main current path. Relaxation of this nonequilibrium spin distribution occurs on a spin-relaxation length scale related to the spin-relaxation time τ_s. Nonequilibrium distribution functions in superconductors come in two sets. The first describes nonequilibrium that is the same for particle-like and hole-like excitation branches, and is denoted by distribution functions $n\mathbf{1} + \mathbf{n} \cdot \mathbf{\sigma}$, where $\mathbf{\sigma}$ is the vector of spin Pauli matrices, and $\mathbf{1}$ is the spin unity matrix. The second describes nonequilibrium situations that lead to an imbalance between particle-like and hole-like quasiparticle branches and is denoted by the second type of distribution function, $n^* \mathbf{1} + \mathbf{n}^* \cdot \mathbf{\sigma}$, where the star denotes the quasiparticle branch imbalance. For collinear spin arrangements it is useful to instead introduce $n_\uparrow = n + n_z$, $n_\downarrow = n - n_z$, $n_\uparrow^* = n^* + n_z^*$, and $n_\downarrow^* = n^* - n_z^*$. Spin-relaxation in diffusive materials takes place typically via spin-orbit coupling in conjunction with elastic impurity scattering (Elliot-Yafet mechanism) [134, 135].

Charge imbalance, on the other hand, is typically connected with an imbalance between particle-like and hole-like Bogoliubov quasiparticles and relaxes by conversion to supercurrents [136, 137]. The two length scales for charge imbalance and for spin-relaxation via the Elliot-Yafet mechanism may compete with each other, and are both typically smaller than relaxation lengths related to inelastic scattering [138].

The nonlocal conductance

$$G_{nl} = \frac{\partial I_{det}}{\partial V_{inj}} \tag{16.54}$$

for a setup where the detector and the injection contacts are a distance L away from each other, is then determined by the nonequilibrium distribution functions $n_\uparrow(E,x)$ and $n_\downarrow(E,x)$ in the superconductor. It is useful to define the spin polarizations P_{det} and P_{inj} via

$$P_{inj} = \frac{G_{inj,\uparrow} - G_{inj,\downarrow}}{G_{inj,\uparrow} + G_{inj,\downarrow}}, \quad P_{det} = \frac{G_{det,\uparrow} - G_{det,\downarrow}}{G_{det,\uparrow} + G_{det,\downarrow}}, \tag{16.55}$$

where, e.g., $G_{det,\uparrow}$ and $G_{det,\downarrow}$ are the spin-resolved detector junction conductances, and the conductances $G_{inj} = G_{inj,\uparrow} + G_{inj,\downarrow}$ and $G_{det} = G_{det,\uparrow} + G_{det,\downarrow}$. For the charge current flowing into the superconductor (electron current into the detector/injector)† the nonlocal conductance is

$$G_{nl} = -G_{det}\left(P_{det}\frac{\partial S(x_{det})}{\partial(eV_{inj})} + \frac{\partial Q^*(x_{det})}{\partial(eV_{inj})} \right), \qquad (16.56)$$

where the spin accumulation, normalized to $\hbar/2$ times the density-of-states in the superconductor in the normal state, N_0, is given by [139, 140]

$$S(x) = \frac{1}{2}\int\limits_{-\infty}^{\infty} dE\, v(E,\Delta)\left[n_\uparrow\left(E + \mu_B B, x\right) - n_\downarrow\left(E - \mu_B B, x\right) \right], \qquad (16.57)$$

with $v(E,\Delta)$ the reduced density-of-states in the superconductor in the absence of Zeeman splitting,‡ and where the charge imbalance Q^*, normalized to eN_0,§ is

$$Q^*(x) = \frac{1}{2}\int\limits_{-\infty}^{\infty} dE\, v(E,\Delta)q(E,\Delta)\left[n_\uparrow^*\left(E + \mu_B B, x\right) + n_\downarrow^*\left(E - \mu_B B, x\right) \right], \qquad (16.58)$$

with $q(E,\Delta)$ the effective quasiparticle charge. At zero-field, the relation $v(E,\Delta)q(E,\Delta) = 1$ holds for $E \geq 0$. In writing Equation 16.56, the simplifying assumption was made that the dependence of the superconducting gap function Δ at the detector on the injection voltage V_{inj} could be neglected. In principle, a suppression of Δ at the injector contact slightly depends on V_{inj}, and this suppression can relax on a certain length scale, in which case the magnitude of the gap at the detector contact can slightly depend on V_{inj} too. Here these effects are neglected for simplicity.

In thin films with a magnetic field applied within the plane of the film, orbital depairing takes place [21, 141, 142]. The reduced density-of-states for negligible orbital depairing effects is given by the BCS form $\text{Re}\left[|E| / \sqrt{E^2 - \Delta^2} \right]$. Otherwise, a pair-breaking parameter $\Gamma = \frac{1}{2}\Delta_0(B/B_c)^2\sqrt{1 - (2\mu_B B_c / \Delta_0)^2}$ determines the gap Δ at zero temperature in terms of its zero-field value Δ_0 via the relation $\ln(\Delta/\Delta_0) = -\pi\Gamma/(4\Delta)$, which holds for $B \leq B_1$ where B_1 is slightly below the critical field B_c, and between B_1 and B_c the

---

† $e < 0$ is the charge of the electron.

‡ The Zeeman terms have been moved into the distribution functions $n_\uparrow$ and $n_\downarrow$ by a variable transformation in the integral.

§ The spin accumulation is $s(x) = \frac{\hbar}{2}N_0 S(x)$, and the charge imbalance is $q^* = eN_0 Q^*$.

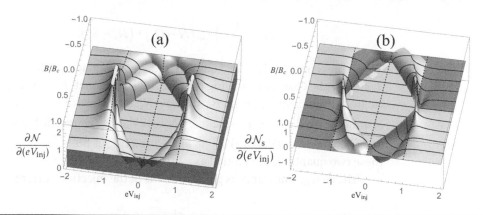

FIGURE 16.25 Plot of (a) $\partial N / \partial (eV_{inj})$ and (b) $\partial N_s / \partial (eV_{inj})$ as a function of injection voltage eV_{inj} and B/B_c, where B_c is the orbital critical field of the thin superconducting film. The Zeeman interaction $\mu_B B$ leads to a marked spin-dependence. In these plots $\mu_B B_c = 0.35\Delta_0$ was assumed. Experimental data confirming these plots have been published in Ref. [130].

superconductor loses its spectral gap[†] [143, 144, 145]. At finite temperatures, the gap is further reduced. The complex solution of the nonlinear equation $E = \left(\Delta - \Gamma / \sqrt{1-u^2}\right) u$ determines the reduced density-of-states for finite pair breaking, $v(E, \Delta) = \mathrm{Re}\sqrt{u^2 / (u^2 - 1)}$ [143, 144].

In order to study the spin accumulation at the injector contact, the spin-resolved injector current is needed, which is given by

$$I_{inj,\sigma} = \frac{G_{inj,\sigma}}{e}\left(N_\sigma - \sigma S(x_{inj}) - Q^*(x_{inj})\right) \tag{16.59}$$

where $\sigma = 1$ corresponds to $\uparrow$ and $\sigma = -1$ to $\downarrow$, and where

$$N_\sigma = \int_0^\infty dE\, v(E, \Delta)\left[n_0\left(E + \sigma\,\mu_B B - eV_{inj}\right) - n_0\left(E - \sigma\,\mu_B B + eV_{inj}\right)\right], \tag{16.60}$$

with $n_0(E) = [\exp(E/k_B T) + 1]^{-1}$ the equilibrium Fermi distribution function. At zero temperature, one obtains from Equation 16.60

$$\frac{\partial N_\sigma}{\partial (eV_{inj})} = v(eV_{inj} - \sigma\,\mu_B B, \Delta). \tag{16.61}$$

Writing $N = (N_\uparrow + N_\downarrow)/2$ and $N_s = (N_\uparrow - N_\downarrow)/2$, one obtains the total current $I_{inj} = I_{inj,\uparrow} + I_{inj,\downarrow}$ and the spin-antisymmetric current $I_{inj,s} = I_{inj,\uparrow} - I_{inj,\downarrow}$,

$$I_{inj} = \frac{G_{inj}}{e}\left[\left(N - Q^*\right) + P_{inj}(N_s - S)\right] \tag{16.62}$$

[†] In the gapless region one first obtains the complex quantity w from $\mu_B B = \left(\Delta - \Gamma / \sqrt{1-w^2}\right) w$, and then Δ via $\ln(\Delta / \Delta_0) = -\pi\Gamma / (4\Delta) - \mathrm{Im}\left\{\arcsin(w) + \left[\Gamma / (4\Delta)\right]\ln\left[(1-w)/(1+w)\right]\right.$ $\left. - w\Gamma / \left[2\Delta(1-w^2)\right]\right\}$

$$I_{\text{inj},s} = \frac{G_{\text{inj}}}{e}\Big[P_{\text{inj}}\big(N - Q^{*}\big) + \big(N_s - S\big)\Big]. \tag{16.63}$$

The expression $G_{\text{inj}}N/e$ in Equation 16.62 corresponds to the expression I_{Giaver} for the Giaever tunneling current [146]. Note that N is antisymmetric in V_{inj} and symmetric in B, whereas N_s is symmetric in V_{inj} and antisymmetric in B. As a consequence, $\partial N/\partial(eV_{\text{inj}})$ is symmetric both in eV_{inj} and B, whereas $\partial N_s/\partial(eV_{\text{inj}})$ is antisymmetric in both. In Figure 16.25, these two quantities are plotted as a function of eV_{inj} and B.

The charge imbalance is given in terms of the injection current by

$$Q^{*}\big(x_{\text{inj}}\big) = N - \frac{eI_{\text{inj}}}{G_{\text{inj}}} + P_{\text{inj}}\big(N_s - S\big). \tag{16.64}$$

The first two terms in Equation 16.64 can then be written as $e\big(I_{\text{Giaever}} - I_{\text{inj}}\big)/G_{\text{inj}}$. The spin-antisymmetric current in terms of the injection current is

$$I_{\text{inj},s} = P_{\text{inj}}I_{\text{inj}} + \frac{G_{\text{inj}}}{e}\big(1 - P_{\text{inj}}^2\big)\big(N_s - S\big). \tag{16.65}$$

Within a relaxation time approximation, the spin accumulation $S\big(x_{\text{inj}}\big)$ is proportional to $I_{\text{inj},s}$. This leads to the following situations for the two setups shown in Figure 16.24.

For the setup in Figure 16.24a there is no externally applied field, i.e., $B = 0$. In this case, $N_s = 0$, and the spin accumulation is created at the injector contact due to the spin-polarization P_{inj}. In this case, one has

$$S\big(x_{\text{inj}}\big) \sim P_{\text{inj}}I_{\text{inj}}\big(eV_{\text{inj}}\big). \tag{16.66}$$

Thus, for Figure 16.24(a), the spin accumulation is antisymmetric in V_{inj}, just like the charge imbalance (the latter follows from Equation 16.64, as I_{inj}, N, and S are all antisymmetric in V_{inj}, and N_s is zero). However, the two contributions can be separated as the contribution to G_{nl} from the spin accumulation in Equation 16.56 is proportional to $P_{\text{inj}}P_{\text{det}}$, whereas the contribution from the charge imbalance is insensitive to the relative sign of P_{inj} and P_{det}. Thus, switching the two ferromagnets from parallel to antiparallel allows for separating of the two contributions.

For the case of Figure 16.24b, where a Zeeman splitting of the superconducting density-of-states appears due to an externally applied magnetic field, an injection current polarization of nearly 100% can be achieved in the region of $|eV| = \Delta(B) \pm \mu_B B$, between the upper and lower Zeeman band. Here, the spin accumulation appears due to the spin-split density-of-states in the superconductor. In this case, as $P_{\text{inj}} = 0$, one obtains

$$S\big(x_{\text{inj}}\big) \sim \frac{G_{\text{inj}}}{e} N_s\big(eV_{\text{inj}}\big). \tag{16.67}$$

Thus, for Figure 16.24 b, the spin accumulation S is symmetric with respect to V_{inj} and antisymmetric with respect to B. This allows for the separating of this contribution from Q^*, as the latter has opposite symmetries (this follows from Equation 16.64, as I_{inj} and N are symmetric in B and antisymmetric in V_{inj}, and the last term involving P_{inj} is zero).

The two quantities $S(x_{det})$ and $S(x_{inj})$ are connected by a spin diffusion process from x_{inj} to x_{det}, which gives rise to a typical decay length scale over which $S(x)$ decays. The details of these diffusion processes are obtained by solving transport equations in a nonequilibrium situation [147–151]. In the high-field limit, it is possible to choose the distance between the two contacts in Figure 16.24 shorter than this length scale, however, longer than the length scale governing the decay of charge imbalance.

16.8 FUTURE DIRECTIONS

16.8.1 THERMOELECTRIC EFFECTS

The possibilities to couple charge and energy transport are enhanced in superconducting spintronics by the fact that the density-of-states spectrum in superconductors shows a strong energy dependence, which in conjunction with spin-dependent Zeeman shifts or due to spin-dependent interfacial scattering can be strongly particle-hole asymmetric for each spin direction. This allows for very large enhancements of thermoelectric effects compared to normal state metallic devices. Peltier cooler devices with considerable thermoelectric figures of merit have been theoretically suggested [152, 153] and large thermoelectric currents have been experimentally observed in superconductor-ferromagnet tunnel devices [154]. Microrefrigeration and thermometry are other important aspects for which superconducting spintronics might become important [155]. The controlled manipulation and management of energy currents and local and nonlocal heating and cooling effects in superconducting nanodevices are just at the beginning of being explored in detail. Clearly here lies a great potential of superconducting spintronics.

16.8.2 SPIN TORQUE

Central notions in spintronics, like nonequilibrium spin torque and spin-transfer torque [156–159], play an equally important role in S-F structures and are discussed in Chapters 7–9, Volume 1. For clean materials in equilibrium, the torque can be related to the spectrum of spin-polarized Andreev bound states, while for small bias voltages an ac component develops that is determined by the nearly adiabatic dynamics of the Andreev bound states [160, 161]. The equilibrium spin-transfer torque τ_{eq} in an S-FI-N-FI'-S' structure is related to the Josephson current I_c, the phase difference between S and S', $\Delta\chi$, and the angle $\Delta\varphi$ between FI and FI', by [162]

$$\hbar\frac{\partial I_c}{\partial \Delta\varphi} = 2e\frac{\partial \tau_{eq}}{\partial \Delta\chi}.$$

(16.68)

In ballistic Josephson junctions, the dispersion of an Andreev bound state with superconducting phase difference $\Delta\chi$ yields the bound-state contribution to the Josephson current. Similarly, the dispersion of an Andreev state with $\Delta\varphi$ yields the contribution of this state to the spin current for spin-polarization in direction of the spin torque.

In an S-FI-N-FI'-S' structure, the dc spin current shows subharmonic gap features due to multiple Andreev reflections (MAR), similar as the charge current shows in voltage biased Josephson junctions [163, 164]. For high transmission junctions, the main contribution to the dc spin current comes from consecutive spin rotations due to spin-dependent interface scattering when electrons and holes undergo MAR [161].

For a junction with a bias voltage $eV \ll \Delta$, the time evolution of spin-transfer torque is governed by the nearly adiabatic dynamics of the Andreev bound states. However, the dynamics of the bound-state spectrum leads to a nonequilibrium population of the Andreev bound states, for which reason the spin-transfer torque does not assume its instantaneous equilibrium value [161]. For the occupation to change, the bound-state energy must evolve in time to the continuum gap edges, where it can rapidly equilibrate with the quasiparticles, similar as in the adiabatic limit of ac Josephson junctions [165].

16.8.3 NONEQUILIBRIUM MAGNETIZATION DYNAMICS

Recent experiments concentrate on the possibility of exciting spin-waves in superconductor-ferromagnet devices via ferromagnetic resonance and studying the resulting effects of the spin pumping on the superconducting properties of the structure or, inversely, the effect of superconductivity on the magnetization dynamics in the ferromagnet. In particular, the option of generating spin currents via ferromagnetic resonance in order to create a nonequilibrium spin-polarization in superconductors has recently attracted attention. Precession of the ferromagnetic magnetization leads to pumping of the spin angular momentum [166]. The effective Gilbert damping of the spin precession in superconductor/ferromagnet heterostructures is affected by a reduction of losses when the device switches from the normal-conducting into the superconducting state [167, 168]. In the presence of a finite applied bias voltage, a transverse spin current pumped by ferromagnetic resonance strongly depends on Andreev reflection at the interfaces between the superconductor and the ferromagnet and on the mixing conductance [169].

In ferromagnetic Josephson junction devices, it has been shown that the coupled dynamics of magnetization and Josephson phase leads to Josephson plasma waves coupled to oscillations of the magnetization, affecting the form of the current-voltage characteristics in weak magnetic fields [170, 171]. A precessing magnetization can in Josephson junctions lead to long-range triplet amplitudes [172], to the nonequilibrium population of Andreev sidebands, and to dynamical spin currents that feed back to the precessing spins via spin-transfer torques [173].

Quasiparticle spin resonance has been applied recently in order to study the spin coherence time using on-chip microwave detection techniques [174].

16.8.4 Spin-Orbit Coupling

Spin-orbit effects play an important role whenever the inversion symmetry is broken, either in the bulk material (when it is lacking a center of inversion) or at interfaces and surfaces [175]. A prominent example of the spin-orbit effects at surfaces is the Rashba-Bychkov spin-orbit coupling [176, 177], and in the bulk, an example is the Dresselhaus coupling due to bulk-inversion asymmetry [178]. Interesting effects related to the spin-orbit interaction include the spin-Hanle effect [179–182], which describes the coherent rotation of a spin in an external magnetic field, and the Aharonov-Casher effect [183], leading to a phase on the Josephson current through a semiconducting ring attached to superconducting leads [184]. The Aharonov-Casher effect was, e.g., observed experimentally in the ring structures of HgTe/HgCdTe quantum wells [185] and in superconducting Josephson structures [186, 184, 187, 188]. The effect allows for control of the Josephson current through the control of the Aharonov-Casher phase by the gate voltage. Thus, this effect is a promising candidate for realizing new types of controllable devices in superconducting spintronics based on geometric phases.

Andreev reflection can be controlled by changing the magnetization orientation in spin-polarized transport across F/S junctions in the presence of Rashba and Dresselhaus interfacial spin-orbit fields [189]. This opens an avenue toward sensitive probing of interfacial spin-orbit fields in F/S junctions via Andreev reflection spectroscopy.

Another important consequence of spin-orbit coupling is the spin-Hall effect and the inverse spin-Hall effect [190–193], discussed further in Chapter 8, Volume 2 and Chapters 7–9, Volume 3. The spin-Hall effect is the generation of a spin current from a charge current as a result of spin-dependent asymmetric scattering in the presence of spin-orbit interaction. The inverse spin-Hall effect describes the reciprocal effect of a conversion of a spin current into a charge current. In superconductors, such as NbN, these effects are mediated by quasiparticle excitations above the superconducting gap, leading to a strong enhancement of the effects in the superconducting state [194].

Finally, the wide spectrum of phenomena in hybrid systems composed of superconductors and topological insulators (see, e.g., Refs. [195–198) constitute a prime example of how the field can contribute new insights to fundamental research and enable emerging applications in topologically protected quantum computing, discussed in Chapter 15, Volume 2, based on Marjona fermions. Instead of a strong spin-orbit coupling inherent to topological insulators, synthetic spin-orbit coupling generated through magnetic textures could also enable the formation of such Marjona fermions [199].

16.9 CONCLUDING REMARKS

The first two decades of the new millennium have seen a series of pivotal predictions and experimental discoveries relating to the interplay between superconductivity and ferromagnetism. This resulted in new synergies between the fields of superconductivity and spintronics. Superconducting spintronics has the potential and flexibility to become one of the major players in meeting the challenge of energy-efficient large-scale computing and data management, with the cost for cooling being largely outweighed by the gain in energy efficiency. The building blocks of superconducting spintronics are equal-spin Cooper pairs, which are generated at the interface between superconducting and magnetic materials in the presence of noncollinear magnetism. Such spin-polarized Cooper pairs carry spin-supercurrents in ferromagnets and thus contribute to spin transport and spin control. Geometric Berry phases appear in structures with a noncoplanar magnetization profile, enhancing functionality. Spin-orbit locking in noncentrosymmetric materials may lead to topological superconductivity. Thus, the hitherto notorious incompatibility of superconductivity and ferromagnetism has been not only overcome, but turned synergistic, providing an extraordinary potential for applications.

ACKNOWLEDGMENTS

I acknowledge helpful discussions and communications with Mikael Fogelström regarding Section 16.4, with Yakov Fominov, Pavel Leksin, and Bernd Büchner regarding Section 16.5, with Detlef Beckmann regarding Section 16.7, and with Lesley Cohen, Mark Blamire, Jason Robinson, Kun-Rok Jeon, and Hidekazu Kurebayashi concerning new experimental activities and devices. This work was supported by EPSRC grant EP/N017242/1.

REFERENCES

1. P. M. Tedrow and R. Meservey, Spin-dependent tunneling into ferromagnetic nickel, *Phys. Rev. Lett.* **26**, 192–195 (1971).
2. R. Meservey and P. M. Tedrow, Spin-polarized electron tunneling, *Phys. Rep.* **238**, 173–243 (1994).
3. V. L. Berezinskiĭ, Новая модель анизотропной фазы сверхтекучего He³, *Pis'ma Zh. Eksp. Teor. Fiz.* **20**, 628–631 (1974), Engl. transl.: *New model of the anisotropic phase of superfluid He³*, *JETP Lett.* **20**, 287–289 (1974).
4. B. S. Chandrasekhar, A note on the maximum critical field of high-field superconductors, *Appl. Phys. Lett.* **1**, 7–8 (1962).
5. A. M. Clogston, Upper limit for the critical field in hard superconductors, *Phys. Rev. Lett.* **9**, 266–267 (1962).
6. P. Fulde and R. A. Ferrell, Superconductivity in a strong spin-exchange field, *Phys. Rev.* **135**, A550–A563 (1964).
7. A. I. Larkin and Yu. N. Ovchinnikov, Неравномерное состояние сверхпроводников, *Zh. Eksp. Teor. Fiz.* **47**, 1136–1146 (1964), Engl. transl.: Nonuniform state of superconductors, *Sov. Phys. JETP* **20**, 762–769 (1965).
8. V. V. Ryazanov, V. A. Oboznov, A. Yu. Rusanov, A. V. Veretennikov, A. A. Golubov, and J. Aarts, Coupling of two superconductors through a ferromagnet: Evidence for a π junction, *Phys. Rev. Lett.* **86**, 2427–2430 (2001).
9. V. A. Oboznov, V. V. Bol'ginov, A. K. Feofanov, V. V. Ryazanov, and A. I. Buzdin, Thickness dependence of the Josephson ground states of superconductor-ferromagnet-superconductor junctions, *Phys. Rev. Lett.* **96**, 197003 (2006).
10. T. Tokuyasu, J. A. Sauls, and D. Rainer, Proximity effect of a ferromagnetic insulator in contact with a superconductor, *Phys. Rev. B* **38**, 8823–8833 (1988).
11. M. Eschrig, J. Kopu, J. C. Cuevas, and G. Schön, Theory of half-metal/superconductor heterostructures, *Phys. Rev. Lett.* **90**, 137003 (2003).

12. A. Cottet, D. Huertas-Hernando, W. Belzig, and Yu. V. Nazarov, Spin-dependent boundary conditions for isotropic superconducting Green's functions, *Phys. Rev. B* **80**, 184511 (2009) [Erratum: *Phys. Rev. B* **83**, 139901 (2011)].

13. M. Eschrig, A. Cottet, W. Belzig, and J. Linder, General boundary conditions for quasiclassical theory of superconductivity in the diffusive limit: Application to strongly spin-polarized systems, *New J. Phys.* **17**, 083037 (2015).

14. A. Brataas, Yu. V. Nazarov, G. E. W. Bauer, Finite-element theory of transport in ferromagnet-normal metal systems, *Phys. Rev. Lett.* **84**, 2481–2484 (2000).

15. M. Fogelström, Josephson currents through spin-active interfaces, *Phys. Rev. B* **62**, 11812–11819 (2000).

16. F. S. Bergeret, A. F. Volkov, and K. B. Efetov, Long-range proximity effects in superconductor-ferromagnet structures, *Phys. Rev. Lett.* **86**, 4096–4099 (2001).

17. M. Eschrig, T. Löfwander, T. Champel, J. C. Cuevas, J. Kopu, and G. Schön, Symmetries of pairing correlations in superconductor-ferromagnet nano-structures, *J. Low Temp. Phys.* **147**, 457–476 (2007).

18. Y. Tanaka and A. A. Golubov, Theory of the proximity effect in junctions with unconventional superconductors, *Phys. Rev. Lett.* **98**, 037003 (2007).

19. Z. Y. Chen, A. Biswas, I. Žutić et al., Spin-polarized transport across a $La_{0.7}Sr_{0.3}MnO_3/YBa_2Cu_3O_{7-x}$ interface: Role of Andreev bound states, *Phys. Rev. B* **63**, 212508 (2001).

20. F. Hübler, M. J. Wolf, T. Scherer, D. Wang, D. Beckmann, and H. v. Löhneysen, Observation of Andreev bound states at spin-active interfaces, *Phys. Rev. Lett.* **109**, 087004 (2012).

21. F. Pérez-Willard, J. C. Cuevas, C. Sürgers, P. Pfundstein, J. Kopu, M. Eschrig, and H. v. Löhneysen, Determining the current polarization in Al/Co nano-structured point contacts, *Phys. Rev. B* **69**, 140502 (2004).

22. T. Löfwander, R. Grein, and M. Eschrig, Is CrO_2 fully spin polarized? analysis of Andreev Spectra and excess current, *Phys. Rev. Lett.* **105**, 207001 (2010).

23. R. Grein, T. Löfwander, G. Metalidis, and M. Eschrig, Theory of superconductor-ferromagnet point-contact spectra: The case of strong spin polarization, *Phys. Rev. B* **81**, 094508 (2010).

24. F. B. Wilken and P. W. Brouwer, Impurity-assisted Andreev reflection at a spin-active half metal-superconductor interface, *Phys. Rev. B* **85**, 134531 (2012).

25. K. A. Yates, M. S. Anwar, J. Aarts et al., Andreev spectroscopy of CrO_2 thin films on TiO_2 and Al_2O_2, *Europhys. Lett.* **103**, 67005 (2013).

26. R. J. Soulen Jr., J. M. Byers, M. S. Osofsky et al., Measuring the spin polarization of a metal with a superconducting point contact, *Science* **282**, 85–88 (1998).

27. S. K. Upadhyay, A. Palanisami, R. N. Louie, and R. A. Buhrman, Probing ferromagnets with Andreev reflection, *Phys. Rev. Lett.* **81**, 3247–3250 (1998).

28. I. Žutić and S. Das Sarma, Spin-polarized transport and Andreev reflection in semiconductor/superconductor hybrid structures, *Phys. Rev. B* **60**, R16322–R16325 (1999).

29. C. Byers and T. Flatte, Probing spatial correlations with nanoscale two-contact tunneling, *Phys. Rev. Lett.* **74**, 306–309 (1995).

30. G. Deutscher and D. Feinberg, Coupling superconducting-ferromagnetic point contacts by Andreev reflections, *Appl. Phys. Lett.* **76**, 487–489 (2000).

31. D. Beckmann, H. B. Weber, and H. von Löhneysen, Evidence for crossed Andreev reflection in superconductor-ferromagnet hybrid structures, *Phys. Rev. Lett.* **93**, 197003 (2004).

32. G. Metalidis, M. Eschrig, R. Grein, and G. Schön, Nonlocal conductance via overlapping Andreev bound states in ferromagnet-superconductor hetero-structures, *Phys. Rev. B.* **82**, 180503(R) (2010).

33. J. C. Cuevas and M. Fogelström, Quasiclassical description of transport through superconducting contacts, *Phys. Rev. B.* **64**, 104502 (2001).

34. Yu. S. Barash and I. V. Bobkova, Interplay of spin-discriminated Andreev bound states forming the 0-π transition in superconductor-ferromagnet-superconductor junctions, *Phys. Rev. B* **65**, 144502 (2002).

35. A. A. Golubov, M. Yu. Kupriyanov, and Ya. V. Fominov, *Pis'ma Zh. Eksp. Teor. Fiz.* **75**, 709–713 (2002); Nonsinusoidal current-phase relations in SFS Josephson junctions, *JETP Lett.* **75**, 588–592 (2002).

36. V. Ambegaokar and A. Baratoff, Tunneling between superconductors, *Phys. Rev. Lett.* **10**, 486–489 (1963) [Erratum *Phys. Rev. Lett.* **11**, 104 (1963)].

37. I. O. Kulik and A. N. Omel'yanchuk, Свойства сверхпроводящих микром остов в чистом пределе, *Fiz. Nizk. Temp.* **3**, 945–948 (1977), Engl. transl.: Properties of superconducting microbridges in the pure limit, *Sov. J. Low Temp. Phys.* **3**, 459–461 (1977).

38. L. N. Bulaevskiĭ, V. V. Kuziĭ, and A. A. Sobyanin, Сверхпроводящая система со слабой связью с током в основном состоянии, *Pis'ma Zh. Eksp. Teor. Fiz.* **25**, 314–318 (1977), Engl. transl.: Superconducting system with weak coupling to the current in the ground state, *JETP Lett.* **25**, 290–294 (1977).

39. I. O. Kulik, *Zh. Eksp. Teor. Fiz.* **49**, 1211–1214 (1965), Engl. transl.: Magnitude of the critical Josephson tunnel current, *Sov. Phys. JETP* **22**, 841–843 (1966).

40. A. I. Buzdin, L. N. Bulaevskiĭ, and S. V. Panyukov, Критические колебания тока в зависимости от обменного поля и толщины ферромагнитного металла (F) в джозефсоновском переходе S-F-S, *Pis'ma Zh. Eksp. Teor. Fiz.* **35**, 147–152 (1982), Engl. transl.: Critical-current oscillations as a function of the exchange field and thickness of the ferromagnetic metal (F) in an S-F-S Josephson junction, *JETP Lett.* **35**, 178–180 (1982).

41. A. V. Andreev, A. I. Buzdin, and R. M. Osgood, π phase in magnetic-layered superconductors, *Phys. Rev. B* **43**, 10124–10131 (1991).

42. A. I. Buzdin and M. Yu. Kupriyanov, Джозефсоновское соединение с ферромагнитным слоем, *Pis'ma Zh. Eksp. Teor. Phys.* **53**, 308–312 (1991), Engl. trans.: Josephson junction with a ferromagnetic layer, *JETP Lett.* **53**, 321–326 (1991).

43. N. M. Chtchelkatchev, W. Belzig, Yu. V. Nazarov, and C. Bruder, *Pis'ma Zh. Eksp. Teor. Fiz.* **74**, 357–361 (2001), Reprinted in π–0 transition in supercondu ctor-ferromagnet-superconductor junctions. *JETP Lett.* **74**, 323–327 (2001).

44. A. I. Buzdin, Proximity effects in superconductor-ferromagnet heterostructures, *Rev. Mod. Phys.* **77**, 935–976 (2005).

45. A. V. Veretennikov, V. V. Ryazanov, V. A. Oboznov, A. Y. Rusanov, V. A. Larkin, and J. Aarts, Supercurrents through the superconductor-ferromagnet-sup erconductor (SFS) junctions, *Physica B.* **284–288**, 495–496 (2000).

46. T. Kontos, M. Aprili, J. Lesueur, G. Genêt, B. Stephanidis, and R. Boursier, Josephson junction through a thin ferromagnetic layer: Negative coupling, *Phys. Rev. Lett.* **89**, 137007 (2002).

47. H. Sellier, C. Baraduc, F. Lefloch, and R. Calemczuk, Temperature-induced crossover between 0 and π states in S/F/S junctions, *Phys. Rev. B* **68**, 054531 (2003).

48. V. V. Ryazanov, V. A. Oboznov, A. S. Prokofiev, V. V. Bolginov, and A. K. Feofanov, Superconductor-ferromagnet-superconductor π-junctions, *J. Low Temp. Phys.* **136**, 385–400 (2004).

49. G. Eilenberger, Transformation of Gor'kov's equation for type II superconductors into transport-like equations, *Z. Phys.* **214**, 195–213 (1968).

50. A. I. Larkin and Yu. N. Ovchinnikov, Квазиклассический метод в теории сверхпроводимости, *Zh. Eksp. Teor. Fiz.* **55**, 2262–2272 (1968), Engl. transl.: Quasiclassical method in the theory of superconductivity, *Sov. Phys. JETP* **28**, 1200–1205 (1969).

51. F. S. Bergeret, A. F. Volkov, and K. B. Efetov, Josephson current in superconductor-ferromagnet structures with a nonhomogeneous magnetization. [Note: after Eq. (2) it should read γ = $T(\mu)/2$.], *Phys. Rev. B* **64**, 134506 (2001).

52. N. G. Pugach, M. Yu. Kupriyanov, E. Goldobin, R. Kleiner, and D. Koelle, Super conductor-insulator-ferromagnet-superconductor Josephson junction: From the dirty to the clean limit. [Note: in Eq. (10) it should read $D(\mu_f)/2$ instead of $D(\mu_f)$.], *Phys. Rev. B* **84**, 144513 (2011).

53. Y. Blum, A. Tsukernik, M. Karpovski, and A. Palevski, Oscillations of the superconducting critical current in Nb-Cu-Ni-Cu-Nb junctions, *Phys. Rev. Lett.* **89**, 187004 (2002).

54. V. Shelukhin, A. Tsukernik, M. Karpovski et al., Observation of periodic π-phase shifts in ferromagnet-superconductor multilayers, *Phys. Rev. B* **73**, 174506 (2006).

55. A. A. Bannykh, J. Pfeiffer, V. S. Stolyarov, I. E. Batov, V. V. Ryazanov, and M. Weides, Josephson tunnel junctions with a strong ferromagnetic interlayer, *Phys. Rev. B* **79**, 054501 (2009).

56. M.A. Khasawneh, W. P. Pratt, and N. O. Birge, Josephson Junctions with a synthetic antiferromagnetic interlayer, *Phys. Rev. B* **80**, 020506(R) (2009).

57. D. Sprungmann, K. Westerholt, H. Zabel, M. Weides, and H. Kohlstedt, Josephson tunnel junctions with ferromagnetic $Fe_{0.75}Co_{0.25}$ barriers, *J. Phys. D: Appl. Phys.* **42**, 075005 (2009).

58. J. W. A. Robinson, S. Piano, G. Burnell, C. Bell, and M. G. Blamire, Critical current oscillations in strong ferromagnetic Pi-junctions, *Phys. Rev. Lett.* **97**, 177003 (2006).

59. A. K. Feofanov, V. A. Oboznov, V. V. Bol'ginov et al., Implementation of super conductor/ferromagnet/superconductor π-shifters in superconducting digital and quantum circuits, *Nat. Phys.* **6**, 593–597 (2010).

60. T. Yu. Karminskaya and M. Yu. Kupriyanov, Эффективное уменьшение энергии обмена в S (FN) S джозефсоновских структурах, *Pis'ma Zh. Eksp, Teor. Fiz.* **85**, 343–348 (2007), Engl. transl.: Effective decrease in the exchange energy in S–(FN)–S Josephson structures, *JETP Lett.* **85**, 286–281 (2007).

61. T. Yu. Karminskaya and M. Yu. Kupriyanov, Переход из состояния 0 в состояние π в структурах Джозефсона S-FNF-S, *Pis'ma Zh. Eksp, Teor. Fiz.* **86**, 65–70 (2007), Engl. transl.: Transition from the 0 state to the π state in S–FNF–S Josephson structures, *JETP Lett.* **86**, 61–66 (2007).

62. T. E. Golikova, F. Hübler, D. Beckmann, I. E. Batov, T. Yu. Karminskaya, M. Yu. Kupriyanov, A. A. Golubov, and V. V. Ryazanov, Double proximity effect in hybrid planar superconductor-(normal metal/ferromagnet)-superconductor structures, *Phys. Rev. B* **86**, 064416 (2012).

63. J. Gelhausen and M. Eschrig, Theory of a weak-link superconductor-ferromagnet Josephson structure, *Phys. Rev. B* **94**, 104502 (2016).

64. H. K. Wong, B. Y. Jin, H. Q. Yang, J. B. Ketterson, and J. E. Hilliard, Superconducting properties of V/Fe superlattices, *J. Low Temp. Phys.* **63**, 307–315 (1986).

65. A. I. Buzdin and M. Yu. Kupriyanov, Температура перехода сверхрешетки сверхпроводник-ферромагнетик, *Pis'ma Zh. Eksp. Teor. Fiz.* **52**, 1089–1091 (1990), Engl. transl.: Transition temperature of a superconductor-ferromagnet superlattice, *JETP Lett.* **52**, 487–491 (1990).

66. Z. Radović, M. Ledvij, L. Dobrosavljević-Grujić, A. I. Buzdin, and J. R. Clem, Transition temperatures of superconductor-ferromagnet superlattices, *Phys. Rev. B* **44**, 759–764 (1991).

67. L. R. Tagirov, Proximity effect and superconducting transition temperature in superconductor/ferromagnet sandwiches, *Physica C* **307**, 145–163 (1998).

68. L. R. Tagirov, Low-field superconducting spin switch based on a superconductor /ferromagnet multilayer, *Phys. Rev. Lett.* **83**, 2058–2061 (1999).

69. Ya. V. Fominov, N. M. Chtchelkatchev, and A. A. Golubov, *Pis'ma Zh. Eksp. Teor. Fiz.* **74**, 101–104 (2001), Critical temperature of superconductor/ferromagnet bilayers, *JETP Lett.* **74**, 96–99 (2001); Nonmonotonic critical temperature in superconductor/ferromagnet bilayers, *Phys. Rev. B.* **66**, 014507 (2002).

70. C. Strunk, C. Sürgers, U. Paschen, and H.v. Löhneysen, Superconductivity in layered Nb/Gd films, *Phys. Rev. B* **49**, 4053–4063 (1994).

71. J. S. Jiang, D. Davidović, D. H. Reich, and C. L. Chien, Oscillatory superconducting transition temperature in Nb/Gd multilayers, *Phys. Rev. Lett.* **74**, 314–317 (1995).

72. Th. Mühge, N. N. Garif'yanov, Yu. V. Goryunov, G. G. Khaliullin, L. R. Tagirov, K. Westerholt, I. A. Garifullin, and H. Zabel, Possible origin for oscillatory superconducting transition temperature in superconductor/ferromagnet multilayers, *Phys. Rev. Lett.* **77**, 1857–1860 (1996).

73. J. Aarts, J. M. E. Geers, E. Brück, A. A. Golubov, and R. Coehoorn, Interface transparency of superconductor/ferromagnetic multilayers, *Phys. Rev. B* **56**, 2779–2787 (1997).

74. L. Lazar, K. Westerholt, H. Zabel, L. R. Tagirov, Yu. V. Goryunov, N. N. Garif'yanov, and I. A. Garifullin, Superconductor/ferromagnet proximity effect in Fe/Pb/Fe trilayers, *Phys. Rev. B* **61**, 3711–3722 (2000).

75. I. A. Garifullin, D. A. Tikhonov, N. N. Garif'yanov et al., Re-entrant superconductivity in the superconductor/ferromagnet V/Fe layered system, *Phys. Rev. B* **66**, 020505(R) (2002).

76. V. Zdravkov, A. Sidorenko, G. Obermeier et al., Reentrant superconductivity in Nb/Cu$_{1-x}$Ni$_x$ bilayers, *Phys. Rev. Lett.* **97**, 057004 (2006).

77. V. I. Zdravkov, J. Kehrle, G. Obermeier et al., Reentrant superconductivity in superconductor/ferromagnetic-alloy bilayers, *Phys. Rev. B* **82**, 054517 (2010).

78. M. Eschrig, Spin-polarized supercurrents for spintronics, *Phys. Today* **64**, 43–49 (2011).

79. F. S. Bergeret, A. F. Volkov, and K. B. Efetov, Odd triplet superconductivity and related phenomena in superconductor-ferromagnet structures, *Rev. Mod. Phys.* **77**, 1321–1373 (2005).

80. M. Eschrig and T. Löfwander, Triplet supercurrents in clean and disordered half-metallic ferromagnets, *Nat. Phys.* **4**, 138–143 (2008); arXiv:cond-mat/0612533 (2006).

81. M. Eschrig, Spin-polarized supercurrents for spintronics: A review of current progress, *Rep. Prog. Phys.* **78**, 104501 (2015).

82. M. Houzet and A. I. Buzdin, Long range triplet Josephson effect through a ferromagnetic trilayer, *Phys. Rev. B* **76**, 060504(R) (2007).

83. V. Peña, Z. Sefrioui, D. Arias et al., Coupling of superconductors through a half-metallic ferromagnet: Evidence for a long-range proximity effect, *Phys. Rev. B* **69**, 224502 (2004).

84. R. S. Keizer, S. T. B. Goennenwein, T. M. Klapwijk, G. Miao, G. Xiao, and A. Gupta, A spin triplet supercurrent through the half-metallic ferromagnet CrO$_2$, *Nature (London)* **439**, 825–827 (2006).

85. M. S. Anwar, F. Czeschka, M. Hesselberth, M. Porcu, and J. Aarts, Long-range supercurrents through half-metallic ferromagnetic CrO$_2$, *Phys. Rev. B* **82**, 100501(R) (2010).

86. D. Sprungmann, K. Westerhold, H. Zabel, M. Weides, and H. Kohlstedt, Evidence for triplet superconductivity in Josephson junctions with barriers of the ferromagnetic Heusler alloy Cu$_2$MnAl, *Phys. Rev. B* **82**, 060505(R) (2010).

87. C. S. Turel, I. J. Guilaran, P. Xiong, and J. Y. T. Wei, Andreev nanoprobe of half-metallic CrO$_2$ films using superconducting cuprate tips, *Appl. Phys. Lett.* **99**, 192508 (2011).

88. Y. Kalcheim, O. Millo, M. Egilmez, J. W. A. Robinson, and M. G. Blamire, Evidence for anisotropic triplet superconductor order parameter in half-metallic ferromagnetic La$_{0.7}$Ca$_{0.3}$Mn$_3$O proximity coupled to superconducting Pr$_{1.85}$Ce$_{0.15}$CuO$_4$, *Phys. Rev. B* **85**, 104504 (2012).

89. C. Visani, Z. Sefrioui, J. Tornos et al., Villegas, Equal-spin Andreev reflection and long-range coherent transport in high-temperature superconductor/half-metallic ferromagnet junctions, *Nat. Phys.* **8**, 539–543 (2012).

90. I. Sosnin, H. Cho, V. T. Petrashov, and A. F. Volkov, Superconducting phase coherent electron transport in proximity conical ferromagnets, *Phys. Rev. Lett.* **96**, 157002 (2006).

91. J. W. A. Robinson, J. D. S. Witt, and M. G. Blamire, Controlled injection of spin-triplet supercurrents into a strong ferromagnet, *Science* **329**, 59–61 (2010).

92. T. S. Khaire, M. A. Khasawneh, W. P. Pratt Jr., and N. O. Birge, Observation of spin-triplet superconductivity in Co-based Josephson junctions, *Phys. Rev. Lett.* **104**, 137002 (2010).

93. R. Grein, M. Eschrig, G. Metalidis, and G. Schön, Spin-dependent cooper pair phase and pure spin supercurrents in strongly polarized ferromagnets, *Phys. Rev. Lett.* **102**, 227005 (2009).

94. Y. Asano, A. A. Golubov, Ya. V. Fominov, and Y. Tanaka, Unconventional surface impedance of a normal-metal film covering a spin-triplet superconductor due to odd-frequency cooper pairs, *Phys. Rev. Lett.* **107**, 087001 (2011).

95. Ya. V. Fominov, Y. Tanaka, Y. Asano, and M. Eschrig, Odd-frequency superconducting states with different types of Meissner response: Problem of coexistence, *Phys. Rev. B* **91**, 144514 (2015).

96. S. Oh, D. Youm, and M. R. Beasley, A superconductive magnetoresistive memory element using controlled exchange interaction, *Appl. Phys. Lett.* **71**, 2376–2378 (1997).

97. A. I. Buzdin, A. V. Vedyaev, N. N. Ryzhanova, Spin-orientation-dependent superconductivity in F/S/F structures, *Europhys. Lett.* **48**, 686–691 (1999).

98. T. Löfwander, T. Champel, and M. Eschrig, Phase diagrams of ferromagnet-superconductor multilayers with misaligned exchange fields, *Phys. Rev. B* **75**, 014512 (2007).

99. Ya. V. Fominov, A. A. Golubov, T. Yu. Karminskaya, M. Yu. Kupriyanov, R. G. Deminov, and L. R. Tagirov, *Pis'ma Zh. Eksp. Teor. Fiz.* **91**, 329–333 (2010), Reprinted: Superconducting triplet spin valve, *JETP Lett.* **91**, 308–313 (2010).

100. Ch.-Te Wu, O. T. Valls, and K. Haltermann, Proximity effects and triplet correlations in ferromagnet/ferromagnet/superconductor nanostructures, *Phys. Rev. B* **86**, 014523 (2012).

101. P. V. Leksin, N. N. Garif'yanov, A. A. Kamashev et al., Isolation of proximity-induced triplet pairing channel in a superconductor/ferromagnet spin valve, *Phys. Rev. B* **93**, 100502(R) (2016).

102. Ya. V. Fominov, A. A. Golubov, M. Yu. Kupriyanov, *Pis'ma Zh. Eksp. Teor. Fiz.* **77**, 609–614 (2003), Reprinted: Triplet proximity effects in FSF trilayers, *JETP Lett.* **77**, 510–515 (2003).

103. M. Božović and Z. Radović, Ferromagnet-superconductor proximity effect: The clean limit, *Europhys. Lett.* **70**, 513–519 (2005).

104. K. Halterman and O. T. Valls, Nanoscale ferromagnet-superconductor-ferromagnet switches controlled by magnetization orientation, *Phys. Rev. B* **72**, 060514(R) (2005).

105. G. Nowak, H. Zabel, K. Westerholt et al., Superconducting spin valves based on epitaxial Fe/V superlattices, *Phys. Rev. B.* **78**, 134520 (2008).

106. J. Y. Gu, C.-Y. You, J. S. Jiang, J. Pearson, Ya. B. Bazaliy, and S. D. Bader, Magnetization-orientation dependence of the superconducting transition temperature in the ferromagnet-superconductor-ferromagnet system: CuNi/Nb/CuNi, *Phys. Rev. Lett.* **89**, 267001 (2002).

107. A. Potenza and C. H. Marrows, Superconductor-ferromagnet CuNi/Nb/CuNi trilayers as superconducting spin-valve core structures, *Phys. Rev. B* **71**, 180503(R) (2005) [Erratum: *Phys. Rev. B.* **72**, 069901(E) (2005)].

108. K. Westerholt, D. Sprungmann, H. Zabel et al., Superconducting spin valve effect of a V layer coupled to an antiferromagnetic [Fe/V] superlattice, *Phys. Rev. Lett.* **95**, 097003 (2005).

109. I. C. Moraru, W. P. Pratt, Jr., and N. O. Birge, Magnetization-dependent T_c shift in ferromagnet/superconductor/ferromagnet trilayers with a strong ferromagnet, *Phys. Rev. Lett.* **96**, 037004 (2006).

110. J. Zhu, I. N. Krivorotov, K. Halterman, and O. Valls, Angular dependence of the superconducting transition temperature in ferromagnet-superconductor-ferromagnet trilayers, *Phys. Rev. Lett.* **105**, 207002 (2010).

111. P. V. Leksin, N. N. Garif'yanov, I. A. Garifullin, J. Schumann, V. Kataev, O. G. Schmidt, and B. Büchner, Manifestation of new interference effects in a superconductor-ferromagnet spin valve, *Phys. Rev. Lett.* **106**, 067005 (2011).

112. P. V. Leksin, N. N. Garif'yanov, I. A. Garifullin et al., Evidence for triplet superconductivity in a superconductor-ferromagnet spin valve, *Phys. Rev. Lett.* **109**, 057005 (2012).

113. A. Yu. Rusanov, S. Habraken, and J. Aarts, Inverse spin switch effects in ferromagnet-superconductor-ferromagnet trilayers with strong ferromagnets, *Phys. Rev. B* **73**, 060505(R) (2006).

114. R. Steiner and P. Ziemann, Magnetic switching of the superconducting transition temperature in layered ferromagnetic/superconducting hybrids: Spin switch versus stray field effects, *Phys. Rev. B* **74**, 094504 (2006).

115. D. Stamopoulos, E. Manios, and M. Pissas, Enhancement of superconductivity by exchange bias, *Phys. Rev. B* **75**, 014501 (2007).

116. A. Singh, C. Sürgers, R. Hoffmann et al., Spin-polarized current versus stray field in a perpendicularly magnetized superconducting spin switch, *Appl. Phys. Lett.* **91**, 152504 (2007).

117. N. Banerjee, C. B. Smiet, R. G. J. Smits et al., Evidence for spin selectivity of triplet pairs in superconducting spin valves, *Nat. Commun.* **5**, 3048 (2014).

118. V. I. Zdravkov, J. Kehrle, G. Obermeier et al., Experimental observation of the triplet spin-valve effect in a superconductor-ferromagnet heterostructure, *Phys. Rev. B* **87**, 144507 (2013).

119. A. A. Jara, Ch. Safranski, I. N. Krivorotov et al., Angular dependence of superconductivity in superconductor/spin-valve heterostructures, *Phys. Rev. B.* **89**, 184502 (2014).

120. M. G. Flokstra, T. C. Cunningham, J. Kim et al., Controlled suppression of superconductivity by the generation of polarized Cooper pairs in spin-valve structures, *Phys. Rev. B.* **91**, 060501(R) (2015).

121. A. Singh, S. Voltan, K. Lahabi, and J. Aarts, Colossal proximity effect in a superconducting triplet spin valve based on the half-metallic ferromagnet CrO_2, *Phys. Rev. X* **5**, 021019 (2015).

122. T. Löfwander, T. Champel, J. Durst, and M. Eschrig, Interplay of magnetic and superconducting proximity effects in ferromagnet-superconductor-ferromagnet trilayers, *Phys. Rev. Lett.* **95**, 187003 (2005).

123. F. S. Bergeret, A. F. Volkov, and K. B. Efetov, Enhancement of the Josephson current by an exchange field in superconductor-ferromagnet structures, *Phys. Rev. Lett.* **86**, 3140–3143 (2001).

124. A. Iovan, T. Golod, and V. M. Krasnov, Controllable generation of a spin-triplet supercurrent in a Josephson spin valve, *Phys. Rev. B* **90**, 134514 (2014).

125. T. Champel, T. Löfwander, and M. Eschrig, 0–π transitions in a superconductor/chiral ferromagnet/superconductor junction induced by a homogeneous cycloidal spiral, *Phys. Rev. Lett.* **100**, 077003 (2008).

126. A. F. Volkov, Ya. V. Fominov, and K. B. Efetov, Long-range odd triplet superconductivity in superconductor-ferromagnet structures with Néel walls, *Phys. Rev. B* **72**, 184504 (2005).

127. Ya. V. Fominov, A. F. Volkov, and K. B. Efetov, Josephson effect due to the long-range odd-frequency triplet superconductivity in SFS junctions with Néel domain walls, *Phys. Rev. B* **75**, 104509 (2007).

128. N. G. Pugach, M. Safonchik, T. Champel et al., Superconducting spin valves controlled by spiral re-orientation in B20-family magnets, *Appl. Phys. Lett.* **111**, 162601 (2017).

129. N. Poli, J. P. Morten, M. Urech, A. Brataas, D. B. Haviland, and V. Korenivski, Spin injection and relaxation in a mesoscopic superconductor, *Phys. Rev. Lett.* **100**, 136601 (2008).

130. F. Hübler, M. J. Wolf, D. Beckmann, and H. V. Löhneysen, Long-range spin-polarized quasiparticle transport in mesoscopic Al superconductors with a Zeeman splitting, *Phys. Rev. Lett.* **109**, 207001 (2012).

131. C. H. L. Quay, D. Chevallier, C. Bena, and M. Aprili, Spin imbalance and spin-charge separation in a mesoscopic superconductor, *Nat. Phys.* **9**, 84–88 (2013).

132. M. J. Wolf, F. Hübler, S. Kolenda, H. V. Löhneysen, and D. Beckmann, Spin injection from a normal metal into a mesoscopic superconductor, *Phys. Rev. B* **87**, 024517 (2013).

133. M. J. Wolf, C. Sürgers, G. Fischer, and D. Beckmann, Spin-polarized quasiparticle transport in exchange-split superconducting aluminum on europium sulfide, *Phys. Rev. B* **90**, 144509 (2014).

134. R. J. Elliot, Theory of the effect of spin-orbit coupling on magnetic resonance in some semiconductors, *Phys. Rev.* **96**, 266–279 (1954).

135. Y. Yafet, g factors and spin-lattice relaxation of conduction electrons, *Solid State Phys.* **14**, 1–98 (1963), in *Advances in Research and Applications*, F. Seitz and D. Turnbull (Eds.), Academic Press, New York, 1963, doi:10.1016/S0081-1947(08)60259-3.

136. J. Clarke, Experimental observation of pair-quasiparticle potential difference in nonequilibrium superconductors, *Phys. Rev. Lett.* **28**, 1363–1366 (1972).

137. M. Tinkham and J. Clarke, Theory of pair-quasiparticle potential difference in nonequilibrium superconductors, *Phys. Rev. Lett.* **28**, 1366–1369 (1972).

138. D. Beckmann, Spin manipulation in nanoscale superconductors, *J. Phys.: Condens. Matt.* **28**, 163001 (2016).

139. H. L. Zhao and S. Hershfield, Tunneling, relaxation of spin-polarized quasiparticles, and spin-charge separation in superconductors, *Phys. Rev. B* **52**, 3632–3638 (1995).

140. S. Takahashi, H. Imamura, and S. Maekawa, Spin imbalance and magnetoresistance in ferromagnet/superconductor/ferromagnet double tunnel junctions, *Phys. Rev. Lett.* **82**, 3911–3914 (1999).

141. P. G. de Gennes, *Superconductivity of Metals an Alloys*, W. A. Benjamin Inc., New York, 1966, Addison-Wesley, Reading, MA (1989).

142. K. Maki, Gapless superconductivity, in *Superconductivity*, R. D. Parks (Ed.), Chapter **18**, Dekker, New York, pp. 1035–1105, 1969.

143. A. A. Abrikosov and L. P. Gor'kov, Вклад в теорию сверхпроводящих сплавов с парамагнитными примесями, *Zh. Eksp. Teor. Fiz.* **39**, 1781–1796 (1960), Engl. transl.: Contribution to the theory of superconducting alloys with paramagnetic impurities, *Sov. Phys. JETP* **12**, 1243–1253 (1961).

144. K. Maki, The behavior of superconducting thin films in the presence of magnetic fields and currents, *Prog. Theor. Phys.* **31**, 731–741 (1964).

145. K. Maki, Pauli paramagnetism and superconducting state. II, *Prog. Theor. Phys.* **32**, 29–36 (1964).

146. I. Giaever, Energy gap in superconductors measured by electron tunneling. *Phys. Rev. Lett.* **5**, 147–148 (1960).

147. O. Shevtsov and T. Löfwander, Spin imbalance in hybrid superconducting structures with spin-active interfaces, *Phys. Rev. B* **90**, 085432 (2014).

148. I. V. Bobkova and A. M. Bobkov, Long-range spin imbalance in mesoscopic superconductors under Zeeman splitting, *JETP Lett.* **101**, 118–124 (2015).

149. M. Silaev, P. Virtanen, F. S. Bergeret, and T. T. Heikkliä, Long-range spin accumulation from heat injection in mesoscopic superconductors with Zeeman splitting, *Phys. Rev. Lett.* **114**, 167002 (2015).

150. T. Krishtop, M. Houzet, and J. S. Meyer, Nonequilibrium spin transport in Zeeman-split superconductors, *Phys. Rev. B.* **91**, 121407(R) (2015).

151. I. V. Bobkova and A. M. Bobkov, Injection of nonequilibrium quasiparticles into Zeeman-split superconductors: A way to create long-range spin imbalance, *Phys. Rev. B* **93**, 024513 (2016).

152. P. Machon, M. Eschrig, and W. Belzig, Nonlocal thermoelectric effects and nonlocal Onsager relations in a three-terminal proximity-coupled superconductor-ferromagnet device, *Phys. Rev. Lett.* **110**, 047002 (2013).

153. A. Ozaeta, P. Virtanen, F. S. Bergeret, T. T. Heikkilä, Predicted very large thermoelectric effect in ferromagnet-superconductor junctions in the presence of a spin-splitting magnetic field, *Phys. Rev. Lett.* **112**, 057001 (2014).

154. S. Kolenda, M. J. Wolf, D. Beckmann, Observation of thermoelectric currents in high-field superconductor-ferromagnet tunnel junctions, *Phys. Rev. Lett.* **116**, 097001 (2016).

155. F. Giazotto, T. T. Heikkilä, A. Luukanen, A. M. Savin, and J. P. Pekola, Opportunities for mesoscopics in thermometry and refrigeration: Physics and applications, *Rev. Mod. Phys.* **78**, 217–274 (2006).

156. J. C. Slonczewski, Current-driven excitation of magnetic multilayers, *J. Magn. Magn. Mater.* **159**, L1–L7 (1996).

157. D. C. Ralph and M. D. Stiles, Spin transfer torques, *J. Magn. Magn. Mat.* **320**, 1190–1216 (2008).

158. A. Brataas, A. D. Kent, and H. Ohno, Current-induced torques in magnetic materials, *Nat. Mater.* **11**, 372–381 (2012).

159. N. Locatelli, V. Cros, and J. Grollier, Spin-torque building blocks, *Nat. Mater.* **13**, 11–20 (2014).

160. E. Zhao and J. A. Sauls, Dynamics of spin transport in voltage-biased Josephson junctions, *Phys. Rev. Lett.* **98**, 206601 (2007).

161. E. Zhao and J. A. Sauls, Theory of nonequilibrium spin transport and spin-transfer torque in superconducting-ferromagnetic nanostructures, *Phys. Rev. B* **78**, 174511 (2008).

162. X. Waintal and P. W. Brouwer, Magnetic exchange interaction induced by a Josephson current, *Phys. Rev. B* **65**, 054407 (2002).

163. R. Kümmel and W. Senftinger, Andreev-reflected wave packets in voltage-biased superconducting quantum wells, *Z. Phys. B – Condensed Matter* **59**, 275–281 (1985).

164. G. B. Arnold, Superconducting tunneling without the tunneling Hamiltonian. II. Subgap harmonic structure, *J. Low Temp. Phys.* **68**, 1–27 (1987).

165. D. V. Averin and A. Bardas, ac Josephson effect in a single quantum channel, *Phys. Rev. Lett.* **75**, 1831–1834 (1995).

166. Y. Tserkovnyak, A. Brataas, G. E. W. Bauer, and B. I. Halperin, Nonlocal magnetization dynamics in ferromagnetic heterostructures, *Rev. Mod. Phys.* **77**, 1375–1421 (2005).

167. C. Bell, S. Milikisyants, M. Huber, and J. Aarts, Spin dynamics in a superconductor-ferromagnet proximity system, *Phys. Rev. Lett.* **100**, 047002 (2008).

168. J. P. Morten, A. Brataas, G. E. W. Bauer, W. Belzig, and Y. Tserkovnyak, Proximity effect-assisted decay of spin currents in superconductors, *Eur. Phys. Lett.* **84**, 57008 (2008).

169. H. J. Skadsem, A. Brataas, J. Martinek, and Y. Tserkovnyak, Ferromagnetic resonance and voltage-induced transport in normal metal-ferromagnet-superconductor trilayer, *Phys. Rev. B* **84**, 104420 (2011).

170. S. Mai, E. Kandelaki, A. F. Volkov, and K. B. Efetov, Interaction of Josephson and magnetic oscillations in Josephson tunnel junctions with a ferromagnetic layer, *Phys. Rev. B* **84**, 144519 (2011).

171. S. Hikino, M. Mori, S. Takahashi, and S. Maekawa, Microwave-induced supercurrent in a ferromagnetic Josephson junction, *Supercond. Sci. Technol.* **24**, 024008 (2011).

172. M. Houzet, Ferromagnetic Josephson junction with precessing magnetization, *Phys. Rev. Lett.* **101**, 057009 (2008).

173. C. Holmqvist, W. Belzig, and M. Fogelström, Spin-precession-assisted supercurrent in a superconducting quantum point contact coupled to a single-molecule magnet, *Phys. Rev. B* **86**, 054519 (2012).

174. C. H. L. Quay, M. Weideneder, Y. Chiffaudel, C. Strunk, and M. Aprili, Quasiparticle spin resonance and coherence in superconducting aluminium, *Nat. Comm.* **6**, 8660 (2015), doi:10.1038/ncomms9660.

175. K. V. Samokhin, Spin-orbit coupling and semiclassical electron dynamics in noncentrosymmetric metals, *Ann. Phys. (N.Y.)* **324**, 2385–2407 (2009).

176. E. I. Rashba, Свойства полупроводников с контуром экстремума. Циклотрон и комбинационный резонанс в магнитном поле, перпендикулярном плоскости петли, *Fizika Tverd. Tela* **2**(6), 1224–1238 (1960), Engl. transl.: Properties of semiconductors with an extremum loop.1. Cyclotron and combinational resonance in a magnetic field perpendicular to the plane of the loop, *Sov. Phys. Solid State* **2**, 1109–1122 (1960).

177. Yu. L. Bychkov and E. I. Rashba, Oscillatory effects and the magnetic suscepti-bility of carriers in inversion layers, *J. Phys. C.* **17**, 6039–6045 (1984).

178. G. Dresselhaus, Spin-orbit coupling effects in zinc blended structures, *Phys. Rev.* **100**, 580–586 (1955).

179. W. Hanle, Über den Zeemaneffekt bei der Resonanzfluoreszenz (On the Zeeman effect in resonance flourescence), *Naturwissenschaften* **11**, 690–691 (1923).

180. I. Žutić, J. Fabian, and S. Das Sarma, Spintronics: Fundamentals and applica-tions, *Rev. Mod. Phys.* **76**, 323–410 (2004).

181. D. D. Awschalom and M. E. Flatté, Challenges for semiconductor spintronics, *Nat. Phys.* **3**, 153–159 (2007).

182. M. Silaev, P. Virtanen, T. T. Heikkilä, and F. S. Bergeret, Spin Hanle effect in mesoscopic superconductors, *Phys. Rev. B.* **91**, 024506 (2015).

183. Y. Aharonov and A. Casher, Topological quantum effects for neutral particles, *Phys. Rev. Lett.* **53**, 319–321 (1984).

184. Xin Liu, M. F. Borunda, Xiong-Jun Liu, and J. Sinova, Control of Josephson current by Aharonov-Casher phase in a Rashba ring, *Phys. Rev. B.* **80**, 174524 (2009).

185. M. König, A. Tschetschetkin, E. M. Hankiewicz et al., Direct observation of the Aharonov-Casher phase, *Phys. Rev. Lett.* **96**, 076804 (2006).

186. K. A. Matveev, A. I. Larkin, and L. I. Glazman, Persistent current in supercon-ducting nanorings, *Phys. Rev. Lett.* **89**, 096802 (2002).

187. I. M. Pop, B. Doucot, L. Ioffe et al., Experimental demonstration of Aharonov-Casher interference in a Josephson junction circuit, *Phys. Rev. B* **85**, 094503 (2012).

188. M. T. Bell, W. Zhang, L. B. Ioffe, and M. E. Gershenson, Spectroscopic evi-dence of the Aharonov-Casher effect in a Cooper pair box, *Phys. Rev. Lett.* **116**, 107002 (2016).

189. P. Högl, A. Matos-Abiague, I. Žutić, and J. Fabian, Magnetoanisotropic Andreev reflection in ferromagnet/superconductor junctions, *Phys. Rev. Lett.* **115**, 116601 (2015).

190. S. Takahashi and S. Maekawa, Hall effect induced by a spin-polarized current in superconductors, *Phys. Rev. Lett.* **88**, 116601 (1999).

191. H. Kontani, J. Goryo, and D. S. Hirashima, Intrinsic spin Hall effect in the *s*-wave superconducting state: Analysis of the Rashba model, *Phys. Rev. Lett.* **102**, 086602 (2009).

192. S. Takahashi and S. Maekawa, Spin Hall effect in superconductors, *J. Appl. Phys.* **51**, 010110 (2012).

193. T. Wakamura, N. Hasegawa, K. Ohnishi, Y. Niimi, and Y. Otani, Spin injection into a superconductor with strong spin-orbit coupling, *Phys. Rev. Lett.* **112**, 036602 (2014).

194. T. Wakamura, H. Akaike, Y. Omori, Y. Niimi, S. Takahashi, A. Fujimaki, S. Maekawa, and Y. Otani, Quasiparticle-mediated spin Hall effect in a supercon-ductor, *Nat. Mater.* **14**, 675–679 (2015).

195. L. Fu and C. L. Kane, Superconducting proximity effect and Majorana fermions at the surface of a topological insulator, *Phys. Rev. Lett.* **100**, 096407 (2008).

196. X.-L. Qi, T. L. Hughes, S. Raghu, and S.-C. Zhang, Time-reversal-invariant topological superconductors and superfluids in two and three dimensions, *Phys. Rev. Lett.* **102**, 187001 (2009).

197. Y. Tanaka, M. Sato, and N. Nagaosa, Symmetry and topology in superconductors -odd-frequency pairing and edge states, *J. Phys. Soc. Jpn.* **81**, 011013 (2012).
198. J. Alicea, New directions in the pursuit of Majorana fermions in solid state systems, *Rep. Prog. Phys.* **75** 076501 (2012).
199. G. L. Fatin, A. Matos-Abiague, B. Scharf, and I. Žutić, Wireless majorana fermions: From magnetic tunability to braiding, *Phys. Rev. Lett.* **117**, 077002 (2016).

Multiferroic Tunnel Junctions

Manuel Bibes and Agnès Barthélémy

E lectron tunneling through a vacuum or an ultrathin insulating barrier (1 to about 5 nm) is a well-known phenomenon that dates back to the development of quantum mechanics [1] and is described in all quantum mechanics textbooks. It recently attracted a lot of attention not only from a fundamental point of view but also for its device perspectives. This is exemplified by the work performed on superconducting [2] and magnetic tunnel junctions (see Chapters 11 through 13, Volume 1). These latter junctions are composed of a thin dielectric barrier sandwiched between two ferromagnetic electrodes. Upon changing the relative orientation of the magnetizations of the two electrodes, the conductance of the junctions changes, leading to a tunnel magnetoresistance (TMR) effect. This TMR is nowadays exploited for the realization of a new type of nonvolatile memory called magnetoresistive random access memory (MRAM) (see Chapter 13, Volume 3) in which the information is stored by the parallel or antiparallel orientation of the magnetizations of the electrodes.

Until now, very little work has been reported on junctions with a ferroelectric (FE) tunnel barrier despite the fact that the basic idea was reported by Esaki almost 40 years ago [3]. This may be because the growth of ultrathin films of FE materials is challenging and also because of the collective nature of ferroelectricity that is not always conserved at the nanometer scale required for operating in the tunneling regime. Nevertheless, recent theoretical and experimental advances on perovskite FE oxide thin films demonstrate clearly that in some conditions ferroelectricity persists down to at least 1 nm (see Section 17.2). This makes the conception of ferroelectric tunnel junctions (FTJs) composed of an FE tunnel barrier and two metallic electrodes feasible (see Section 17.2). Furthermore, if the two electrodes are ferromagnetic, or if the barrier on its own is composed of a multiferroic material (see Section 17.1), the junction becomes a multiferroic tunnel junction (MFTJ) (see Sections 17.3 and 17.4), in which the interplay between ferroelectricity and magnetism not only holds great promises for applications [4] but also is a fantastic playground for physicists.

17.1 INTRODUCTION TO MULTIFERROICS

Multiferroics are a relatively rare class of multifunctional materials that simultaneously exhibit several ferroic orders among ferromagnetic, ferroelectric, and ferroelastic (ferrotoroidic ordering [5] is also sometimes included) [6, 7]. Given the scarcity of compounds that present two or more strictly ferroic orders, antiferroic orders (e.g., antiferromagnetic) are usually considered. Most of the currently investigated multiferroics are generally magnetic and FE, very few showing a finite large moment (corresponding to ferro- or ferrimagnetic ordering). Practically, the vast majority of multiferroics are thus FE antiferromagnets or weak ferromagnets.

While the coexistence of several ferroic orders in the same material has been thought, since the 1970s, to be an interesting way to realize multiple-state memories [8], what looks more appealing today is the coupling that often exists between the ferroic orders. As sketched in Figure 17.1, this coupling may enable the manipulation of a given ferroic order parameter (FE polarization, magnetization, and strain) by an external stimulus different from the usual one (electric field, magnetic field, and stress, respectively). In the context of spintronics, a promising type of coupling is the so-called magnetoelectric coupling [9] that links the FE polarization and the magnetization, and potentially allows the manipulation of polarization by a magnetic field, and most importantly, of magnetization by an electric field.

Single-phase FE-magnetic multiferroics (simply referred to as "multiferroics" henceforth) are scarce and many are oxides [10, 11]. The technological

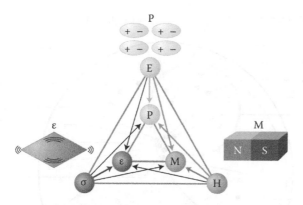

FIGURE 17.1 Phase control in ferroics and multiferroics. The electric field E, magnetic field H, and stress σ controls the electric polarization P, magnetization M, and strain ε, respectively. In a ferroic material, P, M, or ε are spontaneously formed to produce ferromagnetism, ferroelectricity, or ferroelasticity, respectively. In a multiferroic, the coexistence of at least two ferroic forms of ordering leads to additional interactions. In a magnetoelectric multiferroic, a magnetic field may control P or an electric field may control M. (After Spaldin, N.A. et al., *Science* 309, 391, 2005. With permission.)

advances in oxide thin film growth in the 1990s naturally led researchers to favor oxide materials to explore the physics and device potential of multiferroics in the thin film form. Figure 17.2 shows a Venn diagram classifying different types of insulating ferroic oxides according to their FE and/or magnetic properties. While many FE or ferri/ferromagnetic insulating oxides do exist, only a handful of materials are believed to possess both properties simultaneously (experimental indications, somewhat controversial, do exist for $La_{0.1}Bi_{0.9}MnO_3$ [12], $CoCr_2O_4$ [13], and Bi_2FeCrO_6 [14]), and only at a low temperature. Many more are FE and antiferromagnetic, and remarkably most of these show a magnetoelectric coupling. So far, $BiFeO_3$ is perhaps the only room temperature magnetoelectric multiferroic [15], with an FE Curie temperature of 1100 K [16] and a Néel temperature of 640 K [17]. From the inspection of Figure 17.2, it clearly appears that many FE, ferromagnetic, and multiferroic materials crystallize in a reduced number of structural families, such as the perovskite structure. Accordingly, they may be combined into epitaxial heterostructures in which their properties can be exploited independently or through couplings appearing at interfaces. In particular, so-called artificial multiferroics can be engineered by the combination of, for example, ferroelectrics with magnetic materials [18]. As we will see in Sections 17.3 and 17.4, tunnel junctions may be defined using either single-phase or artificial multiferroics.

Since the renaissance of multiferroics in 2003, much effort has been made to decipher the complex physical mechanisms at the origin of the magnetoelectric coupling. More practically, various strategies have also been deployed to achieve an electrical control of magnetization. Approaches based on single-phase multiferroics have had limited success. In $HoMnO_3$, it was shown that an electric field could modify the type of magnetic ordering and generate a large finite magnetic moment, but only at low temperature [19].

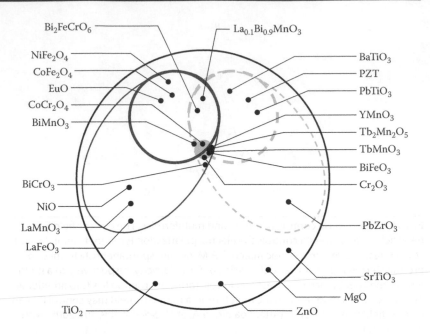

FIGURE 17.2   Classification of insulating oxides. The largest circle represents all insulating oxides among which one finds electrically polarizable materials (light gray, dotted ellipse) and magnetically polarizable materials (dark gray ellipse). Within each ellipse, the circle represents materials with a finite polarization (ferroelectrics) and/or a finite magnetization (ferro- and ferrimagnets). Depending on the definition, multiferroics correspond to the intersection between the ellipses or the circles. The small disk in the middle denotes systems exhibiting a magnetoelectric coupling. (After Béa, H. et al., *J. Phys.: Condens. Matter* 20, 434221, 2008. With permission; inspired by Eerenstein, W. et al., *Nature* 442, 759, 2006.)

The lack of room temperature FE-ferromagnets has driven the research to bilayer systems combining either an FE-antiferromagnet and a soft ferromagnet coupled by exchange [20–23], or a piezoelectric–FE with a magnetostrictive ferromagnet [24–27]. In these so-called artificial multiferroics, an "extrinsic" magnetoelectric coupling may be engineered, with coupling coefficients much larger than that of the "intrinsic" magnetoelectric coupling present in single-phase multiferroics.

Using this bilayer approach, the local domain structure of microstructured ferromagnetic pads was shown to be efficiently manipulated by an electric field. In Figure 17.3a, we present the results of Chu et al. on $BiFeO_3$/CoFe architectures [28]. The ferromagnetic domains in the CoFe are exchange-coupled to the antiferromagnetic–FE domains in the $BiFeO_3$ (BFO). Applying an in-plane voltage across the BFO modifies the antiferromagnetic–FE domain structure, and consequently, the ferromagnetic domain structure of the CoFe dot probed by XMCD-PEEM. Figure 17.3b illustrates the complementary study of Chung et al. on an architecture combining a thick piezoelectric $Pb(Zr,Ti)O_3$ (PZT) layer with Ni nanobars [26]. Due to the converse piezoelectric effect, when a voltage is applied to the PZT, it deforms and so does the other layers grown on top of it, including the Ni elements. The resulting magnetoelastic anisotropy induces changes

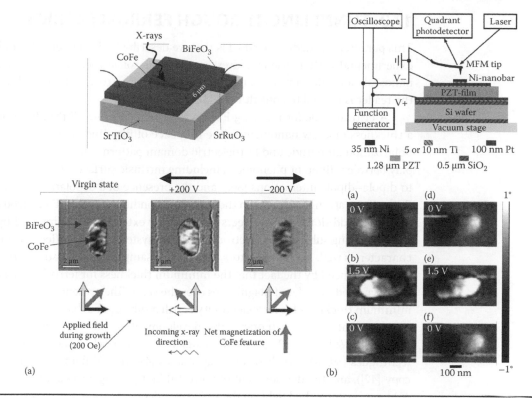

FIGURE 17.3 Magnetoelectric control of magnetization. (a) Top: Sketch of a device consisting of a micronic CoFe dot deposited onto a $BiFeO_3$ film equipped with $SrRuO_3$ planar electrodes. The ferromagnetic domain structure of the CoFe dot is probed by x-ray photoemission electron microscopy (bottom) and the domain patterned is reversibly switched by applying an in-plane voltage of ± 200 V across the $BiFeO_3$. (After Chu, Y.H. et al., *Nat. Mater.* 7, 478, 2008. With permission.) (b) Top: Sketch of the magnetoelectric Ni nanobar/PZT film and associated experimental setup. Bottom: Magnetic force microscopy images of the Ni nanobar for various voltages applied across the PZT. (After Chung, T.-K. et al., *Appl. Phys. Lett.* 94, 132501, 2009. With permission.)

in the ferromagnetic domain structure in Ni, as visible from magnetic force microscopy (MFM) images. Prospective developments beyond these promising results include the magnetoelectric control of the response of a spintronic device [4]. This would pave the way toward MRAMs in which a spin-based binary information would be written by the application of a voltage across a multiferroic, that is, at a very low energy cost compared to even the best state-of-the-art approaches, such as spin-transfer (see Chapters 7, 8, and 5, Volume 1).

En route to the grand objective of a magnetoelectrically controllable spintronics device, a lot has been learned on the physical properties of multiferroics, especially in thin film form. It is now possible to engineer BFO films with determined multiferroic domain structures [29], to tailor their polarization and critical temperatures by epitaxial strain [30], etc. However, the behavior of multiferroic films at thicknesses compatible with tunneling (typically 1–5 nm) is not well known. In the following paragraphs, we discuss important related issues currently under investigation, namely, the physics of tunneling through a ferroelectric and the problematic depression of ferroelectricity often observed in ultrathin films.

17.2 TUNNELING THROUGH FERROELECTRICS

This paragraph is dedicated to FTJs that we describe both theoretically and experimentally after briefly addressing the issue of ferroelectricity at the nanoscale level (for a more complete description of ferroelectric thin films, the reader is invited to consult Refs. [31–35]).

A critical issue for realizing FTJs is to be able to grow FE thin films at a thickness of a few nanometers. The influence of film thickness on the FE polarization amplitude and ferroelectric domain pattern is due to the interplay between different phenomena including intrinsic surface effects related to dipole–dipole interactions [36],* and the presence of a depolarizing field. Both effects strongly depend on the boundary conditions at the FE/electrode interface. Additional strain effects produced by external stress imposed by the underlying substrate or electrode/substrate system strongly affect the FE character as well. Extrinsic factors, such as low sample quality, also contribute as evidenced by the fact that the minimum thickness for ferroelectricity has decreased by order of magnitudes over the years. The presence or not of a minimum thickness for ferroelectricity is still a subject of debate, but recent experimental advances (notably the development of piezoresponse force microscopy (PFM) [37, 38], improvements in x-ray photoelectron diffraction techniques [39, 40], synchrotron x-ray studies [41], or UV Raman spectroscopy [42]), and the development of powerful first principle calculations [43, 44] have recently shed light on these FE size effects.

The screening of the depolarizing field appears as the key parameter that sets the critical thickness. The effect of the depolarizing field increases as the film thickness decreases [44] and may lead to the suppression of ferroelectricity and the concomitant reduction of the tetragonality of the unit cell. Other ways to reduce this huge electrostatic energy at low thickness are by the formation of domains with FE polarization pointing in opposite directions but also by electrical conduction within the film. This depolarizing field arises from uncompensated charges appearing at the surfaces of the FE thin film. In an ideal FE capacitor, composed of perfectly conducting plates sandwiching the FE film, the screening of charges located at the electrode/FE interface exactly compensate the surface charges related to the FE polarization. However, in real FE capacitors, the screening charges extend over a finite region in the metallic electrodes resulting in the presence of finite interface dipoles and a subsequent nonzero depolarizing field. The spatial extension of this region, that is, the effective screening length, depends strongly on the material used as electrodes (e.g., 0.2–1.9 nm for $La_{2/3}Sr_{1/3}MnO_3$ [45, 46], ~0.5 nm for $SrRuO_3$ [44], and <0.1 nm for Au).

---

* Ferroelectricity has long been viewed as a collective phenomenon resulting from the alignment of localized dipoles (resulting in a spontaneous polarization) within a correlation volume that extend for many ferroelectrics over a distance of the order of 10–50 nm along the polarization axis. In this picture, the occurrence of ferroelectricity is allowed only for films thicker than this correlation length which in the case of PZT for example is 20 nm. However, with appropriate boundary conditions, recent ab initio calculations as well as experimental results reveal that ferroelectricity can be preserved at the nanometer scale.

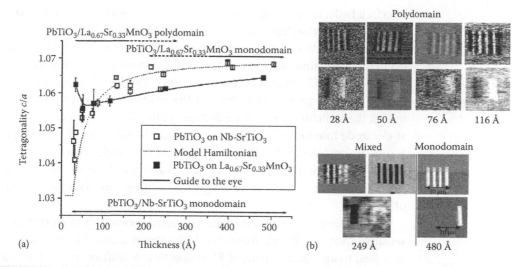

FIGURE 17.4 (a) Evolution of the c/a ratio with the $PbTiO_3$ film thickness grown on Nb-doped $SrTiO_3$ and $La_{2/3}Sr_{1/3}MnO_3$ buffered $SrTiO_3$. Very different behaviors are obtained due to the different screening conditions. (b) The recovery of the c/a ratio at low thickness is accompanied by a change from a monodomain to a polydomain configuration of the polarization, as shown by piezoresponse atomic force microscopy measurements. (After Lichtensteiger, C. et al., *Appl. Phys. Lett.* 90, 052907, 2007. With permission.)

Through its influence on the screening length of the depolarizing field, the choice of the electrodes has important effects on the critical thickness and/ or on the formation of domains. For example, using high-resolution x-ray to study single-domain $PbTiO_3$ deposited on conductive Nb-doped $SrTiO_3$ substrates, Lichtensteiger et al. [47] pointed out in 2005, the systematic decrease of the c-axis lattice parameter with decreasing film thickness below 20 nm, reflecting the decrease of the FE polarization (see Figure 17.4a). This reduction of the polarization is attributed to the presence of a residual unscreened depolarizing field. When grown on $La_{2/3}Sr_{1/3}MnO_3$ electrodes, the same $PbTiO_3$ thin films first show a decrease of c/a with decreasing film thickness followed by a recovery of c/a at small thicknesses (Figure 17.4a). This recovery is accompanied by a change from a single domain to a polydomain configuration of the polarization, as was directly shown by piezoresponse atomic force microscopy (AFM) measurements (Figure 17.4b). Surprisingly, the LSMO electrodes do not seem to screen as well as Nb-doped $SrTiO_3$ electrodes despite their better bulk metallicity. More recent x-ray scattering measurements and ab initio calculations performed by Wang et al. evidenced how the chemical environment can control the polarization orientation in a $PbTiO_3$ FE thin film [49]. Interestingly, recent theoretical predictions from first principle calculations of ultrathin $Au/BaTiO_3/Au$ FE capacitors predict that the covalent bonding mechanism at the interface between Ba-O-terminated films and the simple metal can lead to an overall enhancement of the driving force of the film toward a polar state rather than a suppression [50]. This result not only evidences the limitation of the simple Thomas–Fermi screening approach to understanding the existence of a critical thickness, and that a microscopic analysis will generally be necessary to describe interfacial effects in FE capacitors, but also the opportunity to improve the FE properties of thin films by selecting the appropriate electrode.

Additional effects, such as polarization fatigue, have not yet been studied in ultrathin films but are well known for thin films in the 100 nm range, usually used in FE random access memories (FeRAMs). Fatigue is generally related to the migration of oxygen to the electrode–FE interface that can inhibit the switching by pinning domain walls [51], the formation of layers at electrode interfaces that effectively reduce the total electric field applied across the device, or inhibit the nucleation of oppositely polarized domains [52, 53]. These mechanisms modify the dynamics of FE domains in applied electric fields and should have a significant impact on the characteristics of FTJs.

Interestingly, in perovskite materials, ferroelectricity is also often highly sensitive to strain. If judiciously used, the strain induced by the substrate may lead to an enhancement of FE properties. A shift of the FE transition temperature (Tc) and an increase of the remnant polarization (Pr) with strain have been predicted [54–58] and observed [59–62]. A particularly impressive example of the power of such strain effects is the demonstration of strain-induced ferroelectricity at room temperature in $SrTiO_3$ [63]. Another remarkable illustration is the large increase of the remnant polarization (by at least 250%) as well as of the transition temperatures (by nearly 500°C) in $BaTiO_3$ thin films grown on rare-earth scandates substrates inducing a biaxial compressive strain [62]. Strain engineering may thereby counteract the intrinsic thickness effect and restore ferroelectricity at low thicknesses [64].

All ferroelectrics are nevertheless not as sensitive to strain. $BiFeO_3$, for example, in which the FE character is driven by the Bi lone pair, has been predicted to be virtually strain insensitive [65]. Experimentally, a small variation of the remnant polarization was found under biaxial strain and attributed to the polarization rotation inside the (110) plane [66]. Finally, it has to be noticed that the absence of critical thickness has been predicted in improper ferroelectrics [67].

Characterizing the FE behavior at the nanometer scale is a difficult task. Indeed, for films as thin as a few nanometers, the tunneling current impedes the characterization through standard polarization versus electric field loops (P(E) loops). Noticeable exceptions (see Figure 17.5a and b) are P(E) loop measurements realized on fully strained 3.5 nm at 77 K [68] and 5 nm at room temperature [69] in $BaTiO_3$ capacitors with $SrRuO_3$ electrodes deposited on $SrTiO_3$. As shown by the current–voltage response of the capacitor (Figure 17.5a), a large contribution of the leakage to the total current is observed in this tunneling regime. PFM appears as the technique of choice for probing ferroelectricity in ultrathin films. Figure 17.5c and d present two examples of the PFM characterization of nanometer-thick films. In the first example (Figure 17.5c), stripes with a polarization either up or down have been written on a $BiFeO_3$ thin film deposited on $La_{2/3}Sr_{1/3}MnO_3$-buffered $SrTiO_3$ using a conductive tip AFM (CT-AFM); PFM is then used to read the signal. The clear and reversible out-of-plane phase contrast is indicative of the FE character of this 2 nm $BiFeO_3$ thin film [70]. In the following example

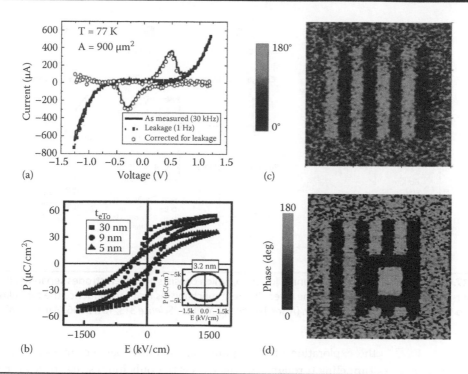

FIGURE 17.5 (a) Current versus voltage response of the 3.5 nm $BaTiO_3$ capacitor with $SrRuO_3$ electrodes measured at 77 K. The black line shows the response measured at the excitation signal frequency of 30 kHz. The measured current is the sum of three contributions: the FE displacement current caused by the switching of the spontaneous polarization and leakage current very important in the tunneling regime. The square dots denote the current response measured at a frequency of 1 Hz at which the leakage only contributes to the signal. The circles show the difference between these two current–voltage curves, which represents the leakage-compensated current response. (After Petraru, A. et al., *Appl. Phys. Lett.* 93, 072902, 2008. With permission.) (b) Polarization versus electric field loops at room temperature for 30, 9, and 5 nm thick $BaTiO_3$ sandwiched between two SrRuO3 electrodes. A clear FE loop is measured for the 5 nm thick film, whereas for the 3.2 nm one, leakage current dominates. (After Kim, Y.S. et al., *Appl. Phys. Lett.* 86, 102907, 2005. With permission.) (c) PFM out of plane phase image of a 2 nm $BiFeO_3$ film deposited on $La_{2/3}Sr_{1/3}MnO_3$ buffered $SrTiO_3$ after poling the $BiFeO_3$ film with alternately positive and negative DC voltage along 500 nm wide stripes. (After Béa, H. et al., *Jpn. J. Appl. Phys.* 45, L187, 2006. With Permission.) (d) PFM out-of-plane phase image of a 1 nm $BaTiO_3$ film grown on $La_{2/3}Sr_{1/3}MnO_3$ buffered $NdGaO_3$ substrate after writing 4 × 0.5 μm^2 stripes with a polarization either up (bright) or down (dark). The polarization in these domains can be switched back and forth by applying positive or negative bias. (After Garcia, V. et al., *Nature* 460, 81, 2009. With permission.)

(Figure 17.5d), $BaTiO_3$ has been deposited on $La_{2/3}Sr_{1/3}MnO_3$-buffered $NdGaO_3$ that imposes a large biaxial strain in order to stabilize its FE behavior [62]. The clear PFM contrast between written stripe-shaped domains with a polarization up or down directly evidences the FE character of this 1 nm thin film. Subsequent square domains have been rewritten to assess the reversibility of the writing process [64].

Ultrathin films deposited on different substrates and electrodes, inducing different boundary conditions and different strains, have been reported to be FE, experimentally. Films as thin as 4 nm of $Pb(Zr_{0.2}Ti_{0.8})O_3$ [37], 2 nm of $BiFeO_3$ [70, 71] or $(La,Bi)MnO_3$ [12], and ~1 nm of $PbTiO_3$ [41], P(VDF-TrFE) [72], and $BaTiO_3$ [64] have been reported to be FE, setting the upper limit for the effective critical thickness of these materials and allowing

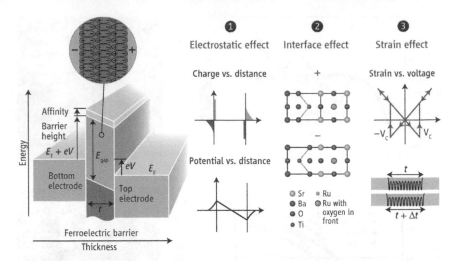

FIGURE 17.6 Schematic diagram of a tunnel junction, which consists of two electrodes separated by a nanometer-thick FE barrier layer and of the three different mechanisms that can influence the tunneling process due to the FE character of the barrier. (After Tsymbal, E.Y. et al., *Science* 313, 181, 2006. With permission.)

the exploration of their potential as tunnel barriers in FTJs and MFTJs. Tunneling through ferroelectrics is not only interesting from a fundamental point of view but it could also be of great potential for applications in the field of data storage, based on heterostructures in which the information is encoded by the direction of the FE polarization, and read nondestructively. This would represent a major advance over existing FeRAMs in which data readout is destructive, so that the initial orientation of the polarization has to be restored after reading, which is time and power consuming. Very few theoretical and experimental works have been reported on the topic of FTJs, but the recent advances in FE ultrathin films have paved the way toward more developments. Experimental and theoretical work in this nascent field* has been initiated by the group of Hermann Kohlstedt in Jülich [74–76].

From the theoretical point of view, Tsymbal and Kohlstedt have recently summarized three possible mechanisms through which the tunnel current would be modulated by the reversal of polarization in the FE barrier [77] (see Figure 17.6). The first mechanism, described in detail in Refs. [78–80], is an electrostatic effect that yields an asymmetric deformation of the potential profile of the barrier owing to the presence of charges at the barrier–electrode interfaces that are differently screened in the two different electrodes. An interesting way to increase this effect, based on composite tunnel barriers combining an FE film with a nonpolar dielectric material, has recently been proposed [81]; the dielectric layer serves as a switch that changes its barrier height from a low to a high value when the polarization of the FE is reversed. In the second mechanism, the so-called

---

* The concept of ferroelectric tunnel junctions, first proposed by L. Esaki in 1971, was the object of a patent deposited by Phillips Corp. in 1996 [73].

interface effect, the interfacial density of states of the electrodes is modified according to the FE-polarization-direction-dependent position of the ions in the last atomic layer in the FE, which affects the tunnel current [82, 83]. The third mechanism is related to the converse piezoelectric effect through which the tunnel barrier width would be changed upon switching the polarization direction. Since the tunnel current depends exponentially on the barrier width, a substantial modulation of the current can indeed be expected [78–80]. All these three mechanisms may contribute to the modulation of the tunneling current by the FE character of the barrier and give rise to large tunnel electroresistance (TER) phenomena. More sophisticated models based on first principle calculations predict sizable changes in the transmission probability across the Pt/BaTiO$_3$ interface as well as in the complex band structure of BaTiO$_3$ with polarization reversal, two ingredients controlling the electron tunneling rates through epitaxial tunnel junctions [82].

Experimentally, it is a challenge to unambiguously interpret the observed resistive switchings in terms of a modulation of the tunneling current by ferroelectricity. Indeed, bias-induced resistive switches have been observed in nonFE oxides [84]. The first breakthrough in FTJs was the demonstration of hysteretic I(V) curves in the SrRuO$_3$/BaTiO$_3$(6 nm)/SrRuO$_3$ [74] and SrRuO$_3$/PZT(6 nm)/SrRuO$_3$ [76] tunnel junctions with, in some cases, clear switchings from a high-resistance state to a low-resistance state giving rise to large TER phenomena (see Figure 17.7a). Nevertheless, as mentioned by the authors of these articles, the correlation between ferroelectricity and electroresistance could not be demonstrated unambiguously and additional effects, such as electromigration could not be refuted. However, a direct relationship between FE polarization reversal and resistance switching has been observed in more recent in situ scanning tunneling microscopy imaging and spectroscopy experiments, which evidence hysteretic features consistent with the FE switching process in the I(V) curves of 4–10 monolayer BaTiO3 thin films deposited on SrRuO$_3$ buffered SrTiO$_3$ [85].

Recently, following a procedure defined by Kohlstedt et al. [86] to unambiguously demonstrate the ferroelectricity-related nature of the resistive switch, current–voltage curves and piezoresponse measurements performed using CT-AFM in ultrahigh vacuum were collected simultaneously on 30 nm thick Pb(Zr$_{0.2}$Ti$_{0.8}$)O$_3$ films deposited on an La$_{2/3}$Sr$_{1/3}$MnO$_3$ electrode on a SrTiO$_3$(001) substrate. These authors evidenced a very large modulation of the tunneling current by a factor of about 200 and its tunability by the FE nature of the film [87]. In these experiments, electron tunneling is induced by injecting electrons from the metal tip into the oxide in the Fowler–Nordheim regime. Injection through the ferroelectrically modulated Schottky barrier at the interface was invoked to interpret the data. Additionally, a smaller modulation of the transport properties was reported using a 5 nm BiFeO$_3$ ultrathin FE tunneling barrier grown on an La$_{2/3}$Sr$_{1/3}$MnO$_3$ -electrode [87].

A definite correlation between the FE character and the electroresistance phenomena in the standard tunneling regime has been recently

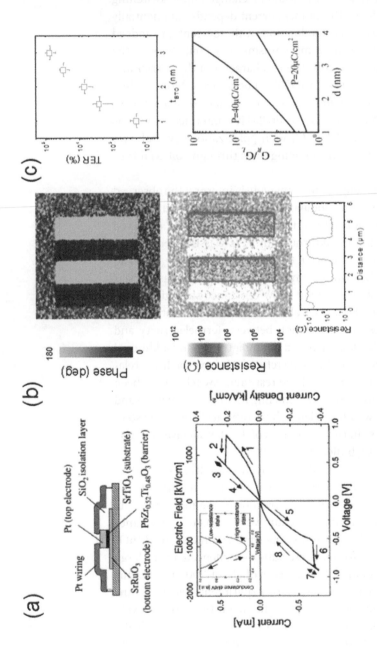

FIGURE 17.7   (a) Sketch (top) and current versus voltage behavior (bottom) of the FTJs of Rodriguez Contreras et al. [76]; the inset shows the dynamic conductance plotted versus voltage for the low- and high-resistance states. (b) Parallel PFM (top) and conductive-tip AFM (middle) mappings of a 3 nm BaTiO₃ film grown on La₂/₃Sr₁/₃MnO₃/NdGaO₃(001) after poling the BaTiO₃ film along stripes with positive or negative DC voltage. The resistance difference between negatively and positively poled areas defines a tunnel electroresistance (TER) of ~75,000%. The bottom curve is a resistance profile across the series of stripes in the middle figure. (After Garcia, V. et al., *Nature* 460, 81, 2009. With permission.) (c) Thickness dependence of the TER across FE BaTiO₃ tunnel barriers measured by Garcia et al. (Top, after Garcia, V. et al., *Nature* 460, 81, 2009. With permission.) Calculated by Zhuravlev et al. (Bottom, after Zhuravlev, M.Y. et al., *Phys. Rev. Lett.* 94, 246802, 2005. With permission.)

brought out by combining PFM and CT-AFM experiments at room temperature [64]. Using PFM, Garcia et al. wrote stripes with an FE polarization pointing either up or down in highly strained $BaTiO_3$ thin films (1–3 nm) deposited on buffered $NdGaO_3$ (see the top of Figure 17.7b). Subsequently, using CT-AFM, they collected resistance maps of the poled domains (see the middle of Figure 17.7b), taking advantage of the finite tunnel current at these low barrier thicknesses. The resistance contrast between positively and negatively poled regions corresponds to TER effects of 200%, 10,000%, and 75,000% for 1, 2, and 3 nm (see the bottom of Figure 17.7b), respectively. From the thickness dependence of the TER effect (see the top of Figure 17.7c), the authors conclude that the predominant mechanism at the origin of this TER effect is mainly the modulation of the barrier profile by the FE character of the barrier as predicted by Zhuravlev et al. [79] and Kohlstedt et al. [78] (see the bottom of Figure 17.7c). The scalability down to 70 nm, corresponding to densities greater than 16 Gbit in.$^{-2}$, was also established. This has been confirmed more recently on 4.8 nm $BaTiO_3$ ultrathin films deposited on $SrRuO_3$-buffered $SrTiO_3$ substrates inducing a smaller strain [88]. The more modest TER effect may result from a smaller strain that induces a less significant polarization than in the case of Garcia et al. [64]. This demonstrates that in FTJs, the direction of the polarization can be determined without altering it by exploiting the polarization-dependent leakage current in the tunneling regime. This could be of great promise for next-generation FeRAMs with nondestructive readouts [89]. Their realization in the solid-state form with top electrodes and standard row and column addressing, as well as their integration into CMOS technology will have to be demonstrated before their real potential as memory devices can be assessed [89]. Additional effects related to the piezoelectric character of the barrier, which is generally impeded in thin films due to the clamping by the substrate, could be expected if these devices are patterned with lateral dimensions of the order or below 100 nm [90]. This may enlarge [78–80] the parameters playing a key role in the ferroelectricity-induced modulation of tunneling current in FTJs.

The functionality of FTJs can be further enriched by replacing the normal metal electrodes by a ferromagnetic one or by using an FE–ferromagnetic material as a tunnel barrier, thereby defining multiferroic (FE and ferromagnetic) tunnel junctions (MFTJs) described in the following sections.

17.3 INTRINSIC MULTIFERROIC TUNNEL JUNCTIONS

Taking advantage of the multifunctional character of $La_{0.1}Bi_{0.9}MnO_3$, one of the only ferromagnetic–FE multiferroic compounds, Gajek et al. designed an MFTJ with a ferromagnetic–FE barrier [12]. $La_{2/3}Sr_{1/3}MnO_3$ and Au were used as electrodes. As expected from the FE nature of the $La_{0.1}Bi_{0.9}MnO_3$

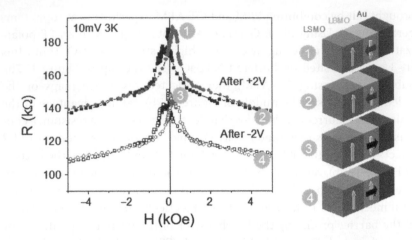

FIGURE 17.8 Variation of the resistance of an Au/LBMO(2 nm)/LSMO junction as a function of the magnetic field and previously applied bias displaying four resistance states whose magnetic and electric configurations are represented on the right-hand side. (After Velev, J.P. et al., *Phys. Rev. Lett.* 95, 216601, 2005; Bowen, M. et al., *Phys. Rev. B* 73, 140408(R), 2006. With permission.)

compound,* a TER effect of about 20% was observed upon reversing the FE polarization of the barrier. An additional spin-filtering effect and the subsequent TMR arise from the ferromagnetic character of the barrier [91, 92]. This spin-filtering effect, observed for both $BiMnO_3$ and $La_{0.1}Bi_{0.9}MnO_3$, is related to the fact that the barrier height is spin dependent because the bottom level of the conduction band is spin-split by exchange interaction in a ferromagnetic material [93], see also Chapter 14, Volume 1. This allows the tunneling electrons to be efficiently filtered according to their spin, and results in a highly spin-polarized current, thus leading to a large tunnel magnetoresistance (TMR, defined as the difference between the junction resistance in the antiparallel (AP) and parallel (P) configurations of the $La_{2/3}Sr_{1/3}MnO_3$ and $La_{0.1}Bi_{0.9}MnO_3$ magnetizations, that is, the TMR = (RAP − RP)/RP) effect when one of the electrodes (in that case, $La_{2/3}Sr_{1/3}MnO_3$) is ferromagnetic. TMR effects of ~50% and ~90% have been observed for $BiMnO_3$ and $La_{0.1}Bi_{0.9}MnO_3$, respectively, corresponding to a spin-filtering efficiency of about 22% and 35%, respectively [92, 94]. Using the FE and ferromagnetic character of a 2 nm $La_{0.1}Bi_{0.9}MnO_3$ tunnel barrier, Gajek et al. were consequently able to define a four-resistance state system (see Figure 17.8 [11]). It was proposed that these four resistance states are encoded by the two-order parameters in this multiferroic barrier—magnetization and polarization—through the mechanisms of spin filtering and the modulation of the barrier profile by the FE nature of the compound [12]. As expected from the ferromagnetic Curie temperature of the compound (95 K, for 10% La doping), the TMR effect disappears around 90 K, whereas the TER effect

---

* It has to be noticed that the ferroelectric character of $BiMnO_3$ thin films remains questionable due to the high leakage current that prevent poling of the sample. Only in $La_{0.1}Bi_{0.9}MnO_3$ films, even as thin as 2 nm, has a clear ferroelectric behavior has been observed.

is still observed at room temperature (in agreement with the estimated FE TC of the compound, 450 K [95]).

Electron transport through these multiferroic tunneling junctions has been investigated theoretically by Ju et al. [96] taking into account the screening of the polarization charges in the metallic electrodes as well as the exchange splitting of the barrier. The TMR effect was found to depend strongly on the FE polarization of the barrier and on the contrast between the screening length of the two electrodes that are the two parameters at the origin of the TER effect in this simple model [78, 79]. This could explain the quite small TMR effect observed in the $Au/La_{0.1}Bi_{0.9}MnO_3(2$ nm$)/La_{2/3}Sr_{1/3}MnO_3$ multiferroic junctions for which the screening lengths of the two electrodes are very dissimilar [12]. First principle calculations of the systems, such as those performed for the $BaTiO_3$ tunnel barriers [83, 97], taking into account the symmetry filtering effect affecting the transmission coefficient and the coupling of the wave function of the electrodes with the evanescent wave functions in the barrier, are however necessary to fully understand the transport properties in these epitaxial tunnel junctions. Nevertheless, these predictions [96] underline the complexity to optimize the TER and TMR effects in these multiferroic junctions based on an FE–ferromagnetic single material. Solutions can be found in extrinsic MFTJs in which the different functionalities underlying the operation of these devices—ferromagnetism and -ferroelectricity—may be optimized independently (see Chapter 16, Volume 1).

17.4 ARTIFICIAL MULTIFERROIC TUNNEL JUNCTIONS

In intrinsic MFTJs, the two-order parameters (magnetization and polarization) of the multiferroic tunnel barrier determine the transport properties. However, in artificial MFTJs, the spin-dependent and ferroelectricity-related contribution to the transport properties are, in first approximation, physically separated as the barrier is made of a nonferromagnetic FE material, and the ferromagnetic order is only present in the electrodes. These junctions are thus expected to show four resistance states due to both TMR (the same as in conventional MTJs) and to TER (as in FTJs with nonmagnetic electrodes) [98]. Additionally, artificial MFTJs allow more flexibility in the choice of materials and in the optimization of the FE and ferromagnetic properties. As will be discussed in the following paragraphs, the interplay between ferroelectricity and ferromagnetism at the interfaces between the barrier and the electrodes can also result in a giant modulation of TMR (and, therefore, of the tunnel current spin polarization) induced by FE polarization reversal.

This ferroelectricity-induced modulation of spin polarization is related to the second mechanism responsible for TER in FTJs, put forward by Tsymbal and Kohlstedt [77] (see Figure 17.6), namely, the modification of the interfacial density of states due to different orbital bonding depending on the direction of the polarization in the FE barrier. If the electrode is made of a ferromagnetic material with a spin-polarized density of states, the FE -polarization-direction-dependent modulation of the density of states will be different for spin-up and spin-down, resulting in a modulation of the interfacial spin polarization and of magnetic properties depending on the electron density close to the Fermi level EF (e.g., the magnetic moment or the magnetocrystalline anisotropy). This effect is thus reminiscent of the predicted [99, 100] and observed [101, 102] influence of a large external electric field on the magnetic anisotropy of ferromagnetic ultrathin layers due to changes in the spin-dependent electron density at the EF. An important benefit of using an FE rather than applying an external electric field to induce changes in the spin-dependent density of states close to EF is that those changes will be driven by the direction of the FE polarization and will thus be nonvolatile.

Ab initio calculations of the interfacial electronic structure of FE/ferromagnet heterostructures have been performed for several systems including $BaTiO_3$/Fe [82, 103–105], $PbTiO_3$/Fe [103, 105], $BaTiO_3/Co_2MnSi$ [106], $BaTiO_3/SrRuO_3$ [97], and $BaTiO_3/Fe_3O_4$ [107]. In Figure 17.9, we show typical results for the case of $BaTiO_3$/Fe. In Figure 17.9a, the orbital-resolved density of states profiles for interfacial atoms are displayed [82]. Clearly, the density of states at the interface (dotted and solid curves) is modified compared to the bulk (shaded gray) in both $BaTiO_3$ and Fe, indicating some substantial redistribution of charge. Importantly, this redistribution and accordingly, the spin-dependent density of states is different depending on the direction of the FE polarization (for the solid and dotted curves, the FE polarization of the BTO is pointing toward and away from the Fe layer, respectively). For the type of interface structure considered here, the spin polarization is found more negative when the FE polarization is pointing away from the Fe. Figure 17.9b reproduces the results of calculations by Fechner et al. for one monolayer of Fe on top of $BaTiO_3$ [103]. Consistent with the prediction of Duan et al. [82] (Figure 17.9a), they also found a globally negative spin polarization for the Fe layer at the interface with $BaTiO_3$ that increases in absolute value when the FE polarization points away from the Fe. However, the precise influence of the FE polarization direction on the interfacial spin polarization is expected to depend on the details of the interface structure [108]. We emphasize that while those calculations predict relatively modest changes in the local magnetic moments induced by FE polarization reversal, the expected effect on the interfacial spin-dependent density of states at EF is usually much larger making tunneling (with its extreme sensitivity to interfacial effects) a particularly well-suited probe of these phenomena.

Although the amplitude of the TMR may in simple cases be estimated from inspecting the spin-dependent density of states of the electrodes (using the so-called Jullière model [109]), it is now well accepted that for epitaxial systems, the complex band structure of the barrier material plays a key role (see Chapter 13, Volume 1). This is epitomized by the model Fe/MgO/Fe system [110, 111] in which the complex band structure of MgO selectively transmits the electronic wave functions of the Δ_1 symmetry that have a very high spin polarization compared to that of the density of states (integrated over the full Brillouin zone, and thus, summing up the contribution of wave functions of all possible symmetries). Calculations of the complex electronic band structure of perovskite oxides have been carried out [112,

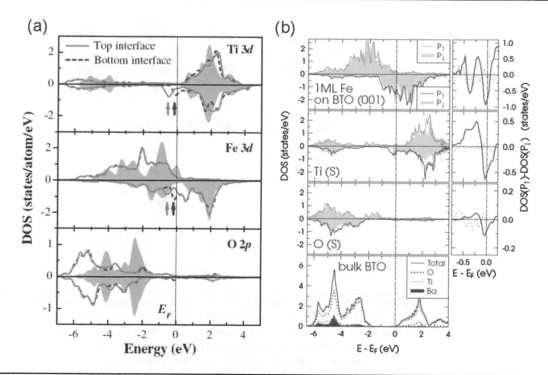

FIGURE 17.9 (a) Orbital-resolved DOS for interfacial atoms in an Fe=BaTiO$_3$ multilayer for m = 4: (top) Ti 3d, (middle) Fe 3d, and (bottom) O 2p. Majority- and minority-spin DOS are shown in the upper and lower panels, respectively. The solid and dashed curves correspond to the DOS of atoms at the top and bottom interfaces, respectively. The shaded plots are the DOS of atoms in the central monolayer of (middle) Fe or (top), (bottom) TiO2 that can be regarded as bulk. The vertical line indicates the Fermi energy (EF). (After Duan, C.G. et al., *Phys. Rev. Lett.* 97, 047201, 2006. With permission.) (b) Main panels on the left: electronic structure at the surface of Fe/BaTiO$_3$(001): spin-resolved density of states (DOS) of Fe in layer S+1 (top) as well as of Ti (second from top) and O (third from top) in layer S for FE polarization pointing up (lines) and down (gray). Bottom: total and partial DOS of bulk BaTiO$_3$, with the bottom of the conduction band taken as energy reference. The small panels on the right show the spin-resolved difference of the DOS for FE polarization pointing up and down of Fe, Ti, and O (majority: dotted; minority: solid).

113] and Velev et al. have even calculated the influence of ferroelectricity on that of BaTiO$_3$ [83]. More recently, full ab initio calculations of the spin-dependent transport properties of SrRuO$_3$/BaTiO$_3$/SrRuO$_3$ artificial MFTJs have been reported [97] (note that despite the use of two SrRuO$_3$ electrodes, the two barrier/electrodes interfaces are asymmetric due to different stacking sequences of the perovskite blocks). Not only are four different resistance states predicted, but also TMR ratios of different amplitude depending on the direction of the FE polarization in the BaTiO$_3$ tunnel barrier.

On the experimental side, there have been so far very few efforts to explore the physics and the potential of artificial MFTJs. Garcia et al. have fabricated La$_{2/3}$Sr$_{1/3}$MnO$_3$/BaTiO$_3$/Fe tunnel junctions in which the ultrathin (1 nm) BaTiO$_3$ film used as the barrier was shown to be FE at room temperature [64]. In these junctions, a negative TMR was measured, as expected (within Jullière's model [109]) from the positive spin polarization of La$_{2/3}$Sr$_{1/3}$MnO$_3$ and the calculated negative spin polarization of the density of states of Fe at the interface with BaTiO$_3$ [82, 103] (see Figure 17.10). Remarkably, the

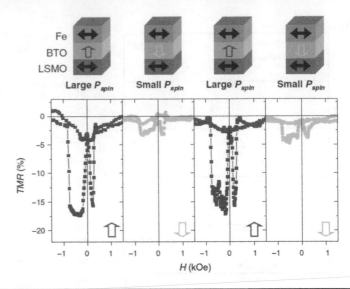

FIGURE 17.10 Tunnel magnetoresistance (TMR) versus magnetic field curves for an Fe/BaTiO$_3$(1 nm)/La$_{2/3}$Sr$_{1/3}$MnO$_3$ tunnel junction (VDC = −50 mV, T = 4 K) after poling the FE BaTiO$_3$ tunnel barrier down (VP+), up (VP−), down (VP+), and up (VP−). A clear modulation of the TMR with FE polarization orientation is achieved. (After Garcia, V. et al., *Science* 327, 1106, 2010. With permission.)

amplitude of the TMR was found to vary reversibly with poling voltage pulses of ±1 V used to orient the FE polarization in the BaTiO$_3$ barrier toward or away from the Fe electrode [114]. A larger TMR, corresponding to a larger (in absolute value) spin polarization for the Fe, was found when the FE polarization was pointing toward the Fe.

While very much remains to be done in the exciting novel field of artificial MFTJs (and more generally on the electric field and/or FE control of interfacial magnetism), these first theoretical and experimental results look promising to propose alternative solutions for the local manipulation of spin-based properties at a low energetic cost. In the future, the FE control of spin polarization predicted and observed at FE/ferromagnet interfaces may be useful to control the response of lateral spin-injection devices, such as spin transistors [98, 115].

17.5 CONCLUSIONS AND PERSPECTIVES

Although the research on MFTJs is still in its infancy, very important advances have been achieved in the past three or four years. The field is at the confluence of several very active areas, such as spintronics, oxide interfaces, and nanoscale ferroelectrics; and advances are thus partly fueled by the fast progress made on these topics. From a practical perspective, MFTJs are some sort of a merger of magnetic and FTJs, with the potential not only to address issues in MRAMs (by reducing the writing energy) and in FeRAMs (by providing a simple and nondestructive readout scheme), but also to combine the best of both technologies (magnetic storage, simple electrical readout) to propose novel storage concepts. Interestingly, there is great synergy between the theoretical and experimental work in this field, with several outstanding achievements, especially in FTJs (note, for instance, the remarkable agreement between the predicted [79] and observed [64] exponential increase of TER with barrier thickness). This joint effort from first-principles specialists on one side and experimentalists on the other side will be decisive in solving the numerous remaining issues, such as fatigue in FE tunnel barriers, FE nanoscale domain dynamics in the GHz regime, and interface-driven magnetoelectric effects.

ACKNOWLEDGMENTS

The authors wish to thank V. Garcia for his critical reading of the manuscript. Partial support from the French ANR Programme Blanc ("Oxitronics") and P-Nano ("Méloic"), and the Triangle de la Physique is acknowledged.

REFERENCES

1. J. Frenkel, On the electrical resistance of contacts between solid conductors, *Phys. Rev.* **36**, 1604–1618 (1930).
2. I. Giaever, Energy gap in superconductors measured by electron tunneling, *Phys. Rev. Lett.* **5**, 147–148 (1960).
3. L. Esaki, R. B. Laibowitz, and P. J. Stiles, Polar switch, *IBM Tech. Disclos. Bull.* **13**, 2161 (1971).
4. M. Bibes and A. Barthélémy, Towards a magnetoelectric memory, *Nat. Mater.* **7**, 425–426 (2008).
5. B. B. Van Aken, J.-P. Rivera, H. Schmid, and M. Fiebig, Observation of ferrotoroidic domains, *Nature* **449**, 702–705 (2007).
6. W. Eerenstein, N. D. Mathur, and J. F. Scott, Multiferroic and magnetoelectric materials, *Nature* **442**, 759–765 (2006).
7. N. A. Spaldin and M. Fiebig, The renaissance of magnetoelectric multiferroics, *Science* **309**, 391–392 (2005).
8. J. F. Scott, Multiferroic memories, *Nat. Mater.* **6**, 256–257 (2007).
9. M. Fiebig, Revival of the magnetoelectric effect, *J. Phys. D: Appl. Phys.* **38**, R123–R152 (2005).
10. M. Bibes and A. Barthélémy, Oxide spintronics, *IEEE Trans. Electron. Dev.* **54**, 1003–1023 (2007).
11. H. Béa, M. Gajek, M. Bibes, and A. Barthélémy, Spintronics with multiferroics, *J. Phys.: Condens. Matter* **20**, 434221 (2008).
12. M. Gajek, M. Bibes, S. Fusil et al., Tunnel junctions with multiferroic barriers, *Nat. Mater.* **6**, 296–302 (2007).
13. Y. Yamasaki, S. Miasaka, Y. Kanko, J.-P. He, T. Arima, and Y. Tokura, Magnetic reversal of the ferroelectric polarization in a multiferroic spinel oxide, *Phys. Rev. Lett.* **96**, 207204 (2006).
14. R. Nechache and C. Harnagea, A. Pignolet et al., Growth, structure and properties of epitaxial thin films of first-principles predicted multiferroic Bi_2FeCrO_6, *Appl. Phys. Lett.* **89**, 102902 (2006).
15. G. Catalan and J. F. Scott, Physics and applications of bismuth ferrite, *Adv. Mater.* **21**, 2463–2485 (2009).
16. J. R. Teague, R. Gerson, and W. J. James, Dielectric hysteresis in single crystal $BiFeO_3$, *Solid State Commun.* **8**, 1073–1074 (1970).
17. P. Fischer, M. Polomska, I. Sosnowska, and M. Szymanski, Temperature dependence of the crystal and magnetic structures of $BiFeO_3$, *J. Phys. C: Solid State Phys.* **13**, 1931–1940 (1980).
18. H. Zheng, J. Wang, S. E. Lofland et al., Multiferroic $BaTiO_3$–$CoFe_2O_4$ nanostructures, *Science* **303**, 661–663 (2004).

19. T. Lottermoser, T. Lonkai, U. Amann, D. Holhwein, J. Ihringer, and M. Fiebig, Magnetic phase control by an electric field, *Nature* **430**, 541–544 (2004).

20. J. Dho, X. Qi, H. Kim, J. L. MacManus-Driscoll, and M. G. Blamire, Large electric polarization and exchange bias in multiferroic $BiFeO_3$, *Adv. Mater.* **18**, 1445–1448 (2006).

21. H. Béa, M. Bibes, S. Cherifi et al., Tunnel magnetoresistance and robust room temperature exchange bias with multiferroic $BiFeO_3$ epitaxial thin films, *Appl. Phys. Lett.* **89**, 242114 (2006).

22. H. Béa, M. Bibes, F. Ott et al., Mechanisms of exchange bias with multiferroic $BiFeO_3$ epitaxial thin films, *Phys. Rev. Lett.* **100**, 017204 (2008).

23. V. Laukhin, V. Skumryev, X. Martí et al., Electric-field control of exchange bias in multiferroic epitaxial heterostructures, *Phys. Rev. Lett.* **97**, 227201 (2006).

24. W. Eerenstein, W. Wiora, J.-L. Prieto, J. F. Scott, and N. D. Mathur, Giant sharp and persistent converse magnetoelectric effects in multiferroic epitaxial heterostructures, *Nat. Mater.* **6**, 348–351 (2007).

25. C. Thiele, K. Dörr, O. Bilani, J. Rödel, and L. Schultz, Influence of strain on the magnetization and magnetoelectric effect in $La_{0.7}A_{0.3}MnO_3$/PMN-PT(001) (A=Sr,Ca), *Phys. Rev. B* **75**, 054408 (2007).

26. T.-K. Chung, S. Keller and G. P. Carman, Electric-field-induced reversible magnetic single-domain evolution in a magnetoelectric thin film, *Appl. Phys. Lett.* **94**, 132501 (2009).

27. M. Weiler, A. Brandlmaier, S. Geprägs et al., Voltage controlled inversion of magnetic anisotropy in a ferromagnetic thin film at room temperature, *New J. Phys.* **11**, 013021 (2009).

28. Y. H. Chu, L. W. Martin, M. B. Holcomb et al., Electric-field control of local ferromagnetism using a magnetoelectric multiferroic, *Nat. Mater.* **7**, 478–482 (2008).

29. Y. H. Chu, Q. He, C.-H. Yang et al., Nanoscale control of domain architecture in $BiFeO_3$ thin films, *Nanoletters* **9**, 1726–1730 (2009).

30. C. J. Fennie and K. M. Rabe, Magnetic and electric phase control in epitaxial $EuTiO_3$ from first principles, *Phys. Rev. Lett.* **97**, 267602 (2006).

31. C. H. Ahn, K. M. Rabe, and J.-M. Triscone, Ferroelectricity at the nanoscale: Local polarization in oxide thin films and heterostructures, *Science* **303**, 488–491 (2004).

32. K. Rabe, C. H. Ahn, and J.-M. Triscone, *Physics of Ferroelectrics, Topics in Applied Physics*, Vol. 105, Springer Verlag, Berlin, Heidelberg, 2007.

33. N. Setter, N. Damjanovic, L. Eng et al., Ferroelectric thin films: Review of materials, properties and applications, *J. Appl. Phys.* **100**, 051606 (2006).

34. M. Dawber, K. M. Rabe, and J. F. Scott, Physics of ferroelectric oxides, *Rev. Mod. Phys.* **77**, 1083–1130 (2005).

35. M. Dawber, N. Stucki, C. Lichtensteiger, S. Gariglio, and J.-M. Triscone, New phenomena at the interfaces of very thin ferroelectric oxides, *J. Phys.: Condens. Matter* **20**, 264015 (2008).

36. M. E. Lines and A. M. Glass, *Principles and Applications of Ferroelectrics and Related Materials*, Oxford University Press, Oxford, 1979; S. Li, J. A. Easterman, J. M. Vetrone, C. M. Foster, R. E. Newnham, and L. E. Cross, Dimension and size effects in ferroelectrics, *Jpn. J. Appl. Phys. Part 1* **36**, 5169–5174 (1977); N. Sai, A. M. Kolpak and A. M. Rappe, Ferroelectricity in ultrathin perovskite films, *Phys. Rev. B* **72**, 020101(R) (2005).

37. T. Tybell, C. H. Ahn, and J.-M. Triscone, Ferroelectricity in thin perovskite films, *Appl. Phys. Lett.* **75**, 856–858 (1999).

38. S. Kalinin and A. Gruverman, *Scanning Probe Microscopy: Electrical and Electromechanical Phenomena at the Nanoscale*, Springer, Berlin, 2006.

39. L. Despont, C. Koitzsch, F. Clerc et al., Direct evidence for ferroelectric polar distortion in ultrathin lead titanate perovskite films, *Phys. Rev. B* **73**, 094110 (2006).

40. W. F. Egelhoff Jr., X-ray photoelectron and Auger electron forward scattering: A new tool for surface crystallography, *Crit. Rev. Solid State Mater. Sci.* **16**, 213–235 (1990).

41. D. D. Fong, G. B. Stephenson, S. K. Streiffer et al., Ferroelectricity in ultrathin perovskite films, *Science* **304**, 1650–1653 (2004).

42. D. A. Tenne, A. Bruchhausen, N. D. Lanzillotti-Kimura et al., Probing nanoscale ferroelectricity by ultraviolet Raman spectroscopy, *Science* **313**, 1614–1616 (2006).

43. J. Junquera and P. Ghosez, Critical thickness for ferroelectricity in perovskite ultrathin films, *Nature* **422**, 506–509 (2003).

44. P. Ghosez and J. Junquera, First-principles modelling of ferroelectric oxides nanostructures, in *Handbook of Theoretical and Computational Nanotechnology*, M. Reith and W. Schommers (Eds.), Vol. 9, American Scientific Publishers, Stevenson Ranch, CA, pp. 623–728, 2006.

45. M. Dzero, L. P. Gor'kov, and V. Z. Kresin, On magnetoconductivity of metallic manganite phases and heterostructure, *Int. J. Mod. Phys. B* **10**, 2095–2115 (2003).

46. X. Hong, A. Posadas, C. H. Ahn, Examining the screening limit of field effect devices via the metal-insulator transition, *Appl. Phys. Lett.* **86**, 142501 (2005).

47. C. Lichtensteiger, J.-M. Triscone, J. Junquera, and P. Ghosez, Ferroelectricity and tetragonality in ultrathin $PbTiO_3$ films, *Phys. Rev. Lett.* **94**, 047603 (2005).

48. C. Lichtensteiger, M. Dawber, N. Stucki et al., Monodomain to polydomain transition in ferroelectric $PbTiO_3$ thin films with $La_{0.67}Sr_{0.33}MnO_3$ electrodes, *Appl. Phys. Lett.* **90**, 052907 (2007).

49. R. V. Wang, D. D. Fong, F. Jiang et al., Reversible chemical switching of a ferroelectric film, *Phys. Rev. Lett.* **102**, 047601 (2009).

50. M. Stengel, D. Vanderbilt, and N. A. Spaldin, Enhancement of ferroelectricity at metal-oxide interfaces, *Nat. Mater.* **8**, 392–397 (2009).

51. M. Dawber and J. F. Scott, A model for fatigue in ferroelectric perovskite thin films, *Appl. Phys. Lett.* **76**, 1060–1062 (2000).

52. A. K. Tagantsev, I. Stolichnov, E. L. Colla, and N. Setter, Polarization fatigue in ferroelectric films: Basic experimental findings, phenomenological scenarios, and microscopic features, *J. Appl. Phys.* **90**, 1387–1402 (2001).

53. P. K. Larsen, G. J. M. Dormans, D. J. Taylor, and P. J. van Veldhoven, Ferroelectric properties and fatigue of $PbZr_{0.51}Ti_{0.49}O_3$ thin films of varying thickness: Blocking layer model, *J. Appl. Phys.* **76**, 2405–2413 (1994).

54. A. F. Devonshire, Theory of barium titanate, *Phil. Mag.* **42**, 1065–1079 (1951).

55. N. A. Pertsev, A. G. Zembilgotov, and A. K. Tagantsev, Effect of mechanical boundary conditions on phase diagrams of epitaxial ferroelectric thin films, *Phys. Rev. Lett.* **80**, 1988–1991 (1998).

56. Y. L. Li, S. Y. Hu, Z. K. Liu, and L. Q. Chen, Phase-field model of domain structures in ferroelectric thin films, *Appl. Phys. Lett.* **78**, 3878–3880 (2001).

57. M. Sepliarsky, S. R. Phillpot, M. G. Stachiotti, and R. L. Migoni, Ferroelectric phase transitions and dynamical behavior in $KNbO_3/KTaO_3$ superlattices by molecular-dynamics simulation, *J. Appl. Phys.* **91**, 3165–3171 (2002).

58. J. B. Neaton and K. M. Rabe, Theory of polarization enhancement in epitaxial $BaTiO_3/SrTiO_3$ superlattices, *Appl. Phys. Lett.* **82**, 1586–1588 (2003).

59. E. D. Specht, H.-M. Christen, D. P. Norton, and L. A. Boatner, X-ray diffraction measurement of the effect of layer thickness on the ferroelectric transition in epitaxial $KTaO_3/KNbO_3$ multilayers, *Phys. Rev. Lett.* **80**, 4317–4320 (1998).

60. N. Yanase, K. Abe, N. Fukushima, and T. Kawakubo, Thickness dependence of ferroelectricity in heteroepitaxial $BaTiO_3$ thin film capacitors, *Jpn. J. Appl. Phys.* **38**, 5305–5308 (1999).

61. S. K. Streiffer, J. A. Eastman, D. D. Fong et al., Observation of nanoscale 180° stripe domains in ferroelectric $PbTiO_3$ thin films, *Phys. Rev. Lett.* **89**, 067601 (2002).

62. K. J. Choi, M. Biegalski, Y. L. Li et al., Enhancement of ferroelectricity in strained $BaTiO_3$ thin films, *Science* **306**, 1005–1009 (2004).

63. H. Haeni, P. Irvin, W. Chang et al., Room-temperature ferroelectricity in strained $SrTiO_3$, *Nature* **430**, 758–761 (2004).

64. V. Garcia, S. Fusil, K. Bouzehouane et al., Giant tunnel electroresistance for non-destructive readout of ferroelectric states, *Nature* **460**, 81–84 (2009).

65. C. Ederer and N. A. Spaldin, Effect of epitaxial strain on the spontaneous polarization of thin film ferroelectrics, *Phys. Rev. Lett.* **95**, 25760 (2005).

66. H. W. Jang, S. H. Baek, D. Ortiz et al., Strain-induced polarization rotation in epitaxial (001) $BiFeO_3$ thin films, *Phys. Rev. Lett.* **101**, 107602 (2008).

67. N. Sai, C. J. Fennie, and A. A. Demkov, Absence of critical thickness in an ultrathin improper ferroelectric film, *Phys. Rev. Lett.* **102**, 107601 (2009).

68. A. Petraru, H. Kohlstedt, U. Poppe et al., Wedgelike ultrathin epitaxial $BaTiO_3$ films for studies of scaling effects in ferroelectrics, *Appl. Phys. Lett.* **93**, 072902 (2008).

69. Y. S. Kim, D. H. Kim, J. D. Kim et al., Critical thickness of ultrathin ferroelectric $BaTiO_3$ films, *Appl. Phys. Lett.* **86**, 102907 (2005).

70. H. Béa, S. Fusil, K. Bouzehouane et al., Ferroelectricity down to at least 2 nm in multiferroic $BiFeO_3$ epitaxial thin films, *Jpn. J. Appl. Phys.* **45**, L187–L189 (2006).

71. Y. H. Chu, T. Zhao, M. P. Cruz et al., Ferroelectric size effects in multiferroic $BiFeO_3$ thin films, *Appl. Phys. Lett.* **90**, 252906 (2007).

72. A. V. Bune, V. M. Fridkin, S. Ducharme et al., Two-dimensional ferroelectric films, *Nature* **391**, 874–877 (1998).

73. R. Wolf, P. W. M. Blom, and M. P. C. Krijn, U.S. Patent, Patent number 5,541,422, July 30, 1996.

74. H. Kohlstedt, N. A. Pertsev, and R. Waser, Size effects on polarization in epitaxial ferroelectric films and the concept of ferroelectric tunnel junctions including first results, *Mater. Res. Symp. Proc.* **688**, C6.5.1 (2002).

75. J. Rodríguez Contreras, J. Schubert, U. Poppe et al., Structural and ferroelectric properties of epitaxial $PbZr_{0.52}Ti_{0.48}O_3$ and $BaTiO_3$ thin films prepared on $SrRuO_3/SrTiO_3$(100) substrates, *Mater. Res. Symp. Proc.* **688**, C8.10.1 (2002).

76. J. Rodríguez Contreras, H. Kohlstedt, U. Poppe, R. Waser, C. Buchal, and N. A. Pertsev, Resistive switching in metal--ferroelectric-metal junctions, *Appl. Phys. Lett.* **83**, 4595–4597 (2003).

77. E. Y. Tsymbal and H. Kohlstedt, Tunneling across a ferroelectric, *Science* **313**, 181–183 (2006).

78. H. Kohlstedt, N. A. Pertsev, J. Rodríguez-Contreras, and R. Waser, Theoretical current-voltage characteristics of ferroelectric tunnel junctions, *Phys. Rev. B* **72**, 125341 (2005).

79. M. Y. Zhuravlev, R. F. Sabirianov, S. S. Jaswal, and E. Y. Tsymbal, Giant electroresistance in ferroelectric tunnel junctions, *Phys. Rev. Lett.* **94**, 246802 (2005).

80. K. M. Indlekofer and H. Kohlstedt, Simulation of quantum dead-layers in nanoscale ferroelectric tunnel junctions, *Europhys. Lett.* **72**, 282–286 (2005).

81. M. Ye. Zhuravlev, Y. Wang, S. Maekawa, and E. Y. Tsymbal, Tunneling electroresistance in ferroelectric tunnel junctions with a composite barrier, *Appl. Phys. Lett.* **95**, 052902 (2009).

82. C. G. Duan, S. S. Jaswal, and E. Y. Tsymbal, Predicted magnetoelectric effect in $Fe/BaTiO_3$ multilayers: Ferroelectric control of magnetism, *Phys. Rev. Lett.* **97**, 047201 (2006).

83. J. P. Velev, C.-G. Duan, K. D. Belaschenko, S. S. Jaswal, and E. Y. Tsymbal, Effect of ferroelectricity on electron transport in $Pt/BaTiO_3/Pt$ tunnel junctions, *Phys. Rev. Lett.* **98**, 137201 (2007).

84. R. Waser and M. Aono, Nanoionics-based resistive switching memories, *Nat. Mater.* **6**, 833–840 (2007).

85. S. Shin, S. V. Kalinin, E. W. Plummer, and A. P. Baddorf, Electronic transport through in-situ grown ultrathin $BaTiO_3$ films, *Appl. Phys. Lett.* **95**, 03290 (2009).

86. H. Kohlstedt, A. Petraru, K. Szot et al., Method to distinguish ferroelectric from nonferroelectric origin in case of resistive switching in ferroelectric capacitors, *Appl. Phys. Lett.* **92**, 062907 (2008).

87. P. Maksymovych, S. Jesse, P. Yu, R. Ramesh, A. P. Baddorf, and S. V. Kalini, Polarization control of electron tunneling into ferroelectric surfaces, *Science* **324**, 1421–1425 (2009).

88. A. Gruverman, D. Wu, H. Lu et al., Tunneling electroresistance in ferroelectric tunnel junctions at the nanoscale, *Nano Lett.* **9**, 3539–3543 (2009).

89. P. Zubko and J.-M. Triscone, A leak of information, *Nature* **460**, 45–46 (2009).

90. S. Bühlmann, B. Dwir, J. Baborowski, and P. Muralt, Size effect in mesoscopic epitaxial ferroelectric structures: Increase of piezoelectric response with decreasing feature size, *Appl. Phys. Lett.* **80**, 3195–3197 (2002).

91. M. Gajek, M. Bibes, A. Barthélémy, M. Varela, and J. Fontcuberta, Perovskite-based heterostructures integrating ferromagnetic-insulating $La_{0.1}Bi_{0.9}MnO_3$, *J. Appl. Phys.* **97**, 103909 (2005).

92. M. Gajek, M. Bibes, A. Barthélémy et al., Spin filtering through ferromagnetic $BiMnO_3$ tunnel barriers, *Phys. Rev. B* **72**, 020406(R) (2005).

93. J. S. Moodera, T. S. Santos, and T. Nagahama, The phenomena of spin-filter tunnelling, *J. Phys. Condens. Matter* **19**, 165202 (2007).

94. M. Gajek, M. Bibes, M. Varela et al., $La_{2/3}Sr_{1/3}MnO_3$–$La_{0.1}Bi_{0.9}MnO_3$ heterostructures for spin filtering, *J. Appl. Phys.* **99**, 08E504 (2006).

95. Z. H. Chi, C. J. Xiao, S. M. Feng et al., Manifestation of ferroelectromagnetism in multiferroic $BiMnO_3$, *J. Appl. Phys.* **98**, 103519 (2005).

96. S. Ju, T. Y. Cai, G.-Y. Guo, and Z.-Y. Li, Electrically controllable spin filtering and switching in multiferroic tunneling junctions, *Phys. Rev. B* **75**, 064419 (2007).

97. J. Velev, C.-G. Duan, J. D. Burton et al., Magnetic tunnel junctions with ferroelectric barriers: Prediction of four resistance states from first principles, *Nano Lett.* **9**, 427–432 (2009).

98. M. Y. Zhuravlev, S. S. Jaswal, E. Y. Tsymbal, and R. F. Sabirianov, Ferroelectric switch for spin injection, *Appl. Phys. Lett.* **87**, 222114 (2005).

99. C.-G. Duan, J. P. Velev, R. F. Sabirianov et al., Surface magnetoelectric effect in ferromagnetic metal films, *Phys. Rev. Lett.* **101**, 137201 (2008).

100. K. Nakamura, R. Shimabukuro, Y. Fujiwara, T. Akiyama, and T. Ito, Giant modification of the magnetocrystalline anisotropy in transition-metal monolayers by an external electric field, *Phys. Rev. Lett.* **102**, 187201 (2009).

101. M. Weisheit, S. Fähler, A. Marty, Y. Souche, C. Poinsignon, and D. Givord, Electric-field-induced modification of magnetism in thin-film ferromagnets, *Science* **315**, 349–351 (2007).

102. T. Maruyama, Y. Shiota, T. Nozaki et al., Large voltage-induced magnetic anisotropy change in a few atomic layers of iron, *Nat. Nanotechnol.* **4**, 158–161 (2009).

103. M. Fechner, I. V. Maznichenko, S. Ostanin et al., Magnetic phase transition in two-phase multiferroics predicted from first-principles, *Phys. Rev. B* **78**, 212406 (2008).

104. J. P. Velev, C.-G. Duan, K. D. Belashchenko, S. S. Jaswal, and E. Y. Tsymbal, Effects of ferroelectricity and magnetism on electron and spin transport in Fe/$BaTiO_3$/Fe multiferroic tunnel junctions, *J. Appl. Phys.* **103**, 07A701 (2008).

105. J. Lee, N. Sai, T. Cai, Q. Niu, and A. A. Demkov, Interfacial magnetoelectric coupling in tri-component superlattices, *Phys. Rev. B* **81**, 144425 (2010).

106. K. Yamauchi, B. Sanyal, and S. Picozzi, Interface effects at a half-metal/ferroelectric junction, *Appl. Phys. Lett.* **91**, 062506 (2007).

107. M. K. Niranjan, J. P. Velev, C.-G. Duan, S. S. Jaswal, and E. Y. Tsymbal, Magnetoelectric effect at the Fe_3O_4/$BaTiO_3$ interface: A first-principles study, *Phys. Rev. B* **78**, 104405 (2008).

108. A. Gloter et al., unpublished.

109. M. Jullière, Tunneling between ferromagnetic films, *Phys. Lett. A* **54**, 225–226 (1975).

110. W. H. Butler, X.-G. Zhang, T. C. Schulthness, and J. M. MacLaren, Spin-dependent tunneling conductance of Fe|MgO|Fe sandwiches, *Phys. Rev. B* **63**, 054416 (2001).

111. J. Mathon and A. Umerski, Theory of tunneling magnetoresistance of an epitaxial Fe/MgO/Fe(001) junction, *Phys. Rev. B* **63**, 220403(R) (2001).

112. J. P. Velev, K. D. Belashchenko, D. A. Stewart, M. van Schilfgaarde, S. S. Jaswal, and E. Y. Tsymbal, Negative spin polarization and large tunneling magnetoresistance in epitaxial Co|SrTiO₃|Co magnetic tunnel junctions, *Phys. Rev. Lett.* **95**, 216601 (2005).

113. M. Bowen, A. Barthélémy, V. Bellini et al., Observation of Fowler-Nordheim hole tunneling across an electron tunnel junctions due to total symmetry filtering, *Phys. Rev. B* **73**, 140408(R) (2006).

114. V. Garcia, M. Bibes, L. Bocher et al., Ferroelectric control of spin polarization, *Science* **327**, 1106–1110 (2010).

115. J. Wang and Z. Y. Li, Multiple switching of spin polarization injected into a semiconductor by a multiferroic tunneling junction, *J. Appl. Phys.* **104**, 033908 (2008).

Index

Printed and bound by CPI Group (UK) Ltd, Croydon, CR0 4YY

24/10/2024

01778295-0016